563

STRATEGIES FOR MARS:
A GUIDE TO HUMAN EXPLORATION

AAS PRESIDENT Dr. Arnauld E. Nicogossian	NASA Headquarters
VICE PRESIDENT—PUBLICATIONS Paula Korn	Brown University

EDITORS
    Dr. Carol R. Stoker
    Carter Emmart
	NASA Ames Research Center
	National Center for Atmospheric Research

SERIES EDITOR Robert H. Jacobs	Univelt, Incorporated
ASSISTANT EDITOR Kelly Snook	
ARTWORK Carter Emmart	National Center for Atmospheric Research

Partial funding for the preparation of this book was provided by Carl Pilcher at the NASA Headquarters Office of Exploration.

Front Cover Illustration:
    In 1984, Thomas Paine, former NASA Administrator, Chairman of the National Commission on Space, and Martian pioneer, designed the Mars flag. His hope was that the flag would provide a symbol around which to rally support for human exploration of Mars. His flag was conceived as an award to be given to those who continue to forge a Mars-bound future and is now presented annually to the winner of the Thomas Paine Memorial Award. The cover painting depicts two astronauts erecting this flag on Mars. One of the astronauts, the young Thomas Paine, was named after his grandfather. This book is dedicated to the memory of Tom Paine, and to planting his flag on Mars.

A Few Thoughts for a Fellow Traveler

It was not quite sunrise on Tranquility this go 'round
when I took out the 'scope
just past sundown.

Earthshine was particularly bright that dusk
Beautiful,
I noticed it on the thin crescent as I talked to some friends
"Something special about tonight's Moon," I thought.

Next night I heard the news
that you left us yesterday
bound for where we're all going
leaving us a flag,
. . . and the torch.

Through you, Tom,
we lit new soil with human footprints
that will endure longer than humans
have dreamed the dream
or walked the Earth.

If goodness comes wrapped in kindness,
then the life you worked so hard to offer the Universe
IS just.
A "should"
more than a "will."

So much
seems to twirl beneath the Sun,
so many possibilities. . .
so much to do.

— **Carter Emmart**

STRATEGIES FOR MARS: A GUIDE TO HUMAN EXPLORATION

Edited by
Carol R. Stoker
Carter Emmart

Volume 86
SCIENCE AND TECHNOLOGY SERIES
A Supplement to Advances in the Astronautical Sciences

615044

Published for the American Astronautical Society by
Univelt, Incorporated, P.O. Box 28130, San Diego, California 92198

Copyright 1996

by

AMERICAN ASTRONAUTICAL SOCIETY

AAS Publications Office
P.O. Box 28130
San Diego, California 92198

Affiliated with the American Association for the Advancement of Science
Member of the International Astronautical Federation

First Printing 1996

ISSN 0278-4017

ISBN 0-87703-405-2 (Hard Cover)
ISBN 0-87703-406-0 (Soft Cover)

*Published for the American Astronautical Society
by Univelt, Incorporated, P.O. Box 28130, San Diego, California 92198*

Printed and Bound in the U.S.A.

Foreword

Over the last two decades, a consensus has developed that human exploration of Mars should be the long term goal of the American space program. A strategy for Mars exploration has emerged which posits that scientific exploration should be a long-term endeavor that lays the groundwork for Mars settlement and that Mars explorers should "live off the land" using Martian resources. One of the key forums for developing the strategy for Mars exploration has been the Case for Mars conferences, organized by a group of Mars enthusiasts known as the Mars Underground, and cosponsored by the American Astronautical Society. This book had its genesis in the many ideas presented and discussions held at Case for Mars over the last 15 years.

Strategies for Mars: A Guide to Human Exploration is intended to provide a broad background on the wide range of topics relevant to human exploration of Mars. Since human exploration is an intrinsically interdisciplinary endeavor, one needs to have at least a general understanding of a broad range of subjects in order to contribute to the work of preparing to explore Mars, or to make informed decisions about it. While the literature on each of these subjects is available, it is an onerous job to collect and digest it. Consequently, until now, few individuals have grokked[*] all the relevant issues. To produce an accessible reference book on human exploration of Mars, the editors solicited review articles from leading authorities in each major subject area. We asked the authors to keep each review on a fairly general level, so that the book could be understood by a broad audience.

Carter Emmart's illustrations, throughout the book, visualize some of the wide array of concepts envisioned here for Mars missions. Particular visual ideas were reached while editing the volume. While some of the drawings, such as the initial Mars base, serve to illustrate a specific concept, others were meant to convey more general aspects of the daily life of those who will eventually work and live on Mars.

The order in which the subjects are reviewed follows the progression of a Mars exploration program as I briefly preview below.

The biggest hurdle to human exploration of Mars is developing the political and popular will to go. At the time of the first footfall on the Moon, the first Mars landing was planned for 1984. If the pace of American activities in space during the 1960s had been maintained until now, there is no doubt that there would be human settlements on Mars today. But America lacked the political will to sustain that program and very little progress has been made since the termination of the Apollo program in the early 1970s. Most researchers believe that a

---

[*] I have borrowed the word "grok" from Robert Heinlen's classic science fiction Stranger in a Strange Land. It is a Martian word meaning in essence to understand so fully as to incorporate the concept into one's being.

human landing on Mars could be accomplished within 10 years of the decision to start, provided a large-scale effort. A more moderately paced program will take even longer. Thus, it will be necessary to first attain a public consensus that the activity is worthwhile, and to maintain that consensus over a period of more than a decade. To build that consensus, there must be a clear presentation of the rationale behind Mars exploration. The first section of the book presents some different views of that rationale (Chapters 1-3), as well as a look at what will be required to build and maintain public support for the activity (Chapters 3 and 4). Finally, no activity of the scale of Mars exploration will succeed unless it is properly managed. Chapter 6 presents a strategy for managing the program.

Mars missions naturally divide into two major phases: **getting there** and **being there**. Getting there involves the design of the interplanetary transportation system, a problem which has attracted the best minds in rocket science for many decades. Wernher von Braun began his studies of Mars missions in the 1940s, long before any rocket-launched vehicle had reached Earth orbit. The von Braun era Mars mission studies (Chapter 7) form a foundation for the design of Mars missions.

Orbital mechanics dictates some of the most important characteristics of Mars missions—namely the mass that must be launched off the Earth, and the total mission duration. Chapter 8 reviews these issues. Because of the phasing of the orbits of Earth and Mars, missions which have the shortest interplanetary transit times also have the longest surface stay times. The overall duration of a Mars mission is two years or greater. Mission duration leads to many of the challenges to the crew discussed in subsequent chapters.

Since the von Braun era, many "point"[*] designs of Mars missions have been presented. Chapter 9 presents a framework for comparing Mars mission transportation systems using some of the recent design studies as examples.

One consequence of the long interplanetary trip time to Mars (6-18 months depending on mission profile chosen) is that something must be done to reduce the physiological consequences of extended travel in 0 g. One possible approach, artificial gravity produced by rotating the spacecraft, is reviewed in Chapter 10.

Ultimately, routine access to Mars will depend on developing advanced propulsion technology to reduce both payload mass and trip time to Mars. During the 1960s nuclear propulsion technology was developed to a fairly high level, but the technology programs were canceled at the end of the Apollo program. The two major types of nuclear propulsion (thermal and electric) are reviewed in Chapters 11 and 12.

Keeping a human crew physically and emotionally healthy during the long duration of a Mars mission is a key consideration. Rescue from Earth will be impossible in the event of a health emergency. The biomedical issues associated with the interplanetary transportation phase of the mission are reviewed in Chapter 13. Just as important is maintaining the

---

[*] By point design I mean an end-to-end mission design which results from making specific choices from a range of options for the major mission features such as crew size, mission duration, propulsion system, etc.

psychological health of the crew during these long duration missions, as reviewed in Chapter 14.

Some important lessons in how to conduct Mars missions can be learned by studying relevant analogs. Exploration of the Earth in the 17th-19th centuries involved long sea voyages that were good analogs to Mars missions. Chapter 15 presents a history of the great sea voyages of exploration and the conflicts that arose in trying to achieve scientific objectives without compromising other aspects of the missions.

The crew of a Mars mission will be exposed to high levels of radiation both during the interplanetary transit and on the surface of Mars. Chapter 16 reviews this environment and how to shield from it.

Compared to "**getting there**" relatively little work has been done on the "**being there**" aspect of Mars exploration. Sustaining humans on Mars is a different problem than sustaining them in space. Size and volume is less of a problem because habitats on Mars can be built to arbitrary size. Plants can be grown to both provide food and assist with recycling of breathing gases. Highly closed systems are not required because Martian materials such as air and soil can be used to supply consumables. Chapter 17 reviews life support strategies on the Martian surface. Biosphere II was created as an analog environment for a closed life support system on Mars, although it is vastly larger than anything to be expected on an early Mars mission. To simulate a Mars mission, a crew of eight people spent two years in the closed environment of the Biosphere II. Some of the key results of that closure experiment are reviewed in Chapter 18.

The long supply lines, and the cost of transportation from Earth, argue for using Martian resources to the largest extent possible to support humans on Mars. Finding ways to use Martian resources will certainly be a key step toward human settlement of Mars. Chapter 19 reviews the concepts for using Martian resources.

Mobility will be a key requirement for Mars exploration, both for scientific exploration and for operations at a Mars base. Chapter 20 presents concepts for providing human mobility.

All of the above considerations for "being there" fold into the design of a Mars base. Chapter 21 discusses a design and habitation strategy for a first base on the Martian surface.

The key activity on the Martian surface will be scientific exploration, and the mission must be designed to enable science. The key scientific questions on Mars, and the approach to doing science from a Martian base are reviewed in Chapters 22 and 23.

The cost of Mars exploration is not trivial. Neither is it exorbitant for an endeavor of such unprecedented historical significance. however, as reviewed in Chapter 24, estimates of mission cost are based on cost forecasting models which have historically high costs built in to them. Substantial cultural change at NASA could lower projected mission costs considerably.

The final chapter of the book is a fantasy forecast of the next hundred years of Martian history, written by the late Tom Paine, visionary and spiritual father of the Mars Underground. This book is dedicated to his memory and its objective is to further his last wish: to carry the Mars flag to Mars. Just months before his death, in 1992, Tom Paine directed the illustration accompanying his chapter.

Human settlement of Mars is an event unparalleled in human history. Compared to opening up a new world to humanity, the European colonization of the Americas was a historical footnote. The only evolutionary parallel on Earth is the colonization of the newly formed continents by the creatures of the sea. The effect on evolution, on the future, is that profound. Moreover, as a species, we turn away from it at our peril. From the geologic record we have learned that life on Earth is periodically pummeled to near extinction by the impact of giant meteorites. Some species survive these giant impact events, but the majority do not. We may have 100 million years before this happens, or it could happen next year, but it is certain that it will happen. The choice is ours. We must take this evolutionary step off our planet and into the future, or be as dead as the dinosaurs. We stand at the threshold of the future. Mars awaits us.

Carol R. Stoker, Volume Editor
Carter Emmart, Volume Editor

Prologue

STEPS TO MARS[*]

Daniel S. Goldin[†]

Today, I want to paint a picture of some of the possibilities for humankind in space. I want to take you on a fantasy journey to Mars, where we'll see robots building the first cityscape on Mars and astronauts searching for signs of early life.

Before I talk about going to Mars, I want to take you on a trip back in time. Let's go back to an ancient, violent age, four and one-half billion years ago. The Earth and Mars are forming in similar ways. They are red hot, glowing with a boiling lava surface. For over a half billion years, they are bombarded with asteroids composed of rocks and metals. There is no life. No living things could withstand the heat and the pounding these planets are taking. Finally, after 600 million years, 3.9 billion years ago, the bombardment stops. The noise and the fury cease. There is just an occasional splashdown of one of the leftover small planetoids. Both the Earth and Mars have cooled down. Their surface temperatures are below the boiling point of water. Polar caps are beginning to form. Rain falls continually as the steam in their warm atmospheres begins to condense. This leaves behind thick atmospheres rich in carbon dioxide and nitrogen. The Earth and Mars look very much alike. There are large bodies of water on both planets, formed from water steaming out of volcanoes and from impacting comets. During this period, geological features are developing on Mars that will be important billions of years later.

Large impact craters and basins are forming where sediments will later be deposited and preserved. These areas might later harbor traces or the very essence of life itself. Volcanoes are erupting and emitting water vapor. The heat and water are forming hydrothermal circulation systems that could later generate mineral deposits. These systems could produce environments to sustain life. The intense bombardment by meteorites is forming a thick, loose rock layer, a regolith. In the future, it might be the storehouse of subsurface water or ice. Soon, the Earth will produce the first self-replicating molecule. These molecules will assemble themselves into the first single-cell organism. The long march toward intelligent life on Earth is beginning. The same process on Mars may be going on, but the result may be different. We don't know. The result may have been a life form unlike anything on Earth.

---

[*] This paper is the transcript (with minor editing) of oral remarks presented at the Steps to Mars Conference, July 15, 1995, Washington D.C. The speech provides an excellent framework for and overview of the concepts presented in the book.

[†] NASA Administrator, l991-present, Washington, D.C. 20546.

Now let's move forward from 3.9 billion years ago to 3.5 billion years ago. Earth and Mars are beginning to diverge. The evolution of the Earth's surface and the atmosphere is largely affected by the fact that life has taken a foothold on our home planet. Mars is on a different path. It is just beginning to diverge from Earth. Its evolution is being shaped almost exclusively by geophysical, not biological, processes. Life, if it managed a foothold on Mars, couldn't survive on the surface because it couldn't have withstood the change in climate on Mars. Mars is too small a planet to trap the heat needed to move its huge crustal plates, and that slows down the renewal of the atmosphere through volcanic action. So the Mars atmosphere begins to thin. The planet becomes dryer and colder. Ice is now the dominant form of water on the surface of Mars and below the surface. Life languishes in shrinking pockets of warmth and the remaining small pockets of liquid water.

Not so on Earth. Life flourishes on its warm, wet surface. On Earth, our fossil records go back 3.5 billion years to fossils of algae that were found in Australia. The earliest beginnings of life before that are still a mystery to us. We can't go back any further than 3.5 billion years. Earlier relics of life were obliterated by the constant change in the Earth's surface as a result of the tectonic activity. But this may not be the case on Mars. The fossils of chemical evolution, if they exist, could still be there, untouched. Because of the cold, arid climate on Mars, there is little erosion. Mars also appears to lack plate tectonics. So, much of the planet's ancient crust is probably preserved. Early life on Mars could have fossilized quickly after dying, just like on Earth. Even when the surface of Mars became hostile to life, it may have become benign to fossils. If a fossil formed on Mars 3.5 billion years ago, it is probably still sitting there today. It may be encased in sedimentary material, just waiting for us to find it. When we begin to dig on Mars, when we look through the layers of its crust, we'll look back through a time machine.

It is widely believed that every living thing on Earth shares a common ancestor—apparently thermophylic, or heat-loving, bacteria. Who knows what we will find on Mars? We might discover traces of whole new kingdoms of early life. Maybe we will learn that the same building blocks of life washed over both planets simultaneously. We could find fossils of cells with elements of proteins similar to those here on Earth. We might find the one fossilized cell that is the missing link between the planets. We might find actual life—imagine that! We might find extant life in some specialized environment. It may be just below a few grains in translucent sandstone—a microscopic greenhouse, as in Antarctica. We may find life buried deep in some of the ancient sediments on Mars. Any of these possibilities would profoundly affect how we think about who we are and why we are here.

Mars may be our next destination in space. Its secrets, and what it can tell us about our own planet, are intriguing. We have learned from our robotic space travels that it is the most likely planet to have developed life. We are in the life zone of our solar system. Mars has surface conditions the most like our own here on Earth. If Mars is humankind's next destination, the next launch date could be 2018. That is when it would take the least amount of energy to launch to Mars in a 20-year period. We have time to plan. We have time to do it right. We have time to figure out how humans can live and work safely and efficiently in space on the international Space Station. We have time to figure out how to perform the mission for an affordable price that will allow a sustained presence in the solar system and not be a one-shot firecracker.

Final preparations need not start in earnest until the end of the first decade of the next century. But we can still make a landing in 2018 if we begin laying the groundwork now. Our goal should be a sustained presence on Mars and in the solar system and not a one-shot, feel-good spectacular mission. We are interested in emigration, not invasion. During the twenty-fifth anniversary of Apollo, reporters asked me if I, or the President, were going to announce a mission to Mars, make a big splash. I said, "That's too easy. A one-shot spectacular is not what we're after."

Now, Apollo was a one-shot sprint mission, but Apollo also met a national need—beating the Soviet Union to the Moon. We did that, and I'm proud of it. I participated. But it was brute force. We spent about 5% of the national budget to make this happen, but we had no vision beyond landing on the Moon so we went into the wilderness for 25 years. The program spent 5% of the national budget in today's terms and we did it in 8 years, which would be the equivalent today of spending 75 billion dollars a year for 8 years. This is not what we're about. People have this feel-good sense about Apollo and say, "Dan, why can't we do it again?" Well, now we are investing eight-tenths of one percent of the federal budget—not 5%—and we're into a marathon, not a sprint. We are into sustaining a presence in the solar system and in space, not a feel-good thing because we no longer have to beat the Soviet Union. We are going to partner with the Russians, the French, Germans, Japanese and others. We don't have that kind of money any more, and that's not all that bad.

The bad news is our budget is cut. The good news is it is forcing us to do things differently, more innovatively, more imaginatively, more cooperatively. We're going to use a new-think on our next piloted missions. We will leverage 21st century technology. We have come a long way since Apollo, and we don't need brute force any more. We will work with international partners, industry, government and academia. We will take the time to develop and apply new technology before building the Mars rockets and equipment. We don't need to have a jobs program to go to Mars, we need to go to Mars. We will create economic opportunity as we do. We will develop space. Our rockets won't just take us to Mars, they will open the space frontier. Unlike Apollo, we will live off the land. The most successful expeditions in human history didn't try to carry all their supplies with them. Lewis and Clark lived off the land, and that is exactly what we will do. One set of choices is brute force versus technological finesse. We choose technological finesse. Rushing to Mars would be brute force. The use of brawn would drive the cost beyond anything tolerable.

Our other choice is this—we in the United States can lead the world in this noble venture or we can buy a ticket and just watch. I submit that America is proactive. We should lead and not sit in the grandstands. If we lead the world, we will shape humankind's boldest adventure. But we have to start now. There is a lot of work to do. We should be pursuing two parallel paths right now, and that's exactly what we are doing. These two parallel paths are robotics precursor missions and fundamental missions to understand the rigors of human space flight.

Let me start with the robots. We are entering the second era of scientific, robotic Mars exploration with the Mars Surveyor program. This isn't a one-shot program. We will take advantage of every launch window to Mars to send a spacecraft. It is a continuing series of reconnaissance missions to fully explore the planet Mars. We are going to explore Mars carefully, deliberately and systematically. We will be bold, but not brash. This won't be a sprint mission, this is a marathon. Instead of rushing up to grab the first rock or handful of soil we can, we will get a better understanding of the planet. We will figure out where it makes sense to take samples. We will want to be smart before we go. That way, the samples we do bring back will tell us the most about Martian evolution, climate, water, and other resources.

Our first job is to map the planet from orbit. Towards this end, we will send small, low-cost orbiters to Mars in '96, '98 and 2001. The first Global Surveyor will be launched in '96 to map the geology and topography of Mars. I'm not really thrilled with the spatial resolution yet, but it will give us a broad picture. As time goes on, the resolution will improve. It will give us a start on the global geochemistry pattern. The second Global Surveyor in 1998 and the third in 2001 will give us that map and, in addition, examine the atmosphere and climate. They will also search for water in the atmosphere, on the surface and just below the surface. These two orbiters will also complete the orbital communications network of three orbiters going around Mars. Our landed missions will use the network to communicate back to Earth.

Building global maps from orbit is our first task. Our second task is to scout selected areas on the Martian surface with landers and rovers using information we obtained from the orbiters. We will need to examine the local conditions, rock and soil types, the potential for water and other resources, and signs of the planet's history. We will send a series of small, low-cost landers to many different areas of Mars. Then we will select the one or two sites to actually collect samples and bring them back to Earth. The first of our landed missions is Mars Pathfinder '96, which is now called Sojourner. It was named by a young lady who was selected out of 3,500 people that submitted suggestions. It was wonderful to hear her essay presented at the Steps to Mars Conference.

Let me talk about the first of our landed missions. There will be a micro-rover to provide mobility on the surface. Pathfinder is only the first. We will send similar landed missions in '98, 2001 and 2003. (I might point out that the Russians are also sanding a robot in '96. We have a payload on their lander in '96 and they have a payload on our lander in '98.) The landed missions planned for '98 and 2001 will go to places as diverse as ancient cratered highlands and the planet's poles. Landers will analyze soil and rocks. One of the things that this will give us is ground truths for orbital remote sensing measurements. The landers may also drill below the surface to better understand the geological history of Mars and search for permafrost. When this phase in our robotics mission is complete, look at what we will have done! We will have embarked on the search for clues to Martian ancient history and its present environment. We will have searched for signs of past water and geological activities. We will have looked for signs of minerals that will tell us the story of ancient water and life on Mars. We will send more landers in 2003 that will be part of a large international effort to place a network of landers on the surface of Mars. This includes a network of seismometers to understand the interior of Mars and a network of weather stations to understand Martian weather and climate. Imagine getting a weather report from the polar station on Mars on the 11 o'clock news every night.

We want to pave the way to go to Mars. After we finish scouting the global diversity of the Mars surface with small landers, we will be ready for a sample return mission. That could happen as early as 2005, maybe even as early as 2003. Also by then, we hope the Mars Surveyor program will involve Europe, Japan, France, Russia, Germany and other countries on planet Earth so we could all go to Mars together. A sample return is just the beginning. We will need ways to move around on the planet. We are looking at a range of things from micro-rovers like the Pathfinder to larger ones built with the Russians for larger area coverage. We are looking at super pressure balloons and specialized—I don't want to use the word "aircraft" because the air is different on Mars. What I mean is power flight instead of just a balloon. Another possibility is a hopper technique with rockets. We are also interested in micro-landers weighing only a few kilograms each and costing less than a million dollars each. They could be deployed on approach to Mars or from orbit. They could be sharp-pointed to penetrate the soil upon landing and could carry micro-cameras and put micro seismometers and weather stations on a chip on the surface of Mars. The possibilities are fascinating.

However we do this phase of exploration, one thing is certain. We should involve many more people than the scientists and engineers on the payroll. We should and will involve the American public. People will see on their televisions what our rovers see on the surface of Mars. People of all ages and backgrounds, all over the world, will fly with a balloon across the Martian surface. They will look out over the Martian landscape like they were in a jet liner flying over Earth. But better than that, we are going to be doing scientific analysis in high schools and colleges. Science will not be only for NASA employees and our contractors, but it will be for students around the world. I want to tell you these young kids want this. This is the beginning of the future. We are already doing it now in high schools around the country. They are helping us with Mission to Planet Earth in Inglewood, California.

The pinnacle of the Mars robotics exploration program will be sample return. We will do this all efficiently and cheaply. The Mars Pathfinder is about 9% of the cost of Viking. In today's dollars, Viking would be over 3 billion dollars. Pathfinder will cost 270 million dollars. The Mars Global Surveyor will cost about 250 million, not the nearly one billion dollars the Mars Observer cost. But we are not stopping there. The President's budget for 1996 proposed the New Millennium program. It is the object of this program, by the turn of the century, to make spacecraft 10 times cheaper, 10 times faster, 10 times better. We can do it!

Now, let's fast forward into the next century and see what robots could be doing in future times. The most exciting thing they could do is search for the areas where life or traces of it might have existed. These fossils could be hidden within certain geological formations. The robots will have to seek out the right outcrops, guided by the mapping done by earlier orbiters. They could also do things to prepare humans for landing on Mars. We could send a robot fuel production plant to Mars. We could send it as early as 2005. The international **Mars I** crew could use the fuel made on Mars to explore the planet. The fuel production plant sucks in Martian atmosphere and spews out rocket fuel. The fuel would be stored in tanks on the surface of Mars. We could also plant a radio beacon right next to the tanks for the astronauts to home in on as they approach Mars (to help land close to the fuel tanks). Robots could be doing many more things to prepare for the astronauts. They could be building a water and wastewater recycling station the size of a barn. They could begin to build the first infrastructure, the first cityscape on Mars. A convoy of robots could map Mars. They would drill in high probability areas to find water, metals, minerals and other resources to support humans and get ready to zero in on these locations. Let me say a word about the search for water. We already know we can obtain limited quantities of water on Mars. We can get it from the atmosphere or at the polar ice caps, but there may be—just maybe—aquifers below the surface with liquid water or ice.

The robots could also carry a mobile laboratory where they could deposit soil or rock samples for analysis. Slowly, the profile for the area where the robots are working would emerge. Computers would use the data to build a three-dimensional tomographic or thermal map of the subsurface of Mars. The robots would develop a picture of the subsurface structure—layers of rock, the presence of water or ice, the thickness of sediments. Through chemical and seismic and mineralogical analysis, they could determine what the crust is made of. Why are certain types of rocks and minerals present? And why are others that we find on Earth not there?

As we move toward launching the human mission in 2018, here is what we will have accomplished. Our scientific exploration will have done detailed surveys of sites we want humans to examine for subtle, hard-to-find evidence of life. We will have scoured the surface to find out where water and other resources are most accessible. When we have found the best sites and studied them with more robots, the world—and I emphasize the world—will send the first human beings to the planet Mars. We will also have figured out how human beings can live and work in space for long periods, and do many other functions. We will learn much of this aboard the international Space Station. In the process, we will benefit the people of Earth. That is part of the new-think at NASA.

Let's fast-forward a little further ahead. It is 2018, just moments before the **Mars I** takes off on its two-year mission. It will take them six or seven months to get to Mars. They will spend 30 days on the surface, and they will take 14 or 15 months to get back. Right now, they are on the launch pad. When they get to low-Earth orbit, they will transfer to the Mars vehicle. Once they are aboard, they will talk to Mission Control in Houston. They will also talk to Operation Centers in Kaliningrad, Bavaria and Tsukuba. Before they blast off, let's take a moment. Let's look at what it has taken to get them to this point.

Before we send people into space for several years, we will have had to develop affective countermeasures to micro- and partial gravity. This would have tremendous benefits for the elderly people suffering from diseases such as osteoporosis on Earth. We will also have developed pre-screening procedures to try to ensure a healthy crew. By the year 2018, that could have led to the ability to identify medical conditions long before the symptoms become overt, so preventative measures can be used. Think about the effects of this on Earth. Think about what it would mean if we could identify and counteract a predisposition for disease. Clearly, that will not solve the whole problem. But before going to Mars we need to do this work and learn.

What if our astronauts do get sick on their trip? They won't be able to take hospitals or doctors with them in huge quantities. So we will have had to develop chemical surgery techniques that can heal without scalpels or incisions. We could put micro-machines into their bodies. Doctors could manipulate them from the ground. Back in the 1990s, these little machines were only a concept in the minds of engineers and scientists at NASA and NIH. In 2018, they could monitor what is happening in the body. They could carry antibodies directly to a certain part of the body. They could go to where a problem is and then begin to fix it. That will have made health care on Earth more accessible, cheaper, and less intrusive than it has ever been. The same kind of vital sensors we inject under the skin of astronauts to monitor, diagnose and treat them from the ground could be used throughout the world. They could be used by surgeons to make prenatal corrections. They could be used to continually monitor the elderly. If something goes wrong, a doctor could be on the phone or another interactive system within minutes. People in rural areas in the United States and remote villages around the world could have access to low cost, high-quality health care. We could be diagnosed and treated in our homes. A hospital stay would be the exception.

And, of course, we will have a new rocket. It will be a single-stage, not a three-stage rocket. Our single-stage, reusable vehicle will open up the space frontier. NASA won't have developed it and won't run it—industry will. Together, industry, government and academia will be opening the space frontier. Entrepreneurs will be building industrial parks in space. The international Space Station won't be alone. It will be just one in a constellation, along with, perhaps, some stations on the surface of the Moon. Our spaceship will be self-diagnosing and self-repairing. Its systems will incorporate micro- and nano-electrical, electronic, and mechanical technology. It will use the most advanced expert decision-making systems to detect and report problems and do simple fixes. Our astronauts are taking along robotic assistants. They are capable of autonomous operation for exploration, science, and rescue tasks. They are controlled by advanced interfaces like voice recognition. They can see, think and hear.

By 2018, we will have done a much better job of integrating people and machines. We will have had to compensate for humans' limitations in absorbing and monitoring information. The Mars crew will need this kind of help because people on the ground won't be able to talk to them in real time. There is more than 10-minutes of time lapse in getting a message from Mars to Earth and back. We won't have thousands in Mission Control like we did on the Moon Shot. It will just be a few astronauts onboard that spacecraft, integrated with the expert decision-making machines. Our crew will have to be able to absorb complex information quickly and make instant decisions without communicating with Mission Control.

I have taken you from billions of years back in time to 2018, where **Mars I** sits on the launch pad, awaiting take-off. Let's take one more jump. Let's look at what the astronauts could be doing once they arrive on Mars. They have three basic missions to perform. One, they will do things connected with finding past life on Mars. Two, they will be doing things to support life in the present. And three, they will be doing things for life on Mars in the future.

Let's start with life in the past. Suppose that robots have located areas that appear highly probable for containing fossil cell life. The next step is for humans to confirm that revolutionary finding. A team of astronauts could explore the outcrop where the fossil was found. They would carefully select the most promising rocks to bring back to the base for study. Several of our astronauts could go down into an ancient, dried-out lake bed in the northern plateau to do a core drilling. They will drill around the edges and into the center then analyze the layers for clues to many things. They may find clues to what the atmosphere was like on Mars billions of years ago or more clues to the earliest forms of life.

The **Mars I** crew will also worry about life in the present. They will need to make a place on Mars where humans can survive and grow and do research. They will also look for Martian life that has survived from ancient times in a subsurface micro oasis. They will also look for water, more precious than gold. Thanks to the robotics precursor missions they will have a good idea where to start. They will use subsurface, ground-penetrating radar—kind of a high-tech divining rod. As scientists, they know that finding water would be important in understanding the past climate on Mars. As pioneers, they know it is critical to human survival. The astronauts could also build hothouses. Humans won't go to Mars alone. They will go as a community of life forms, bringing with them seeds and plants grown on the trip. The studies done at the turn of the century in gravitropism—the way plants react to gravity—were an important step in that process. NASA began thinking about that in the '90s. Eventually, we will develop plants that can grow out in the open. The astronauts also could bring along additional production plants to break down the Martian water into oxygen for breathing and hydrogen to use as fuel.

The third thing our crew could do is look toward life in the future. They will begin to assess whether local or broad-range terraforming is feasible—whether it is possible for human beings to colonize Mars. They might take an inventory of the planet, much like the U.S. Geological Survey did in the Old West in the late 1800s. The Viking Mission of 1976 literally only scratched the surface of Mars. There is a lot we don't know about Mars. So they could take inventory, building on what the robots have already done. Is all the water that once was on Mars still there, or did some of it or most of it escape into space? If most of it escaped, reconstructing a habitable planet is going to be much more difficult. If it is ice, or trapped in the ground like oil is trapped on Earth, it will be a lot easier. They could run studies and take measurement to try to determine whether people could live on Mars for long periods. Can people live in one-third gravity with no ill effects? There is no magnetic field on Mars—is that a barrier to sustaining human presence?

They will also forward the search for planets around other stars. Our Solar System contains four Earth-like planets—Venus, Earth, Mercury and Mars. By better understanding the evolution of Mars, we will get a better understanding of how planets in general form and evolve. This will help us in our search for solar systems around other stars. As we search for the nearest stars over the next 10-25 years, we may find another Earth-sized planet. We will be better able to know whether we have found another Venus, Earth or Mars because of what the **Mars I** astronauts find.

The next human crew to Mars will stay longer. They could stay for nearly two years. They will develop permanent habitats and grow a lot of their own food. They will bring "hopper rockets" that will let them get anywhere on the planet. Someday, there will be a colony on Mars. Humans will live and work on the Red Planet. Someday after that, humankind will hear a brand now sound. A newborn baby's thin wail of life—the fragile voice of the very first Martian.

Acknowledgment

Special thanks to the talented people who helped give me a vision for this fantasy trip to Mars. One of them is Chris McKay, who is now in the Arctic doing relative planetology experiments to better understand what happened on Mars and what it means to the evolution of our own planet. I also want to thank France Cordova, Harry Holloway, Wes Huntress, Jeff Plescia, John Kerridge, Michael Meyer, Bob Zubrin and Diane Ballard.

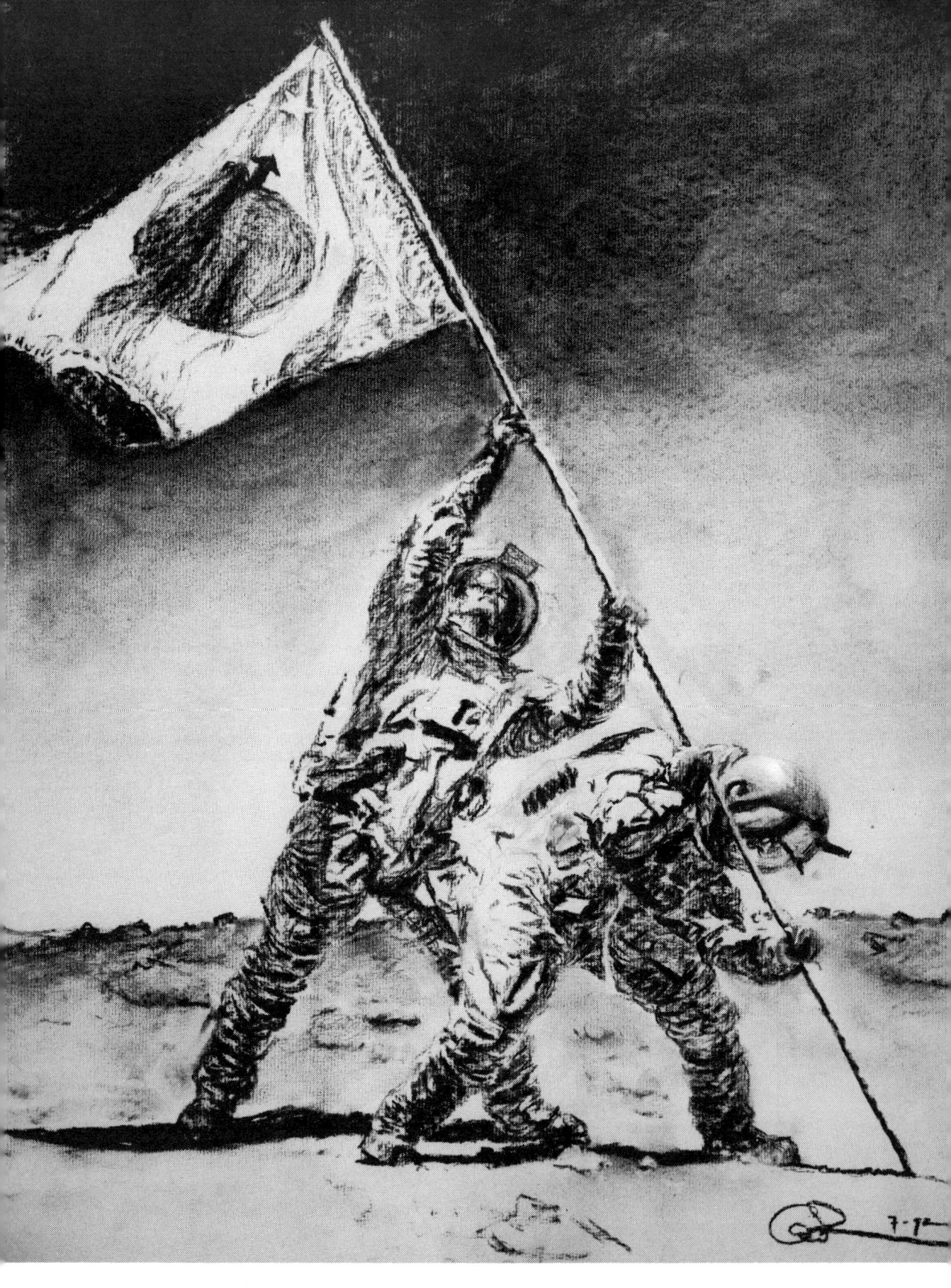

Contents

	Page
Fellow Traveler - Poem and Book Dedication to Thomas O. Paine	iv
Foreword	vii
Prologue. Steps to Mars Daniel S. Goldin	xi

SECTION I - MAKING THE CASE FOR MARS   1

Chapter 1. Why Should Humans Explore Space? (AAS 95-471)
    Lawrence G. Lemke 3

Chapter 2. The Significance of the Martian Frontier (AAS 95-472)
    Robert M. Zubrin 13

Chapter 3. The Millennium Project (AAS 95-473)
    Harrison H. Schmitt 27

Chapter 4. Mars: The Media. . . the Masses. . . and the Message (AAS 95-474)
    Leonard David 41

Chapter 5. Strategic Communications Planning and the Case for Mars (AAS 95-475)
    Frank White 51

Chapter 6. Managing the Exploration of the Moon and Mars (AAS 95-476)
    Michael D. Griffin 59

SECTION II - GETTING THERE: INTERPLANETARY TRANSPORTATION ISSUES   67

Chapter 7. Mars Mission Concepts: The von Braun Era (AAS 95-477)
    Frederick I. Ordway III 69

Chapter 8. Pathways to Mars: An Overview of Flight Profiles and Staging Options for Mars Missions (AAS 95-478)
    John C. Niehoff and Stephen J. Hoffman 99

Chapter 9. Mars Mission Designs: Comparing the Near Term Options (AAS 95-479)
    Malcolm A. LeCompt and Julie P. Stets 127

	Page
Chapter 10. Artificial Gravity: Design Implications for Mars Vehicles (AAS 95-480) Lawrence G. Lemke	153
Chapter 11. Nuclear Rockets: High-Performance Propulsion for Mars (AAS 95-481) Clayton W. Watson	167
Chapter 12. Nuclear Electric Propulsion for Human Mars Missions (AAS 95-482) Ernst Stuhlinger	193

SECTION III - LIVING IN SPACE: THE HUMAN ELEMENT 223

Chapter 13. Biomedical Issues in the Exploration of Mars (AAS 95-483)
Rosalind A. Grymes, Charles E. Wade and Joan Vernikos 225

Chapter 14. The Human Side of Mars Flight: A Review of Human Factors Issues (AAS 95-484)
Mary M. Connors and Albert A. Harrison 241

Chapter 15. From the Great Voyages of Exploration to Missions to Mars (AAS 95-485)
Ben Finny 267

Chapter 16. The Interplanetary Radiation Environment and Methods to Shield from it (AAS 95-486)
Lawrence W. Townsend and John W. Wilson 283

SECTION IV - BEING THERE: LIVING AND WORKING ON MARS 325

Chapter 17. Moving in on Mars: The Hitchhiker's Guide to Martian Life Support (AAS 95-487)
Penelope J. Boston 327

Chapter 18. Living in Space: Results from Biosphere 2's Initial Closure, an Early Testbed for Closed Ecological Systems on Mars (AAS 95-488)
Mark Nelson and William F. Dempster 363

Chapter 19. Using the Resources of Mars for Human Settlement (AAS 95-489)
Thomas R. Meyer and Christopher P. McKay 393

Chapter 20. Mars Rovers (AAS 95-490)
Benton C. Clark 445

Chapter 21. First Mars Outpost Habitation Strategy (AAS 95-491)
Marc M. Cohen 465

SECTION V - SCIENCE ON MARS — 513

Chapter 22. Scientific Objectives of Human Exploration of Mars (AAS 95-492)
    Michael H. Carr 515

Chapter 23. Science Strategy for Human Exploration of Mars (AAS 95-493)
    Carol R. Stoker 537

SECTION VI - COSTS AND BENEFITS OF MARS EXPLORATION — 561

Chapter 24. The Cost of Sending Humans to Mars (AAS 95-494)
    Humbolt C. Mandell, Jr. 563

Chapter 25. Mars Colonization: Technically Feasible, Affordable, and a Universal Human Drive (AAS 95-495)
    Thomas O. Paine 579

EPILOGUE — 593

Chapter 26. Beyond Mars... Into the Universe at Large (AAS 95-496)
    Leonard David 595

APPENDIX — 599

Publications of the American Astronautical Society 601

INDEX — 615

Numerical Index 617

Author Index 619

Section I

Making the Case for Mars

The Apollo program is the hero's journey.

Chapter 1

WHY SHOULD HUMANS EXPLORE SPACE ?

Lawrence. G. Lemke[*]

The argument over the reason for humans-in-space and, by extension, how much the nation should spend on it, is as old as the program itself. Here, it is argued that the humans-in-space program is properly a transformational, rather than a utilitarian activity. Observations from the perspectives of history, psychology, and cultural anthropology lead to the conclusion that humans-in-space is a drama in which the heroes act out the plot of humankind seeking other homes for life. This conclusion raises the central dilemma for NASA and, indeed, the nation; should the federal government be in the cultural transformation business at all? Can it not be? It is assumed that there will continue to be a human space flight program and there will continue to be controversy about it. Tactically, the human space flight program can maximize its chances for success by having the drama appear to be as compelling as possible. Since the human space flight program will be judged aesthetically, as a piece of dramatic art, it should be constructed as such. The plot of seeking other homes for life is best portrayed on Mars, which should therefore be the centerpiece. For the foreseeable future, other plot elements such as space stations and the Moon may play supporting roles. An essential first step is to make human operations in Earth orbit much less heroic. Fortunately, a concrete step in this direction (reducing dependence on the Shuttle) can be taken while decreasing, not increasing, short-term net expenditures for humans-in-space.

INTRODUCTION

"Why have humans in space?" has, of course, been a perennial question ever since the nation started spending significant amounts of money on it:

"As I look back upon these two years of involvement in this exciting activity, I find myself wishing that we could have been operating in support of more clearly understood and nationally accepted goals or purposes... How can we decide how important it is to spend, on an urgent basis, the very large sums of money required to put a man into orbit or to explore the atmosphere and surface of Mars or Venus unless we have a pretty firm grasp of what the purpose behind the whole space effort really is?

---

[*] Chief, Advanced Space Projects Branch, NASA Ames Research Center, Moffett Field, California 94087.

And yet, who knows the answers to this and many similar questions today? Who is thinking about them and doing something about developing some answers?"

— T. Keith Glennan, first NASA Administrator

Thus opens the 1964 book, *Pride and Power*, by Vernon Van Dyke. This is one of the first scholarly attempts to examine the rationale for the then nascent space program, from the political science perspective. It is a reasonable point of departure in considering the question today because it illustrates several important points.

The first point is that this is not a new issue for NASA administrators and others with a professional interest in the space program. The Apollo years are sometimes looked upon from within the space program as the fabled era in which there was a pure unanimity of societal approval for the humans-in-space movement. The date of publication of *Pride and Power* shows that this is not true. It was written after president Kennedy had committed the nation to put humans on the Moon, but before the nation had achieved that goal.

The second point is that it is relatively straightforward to answer the question—at least in an intellectual sense—through diligent application of the techniques and insights of the various relevant disciplines, as professor Van Dyke has illustrated by example. Unfortunately, the question doesn't seem to stay answered.

Nevertheless, I believe, as was stated by author Frank White in the 1989 TV documentary program *Space Workers*, ". . . that a society can not persist in an activity and not know why it is doing it." For those of us who want society to persist in space exploration, it is therefore important to seek to continually reanswer the question.

WHAT IS HUMANS-IN-SPACE NOT ABOUT?

Humans-in-space is not about doing science and technology, as conventionally defined (meaning, to deliver the largest number of citations in refereed journals or the largest number of patents, per program dollar). This is why the non-aerospace professional science and technology community consistently opposes or refuses to support large human space flight projects whenever they are proposed (e.g., Apollo, Shuttle, Station, SEI).

Humans-in-space is not about making money, as conventionally defined (meaning, to deliver a reasonably predictable rate of return on investment within a foreseeable period of time). This is why humans-in-space is not yet a private-sector activity.

Humans-in-space is not about national defense, as conventionally defined (meaning, to provide a valuable war-fighting capability that can't be provided more cheaply or reliably by alternative means). This is why the U.S. military can not agree on a compelling mission for humans in space, even after multiple attempts.

Finally, humans-in-space is no longer about politically defeating a doctrinal adversary on the international stage, as in the Moon race. Our primary doctrinal adversary has conceded defeat and seeks to become a partner.

In a public address at the *Pathway to the Planets* conference sponsored by NASA's Office of Exploration in 1989, Mark Albrecht, the then executive secretary of the National Space Council, summed up this line of reasoning very succinctly when he said, "Humans in space does not emerge from the bottom up as the most cost-effective solution to any problem."

Goal-directed human activities can be logically divided into two categories, utilitarian and transformational. Utilitarian activities, such as providing food, clothing, and shelter, are intended to maintain existence within a *status quo*. Transformational activities, such as art, religion, and politics, are intended to create a new *status quo*. Humans-in-space certainly appears to be goal-directed, and we have just asserted that it is not utilitarian. What kind of transformational activity is it?

THE POLITICAL SCIENCE VIEW

Most citizens believe that government programs are rightly judged by how faithfully they perform according to the internal logic of their ostensible goals. For example, a transportation program should be primarily judged by how well it builds roads, airports, etc., and an education program should be judged by how well it educates students.

Many citizens who live inside the Washington D.C. beltway have a contrary view. They believe that the most important attribute of a government program is which politicians win or lose and which agendas are strengthened or weakened by its existence. In this view, all government programs are really about politics, regardless of their ostensible goals.

Professor John Logsdon, of George Washington University, for example, is an able and thoughtful proponent of this view, and in personal conversations has forcefully argued that the *sine qua non* of any program—including the space program—is that it must be structured to provide sufficient political benefit to its advocates so that it will contribute to their survival. Nevertheless, he goes on to argue, in his essay, *A Sustainable Rationale for Manned Space Flight*, that political viability is a necessary, but not a sufficient condition for existence of a public program. He suggests that, "The search for a convincing rationale for manned space flight ultimately leads to perhaps the most ambitious conceivable objective: permanent movement of some humans off of this planet to other locations in the solar system. . . . Preparing to leave Earth is an appropriate rationale for human activities in space, if the broad sweep of history, rather than the fiscal and political constraints of the time, are taken into account."

THE HISTORICAL VIEW

To remind ourselves what a program is about, it is always illuminating to revisit the body of thought that existed before people got paid for it. This body of thought was perhaps best evolved in the give-and-take of the amateur space flight clubs that came into existence in various nations during the 1930's.

In his classic, *Rockets, Missiles, and Space Travel* (1951 edition), Willy Ley writes primarily from the perspective of the German *Verein für Raumschiffart*, or VfR (society for space flight), and makes it clear that humans-in-space is about an idea—the idea that there are other homes for life than Earth. This idea has two components, and is therefore slightly more inclusive than the one quoted from professor Logsdon, above. One is indeed the notion that Earth life—and humans, in particular—can go to other places to live, and the other component is the notion that alien life might exist other places and can interact with Earth life. Note that the operable term is "idea," since both of these component notions are essentially without existence proofs, even today, and thus require an active suspension of disbelief from their adherents.

Willy Ley shows how the evolution of this idea paralleled in works of fiction the growth of scientific knowledge about the cosmos. Whenever new facts were discovered about potential "other places" (i.e., moons, planets), human imagination wove tales about what it would be like to transport Earthly life to those destinations, or conversely, what it would be like to interact with alien life from those places.

An important fact about these fictional tales is that they are distributed across all cultures and all times. The first such tale recorded in western literature is *True History*, by Lucian of Samosata, Greece, written in the year 80 BC. It purports to be a previously "lost" or "missing" story of Odysseus' voyage to the Moon. By AD 1634, no less a figure than Johannes Kepler wrote again on the theme of human travel to the Moon, this time attempting to make the fiction more scientific by invoking a primitive notion of gravity to explain the transport of characters to and from the Moon.

And so on, to the present time. In 1988, students of the inaugural class of the International Space University took it upon themselves to write a notional *Philosophical Preamble for Space Development* as part of a lunar base design study, drawing only upon the thoughts and feelings they brought with them from their native cultures. They wrote, in part,

> "The purpose of this preamble is to describe our philosophical perspective and to provide a framework of principles regarding why and how we should pursue space activities. Because we have faith in the human spirit, in our future and in our survival, we believe that space settlement is inevitable. Our first dream of going into space was our first step towards it. . . . By both creating livable habitats and physically adapting to the new environments we encounter, the exploration of space offers us the possibility of a cosmic immortality of life. . . .," etc.

The most important feature of this preamble is the list of signatories, which contains names from every continent on Earth, and many from countries that don't even have space programs.

Not everyone is captivated by the idea of other homes for life, but history shows that those who are, are widely distributed across time, geography, and culture. The only tenable explanation is that this theme is tapping into deep unconscious motivations that seem to be stable on the time scale of centuries. Thus, one answer to the question of what humans-in-space is about is, "It's what it's always been about—other homes for life." We in the U.S. space program are simply the most recent custodians of this idea. In fact, the major elements of the U.S. human space flight program (which accounts for the majority of civil space dollars expended) are an attempt to follow the "classic agenda" that was first enunciated by the amateur rocket societies beginning in the 1930's and discussed more recently by Michael Michaud in *Reaching for the High Frontier: the American Pro-space movement, 1972-84*.

THE CULTURAL-ANTHROPOLOGICAL VIEW

A comparative ethnologist would not have difficulty identifying what kind of activity human space flight is; it is a ritual, presided over by its shamans, astronauts and cosmonauts.

In 1949, Joseph Campbell, published his seminal book, *The Hero With a Thousand Faces*. This book represented more than a decade's worth of work in exploring and interpreting the great belief systems of the world's cultures—in Europe, Asia, Africa, the Americas, and Polynesia. The title derives from his discovery that all belief systems contain stories, either oral or written, that instruct individuals about the nature of physical reality, the relation of the individual to society, and the relation of the individual to the metaphysical. Campbell's first contribution was to realize and then document the fact that all these stories are variations of a single structure (the "monomyth").

At the center of the monomyth is the hero's journey, which has among its essential elements the call to adventure, the crossing of the threshold into a realm of personal rebirth, the mastery of two worlds, the peak experience, and the return to the normal world with a prize. Jason and the Argonauts is a mythic adventure with this structure, so are the legends of King Arthur, so are the *Star Wars* films, and so was the Apollo program (see frontispiece for this chapter).

Another of Joseph Campbell's contributions was to recognize that belief systems, like other complex phenomena, have a developmental history. Their evolution is driven by how well or poorly they address the important concerns of the societies in which they must exist. Thus, when humans lived as hunter-gatherers, belief systems were concerned with elemental forces and animistic themes. After the rise of cities, literature, and empires, belief systems became more appropriately concerned with law, conflict, and religious salvation.

In the 1971 book, *Myths To Live By*, Joseph Campbell discusses the impact of science on myth and the significance of the Apollo Moon walk. His thesis is that humanity is now at a developmental stage where traditional belief systems are no longer functional and a new one must emerge. According to Campbell, the new and powerful factors that traditional belief systems inadequately address are: modern empirical knowledge about the origin, processes, and destiny of the cosmos, the power of applied science and technology to transform the human experience within the lifetime of the individual, recognition of the finite nature of the Earth, and the emergence of a global community. I use the shorthand statement that the new belief system will be about growing up as a species, and assert that space exploration is the most concentrated and potent naturally occurring mixture of these themes.

To summarize: Humankind (especially the part that is most intimately affected by science and technology, the finite boundaries of the Earth, and global interconnectedness) needs to work out a new belief system because the traditional ones are becoming progressively irrelevant. The importance of a relevant belief system is what it has always been—to explain the natural world, to make lawful and systematic the relationship of individuals to society, and to provide meaning to life by connecting the individual to something larger. Space exploration is the single best metaphor that encompasses all the critical elements of growing up as a species.

Human space flight is the passion play that can motivate participation in this effort. Humans in space are heroes, in the anthropological sense. As photons are the quanta that carry an electromagnetic field, heroes are the quanta that carry a mythic field. They are the actors that make the belief system accessible to the 99+% of the population that can not be primary intellectual contributors. This function is even more important now than in the past because a modern belief system will necessarily be more complex, and therefore inaccessible, than previous ones. In short, human space flight is the heroic tale appropriate to the mythology of our time.

HOW MUCH SHOULD WE SPEND ON HUMANS-IN-SPACE?

The argument up to this point is that putting humans in space is a kind of theater about an idea that is central to an emerging belief system. Obviously, this poses a dilemma for NASA; the problem is not in discovering what humans-in-space is about, the problem is in deciding how open and honest to be about it. On the one hand, if NASA does not say what humans-in-space is really about, it has to either say nothing at all in defense of the centerpiece of its program, or make up preposterous lies, such as that the purpose of a space station is to produce perfect ball bearings. Either way, the resulting program will drift into progressive incoherence as entropy overtakes it and renders it eventually insupportable by even its most cynical political patrons. I assert that this is a reasonable description of the human space flight program since the end of Apollo.

On the other hand, NASA can be more open about what it is really up to and risk outright extermination, which nearly happened when vice president Spiro Agnew's Space Task Group made its recommendations in 1969. There are several inherent qualities of the real humans-in-space agenda that obviously render it vulnerable to attack. First, it is idealistic, meaning that there is no quantifiable relationship between dollars in and value out. Second, it is altruistic, meaning that the presumed benefits, intangible as they are, will not necessarily accrue to the same individuals who pay for the program. Third, it is quasi-religious, which can be opposed both because people subscribe to other belief systems, or because of our nation's principle of separation of church and state.

This last point, I believe, is the crux of the issue, the main point of this essay, and a real dilemma at the national level. Is it appropriate for the U.S. government to be in the idea and belief business at all? On the one hand, the principle of preventing our belief system from being manipulated by our political system by separating church and state has proven to be profoundly beneficial in its overall effect on the nation and is irrevocably rooted in our culture, in any event. Moreover, even if our philosophy did not oppose it, federal bureaucrats would not necessarily be the instruments of choice for propagation of a belief system, simply on the grounds of efficiency and effectiveness.

On the other hand, the idea that organizations, up to and including nations, need a shared vision to maintain health and survival is also self-evident. Moreover, if it is true that the modern belief system is to be about growing up as a species, this is a theme that ultimately must be addressed at least at the national level, and the federal government is the only instrument this nation has for addressing issues at such a level. So, the federal government does not formally have the charter to be concerned about the nation's belief system, but it also has to be if it is truly being responsible about cultural health. And, we also have the evidence that the last time the government did act strongly along these lines (in the Apollo program) the nation felt very positive about it. Like any true dilemma, this one does not have a clean and final solution, it will simply have an ebb and flow of controversy for the foreseeable future.

Therefore, it would be inconsistent to propose a closed-form solution to the question posed by Mr. Glennan, in the first paragraph. The nation should spend on space exploration whatever it feels comfortable with. That amount will vary depending on the perceived quality of the program and the overall state of the nation at any particular time. How much public candor the program can stand changes from time to time and is a function of the individual personalities involved, as much as anything else. However, the one factor that will always maximize the probability of public acceptance is to have an obviously well-crafted program. Fortunately, because it is fundamentally heroic, the humans-in-space program can always speak most eloquently through its deeds.

If humans-in-space is a play about an idea, then it follows that it should be judged on aesthetic grounds as a work of art. I suggest that is essentially what the public does (perhaps not consciously) and is what NASA should do in planning the program. Modern media constantly bombards the public with professionally crafted bits of theater, whether for the purpose of entertainment, the selling of goods and services, or the promotion of a political agenda. This has had the effect of setting the standards for the production of images that hope to compete for attention in the public consciousness. Humans-in-space must be a dynamic, compelling story written by and for those who can suspend disbelief in its premise and are intellectually capable of understanding its subtlety, not by and for those who merely profit from it economically. It also must be produced with high standards so that it remains true to the concept.

What should the plot be? For the foreseeable future, the organizing goal should be exploring Mars as another home for life, past, present, or future. The reason is that it has the best total combination of thematic elements. Recall that the idea of other homes for life has two subcomponents, transporting Earth life to alien locations, and alien life encountering Earth life. It is certainly possible to transport Earth life to the Moon or to space stations, but Mars is the only credible location that encompasses both subcomponents, since it has a non-trivial possibility of having once harbored life and of being terraformed in the future. The path to Mars can certainly be made to go through an orbital station or the Moon, but it should be clear to all what the ultimate objective is and why. Also, Mars is the most distant attainable objective, and therefore the most heroic. The stage of heroic action should be set as far out as seems feasibly attainable.

The corollary to this is that human actions at intermediate locations should actually be, and should be seen to be, progressively less heroic. This is, of course, opposite to the direction the humans-in-space program has been going for the last 20 years with the Shuttle program. It follows that this trend needs to be reversed. The Shuttle was publicly sold on the basis of utilitarianism (which turned out to be fraudulent) but was in fact desired and is maintained as a stage for the humans-in-space play. The proper objection to the Shuttle is not that it is a stage for the humans-in-space play, but that it is such a *poor* stage, because the range of drama it allows is so small compared to the Apollo drama of more than 20 years ago. Undoubtedly the Shuttle program seemed like a good political bargain in 1969 (compared to having no human space flight program at all), but in fact it has become a block to further progress. It is neither transformational enough (compared to Apollo launches, which were of approximately the same magnitude) to be justified on that basis nor utilitarian enough to be justified on that basis. As the only available example of humans-in-space in this country; when it extrapolated to Moon-Mars exploration, it makes that program look prohibitively expensive.

Probably the biggest and most immediate challenge that must be faced by the humans-in-space program is making human operations in Earth orbit much less expensive and much more utilitarian so that the focus can once again move outward. We will know we have succeeded when going into Earth orbit no longer qualifies one to become NASA Administrator. This challenge is not primarily technical, it is political, economic, and psychological, and can probably be accomplished without a net budget increase. Without decisive movement in this direction, the play will likely have no next act.

REFERENCES

Campbell, J., *The Hero With a Thousand Faces*, Princeton, NJ, Princeton University Press, 1949.

Campbell, J., *Myths to Live By*, New York, NY, Bantam Books, 1973.

Ley, W., *Rockets, Missiles, and Space Travel*, New York, NY, Viking Press, 1954.

Logsdon, J., "A sustainable rationale for manned space flight," *Space Policy*, 5, February, 1989.

Michaud, M.A.G., *Reaching for the High Frontier: The American Pro-Space Movement, 1972-84*, New York, NY, Praeger, 1986.

Van Dyke, V., *Pride and Power*, Urbana, IL, University of Illinois Press, 1964.

White, F., *The Overview Effect*, Boston, MA, Houghton Mifflin, 1987.

Crew habitat aerobraking over Vallis Marineris.

AAS 95-472

Chapter 2

THE SIGNIFICANCE OF THE MARTIAN FRONTIER

Robert Zubrin[*]

As the great historian Frederick Jackson Turner pointed out 100 years ago, Western humanist civilization, as we know and value it today, was born in expansion, grew in expansion, and can only exist in a dynamic expanding state. While some form of human society might persist in a non-expanding world, that society will not feature freedom, creativity, individuality, or progress, and placing no value on those aspects of humanity that differentiate us from animals it will place no value on human rights or human life as well. Such a dismal future might seem an outrageous prediction, except for the fact that for nearly all of its history most of humanity has been forced to endure static modes of social organization, and the experience has not been a happy one. Free societies are the exception in human history, they have only existed during the four centuries of frontier expansion of the West. That history is now over, the frontier that was opened by the voyage of Christopher Columbus is now closed. If the era of Western humanist society is not to be seen by future historians as some kind of transitory golden age, a brief shining moment in an otherwise endless chronicle of human misery, then a new frontier must be opened.

An open frontier on Mars will allow for the preservation of cultural diversity which must vanish within the single global society that is being created on Earth. The necessities of life on Mars will create a strong driver for technological progress that will produce a flood of innovations that will upset any tendency towards technological stagnation on the mother planet. The labor shortage that will exist on Mars will function in much the same way as the labor shortage did in 19th century America; driving not only technological but social innovation, increasing pay and public education, and in every way setting a new standard for a higher form of humanist civilization. Martian settlers, building new cities, defining new laws and customs, and ultimately transforming their planet will know, and prove to all outside observers, that human beings are the makers of their world, and not merely its inhabitants. By doing so they will reaffirm in the most powerful way possible the humanist notion of the dignity and value of mankind.

---

[*]   Martin Marietta Astronautics Division, P.O. Box 179, Denver, Colorado 80201.

INTRODUCTION: THE AMERICAN FRONTIER

It was 100 years ago, 1893, at the annual conference of the American Historical Association, that a young professor of history from the then relatively obscure University of Wisconsin got up to speak. Frederick Jackson Turner's talk was scheduled as the last one in the evening session, preceded by a series of excruciatingly boring papers on topics so obscure that kindness forbids even the reprinting of their titles. Nevertheless, for some unexplained reason, the majority of the conference participants stayed up to hear him. Perhaps somehow a rumor had gotten afoot that something important was about to be said, if so it was correct, for in one bold sweep of brilliant insight Turner laid bare the source of the American soul. It was not legal theories, precedents, traditions, national or racial stock that was the source of the egalitarian democracy, individualism, and spirit of innovation that characterize America, it was the existence of the frontier.

". . . to the frontier the American intellect owes its striking characteristics. That coarseness of strength combined with acuteness and inquisitiveness; that practical, inventive turn of mind, quick to find expedients; that masterful grasp of material things, lacking in the artistic but powerful to effect great ends; that restless, nervous energy; that dominant individualism, working for good and evil, and withal that buoyancy and exuberance that comes from freedom - these are the traits of the frontier, or traits called out elsewhere because of the existence of the frontier. Since the days when the fleets of Columbus sailed into the waters of the New World, America has been another name for opportunity, and the people of the United States have taken their tone from the incessant expansion which has not only been open but has even been forced upon them. He would be a rash prophet who should assert that the expansive character of American life has now entirely ceased. Movement has been its dominant fact, and, unless this training has no effect upon a people, the American energy will continually demand a wider field for its exercise. But never again will such gifts of free lands offer themselves. For a moment, at the frontier, the bonds of custom are broken and unrestraint is triumphant. There is not tabula rasa. The stubborn American environment is there with its imperious summons to accept its conditions; the inherited ways of doing things are also there; and yet, in spite of the environment, and in spite of custom, each frontier did indeed furnish a new opportunity, a gate of escape from the bondage of the past; and freshness, and confidence, and scorn of older society, impatience of its restraints and its ideas, and indifference to its lessons, have accompanied the frontier. What the Mediterranean Sea was to the Greeks, breaking the bonds of custom, offering new experiences, calling out new institutions and activities, that, and more, the ever retreating frontier has been to the United States directly, and to the nations of Europe more remotely. And now, four centuries from the discovery of America, at the end of a hundred years of life under the Constitution, the frontier has gone, and with its going has closed the first period of American history."

Turner was unstoppable. America's greatest leaders were all men of the frontier—Washington, Jefferson, Jackson, and Lincoln, he said, and the great struggles of American history have all hinged ultimately upon the fate of frontier. The attempt by the

British crown to close the frontier drove the revolution. The Civil War began in the frontier territories and it was the fight for the future of the frontier, not abstract issues of State's Rights or morality, that sent tens of thousands marching into battle at Shiloh and Gettysburg. Most importantly, Turner showed how the very character of Americans, their philosophical outlook, and their society are all based upon the frontier. The frontier creates a perpetual labor shortage in the settled areas, which drives up wages and thus technological innovation. With people in short supply, each one is valued more preciously, putting a premium on popular education and elevating the general estimate of the human dignity of the common man. So long as the frontier exists, the factory worker back East always has another option in the West, and even if he does not choose to exercise it, he has to be treated with the respect due to someone who can quit—who works by choice, not duress. So long as the frontier is open and new fortunes can be made, the establishment of a closed aristocracy is impossible.

The Turner thesis was a bombshell, which within a few years created an entire school of historians who proceeded to demonstrate that not only American culture, but the entire worldwide Western progressive humanist civilization that America has generally represented in its most distilled form was the result of the Great Frontier of global settlement opened to Europe by the Age of Exploration. It was the Great Frontier that shattered the static, stultifying, irrational, dogmatic, and completely stratified world of medieval Christendom, unchaining thought, hope, and imagination to revolutionize the world.

Then the question arises: with the end of the frontier, what happens to America and all it has stood for? Can a free, egalitarian, democratic, innovating society with a can-do spirit be preserved in the absence of room to grow? Maybe the question was premature in Turner's time, after all, even with the vanishing of the line of settlement, most of the country was still empty. In any case, a popular culture based on 400 years of frontier individualism does not die instantly, and the children of America's last generation of pioneers could take America through World War II and on to the Moon. But what of now? What do we see around us now but an ever more apparent loss of vigor of American society, increasing fixity of the power structure and bureaucratization of all levels of society, impotence of political institutions to carry off great projects, the cancerous proliferation of regulations affecting all aspects of public, private and commercial life, the spread of irrationalism, the banalization of popular culture, the loss of willingness by individuals to take risks, to fend for themselves or think for themselves, economic stagnation and decline, the deceleration of the rate of technological innovation and a loss of belief in the idea of progress itself. Everywhere you look, the writing is on the wall. Without a frontier from which to breathe life, the spirit that gave rise to the progressive humanistic culture that America for the past several centuries has offered to the world is fading. Once again, the issue is not just one of national loss. Human progress needs a vanguard, and no replacement is in sight.

The creation of a new frontier thus presents itself as America's and humanity's greatest social need. Nothing is more important because, apply what palliatives you will, without a frontier to grow in, not only American society, but the entire global civilization based upon Western enlightenment values of humanism, reason, science, and progress will ultimately die.

MARS: A NEW FRONTIER

I believe that humanity's new frontier can only be on Mars. Why is this the case? Why for example can it not be on Earth, on or under the oceans, or perhaps in such remote regions as Antarctica? And if it must be in space, why on Mars? Why not on the Moon or in artificial satellites in orbit about the Earth?

It is true that settlements on or under the sea or in Antarctica are entirely possible, and their establishment and access would be much easier than that of Martian colonies. Nevertheless, the fact of the matter is that, at this point in history, such terrestrial developments cannot meet an essential requirement for a frontier, to wit, they are insufficiently remote to allow for the free development of a new society. Put simply, in this day and age, with modern terrestrial communication and transportation systems, anywhere on Earth the cops are too close. If people are to have the dignity that comes with making their own world, they must be free of the old.

Why then, not the Moon? The answer is because there's not enough there. True, the Moon has a copious supply of most metals and oxygen, in the form of oxidized rock, and a fair supply of solar energy, but that's about it. For all intents and purposes, the Moon has no hydrogen, nitrogen, or carbon (they are present in the Lunar soil in parts per million quantities, somewhat like gold in sea water—if there were concrete on the Moon, Lunar colonists would mine it for water), and these are three of the four elements most necessary for life. You could bring seeds to the Moon and grow plants in enclosed greenhouses there, but nearly every atom of carbon, nitrogen, and hydrogen that goes into making those plants would have to be imported from another planet. While sustaining a Lunar scientific base under such conditions is relatively straightforward, growing a civilization there would be impossible. The difficulties supporting significant populations in artificial orbiting space colonies would be even greater.

Mars has what it takes. It's far enough away to free its colonists from intellectual, legal, or cultural domination by the old world, and rich enough in resources to give birth to a new. The Red Planet may appear at first glance to be a desert, but beneath its sands are oceans of water in the form of permafrost, enough in fact, if it were melted and Mars' terrain were smoothed out, to cover the entire planet with an ocean several hundred meters deep. Mars' atmosphere is mostly carbon-dioxide, providing enormous supplies of the two most important biological elements in a chemical form from which they can be directly taken up and incorporated into plant life. Mars has nitrogen too, both as a minority constituent (3%) in its atmosphere and probably as nitrate beds in its soil as well. For the rest, all the metals, silicon, sulfur, phosphorus, inert gases, and other raw materials needed to create not only life but an advanced technological civilization, can readily be found on Mars.

The United States has, *today*, all the technology needed to send humans to Mars. If a "travel light and live off the land" strategy were adopted, then the first human exploration mission could be launched within ten years at a cost less than 20% of NASA's existing budget. Once humans have reached Mars, bases could rapidly be established to support not only exploration, but experimentation to develop the broad range of civil, agricultural, chemical and industrial engineering techniques required to turn the raw materials of Mars into food, propellant, ceramics, plastics, metals, wires, structures, habitats, etc. As these techniques are mastered, Mars will become capable of supporting an ever increasing population, with an expanding division of labor, capable of mounting engineering efforts on an exponentially increasing scale. Once the production infrastructure is in place, populating Mars will not be a problem—under current medical conditions an immigration rate of 100 people per year would produce population growth on Mars in the 21st century comparable to that which occurred in colonial America in the 17th century. Within a century, an engineering capability could be created on Mars with the capability to literally transform the planet, if not to a fully Earth-like environment at least to the warm, wet conditions of Mars' primitive past, making a desert world into a new home for a new spectrum of descendants of terrestrial life.

Mars can be settled, and the fact that Mars can be thus settled and altered defines it as the New World that can create the basis for a positive future for *terrestrial* humanity for the next several centuries.

WHY HUMANITY NEEDS MARS

"We hold these truths to be self evident, that all men are created equal, and endowed by their creator with certain inalienable rights, among them life, liberty, and the pursuit of happiness. . ."
Declaration of Independence, 1776

"Everything has tended to regenerate them; new laws, a new mode of living, a new social system; here they are become men."
**Jean de Crevecoeur,
"Letters from an American Farmer," 1782**

To see best why 21st century humanity will desperately need an open frontier on Mars, we need to look at modern Western humanist culture and see what in it makes it so much more desirable a mode of society than anything that has ever existed before. Then we need to see how everything we hold dear will be wiped out if the frontier remains closed.

The essence of humanist society is that human beings are valued, that human life and human rights are held precious beyond price. Such notions have been for several thousand years the core philosophical values of Western civilization, dating back to the Greeks and the Judeo-Christian ideas of the divine nature of the human spirit. Yet they could never be implemented as a practical basis for the organization of society until the great explorers of the age of discovery threw open a New World in which the dormant seed of medieval Christendom could grow and blossom forth into something the likes of which the world had never seen before; something so wonderful that for 400 years millions of men and women all over the world have abandoned everything they had, traveled thousands of miles, braving incredible dangers and hardships to make themselves parts of it, and millions of others have conspired and fought, often against tremendous odds, to bring it to their homelands.

The problem with Christendom was that it was *fixed*, it was a play for which the script had been written and the leading roles both chosen and assigned. The problem was not that there were insufficient natural resources to go around—medieval Europe was not heavily populated, there were plenty of forests and other wild areas—the problem was that all the resources were *owned*. A ruling class had been selected and a set of ruling institutions, ideas and customs had been selected, and by the law of "Survival of the Firstest," none of these could be displaced. Furthermore, not only the leading roles had been chosen, but also those of the supporting cast and chorus, and there were only so many such parts to go around. If you wanted to keep your part, you had to keep your place, and there was no place for someone without a place.

The New World changed all that by supplying a place in which there were no established ruling institutions, an improvisational theater big enough to welcome all comers with no parts assigned. On such a stage, the players are not limited to the conventional role of actors, they become playwrights and directors as well. The unleashing of creative talent that such a novel situation allows is not only a great deal of fun for those lucky enough to be involved, it changes the view of the spectators as to the capabilities of actors in general. People who had no role in the old society could define their role in the new. People who did not "fit in" in the old world could discover and demonstrate that, far from being worthless, they were invaluable in the new, *whether they went there or not.*

The New World destroyed the basis of aristocracy and created the basis of democracy, it allowed the development of diversity by allowing escape from those institutions that were imposing uniformity, it destroyed a closed intellectual world by importing unsanctioned data and experience, it allowed progress by escaping the hold of those institutions whose continued rule required continued stagnation, and it drove progress by defining a situation in which innovation to maximize the capabilities of the limited population available was desperately needed. It raised the dignity of man by raising the price of labor and by demonstrating for all to see that human beings can be the creators of their world, and not just its inhabitants.

Now consider the probable fate of humanity in the 21st century under two conditions, with a Martian frontier and without it.

In the 21st century, without a Martian frontier, there is no question that human diversity will decline severely. Already, in the late 20th century, advanced communication and transportation technologies have been eroding the healthy diversity of human cultures on Earth, and this tendency can only accelerate in the 21st. On the other hand, if the Martian frontier is opened, then this same process of technological advance will also enable us to establish a new branch of human culture on Mars and eventually worlds beyond. The precious diversity of humanity can thus be preserved on a broader field, but only on a broader field. One world will be just too small a domain to allow the preservation of the diversity that is needed not just to keep life interesting, but to assure the survival of the human race.

Without the opening of a new frontier on Mars, continued Western civilization faces the risk of technological stagnation. To some this may appear to be an outrageous statement, as the present age is frequently cited as one of technological wonders. In fact, however, the rate of progress within our society has been decreasing, and at an alarming rate. To see this, it is only necessary to step back and compare the changes which have occurred in the past 30 years with those that occurred in the 30 years preceding and the 30 years before that. Between 1903 and 1933 the world was revolutionized; cities were electrified, telephones and broadcast radio became common, talking motion pictures appeared, automobiles became practical, and aviation progressed from the Wright Flyer to the DC-3 and Hawker Hurricane. Between 1933 and 1963 the world changed again, with the introduction of color television, communication satellites and interplanetary spacecraft, computers, antibiotics, SCUBA gear, nuclear power, Atlas, Titan, and Saturn rockets, Boeing 727's and SR-71's. Compared to these changes, the technological innovations from 1963 to the present are insignificant. Immense changes should have occurred during this period, but did not. Had we been following the previous 60 years technological trajectory, we today would have videotelephones, solar powered cars, maglev trains, fusion reactors, hypersonic intercontinental travel and regular passenger transportation to orbit, undersea cities, open-sea mariculture, and human settlements on the Moon and Mars. Even more indicative of technological decadence than the nonappearance of these innovations, is the fact that a fundamental advance in technology in an area basic to the total process of production that was already emerging in 1963, namely nuclear power, has been blocked in its implementation by political forces dedicated to preserving the technological status quo, in the process raising technophobia to the status of a fashionable political and philosophical creed.

It is important to understand this. The widespread introduction of commercial nuclear power in the Western world was not stopped by the small groups that duel with the industry in public hearings and courtrooms. Whatever one might think of the pluses and minuses of nuclear power, the fact remains that the anti-nuclear activists have only been allowed to have their way with commercial nuclear industry because the world's dominant financial institutions currently hold the mortgages on literally trillions of dollars worth of coal, oil and gas reserves, all of which would be severely devalued should a replacement source of energy come on line. Such investments have caused these financial institutions and their governmental allies to develop a preference for stagnation in energy technology that is extremely difficult to overcome. Indeed, nuclear technology is only supported by the powerful in the Western world today for the decisive military applications of nuclear weapons and nuclear submarines—in the case of deployment of such instruments no advice from Sierra Club is requested. Analogous paradigms hold true in other important areas of the economy, so that practically the only areas where notable technological progress is occurring currently is in products such as home computers that do not compete directly with previously well-established industries. As the interlocking of terrestrial institutions of political and economic power becomes ever more intimate and incestuous in the 21st century, this trend towards technological stagnation can only deepen.

Unless of course, there is an alternative uncontrolled domain that drives progress from the outside, and this is what the Martian frontier will provide. Consider a nascent Martian civilization: its future will depend critically upon the progress of science and technology to which the colonists will therefore enthusiastically contribute. Thus just as the inventions of produced by the "Yankee Ingenuity" of frontier America were a powerful driving force on world-wide human progress in the 19th century, so the "Martian Ingenuity" born in a culture that puts the utmost premium on intelligence, practical education, and the determination required to make real contributions will make much more than its fair share of the scientific and technological breakthroughs that will dramatically advance the human condition in the 21st. A prime example of where this is likely to occur is energy production. Mars does have one major energy resource that we do currently know about; deuterium, which can be used as the fuel in nearly waste-free thermonuclear fusion reactors. Earth has large amounts of deuterium too, but with all of its existing investments in other, more polluting, forms of energy production, the research that would make possible practical fusion power reactors has been allowed to stagnate. The Martian colonists are certain to be much more determined to get fusion on-line, and in doing so will massively benefit the mother planet as well. Fusion power will also lead to fusion propulsion, making possible spaceships that will carry hundreds of passengers and thousands of tons of payload rapidly back and forth between Earth and Mars, thus accelerating the rate of colonization and opening up the possibility of emigration to Mars to more and more people. Not only would such technology cause travel times between Earth and Mars to shrink from months to weeks, but travel times to the outer solar system would be reduced from years to months, and even voyages to the stars could become possible on a time scale of decades instead of millennia. Thus by acting as a driver on technology, the Martian frontier can become a gateway to the practically infinite hinterland that lies beyond.

The parallel between the Martian frontier and that of 19th century America as technology drivers is, if anything, vastly understated. America drove technological progress in the last century because its Western frontier created a perpetual labor shortage in the east, thus forcing the development of labor-saving machinery and providing a strong incentive for improvement of public education so that the skills of the limited labor force available could be maximized. This condition no longer holds true in America, in fact far from prizing each additional citizen, immigrants are no longer welcome here and a vast "service sector" of bureaucrats and menials has been created to absorb the energies of the majority of the population which is excluded from the productive parts of the economy. Thus in the late 20th century, and increasingly in the 21st, each additional citizen is and will be regarded as a burden. On 21st century Mars, on the other hand, conditions of labor shortage will apply with a vengeance. Indeed, it can be safely said that no commodity on 21st century Mars will be more precious, more highly valued, and more dearly paid for than human labor time. Pay rates will be higher on Mars, workers will be treated better, and public education will be driven much harder than ever was the case on Earth. Just as the example of 19th century America changed the way the common man was regarded and treated in Europe, so the impact of progressive Martian social conditions will not only be felt on Mars. Put simply, a new standard will be set for a higher form of humanist civilization on Mars, and viewing it from afar, the citizens of Earth will rightly demand nothing less for themselves.

The frontier drove the development of democracy in America by creating a self-reliant population which insisted on the right to self-government. It is doubtful that democracy can persist without such people. True, the trappings of democracy exist in abundance in America today, but meaningful public participation in the process has all but disappeared. No representative of a new political party has been elected president of the U.S. since 1860, neighborhood political clubs and ward structures that allowed citizen participation in party deliberations are gone, as are the camp meetings and torchlight election parades. With a re-election rate of 95%, the U.S. Congress is about as susceptible to the people's will as the British House of Lords, and regardless of the will of Congress, the real laws, covering ever broader areas of economic and social life, are increasingly being made by a plethora of regulatory agencies whose officials do not even pretend to have been elected by anyone. Judges are still elected in many places, but the elections generally feature little public involvement, so that rather than representing any concept of justice as understood by the public, the judicial system has come to function largely as an autonomous legal caste. Clearly, if it is not to continue its ongoing degeneration into sham, democracy in America and elsewhere in Western civilization needs a shot in the arm. That boost can only come from the example of a frontier people whose civilization incorporates the ethos that breathed the spirit into democracy in America in the first place. As Americans showed Europe in the last century, so in the next the Martians can show us the way away from oligarchy.

There are greater threats that a humanist society faces in a closed world than the return of oligarchy, and if the frontier remains closed in the 21st century we are certain to face them. These threats are the spread of various sorts of anti-human ideologies and the development of political institutions that incorporate the notions that spring from them as a basis of operation. At the top of the list of such pathological ideas which tend to spread naturally in a closed society is the Malthus theory, which holds that since the world's resources are more or less fixed, population growth must be restricted or all of us will descend into bottomless misery. Malthusianism is scientifically bankrupt and all predictions made upon it have been wrong, because human beings are not mere consumers of resources. Rather we create resources by the development of new technologies that find use for them. The more people, the faster the rate of innovation, and this is why contrary to Malthus, as the world's population has increased, the standard of living has increased, and at an accelerating rate. Nevertheless, in a closed society Malthusianism has the appearance of self-evident truth, and herein lies the danger. Because if the idea is accepted that the world's resources are fixed, then each person is ultimately the enemy of every other person, and each race or nation is the enemy of every other race or nation. The inevitable result is the creation of tyrannical regimes to restrict population growth, such as that now prevailing in China, or worse, the development of Nazi style genocidal governments as various populations become convinced that their vital self interest requires the elimination of those other races that are allegedly competing with them for the world's finite resources. Only in a universe of unlimited resources can all men be brothers.

It is not enough to argue against Malthusianism in the abstract, such debates are not settled in academic journals. Unless people can *see* broad vistas of unused resources in front of them, the belief in limited resources tends to follow as a matter of course. Unless the frontier is re-opened, the probability is high that humanity will create hell for itself in the 21st century.

Is the world that humans live in changeable or is it fixed? Are we the makers of our world or just its inhabitants? In a society that is growing into a frontier, the creative role of humans is self-evident, and the dignity of man is raised accordingly. Nineteenth century Americans, building cities, draining swamps, and digging canals could have no doubt as to humanities role as improvers of creation. Today much of what they saw as progress is cited by many as environmental destruction. Despite abundant scientific evidence that evolution is intrinsic to nature, a belief is spreading that nature as it is at the moment is sacrosanct, and that humans should not have the right to change it. An open frontier on Mars would not merely restore the 19th century American humanist views in such matters, it would raise it to unprecedented heights, because in the process of terraforming Mars we will not merely be taming a wild world, but bringing a dead one to life. What greater affirmation of the positive nature of the human creative spirit could there be?

THE NEVER ENDING RENAISSANCE

"We have come recently to boast of a global economy without thinking of its implications, of how unfortunate we are in finding it. It would be more cheering if news should come that by some freak of the solar system another world had swung gently into our orbit and moved so close that a bridge could be built over which people could pass to new continents untenanted and new seas uncharted. Would those eager immigrants repeat the process they followed when they had that opportunity, or would they redress the grievances of the old Earth by a new bill of rights. . .? The availability of such a new planet, at any rate, would prolong, if it did not save, a civilization based on dynamism, and in the prolongation the individual would again enjoy a spell of freedom. . . .

"It would be very interesting to speculate on what the human imagination is going to do with a frontierless world where it must seek its inspiration in uniformity rather than variety, in sameness rather than contrast, in safety rather than peril, in probing the harmless nuances of the known rather than the thundering uncertainties of unknown seas or continents. The dreamers, the poets, and the philosophers are after all but instruments which make vocal and articulate the hopes and aspirations and the fears of a people.

"The people are going to miss the frontier more than words can express. For four centuries they heard its call, listened to its promises, and bet their lives and fortunes on its outcome. It calls no more. . . ."

Walter Prescott Webb, "The Great Frontier," 1951.

Western humanist civilization as we know and value it today was born in expansion, grew in expansion, and can only exist in a dynamic expanding state. While some form of human society might persist in a non-expanding world, that society will not feature freedom, creativity, individuality, or progress, and placing no value on those aspects of humanity that differentiate us from animals it will place no value on human rights or human life as well. Such a dismal future might seem an outrageous prediction, except for the fact that for nearly all of its history most of humanity has been forced to endure static modes of social organization, and the experience has not been a happy one. Free societies are the exception in human history, they have only existed during the four centuries of frontier expansion of the West. That history is now over, the frontier that was opened by the voyage of Christopher Columbus is now closed. If the era of Western humanist society is not to be seen by future historians as some kind of transitory golden age, a brief shining moment in an otherwise endless chronicle of human misery, then a new frontier must be opened. Mars beckons.

But Mars is only one planet, and with humanity's power over nature rising exponentially as they would in an age of progress that an open Martian frontier portends, the job of transforming and settling it is unlikely to occupy our energies for more than three or four centuries. Does the settling of Mars then simply represent an opportunity to "prolong but not save a civilization based upon dynamism?" Isn't it the case that humanist civilization is ultimately doomed anyway? I think not. The universe is vast; its resources, if we can access them, are truly infinite. During the four centuries of the open frontier on Earth, science and technology have advanced at an astonishing pace. The technological capabilities achieved during the 20th century would dwarf the expectations of any observer from the 19th, the dreams of one from the 18th, and seem outright magical to someone from the 17th. The nearest stars are incredibly distant, about 100,000 times as far away as Mars; yet Mars itself is about 100,000 times as far from Earth as America is from Europe. If the past four centuries of progress have multiplied our reach by so great a ratio, might not four more centuries of freedom do the same again? There is ample reason to believe that they would. Terraforming Mars will drive the development of new and more powerful sources of energy; settling the Red Planet will drive the development of ever faster modes of space transportation. Both of these capabilities, in turn, will open up new frontiers ever deeper into the outer solar system, and the harder challenges posed by these new environments will drive the two key technologies of power and propulsion ever more forcefully. The key thing is not to let the process stop, for if it is allowed to stop for any length of time society will crystallize into a static form that is inimical to the resumption of progress. That is what defines the present age as one of crisis. Our old frontier is closed, the first signs of social crystallization are clearly visible. Yet, progress, while slowing, is still extant; our people still believe in it and our ruling institutions are not yet incompatible with it. We still possess the greatest gift of the inheritance of a four hundred year long Renaissance, to wit, the capacity to initiate another by opening the Martian frontier. If we fail to do so, our culture will not have that capacity long. Mars is harsh, the people who settle it will need not only technology, but the scientific outlook, creativity, and free-thinking individualistic inventiveness that stand behind it. Mars will not allow itself to be settled by people from a static society; those people won't have what it takes. We still do. Mars today waits for the children of the old frontier, but Mars will not wait forever.

Like an aircraft moving down a runway, Western civilization used the freedom afforded by the open frontier to accelerate itself to takeoff speed. The end of the runway has now been reached. If our journey is to continue, we must now take courage and fly.

Sun-facing crater walls are used to lay out solar cells.

Chapter 3

THE MILLENNIUM PROJECT

Harrison H. Schmitt[*]

Humankind's project for the Third Millennium could be the establishment of a permanent outpost on Mars by the year 2010. The primary and ultimate outcome of this outpost would be human settlement. An important related goal could be development of helium-3 fusion, using lunar helium-3, as an environmentally acceptable replacement for fossil fuels on Earth.

The project's first Mars expedition could be on its way by the end of the first decade of the Third Millennium. A true Martian settlement could be functional by 2015, but a permanently occupied outpost, resupplied with lunar derived consumables, would be possible within a few years of the establishment of commercial helium-3 production at a permanent lunar base.

INTERMARS, a management concept for participant based international space cooperation in this Millennium Project, satisfies all the established constraints of space law and is consistent with the principles of free enterprise. From the beginning of INTERMARS, a clear mechanism should be created by which permanent settlers can be represented in its organizational entities and can have majority control of INTERMARS at an appropriate level of population.

A commitment to the Millennium Project would place humankind on the threshold of the next great adventure, the settlement of Mars. The Project can be for the children of the world what Space Shuttles have been to their parents and what Apollo represented to their grandparents: the embodiment of the best in the human spirit.

THE CHALLENGE

Humankind sought and attained greatness with the first explorations of the Moon between 1967 and 1973. During these momentous years, our species also took its first clear steps of evolution into the solar system. Eventually those steps lead into the galaxy. Clearly now, as the Pueblo Indians of America relate the lesson of their ancestors, "We walk on the Earth, but we live in the sky."

---

[*] P.O. Box 14338, Albuquerque, New Mexico 87191.

How should we react to this new, uncertain, but undeniably exciting direction into the future? What next step would provide both benefits to humankind on Earth and a logical progression for our species in space?

Doing nothing does not appear to be an acceptable alternative. Standing still would court disaster in the face of mounting international challenges posed by population growth, threats to the global environment, and the loss of generations of young people to poverty, disease, poor education, drugs, and crime.

One commitment might be made to a far reaching project focused on the turn of the Third Millennium. This rare, motivating milestone looms only a few years away, and the technological and psychological potential has grown for a Millennium Project that will match both society's needs to advance and its capabilities to achieve.

Such a Millennium Project could be the establishment of a permanent human outpost on Mars by the year 2010. The primary and ultimate outcome of this outpost might well be the human settlement of that planet [Schmitt, 1978 and 1985].

An equally important related goal for this Millennium Project could be the provision of an environmentally and economically acceptable replacement for fossil fuels on Earth. Easily extracted solar wind volatiles, trapped in Lunar soil, include a light isotope of helium—helium-3 (^{3}He). Future fusion power plants, fueled by lunar helium-3, could provide essentially unlimited as well as environmentally acceptable electrical power for increasing billions of human beings on Earth [Kulcinski and Schmitt, 1987]. This goal requires the coordinated development of both terrestrial fusion technology and the technology to deliver lunar helium-3 to Earth.

Although Mars ultimately should supply all or most of the materials necessary to support its inhabitants, by-products of the production of lunar helium-3 could provide the lowest cost source of consumables, including hydrogen, oxygen, and water, critical to sustaining the early Martian settlers. Food produced on the Moon also may be imported in the early years on Mars.

Thus, with a vision of a outpost of human civilization on Mars and access to lunar resources, humankind stands on the threshold of the next great adventure, the settlement of Mars. Education, environment, technology, science, civilization, and evolution all will benefit from this adventure as will the purely human desire to "be there" when great events unfold.

The benefits for education will be the rejuvenation of the motivation, teaching incentives, technological foundations, and learning choices necessary to revitalize the knowledge and skills of young people.

The benefits for environmental preservation will be the development of abundant fusion fuel as an acceptable global substitute for fossil fuels.

The benefits to technology will be the creation of a vastly enhanced reservoir of know-how as humans meet and overcome new and unknown technical and personal challenges. The humanitarian as well as the economic value of this reservoir will be as immeasurable as that which resulted from the discoveries of Columbus, the opportunities of the Louisiana Purchase, or the challenges of Apollo.

The benefits to science will come from the detailed search for fossilized life forms that may have evolved by three and one half billion years ago in the oceans of early Mars, contemporaneously with early single-celled life already known as fossils on Earth. If found on Mars, such fossils will give the answer to one of the deepest of philosophical questions, that is, life on Earth is not just some cosmic accident.

The benefits for human civilization will be the perpetuation of the institutions of human freedom. The challenge of new frontiers in space should once again reinforce and enhance these institutions.

Possibly most significantly, the benefits to human evolution and species survival will be the establishment of self-sufficient enclaves on the first stepping stones into the universe. These planetary harbors can lead to permanence as a species whether or not life exists by cosmic accident or by cosmic design.

THE MILLENNIUM PROJECT

Philosophical and psychological momentum for returning to deep space and specifically for going to Mars continues to build among the young people of the Earth. Contacts with students and teachers during the last two decades and the enthusiasm greeting the formation of an International Space University [Schmitt, 1975, and Hawley, 1991] confirm this conclusion. In fact, rather than thinking much about going back to the Moon, many young people now have their eyes only on Mars. Ultimately, they or people like them will go to Mars. Like most of our ancestors, they will never be satisfied with either the comforts or the restrictions of home and Earth. Among them will be found the parents of the first Martians.

New technological developments required for early Martian settlement consist largely of those related to long duration interplanetary travel and habitation. The three most needed categories of new technology involve continuously accelerating and decelerating interplanetary rocket engines, space and planetary facilities of indefinite maintainability, and countermeasures against any adverse physical effects of space adaptation and re-adaptation (see The Human Factor, below).

Other technologies lie closer at hand. For example, heavy lift launch systems constitute essential elements for implementation of the Millennium Project, however, the technology for such systems exists. We lack only a decision to proceed with their development or appropriate access and testing, as in the case of the Russian Energia rocket. Basic spacecraft technologies, including those for interplanetary craft, landers, and rovers, also require only detailed consideration of the consequences of solar particle events, prolonged flight and weightlessness, and the Martian environment.

Development of interplanetary propulsion systems, on the other hand, must have both a high priority and an open mind. Comparisons of several potential candidates should be made, including those based on chemical, fission, fusion, and solar energy systems. Tradeoff studies should consider cost, reliability, transit time, and schedule. In addition, such studies should look at potential synergism between each approach and other technical and operational issues, such as electrical power generation, radiation protection, space adaptation, and the terrestrial dual use of the technology.

An operating Earth orbit space station has a clear and unique role to play in the technological and scientific maturation of the Millennium Project and should be made available on this rationale alone. First, the design and operation of a space station should be constrained by the need to learn how to build, operate, and maintain long duration facilities in space. In such a station, for example, issues of maintainability take precedence over issues of redundancy. Second, as discussed below (The Human Factor), the space station permits the development of both the scientific and the operational basis for long term occupational medicine in space. A space station can serve many other roles, but these two make it a contributing element of the Millennium Project.

With appropriate technologies and equipment available, implementation of the Millennium Project could proceed. Well within an annual budget allocation of one percent of the Gross National Product of the United States, the Project's first Mars expedition could be on its way by the end of the first decade of the Third Millennium if not sooner. A true Martian settlement could be functional by 2015, but a permanently occupied outpost, resupplied with lunar derived consumables, would be possible within a year or two of the establishment of commercial helium-3 production at a permanent lunar settlement. With the experience of Apollo already behind us, a permanent lunar settlement could be in place before the turn of the Millennium.

[Neal et al., 1989] describe a Mars exploration mission strategy in which three or four human precursor expeditions would evaluate potential sites for a permanent base. The crews of the first two or three of such missions would each land in, explore, and evaluate areas considered favorable candidates for a permanent settlement or areas where scientific data could be gathered that would support outpost selection decisions. Crews and sensors in orbit during surface activities would conduct a detailed reconnaissance of the rest of the Martian surface and of its moons, Phobos and Deimos. Upon departure of each lander from the surface, rovers would be configured for remotely controlled exploration and sampling operations between the area just explored and one of the next sites to be evaluated.

Using the findings of earlier missions, the third or fourth expedition would land at the site selected for the first permanent outpost. The payload would include the capability to begin the emplacement of the basic infrastructure for eventual settlement. From that point on, the first outpost site could be permanently occupied with actual settlers as soon as sufficient stores of consumables existed.

Each early expedition should leave the vicinity of the Earth about every two years, depending on available propulsion technology, and stay at Mars about eighteen months. This initial frequency of flights has many advantages in developing and maintaining an Earth-Moon-Mars servicing infrastructure as well as satisfying the emotional urges of a generation of Earthlings who want to "press on."

Orbital reconnaissance of potential landing sites as well as of the Martian moons could be performed by orbiting crew members on all the early expeditions. However, deep Martian valleys in regions that show strong photogeologic evidence of being underlain by water rich permafrost [Carr *et al.*, 1984] make likely locations for the first permanent outpost.

Ideally, crew members left in orbit during each mission's initial surface activities would rotate down to the surface for the examination of a second site. Not only would such rotation be psychologically and physiologically appropriate for crews who have flown all the way to Mars, but the technical capability for rotation would add important margins for success to each mission. Rotation of crews also would permit the evaluation of two potential outpost sites on each precursor expedition, and the timetable for a permanent outpost could be moved forward.

A great deal of autonomy in decision making and data evaluation will lie with Mars expedition crews due to communication time delays and occasional total lack of communications between those crews and Earth. For the most part, the Mission Control Center will be in Martian orbit with the crew assisted by sophisticated data processing and data fusion systems. Mission operations on Earth will be relegated to the role of long term planning and analysis rather than real-time control.

Dust will be the most challenging technical aspect of equipment design and in living and working on Mars. Unlike the Earth's Moon, which has no atmosphere, fine dust on Mars blows around in great global storms and settles very slowly. These storms may require the temporary confinement of the explorers and eventual settlers to shelters much like during winter at year-round Antarctic bases.

Chemical and agricultural characteristics of Martian soil constitute one of the major planning uncertainties for the Millennium Project. Provisions to evaluate and protect against any adverse properties will be necessary. Like the lunar soil and new volcanic ash on Earth, however, Martian soils probably will be very fertile. In fact, particles of clay as well as impact and volcanic glass may be important soil constituents. However, the existence of a Martian atmosphere, its oxidizing nature, and the possible presence of sulfur in local soil crusts may mean that some treatment of the soil will be required before it can be farmed productively. An alternative method for growing food could be hydroponics, depending on the availability of sufficient quantities of water. Imported food from the Moon may be required before any settlement reaches agricultural self-sufficiency.

Much of the excitement felt for Mars arises from the scientific insights we expect from exploration and study of Martian rocks and soils [Carr *et al.*, 1984, and Reiber, 1988]. Indeed, one of the major criteria for selection of the site of the first outpost will be its value as a scientific base as well as suitability for permanent habitation.

Scientific interest in Mars stems in great degree from insights gained from exploration on the Moon [Schmitt, 1991]. The extraordinary intensity of the impact history of the first 500 million years of lunar evolution strongly suggests a similar history for the Earth and Mars [Schmitt, 1975]. The presence of clay minerals and the continuous supply of energy and complex hydrocarbon molecules from debris falling into the Earth's warm atmosphere and seas during this period may have been critical to the beginning of terrestrial life. The crystal structures of clay minerals may have been the basis for the creation of the first very complex organic chemicals which could "evolve" and "reproduce" either on their clay templates [Cairns-Smith, 1985] or as independent complex organic compounds.

Whether clay structures were involved or not, the 3.5 billion year old bacterial fossils in Australia and Africa [Schopf, 1983, and Walsh and Lowe, 1985], and significantly more highly evolved life forms found in the younger but still ancient rocks at Spitzbergen [Knoll, 1991], suggest the possibility of very early, albeit temporary, organic evolution on some of the terrestrial planets besides the Earth.

On Mars, the process of terrestrial biological evolution may have been arrested in various phases as the Martian surface environment became too cold and the atmosphere too thin to support liquid water. Thus, the possibility of fossil evidence of early life on Mars adds extraordinary excitement to future paleontological exploration of its layered rock sequences.

In spite of immediate emotional and scientific interest in Mars, the fortuitous existence of the Moon and its resources so near the Earth, but still outside Earth's gravity well, give a significant opportunity to accelerate the timetable for implementing the Millennium Project. Establishment of a permanent settlement on the Moon, based on the commercial production of helium-3 for use as a fusion energy fuel on Earth [Kearny, 1989, and Schmitt, 1993], fully supports the desire to live on Mars as soon as possible.

First of all, most of the technology needed for the creation of a permanent lunar settlement, with a resources production economy, will support the technological requirements for establishing a Martian outpost. The compatible technologies include systems for heavy lift launch vehicles, surface habitats and mobility, resource production, regular and routine work outside habitats, and new concepts in equipment automation, reliability, longevity, and maintainability.

Second, the direct and indirect by-products of helium-3 production from lunar surface materials will provide a ready source of necessary consumables for Martian inhabitants prior to the creation of their own consumables industry. These lunar produced consumables include hydrogen, oxygen, water, nitrogen and carbon compounds, and food.

Third, the integration of space development with the advancement of helium-3 fusion technology on Earth provides the foundation for efficient Earth-Mars and Moon-Mars propulsion systems based on that same fusion technology. Decreased transit time to and from Mars represents a desirable long term objective for the Millennium Project. A fusion based propulsion technology also opens up the possibility of travel to near-by stars.

General technical assurance in discussing the establishment of a permanent outpost on Mars as a Millennium Project comes from four areas of experience: the confidence and knowledge gained from the Apollo expeditions to the Moon, the spectacular and detailed data returned about Mars by the Viking landers and orbiters, the empirical success of long term space flight by humans and systems in the Soviet, Russian, and U.S. space programs, and the historical success of human beings that undertook difficult and stressful exploration and settlement on Earth. In fact, there exists as much or more knowledge about Mars as a planet and as an exploration objective, and about our own ability to go there, than existed about the Moon and space travel before Armstrong and Aldrin landed at Tranquillity Base in July of 1969.

THE HUMAN FACTOR

Many of the major debates about exploration of space exist almost entirely because of the desire for humans to participate directly in that exploration. The initiation of the Space Age saw little questioning of what then seemed to be an essential presence of humans as explorers and national symbols at the space frontiers. With success, however, came the luxury of doubts. There have been increasingly detailed and complex arguments concerning the relative "cost-benefits" of human versus automated space activities as if they are mutually exclusive. In addition, there have been arguments concerning the nature of space exploration relative to "human needs" on Earth as if, again, they exist as totally separate and distinct issues.

In the debate on the value of humans in space, both sides may have lost sight of the essential fact that human beings have yet to be, and probably will never be, emotionally satisfied with an indirect and passive role in the making of history. The mass of humanity identifies with the physical presence of individuals at the frontiers of human endeavors. This personal identification has been communicated on a worldwide basis since the flight of Lindbergh.

For the most part, people do not identify with "scientific results" or with robots no matter how exciting the consequences may be to scientists and engineers. The excitement, motivation, and dedication generated by Project Apollo was not the consequence of the promise of science. It was, rather, the consequence of watching people strive to surpass themselves under conditions that inherently carry a high level of risk, unknowns, and competition. We go to surprising lengths at times to deny or to quantify this seemingly natural psychological drive.

Once on the way to Mars, in orbit around it, and on its surface, human beings will do what they do better than robots anywhere—analyze situations on the spot, predict consequences, devise alternative strategies, select from those strategies, test the results, and revise plans and actions as necessary based on experience and intuition. In the resolution of the unexpected opportunity, condition, or problem, humans clearly demonstrate superiority over the most advanced artificially intelligent robots. This superiority may be lost some day, but that day has not appeared on the horizon as yet.

In a real sense, humans exhibit instant and automatic reprogramming in the qualitative response to stimuli, such as those represented by emergencies, observations, and opportunities. The ability of men and women to improvise in the face of such stimuli will be as essential to the success of a complex enterprise like The Millennium Project as will be the automation of repetitive and quantitative tasks.

If human beings constitute essential operational and psychological elements of The Millennium Project, it is appropriate to ask, "Can they survive and function effectively in the environment of space for the time required for travel to and from Mars?" The empirical answer appears to be, "yes," however, neither the American nor the Soviet and Russian space programs to date have provided the rigorous scientific proof of this answer.

Ignorance rather than knowledge best describes the current state of investigations concerning human adaptation and readaptation to various space environments [Schmitt, 1990]. On the one hand, obvious physiological adaptive responses to changes in the body's sensory, cardiopulmonary, and autonomic nervous systems take place when human beings become weightless. As far as has been shown to date, these responses reverse upon re-exposure to a gravitational field, requiring lengths of time roughly proportional to a modest fraction of the time exposed to weightlessness.

Obvious neurological symptoms in the first stage of space adaptation, such as headaches and nausea, vary greatly in severity from individual to individual, but the process reaches completion within three to four days in all but a very few cases. No correlation has yet been found between the severity of individual neurological symptoms in space and similar symptoms associated with terrestrial motion sickness.

The more subtle effects of full multisensory conflict, autoregulatory dysfunction, and hemodynamic alteration appear to stabilize after several weeks of exposure to weightlessness. Bone mineral balance, muscular conditioning, and long-term vestibular response comprise the important known exceptions, all of which appear to have shown continuous adverse change during Soviet space flights of durations approaching those necessary for missions to Mars [Soviet Technical Presentations, 1986].

The basic causative mechanisms of space adaptation symptoms remain unknown. Vestibular agitation (head motion) aggravates the severity of neurological symptoms in the few days that the first stage of adaptation persists, but such agitation does not prevent initial neurological adaptation. The many potential countermeasures against space adaptation symptoms, unfortunately, have not been systematically evaluated or compared. Further, the possibility of artificially induced readaptation during weightlessness or accelerated readaptation on a planetary surface has not been investigated adequately.

Obviously, many physiological and psychological unknowns remain relative to long duration missions to Mars and to subsequent activities on the Martian surface. The Space Shuttle, Space Station *Freedom*, and Mir, however, have great potential for systematic clinical studies and clinical tests of countermeasures.

The development of a systematic clinical characterization protocol, the use of that protocol on an adequately sized cadre of physician-observer test subjects, and the repeated re-flight of that cadre to investigate countermeasures would seem to be the only truly scientific means to gain both understanding and believable medical intervention regimes. The application of this protocol could begin with full adaptation to slightly elevated rotational g levels on Earth, followed by full readaptation to one g. Pairs from the physician cadre then could be flown on each of the relatively short Space Shuttle flights and re-flown for longer durations on Space Station *Freedom* or Mir.

Without this approach, very conservative design parameters and much delay will be imposed on the early years of the Millennium Project. Although the empirical Soviet and Russian experience with long duration flight suggests that no insurmountable physiological problems exist, neither they nor we have committed fully to gathering the data necessary to optimize missions to Mars. These data are essential if we expect to provide appropriate space occupational medical services and accelerate the time when the settlement of space away from Earth can begin.

MANAGEMENT OF THE VISION

People throughout the world want space to be a frontier for human cooperation as well as a frontier of freedom and achievement. Unfortunately, narrow political designs as well as the legitimate national interests of nations make broad cooperation in any endeavor very difficult.

Cooperation, however, is not impossible. The success of the Apollo-Soyuz mission in 1974 and various Soviet-French efforts shows the possibility of successful joint endeavors involving international adversaries if objectives remain relatively limited and well defined. Spacelab, developed by the European Space Agency and flown successfully by the United States on the Space Shuttle, and the Mir space station program of the Soviet Union, demonstrate the potential for close cooperation between diverse nations. Most importantly, the INTELSAT and INMARSAT telecommunication organizations, provide examples where nations of all levels of economic development and all varieties of political persuasion have found it in their self interest to work cooperatively on space related projects.

Broadly based international cooperation in space ultimately requires a commitment to the rule of a body of space law. Although not perfect and certainly not complete, currently recognized and internationally sanctioned tenets of space law provide a workable foundation for future cooperation [Joyner and Schmitt, 1984, and Bilder *et al.*, 1989]. The world, however, must be very cautious about agreeing to any legal framework for space that either limits rational free enterprise activities or allows one-nation one-vote control.

The "Moon Treaty" and "Law of the Sea Convention" developed by the United Nations illustrate the dangers of ill-conceived international frameworks incorporating both one nation-one vote control and unworkable applications of the otherwise acceptable notion that the Moon and the sea constitute the "common heritage of mankind." Fortunately, few developed nations have ratified these documents, nor should they be ratified without major amendment.

Another concept for participant based international space cooperation incorporates the approaches suggested both in "INTERLUNE" [Joyner and Schmitt, 1984, and Bilder *et al.*, 1989] and in "INTERMARS" [Schmitt, 1986]. These concepts attempt to satisfy all the established constraints of space law as well as be consistent with the principles of free enterprise. Most importantly, a concept like INTERMARS would bring into the management of the Millennium Project those nations and other interests with the greatest motivations for insuring its successful implementation.

INTERMARS derives from the international culture of the space age and from the recognition that space resources make up a common heritage of the spaceship Earth. The INTERMARS concept does not require that territorial sovereignty be given up in space, nor does it require that free enterprise opportunities be abandoned in space. It merely requires that sovereignty and opportunity be shared.

The theoretical advantages of a participant based international organization, such as INTERMARS, will only be realized if the actual institutional structure provides both an equitable system for various interests to exert influence and control and for efficient and proper management of the outpost. Two distinct mechanisms exist for nations, users, and investors to be involved in INTERMARS. The first relates to the creation and operation of a permanent Martian outpost and would attract those nations that contribute directly and substantively to the activities required to establish and stabilize initial operations. The second mechanism relates to the use and the terms and conditions for use of the outpost, its accessible resources, and the proprietary technologies required to establish it. This second mechanism would attract those nations, users, and investors who contract with or invest in INTERMARS in order to benefit from its activities [see Schmitt, 1992, for organizational details of the similar INTERLUNE concept].

The concept of establishing a permanent Martian base includes the high probability that such a base would ultimately become a settlement of permanent residents. As our history on Earth indicates, permanent residents will eventually desire a controlling voice in the governing of their activities. We should take this possibility into account in the initial structure of INTERMARS so as to avoid the conflicts that have plagued colonial establishments in the past. From the beginning of INTERMARS, a clear mechanism should be created by which permanent settlers can be represented in its organizational entities and can have majority control of INTERMARS at an appropriate level of population.

Once a reality, and once clearly successful, INTERMARS would attract many of those nations that may at first be reluctant to participate. Although conceived as an international self regulating monopoly, INTERMARS should always be open to new members and investors. Thus, INTERMARS could achieve its broad humanistic goals as well as its technical and economic purposes as the implementing arm of The Millennium Project.

CONCLUSION

Dreams and visions set humankind apart among life on Earth. Deeply ingrained speculations about space life and space travel have permeated the consciousness of people for thousands of years and, indeed, probably have their roots in early human fascination with the Sun, Moon, and heavens. As a result, there developed a worldwide, personal identification by hundreds of millions of people with the excitement and promise of the Apollo explorations on the Moon.

In part because of their awareness of achievements in space, young people alive today expect great advancements in the benefits and quality of civilization in their lifetimes. Because of these aspirations, those who believe in the importance of the extension of civilization into space have the need as well as the moral obligation to continually touch upon the lives and imagination of these young people.

Why undertake a Millennium Project that stretches human technological and psychological reach to the limit? The answers are forming in the minds of those young people who will lead humankind into the Third Millennium—the generations now in school, now playing around our homes, now driving us to distraction as they struggle toward adulthood. They will settle the Moon and then Mars. They will do this because they want to "be there." Our role has become merely one of staying out of the way while preserving and expanding their opportunities and challenges. Opportunity and challenge turns young dreamers and visionaries into doers.

"Being there" remains the essential human ingredient in life's meaningful experiences. Standing on the rim of the Grand Canyon, reaching the summit of Everest, true love, a religious awakening, or the birth of a child have been such experiences for many. Now, "being there" in the middle of the majestic plains and valleys of the Moon and "being there" in orbit above this remarkable planet called Earth stand among the uniquely meaningful experiences of humankind.

"Being there" adds the human element to life's events. The desire to "be there" will drive our young people away from the established paths of history on Earth and to the planets and the stars. Video pictures and data streams from robots on Mars, no matter how good or how complete, will never be enough for the parents of the first Martians.

The Millennium Project can be for the children of the world what Space Shuttles and Mir have been to their parents and what Apollo and Soyuz represented to their grandparents: the embodiment of the best in the human spirit. Maybe most importantly, if our determination becomes unequivocal and if the United States proceeds with a sense of historical urgency, astronauts and cosmonauts from the world over may be able to join hands in this great adventure.

Few technical unknowns span the distance between us today and the realization of the Millennium Project. Certainly, little needs to be done compared to the task that faced Americans and Soviets when they entered the race to the Moon. Whatever technical options may turn out to be appropriate, the time to create those options has arrived. The certainty we can have of success in undertaking the establishment of a permanent outpost on Mars comes from two predominant sources: the confidence and knowledge gained from the Apollo expeditions to the Moon and the spectacular and detailed data returned by the Viking landers and orbiters of Mars. We know more about Mars as a planet than we did about the Moon before Armstrong and Aldrin landed in 1969.

In addition, the reports of two major studies, that of the Advisory Committee on the Future of the U.S. Space Program [Augustine, 1990] and the Synthesis Group on America's Space Exploration Initiative [Stafford, 1991] fully document and support a commitment to our future in space. They develop clear rationales and options for Missions to Planet Earth and from Planet Earth with "the long term magnet" being the planet Mars and the best way to get there being by way of a return to the Moon to stay. The only concern one can have with these two excellent reports, and indeed in policy positions of the federal government, lies in their lack of urgency. The Millennium Project and its tie to critical problems of today, particularly in education, global environment, and international motivation, would restore a sense of immediacy to large scale space endeavors.

However, space activities will be sustained less by the needs of technology or science or politics than by emotions—the emotions of young Americans and young people the world over. Indeed, many of those living under the yoke of oppression, or just emerging from under that yoke, also must look to space as the Earth's frontier. As with our ancestors, their real freedom may lie across a new ocean—the new ocean of space.

REFERENCES

Augustine, N.R., Chairman, Report of the advisory committee on the future of the U.S. Space, *U.S. Government Printing Office*, Washington, D.C., 1990.

Bilder, R.B., E.N. Cameron, G.L. Kulcinski, and H.H. Schmitt, Legal regimes for the mining of Helium-3 from the Moon, *WCSAR-TR-AR3-8901-1*, University Wisconsin, Madison, WI, 1989.

Cairns-Smith, A.G., *Scientific American*, 252, 6, pp. 90-100, 1985.

Carr, M.H., R.S. Saunders, R.G. Storm, and D.E. Wilhelms, The geology of the terrestrial planets, *NASA SP-469*, Washington, D.C., 1984.

Kearny, J.J., Chairman, Report of NASA lunar energy enterprise case study task force, *NASA Technical Memorandum 101652*, 1989.

Knoll, A.H., End of the proterozoic eon, *Scientific American*, 265, 4, pp. 64-73, 1991.

Kulcinski and H.H. Schmitt, The Moon: An abundant source of clean and safe fusion fuel for the 21st century, *11th International Scientific Forum on Fueling the 21st Century*, September 29-October 6, Moscow, USSR, 1987.

Hawley, T.B., Space cadets, *Ad Astra*, May, 1991.

Joyner, C.C. and H.H. Schmitt, Extraterrestrial law and lunar bases: general legal principles and a particular regime proposal (INTERLUNE), in *lunar Bases and Space Activities of the 21st Century*, W.W. Mendell, ed., lunar and Planetary Institute, Houston, 1984.

Neal, V., N. Shields, Jr., G.P. Carr, W. Pogue, H.H. Schmitt, and A.E. Schulze, Extravehicular activity in Mars surface exploration, in *Advanced Extravehicular Activity Systems Requirements Definition Study, NASA-17779-Phase III*, 1989.

Reiber, D.B., ed., *The NASA Mars Conference*, American Astronautical Society *Science and Technology Series*, no. 71, 1988.

Schmitt, H.H., The progeny of Skylab, *Progress In Astronautics and Aeronautics Series*, AIAA/AGU Conference of Scientific Experiments of Skylab, October 30, 1974.

Schmitt, H.H., Evolution of the Moon: The 1974 model, *Space Science Reviews*, 18, pp. 259-279, 1975.

Schmitt, H.H., The Chronicles Plan, *Congressional Record*, 124, no. 167-Part IV, Oct. 13, 1978.

Schmitt, H.H., The Millennium Project: Mars 2000, in *lunar Bases and Space Activities of the 21st Century*, lunar and Planetary Institute, Houston, 1985.

Schmitt, H.H., INTERMARS; in *Manned Mars Missions Working Group Papers*, Volume II, NASA M002, MSFC, Huntsville, AL, June 1986.

Schmitt, H.H., Something funny may happen on the way to Mars, *Aviation, Space, and Environmental Medicine*, January 1990, pp. 79-81, 1990.

Schmitt, H.H., Evolution of the Moon: Apollo model, *American Mineralogist*, 76, pp. 773-784, 1991.

Schmitt, H.H., INTERLUNE concept for helium-3 fusion development, in *Space 92, The Third International Conference on Engineering, Construction, and Operations in Space*, Denver, CO, May 31-June 4, 1992.

Schmitt, H.H., Helium-3 fusion: A business approach to commercialization, presented to the *Earth Resources Consortium Envirosciences Expo '93*, Denver, CO, October 8, 1993.

Schopf, J. W., ed., *Earth's Earliest Biosphere*, Princeton University Press, pp. 214-238, 1983.

Soviet Technical Presentations, *Space Adaptation Syndrome Conference*, Baylor University and University Space Research Association, Houston TX, Feb. 2-12, 1986.

Stafford, T.P., Chairman, *America at the Threshold*, Report of the Synthesis Group on America's Space Exploration Initiative, U.S. Government Printing Office, Washington, D.C., 1991.

Walsh, M.M., and Lowe, F., *Nature*, 314, pp. 530-532, 1985.

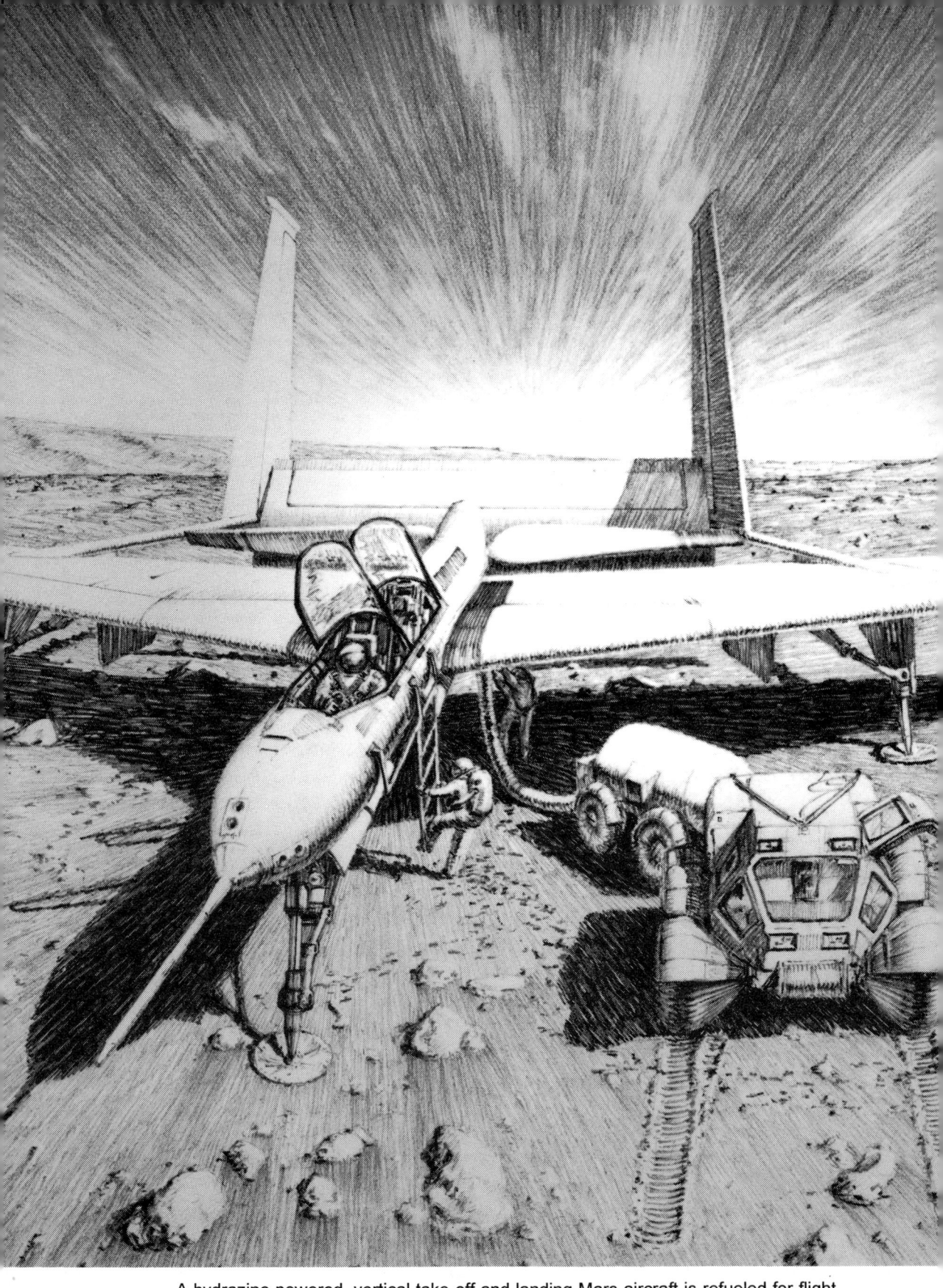
A hydrazine-powered, vertical take-off and landing Mars aircraft is refueled for flight.

AAS 95-474

Chapter 4

MARS: THE MEDIA . . . THE MASSES . . . AND THE MESSAGE

Leonard David[*]

Public support for adventurous travel to Mars will be largely shaped by the media—for better or for worse. Surveying the past 30+ years of space exploration as reported on by electronic and print media shows a roller coaster response in terms of "good" and "bad" press. This coverage has provided a subliminal backdrop for the public, from which it draws its support and contempt for space exploration. Human excursions to Mars, in terms of public interest for such missions, will find stronger support if, indeed, there is broader interest in space, be that the space station, a lunar base, or other human and robotic space activity.

INTRODUCTION

"Are the shuttle astronauts still drinking Tang? They train for years. It's hard work. It's low pay. Give them fresh-squeezed . . . how much more can it cost?"
— **Jay Leno, comedian**

"Television is chewing gum for the eyes."
— **Frank Lloyd Wright, American Architect**

If all goes according to presidential rhetoric, the first foot of an astronaut to settle on the surface of Mars will occur no later than 4:17:43 P.M. EDT on July 20, 2019—five decades to the second after Apollo 11 astronauts Neil Armstrong and Buzz Aldrin set their landing vehicle down on the Moon's Sea of Tranquility. Speaking before a Texas A&I University gathering in May 1990, President George Bush put a firm date on the first American footfall on Mars. "I believe that before Apollo celebrates the fiftieth anniversary of its landing on the Moon, the American flag should be placed on Mars," Bush said. The pronouncement followed yet another space directive from Bush some ten months earlier. Standing on the steps of the National Air and Space Museum in a public celebration of the twentieth anniversary of humankind's first step on our neighboring Moon, Bush placed the nation on three astronautical headings. "First, for the coming decade, for the 1990s, space station *Freedom*—our critical next step in all our space endeavors. And for the next century—back to the Moon . . . and this time, back to stay. And then . . . a journey into tomorrow—a journey to another planet—a manned mission to Mars," Bush stated. As for why the Moon? . . . why Mars?, Bush

---

[*] Space Data Resources & Information, P.O. Box 23883, Washington, D.C. 20026.

continued by saying "it is humanity's destiny to strive, to seek, to find. And because it is America's destiny to lead," he exclaimed. So much for the rhetoric of rocketry.

The fact is that, the very moment Bush was casting the nation off to distant celestial shores, opinion poll takers of the Gallup Organization in New Jersey were busy pulsing by phone the thoughts of the American public regarding the nation's space endeavors. The pollsters found that support for the space program was "lukewarm at best" and not a national priority. Moreover, the Gallup poll takers found that the American public favors decreasing NASA's budget. In addition, those willing to support an increase in spending on space is on the decline, compared with past Gallup surveys.

These trends were more recently fortified by Jon Miller, Director of the Public Opinion Laboratory at the Northern Illinois University in Dekalb, Illinois. In work supported by the National Science Foundation [NSB, 1991], Miller found that public assessments of space exploration are changing—and in a negative direction. Between 1985 and 1990, his findings show that the proportion of public reporting "benefits exceed costs" fell from 53 percent to 42 percent, and the proportion of citizenry more concerned with costs outweighing benefits grew from 38 to 47 percent. Overall, Miller's study indicates that space exploration, while highly visible, has relatively low political saliency to citizens not attentive to space exploration.

In concert with this finding is Peter Clarke, Director of the Annenberg School for Communication at the University of California. His analysis of public views concerning space, by way of special focus groups, provides a "loud and clear" message [Clarke, 1991]. "People are convinced that space exploration is frivolous, a luxury, a hobby to be indulged only after other societal tasks are well in hand. The environment, housing, transportation, and health problems strike folks as far more vital—especially in a tight economy and in a nation burdened by debt and unfavorable balances of world trade," Clarke reports.

Lastly, in a poll of 1,000 individuals in February 1992 by Yankelovich, Skelly and White/Clancy, Shulman Inc. of Newport Beach, California [Lawler, 1992], a confusing mix of responses by the public was recorded concerning the proper focus of the U.S. space effort. However, by far the highest public support centered on NASA programs that monitor the Earth's environment—the issue of the moment in the public mind.

It is clear that the days of unquestionable public support for space exploration has long since waned. The U.S. space effort is increasingly vying for funding with a number of pressing social concerns: public education, housing, and the environment, among them. The public, suggests the polls, would like to see an American space agenda that shores up the country's educational system, augments its technical competitiveness in the community of nations, and is responsive in helping to improve the environmental quality of life. Gone are the yesterdays when America jumped into the future and sent astronauts bounding across the Moon's cratered terrain, propelled there as a product of Soviet and American space rivalry. Project Apollo—as great of an achievement as it was—has been reduced to playback mode on the home video recorder.

THE ROLE OF THE MEDIA

Thus, the questions that remain are whether a rationale for dispatching crews to Mars can fit within the public's wants and needs as now expressed? And what role will the media play in crafting public interest and disinterest in planting footprints on the Red Planet sometime in the early decades of the 21st century? Partial answers to these questions are anchored in the past three decades of human spaceflight.

There is no doubt that the media—electronic and print—has proven its ability to both help *and* hinder the shaping of public attitudes about space. The roots of media involvement with the nation's space program has had a long history. The National Aeronautics and Space Act of 1958, as amended, calls upon NASA, in order to carry out the purpose of the Act, to "provide for the widest practicable and appropriate dissemination of information concerning its activities and the results thereof."

An important hallmark of NASA communications is continued openness of the space program to public scrutiny, a policy that has garnered world-wide admiration, respect and emulation by other countries. A decision to maintain an "open" space program was made early in NASA's history. That decision was largely based on the fact that taxpayer monies were being used for the U.S. civilian thrust into space, and that the public was entitled to fully understand America's role in exploring the frontier of space. Although an approach of space agency openness was not unanimously held in the late 1950s, NASA's first administrator, the late James Webb, organized the day-to-day workings of the space agency to be visible to the public, and structure NASA in such a way as to become a trusted source of information. It was Webb's philosophy [Webb, 1969] that "society needs as much assurance of success as it can have when it commits its resources in large amounts. It needs assurance that the leaders to whom those large resources are entrusted will use them to strengthen, and not weaken, valuable existing institutions and groups in the nation's economic, social, and political structure." This policy was strongly reflected in America's reach for the Moon [Logsdon, 1970]. The first steps to the lunar surface began with simple and crude flights of the Mercury, one-seater, space capsule. The first manned test of Mercury on May 5, 1961—the 15 minute suborbital hop into the ocean of Alan Shepard—provided anxious moments for the public, Congress and for the President of the United States—John F. Kennedy. It was his decision of openness that was perpetuated in the Gemini and the Apollo program, continued through the Skylab, Apollo-Soyuz projects, and now is resident in the era of the Space Shuttle.

Those early days of pioneering the space frontier—one could call it barnstorming—was high drama for the media, particularly television which was also in its infancy. Notes media historian, Barbara Matusow [Matusow, 1983], "Space was intrinsically a wonderful television story. . . A magnificent pictorial display, overlaid with such appealing themes as human courage, American ingenuity, and the charting of the unknown, a space shot had a structure that worked on television—a breathtaking opening, a suspenseful middle, and a dramatic conclusion. And everything occurred on a fixed schedule, so the cameras could be situated in the right place. And space was one of the few positive stories of the tumultuous '60's. . ."

Compare Matusow's thoughts with those of CBS News correspondent, Eric Severeid, who remarked on the eve of Apollo 11's liftoff toward the Moon [CBS, 1970], "We are a people who hate failure. It's un-American. It is a fair guess that the failure of Apollo 11 would not curtail future space programs but re-energize them. Success may well curtail them because for a long time to come future flights will seem anti-climactic, chiefly of laboratory, not popular interest, and the pressure to divert these great sums of money to inner space, *terra firma* and inner man, will steadily grow."

The moment when Neil Armstrong set foot on the lunar surface, 94 percent of homes with televisions were tuned to the space event [Bliss, 1991]. A billion people on Earth heard Armstrong's first words from the Moon. And as Severeid correctly predicted, follow-on Apollo missions met with less and less media and public attention. Similarly, the maiden flight of the first space shuttle in April 1981, dominated the news . . . with repeated shuttle missions spurring scant coverage; only Cable News Network and NBC's West Coast TV covered live the *Challenger's* last liftoff on that fateful day in January 1986.

Severeid's prescient thought as to public reaction to space failure held true following the *Challenger* disaster. The accident did not diminish public support for the space shuttle program [NSB, 1987]. Immediately after the accident, only a minority thought the explosion was a major setback for the shuttle program. One-half were willing to increase federal funding for the program if that were needed to get it back on schedule. Nearly everyone believed that human flights would be resumed and that the shuttle is still an outstanding example of American technology.

However, following months of open hearings by the Rogers Commission that investigated the accident, much of it presented night after night on the evening news, the public belief that there was a basic design flaw in the shuttle resulted in a majority willing to characterize the accident as a major setback for the shuttle program. While a very large majority continued to believe that the shuttle is an outstanding example of U.S. engineering prowess, the number who strongly agreed with this had dropped significantly.

It is worth noting that a self-assessment by reporters following the *Challenger* explosion found that the media, itself, was a "contributing factor" in the tragedy [Easterbrook, 1991]. Too much "boosterism" and not enough critical reporting by dazzle-eyed journalists helped to bring the *Challenger* down, complained one reporter. Yet the aftermath of the event also put a powder keg with a lit fuze underneath the media elite to become ever more vigil as to the inner workings of NASA. "I thought my job was to discover the cause, fix it, ensure future safety and reliability, and return the space shuttle to safe flight. I quickly discovered, however, that I was embroiled in politics, budgets, and a critical reexamination of NASA, *all surrounded by a media zoo*," (my emphasis) recalled NASA administrator, Richard Truly [Truly, 1992].

In the case of Apollo's multiple landings on the Moon, and now the shuttle-after-shuttle liftoffs, the signal here is that repetition stirs media and public disinterest. On the other hand, the fact that humans are involved in exploring space—not just space surveying by robots—has proven time and time again crucial in stimulating public and media attention to space; that attention is divided between how much it costs for human spaceflight as well as the adventure, excitement and daring of astronaut exploits. As often cited by Congress: "No Buck Rogers—no bucks . . . No bucks—no Buck Rogers."

Observed political columnist William Saphire, questioning the lack of public attendance at the launch of a *robot* spacecraft to Jupiter [Saphire, 1980]: "The answer, of course, is that there are no human beings aboard and there can be no adventure without danger. But one day there will be men aboard (and women, and blacks, and young people, and ethnics) and this whole world will hold its breath as the human spirit reaches up and touches another whole world." A cynic would say, don't hold your breath too long for that day, particularly when one glances at a spate of news magazines with titles that read:

THE SPACE PROGRAM'S MID-LIFE CRISIS
NASA strives to avoid collapse and to chart its course for the 1990s
— *Government Executive*, October 1988

20 YEARS AFTER APOLLO IS THE U.S. LOST IN SPACE?
— *Popular Science*, July 1989

STAR CROSSED - THE HUBBLE TELESCOPE
NASA's $1.5 Billion Blunder
— *Newsweek*, July 9, 1990

THE CASE AGAINST NASA
Nowhere to go but up
— *The New Republic*, July 8, 1991

WHO KILLED THE AMERICAN SPACE PROGRAM?
How lies, wishful thinking and shortsightedness ruined NASA
— *Florida Trend*, February 1992

These are, of course, juxtaposed with numbers of other articles who promote rapture of the deep (space) in the public mind by showing the latest Hubble images, Magellan Venus photos, or Earth pictures taken by satellites and shuttle crews. As example:

ONWARD TO MARS
— *Time*, July 18, 1988

OUR PLANET
Breathtaking Views from Space by Those Who Have Been There
— *Life*, November 1988

THE BLUE PLANET
A Close Encounter With Neptune
— Newsweek, September 4, 1989

BEHOLD THE EARTH
Startling new pictures show our planet as we've never seen it before
— Life, April 1992

Arguably, this mix of mixed messages plays havoc with the public mind. Space is exciting . . . space is expensive . . . NASA does great things . . . NASA can foul up. All these media-driven images meld together, thus contributing to the general reservoir of public attitude about space from which that public draws its support as well as contempt for space exploration. For a large body of public, these images form little more than a subliminal backdrop. Also, I would contend that the public is unable to discern "space" exploration from a more focused rationale for just Mars exploration.

The popularization of space owes a great deal of credit to science fiction. Even the countdown was a product of science fiction [Green, 1969]. As motion picture director Fritz Lang decided for the first science-fiction feature, "By Rocket to the Moon," that counting down made more sense than counting up. Explained Lang: "When you count, 'One, two, three, four, five, six,' no one knows when you are finished."

Throughout history, by way of the pages of science fiction literature, Mars has been peopled many times. Most notable of these was the work of novelist H. G. Wells who envisioned Mars as a threat to Earth in his turn-of-the-century classic, *War of the Worlds*. The writer scripted Martian forces using advanced weapons attacking our planet, but who, in the end, were unable to resist lowly forms of bacteria here on Earth. Broadcaster Orson Wells created a public panic by adapting *War of the Worlds* into a highly realistic radio newscast on Halloween night. That resulted in hundreds of terrorized people fleeing from the vanguard of Martian armies who had supposedly touched down in the farmlands of New Jersey. Radio listeners ran into churches. Some even wrapped their heads in wet towels to counter the poison gas sprayed into the air by advancing Martian troops! More recently, public access to Mars was made possible through the movie, *Total Recall*, staring Arnold Schwarzenegger. Theater audiences were given a crash course on a "shake and bake" method to terraform Mars, with Schwarzenegger taking lead duty as a planetary chef.

As with science fiction movies that revolve around space themes, advertising offers an even softer subliminal message. Be it lost luggage on the Moon; an electronic manufacturer that claims its new television model can automatically capture all the shades of Jupiter; or an airline asking potential passengers to "ride a spaceship to a place that's out of this world" as a promotional pitch for their new widebody DC-10 route to Hawaii—these advertising messages help sensitize the public to space exploration and human space travel. Similarly, other forms of communication with a space theme, be it music, poetry, dance, or a comedian's joke, all play a role in acclimating the public to space travel, including long sojourns to the distant dunes of Mars.

Public acceptability of human space travel will depend greatly on access of the media to space events. While some may sense that staged "media events" trivialize space, efforts should be made to expand such activity. For instance, on March 30, 1992, the space shuttle crew aboard *Atlantis* participated in the 64th Annual Academy Awards festivities, viewed by millions, by carrying an Oscar into space, thereby saluting the film making work of George Lucas, director of the popular *Star Wars* movie.

Likewise, on June 29, 1992, viewers of ABC's Good Morning America show were treated to "breakfast with the astronauts"—a live interview with crewmembers of space shuttle *Columbia* as they circuited the globe. Millions of viewers could see firsthand the mix of crew, types of experiments—including AIDS and osteoporosis research—and how human beings cope with living and working in microgravity.

Humans watching humans doing human things in space—television has proven the key medium in this regard, be it capture and repair of an errant communications satellite, the building of a space station, or a first step on to Mars. Not to discount the value of nuts and bolts over flesh and bone, new generations of Mars robotic landers and surface rovers will obviously heighten interest in the Red Planet. As always, the prospects for finding life on Mars is likely to fuel public and media excitement. Other space events are likely to play a role in bolstering interest in Mars, particularly cosmological findings from the Hubble Space Telescope or a successful search for extraterrestrial intelligence.

It is difficult to project what set of conditions will align to enable the human exploration of Mars: scientific knowledge; social enrichment; the sharpening of technological skills; a political statement; international cooperation; economic augmentation; or for environmental/survival purposes. Given the Apollo experience, how to sustain a Mars exploration program beyond initial missions is likely to prove challenging, despite the post-Vietnam-like slogan often heard from Mars proponents of "Not another Apollo!"

What ever the rationale, in view of today's social climate and public/media interest in space, Mars exploration cannot be divorced from a general need to cultivate and involve the public in other modes of space travel: the shuttle program, the space station, a lunar base, etc. Only through the maturation of public and media interest in the broader context of space exploration, can we hope to garner the support, monies, talent, and technological wherewithal to build a sustainable human link between the third and fourth planets.

Until then, we will be doing little more than writing one more chapter in the Wait Watcher's Guide to Space Exploration.

REFERENCES

Bliss, Jr., E., *Now The News - The Story of Broadcast Journalism*, 367 pp., Columbia University Press, New York, 1991.

CBS, 10:56:20 PM EDT 7/20/69 - The historic conquest of the moon as reported to the American people by CBS News over the CBS Television Network, *CBS News Limited Edition*, 10 pp., Columbia Broadcasting System, Inc., 1970.

Clarke, P., Bringing space home to the American people, delivered to The Seventh Annual National Space Symposium, Colorado Springs, Colorado, April 10, 1991.

Easterbrook, G., The Cast Against NASA -Nowhere to go but up, *The New Republic*, 22 pp., July 8, 1991.

Green, M., *Television News - Anatomy and Process*, 9pp., Wadsworth Publishing, Belmont, California, 1969.

Lawler, A., Poll shows Americans like Earth observation, in *Space News,* 14 pp., March 23-29, 1992.

Logsdon, J. M., *The Decision to Go to the Moon: Project Apollo and the National Interest,* pp. 122-123, The MIT Press, Cambridge, Massachusetts, 1970.

Matusow, B., *The Evening Stars - The Making of the Network News Anchor*, 150 pp., Ballantine Books, New York, 1983.

National Science Board of the National Science Foundation, *Science & Engineering Indicators,* (NSB-87-1), Washington D.C., pp. 162-163, 1987.

National Science Board of the National Science Foundation, *Science & Engineering Indicators*, Tenth Edition, pp. 178-179, (NSB-91-1), Washington D.C., 1991.

Saphire, W., Because it's there, *Saphire's Washington*, pp. 215-217, Time Books, New York, 1980.

Truly, R. H., Remarks delivered at the National Space Club luncheon, 3 pp., Washington, D.C., February 26, 1992.

Webb, J. E., *Space Age Management: The Large-Scale Approach*, 5 pp., McGraw-Hill Book Company, New York, 1969.

Special thanks to Alfred Robert Hogan for making available his *Science on the Set . . . A Critical Review and Analysis of CBS Television News Coverage of Project Apollo*, Independent Study, University of Maryland, August 18, 1987.

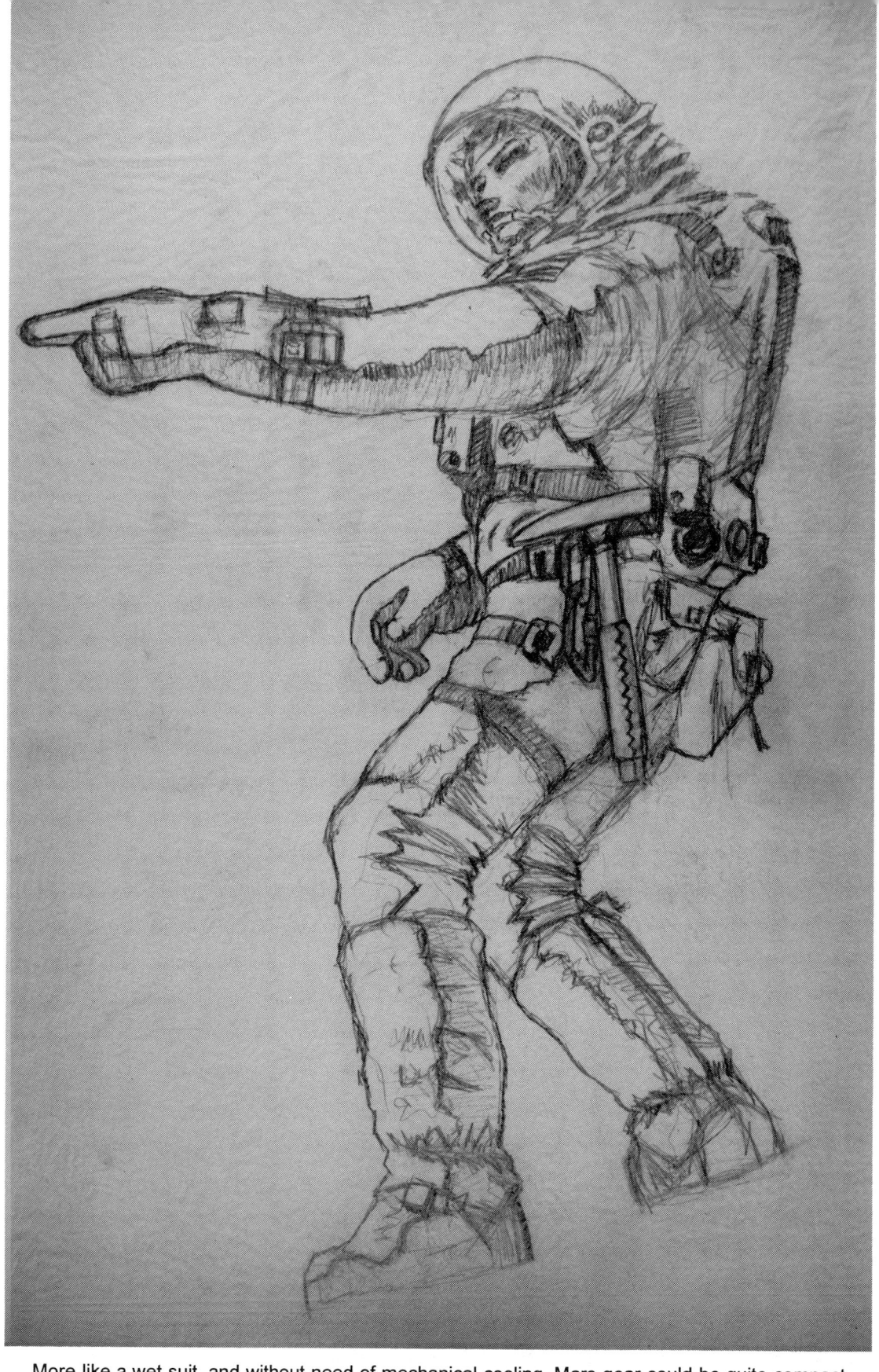

More like a wet suit, and without need of mechanical cooling, Mars gear could be quite compact.

AAS 95-475

Chapter 5

STRATEGIC COMMUNICATIONS AND THE CASE FOR MARS

Frank White[*]

This paper argues that the Case for Mars must be made in the "court of public opinion," and that making the case requires a knowledge of strategic communications. It is through the use of communications that society can be convinced to invest resources in Mars missions rather than other worthy goals. An effective strategic communications plan should begin with positioning, which means determining current perceptions about Mars exploration. Once current perceptions are determined, goals for affecting perceptions can be defined. Positioning leads to messaging for key publics, and then to implementation and evaluation. If the effort succeeds, new behavior will follow the new perceptions, and a new era of Mars exploration can begin.

INTRODUCTION

The term "Case for Mars" has a legalistic tone to it, suggesting that those in favor of exploring Mars must be advocates for it, convincing others of its value.

In fact, it is an idea that is on trial—the vision of human exploration and development of Mars. This trial is not happening in a court of law, but rather in the court of public opinion, an arena in which many of society's most critical decisions are made. As in law, there is also a tradition of advocacy in this domain, a type of communications advocacy known as "strategic communications."

As we rapidly evolve into an information-based society, those who do not understand strategic communications will be less successful in making their case to others, and this applies to the case for Mars as well.

A reasonable percentage of the planet's population may well be in favor of exploring and developing Mars. Among space exploration advocates, Mars has long been the primary target of exploration within the solar system. Mars inspired rocket pioneer Robert Goddard, the early Russian space philosophers, and Wernher von Braun, who popularized his concepts in the early 1950s.

---

[*] Frank White is a strategic communications consultant and author. His books include *The Overview Effect*, *The SETI Factor*, and (with Isaac Asimov) *Think About Space* and *March of the Millennia*. His latest book, *International Space*, is published by Walker and Company, 1993. Address: 1257 Worcester Road, #104, Framingham, Massachusetts 01701.

For a time, interest in Mars was submerged, both literally and figuratively, and only a hardy group of advocates, known as the "Mars Underground" kept the dream alive. Over time, Mars became a more acceptable topic of discussion among the space exploration establishment and the public at large. *Life* magazine even ran an article on terraforming Mars [Darrach *et al.*, 1991], and many pro-space organizations have supported an international mission to the planet.

However, if this support is to be translated into action, substantial public and private resources will have to be devoted to a Mars mission. For that to happen, the public must be strongly in favor of putting those resources behind a Mars mission, rather than other worthy alternatives, such as cleaning up the environment, feeding hungry people, or curing AIDS.

The substantive case for Mars, i.e., why human beings ought to go there, is extremely interesting, but it will be discussed rather briefly here, and seen as the core "message" for a strategic communications effort. More attention will be devoted to how that message might best be delivered to achieve the desired result.

STRATEGIC COMMUNICATIONS

Strategic communications has grown out of public relations, advertising, and related communications disciplines. The essential purpose of strategic communications is to use the transmission of information to key audiences in order to achieve a desired goal. The goal can truly be anything, from electing a political candidate, to selling a product, to gaining support for a space mission.

Every person, organization, and cause has an image in the world, a way that they are perceived by "key publics." This image has also been characterized as a "position," or "share of mind." A successful communications campaign begins with finding out what that baseline image is, then uses the media to change the perception so that it is more aligned with desired goals.

Strategic communications transcends any particular communications discipline because it uses many different communications methods to enhance or transform a public image. The focus is on the result, not the means of achieving it.

Strategic communications also recognizes the fact that in a complex global society where most people get their information second-hand, "perception is reality."

No one has the time to check out everything first-hand, so we all rely on the media for our knowledge of important matters. While we may not consciously realize it, all of that information is to some extent slanted by advocates of one cause or another, even when it is filtered through a supposedly objective media.

Advocates need to realize that the purpose of changing an image is really to change behavior. While the focus is not directly on behavior, once a perception has been changed, behavior change often follows.

A person who believes a mission to Mars is a waste of resources that could be devoted to world peace is unlikely to call his or her representatives in Congress to support funding for the mission. If, however, that person receives new information showing that an international mission would reduce tensions worldwide, they might make that phone call because Mars is now relevant to the person's concerns.

It is very important for the space exploration community to be responsible in working to change perceptions, because the effort is really to get society to commit resources to our goals rather than to other, perhaps equally valuable, activities. Moreover, major shifts in perceptions are possible, and many have occurred in recent memory. The change in U.S./Russian perceptions of one another would, for example, have been unthinkable not so long ago.

The essential question, then, is one of positioning the case for Mars in the minds of key audiences. How is that case perceived now? How do its advocates want it to be perceived? How can strategic communications be used to get from here to there?

POSITIONING THE CASE FOR MARS

Positioning a product, service, or issue means determining what "share of mind" it should have in the awareness of different publics. Today, most citizens are suffering from an overabundance of information. In watching television alone, they are exposed to hundreds of advertising messages each day, and that is only one medium.

As a result, people filter out everything that is deemed irrelevant to them, and they do it quickly and simply. They tag the information, position it within their minds, and then store it there, until something very powerful comes along to change their thinking [Trout and Ries, 1981].

Once a product, service, or issue has taken on a specific position, it is hard to budge it out of that slot, even with strong and convincing facts. It is also hard to move into a new position once you are identified in a specific way.

Here's an example of how positioning works, both positively and negatively: For many years, customers saw IBM as the dominant force in computing, and it was hard for competitors to move IBM out of that position, even when they were able to demonstrate their superiority to IBM. However, IBM's efforts to become a copier company were less successful, because people just did not see IBM as having any expertise in copiers. Xerox, of course, did have copier credibility and was seen as dominant in that market. However, when Xerox in turn wanted to be seen as a computer company, people were equally reluctant to change their perceptions to allow it [Trout and Ries, 1981].

Thus, IBM was stuck with its (positive) positioning as an excellent computer company, even though it wanted to change, and Xerox was stuck with its (positive) positioning as a copier company, even though it also wanted to change.

Perceptions do evolve over time, however. IBM, Xerox, and other companies can re-position themselves by using the right combination of media and messages.

Similarly, the efforts of several different groups have shifted the positioning of the case for Mars. When the original Case for Mars conferences began, they were far outside the mainstream, but eventually Mars exploration achieved acceptance as a reasonable goal.

Even so, the case for going to Mars still must be made, and the required shift in perception needs to be supported, if any missions are actually to occur. Missions to Mars do have a position in the minds of various publics today, but we do not know, accurately, what that positioning is. The right way to do a strategic communications plan would be to conduct research among the key publics, and find how exactly how they perceive the issue. We can then define a desired perception of the case for Mars, and develop a communications campaign to support movement from actual to preferred perceptions.

To do so would also require being more specific about what kinds of activities human exploration of Mars will include. Are we talking about piloted or robotic missions, or a combination of both? Do we plan to go to Mars after establishing a Moon base, or as an alternative to the Moon? Will the missions be primarily government-funded or privately-funded, national or international, etc.?

Looming over all these questions is the most serious issue of all—what is the ultimate goal of human exploration of Mars? There are those who would be fully satisfied with remote robotic research, the primary results being purely scientific. For others, settlement of humans on Mars is the only goal worth pursuing. Terraforming the planet so that it is Earth-like has become an option that is being openly discussed as well.

These questions in turn touch upon an inquiry into the purpose of human exploration of the universe itself, an issue on which there is nothing close to unanimity among the people of this planet.

Even without conducting this initial research, however, we can be fairly certain of one thing—there is and has been a tremendous amount of interest in Mars on the planet Earth. Mars remains the most mysterious and compelling of planets, largely because it is the most Earth-like within our reach. In particular, if life as we know it exists or has existed elsewhere in the solar system, Mars is the most likely place for it. Who wouldn't want to find out what is going on there?

Nevertheless, it is equally likely that very few people would perceive the exploration of Mars as a necessity. Some might openly oppose it, while others might see it as a good or useful thing to do, but not as critical to the future of humanity.

If this assumption is correct, those making the case for Mars have a critical decision before them. The question is whether merely to generate support for existing policy initiatives, or to go further and make the case for Mars a very high priority.

More to the point, if going to Mars is not a high priority, why go at all?

MESSAGES

Perceptions are changed by messages, new information that enables audiences to shift their viewpoints on a given issue. In regard to Mars, advocates probably haven't thought carefully enough about the message to communicate effectively with those who are less concerned with Mars. A good way of knowing whether your message is clear is to ask yourself, "If you had thirty seconds to explain the case for Mars to a total stranger, what would you say?"

The "thirty-second test" is important because people listen to information that affects their own lives, or has life-or-death implications for the species. For example, peace groups were able to bring the global danger of nuclear weapons home to people, even though such weapons might have no obvious impact on their daily existence.

This line of thought seems to lead to an inescapable conclusion: the case for Mars, on its own, may not be strong enough to be seen as a top priority among key audiences unless it is linked to larger messages about space exploration and human evolution.

The best approach may be to create a context that explains why space exploration is important to human and universal evolution, and how Mars fits into that context as an essential component. In my book, *The Overview Effect*, the thesis is developed that the primary benefit of space exploration is that it transforms human consciousness and supports social evolution. Research suggests that different missions have different levels of impact, with a mission to the Moon generally being more significant than a trip into Low Earth Orbit [White, 1987].

Piloted missions to Mars will have a dramatic effect on human awareness of our place in the universe. For the first time, humans will leave the Earth/Moon system and move into a direct awareness of the solar system, reinforcing a movement from a geocentric to heliocentric perspective.

We will also have an opportunity to investigate, on site, the question of whether life has ever existed on Mars, which is crucial in understanding our own place in the universe.

Another theme developed in *The Overview Effect* is that humanity has a choice between evolution and extinction, and that exploring space through a global "Human Space Program" represents the direction of evolution. Mars is seen, in that paradigm, as a critical near-term objective, not for its own sake, but as part of something far grander [White, 1987].

AUDIENCES

One of the most important insights of strategic communications planning and implementation is this: there is not one "public." Instead, there are many publics, and the term "general public" may well be inaccurate. Once we have determined positioning and messages, then, the next step is to determine who needs to receive the messages. In addition, messages must be tailored to each audience specifically.

Ultimately, if the case for Mars is fundamentally the case for an international mission, then the audiences are worldwide. However, it is likely that they still must be approached on a country-by-country basis.

Assuming for the sake of simplicity that we are approaching the United States with these messages, the most important audiences are:

1. The president and the presidential advisors;
2. The space exploration/development community;
3. The public policy community;
4. Influentials within private industry;
5. Influentials within academia;
6. Key analysts in the media;
7. Other publics.

Since the goal of strategic communications is to achieve a result, and resources are always limited, the approach to audiences is obvious: find the audience(s) to whom a delivered message will have the greatest impact, at the lowest cost.

In a democratic political system, this can make for difficult choices. Ultimately, however, the president is the most important audience for macro-projects requiring public funds, because a strongly committed president can make things happen, even when popular support may be uncertain. For example, President Kennedy provided the impetus for Apollo at a time when even the space exploration community was not united on a Moon landing as a goal.

Exploring and developing an entire planet as a global commitment requires the commitment not only of the president and the American people, but also needs political leverage to involve other nations in the endeavor.

If the president and vice president are critical in securing support for the mission, and if the support may have to extend over more than one administration, the issue becomes how to reach these key individuals with messages that will be meaningful over time.

One approach is to reach them through other key audiences. However, this method will be ineffective if going to Mars does not fit into the political philosophy of the administration, and its public policy agenda. The case for Mars needs to be made in terms of the political imperatives of the time.

Within the public policy community, there are other key individuals whose support is essential if missions using government funds are part of the overall plan. In particular, there are members of Congress whose committees control the funds that are necessary to support the venture.

The space exploration community itself cannot be overlooked in terms of strategic communications by Mars advocates. The debate within the community fluctuates among many priorities, and while there is overall agreement on Mars as a goal, the consensus breaks down on timing and other issues—Moon vs. Mars, piloted vs. robotic missions, etc.

Within the space exploration community, the case must be made, as previously mentioned, in the context of a larger rationale for space development itself.

One appropriate communications strategy might be to solidify support within the space exploration community for a specific approach to Mars, with that community then working as message disseminators to other key audiences. This community can be especially valuable in reaching other advocacy groups, such as peace organizations, focusing attention on how Mars exploration supports the goals that they advocate.

MEDIA

The optimum strategic communications strategy integrates advertising, public relations, direct mail, and other media to achieve its purpose.

Advertising is the best way to build awareness among large numbers of people, and it offers maximum control over how a message is delivered. However, it is by far the most expensive vehicle to use, and takes substantial expertise to do it well.

Public relations is the most credible way to disseminate a message, because you must first present your case to the media, who then will decide whether to include it in their news stories and broadcasts. Public relations is a labor intensive communications discipline, and the cost is largely in "people time," which makes it a useful vehicle for volunteer organizations. In terms of public relations, the initial goal should be to integrate the key messages about Mars into existing efforts of organizations such as the National Space Society and The Planetary Society, working outward from there to mainstream media.

Public affairs and lobbying, disciplines that are closely related to public relations, but requiring distinct skills and special backgrounds, should also be integrated into the strategic communications effort.

Finally, direct marketing of products and ideas, usually through direct mail, is an extremely powerful (though often expensive) tool for those with targeted audiences and good mailing lists. Direct mail can certainly be used within the space exploration community, where such lists already exist, and the audience is well-known. It can also be a valuable tool for coalition-building with other groups.

Taken together, an integrated use of these types of media can build up a powerful synergy that eventually can succeed in changing perceptions and achieving desired goals. The final media decisions cannot be made without a strategic communications plan and accompanying budget, but it seems likely that public relations, coupled with direct mail, would bear the brunt of the proposed effort.

ACTION STEPS

Strategic communications may make the difference as to whether there is a major mission to Mars in the near future. However, a strategic communications plan of the type that is needed does not get developed or implemented without having people who are dedicated to making it happen.

The most important next step is for a group of people to come together and create a strategic communications team that can work on this project until the goal is achieved. This effort would have to be funded by organizations and individuals who care whether these missions take place or not. This group, which might call itself Mars Media or MarsCom, would be the space community's focal point for making the case for Mars in the all-important court of public opinion.

Going to Mars will take work and commitment, but there is a good case for doing it. Going to Mars is really a journey, not a mission, and this journey of millions of miles includes many small steps.*

REFERENCES

Darrach, B. and S. Petranek, Our next home, *Life*, 24 pp., May, 1991.

Trout, J. and A. Ries, A., *Positioning: The Battle For Your Mind*, New York: McGraw-Hill, 1981.

White, F., *The Overview Effect: Space Exploration And Human Evolution*, Houghton Mifflin, Boston, p. 100, 1987.

---

* At the 1991 International Space University summer session held in Toulouse, France, all of the students participated in a design project to develop an International Mars Mission (IMM). I worked with a smaller group of students to enunciate the philosophical foundations of the IMM. It was during that work that it became clear to me that going to Mars should be seen as "a journey, not a mission," because it is so much more than one trip.

AAS 95-476

Chapter 6

MANAGING THE EXPLORATION OF THE MOON AND MARS

Michael D. Griffin[*]

This paper discusses the key ingredient for a successful program of exploration and habitation of the Moon and Mars: effective program and project management. There is no technical issue that would preclude establishing a lunar base within five years, or a Mars landing within a decade, if proper management principles were applied. These principles, which have been applied to past successful projects including the space program of the 1960s, include: maintain flat organizations and short chains of command; specify outcome rather than process; select the best people and put them in charge; delegate decision authority down to the lowest possible level. In addition, a successful program will require clearly defined goals, and accepting appropriate risks. The order of priorities must be set among the three basic project parameters: cost, schedule, and performance. Finally, it is essential to select good project and program managers and vest them with the responsibility and authority to manage. A national program office should be put in charge of the program, and it must be able to control the funding for the effort. Space exploration should be the single unifying focus of America's space program and other programs should be judged from this perspective. The exploration and settlement of the Moon and Mars will be so demanding that managing it properly, or not, determines by itself whether the effort is affordable and can succeed. If we cannot agree among ourselves, within government, to direct the program appropriately, then we had best not start.

INTRODUCTION

On 20 July 1989, on the 20th anniversary of the first human landing on the Moon, President Bush announced the centerpiece of the nation's space activity for the remainder of his administration: to return U.S. astronauts to the Moon, "this time to stay," and to begin the exploration of Mars. America had waited twenty years for a succinct definition of its post-Apollo space program. The program became known as the "Space Ex-

---

[*] Space Industries International, 800 Connecticut Ave. N.W., Suite 1111, Washington D.C. 20006. At the time this paper was written, Michael Griffin was the NASA Associate Administrator for the Office of Exploration, the office responsible for implementing the Space Exploration Initiative, a program with the goal of establishing a permanent Lunar base and human landing on Mars.

ploration Initiative."† This is the next logical step in space, as indeed it must be to command the necessary public and congressional support to sustain the effort throughout the current and succeeding administrations. But, so far, such support has been lacking. Why? The Space Exploration Initiative (SEI) commits the U.S. to a permanent mission in space. But it is apparent, across the entire U.S. civil and defense aerospace community, that most space projects cost too much and take too long when judged by any conventional standard. This sets up a classic vicious circle. When project cost is high and funding must be sustained across several administrations, near-perfection in execution is expected. But perfection costs a lot and takes a long time, and so there is great reluctance at the policy making level to approve any large, new national space effort. To be sure, there are no medals awarded for quick, cheap failures. But it is equally true that we cannot sustain expensive projects which benefit subsequent office holders, no matter how successful they may be. Technical as well as political criteria argue for an expeditious approach. This view is espoused most thoughtfully by Freeman Dyson in his essay "Quick is Beautiful," from the book *Infinite in All Directions*. According to Dyson,

> ". . . it seems that major changes come roughly once in a decade. In this situation it makes an enormous difference whether we are able to react to change in three years or in twelve. An industry which is able to react in three years will find the game stimulating and enjoyable, and the people who do the work will experience the pleasant sensation of being able to cope. An industry which takes twelve years to react will be perpetually too late, and the people running the industry will experience sensations of paralysis and demoralization. It seems that the critical time for reaction is about five years. If you can react within five years, with a bit of luck you are in good shape. If you take longer than five years, with a bit of bad luck you are in bad trouble."

The message is clear. To be viable, a project must offer significant returns within a few years of inception. An endeavor requiring a broader effort, such as the human exploration of space, must be structured to offer a continuing series of milestones and decision points. In this context, the SEI program must consist of a set of individually sustainable projects linked by a grander vision. In the wake of initial estimates of $500 B and thirty-plus years to get to Mars, it has been claimed that extensive technology development prior to mission commitment is required. But in reality the technology and much of the infrastructure to enable lunar return is here today. Certainly we need some vehicles and systems not presently available, but these are of types well understood. Technology development is essential over the long haul, but not an impediment to getting started. *There is nothing of a technical nature which precludes a lunar return and the initial establishment of a permanent base within five years from authorization to proceed.* Going to Mars is a tougher problem, but not intractable even with present and near-term technology.

---

† Editor's note: After three years in which congress refused to allocate any funding for SEI, and lacking support from the incoming Clinton administration, the SEI program office was discontinued at NASA. Nevertheless, the principles espoused by this paper will apply to any major space exploration effort, and in particular, for human exploration of Mars.

MANAGEMENT IS THE KEY

Contrary to conventional wisdom, technology is *not* the central issue for SEI. The key problem today is one of management. This is a broad statement, but one which is amply justified. A plethora of studies exist to document the problem. For example, in 1990 Undersecretary of Defense for Acquisition John Betti commissioned a Defense Science Board study to determine how to effect a 50% reduction in the time it takes to acquire new weapons systems. As part of its data gathering effort, the Task Force surveyed 138 acquisition programs for systems from ships to spacecraft. It was found that the current defense systems acquisition process requires over sixteen years from need definition to initial operational capability, longer by approximately 60% than in the 1950s and by 75% than similar commercial programs. The process length is relatively insensitive to program characteristics such as the basis of need, program size, or decision authority. A sufficiently high decision authority can exempt a program from all or part of the process but, once entered, significant "streamlining" of the process is unlikely. The three major factors controlling acquisition timelines were found to be funding instability, excessive oversight, and technical difficulty stemming from the use of immature technology.

The civil aerospace record is similar, reflecting the use of a common laboratory and contractor support base. The Mercury and Gemini projects required five years, and Apollo eight years, from initial concept to achievement of the intended mission. Shuttle needed over twelve years. Comparable results are observed with robotic spacecraft programs.

In summarizing the situation, one cannot do better than to cite the February 1986 testimony of Gen. Bernard A. Schriever, USAF (Ret.), to the Packard Commission:

"There is general consensus that the following situation prevails:

- We have let the acquisition time from laboratory to initial operational capability (IOC) more than double, and costs inflate enormously.

- The defense procurement process has been politicized by a blizzard of legislation.

- We are inhibiting technological innovation, the cornerstone for maintaining qualitatively superior military systems.

- We are not moving rapidly enough in putting new technologies into systems and are no longer taking the prudent risks or using "concurrency" as we did in the 50s.

- We have taken the management authority out of the hands of the Program Director in the wake of a maze of top-down micromanagement.

- We often fail to keep the Program Director and his staff in place long enough for program stability.

- There has evolved an extremely unhealthy adversarial role between the government and industry.

- We have lost public confidence in the government-industry team.

- We have more rules, requirements, documents, people, reviewers, and checkers than ever before involved non-productively in the decision making process."

Gen. Schriever was correct in 1986, and is correct today. Over and over again, the key lessons of project management have been definitively demonstrated: Maintain flat organizations and short chains of command. Specify outcome, not process. Select the best people, and put them in charge. Delegate decision authority and responsibility downward wherever possible. Yet one seeks in vain to observe these principles in action in most modern aerospace programs.

REALISM ABOUT RISK AND FAILURE

We must come to terms with concepts like "risk" and "failure" in a way which has not been popular for at least two decades. To push back the frontier which separates the known from the unknown is an inherently risky proposition. This is true whether the frontier is one of physical limits or intellectual limits. The financial, technical, schedule, and human risks of such activities cannot be eliminated. Indeed, this assertion comprises the single strongest reason for government involvement in research, development, and exploration. Government is the entity which can best amortize the risk of such enterprises. As a society, we tolerate the known inefficiency of government processes and institutions in exchange for the collective benefits of spreading the risk. When an enterprise is no longer unknown and uncertain, we expect the marketplace to be the judge of its worth.

It follows that those who conduct and manage advanced development programs, of which aerospace programs provide virtually the defining example, are in a fundamental way concerned with the business of understanding and managing risk. To be involved in an advanced development program is to accept, either implicitly or explicitly, relatively high odds that things will not go as planned. This is in strong contrast with everyday events which form, obviously, the background of human experience against which we judge the excursions.

Lately we have been too ready to apply the standards of common experience to activities which, by their very nature, lie outside this experience. When a goal is yet to be attained, the path towards it must remain uncertain. The path, if found at all, is found as the result of a trial and error process. When we focus too much on the errors, labeling them as "failures," we prevent ourselves from ever reaching the goal. Not reaching the goal, as opposed to incurring setbacks along the way, constitutes the only true failure.

As in so many things in the world of aerospace, Apollo provides an excellent example of the proper orientation. From the outset, the goal of the program could be summarized in three words: man, Moon, and decade. Considered in the small, the program had a large (but not uncommonly large) number of setbacks, anomalies, and problems, ranging up to the scale of the fire which killed astronauts Grissom, White, and

Chaffee. But the goal was "man, Moon, decade," *not* any given flight test along the way, and so the program survived.

Once a goal at the right level has been set, and the level of risk in reaching it accepted, one must understand the order of priorities among the three basic project parameters: cost, schedule, and performance. Within limits, any two of these can be independently specified, but the third must be allowed to float to some extent when attempting a new task. Any order of priorities can be acceptable, but it is crucial to know what the order is. Failure to choose is not an option. In this country, within the aerospace culture, the default option is *always* performance, schedule, and cost—the expensive, late, success. If a project sponsor wishes any other outcome, he must explicitly state it, and continue to reinforce the statement.

COST, SCHEDULE, AND PERFORMANCE

Following Dyson's philosophy, it may be observed that near-term projects (any effort that has actually moved into the design stage) tend to suffer unless fairly rigorous schedule and cost disciplines are imposed, and the scope of the job adjusted as necessary to meet these parameters. Thus, any *current* project should probably follow guidelines such as, "Do as well as you can, by such-and-such date, for so-and-so money." When a higher level of performance is required, it must be approached in stages, each of which is packaged according to these guidelines. Thus, the Apollo program reached its goal—landing a man upon the Moon and returning him safely to the Earth—on only the fifth manned Apollo flight, eight years after President Kennedy's declaration of that goal. But the operational Apollo spacecraft program was preceded by—in retrospect—the absolutely crucial Gemini program. Gemini's dazzling contributions, gained over the course of ten manned missions placing twenty men aloft in twenty months, built on the lessons of the Mercury program, in which the basic goal was little more than to learn how to put men in space.

Longer term programs often need different priorities. Technology development programs in particular are difficult to accomplish within fixed schedule or budget constraints. The appropriate guideline in this case is usually to specify the desired performance level and the available funding stream, and to let the results mature as they may. Again, understanding the priority order is the crucial step.

PROGRAM MANAGEMENT

It is essential to select good project and program managers. There is no substitute. The availability of good managers and managing engineers at the top level may well be one of the nation's critical limiting resources. Funding without the skill to manage it is of no value.

Having selected the best managers we can find, we must vest them with the authority and responsibility to manage primarily as they deem appropriate. We must judge the program's success based on results rather than process. Executive management must define appropriate milestones and peer review points, and must be sufficiently disciplined not to interfere as long as commitments are met. The extensive review and approval processes with which almost all modern aerospace programs are beset has become a major factor, possibly *the* major factor, in causing the average defense acquisition to require over sixteen years. Today we seem to manage by consensus, and when everyone must agree to a course of action, progress is very slow. The project must become all things to all people. Responsibility and accountability are hopelessly diffused. Lawyers, accountants, and contracting officers who should be working toward a common goal *for* a program manager are instead in a position to levy demands, without responsibility for outcome, *upon* the program manager.

These conclusions concerning effective project management have significant implications at the institutional management level. It is axiomatic that the management structure should exist to facilitate the work to be performed.

The basic structure for managing the Space Exploration Initiative should be that of a National Program Office (NPO) operating under a NASA Associate Administrator for Exploration. The Program Office should have DoD and DoE representation explicitly allocated. This is a logical consequence of the highly interdisciplinary nature of a robust SEI. No single government agency in existence today possesses the necessary fiscal and technical resources. Nonetheless, someone must be clearly in charge, and the logical entity to hold the "first among equals" position of Program Director is the nation's civil space agency.

The Program Office must control the funding for the effort; ideally it would be appropriated in a single line. Obviously, this creates a funding pool that is highly vulnerable to annual congressional examination. However, the SEI will not in any case survive a hostile congressional view of its worth, irrespective of how the funding is allocated. The apparent vulnerability of a single budget line must be offset against the absolute need by the Program Office to control the work for which it is responsible. If NASA, DoD, and DoE receive *ab initio* funding allocations from Congress for the SEI, the Program Office is reduced to specifying the work to be done subject to the constraint of known entitlements. A better arrangement is to allow the Program Office to place the funding where the work can best be done.

Institutional assumptions and priorities must change. As conceived by many in the space community, the SEI requires greatly increased funding *beyond* that earmarked for current programs, which remain untouched. To undertake a space exploration program supposedly requires much additional funding, greatly increased government staff, and more bureaucracy. This is patently absurd. Such an approach cannot succeed politically, nor is it desirable. NASA in FY91 receives about $14 B, easily comparable in constant dollars to the average annual funding for the Apollo years. While some additional funds will be necessary, clearly $14 B can support a substantial program of exploration. What is needed is to reorient the focus of the Agency. Space exploration should be the single unifying focus of America's space program. Other programs must be judged from this perspective. DoD and DoE must similarly adjust their planning. SEI will be a broad national, and likely international, program if it is to exist at all. The DoD and DoE involvement directed by President Bush cannot be limited to helping NASA spend its money.

It is necessary to define carefully the relationship that the Program Office must have with the entities, from whatever agency, which prosecute its work. It is imperative that the creation of additional governmental layers, whether *de facto* or *de jure*, is avoided. Project managers at NASA field centers and DoD or DoE laboratories must report to the Program Office through a *project* chain of command. It must be explicitly stated that the center or Laboratory functional area managers and *institutional* lines of command are not in this reporting chain.

Institutional lines of command are properly involved in many decisions. Institutional managers must decide whether a proposed work package can be accommodated within the available resources of the center. They must match facility and personnel resources with requirements, orchestrate the use of resources among projects according to overall priority, and address current and future needs for staffing and facilities. In brief, the institutional line of command is responsible for the long-term viability of the organization, and for meeting the needs of the various projects housed in its facilities in an efficient manner. It is *not* responsible for directing such work or judging its quality, other than as may be requested by the project line of command.

In brief, it is necessary to adhere more closely than has been the practice in recent years to the classic matrix organization concept. In this concept, project and program managers control money and are responsible for work. Institutional managers have custody of capital assets and personnel, and are responsible for their efficient utilization and continuing growth.

Implicit in these remarks is the recognition that an overly large number of billets at Headquarters is not required to manage SEI. The very existence of Headquarters staff in large numbers precludes effective management. Each person employed must have a function; only a few can have high-level decision authority. The rest can only engage in low-level micromanagement of work being performed in the field. In effect, "staff" people become imbued with "line" authority, but lack accountability for their actions.

The fact is that a major portion of the nation's technical strength lies in the NASA, DoD, and DoE field centers and laboratories. There is very little value which is added, or can be added, by top management in directing the details of work done at these centers. The best technical people are few in number and are rarely found in Washington. There is a fundamental incompatibility, a "rank inversion" in the military parlance, involved in having less capable people in Washington directing, at too low a level, the work of more capable people in the field. It is an unstable arrangement which leads to disregard for the entire system and its processes.

What is needed is a relatively small, but very strong, decision making group at Headquarters, drawn from the best that the nation has to offer *as demonstrated by prior performance in the field* on significant programs. Decision authority on program direction should be vested in these people. The proper function of a management headquarters is to provide unifying program-level guidance, to specify what work is to be done, by whom, for what level of resources. Headquarters may, indeed must, conduct management reviews of how the work is performed, but in general cannot and should not be involved in the details of project execution. Less activity, of greater significance, should be the theme for a management headquarters.

In summary, what is advocated here is a return to management principles which have served the nation well in the past. The history of the Manhattan Project, the ICBM program, the development of the nuclear submarine, the space program of the 1960s, and many other programs provide ample support for this view. Indeed, one can argue that whenever we as a nation are really serious about an enterprise, we manage it as described here. The proposed Space Exploration Initiative will be so demanding that managing it properly, or not, determines by itself whether the effort is affordable and can succeed. If we cannot agree among ourselves, within government, to direct the program appropriately, then we had best not start.

Section II

Getting There:
Interplanetary Transportation Issues

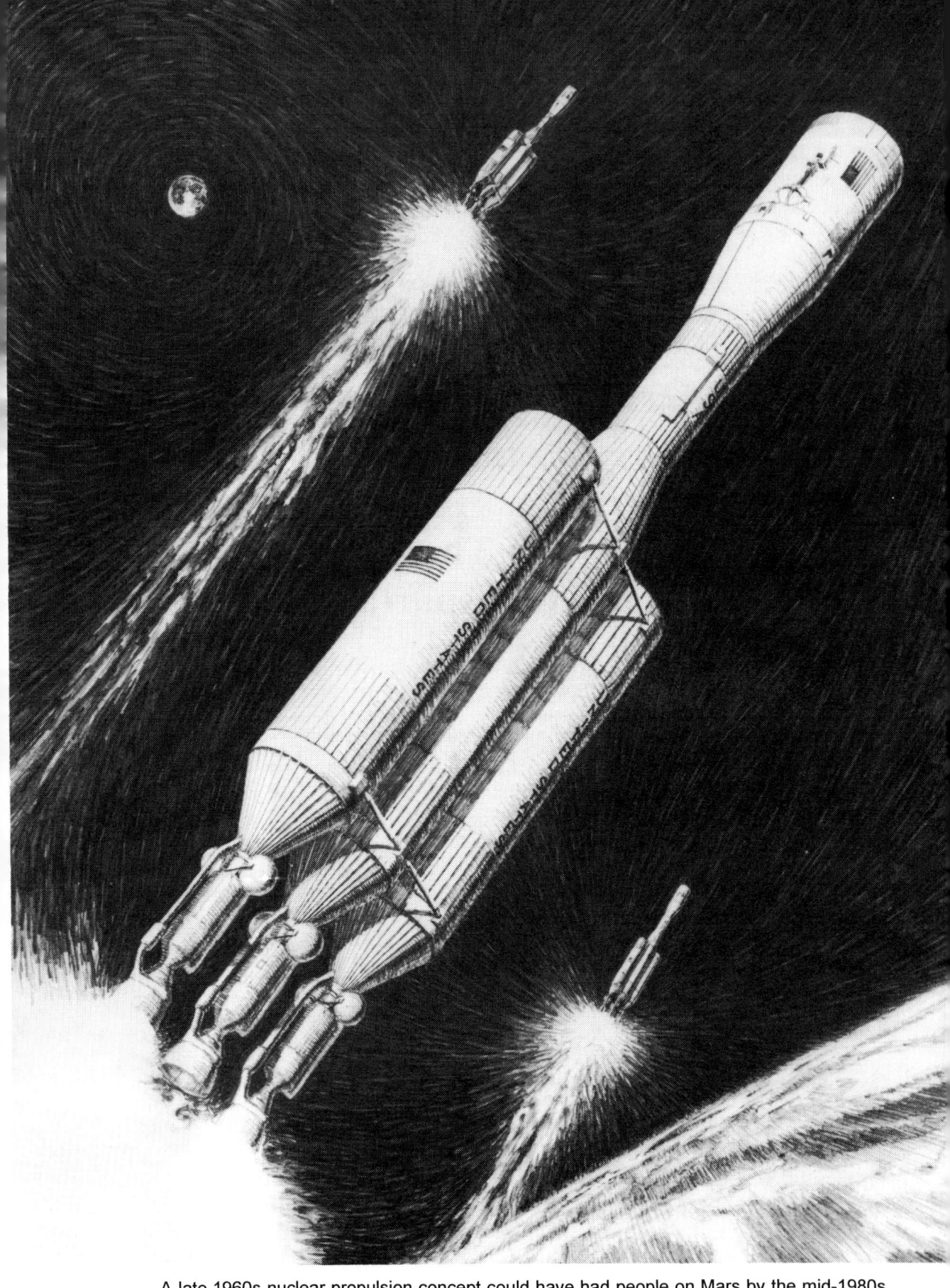
A late 1960s nuclear propulsion concept could have had people on Mars by the mid-1980s.

Chapter 7

MARS MISSION CONCEPTS: THE VON BRAUN ERA

Frederick I. Ordway III[*]

The background of Mars as a world for exploration is reviewed with attention focused on Schiaparelli's detection of "canali" (naturally occurring grooves or channels) and their later interpretation by Lowell as canals constructed by sentient beings to conduct polar meltwater to the planet's parched equatorial regions. Youthful Wernher von Braun's fascination with Lowellian Mars set the stage for the post-World War II publication of *The Mars Project*, which is described. Von Braun subsequently reached out to the public with proposals for Mars exploration in the pages of *Collier's* magazine in 1954 and in the book *The Exploration of Mars* two years later. The paper continues by examining the EMPIRE studies conducted by his NASA-Marshall Space Flight Center team and its contractors in the early 1960s as well as his post-EMPIRE contemplations towards the end of the decade. Von Braun's support of Space Task Group deliberations in 1969 and 1970 are covered as is the Nixon Administration's subsequent rejection of its ambitious manned Mars mission element.

MARS AS A WORLD FOR EXPLORATION

For millennia, Mars has intrigued man. Not only does it move across the skies and thus differ from the apparently "fixed" stars, but it exhibits a distinct reddish color. In *Scipio's Dream* published in 52 B.C. by Marcus Tullius Cicero, the sleeping hero tells of the "seven globes" beneath the sphere of the heavens including Mars "gleaming red and hateful." A thousand years later, Robert Anglicus would describe it as being "of hot and dry nature which consumes by burning."

Like the ancients, modern man is attracted to the planet but for rather different reasons: he wants to *go there*. But why? What is it that makes Mars the target of our ambitions?

For one thing, it is not too far away: astronomically speaking, Mars is the next planet out from the Sun after our Earth. Also, it possesses a thin atmosphere, something our Moon clearly lacks. Moreover, the Martian atmosphere is generally transparent so we can make out surface markings and speculate as to their nature. Although its diameter is only half that of Earth and its surface area a little more than a quarter ours, since

---

[*] U.S. Space & Rocket Center, One Tranquility Base, Huntsville, Alabama 35807.

Mars has no oceans there is actually *more* land area to explore there than on our own world. It is the combination of being relatively close to Earth, possessing an atmosphere of sorts, and revealing a varied, interesting and extensive surface that leads us to look upon Mars with keen exploratory interest.

EMERGING KNOWLEDGE OF THE MARTIAN SURFACE

The nature of the Martian surface has long intrigued astronomers and laymen. The earliest drawings, reproduced in Figure 1, are based on telescopic observations attributed to the Italian observer Francisco Fontana. In 1636, he described "the disk of Mars [as] not uniform in nature, but it appears fiery in the concave part." A couple of years later, he found that the planet exhibited a gibbous phase, characteristic of a planetary body revolving around the Sun as seen from Earth. Then, in 1659, with an improved telescope, Christiaan Huygens discovered a major feature on Mars, the *Syrtis Major*. His drawings included the south polar cap as well. The famed Dutch scientist also demonstrated that Mars rotates around a north-south axis much like the Earth. Several years later, Giovanni Domenico Cassini confirmed the rotation and determined that the Martian day is 24 hours 40 minutes long, remarkably close to the modern value of 24 hours 37 minutes 22.6 seconds; his drawings based on observations made in 1666 appear in Figure 2.

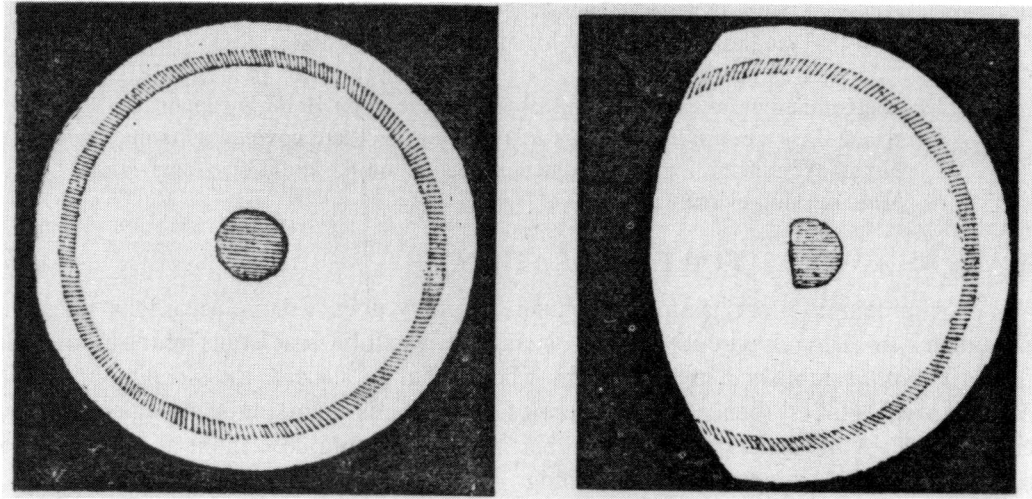

Figure 1 Fontana's observations of Mars made in 1635, left; and 1638, right.

William Herschel repeatedly observed Mars between 1777 and 1783, and worked out—among other things—its axial inclination (23.98 degrees) and the fact that the planet's atmosphere is at best very tenuous. During the following century, remarkable drawings were made by Wilhelm Beer and Johann H. von Mädler in Germany, by the Reverend William R. Dawes in England, and by many others. In 1867, the British astronomer and author Richard A. Procter published a map based on observations made to that date, and provided names for numerous features.

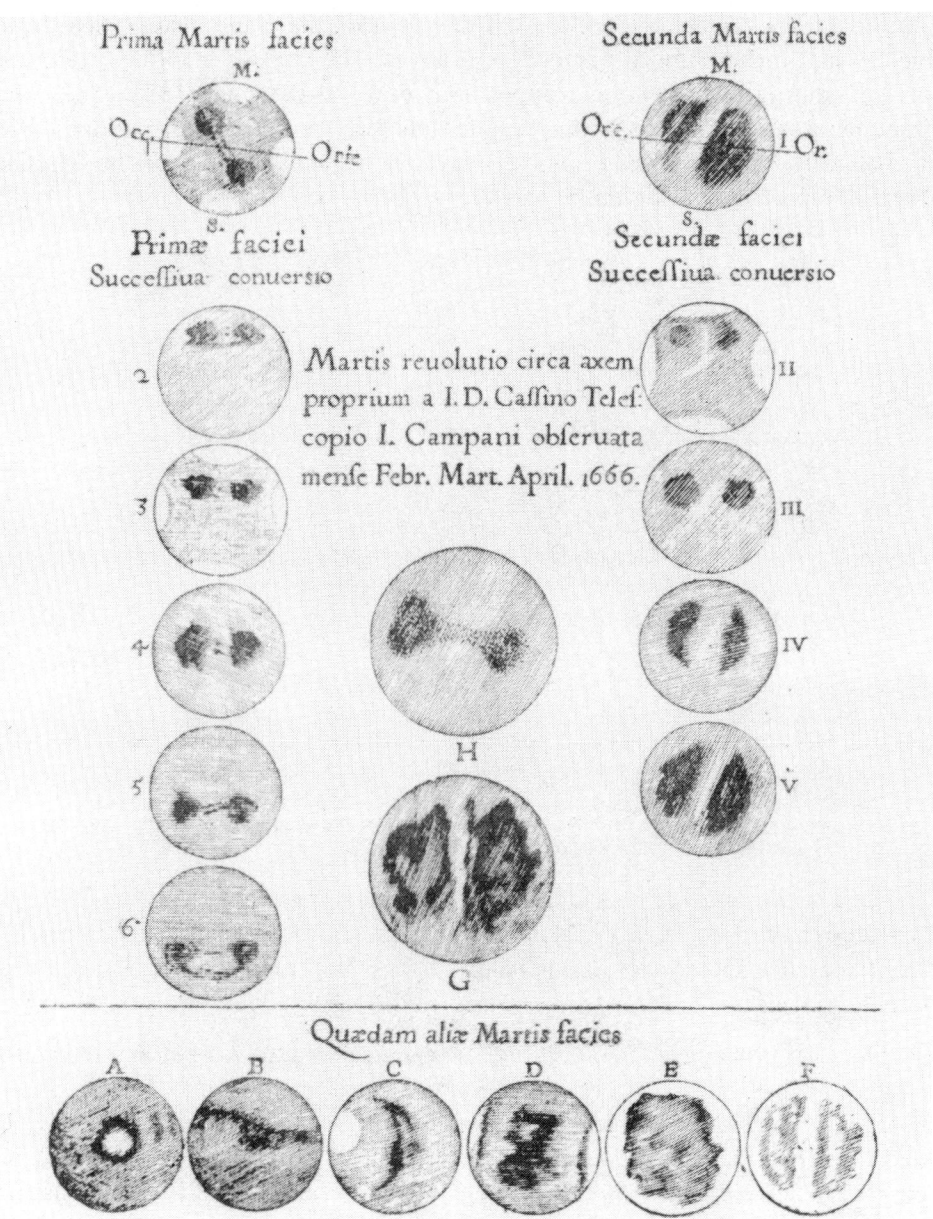

Figure 2 Cassini's observations of Mars made in 1666.

Schiaparelli and the *Canali*

Over a century ago, long-smoldering debates on the possibilities of intelligent life beyond planet Earth began to heat up. They were given impetus from reports published in 1877 by the Italian astronomer Giovanni Virginio Schiaparelli [1878-1899].

When Mars approached Earth during the late summer opposition of that year, the respected observer—using a relatively small 22-centimeter telescope at the Brera Observatory in Milan—detected a network of fine lines on the planet. He termed them *canali*, which to him meant simply channels or grooves. The curious markings were seen—and reported—during the subsequent oppositions of 1879-1880 and 1881-1882. The Italian astronomer studiously avoided any suggestion that the canal-like markings were other than naturally occurring geographic features. Schiaparelli's map of the Martian hemispheres is reproduced as Figure 3.

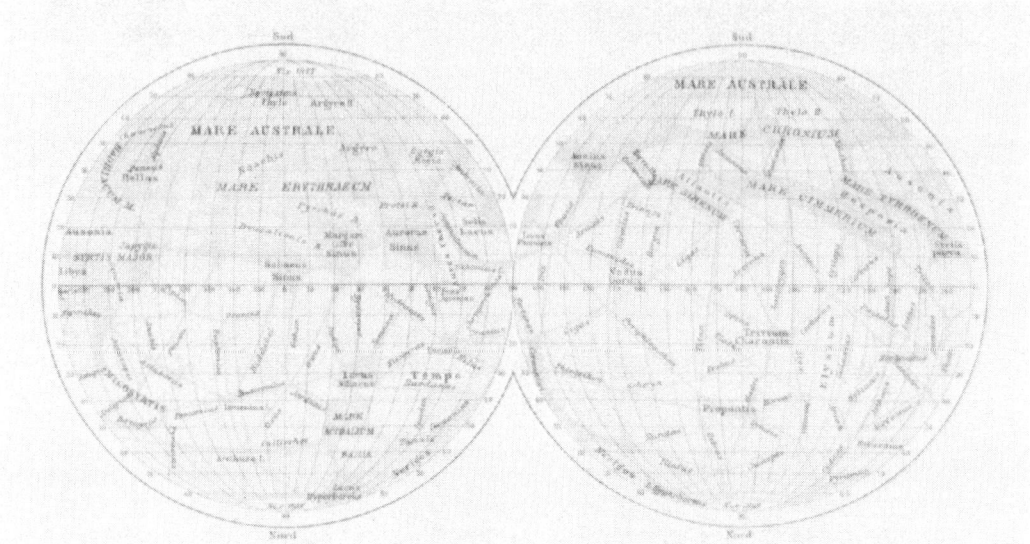

Figure 3 Map of the Martian hemispheres based on observations undertaken by Schiaparelli during oppositions from 1877 to 1888.

For a while, no one else could detect the markings. Then, in the spring of 1886, they were reported by H. C. Wilson at the Cincinnati observatory and Henri J. A. Perrotin of the Nice Observatory. But others, try as they might, never did see the Schiaparellian markings.

One fervent believer in *canali* not as channels or grooves but as the handiwork of intelligent beings was the French astronomer and popularizer Camille Flammarion [1884]. "The considerable variations observed in the network of waterways," he suggested, "testified that this planet is the seat of an energetic vitality" (by which he meant intelligent beings). *Canali* were beginning to be described as artificially constructed *canals*.

Mars as an Abode of Life

All of this attracted the attention of Percival Lowell, member of one of America's most famous and wealthy families and a man well trained in science. During the years following Schiaparelli's discovery of *canali* on Mars, Lowell became increasingly interested in the growing debate as to their true nature. When, during the course of a trip to Japan in 1892-93, he learned that Schiaparelli was giving up observing due to failing eyesight, Lowell resolved to carry out his work.

His first task was to build and outfit a brand new observatory, one dedicated in large part to the study of Mars. Lowell wanted it ready for operation within a year—by October 1884—when the red planet would come into favorable opposition for viewing. He selected a site on the high Coconino Plateau near Flagstaff in Arizona Territory. Then, with the help of associates and equipment from Harvard and elsewhere, he began construction. By late spring 1894, several months before the October opposition, he was ready.

As he trained his telescopes on Mars, *canali* appeared not by the tens but by the hundreds! Some were single, others were double; some long, some not so long. Since the lines crisscrossing Mars were straight, argued Lowell [1895, 1906, 1908], they could not be the result of random geologic processes. Then, too, he found them to be individually of uniform width, something unlikely in nature. He felt that, insofar as is known, physical processes do not give rise to "perfectly regular results; that is, results in which irregularity is not also discernible."

Lowell was especially drawn to what he called oases, which were ". . . not innocent of design. . . For here in the oases we have an end and object for the existence of the canals, and the most natural one in the world, namely that the canals are constructed for the express purpose of fertilizing the oases. . . The canals rendezvous so entirely in defiance of the doctrine of chance because they were constructed to that end. They are not purely natural developments, but cases of assisted nature." Canals and oases are depicted in Figure 4.

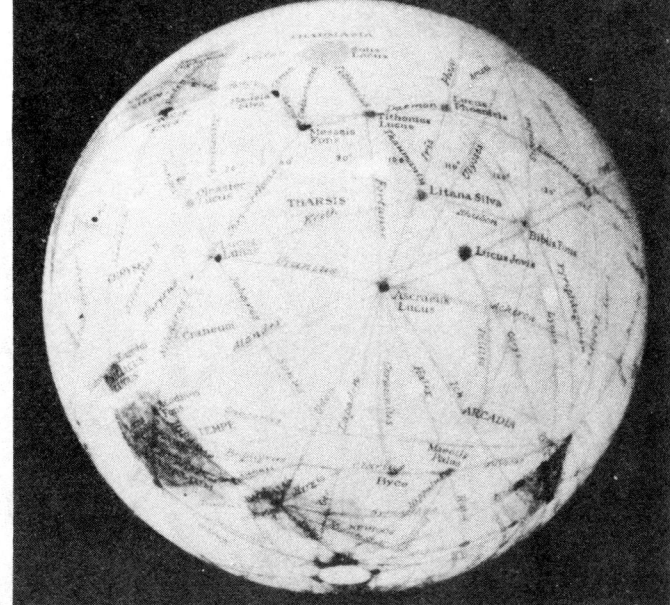

Figure 4 Martian hemisphere based on observations conducted by Percival Lowell featuring the oasis *Ascraeus Lucas* from which many canals are seen to radiate.

Despite Lowell's intriguing theories, with the passing years, doubts about his canals grew. It was not until the advent of the Space Age, however, that their existence was finally shown to be illusory. Though the canal debate may seem to us to have been a waste of energy, there was a positive side to the whole affair, one well expressed by R. L. Waterfield in his *A Hundred Years of Astronomy* [1940]. "Now the story of the 'canals' is a long and sad one," he wrote, "fraught with backbitings and slanders; and many would have preferred that the whole theory of them had never been invented. Yet whatever harm was done was more than outweighed by the tremendous stimulus the theory gave to the study of Mars . . ."

DELIBERATIONS BY WERNHER VON BRAUN

Wernher von Braun was one of many to feel this stimulus. It came to him by reading the scientific literature and press accounts of it, as well as by an occasional foray into science fiction.

In an environment shaped partly by science, partly by exaggerated faith in an idea, and partly by writers of such fiction, he pondered how man might fly to other worlds. When he was brought to the United States from Germany following the end of World War II [Ordway and Sharpe, 1979], he found the opportunity to put his thoughts on paper. During spare time at Fort Bliss in Texas and the nearby White Sands Proving Ground in New Mexico, he developed plans for the human exploration of Mars. The result: *Das Marsprojekt,* which appeared in 1952 in a special issue of the German journal *Welt-raumfahrt* [von Braun, 1952] (Figure 5). The following year, the work came out in English as *The Mars Project* published by the University of Illinois Press [von Braun, 1953]. It was reprinted twice [von Braun, 1962, 1991].

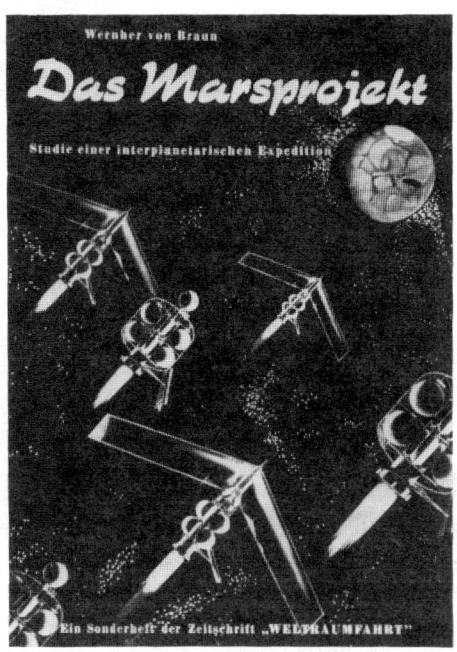 

Figure 5 Cover of *Das Marsprojekt* published in 1952 and that of the most recent edition published in 1991.

Von Braun warned his readers that space travel is an incredibly complex enterprise, one that "can only be achieved by the coordinated might of scientists, technicians, and organizers belonging to very nearly every branch of modern science and industry." He proceeded "to explode once and for all the theory of the solitary space rocket and its little band of bold interplanetary adventurers. No such lonesome, extra-orbital thermos bottle," he predicted, "will ever escape Earth's gravity and drift toward Mars."

Flotilla to Mars

In *The Mars Project* von Braun described a flotilla of 10 spaceships accommodating 70 men. "Each ship," he explained, "will be assembled in a two-hour orbital path around the Earth to which three-stage ferry rockets will deliver all the necessary components such as propellants, structures, and personnel. Once the vessels are assembled, fueled, and 'in all respects ready for space,' they will leave this 'orbit of departure' and begin a voyage that will take them out of the Earth's field of gravity and set them into an elliptical orbit around the Sun."

His spaceships were to be powered by chemical propellants; from the perspective of the early 1950s they offered the only feasible means of traversing interplanetary distances. Although von Braun did "not propose to deny the possibility that nuclear energy may someday propel space vessels," he felt it was unlikely to become available for space application for at least 25 years.

In preparing his readers for the Martian adventure, von Braun emphasized that it must be done "on a grand scale." he went on to explain that "Great numbers of professionals from many walks of life, trained to cooperate unfailingly, must be recruited. Such training will require years before each can fit his special ability into the pattern of the whole. . . The whole expeditionary personnel, together with the inanimate objects required for the fulfillment of their purpose, must be distributed throughout the flotilla of space vessels traveling in close formation so that help may be available in case of trouble of malfunction of a single ship. . ."

Partly because it was published in Germany in a limited circulation journal and in the United States by a university press, *The Mars Project* did not enjoy wide distribution. Also its timing was far from optimum: in the early 1950s, broad interest in space flight had yet to develop.

Outreach to the Public: The *Collier's* Series and Viking

Von Braun was, of course, well aware of this. So, in 1954 with his friend and collaborator Cornelius Ryan, he answered affirmatively in *Collier's* magazine the question "Can we Get to Mars?" [von Braun and Ryan, 1954]. But, he cautioned, ". . . it will be a century or more before he's [man is] ready." (He would change his mind about *that* within the next decade and a half.) The *Collier's* article was essentially a popularized version of *The Mars Project* with the same 10-ship flotilla occupied by 70 scientists and crew members. "What curious information will these first explorers carry back from Mars?" von Braun pondered. "Nobody knows—and it's extremely doubtful that anyone now living will ever know. All that can be said with certainty today," he concluded, "is this: the trip can be made, and will be made . . . someday." Figure 6 illustrates vehicles from the flotilla in orbit around Mars painted by Chesley Bonestell, Figure 7 an original drawing of what von Braun called the "Mars Landing Boat," and Figure 8 Fred Freeman's painting of the winged lander on the polar surface with a tractor train being readied in the foreground for a long trek towards the equator.

Figure 6 Elements of von Braun flotilla arrive at Mars' orbit 600 miles above the surface. Painting by Chesley Bonestell first published in *Collier's* magazine, 30 April 1954.

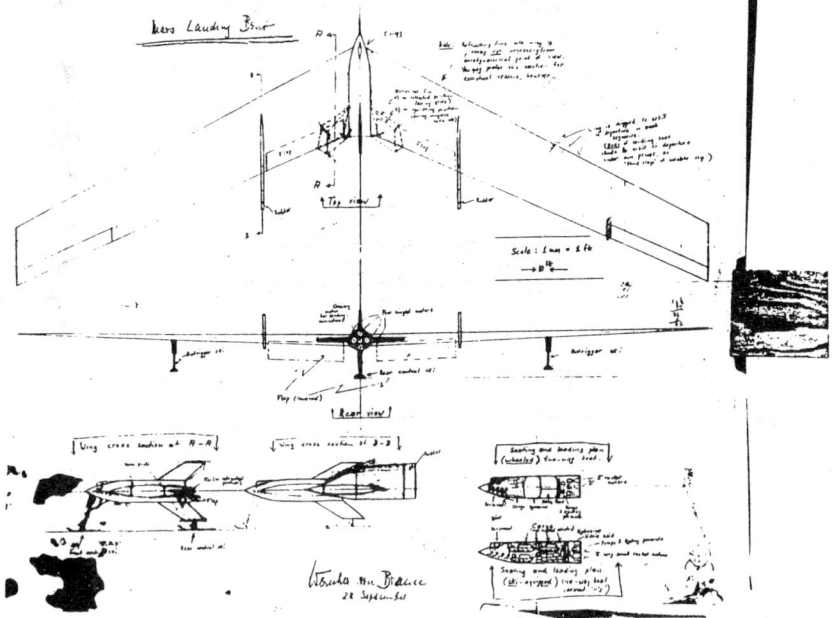

Figure 7 Original drawing of von Braun's "Mars landing boat" or winged lander used by artists Chesley Bonestell and Fred Freeman as basis for final paintings that appeared in *Collier's* magazine in 1954.

Figure 8 Advanced landing party on snow-covered polar surface of Mars painted by Fred Freeman and first published in *Collier's* magazine in 1954. The tractor train in foreground will soon depart for the equatorial regions of the planet.

Despite his realization that expeditions to Mars were not just around the corner, von Braun persisted and in 1956 expanded his ideas in *The Exploration of Mars* [Ley and von Braun, 1956] written with his close associate, Willy Ley. Enhanced with spectacular illustrations, the book included a chapter entitled "Expedition to Mars" that was, in effect, a more developed version of themes earlier introduced in *The Mars Project* and *Collier's*.

"Although," von Braun explained, "it [*The Exploration of Mars*] envisions an expedition of only twelve men traveling in two ships, the total propellant requirement—a good yardstick for the overall logistic effort—is only 10 per cent of that found in *The Mars Project*. This enormous saving is due solely to a superior over-all plan, for the specific construction weight factors have *not* been altered." The following extracts demonstrate von Braun's enthusiasm for his subject.

> At X minus 4 seconds a deep rumble goes through the two ships—the rocket motors are burning at 'ignition stage' . . . The rumble becomes a thunderous roar; the thrust very rapidly builds up to its full value of 396 tons. Ponderously the two large Mars ships begin to move visibly. The thunder and the reverberations of the rocket engines last for a little over 15 minutes. Then, as suddenly as it began, the roar subsides. The pitch of the whining gyros declines, and soon only the rustle of the ventilation blowers remains.

The 260-day coasting flight to Mars has begun. . .

Only ten days remain before the second major power maneuver, the induced capture of the ships by Mars. The distance to the planet . . . has shrunk to 1,400,000 miles. The visible half-disk of Mars glows with an intense orange-red with greenish patches, and the naked eye can easily distinguish the white spot of the southern polar cap melting in the sunlight of the Martian summer. . .

After a thorough survey of the planet from Martian orbit, the men ready themselves for descent onto the surface. Clad in pressurized suits, the first nine human beings to set foot on Mars are grouped around the cabin door. One by one they enter the airlock and listen to the hiss of the escaping air. . . Following their successful landing, the explorers spend 400 days studying the red planet before rejoining the spaceships circling above and making ready for the nearly 9-month return voyage to Earth.

Interlude

During the years that followed publication of *The Exploration of Mars*, von Braun had little time to pursue plans for the exploration of the red planet. More immediately important was the task of developing ballistic missiles; and, beginning in October 1957, of rallying his team at the Army Ballistic Missile Agency in Huntsville, Alabama to respond to the challenge of the Soviet Sputnik triumph [von Braun and Ordway, 1966; Ordway and Liebermann, 1992; Stuhlinger and Ordway, 1994]. A few years later, in the spring of 1961, President Kennedy announced the goal of landing man on the Moon within the decade. For von Braun, whose Army team in the meantime (July 1960) had been transferred by NASA to become the George C. Marshall Space Flight Center, this meant focusing his resources to the giant Saturn series of launch vehicles that were to make the lunar expeditions possible.

Nevertheless, as preparations for the Apollo program gained momentum, von Braun did assign members of his team to continue to study the feasibility of manned Mars exploration. After all, he reasoned, the conquest of the Moon would assuredly pave the way to that more ambitious goal. At about the same time, small groups were conducting similar studies at NASA's Ames Research Center in California; the Langley Research Center in Virginia; the Lewis Research Center in Ohio; and the Manned Spacecraft Center in Texas.

In October 1989, at the 40th International Astronautical Congress in Torremolinos, Spain, Franklin P. Dixon presented a survey paper entitled "Manned Planetary Mission Studies from 1962 to 1968" [Dixon, 1989] in which he noted that "The logical beginning of this emphasis [e.g., planning manned planetary missions] was at the Future Projects Office . . . of the NASA Marshall Space Flight Center . . ." And, indeed, it was there, under Heinz Hermann Koelle and his deputy Harry O. Ruppe, that three contractor studies were soon to be sponsored.

THE EMPIRE STUDIES

To set the stage, in the spring of 1962 a work statement and a procurement request were prepared by NASA laying out tasks to be accomplished to determine the feasibility of manned scientific missions to Mars as well as to Venus during the 1970-72 time period. The ground rules stated that the anticipated contractor-conducted studies would be based on projected state-of-the-art (early 1970's) nuclear propulsion technology, and would consider a number of missions (Mars and Venus flybys, Mars and Venus orbital reconnaissance, and Mars orbital reconnaissance and eventual surface landing).

The total duration of any mission was not to exceed 1-1/2 years and nuclear propulsion was to be used for escape from Earth orbit, for braking into planetary orbit, for escape from planetary orbit, and for any propulsive braking required when returning towards Earth. The studies were only to concern Earth launch vehicles insofar as they were dictated by spacecraft requirements. The crew was to be completely integrated into vehicular systems, and the spacecraft designs were to take into account experience gained in the American Mercury, Gemini, and Apollo programs as well as on-going space station studies. Mars surface landing missions were to be accomplished using a planetary excursion module. Preparations for the Mars initiative were duly reported in the trade literature [Stone, 1962].

Implicit in the initiative, which came to be known as EMPIRE (for Early Manned Planetary-Interplanetary Round-trip Expeditions), was the use of nuclear propulsion. During the study period, the feasibility of such propulsion was being demonstrated by the joint NASA-Atomic Energy Commission NERVA (Nuclear Engine for Rocket Vehicle Application) solid core fission reactor program in which hydrogen was heated to high temperatures yielding exhaust velocities approximately twice those achievable from chemical engines. Building on the success of the Kiwi-A series of reactor tests, in July 1961 the NASA-AEC Space Nuclear Project Office awarded a contract for engine development to the Aerojet-General Corporation; its principal subcontractor was the Westinghouse Astronuclear Laboratory responsible for the flight reactor. During the course of the program thrust levels up to 200,000 pounds were realized. Figure 9 shows a nuclear rocket reactor test at the Nuclear Rocket Development Station at Jackass Flats, Nevada, conducted in June 1968 that lasted for 32 minutes. Figure 10 is the NERVA XE engine undergoing build-up at the Engine Maintenance, Assembly and Disassembly Building.

Building on studies conducted in the late 1950s and early 1960s by government, industrial and individual researchers, in May 1962 NASA's George C. Marshall Space Flight Center selected three contractors to undertake 6-month, 6,000-man-hour EMPIRE studies [NASA, 1962]. The three were the Aeronutronic Division of the Ford Motor Company, General Dynamics/Astronautics, and the Lockheed Missiles and Space Company.

Figures 11, 12 and 13 depict vehicle configurations that emerged from the three studies.

Ruppe, who served as NASA-Marshall study director on the project, recalls that the acronym EMPIRE was suggested by his assistant Jerry Smith and that the enterprise itself originated from discussions held earlier with von Braun and Koelle. "The original requirement," according to Ruppe, "was to study the application of Apollo hardware for early manned planetary missions. Those were limited to fly-by missions. We assumed Saturn 5 (then often referred to as C-5) and Earth orbital operations (without a space station) as Earth launchers. You might remember that at that time in the large assembly building at the Cape [Canaveral] we had a theoretical capability to process several Saturn 5 vehicles in parallel (I do not remember how many, maybe 3). This limitation was part of the input" [Ruppe, 1989].

Figure 9   Nuclear rocket test at Jackass Flats, Nevada's Nuclear Rocket Development Station, 26 June 1968.

Figure 10 Close-up photo of the NERVA XE engine undergoing build-up at the Engine Maintenance, Assembly and Disassembly Building, Jackass Flats, Nevada in 1969.

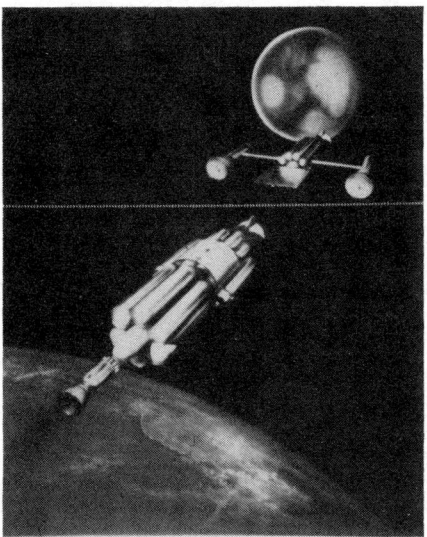

Figure 11 Aeronutronic's spaceship departing Earth under nuclear power with the Mars' arrival craft shown above making the flyby.

Figure 12 General Dynamics/Astronautics' manned spaceship above and cargo spaceship below.

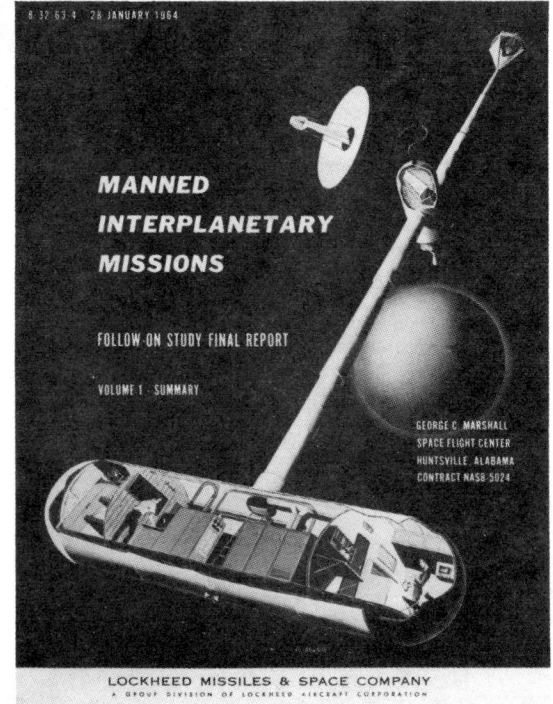

Figure 13 Lockheed's Mars spaceship configuration.

The results of the EMPIRE studies, which were funded at $250,000 each, revealed that manned missions to the two planets appeared feasible during the early 1970s. As Ruppe noted, it was assumed that launch into orbit from the surface of the Earth would be accomplished by the Saturn C-5 (later, 5 or V, possibly with variations) booster configuration though some consideration was also given to the Nova then under study as well as the so-called Super Nova.

The EMPIRE studies [Ordway *et al.*, 1993, 1994, 1996] conducted under the sponsorship of NASA-Marshall's Future Projects Office had reached sufficient maturity by mid-decade to permit von Braun to publish—in November 1965—an article entitled "The Next 20 Years of Interplanetary Exploration" in the journal *Astronautics and Aeronautics* [von Braun, 1965]. Therein, he demonstrated the feasibility of establishing a semi-permanent foothold on Mars by taking maximum advantage of evolving Saturn launch vehicle and Apollo lunar spacecraft hardware and know-how.

POST-EMPIRE APPROACH TO MARS EXPLORATION

Von Braun considered four manned planetary missions: a flyby of Venus in 1975; a flyby of Mars in 1978; a stopover mission to Mars in 1982; and the establishment of a semi-permanent base on that planet with no timetable offered. "All four missions," he wrote, "are supported with a logistics supply system which operates between the surface of the Earth and an orbit around the Earth, and which comes directly out of the Saturn V launch vehicle." He called his logistics vehicle the MLV-3, which, he explained, was one of the many configurations being studied by his team at the Marshall Center. In essence, it was an uprated Saturn V.

> Specifically [von Braun continued], whereas the normal Saturn V, in normal lunar orbital configuration, has a takeoff weight of 6 million lb, the MLV-3 vehicle has been uprated to 7.2 million lb. The thrust of the Saturn, normally 7.5 million lb, has been uprated to 9 million. Whether we get this uprating by uprating the F-1 engine, five of which power the Saturn V first stage, or through the use of solid strap-on boosters, was not specifically investigated, but it has no bearing on the rest of the problem here.
>
> The load-carrying capability of this MLV-3 vehicle is 310,000 lb of net payload into a 495-km orbit at 30-deg inclination. The standard Saturn, for comparison, in a two-stage configuration, lugs about 220,000 lb into a comparable orbit. So the uprating is really quite moderate and entirely within the framework of the time scale for such planetary expeditions...

Von Braun explained that the trip to Mars and return would take 682 days. For his flyby configuration, he planned to use the then-under-development Nerva II nuclear propulsion module which, when fully developed, was to be capable of producing some 230,000 pounds of thrust. His earlier reservations about applying nuclear energy to space travel had, by the mid-1960s, been modified.

> In order to be able to carry out the entire flyby and have enough propellant for the mission [von Braun continued] we have to bring an extra tank of 68,000-lb hydrogen capacity up into orbit. This tank, added to the propulsion module of 207,000-lb capacity, the maximum that one Saturn V could carry, would give us enough velocity

change capability to fly this mission. So, actually, in support of the mission, we have two logistics flights to Earth orbit—one that brings the spacecraft plus the propellant module up, and the other that brings the propulsion module up. . .

To von Braun, the flyby was but an interim mission; the real thing was what he liked to call the "Mars stopover." And by "stopover" he was considering a mission that would last 456 days, with 20 days on Mars—in his example, from 4 to 24 August 1982. The mission required a more complicated vehicle than employed for the flyby, but would still be based on already identified modules. "Our main building block," he proposed, would be "what we call the standard propulsion module, or SPM, which is actually the nuclear-propulsion module." Three SPMs would be needed to lift a total of 2.627 million pounds out of Earth orbit. The so-called "first push booster, or "leave-Earth" stage, would be composed of three parallel-strapped SPMs, each powered by a single Nerva II engine.

Then there was the "arrive-Mars" stage required to execute maneuvers at the Mars end of the interplanetary trajectory. "The booster is dropped, of course, after the injection into the interplanetary ellipse is begun," wrote von Braun, "and an arrive-Mars stage consisting of one standard propulsion module and an extra-large tank will deboost the vehicle into a circular Mars orbit." Following the 20-day visit to Mars, the crew would return to Earth using nuclear power, with a much smaller extra tank of 67,000 pounds of hydrogen content.

The spacecraft contemplated for the mission not only would accommodate the crew during the interplanetary portion of the trip, but would also perform the maneuver of landing on the surface of Mars and later re-ascending into orbit. For what he termed the Mars excursion module, von Braun relied on a lifting-body type of vehicle. This would permit the rather tenuous Martian atmosphere to slow down the craft until its retro-rocket unit took over for the final phase of descent.

After reviewing in considerable detail vehicular design and performance relative to the 20-day stopover mission, von Braun addressed the question of a semi-permanent presence on Mars:

"Obviously," he observed, "this is a far more demanding task. It turns out, and I think this is hardly surprising, that the easiest way of doing it would be to send the first expedition to Mars without a complete return capability. In other words, give this crew all the weight that they would otherwise need in a return flight for extended survival on the planet, and tell them that the Congress willing, we would hope to raise enough money the following year to pick them up again." (A few years later in the film *2001: A Space Odyssey,* Stanley Kubrick and Arthur C. Clarke would introduce this concept for the spaceship "Discovery," which had the capability of traveling one-way to Jupiter. The mission plan called for its crew to be rescued later by a more advanced vehicle with round-trip capability.)

"If we do that," continued von Braun, "then we would have the capability of leaving a group of 12 men on Mars for 1-1/2 years, with all the life-support equipment and ground transportation and everything a man would want there. And this would not weigh any more than the round-trip of the eight men that we mentioned earlier. Then, the following year, we would just have to make one more such trip to Mars to pick them up again." The advantage of this approach would be that the dozen men would not have to take along a departure stage (the nuclear departure stage) on the first flight, but instead would carry along six Mars excursion modules (four to be converted to cargo landers). The men would enjoy the support of 95,000 kilograms (190,000 pounds) of useful payload, half for life support and half for transportation and scientific equipment, etc.

As for timing, if the flight to Mars were to take place in 1984, the retrieval there of the 12 astronauts would take place in 1986. This scheme would ease the overcrowding of the launch facilities at the Cape because, as von Braun pointed out, "the pickup mission would have its own logistic schedule."

Von Braun contemplated establishing "a little village on Mars" to be built up of six MEMs, two of which would be cargo carriers. He suggested that two MEMs be converted into shelters capable of housing five or six men and their life support equipment. MEM shelters would also provide laboratory space in which studies and experiments would be carried out.

Of the six MEMs, only two would be capable of returning to Mars' orbit. This they would do after the twelve explorers had spent a year and a half on the planet. There, the "pickup mission" would rescue them "and boost them with nuclear power back to Earth."

Von Braun pointed out that the cost of such an expedition would be in "the same ballpark" as what was at the time being spent for Apollo. He noted that six Saturn V launch vehicles were then under construction. These he compared to the twelve needed for the Mars mission.

"So," he went on, "while the cost of the manned space program if we envision expeditions like this, may go up somewhat, the fact still remains that the main elements that would support even such ambitious space operations are a byproduct of our present Saturn-Apollo program." As a consequence, he felt that the taxpayer would not have to support a greatly increased financial burden to make the Mars mission a reality. "Actually," von Braun summed up, "it would become possible roughly within the framework of what people are already used to supporting."

Having said this, he posed the inevitable question, "Why do we have to go to Mars?"

> Well [answering his own question], I think that we have to keep trying to expand our sphere of activity. Besides, I have not heard of a good scheme to legislate away the population explosion as yet, so maybe one of these days we'll be starting with an air-conditioned environment and air-conditioned comfort right at the beginning, so we do not have to depend on covered wagons and horse-and-buggy systems.
>
> You will also have gathered from my paper here that I am an optimist with respect to our space program. However, even if those pessimists should prevail with their gloomy predictions that science has brought mankind close to the abyss, and that it's

only a short time before man will blow himself up on this planet, let me suggest that even that seems to indicate that we should pursue this course of going to Mars. In fact, we should indeed hurry, so that we can establish a foothold on a new planet as long as we have one left to take off from.

CHANGING BUDGETARY CLIMATE FOR ADVANCED PROGRAMS

Ironically, as von Braun continued to promote through articles, speeches, and private discussions his plans for the manned exploration of Mars, NASA's budget was peaking and beginning a decline that would last for a decade. From a high of $5.25 billion in fiscal year 1965 it steadily contracted to a low of $3.03 billion in fiscal year 1974. And even after that, when the numbers finally began to rise again, space mission planners had to take into account the ravages wrought by inflation in the intervening years.

Despite the somber budget situation, von Braun did not give up. One more avenue seemed open to him.

As the Apollo program matured in the late 1960 decade, NASA managers began to ponder what to do after the Moon had been conquered. Many proposals were put forth, mostly dealing with the consolidation of man's presence in Earth orbit and on the Moon. But von Braun, with the full support of NASA Administrator Thomas O. Paine and Vice President Spiro T. Agnew, talked of a more distant goal for post-Apollo America: Mars.

Taking advantage of the Apollo 8 and 10 circumlunar and Apollo 11 lunar landing successes of 1968 and 1969, von Braun gave energetic support to a national commitment to manned exploration of the red planet. His persuasive efforts were aimed not only at the NASA management structure, but at individuals within the President's National Aeronautics and Space Council and the President's Science Advisory Committee, various Senate and House committees on Capitol Hill, and elsewhere.

Shortly after the November 1968 election, president-elect Richard M. Nixon established a task force under Science Advisory Committee member and Nobel laureate in physics Charles H. Townes. Its purpose: to provide advice on the nation's post-Apollo space posture. Just before Nixon's inauguration in January 1969, the Townes task force recommended that the nation not embark on costly new space endeavors, including "a manned planetary expedition."

SPACE TASK GROUP DELIBERATIONS

On 13 February shortly after his inauguration, the new president established a Space Task Group charged with developing recommendations for America's future posture in space. Chaired by Vice President Agnew, it had three members (NASA Administrator Thomas O. Paine, Secretary of the Air Force Robert C. Seamans, Jr., and the President's Science Advisor Lee A. Dubridge), and three observers (Robert P. Mayo, Director of the Bureau of the Budget, Under Secretary of State for Political Affairs U. Alexis Johnson, and Chairman of the Atomic Energy Commission Glenn T. Seaborg). The first meeting took place in the vice president's office at the Executive

Office Building at 9 o'clock in the morning, and was attended by Agnew, DeBridge, Seamans, Paine and Mayo. Other meetings followed.

As the STG deliberated, it became clear that the opinion was quite divided, with only Agnew and Paine arguing for a major post-Apollo space program that might include a manned expedition to Mars. Agnew in particular persisted. Paine recalls that during the STG meeting with congressional leaders on 16 May 1969, the vice president "emphasized that the new goals must be simple, attractive, dramatic, and easily comprehended, just as the moon landing goal had been in the first decade of space. In his opinion, space applications would not galvanize this nation into supporting a bold space program, whereas an eventual Mars expedition (open-ended, without a final date) might do so, in concert with other projects in a balanced space program."

While STG pondered the future of space flight, other groups were beginning to offer opinions on what America should be doing in space once the Apollo exploratory expeditions had been completed. Most of them directly or indirectly attempted to influence studies then going on within NASA. At the same time, STG individual members were articulating their positions; thus, on 4 August 1969, Seamans wrote to Vice President Agnew "*I don't believe we should commit this Nation to a manned planetary mission, at least until the feasibility and need are more firmly established.* Experience must be gained in an orbiting space station before manned planetary missions can be planned... A decision to travel to Mars, the only accessible planet, would require new launch vehicle stages and spacecraft modules and a greatly increased annual outlay that would compete with the resources needed to provide immediate benefits from NASA's compatibility."

NASA, meanwhile, was considering options for the future. In early September 1969, it issued a report to the Space Task Group entitled America's Next Decades in Space [NASA, 1969]. Among several possible programs examined by space agency planners, one offered "the first manned expedition [to Mars] in 1986 or 1988."

All the while, the STG was completing its own report for the president. Entitled *The Post-Apollo Space Program: Directions for the Future* [Space Task Group, 1969], it appeared on 15 September. "As a focus for the development of new capability," it said, "we *recommend* [report's emphasis] the United States accept the long-range option or goal of manned planetary exploration with a manned Mars mission before the end of this century as the first target." In his transmittal letter of the same date, the vice president wrote the president:

> The three program options presented in the Space Task Group report for NASA lie between the bounds of a vigorous expansion over our present space activities and an undesirable level of retrenchment which involves termination of manned space flight activities. The three program options recommended for your consideration are Option I, which reaches a maximum expenditure of $9.0 Billion in 1980, and involves a decision to undertake a manned Mars mission in the early 1980's; Option II, which reaches expenditures of about $5.5 Billion in Fiscal Year 1976, and would include launch of a manned mission to Mars about 1986, thereby resulting in peak expendi-

tures of about $8.0 Billion in the early 1980's; and Option III, which defers decision on a Mars manned mission until after 1990.

Agnew went on to note that the STG did not recommend any one program option above the others but did "unanimously recommend that 'the United States accept the long-range option or goal of manned planetary exploration with a manned Mars mission before the end of the century as the first target.'" Personally, Agnew opted for the second option that would result in man on Mars by 1986. Paine took the same view, writing to Nixon on 19 September with the recommendation that "you select Option 2, a balanced and challenging program which includes as major objectives the Earth-orbiting space station, space shuttle and nuclear stage in the 1970's, leading to a manned mission to Mars in the 1980's." Figures 14, 15, 16, 17 and 18 show various aspects of the proposed manned Mars mission based on concepts developed during 1969 and 1970.

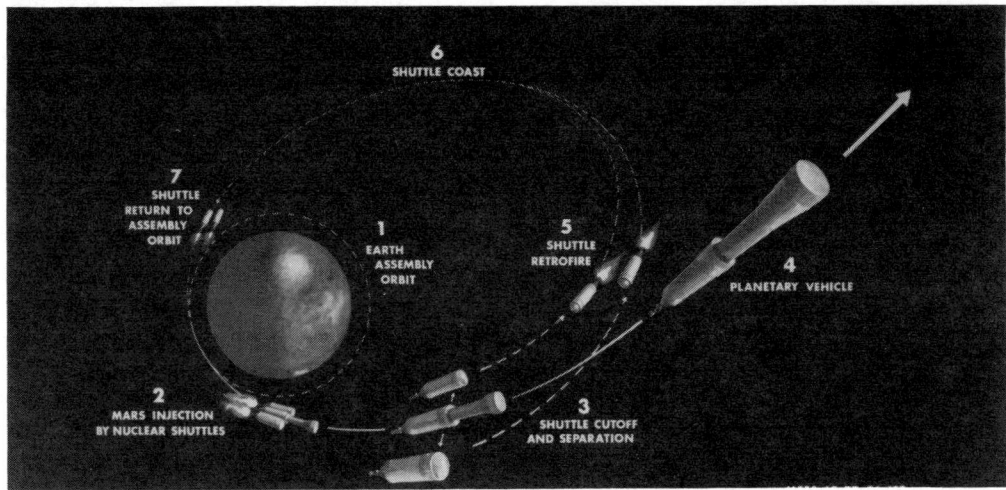

Figure 14 Diagram showing departure maneuvers in Earth orbit of Mars nuclear-powered spaceship. Note separation of nuclear boosters (3) after which the main ship (4) continues along its trans-Martian trajectory.

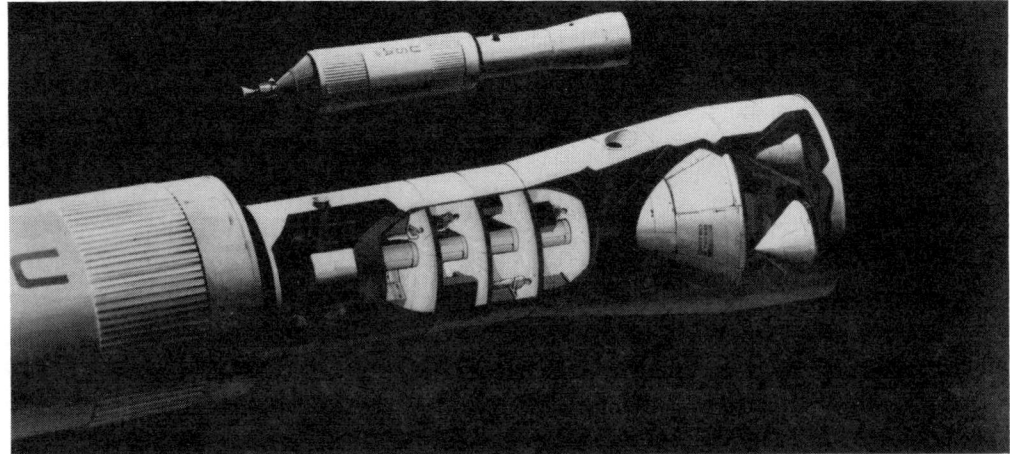

Figure 15 En route configuration of the Mars-bound spaceship.

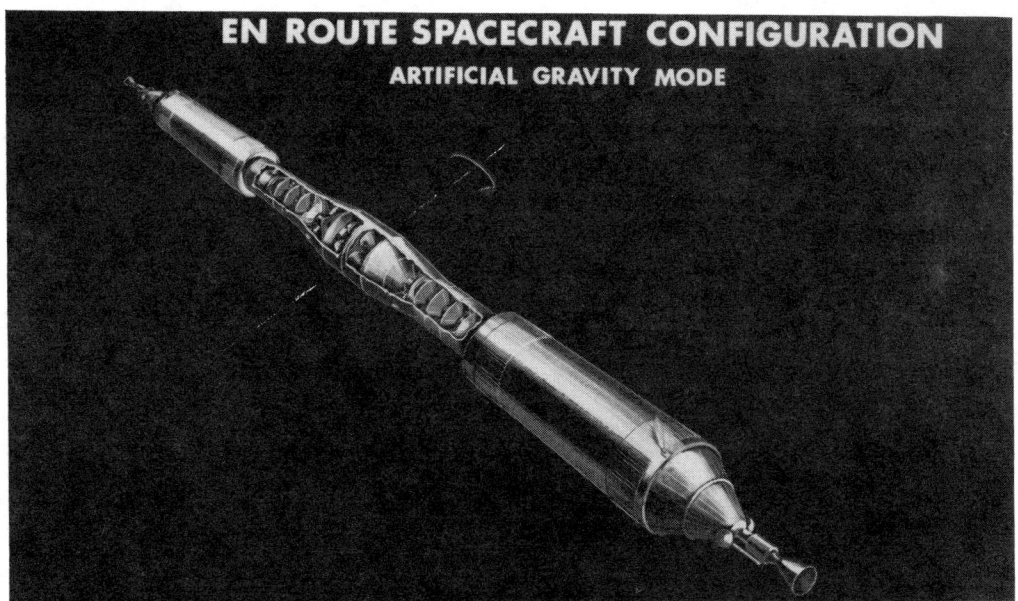

Figure 16 By spinning the spaceship, artificial gravity could be provided for the crew on the long interplanetary voyage.

Figure 17 A Mars Excursion Module has descended onto the Martian surface where preparations are made to begin the exploration phase of the mission.

Figure 18 The Multiple Mars Excursion Modules make up a temporary base on the Martian surface.

Paine also suggested to the president that he make a public statement in which he would outline his views on space. He even suggested the event: the forthcoming dedication of the new Lunar Science Institute at Houston.

But Nixon demurred. The autumn passed and most of the winter. In March 1970, the Space Science and Technology Panel of the President's Science Advisory Committee took up the manned Mars option in its *The Next Decade in Space* report. "Considering the lead times involved," it suggested, "the prerequisites [to manned missions, e.g. demonstration of long-duration manned flight in Earth orbit] could be fulfilled for a manned Mars mission starting somewhere after 1985." But, the panel cautioned, "if unmanned missions show Mars to be less interesting than we expect, the investment for manned Mars exploration could be indefinitely postponed without having lost a larger investment in mission-specific technology."

Into the debate on Mars enthusiastically strode Wernher von Braun, the manned Mars mission's most vigorous champion. His first major appearance was before the Space Task Group on 4 August 1969. He repeated his presentation the very next day at a hearing before the Committee on Aeronautical and Space Sciences of the United States Senate.

Von Braun had enjoyed outlining his thoughts to the STG on promising new post-Apollo space activities, and was pleased that Agnew and Paine were solidly behind him. He was less sure of the other members, but had no doubt whatever where the Bureau of Budget Director Mayo stood. Budgeteers were rarely enthusiastic about space, von Braun knew, and this one was no exception.

The 5th fell on a Tuesday. After a light breakfast at his hotel, he was driven to NASA Headquarters and then, well before the appointed hour of 10, to Capitol Hill. This would not be a routine appearance, for today von Braun would be proposing something out of the ordinary. He'd better be in top form, for there was sure to be some questioning after his prepared remarks.

Arriving at the Old Senate Office Building, von Braun walked briskly to room 235 where the hearing was to take place. He was accompanied by NASA Administrator Paine; Robert F. Allnutt, Assistant Administrator for Legislative Affairs; Dr. George E. Mueller, Associate Administrator for Manned Space Science and Applications; Dr. Homer E. Newell, Associate Administrator; William E. Lilly, Assistant Administrator for Administration; Willis H. Shapely, Associate Deputy Administrator; and members of their staffs. They had to wait for the arrival of a tardy chairman, Senator Clinton P. Anderson.

"I apologize for being late," said the chairman of the Committee on Aeronautical and Space Sciences, "but I was almost here when something came up in the office and I could not get out of it." The New Mexico senator then recited the subjects that Paine and his assembled team would be covering that morning before the assembled senators (Stuart Symington, Missouri; Stephen M. Young, Ohio; Howard W. Cannon, Nevada; Spessard L. Holland, Florida; Margaret Chase Smith, Maine; Mark O. Hatfield, Oregon;

Barry Goldwater, Arizona; William B. Saxbe, Ohio; and Charles McC. Mathias, Jr., Maryland.)

John E. Naugle would lead off by describing the Mariner and Viking unmanned missions to Mars. Von Braun would follow and focus on the same planetary objective—with a staggering difference: his mission would be *manned*. Whereas Naugle would quote something like $700 million for the unmanned Viking experiment then being planned, von Braun would be talking billions. Following von Braun, Mueller would cover a variety of subjects including the on-going Apollo lunar exploration program and such advanced concepts as the Saturn V "workshop" or space laboratory, the space shuttle, the space tug, the space station, the nuclear lunar shuttle, and the establishment in lunar orbit of a space station module.

All the projects and concepts described that morning were exciting, but none challenged the imagination so much as a manned exploring expedition to the red planet. That was the world of von Braun's boyhood dreams, dreams that had never left him. With the triumph of the first Apollo lunar landings fresh in everyone's mind, he reasoned, the time was ripe to secure commitments for the next big thrust into space. To von Braun, momentum was crucial; one should never let up. Moreover, it was difficult to believe that America would wind down its majestic space capability after a few visits to the nearby Moon. There were many beckoning worlds in the solar system. And none was more fascinating than Mars.

Before von Braun got down to the business at hand, NASA Administrator Paine emphasized to the senators that "In no event will it be necessary for us to make . . . a final decision in the next few years for such an expedition [to Mars]. But," he continued, "the question is whether or not, as we make our plans for other components [of NASA's post-Apollo space program], whether such an expedition will be visualized as an eventual possibility, which will then serve somewhat as a focusing requirement on such things as a space station and a space shuttle." As pointed out by space policy historian John M. Logsdon, "Paine thought that, in the wake of the Apollo 11 success, chances for approval of major new space projects were as great as they were ever likely to be. He saw the Mars mission as an 'offer' that NASA should make to the country, to undertake another tremendously challenging but exciting national enterprise like Apollo."

By way of introducing his own testimony, von Braun recalled to the assembled senators and their staffs that while one can fly to the Moon every month, launch opportunities for Mars occur approximately every two years. "It is for this reason," he explained, "that my mission description includes some hard departure and arrival dates, but I ask you not to construe this to mean that this is a hard sales proposal, so to speak. It is only necessary in this case to tie a complete mission analysis to hard departure dates in order to make the story consistent all the way through." He proceeded to describe a now familiar mission involving a 12-man crew traveling in two ships.

> The two ships will make the trip departing from Earth on the 12th of November 1981. The ships increase their initial orbital motion around the Earth to the velocity of departure required to enter a circumsolar ellipse. This will carry them during a

period of 270 days almost half around the Sun to the planet Mars... The expedition will arrive in Mars orbit on August 9, 1982, and will enter an elliptical orbit around Mars with the help of an orbit injection maneuver.

The ships will then stay for a period of 80 days in the elliptical orbit around Mars. From this orbit around Mars they will perform various surface landing operations. Six of the total of 12 crew members will have an opportunity to go down to the surface, three from each ship. The other six men will remain as shipkeepers in the two orbiting vehicles.

In October 1982, 80 days after arrival, the ships will leave Mars orbit again and will make a swingby of the planet Venus on their way home...

The ships, therefore, cross the orbit of the Earth... and they will make a swingby of the planet Venus 123 days after leaving Mars. They will send two unmanned probes into the Venus atmosphere. As the vehicles swing by Venus, they lose some of their circumsolar orbital velocity which will be helpful to reduce the approach velocity to the Earth; 167 days later, on August 14, 1983, the ships will swing back into an orbit around the Earth from which the crew returns to the Earth's surface...

The ships are identical. So we could actually split the whole mission in half and fly only with one ship [and] this would still make a good and respectable expedition. Our proposal to use two ships is based on the thought that ship redundancy will be particularly helpful when the crew is so far away from the Earth that any idea of help provided from the Earth is entirely out of the question. That is why we are going back to the tradition of the old sailing ships. Columbus, as you will remember, took three vessels when he sailed west, and I think the record shows that he would never have returned to report his discovery had he not provided that redundancy in his system.

So if we lose one ship en route, if one ship became incapacitated and unable to return, then its six-man crew could return in the other ship. It will be a little more crowded, but it would still be entirely acceptable to return all 12 men in one ship...

At this point, von Braun spent some time in describing the NERVA nuclear propulsion system needed to make the round-trip flight to Mars, the maneuvers undertaken on entering orbit around Mars, and activities conducted in orbit prior to the initiation of the crucial landing phase. His plan included the dispatch of an unmanned surface return sampler.

Finally, the time arrives for the men to enter their descent craft which, while larger, is not unlike the Lunar Module employed during the Apollo program. But unlike Apollo, the Mars descent craft takes advantage of the thin Martian atmosphere to help brake its downward velocity. A retro-rocket engine provides final deceleration power, permitting the craft to make a Lunar Module-type landing on the surface.

We land on [the] landing gear, open a hatch again, get a vehicle out which provides mobile transportation. We also have quarters here to accommodate the crew for maybe a stay time on the Mars surface of a month or so, so there will be sleeping and cooking facilities and some equipment to take a first glance at samples, maybe some microscopes, maybe some infrared capability, and so forth...

Von Braun next looked at some of the experiments and observations to be undertaken during the mission. These included gravitational and magnetic field measurements, studies of the surface and subsurface composition of the planet, the nature of the atmosphere, and characteristics of any life forms that might be discovered. He also considered the possibility of carrying organisms from Earth to see how they would react to the Martian environment. He recognized the importance of finding water and other useful resources. The latter he considered particularly important; "we have," he emphasized, "to learn to live off the land, so to speak."

Despite support from Vice President Agnew, NASA Administrator Paine, and many others, the reaction to the manned Mars enterprise was cool. At the 5 August hearing, Senator Mark D. Hatfield or Oregon spoke of

> . . . a growing feeling that the space program is one of these optional aspects of our national commitment, and that a polarization is taking place. . . . I do not think there is sufficient communication to impress enough people at this point that what we are doing in space is worth the continued level of expenditures. I think the euphoria of landing men on the Moon is going to fade. I admit it is one of the most magnificent things I have ever witnessed. I think you are all to be commended. But we still have to deal with . . . political realities.

Robert F. Allnutt recalls vividly the sentiments of the time [Allnutt, 1985].

A von Braun appearance on Capitol Hill always posed both an opportunity and a problem for us good gray bureaucrats in NASA headquarters. The opportunity lay, of course, in the fact that Wernher was extremely popular in many quarters and had that celebrity quality that made Members of Congress want to be associated with him and, of course, the background of solid accomplishment that gave him great credibility. The problem lay in the fact that in those days NASA assiduously sought to avoid either the appearance or the reality of seeking support for anything that was not in the President's budget. Since our budgets were under some pressure on the Hill . . . certain of our protectors [there] also had that view, and were anxious not to have the boat rocked. So, for example, while "Tiger" Teague was always after us to be more interesting and imaginative in our appearances, Clinton Anderson was one who very much wanted controlled presentations.

This was particularly true with regard to anything that related to Nerva. Anderson was the person in political life most interested in Nerva. It is almost certainly because of him that he nuclear rocket was included in President Kennedy's famous speech to the joint Session of Congress laying down the Apollo challenge (a fact often overlooked). Nerva was important to New Mexico and the Atomic Energy Commission, over which Anderson exercised great power and which he strongly supported.

The difficulty with Nerva was that it was often linked to a manned Mars mission. Of course, this was the main actual justification for it, although we concocted others so that it would not be tied to any one mission. Anderson was always fearful of the kind of attack that would say 'the only reason we are building Nerva is to go to Mars. A Mars mission will cost $100 billion. Therefore, we can save $100 billion by canceling Nerva,' a two billion dollar project.

This, indeed was the kind of attack sometimes mounted. For, every year in hearings before the Senate Committee on Aeronautical and Space Sciences, the Chairman would ask 'Is there any money in this budget for a manned mission to Mars?' And every year we would respond 'no'.

Allnutt emphasized that von Braun was carefully briefed before his presentation *not* to leave the feeling that he was talking about a project being proposed by NASA or on which the agency was working.

> My recollection of the presentation itself [recalls Allnutt] is that the Committee was spellbound. I don't know if this shows through in the transcribed questions. Sometimes cold print takes all the life out of actual events. But I recall that particularly Chairman Anderson and Margaret Chase Smith were enthralled with the presentation. Wernher had a way of putting things in such a totally exciting and captivating form, and of putting so much enthusiasm into his remarks, that it was difficult even for the skeptics not to be led into some of the enthusiasm...

Dr. Glen P. Wilson, a long-time staff member of the Senate space committee, has similar recollections of von Braun's appearance before the assembled senators and their staffs: he

> was perhaps the most articulate witness we ever had before the committee. He had a way of presenting complicated scientific and technical material in a manner that was both understandable and exciting. One had confidence that all these wonderful and fabulous things could really be done. And for a man whose mother language was not English, he had a mastery of his adopted tongue that was as brilliant as it was astounding [Wilson, 1989].

Despite favorable reaction by individual senators, in general Washington was cool to a major post-Apollo program designed to culminate in manned landings on Mars. Nixon, as we have seen, carefully avoided committing his administration to a major post-Apollo enterprise. To the president, an expensive new program would be unattractive politically; and, in any event, would unlikely be completed during the eight years he expected to be in office. He was also keen on controlling his budget after what he considered the excesses of the Johnson years.

So for the time being Nixon said nothing, despite Paine's plea that "it would be advantageous for you to make a public statement on your view of the nation's future in space." Nixon preferred, it seemed, to follow the advice of his Bureau of Budget chief Robert P. Mayo: maintain silence until the fiscal year 1971 budget had been determined. That budget, it turned out, was far from robust; in fact, it was so lean that Paine had to suspend Saturn V production, stretch out the Apollo lunar landing schedule, and take other austerity moves.

Daily, it became clearer that Nixon was not going to endorse a program built upon manned Mars expeditions as a focusing strategy. Rather, in a space policy statement released from Key Biscayne, Florida on 7 March 1970, he noted that "space expenditures must take their place within a rigorous system of national priorities" and that "we should not try to do everything at once." He outlined three general purposes that should "guide our space program," namely exploration, acquisition of scientific knowledge, and practical application. He then laid out six specific objectives: continue to explore the Moon, "move ahead with the bold exploration of the planets and the universe," reduce substantially the cost of space activities, extend man's capability to operate in the space environment, accelerate the practical applications of space, and encourage increased international cooperation. Buried in the text was a short reference to manned Mars exploration: "There is one major but longer-range goal we should keep in mind as we proceed with our exploration of the planets. As a part of this program we will eventually send men to explore the planet Mars."

The Congress agreed with Nixon's go-slow philosophy. There would be no Apollo-scale initiatives for the 1970s and into the 1980s. The only element of the Space Task Group list of post-Apollo options destined to survive would be the Space Shuttle. On 15 September 1970, Paine resigned as NASA Administrator and with his departure the dream of going to Mars waned. Von Braun, who had transferred from Marshall to NASA Headquarters in February of that year to spearhead planning for the post-Apollo epoch, stayed on for a couple of more years. On 5 January 1972, President Nixon announced that development of the Space Shuttle within the context of a program to make access to near-Earth space routine and economical. In July, at the age of 60, von Braun retired from federal service, his dreams of man on Mars in his lifetime unrealized.

POSTSCRIPT

Proposals for manned expeditions to Mars were few and reserved during the 1970s. Von Braun rarely discussed the subject in speeches, though did devote a page to it in his last book *New Worlds: Discoveries from our Solar System* co-authored by Ordway [von Braun, 1979]. "While other unmanned missions will undoubtedly follow the Vikings. . ." they wrote, "the twenty-first century will most likely witness a manned mission to the red planet." On 16 June 1977, von Braun died in Alexandria, Virginia at the age of only 65.

REFERENCES

Allnutt, R.F., letter to Frederick I. Ordway III, dated 11 February 1985.

Dixon, F.P., "Manned planetary mission studies from 1962 to 1968," *History of Rocketry and Astronautics*, Vol. 17, *AAS History Series*, edited by John Becklake, Univelt, Inc., San Diego, California, 1995 (a volume in the International Academy of Astronautics history series).

Flammarion, C., *La planète Mars et ses conditions d'habitabilité,* Gauthier-Villars, Paris, 1892. See also his *Les Terres du ciel: Voyage astronomique sur les autre mondes,* C. Marpon et E. Flammarion, Paris, 1884.

Ley, W., W. von Braun, *The Exploration of Mars*, Viking, New York, 1956.

Lowell, P., *Mars,* Houghton, Mifflin, Boston, 1895.

Lowell, P., *Mars and its Canals*, Macmillan, New York, 1906.

Lowell, P., *Mars as the Abode of Life*, Macmillan, New York, 1908.

NASA, *America's Next Decades in Space: A Report for the Space Task Group.* Washington, D.C., National Aeronautics and Space Administration, September, 1969.

NASA's Project Empire, *Space Daily,* 29 June 1962.

Ordway, F.I. III and M.R. Sharp, *The Rocket Team*, Heinemann, London, 1979; Thos. Y. Crowell, New York, 1979; MIT Press, Cambridge, Mass., 1982; and Aircraft Designs, Inc., Monterey California, 1991.

Ordway, F.I. III and Liebermann, eds., *Blueprint for Space*, Smithsonian Institution Press, 1992.

Ordway, F.I. III, M.R. Sharpe and R.C. Wakeford, "Early manned planetary-interplanetary roundtrip expeditions," *Journal of the British Interplanetary Society*, Vol. 46 No. 5, May 1993 and Vol. 47 No. 5, May 1994.

Ordway, F.I. III, M.R. Sharpe, and R.C. Wakeford, "EMPIRE: Background and initial dual-planet mission studies," (in press, *History of Rocketry and Astronautics*, Vol. 19, *AAS History Series*, edited by J.D. Hunley, Univelt, Inc., San Diego, California (a volume in the International Academy of Astronautics history series), 1996.

Ruppe, H.O., Letter to Frederick I. Ordway III, dated 10 January 1989.

Schiaparelli, G.V., "Osservazioni astronomiche e fisiche sull 'asse di rotazione e sulla topografia del pianeta Marte...", in *Memoria I-VI*, Reale Academia dei Lincei, Rome, 1878-1899.

Space Task Group, *The Post-Apollo Space Program: Directions for the Future,* Space Task Group Report to the President, Washington, D.C., September 1969. See also von Braun's discussion of this report in "Now that Man has Reached the Moon, What Next?", an essay written in 1969 and previously unpublished, in *Blueprint for Space*, Ordway, F.I. III and R. Liebermann, eds., pp. 166-173, Smithsonian Institution Press, Washington, 1992.

Stone, I., NASA to analyze requirements for manned round trips to planets, *Aviation Week & Space Technology*, 20 April 1962.

Stuhlinger, E. and F.I. Ordway III, *Wernher von Braun Crusader for Space*, Krieger, Melbourne, Florida, 1994 (two volumes).

Von Braun, W., "Das Marsprojekt, Studie einer interplanetarischen Expedition," *Weltraumfahrt*, Unschau, Frankfurt/Main, 1952.

Von Braun, W., *The Mars Project.*, University of Illinois Press, Urbana, 1953.

Von Braun, W., with Cornelius Ryan, "Can we Go to Mars?" in *Collier's*, 30 April 1954.

Von Braun, W., *The Mars Project.*, University of Illinois Press (with a revised preface by von Braun) Urbana, 1962.

Von Braun, W., "The next 20 years of interplanetary exploration," *Astronautics & Aeronautics*, November, 1965.

Von Braun, W., Ordway, F.I. III, *New Worlds: Discoveries From Our Solar System*, Anchor Press/Doubleday, New York, 1979.

Von Braun, W., Ordway, F.I. III, *History of Rocketry and Space Travel*, New York, 1966, 1969 and 1975; Crowell, T.Y; updated with Dave Dooling as *Space Travel: A History*, Harper & Row, New York, 1985.

Von Braun, W., *The Mars Project*.: University of Illinois Press (with forward by Thomas O. Paine.), Urbana, 1991.

Waterfield, R.L., *A Hundred Years of Astronomy*, Macmillan, New York, 1940.

Wilson, G. P., interview by Frederick I. Ordway III, Washington, D.C., 7 December 1989.

A diagram of Buzz Aldrin's concept for cycling spacecraft between Earth and Mars.

AAS 95-478

Chapter 8

PATHWAYS TO MARS: AN OVERVIEW OF FLIGHT PROFILES AND STAGING OPTIONS FOR MARS MISSIONS

John C. Niehoff[*] and Stephen J. Hoffman[†]

This paper discusses possible trajectory and staging scenarios for piloted Mars missions to provide an overview of both traditional and innovative concepts for traveling to and from Mars. The characteristics and requirements of several alternative flight modes are briefly defined, discussed and compared, and the subject matter is addressed from three perspectives: Earth-Mars pathways, planet-centered pathways, and propulsion options. The concepts represent a variety of mission types and trajectories that are primarily defined as ballistic transfers, cyclers, and escalators, and they involve staging options for both Earth and Mars. Decisions about these missions and options will ultimately be affected by a number of evolving factors, such as technology development paths, and should be developed through an active process of study and debate under NASA leadership.

INTRODUCTION

The purpose of this paper is to provide an overview of Mars mission staging options by reviewing both traditional and innovative flight profiles to and from Mars. Flight time and orbital mass performance characteristics associated with these pathway alternatives will be presented to indicate their diversity.

The scope of this paper covers four subjects. First is a discussion of pathways between Earth and Mars followed by a review of some planet-centered strategies. Next is a discussion of propulsion options and their implications on performance as measured by the initial mass of the spacecraft. This paper concludes with a discussion of some of the mission design implications that these pathways suggest.

Mars' Influence on Mission Design

In a broad sense, round trip trajectories between Earth and Mars must involve more than just Earth and Mars. Information about Venus must be incorporated as well. Figure 1 illustrates the orbital relationship of the solar system's three best known inner planets. Earth's orbit is 1 AU (93 million miles) from the Sun, and Mars' somewhat

---

[*] Science Applications International Corp., 1501 Woodfield Road, Suite 202N, Schaumburg, Illinois 60173.
[†] Ph.D., Science Applications International Corp., 17049 El Camino Real, Suite 202, Houston, Texas 77058.

more eccentric orbit is about half again as far (142 million miles, average of the perihelion and aphelion distances). Venus, on the other hand, is about thirty percent (30%) closer to the Sun than Earth.

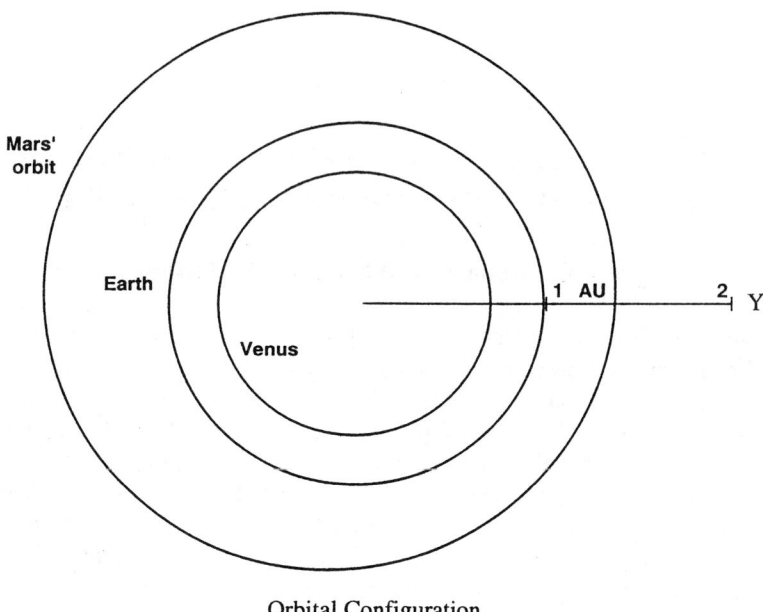

Orbital Configuration

Figure 1 The relative size and shape of the orbits of three inner solar system planets: Venus, Earth, and Mars. The orbits of Venus and the Earth are nearly circular, while the orbit of Mars is slightly elliptical. This non-circular nature of Mars' orbit causes some launch opportunities to be more favorable than others.

Mars' orbit is slightly inclined, at 1.85 degrees, relative to Earth's orbital plane. Its orbit is also somewhat elliptical, coming closest to the Sun at about 1.4 AU before moving out to about 1.6 AU. At this distance, Mars' orbital period is almost twice that of Earth's. Consequently, Earth and Mars move in and out of favorable phasing for transfers to and from each other. If the correct phasing were to occur along the reference line indicated on the illustration, it would take approximately 26 months (roughly two years) before the phasing was again appropriate, and it would have advanced (counter-clockwise) approximately 50 degrees. Consequently, over the span of seven of these intervals, the phasing moves completely around the orbit. Every 15 years (two phase revolutions), a complete cycle of phasing characteristics is repeated. Hence, good launch opportunities to Mars occur every 26 months, and they repeat nearly exactly in character and heliocentric orientation every 15 years.

PATHWAYS: ORBIT TRANSFERS

The discussion of pathways can be divided into three general categories: ballistic, low thrust, and cyclers. Each of these categories can be further subdivided into various classes based on trip time or energy, which are interrelated. Conjunction and Opposition class ballistic trajectories have been considered for piloted missions to Mars for a considerable period of time [Wilson, *et al.*, 1967]. Two of the simplest ballistic cases are illustrated in Figure 2. "Ballistic" means that a spacecraft coasts from one planet to the other, using an impulse provided by a high-energy chemical rocket stage to get started. A spacecraft rocket motor could also be used to achieve Mars orbit insertion (MOI) at arrival and again at departure.

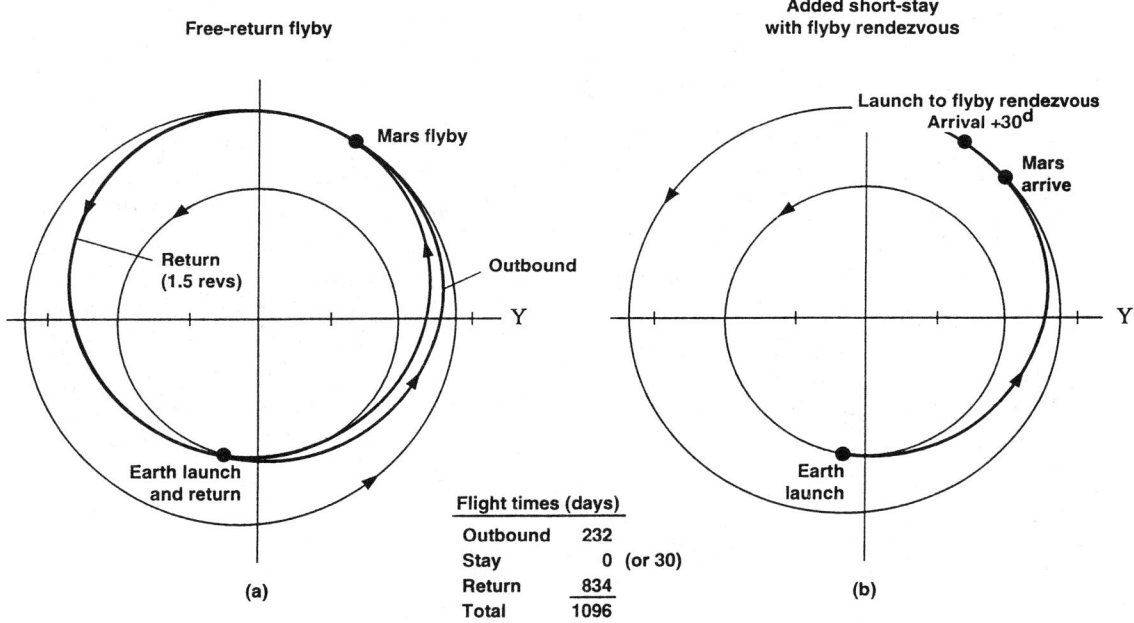

Figure 2 Two illustrative cases of ballistic transfers. Figure 2a illustrates a free return trajectory which will fly by Mars, without stopping, and then continues to orbit the Sun for 1.5 revolutions (for a total of two complete revolutions) before intercepting Earth. Figure 2b illustrates the one-way portion of a two part mission; the second vehicle which returns a crew to Earth uses the same trajectory as illustrated in Figure 2a, but the second vehicle is timed to arrive 30 days after the first, one-way vehicle.

Ballistic Transfer Missions

Free-Return Flyby (Single Spacecraft) — This case is the easiest to perform in terms of velocity change and thus minimizes the amount of propellant required. It is called the free-return flyby. After being launched at Earth, the spacecraft follows a shorter arc to Mars and simply flies by the planet (Figure 2a). This provides only a few hours to observe the planet from what can be a fairly significant distance, and then the spacecraft must continue on to complete 1.5 revolutions about the Sun (due to Earth-Mars phasing) before it returns to Earth. That amounts to about three years of flight time for approximately two hours of optimal viewing at Mars, a relatively unsatisfactory arrangement.

Short-Stay/Flyby-Rendezvous (Two Spacecraft) — Mission productivity can be improved by including a landing on Mars. The Short-Stay/Flyby-Rendezvous mission option involves sending a piloted interplanetary spacecraft (including a lander and an ascent/rendezvous vehicle) timed to arrive and land on Mars thirty (30) days before the anticipated flyby of a second interplanetary spacecraft (Figure 2b). Doing so provides a month of surface operation time before the ascent vehicle must take off to rendezvous with the second interplanetary spacecraft as it flies by, which then proceeds to complete the 1.5 revolutions back to Earth in essentially the same manner as a free-return flyby mission spacecraft (Figure 2a). In this case the second interplanetary vehicle never stops at Mars, making it a fairly efficient concept, since it uses no propellant for capture or escape maneuvers at Mars. But it still provides only thirty days of mission time at Mars as a payback for the rather long three year transfer.

Conjunction Class Mission. The Conjunction Class mission (Figure 3) is perhaps the most traditional concept, as well as being the most frequently studied mission type [Hoffman, *et al.*, 1989, Soldner, 1990]. It is a low-energy mission that has a relatively long Mars stay time. As Figure 3 shows, the trajectory transfer path to Mars spans about 180 degrees. At Mars arrival, Earth is moving into conjunction with Mars, hence the name Conjunction Class for the mission type. The two planets are out of phase for the return, so a long stay time of more than 1.5 years is required. This might be quite appropriate if there are many things to do, as would be the case after an early exploration stage. When the planetary phasing is again correct, the mission is concluded by a relatively short return flight. The long stay time and the shorter transfer times lead to about 2.8 years of total trip time for one of the lowest low-energy cases.

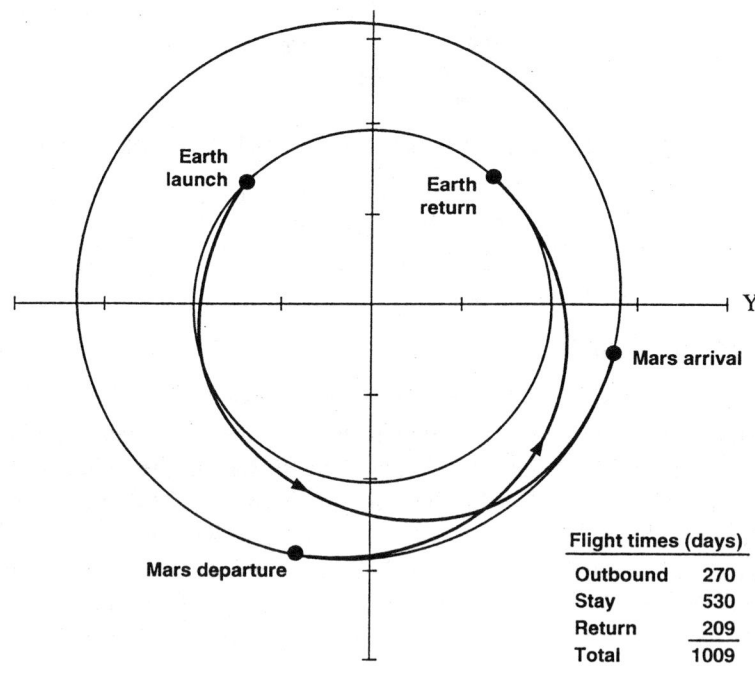

Conjunction Class Mission

Figure 3 A round trip mission using a Conjunction Class trajectory. This option generally requires the lowest amount of energy but the longest total time to complete. The stay time at Mars is dictated by the time it takes for Mars and the Earth to be properly aligned for the return trajectory.

Opposition Class Mission. If a significant improvement in the mission/trip time is desirable, a mission designer can move to the other extreme and use the Opposition Class mission illustrated in Figure 4. This mission type is a very high energy (i.e. a larger velocity change requirement and thus a larger propellant requirement) mission [Hoffman, *et al.*, 1989, Soldner, 1990]. Now an interplanetary spacecraft arrives at Mars as Earth is leaving opposition with Mars hence the identification as an Opposition Class mission. The stay time is relatively short, approximately 20 days, and then the spacecraft must rapidly get back onto a return trajectory to catch up with Earth (which is moving out of phase). To do so, the Earth-return vehicle must fall inside Earth's orbit (i.e. closer to the Sun) to achieve the higher velocity needed. This leads to a situation in which the spacecraft catches Earth at a very high approach velocity, which consequently requires higher velocity changes to become recaptured. The advantage of a high-energy mission of this class is that total flight time is reduced to about 1.6 years, but stay time permits only about two or three weeks on the surface of Mars.

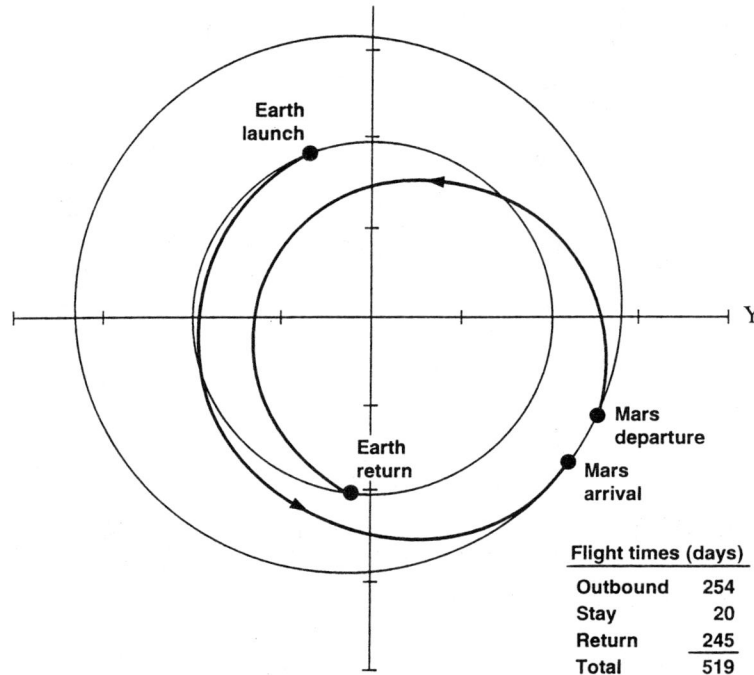

Opposition Class Mission

Figure 4 A round trip mission using an Opposition Class trajectory. This option reduces the total trip time for a mission but at the cost of significantly more energy to complete. The return trajectory must be one that allows the spacecraft to catch up to an Earth that is out of position for a minimum energy return, thus the passage much closer to the Sun than on a Conjunction Class trajectory.

To improve on this type of mission, it has been found that the addition of a Venus swingby makes shorter trip-time missions more favorable from an energy point of view, as reflected in Figure 5. The same type of outbound transfer is used, and a flight crew will now stay at Mars for roughly 60 days (two months instead of three weeks) before again beginning the catch-up maneuver on its return flight. But, on the inward leg for this mission case, a Venus encounter is possible. The swingby at Venus conveniently reshapes the orbit, giving the velocity change required, and also provides a tangential return at Earth—reducing the amount of energy to slow down. Total trip time increases slightly to about 1.9 years (about 0.4 of a year greater than the non-swingby mission), but this mission has a much more favorable velocity change requirement and provides a more reasonable stay time at Mars.

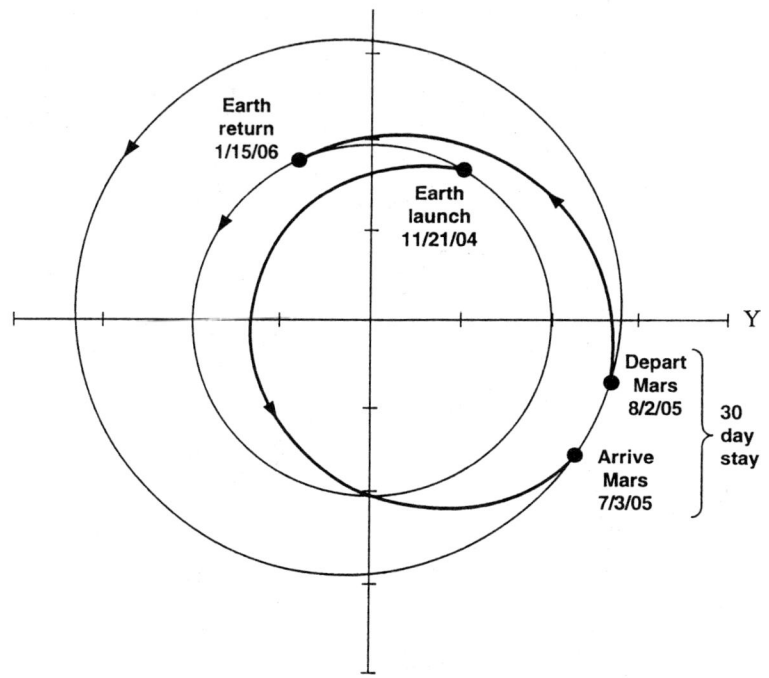

Venus Swingby (Inbound)

Figure 5 Venus can also play a role in Opposition Class trajectories for that part of the trajectory that lies inside Earth's orbit. If the alignments are favorable, Venus can "slingshot" the spacecraft on its way back to Earth, helping to reduce the amount of time and/or propellant that must be used.

Sprint Class Mission. Several years ago a search was conducted for trajectories with flight times similar to the current experience base for flight crews in zero gravity (i.e. one year) in the hope that artificial gravity spacecraft could be avoided [Niehoff and Hoffman, 1989]. This led to the identification of a series of trajectories, dubbed 'Sprints,' with flight times of approximately 400-450 days including a 30 day stop-over at Mars. These trajectories are a subset of the Opposition class with many of the same characteristics, as can be seen in Figure 6. This includes Venus swingbys although these may not be available on every opportunity.

A mirror image of the trajectory shown in Figure 6 was also identified, with the short leg on the outbound transfer and the long leg on the return. The use of the first type of Sprint offers possible psychological advantages in that the crew returns to Earth quickly once the primary mission at Mars is complete. The second type offers possible physiological advantages since the crew arrives at Mars after a relatively short period in zero gravity and thus is physically more capable of conducting the Mars surface mission.

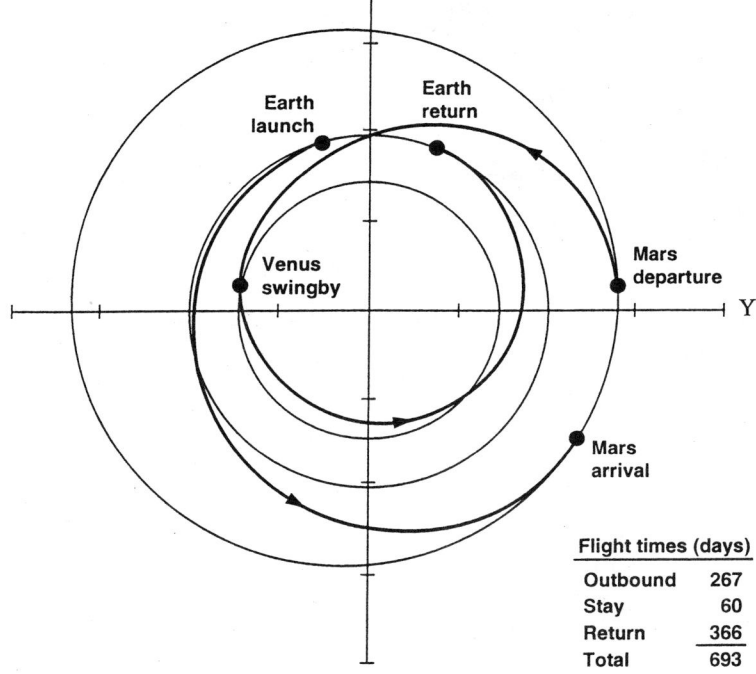

Opposition Class Trajectories: 1:15 Year Sprint

Figure 6 A "Sprint" Class trajectory with a total flight time of a little more than one year—comparable to the currently known ability for humans to survive in weightlessness. Use of "Sprint" trajectories is hoped to eliminate the need for artificial gravity on missions to Mars.

While Sprint trajectories do offer the potential of avoiding artificial gravity spacecraft, this shorter flight time is bought at the expense of higher velocity change requirements and thus larger propellant requirements for each leg of the mission.

Low-Thrust Transfer Missions. The next mission types are called low-thrust trajectory missions and are represented in Figure 7. These use an entirely different propulsion system that continues to provide low thrust over a portion of the flight arc between Earth and Mars. The propulsion systems that can be used for this kind of mission are discussed below.

Low-thrust trajectories differ from the ballistic cases in that they begin with a slow, outward spiraling orbit of the planet, for a considerable period of time after the mission is initiated, before they finally escape Earth's gravity well. They continue to thrust for a portion of this interplanetary flight before the engine is shut down (in the case of the ion propulsion) or before setting the solar sail so it no longer flies against the solar wind. The spacecraft then coasts for a significant portion of the trajectory before again beginning to thrust in order to reshape the spacecraft's orbit to that of Mars. As the spacecraft approaches Mars, it again slowly begins its spiraling flight path, this time inward toward its arrival orbit about Mars.

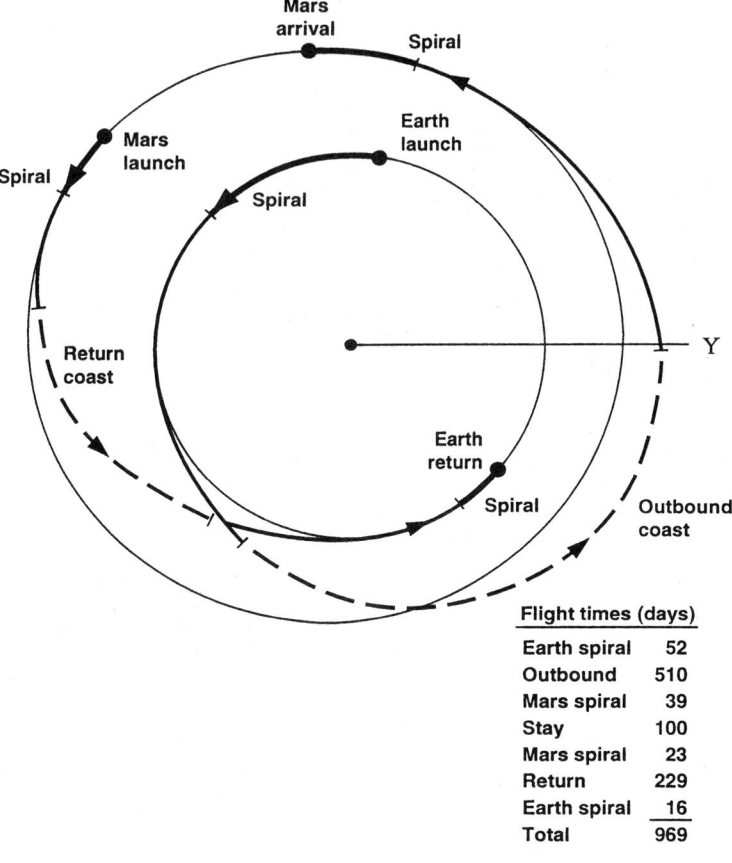

Low-Thrust Transfer

Figure 7 A typical example of a low thrust trajectory, illustrating those periods when the spacecraft must spiral into or out of orbit around either Earth or Mars as well as those periods when it is thrusting or coasting in interplanetary space.

Stay times range from 100 to 200 days; the example mission illustrated in Figure 7 reflects a 100-day stay time. At the end of the "stop-over," the whole process is repeated—first spiraling outward from Mars to begin the return flight to Earth. The total mission time for a low-thrust mission is about 2.5 years. This is somewhat faster than can be achieved with the least energetic ballistic cases, and there are some specific performance advantages with these kinds of propulsion systems. Their current drawback is that the required propulsion systems are currently beyond present capabilities and require extensive technology development.

"Integrated" Missions. Recently, a number of operational considerations have been incorporated into the pathway options for piloted Mars missions. Primary among these considerations are abort options, radiation exposure limits, and zero-gravity expo-

sure limits. One study [Joosten, *et al.*, 1991] examined the implications to an overall mission plan when these three operational factors were taken into account. Quantitative estimates of crew radiation exposure, both in space and on the surface of Mars, were made and compared to NASA standards. Time periods spent in zero-gravity were assessed, taking into account the time spent working in the Mars gravity field. Both of these factors were also taken into account when assessing abort options. The resulting mission profile is made up of various ballistic trajectory types discussed here but the details are beyond the scope of this paper; the reader is referred to Joosten, *et al.* [1991] for additional details.

PATHWAYS: CYCLER-ORBIT MISSIONS

Recently, an old idea has been revived which uses a more innovative approach to flying to and from Mars. This promising concept involves what are called Cycler orbits. These are orbits that are designed to repeatedly re-encounter both Earth and Mars, such that their application is relevant to a potential Mars base which must be supported continuously with both supplies and people. It allows the establishment of an interplanetary infrastructure; a spaceport orbiting in interplanetary rather than Earth space, cycling back and forth like a shuttle between the two planets. As it passes each planet, a shuttle-craft taxi vehicle is launched to rendezvous with the spaceport, carrying supplies and new or replacement crew members. The spaceport itself simply continues to fly along its Cycler orbit [Hollister, 1967, Hollister, 1968].

VISIT (Cycler) Orbits

The first of the Cycler orbits to be discussed are called VISIT (Versatile International Station for Interplanetary Transport) orbits. The most attractive of the two cases illustrated in Figure 8 is the VISIT-1 orbit, which has a 1.25 year period. This orbit is designed to have a commensurability with Earth of four-to-five (4:5), which means that VISIT-1 goes around the Sun four times while the Earth goes around five times. Consequently, the re-encounters at Earth occur once every five years. With Mars, the same orbit has a commensurability of three-to-two (3:2); consequently, VISIT-1 completes three orbits while Mars goes around twice, and re-encounters with Mars occur once every 3.75 years. Working through the Earth and Mars encounters reveals that the VISIT orbits then completely repeat their encounter sequences every fifteen years.

The VISIT-1 orbit has now been verified through computer analysis [Friedlander, *et al.*, 1986]. Figure 9 represents a twenty-year propagation of the orbit, showing that it is quite stable and that the encounters have a slow, regressive-encounter location on both of the planetary orbits. Eventually the orbit must be re-tuned before coasting again for another twenty years. It does not use planetary swingby effects to perturb the station, other than for navigational purposes, being basically a natural orbit in space. With VISIT orbits, one must arrange two or three of the orbits at several different orientations in Earth-Mars space to create more encounters than one every five years at Earth and one every 3.75 years at Mars. Ultimately, then, a network of these orbits could be utilized to establish a total transportation system to support a base on Mars.

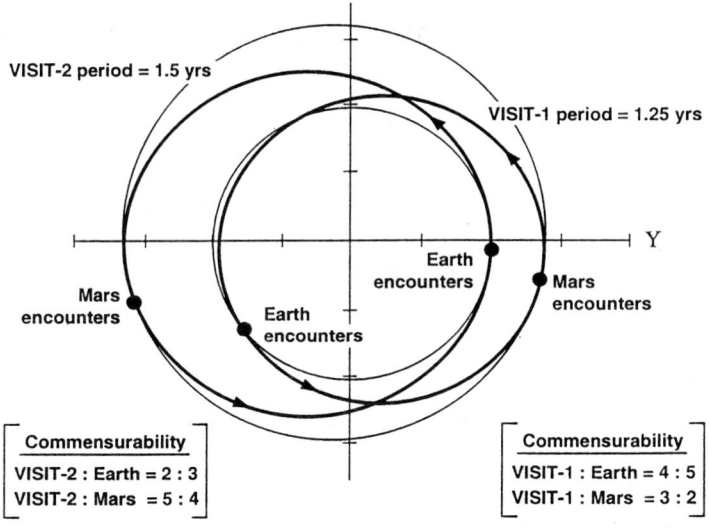

VISIT* Cycler Orbits

Figure 8 VISIT orbits are naturally occurring trajectories between the Earth and Mars. A spacecraft on one of these orbits would cross the orbits of Earth and Mars once every cycle around the Sun, but due to the relative speeds of the spacecraft, Earth, and Mars, an encounter may not occur at every opportunity. Hence the "commensurability" factors for the VISIT-1 and VISIT-2 orbits.

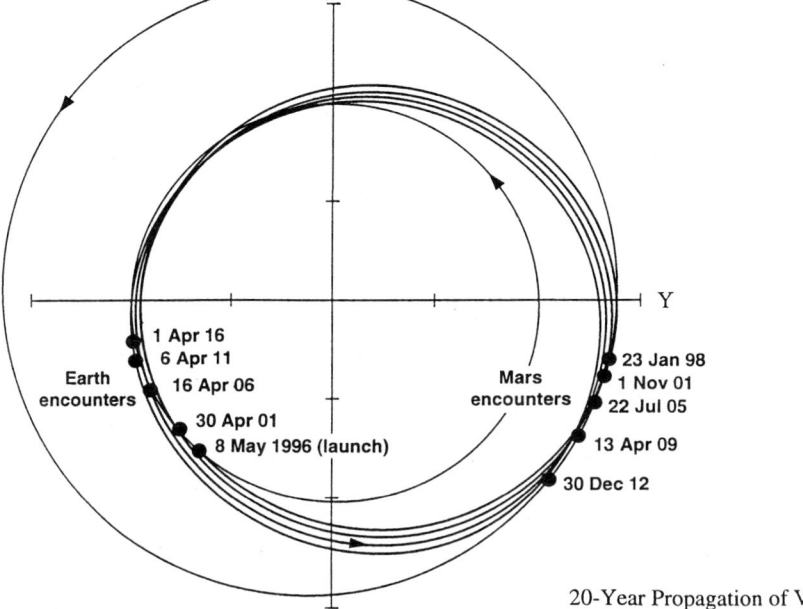

20-Year Propagation of VISIT-1 Orbit

Figure 9 A 20 year propagation of a VISIT-1 orbit indicating that encounters between a spacecraft on one of these orbits and the planets to not always occur in the same place. Thus a spacecraft on one of these orbits must be reset periodically to maintain the desired series of encounters.

Escalator (Cycler) Orbits

An alternative Cycler concept has been proposed by Dr. Buzz Aldrin (Apollo astronaut; personal communication). His concept, illustrated in Figure 10, has come to be known as the UP/DOWN-Escalator orbits. We will first describe how this orbit concept works and then contrast it with the VISIT orbits.

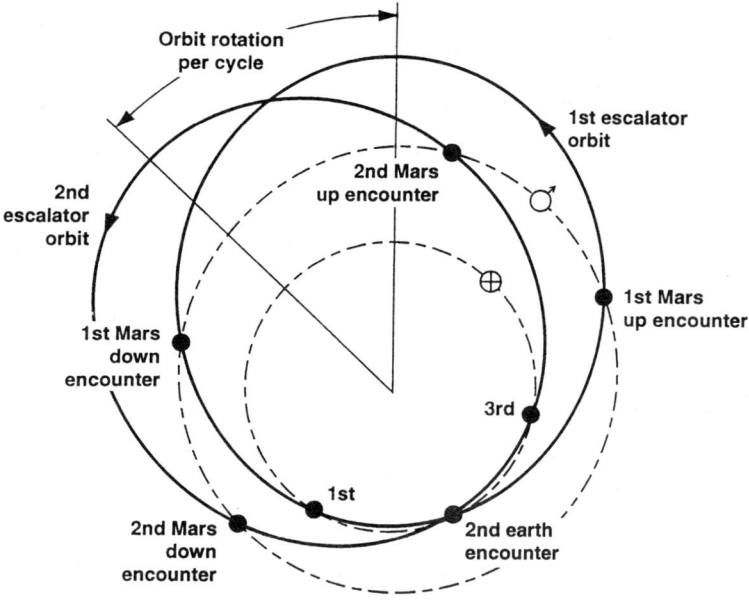

Up/Down Escalator Cycler Orbits

Figure 10 A second example of a Cycler orbit, known as "Up" or "Down Escalator" orbits, in which an encounter with the Earth and with Mars occurs once per orbit. However, the Cycler orbit must be reset after every encounter with Earth to maintain this feature. The Earth flyby can be used to make this change in orientation in most cases, but occasionally a spacecraft on one of these orbits would be required to use some quantity of propellant to maintain the encounter sequence.

The UP component of an Escalator orbit, originating with the first Earth launch, proceeds slightly inside Earth's orbit at first and then outward toward the first Mars encounter. It is essentially a short transfer "up" to Mars. It continues on by Mars for a longer transfer back "down" through Earth's orbit to the second Earth encounter—the DOWN component of the Escalator cycle. At this point, a critical gravity-assist maneuver (provided by Earth) is utilized, rotating the major axis of the orbit. This sets the orbit on a new path for the second Mars encounter and, subsequently, for another long return to the next Earth encounter before again repeating the gravity-assist at Earth. What is actually happening with this orbital process is that the orbiting station is using the Earth-swingby effect to precess itself and keep up with the progressive Earth-Mars phasing orientation (rotating its semimajor axis approximately 50 degrees counterclockwise between successive phases). It continues to do so as this phasing geometry moves around the Sun.

One thing that can be done to enhance this orbital concept is to change the Mars encounter point. If one alters the approach (UP-Escalator trajectory) to re-time the encounter geometry, the Earth-return loop (DOWN-Escalator trajectory) can be significantly shortened. It then becomes, essentially, a mirrored image of the UP-Escalator. The disadvantage of the Escalator orbit is that it results in fairly high encounter velocities at Earth, as well as somewhat higher velocities at the Mars encounters, compared with those of the VISIT orbit. On the other hand, it results in much more regular encounters, which occur once every two years. This suggests that only one or two of these orbits need be utilized rather than the network of three or more VISIT orbit stations.

The Escalator orbit propagation also has been computed [Friedlander, et al., 1986], and Figure 11 illustrates its fifteen-year propagation. It is not necessary to follow and fully understand the orbits illustrated, but it is important to note that they do need occasional realignment using propulsion at certain times. Over a fifteen-year period, propellant to correct an accumulation of about two kilometers per second in small misalignments is needed to keep these Escalator stations in phase and on track with the Earth-Mars geometry.

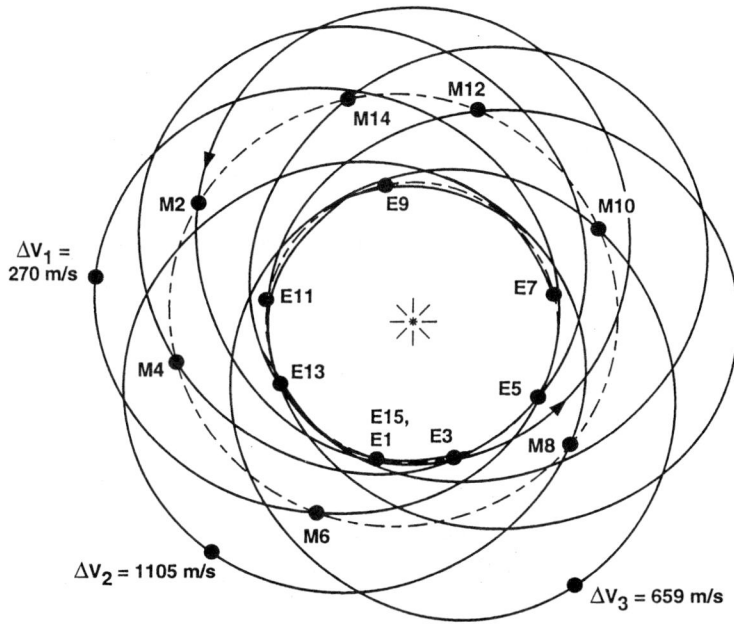

15-Year Propagation of Down Escalator Orbit

Figure 11 A 15 year, seven orbit propagation of the "Down Escalator" orbit, showing those points at which some propellant must be used to change the speed and orientation of the orbit to maintain the encounter sequence.

PLANET-CENTERED STRATEGIES

The planet-centered strategies alluded to earlier are important in developing a piloted Mars mission design. Several representative strategies for operations at both Earth and Mars will be discussed in the next several sections.

Earth Staging

Figure 12 depicts two straight-forward staging strategies for departing Earth, and the most traditional way is to begin, quite simply, in a circular LEO. At the appropriate time (at perigee), an impulse is performed that achieves both the correct direction and speed for transfer to Mars. On the other hand, since the staging base may be supplied with propellants from both Earth and the Moon (perhaps delivering oxygen manufactured on the Moon, and hydrogen, other supplies, and people from Earth), the L1 libration point represents a particularly interesting point at which to stage. This is because it makes possible the opportunity to split the energy required to supply the staging base from both the Moon and Earth.

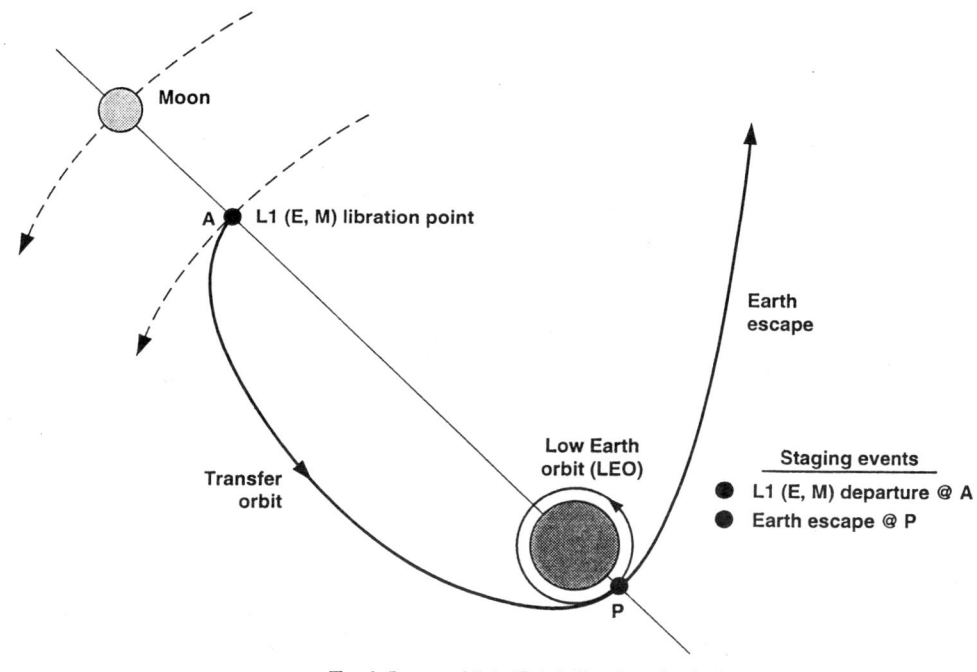

Earth Leo and L1 (E-M) Staging Options

Figure 12 Two staging strategies for departing from Earth: departing from low Earth orbit by increasing speed past escape velocity at point "P" and by departing from the Earth-Moon L1 libration point, first picking up speed by falling in towards the Earth and then increasing speed past escape velocity at point "P."

The escape procedure begins at the libration point with a small, initial impulse, dropping the escape system into a transfer orbit prior to perigee. At perigee, a second impulse is provided to get the system back on the correct escape path. This libration point offers a somewhat lower energy requirement for escape than the conventional LEO, but reaching it from Earth requires a little more energy. The beneficial end result of using this orbit (and the procedures just described), rather than staging at LEO, is that it is more propellant efficient to deliver materials produced on the Moon [Hoffman, 1991a].

Earth-Moon Cycler Orbit — Figure 13 is a rather complex diagram representing a staging base in an Earth-Moon Cycler orbit. This is another of the concepts proposed by Buzz Aldrin (personal communication). The idea is to fashion an orbit that repeatedly goes back and forth between Earth and Moon, picking up at each encounter supplies and system components needed to build the Mars staging base. In this case, the initial orbit incorporates an approach to the Moon that results first in a lunar swingby. It then falls back down to perigee where a very small impulse burn is made, placing the staging base in a phasing orbit. It completes two revolutions in the phasing orbit before another very small burn sends it toward another lunar swingby, such that the process can be repeated again and again.

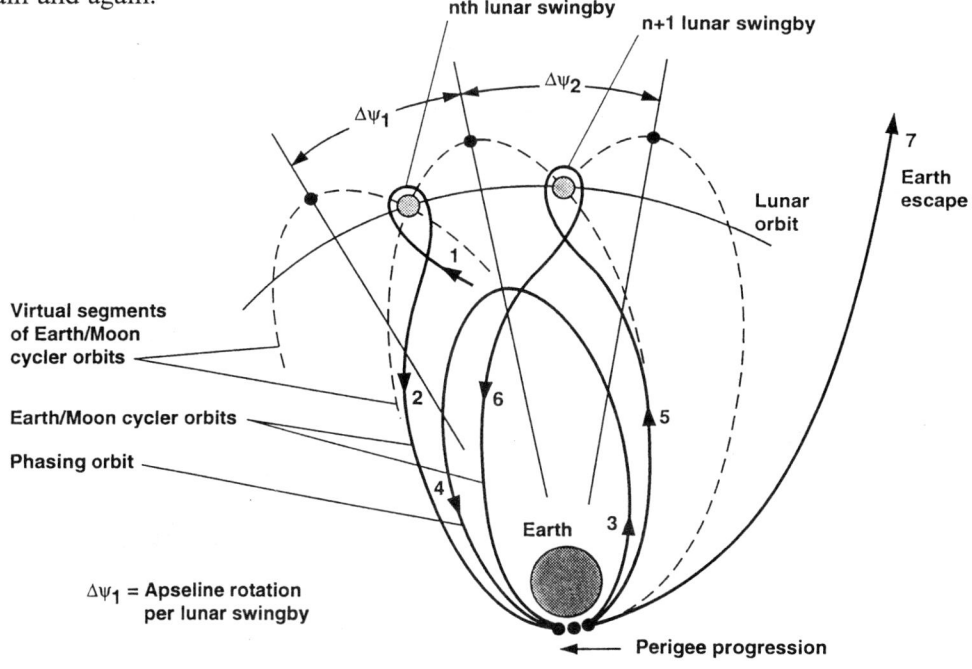

Earth-Moon Cycler Staging Option

Figure 13   An Earth-Moon Cycler orbit, created as a variation of the Earth-Mars "Escalator" orbits. This concept uses the free-return orbits developed for the Apollo missions but adds a "phasing" orbit and propulsive maneuvers at Earth to allow repeated encounters with the Earth and Moon to occur.

With each lunar swingby, the orbit's major axis is rotated a number of degrees in a retrograde fashion, depending on how large the virtual orbit must be (dashed arcs in Figure 13; the spacecraft's orbit if the Moon did not affect it or if no further propulsive maneuvers were made). This is an important characteristic of a Cycler orbit, because one needs to reorient this kind of orbit to match the correct conditions required to escape Earth and achieve a Mars transfer trajectory. This kind of orbit is also a minimum-energy orbit, in the sense that it minimizes the amount of combined energy needed, for example, to put oxygen from the Moon on the Cycler in addition to supplies like hydrogen from Earth. However, it is a relatively high-energy orbit relative to escape, so that the escape maneuver is quite low. Clearly, this is a very advanced concept and it could bear on longer range plans for sustaining a colony on Mars.

Low-Thrust Spiral Departure — Figure 14 illustrates the final Earth-departure staging option to be discussed, low-thrust spiral orbits. Not all of the revolutions of the spiral are shown, but the figure illustrates that much time is spent very close to the planet and much less time is spent at the end of the spiral. This does have a problem associated with it, in that very long periods of time must be spent in the Earth's radiation belts. Consequently, people would not board a low-thrust vehicle departing for Mars until after its spiral orbit emerged from the radiation belts—perhaps at geosynchronous orbit. Using this procedure has an additional advantage in that boarding the spaceship at this later point would reduce the amount of spiral time for the passengers to only 60 of the 163 days that the process requires. This strategy, incidentally, allows the use of the same type of low-thrust propulsion at capture and escape from the planets as is used in interplanetary space, so it is a very efficient system.

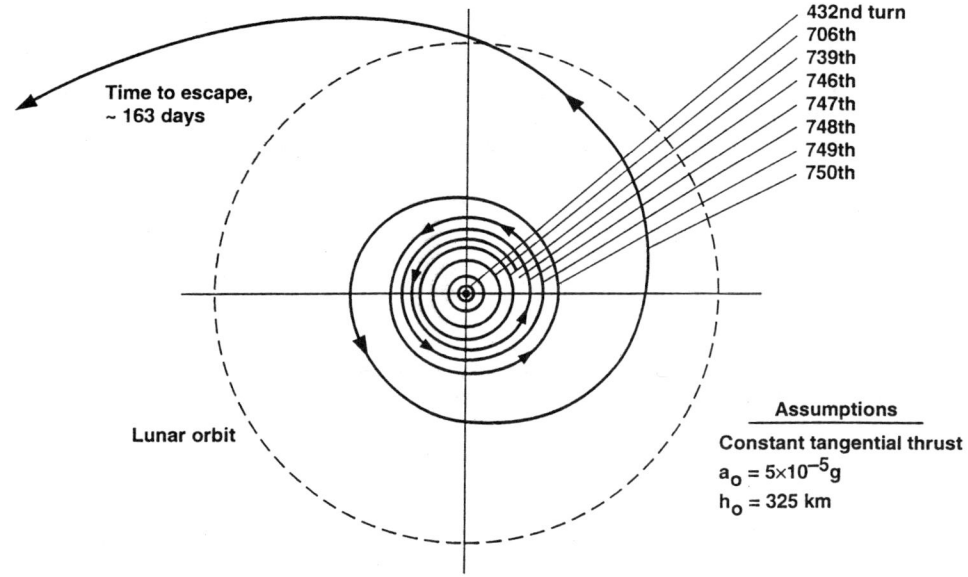

Earth Low-Thrust Spiral Escape Option

Figure 14 A typical low thrust spiral departure from low Earth orbit. The key feature to note here is the large number of orbits spent very close to the Earth.

Mars Staging

Figure 15 outlines several concepts for Mars staging at arrival. First the low-energy, traditional concept in which one impulse is used at periapsis to first place the spacecraft in a loose elliptical orbit. Another small maneuver is performed at the apoapsis point which allows the vehicle to begin a descent trajectory, with entry occurring approximately at Point E (Figure 15).

An alternative, based on the application of somewhat higher energy, uses a much larger burn at the periapsis, placing the spaceship in a low, circular orbit. This alternative concept then requires a somewhat longer burn at Point B in order to achieve the descent trajectory needed to enter the Martian atmosphere [Hoffman, 1991b]. The advantage of the lower orbit is that it reduces phasing problems associated with both the landing site and with the departure, compared with an elliptical orbit. That is, there are typically more opportunities when the orbiting spacecraft passes over the landing site and when that same orbiting spacecraft is aligned for departure on the return trajectory to Earth.

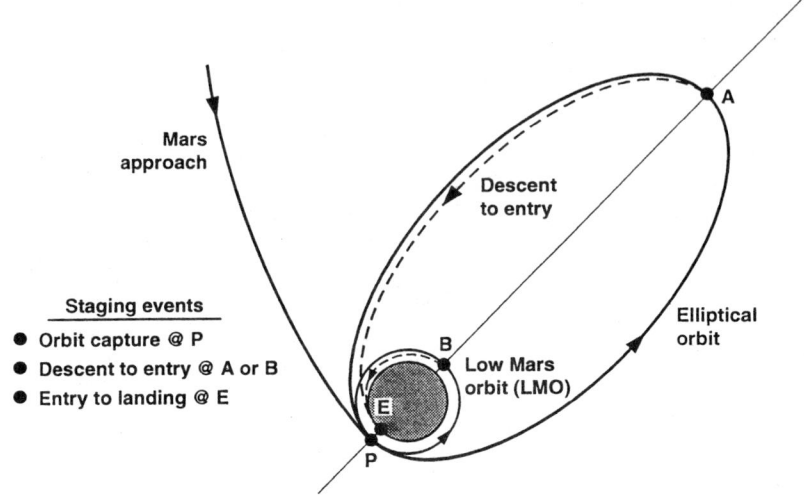

Mars Elliptical and LMO Staging Options

Figure 15 Two alternatives for entering Mars orbit and descending to the surface. The first enters an elliptical orbit and then descends to the surface. The propellant required for this option is relatively small for decent but large for the return from the surface. The second alternative enters a low circular orbit and then descends. The sum of these maneuvers is larger than the first option. However the ascent phase requires less propellant.

For resource production as well as the establishment of a staging base to service a Mars colony (but also suggested as a candidate for the first piloted Mars mission), the option to stage at the inner satellite of Mars (Phobos) has been proposed. Figure 16 illustrates the orbital maneuvering needed to do so. First, capture is achieved at Mars periapsis, followed by a rise to the circular orbit of Phobos. A second burn circularizes the spaceship for rendezvous with Phobos. Phobos is also attractive as a staging base for an early mission not intended to land on Mars as suggested by Dr. Fred Singer [1984]. In this scenario, Phobos could serve as a base from which automated rovers could be sent to and controlled on the surface of Mars.

Phobos is also attractive as a staging base because it may have the composition of a carbonaceous chondrite meteorite. This suggests that it may contain hydrogen and oxygen that could be mined and manufactured on the satellite, thereafter to be used as propellant. This would help reduce the launch mass requirements for piloted Mars missions. Thus, Phobos may be particularly important in shaping piloted missions, at least as a staging base in Mars orbit and possibly as a propellant supply base, as well.

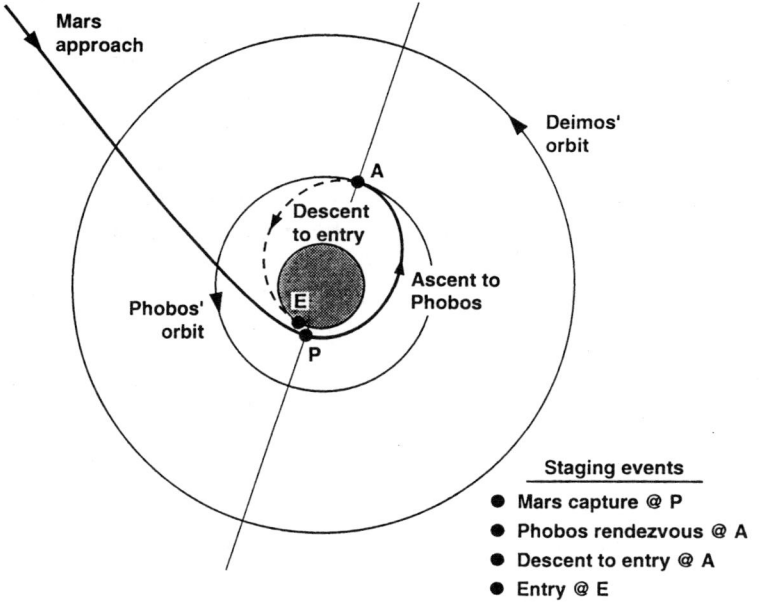

Mars' Phobos Staging Option

Figure 16 The inner moon of Mars, Phobos, could serve as a staging base for operations on the Martian surface. Robots could be teleoperated by humans remaining on this moon or human crews could descend to the surface in small excursion vehicles. Phobos may also be a source of raw materials from which propellants could be made.

MISSION PERFORMANCE COMPARISONS

Propulsion Considerations

One of the multitude of factors that goes into selecting a trajectory, but one of the most closely related, is the choice of systems used to accelerated and decelerate the spacecraft. Ideally one would like a system with high thrust level for a short time or a low to moderate thrust for long durations. In either case, a high system efficiency is desirable to minimize the amount of propellant required for the mission. Table 1 illustrates the range of efficiencies and thrust levels for engines currently available or tested in a simulated environment.

Chemical systems encompass those propulsion devices currently used by all launch vehicles and upper stages. Chemical reactions in solid and/or liquid propellants are used to create a high temperature, high pressure gas which is expanded through a supersonic nozzle to produce thrust. Limiting factors are the maximum temperature and/or pressure which can be sustained by engine components. Nuclear Thermal Rockets (NTRs), ground tested during the NERVA program in the late 1960s and early 1970s [Gunn, 1989], use the heat of radioactive decay rather than chemical reactions to raise the temperature and pressure of propellants. The resulting gases are then expanded through a supersonic nozzle in the same fashion as the chemical system. As can be seen from

Table 1, these systems have a higher efficiency than the chemical systems but with comparable thrust levels. These systems are also limited by the materials used to contain the exhaust gases. Magnetoplasmadynamic (MPD) systems heat propellants via an electric arc struck between two electrodes. This creates a plasma which is accelerated by a magnetic field to produce thrust. These engines have much better efficiencies than either chemical or NTR engines, but at greatly reduced thrust levels. As a consequence, these low thrust engines must remain running for days, weeks, or even months to attain the same change in spacecraft speed accomplished by chemical or NTR engines in just a few minutes. This also changes the type of trajectory followed by the spacecraft from elliptical arcs to spiral segments. Limitations of this engine are related to electrode lifetime (which degrade due to the large current passed through them) and by the size of the power source necessary to generate the electric arc (typically a nuclear reactor). Ion engines, with the highest efficiency of the systems shown in Table 1, produce a stream of charged particles by stripping electrons from individual atoms of the propellant. These ions are accelerated to very high speeds by a magnetic field and are then neutralized by reintroducing the electrons. Due to the process involved, only a relatively small number of atoms are moved through this engine at any one time compared to the other engines discussed, resulting in the lowest thrust level of these systems. Spacecraft using these engines are relegated to the same type of trajectory as the MPD systems and also require a large power source.

Table 1
A QUALITATIVE COMPARISON OF
VARIOUS MAIN ENGINE TYPES SUGGESTED FOR MARS MISSIONS

Specific impulse is an indication of the efficiency with which propellants are accelerated; the higher the number the better. The last two columns, Thrust per engine and runtime, are related in that the lower the thrust level the longer an engine must run to achieve a desired change in velocity.

Engine efficiencies and thrust levels

Engine Type	Specific Impulse (sec)	Thrust per engine (NEWTONS)	Run time/duration
Chemical	150–450	$0.5 - 5 \times 10^6$	Few seconds – 100 of minutes
Nuclear Thermal	825–925	5000–50000	Few minutes – several hours
ARCJET (Electrothermal)	800–1200	1–10	Few seconds – several 100 hours
MPD (Electromagnetic)	2000–5000	10–200	Few seconds – several 100 hours
ION (Electrostatic)	3500–10000	1–10	Few minutes – several days or months

As can be seen, no one system has both desirable characteristics of high thrust and high efficiency, but each can be successfully applied to a mission scenario with the proper choice of trajectory.

At two points on these round trip trajectories, the natural environment can be used to accomplish the same task as the propulsion system. On the outbound trip to Mars and the return trip to Earth, the spacecraft must slow down to enter orbit around the planet. The atmosphere at both planets can be used to slow the spacecraft if proper shielding is included. This is advantageous only if the shielding mass is less than the propulsion system which would accomplish the same result.

Other systems have also been proposed as a means of moving people and equipment between the two planets. Solar sails offer the advantage of using no propellant (and thus have an infinite efficiency) but requires an extremely large and lightweight sail. Tethers, measuring thousands of meters in length, can be used to take advantage of orbit mechanics and gravitational torques to exchange momentum between a spacecraft and some inert mass. This provides another means of placing a spacecraft in a higher (possibly escaping) or a lower orbit without the use of propellants. However, manufacturing a tether with an appropriate strength to weight ratio and a means of controlling this long, flexible structure are two problems which have yet to be solved. Rail guns or mass drivers use a magnetic field to accelerate small, packetized inert masses to change the speed of a spacecraft; laboratory demonstrations of this type of system are promising but full scale systems have yet to be developed. Laser thermal rockets would use a laser (possibly Earth or Moon based) to heat a propellant and expand the hot gases through a supersonic nozzle. Laboratory work on such a system is still under way. Each of these systems have been examined at the theoretical level and in some cases tested with laboratory models. But none have been developed into flight systems.

From this discussion, it can be seen that the choice of propulsion system is intimately involved with the type of trajectory used to travel between Earth and Mars. But as with the trajectory classes available for this mission, there are also significant alternatives to keep this from becoming a bottleneck for mission design.

Having reviewed the pathway concepts, a comparison of the aspects of their flight performance potential (i.e., mass requirements) will be made using the mission example illustrated in Table 2. The requirement posed in this sample problem is the delivery of 75 metric tons of useful payload to the orbit of Mars.

Setting the Stage

To accomplish this task for ballistic round trips, a hypothetical mission will first require a 100-metric-ton spaceship. Following through on that, this mission would have a total flight mass of 175 metric tons departing Earth, of which 75 metric tons would be off-loaded at Mars. The Earth-return flight weight would then be 100 metric tons, the weight of the spaceship itself.

However, if an interplanetary staging base is operating in a Cycler orbit, that 100-metric-ton spaceship would not be needed because much of what it represents is already embodied in the staging base. In fact, the most this hypothetical mission might require is a 10-metric-ton shuttling vehicle to get to and from the Cycler base. The payload is still the same, however, so the total departure mass at Earth in this case will be 85 metric tons—10 metric tons for the transport and 75-metric tons of cargo.

Table 2
PERFORMANCE CHARACTERISTICS AND MISSION ASSUMPTIONS FOR A HYPOTHETICAL MISSION

This mission will provide a common basis to illustrate the propellant mass requirements of trajectory options discussed previously.

Sample problem for performance comparisons

	Round-trip	Cyclers
Payload assumptions		
Leaving earth orbit	175 MT	85 MT
Dropped in Mars orbit	75	75
Returned to earth orbit	100	10
Flight options		
Ballistic round-trips	Flyby/rendezvous, conjunction, opposition, and Venus swingby	
Ballistic cyclers	VISIT-1, up/down escalators	
Low-thrust round-trip	Nuclear-electric, solar sail	
Planet-centered staging		
Earth (ballistic)	Earth-moon cycler	
Earth (low-thrust)	Low circular orbit	
Mars	Phobos	
Propulsion		
Chemical	H_2/O_2 stages (460 sec I_{sp})	
Nuclear-electric propulsion	8 MWe system (MPD thrusters, argon propellant)	
Solar sail	25 km^2 (1 g/m^2)	
Propellant options		
Option 1	From earth/moon	
Option 2	From earth/moon and Mars/Phobos	

The ballistic round trips, as previously defined, are represented in the example by three types of missions: Short-Stay/Flyby-Rendezvous, Conjunction Class, and Opposition Class (including the Venus-swingby case). For the ballistic Cyclers, one should consider both the VISIT-1 and the UP/DOWN-Escalator concepts. Also, as previously discussed, the low-thrust round trips are represented by nuclear-electric (ion) propulsion (NEP) systems and solar-sail propulsion systems.

For planet-centered staging, both the ballistic Earth-Moon Cycler and the low-thrust spiral cases are compared. Note that this puts low-thrust at a bit of a disadvantage because that system must carry itself up from LEO while the Cycler is already almost free of Earth's gravity well. At Mars, the Phobos staging base is chosen in this example.

For chemical propulsion, cryogenic (hydrogen-oxygen) stages are assumed, and an eight-megawatt magnetoplasmadynamic (MPD) thruster (with argon propellant) is the energy source selected for the NEP system. The solar sail considered has a five-kilometer (on a side) surface with a density of one gram per square meter.

Two additional options are worthy of special mention, one in which propellant is produced only at Earth and a second that involves the production of propellant at both Earth and Mars. It should be emphasized that these options apply only to the chemical propulsion cases.

Results

Figure 17 presents the results. Concentrating first on the ballistic cases (with Earth-Moon propellant only), the Opposition Class mission (with a 1.5-year total flight time) requires an enormous amount of Earth-departure mass—3000 metric tons. However, the Venus-swingby case dramatically reduces that requirement, by more than half, and at a penalty of only about six months additional trip time (increasing it to about two years).

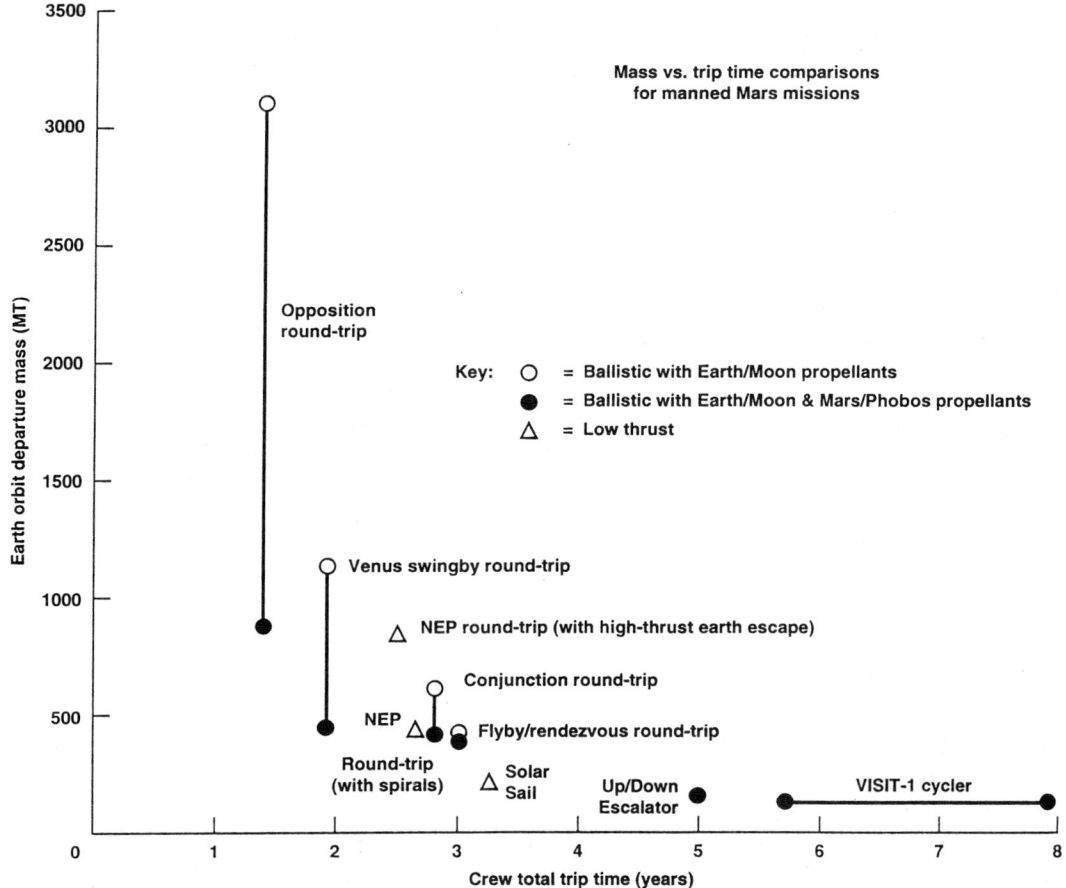

Figure 17 A comparison of the initial mass in low Earth orbit plotted against total trip time to carry out various of the mission options discussed in this section. Shorter trip time generally require larger amounts of mass to accomplish.

When considering the longer Conjunction Class mission, the flight mass is again halved but the trip time increases to three years. And, for a very simple Short-Stay/Flyby Rendezvous mission with stay time of only 30 days, one can again reduce the flight mass to about 500 metric tons. Production of propellant at Mars provides a dramatic reduction in the total mass, but still requires almost 1,000 metric tons with the Opposition Class mission. Again, using a Venus swingby halves this requirement, but further improvements are not significant. In fact, relative to the Short-Stay/Flyby-Rendezvous case (in which the spacecraft doesn't actually stop at Mars with its large mass), propellant production offers almost no improvement at all.

The NEP case is comparable to the best ballistic cases, even though the NEP case must start from LEO (deep in the gravity well) rather than from the more advantageous Cycler orbit. As an additional note, there is really no advantage to using high-energy thrust to escape Earth's gravity quickly, avoiding the spiral, before starting up the electric propulsion system, because the trip time really doesn't improve much and the mass is nearly doubled. The solar sail concept seems quite attractive. It imposes a slightly longer trip time but at only about half the metric tonnage otherwise required.

The UP/DOWN-Escalator case has a total time of five years and represents a very low tonnage when leaving Earth orbit. The VISIT Cyclers are similar but require longer periods of time. It should be remembered, however, that the Cycler orbits cases are (by intention) supporting a Mars base, and a large portion of the associated mission time is spent at the base (by design). That is why these trip times are as long as they are. The UP/DOWN-Escalator actually spends about four years at Mars and only about a year going to and from the planet. With the VISIT-1 cases, time at Mars may be as much as six years while as much as two years may be on the transfer legs in transit.

SUSTAINED MARS BASE OPERATIONS

Table 3 provides a comparison of sortie trip times (flight time plus stay time) for sustained base operations on Mars. It also provides a little better idea of the total transport and expendable-mass requirements for the Cyclers (previously given mass data represented only a shuttlecraft, or taxi, vehicle). This illustration reflects a comparison of the two types of Cyclers, the VISIT orbit and the UP/DOWN-Escalator, plus a simpler Escalator variant—the DOWN-Escalator (just the DOWN component of the initial Escalator concept). Also included are the Conjunction Class mission, used in this context to support a Mars base, and a low-thrust NEP mission.

The flight times for all of the mission types are given in Table 3. One critical aspect that should be noted is the amount of expendable (propellants and supplies) predicted to be consumed over a fifteen-year period [Hoffman, et al., 1986]. It can be seen, for example, that the UP/DOWN-Escalator option requires the highest amount of expendable mass at nearly 33,000 metric tons (MT), but it is nearly matched by the Conjunction Class option (30,200 MT). The VISIT orbits reflects somewhat longer mission times but require only a little more than 20,000 metric tons of expendables, a profile that is similar to that of the DOWN-Escalator. Interestingly, low-thrust NEP is quite impressive when compared with these large masses; its requisite expendable mass is less than 10,000 metric tons. However, this particular NEP concept requires a power plant capable of about twenty megawatts, which implies a requirement for extremely advanced technology.

Table 3
A COMPARISON OF KEY PERFORMANCE INDICATORS
FOR VARIOUS OPTIONS OF SUPPORTING A PERMANENT MARS BASE

The total time indicates the amount of time a crew would be away from Earth. The scale of the supporting operations is indicated by the number of interplanetary spacecraft and magnitude of the expendables (mainly propellant) needed based on the trajectory option.

Flight mode comparisons for sustained Mars base operations

Flight mode	Flight time (years)	Stay time (days)	Total time (years)	Number of interplanetary transfer stations	Total expendables need for 15-year operation (MT)*
Conjunction	1.6	3.2	4.8	2	30,200
Visit-orbit cycler	1.2 – 6.3	1.6 – 5.9	5.7 – 7.9	3	21,500
Up/down escalators	0.9	4.1	5.0	2	32,850
Down escalator	2.1	4.4	6.5	1	23,650
Nuclear-electric low-thrust	2.4	2.4	4.8	2	8,350

* Assumes a 20-man average base size, L1 (E, M) and Phobos staging nodes, and propellant (H_2 and O_2) production at the moon (O_2 only), Phobos and Mars

SUMMARY AND CONCLUSIONS

Several interplanetary trajectory types, associated planet-centered trajectories for arrival and departure, and the mass performance consequences for piloted Mars missions have been reviewed in this paper. For the ballistic mode, this ranged from the least propellant intensive flybys, through conjunction, opposition, and Cycler, to the most propellant intensive Sprint types. Low thrust trajectories, using different propulsion systems, were also discussed. Table 4 summarizes the general characteristics associated with each of these trajectory types. Figure 18 illustrates the progression of these trajectories over time (limited to ballistic types; several other factors must be considered for low thrust) and indicates that an opportunity of one type or another is available almost every year. However, as indicated by the comparative example from the last section, it is not possible to build one spacecraft and propulsion system which can be used for all of the trajectory types. This forces the selection of one specific trajectory type which in turn requires the consideration of other key issues.

Table 4
A SUMMARY OF KEY MISSION PLANNING PERFORMANCE CHARACTERISTICS FOR ROUND TRIP MARS MISSIONS

Flight time indicates the duration that the crew must be supported in the interplanetary environment. Stay time indicates the length of time the crew may be supported on the Martian surface. Payload mass fraction indicates how much of the mass initially in low Earth orbit is actually useful payload and not associated with the propulsion system.

Manned Mars round-trip mission comparisons

Flight Mode	Flight time (years)	Stay time (days)	Total time (years)	Payload mass fraction*
Free-return flyby	3.0	0	3.0	36.8%
Flyby-rendezvous	2.9	30	3.0	13.2%
Conjunction	1.3	530	2.8	11.4%
Opposition	1.4	20	1.4	2.4%
Venus swingby	1.7	60	1.9	5.7%
Nuclear-electric low-thrust	2.4	100	2.6	41.5%

* Assumes Leo departure, Mars staging at Phobos, and aerocapture return
  Mars landing operations are not included

Several of these issues (not necessarily listed in order of importance) include:

Reliability. Longer missions will require higher reliability for hardware and the capability to make repairs (possibly major repairs) while in transit. Short missions could reduce this requirement but at the expense of an increased propellant requirement.

Artificial g. Longer missions also raise the possibility of using artificial gravity to maintain the crew's physical condition. In so doing, the complexity of the spacecraft will probably be increased, compounding the concerns raised above. Again, short trip times could reduce or eliminate this requirement but will increase the propellant required to complete the mission.

Split Mission. In both the hyperbolic rendezvous and split/Sprint mission scenarios, mission success is dependent on the crew's capability to rendezvous with a separate, and hopefully fully functional, spacecraft in the vicinity of Mars. This introduces additional risk beyond that otherwise inherent in this type of mission.

Propulsion. Several different propulsion systems have been identified which could enhance these missions by reducing propellant requirements: nuclear thermal rockets, NEP, or solar sails. However, the technology for each of these systems is still immature compared to liquid propulsion systems currently available.

Aerobraking. An additional enhancement is the use of aerobraking to slow the spacecraft on its arrival at Mars and return to Earth instead of using a propulsion system. In some cases, aerobraking could reduce the total initial mass of the spacecraft by approximately half. But this technology has not been demonstrated for the sizes, entry speeds, and heating rates needed by these missions.

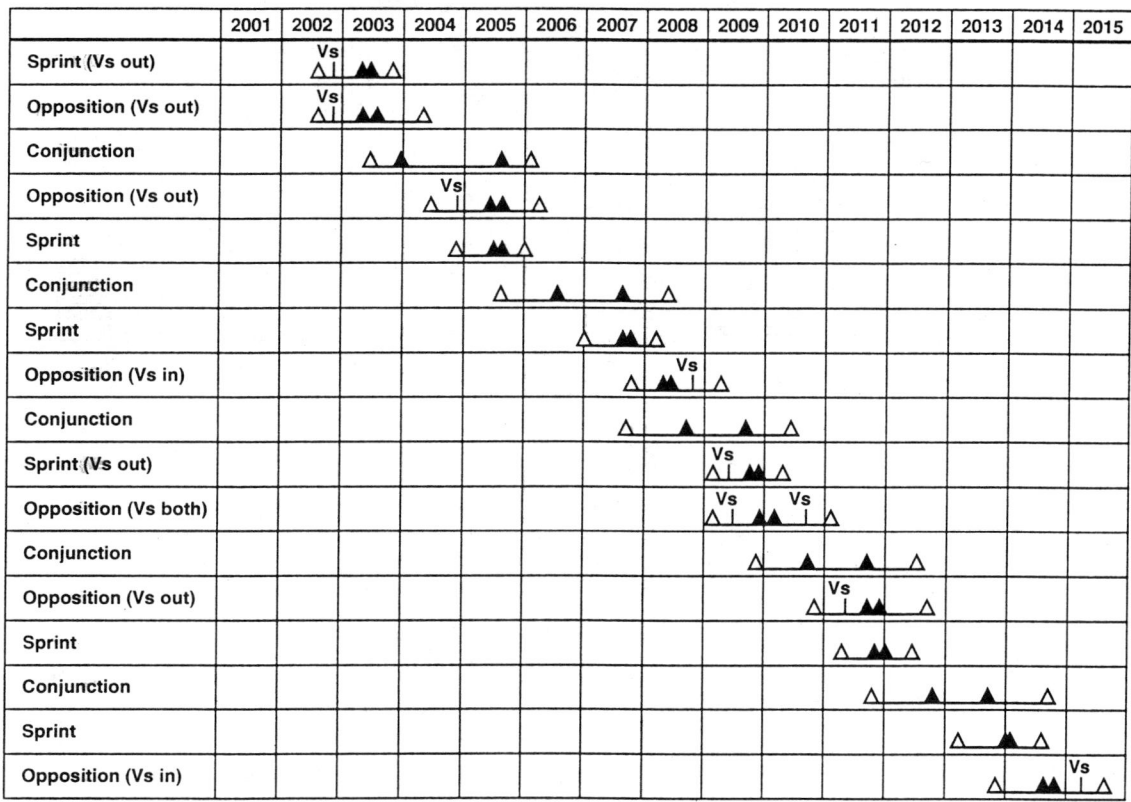

△ Earth departure and arrival   ▲ Mars arrival and departure

Figure 18 This figure illustrates the availability of ballistic trajectories for 15 years beginning early in the next century. An opportunity of one type or another is available on almost an annual basis. However the propellant required for each of these opportunities varies significantly.

Once a specific trajectory type is selected, it may also be necessary to select one specific opportunity (i.e. launch year) around which to design the spacecraft and mission. This is the result of a fact mentioned earlier: Mars' orbit is both elliptical and inclined relative to Earth's orbit. Thus some opportunities will be more attractive than others. For conjunction class trajectories, this variation is relatively small, as is shown in Figure 17. For Sprint class trajectories, Figure 19 indicates that the variation is much more severe. This type of variation places additional pressure on program management to meet a specific schedule in an all-or-nothing environment or to over design the spacecraft allowing launch in more than one opportunity but at the expense of potentially unnecessary hardware.

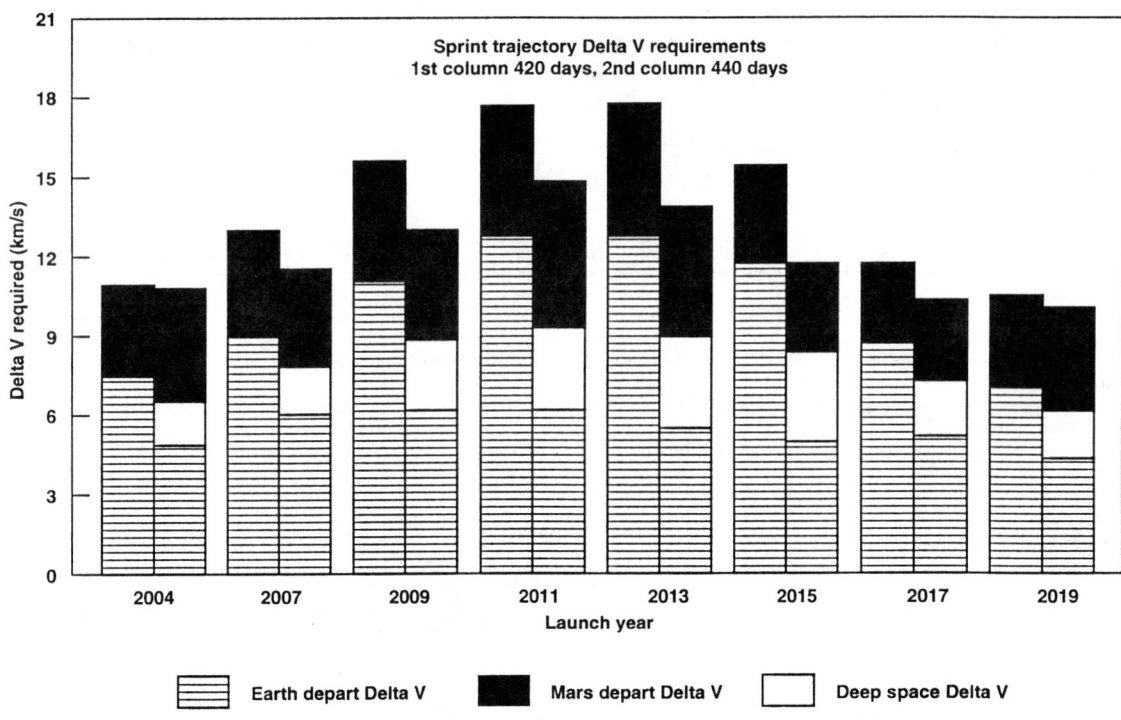

Figure 19 The relative velocity change requirements for Sprint-type trajectories over one 15 year cycle. The elliptical shape of Mars' orbit is the cause of this variation, with poor opportunities occurring when the Mars interception occurs farthest from the Sun.

In summary, it should be apparent that there are many pathways for human flight to Mars. Choices will be dictated by program maturity (exploratory vs. sustained-based phase), technology readiness, risk tolerance, and the desired pace of exploration progress. These decisions will also be affected by space program activity in other areas, such as the Earth Sciences, astrophysics, robotic solar system exploration, and human activity in Earth-Moon space. The ultimate course and pace of a Mars exploration program should evolve from an active national/international process of studies, debate and policy development led by NASA.

REFERENCES

Gunn, S.V., "Development of nuclear rocket engine technology," *AIAA-89-2386*, AIAA/ASME/SAE/ASEE 25th Joint Propulsion Conference, Monterey, CA., July 10-12, 1989.

Friedlander, A.L., J.C. Niehoff, D.V. Byrnes, and J.M. Longuski, J.M., "Circulating transportation orbits between Earth and Mars," *AIAA 86-2009-CP*, AIAA/AAS Astrodynamics Conference, Williamsburg, VA, August 18-20, 1986.

Hollister, W.M., "Castles in space," *Astronautica Acta*, January, 1967.

Hollister, W.M., "Periodic orbits for interplanetary flight," *AAS 68-102*, American Astronautical Society, *AAS Microfiche Series*, Volume 7, 1969 (AAS/AIAA Astrodynamics Specialist Conference, Jackson, WY, September, 1968).

Hoffman, S.J., A.L. Friedlander, and K.T. Nock, "Transportation mode performance comparison for a sustained manned Mars base," *AIAA 86-2016-CP*, AIAA/AAS Astrodynamics Conference, Williamsburg, VA, August 18-20, 1986.

Hoffman, S.J., J.V. McAdams, and J.C. Niehoff, "Round trip trajectories for human exploration of Mars," *AAS 89-201, Orbital Mechanics and Mission Design*, American Astronautical Society, *Advances in the Astronautical Sciences*, Volume 69, pp. 663-679, 1989.

Hoffman, S.J., "Mass performance characteristics of assembly and departure orbits for piloted Mars missions," *AAS 91-122, Spaceflight Mechanics 1991*, American Astronautical Society, *Advances in the Astronautical Sciences*, Volume 75-I, pp. 303-316, 1991 (AAS/AIAA Space Flight Mechanics Meeting, Houston, TX, February 11-13, 1991a).

Hoffman, S.J., "Mass performance implications of mars parking orbit selection for piloted Mars missions," *AAS 91-440, Astrodynamics 1991*, American Astronautical Society, *Advances in the Astronautical Sciences*, Volume 76-III, pp. 1949-1964, 1992 (AAS/AIAA Astrodynamics Specialists Conference, Durango, CO, August 19-22, 1991b).

Joosten, B.K., B.G. Drake, D.B. Weaver, and J.K. Soldner, "Mission design strategies for the human exploration of Mars," *IAF-91-336*, 42nd Congress of the International Astronautical Federation, Montreal, Canada, October 5-11, 1991.

Niehoff, J.C., and S.J. Hoffman, "Piloted sprint missions to Mars," *AAS 87-202, The Case for Mars III*, American Astronautical Society, *Science and Technology Series*, Volume 74, pp. 309-324, 1989.

Singer, S.F., "The Ph-D proposal: Manned mission to Phobos and Deimos," *AAS 81-231, The Case for Mars*, American Astronautical Society, *Science and Technology Series*, Volume 57, p. 39, 1984.

Soldner, J.K., "Round-trip Mars trajectories; new variations on classic mission profiles," *AIAA-90-2932-CP*, AIAA/AAS Astrodynamics Conference, Portland, OR, August 20-22, 1990.

Wilson, S.W., and V.A. Lee, A survey of ballistic Mars-mission profiles, *Journal of Spacecraft and Rockets*, 4, 2, 129-142, February, 1967.

A landing party examines its fore running site verification probe.

AAS 95-479

Chapter 9

MARS MISSION DESIGNS: COMPARING THE NEAR TERM OPTIONS

Malcolm A. LeCompte and Julie P. Stets[*]

Different Mars mission designs recently proposed are examined to illustrate their salient features. The benefits to be derived from manufacturing chemical rocket propellants from native resources on the Martian surface are so great that it will almost certainly be a component of any future human Mars exploration program. A simple analytic tool has been developed to compare different mission approaches including the impact of *in situ* propellant production. Quantitative criteria to evaluate competing approaches to achieve global Mars exploration are proposed. The immediate advantages of each design option, and its impact for long term programs are compared. The prospects for, and benefits to be derived from, an evolutionary program of propellant manufacture are discussed.

POSSIBLE MISSION ARCHITECTURES

There are many ways to transport humans to Mars. The architecture and cost of future Mars missions will depend on the level of technology applied to the task and on programmatic goals. If advanced technologies are used and extensive goals defined, Mars exploration will be expensive. Comprehensive exploration of an entire planet will require an efficient transportation system to reduce costs. However, the technologies for the most efficient propulsion systems are not yet fully developed. Without a major investment to develop advanced propulsion technologies, the most immediately available alternatives are liquid chemical and nuclear thermal engines.

Liquid chemical rocket engines are generally high thrust, but fuel inefficient. Nuclear-thermal rockets are also high thrust and somewhat more efficient, but require some development and have associated safety concerns. Either chemical or nuclear-thermal systems are most frequently associated with near term (within the next 30 years) human Mars exploration. In either case, chemical rockets are most likely to be used for surface launch (ascender) vehicles [French, 1989]. If the propellant for these rockets must be carried from Earth, it replaces valuable cargo tonnage and severely handicaps the expedition's exploratory capability.

---

[*] Aerodyne Research Inc., 45 Manning Rd. Ballerica, Massachusetts 01821-3976.

On the other hand, innovative approaches to Mars exploration can minimize development costs while reducing programmatic cost and complexity. Mars harbors accessible and abundant reservoirs of volatile resources for the manufacture of ascender propellant on the Martian surface [Clark, 1979; Meyer & McKay, 1989 and 1989]. *In situ* ascender propellant production (ISPP) increases the payload fraction of the expedition's initial mass in low Earth orbit (IMLEO) and provides greater exploration capability for any level of effort [Ash *et al.*, 1989, Ramohalli *et al.*, 1987, Ramohalli *et al.*, 1989, Baker & Zubrin, 1990]. Manufacturing propellant on the surface of Mars enables transportation systems based on existing chemical propulsion technology to perform with the efficiency of more advanced systems. It would also enhance the performance of nuclear thermal or more advanced systems as they become available [LeCompte and Stets, 1991].

Spacecraft with high thrust, chemical and nuclear thermal engines use impulsive trajectories when moving between different planetary or solar orbits. Desirable trajectories may minimize flight time or energy requirements, or strike an optimum balance between the two. The set of impulsive maneuvers define the flight path. The energy requirements for initial departure from low Earth orbit trans-Mars injection (TMI) will vary significantly with the solar transfer orbit selected.

Conjunction class missions require the least transfer energy, but usually extend the length of a mission. The total elapsed time for conjunction class expeditions may be about 3 years. However, exploration time on the Martian surface can replace less productive time spent in space. Conjunction class missions leaving from low Earth orbit typically require a velocity increment of less than 4.0 km/sec. Launch windows occur at the synodic frequency, or about every 780 days.

Opposition class missions require greater transfer energy and allow surface stays of no more than a few months. The actual flight time for the opposition class missions may be quite similar to a conjunction class mission that remains on the surface for over a year. Transit time between planets is roughly six to ten months. Opposition class missions leaving from low Earth orbit generally require a velocity increment well over 5.0 km/sec. A gravitational assist from Venus may reduce this somewhat.

A large booster rocket may allow missions to be launched directly to Mars from the Earth's surface. This "Mars Direct" approach is operationally simple but, while payload fraction is maximized, payload mass is severely constrained. Launching from Earth orbit relaxes that constraint, allows assembly of larger vehicles, and provides a convenient locale for operational checks prior to launch. Components launched separately to rendezvous in a highly elliptical orbit provides a compromise that yields much of the benefits of both "Mars Direct" and "Low Orbit Assembly" [Lusignan *et al.*, 1991].

On the approach to Mars, spacecraft may exercise one of three trajectory options. They may enter orbit about the planet, and use smaller craft to land and ascend from the planet's surface. This option is often referred to as "Mars Orbit Rendezvous" (MOR) [von Braun, 1962, Cohen, 1989]. The entire space vehicle also may elect to descend from a pre-established orbit or descend directly to a predetermined site where supplies have been prepositioned. This option is called "Mars Surface Rendezvous" (MSR) [Zubrin *et al.*, 1991]. A spacecraft may fly by the planet and avoid being gravitationally captured. In this case it may be possible for the flyby spacecraft to cycle between Earth and Mars in a quasi-stable orbit. During the flyby, the spacecraft lands cargo vehicles to establish a base and lands an ascender which will rendezvous with the mother-ship after a brief stay or at the next flyby. This approach is called "Mars Flyby Rendezvous" (MFR) [French *et al.*, 1990].

Spacecraft arriving at Mars may use either propulsive or aero-braking for capture into orbit or for descent to the surface. A spacecraft entering orbit or landing on the Martian surface requires a major decrease in spacecraft velocity. Using the Martian atmosphere to dissipate most of the spacecraft's speed allows initial propellant mass to be traded for the smaller mass of a heat shield. Soft landing may be accomplished with the aid of rockets and parachutes. Without parachutes, rockets must dissipate about one km/sec of the lander's velocity. Parachutes may reduce landing speeds to less than half this value.

Ascender designs are not only distinct among the three architectures, they are a major differentiating factor. Advantages and disadvantages in each mission architecture largely accrue through the character of their associated landers. An MSR direct-lander/ascender is simple, but the cab should be large to accommodate the crew for many months in space. A large crew cab requires a large ascender which must carry sufficient propellant to achieve escape velocity and return to Earth. However, transporting a large, fully fueled ascender from Earth is prohibitive, so a significant fraction of its propellants *must* be made on the Martian surface. Transporting the crew between planets in a large MOR spacecraft that parks in Mars orbit allows the ascender mass to be minimized. The Apollo missions to the Moon used the MOR approach for this very reason. MOR is an optimal strategy for landing and ascent from a body with neither atmosphere or volatile resources. However, the orbiting component that remains in orbit must carry a large amount of propellant to break out of orbit and return to Earth. This propellant must be carried from the Earth as cargo. An MFR flyby or cycler spacecraft never becomes gravitationally trapped by Mars so it doesn't need to carry a large amount of propellant. Although the crew cab may be small, the ascender must be large enough to achieve escape velocity and "catch" the cycler. Again, some or all of the large ascender's propellants *must* be made on the surface. A separate habitat is needed to support the crew awaiting the return of the cycling spacecraft.

In summary, propellant manufacturing on the Martian surface impacts each architecture differently because each lander/ascender vehicle has different velocity and payload requirements. In the simplest terms, ascender fuel can be provided in one of four general ways:

1. *All* of the *propellants* required for the return voyage to Earth may be *carried as cargo* on the outbound voyage. This has been the traditional MOR type mission.

2. *Some* of the *propellants* for the return voyage may be *carried* and *some manufactured on the Martian surface*. This option is available to the MOR type architecture.

3. *Most* of the *propellants* for the return trip may be *manufactured at Mars*. This option may be utilized by either the MSR or MFR category.

4. *All* of the *propellants* required to return to Earth are *produced on Mars*. This option is available to the MSR type mission.

The Mars orbit rendezvous technique may be employed with either an opposition class or a conjunction class trajectory. MSR and MFR architectures, options that may allow major reductions in the initial mass placed in low Earth orbit, are really only practical for conjunction class missions, and require some degree of propellant manufacture on Mars.

It may seem reasonable to seek the most advantageous of the many mission design possibilities. However, it is difficult to define an objective criteria for making such a judgment between fundamentally different approaches to Mars exploration. Moreover, the design of Mars missions are usually governed by programmatic goals more than technical issues. For example, a single mission to "plant the flag" would probably be very different than a mission that initiated a long program of exploration. Before an architecture can be selected, programmatic goals must be defined.

National or international sponsors with the resources and the will to send humans to Mars must first define the purpose of the venture. The most appropriate means can then be selected from among the many available options. Defining the programmatic goals will make it possible to define the criteria to make this selection. We assume here that the goal of future Mars exploration will be a continuous and long term exploration of the complete Martian planetary environment. This should be a sufficiently specific goal to provide a reasonable basis for defining a viable set of evaluation criteria with which to make a comparison and evaluation of different mission designs.

We will make this evaluation in two ways: by comparing point designs qualitatively, and by a parametric analysis of performance. Comparing point designs based on each architecture option is useful as a qualitative, conceptual survey that illuminates the features and the relative strengths and weaknesses of each. A point design comparison will also illustrate the difficulties with attempting to evaluate specific missions from among a continuum of possibilities.

On the other hand, parametric models can be used to subject each architectural option to an objective analysis. However, this is only effective if common performance criteria can be defined. Parametric models to evaluate different mission options will be addressed later.

COMPARISON OF FOUR POINT DESIGNS

There are four well documented point designs that are good representative examples of their particular architectures. Each has received both public attention and critical review. The National Aeronautical and Space Administration (NASA), in its 90-Day Study, has proposed a very efficient, cryogenically-propelled, opposition class mission. This mission is based upon work by Boeing, Corp., Huntsville, Alabama [Cohen, 1989]. Martin Marietta Space Systems Division of Denver, Colorado, has proposed a cryogenically boosted conjunction class MOR mission [Clark, 1990]. This same group has also been the source of an innovative MSR mission design [Zubrin, et al., 1991]. Finally, a complete exploration program, also using cryogenic propulsion and based on MFR evolved out of four "Case for Mars" conferences held at the University of Colorado over the last 10 years [French et al., 1990]. The characteristics and specifications associated with each mission design are given in Table 1.

All four point designs propose a single, discarded, cryogenic (H_2/O_2) propulsion stage to accomplish trans-Mars injection (TMI) out of Earth orbit, although performance varies among designs. All use partially closed life support systems which recycle air and waste water. All except the Martin Marietta MOR design use aerobrake capture at Mars, however, each design's aerobrake mass fraction is different. In general the details of each are not very comparable and demand a closer examination to reveal their attributes and vices.

NASA/Boeing 90 Day Study - Opposition MOR

The Mars exploration program described in the 90 Day Study (90DS) uses an opposition class trajectory. The surface stay time is limited to about 30 days. The total flight time would exceed 500 days in this "sprint" scenario. The time of flight for the outbound and inbound legs of the journey is about the same as for a conjunction class trajectory. However, the total mission time is about half that of the conjunction class mission. Also, an opposition class mission requires about twice the initial propellant mass in low Earth orbit due to its greater energy requirement. Shorter surface stay time and greater initial mass may appear to be disadvantages yet these penalties are accepted as a means to minimize risk and cost.

Risk is sometimes presented as proportional to mission duration. In this view, halving the duration of a mission also halves its risk. On the other hand, risk is greatest during brief mission events like reentry. The total number of such events is roughly equivalent for both opposition and conjunction class missions. The duration of trans-planetary cruise is about the same for the two trajectories. Therefore, the difference in perceived risk arises from the disparity in the duration of the surface stay, and the nature of the crew's activities while on the surface.

Moreover, operational costs are typically regarded as the greatest expense of a space mission. If this remains true, then cutting the mission duration in half greatly reduces the programmatic expense. However, it is more likely that long term missions will become more autonomous which will alleviate the high cost of ground based operational support.

Table 1
POINT DESIGN COMPARISON
(masses in metric tons)

	NASA/Boeing 90-Day Study MOR	Martin-Marietta Straight Arrow MOR	Martin-Marietta Mars Direct MSR	Case for Mars MFR&
Crew Hab Mass	29	42	15/7%	248
Consumables Mass	6*	19	9/2	45
Crew Size	4	5	4	15
Artificial Gravity	No	Yes	Maybe	Yes
Descent	Retro-rocket	Parachute	Parachute	Parachute
Cargo Mass	25*	23	25	36
Resource Utilization	No	Partial@	Full	Full
Total Mass Descent Stage	84	52	40	135
Ascender Specifications				
Cab Mas	3.5	<1	7	4(x2)
Crew Size	4	3	4	5(x2)
Dry Mass	7	2	20	30(x2)
Wet Mass	23	14	116	174(x2)
Propellants and Specific Impulse (seconds)				
TMI Propulsion	H_2/O_2 475 sec.	H_2/O_2 460 sec.	H_2/O_2 465 sec.	H_2/O_2 460 sec.
Ascender Propulsion	H_2/O_2 475 sec.	N_2H_4/N_2O_4 334 sec.	CH_4/O_2 373 sec.	CO/O_2 260 sec.
TEI Propulsion	H_2/O_2 475 sec.	N_2H_4/N_2O_4 334 sec.	Not Applicable	N_2H_4/N_2O_4 330 sec
Initial Mass Earth Orbit	801	711	242	1620

* Quantities for Opposition class mission, greater for conjunction class mission.

% crew resides in 15 ton surface hab outbound and on surface. Return to Earth is accomplished in 7 ton ascender cab.

& assumes launch of three separate interplanetary spacecraft, one with cargo lander, two with crew lander.

@ Partial Resource Utilization = only Life Support Replenishment. Full Resource Utilization = Ascender Propellant, Life Support

 Ultimately, the short stay time of the opposition class mission may be its most important shortcoming. How much greater will be the programmatic risk and cost of staging an entire mission which will not appreciably advance the purpose of scientific exploration?

 While the question of trajectory selection remains, the NASA MOR architecture remains fundamentally the same for follow-on missions which follow conjunction class trajectories, although either the initial mass in Earth orbit would be significantly less or else more cargo could be accommodated. Figure 1 schematically illustrates the mission profile and step 3 shows the vehicle layout prior to TransMars Injection. The large cryogenic propulsion stage topped by two aerobrakes appears to be nothing so much as a giant clam, opening atop a six-pack of beer. One aerobrake is used to decelerate components bound for the Martian surface. The other is used to place an lab-habitat and Earth-Return-Stage into Martian orbit. The chief characteristics of this mission design are the absence of any provision for creating artificial gravity and the use of cryogenic fuel imported from Earth for all major propulsion systems. Hydrogen boil-off, in space and on the Martian surface, is minimized by innovative insulation schemes.

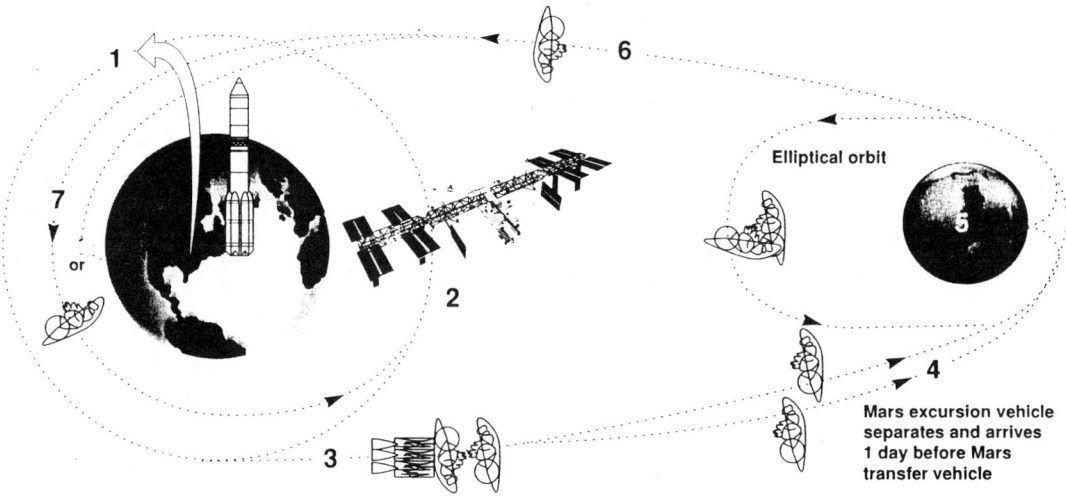

1 Payload delivered to Space Station Freedom
2 Mars transfer vehicle mated with payload at Freedom
3 Trans-Mars phase with Mars transfer vehicle
4 Mars transfer vehicle remains in Mars orbit; Mars excursion vehicle descends to surface
5 Excursion vehicle to/from Mars surface
6 Trans-Earth phase with transfer vehicle
7 Transfer vehicle aerobrake maneuver and return

Figure 1 NASA/Boeing 90 Day Study MOR mission design schematic.

One spacecraft is launched to Mars with a crew of 4. The spacecraft components are mated during interplanetary cruise. The crew can occupy the 29 ton Crew Hab Module outbound, when in Mars orbit and on the return voyage. The crew also has access to the 25 ton Surface Crew Hab outbound and on the surface. The crew habitats, in space and on the surface, are small but endowed with reasonable comfort and allowance for privacy. A radiation shelter is incorporated into the structure of the Crew Hab Module.

For an opposition class mission, the crew may spend about a month on the Martian surface. There are no plans to utilize *in situ* resources to enhance their surface capabilities. They use a small ascender to rendezvous with the orbiting Crew Hab Module for the flight back to Earth. As the spacecraft nears home, return to the surface or low orbit is accomplished using an Apollo style reentry capsule. While the surface habitat on

Mars is available for reuse, virtually all other transportation components are lost. Thus, in terms of hardware, the next mission must be entirely recreated.

A conjunction class mission entails a longer surface stay. Moreover, the smaller launch energy requirement allows the same spacecraft components to deliver more cargo to the Martian surface to accommodate the longer stay. However, very little cargo can be brought off Mars because the small ascender is sized to minimize the mass launched from Earth. Any desire to increase the amount of material brought up from the Martian surface must await development of a larger ascender. The long-duration habitat left in Mars orbit is relatively inaccessible to crew on the surface. Its use as a contingency shelter is therefore somewhat limited. If the crew is forced to retreat to an orbiting shelter it would also be very difficult to advance programmatic goals. MOR missions do not offer graceful degradation.

If the spacecraft components' mass is evenly distributed in a conjunction class mission, it may be possible to create artificial gravity for the outbound leg, after the booster stage is jettisoned. The two aeroshell enshrouded sections may be separated and rotated about a common center of mass with the aid of a flexible connecting cable. Otherwise the crew would spend most of the mission in zero-g and be subjected to very rapid changes in gravity loads between cruise and planetfall.

The 90DS approach to MOR builds on Apollo experience in the most direct way possible: by faithfully reproducing the hardware, planning and management philosophy. It is therefore low risk in a programmatic as well as a technological sense. However, if our original goal had been global and continuous exploration of the Moon, then Apollo would now be regarded as a failed program. Perhaps it would be wise to approach Mars exploration from a fresh perspective, learning from the mistakes of the past.

Martin-Marietta Straight Arrow - Conjunction MOR

A superficially similar approach is proposed by Martin-Marietta in its Straight-Arrow mission design (MMSA). The primary difference between the two concepts resides in the design of the Crew Habitat Module. The MMSA crew hab is roughly 50% larger than its 90DS counterpart. The additional mass is due to larger crew accommodations and the necessary structure to provide artificial gravity. The MMSA spacecraft is seen in Figure 2. It bears some resemblance to an oriental coolie, spinning with arms outstretched, and clutching pop cans.

In transit, the spacecraft is spun at 6 rpm which provides about two thirds Earth gravity (1 g). At some point, the spacecraft might be despun to 4.5 rpm, to provide acclimatization to the Martian environment. At Mars, the spacecraft is propulsively captured into orbit. Three of the crew of 5 descend to the Martian surface to remain for over a year, exploring and developing resource utilization technology. At the end of their stay, these three board their ascent vehicle and rejoin their comrades in the orbiting spacecraft. A cryogenic propulsion system provides the impulse to leave Mars orbit and return to Earth. On the return voyage, the rotation rate can be increased to 7.25 rpm to provide a 1 g environment in preparation for landing. In the vicinity of Earth, the crew boards an Apollo style reentry capsule to complete their journey.

While the MMSA inflight Crew Habitat is larger, its ascender is only about two thirds the size of the 90DS. The lander and ascender use a moderate energy propellant combination, hydrazine and nitrogen tetroxide, based on Space Shuttle technology. It is easier to store, but is significantly less energetic than cryogenic propellants. Since it uses a lower energy propellant than the 90DS, the MMSA ascender must be designed with extremely narrow performance margins. It is therefore even more growth-limited than the 90DS ascender.

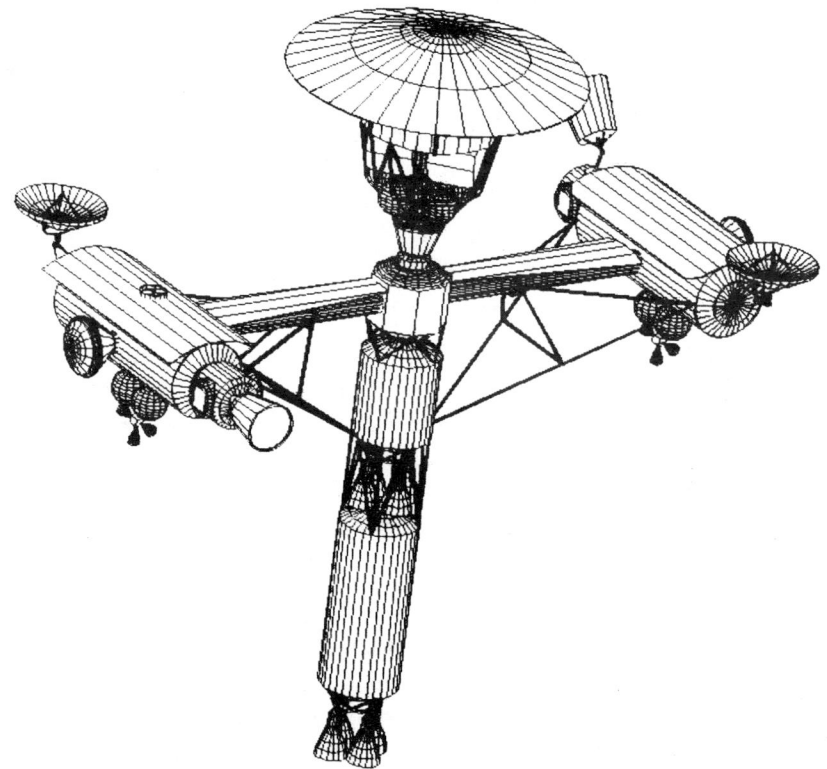

Figure 2 Martin-Marietta Straight Arrow MOR mission design vehicle post trans-Mars Injection and booster stage discarded.

The MMSA design can deposit about 23 tons of cargo to the Martian surface. This is slightly smaller than that delivered by the 90DS mission, and it must house and support a crew of three for a much longer surface stay. In this design, cargo delivery capacity to the surface is sacrificed for crew habitability in space. The lower surface cargo capacity is, at least in part, due to the mass of the additional structure necessary for the spacecraft to create artificial gravity.

Although most of the infrastructure developed for each mission is discarded after use, there are still many attractive features of the MMSA mission design. It retains much of the Apollo/Shuttle legacy but recognizes the physiological stresses imposed by long duration spaceflights punctuated by high levels of gravity. It also makes resource utilization an important part of the exploration process; although it does not realize the full benefits of using *in situ* resources for space transportation needs. Additional gains could clearly be made by locally manufacturing ascender propellant.

Martin Marietta Mars Direct - Conjunction MSR

The innovative approach to Mars exploration taken by Martin Marietta in its Mars Direct design (MMMD) uses the surface rendezvous option to drastically reduce initial mass in low Earth orbit. In fact, this mission design isn't even meant to be launched from Earth orbit. It incorporates components small enough to launch directly to Mars from the Earth's surface. In this, it is by far the most operationally simple design of those considered. The MSR technique was proposed, with little support, for Apollo, as a means of performing the Moon landing quickly and cheaply, albeit with much higher perceived risk. In this approach, the crew travels to the destination planet in a relatively large habitat, which is landed on the surface in proximity to a previously landed return vehicle. On the Moon, with little surface mobility, the consequences of even a minor error could have been catastrophic. Moreover, building a vehicle whose design accommodated landing on the airless, low gravity Moon, as well as a reentry into Earth's dense atmosphere was a major technological challenge to be avoided.

On Mars, the existence of a substantial atmosphere and the possibility of *in situ* propellant production, greatly mitigates these problems. In fact, the MMMD MSR is a mission design particularly suited to volatile rich Mars. Manufacturing the propellant needed to return to Earth at Mars allows the MMMD approach to deliver the greatest payload fraction of any other design. With surface vehicles fueled by native resources, an expedition is not only more fault tolerant to a missed rendezvous, it can explore on a much grander scale.

Unfortunately, while the payload fraction of the initial mass is greatest with MSR, there are practical limits to the magnitude of the payload that can be delivered to the Martian surface. The size of payloads launched directly from the Earth's surface are governed by the performance of heavy-lift boosters. In general, payload limitations may translate to restrictions on surface capability or crew habitability. This is particularly true of the MMMD ascender, in whose small cab the crew must spend about 6 months on the return to Earth.

There are two major flight components in the MMMD design. A surface habitat (SH), which is launched to Mars carrying a crew of 4. Artificial gravity may be created by attaching the SH to the spent upper stage with a cable and rotating the two about their center of gravity. The SH is guided to a landing near an Earth Return Vehicle (ERV), launched 2 years earlier. The ERV has spent the intervening time manufacturing propellant from the Martian atmosphere. The SH will not be launched unless the tanks of the ERV are known to be full. In any event, a spare ERV accompanies the SH. The crew lands on Mars in the SH and uses it as their base of operations. At the completion of their mission, they board the 2-stage ERV and launch directly back to Earth. Figure 3 is an artists conception of the vehicles looking like a pill box (the habitat) near a giant inverted thimble (the ascender).

Figure 3 Martin-Marietta Mars Direct MSR Mission Design vehicles on Martian surface.

Other mission designs create no "capability-in-being" for Mars exploration since their major mission components are dedicated to transportation and expended in that role. The MMMD design obtains double utilization out of its mission components. The Martian base habitat that serves as a bus for transporting the crew to Mars remains on the surface for future use. Neglecting the booster, only the first stage of the ERV is not recoverable or reusable. Thus, the MMMD MSR option creates permanent surface infrastructure on Mars rather than disposable transportation infrastructure.

The requirement for launching vehicles from the Earth's surface makes this scheme operationally simple. It also restricts the permissible mass of the vehicles to the payloads of a single, Saturn-5 class, booster. If the vehicles can be assembled and launched from Earth orbit, this constraint is somewhat relaxed. However, the major rationale for this approach is that it minimizes the required initial mass needed for the mission. We will see that, as the size of the crew cab on the ERV increases, the advantages of this scheme are diminished. The MSR technique is only competitive if some level of crew habitability can be sacrificed for total performance. To give this qualification meaning, we will define a performance criteria in a later section of this paper.

Case for Mars - Conjunction MFR

The MMMD design optimized the creation of surface infrastructure. The design that grew out of the series of "Case for Mars" conferences instead focused on establishing a permanent transportation system able to support long term exploration of Mars. In the Case for Mars design (CfM), reducing the initial mass in low Earth orbit was secondary to establishing an infrastructure with continuity for both transportation and exploration. Continuity in an interplanetary transportation system is achieved if a spacecraft can be recovered at the end of a mission, and reused for subsequent flights. Exploration continuity is established by maximizing the use of *in situ* resources to enhance surface capability and also infrastructure. Communal self-sufficiency is the goal on the surface of Mars, while regularly scheduled service between Earth and Mars is the transportation objective.

Ideally, cargo and human occupied components of the transportation system are treated differently, both in their use and development. Each infrastructure is optimized for its particular purpose.

Figure 4 Case for Mars MFR interplanetary spacecraft en route to Mars.

Three separate vehicles are sent to Mars during the same launch opportunity. Each individual spacecraft is, by itself, launched in the same manner as, and roughly equivalent in size and capability, to a 90DS or MMSA spacecraft assembly. Each carries its own lander/ascender. The three components rendezvous and dock soon after launch. Together, they form a very large craft for travel in relative comfort between planets.

Compared to other options, the CfM vehicle is a space-going hotel. This structure, resembling a giant lawn sprinkler, is shown in Figure 4. It is spun to create artificial gravity. The crew cabin of the lander/ascender is surrounded by propellant and can act as storm cellar to mitigate the radiation hazard. Cargo landers are derived from human occupied lander/ascenders and launched separately for one way delivery of material and use as shelters. However, for comparison, we have treated one of the three available lander/ascenders as a cargo vehicle.

As the craft nears Mars, the lander/ascenders are released and descend to the surface. A base for long term exploration is established and propellant production facilities are installed to create the supply needed to return to the mother ship when it returns some years hence. When a relief ship does arrive, only part of the large crew departs. Their replacements are delivered to the embryonic community on the surface. When the large spacecraft returns to Earth it is captured in a high apogee orbit for future recovery and reuse, after refurbishment.

The major disadvantages of this approach include the absence of an available spacecraft to allow a contingency return to Earth, and the very high initial mass of the conceptual vehicle. However, once the system is established, it becomes a virtual conveyor belt to facilitate exploration and the expansion of the human presence on Mars. Moreover, these vehicles, created primarily for interplanetary travel, may also be useful for missions to the asteroids as well as Mars. To this extent, an infrastructure for solar system exploration is created, not just a means to explore Mars.

Point Design Summary

The foregoing qualitative architecture comparison shows that the design of Mars missions is motivated more by exploration priorities than by available technology. In other words, choosing an architecture is not so much defining a transportation technology as defining programmatic objectives. Since each architecture shares the common goal of long term Martian exploration, comparative evaluations should be based on criteria that serve this end.

As outlined in NASA's 90 Day Study, after the first mission is accomplished, subsequent flights to Mars will use conjunction class trajectories. The disparity apparent in a comparison of opposition and conjunction class missions is therefore a transitory issue pertaining only to the initial mission. However, use of the MOR architecture is a necessary consequence of the opposition class trajectory. Moreover, the space vehicles developed for the first mission, and used for later, conjunction class flights, are only suitable for MOR type missions. Thus, requiring that the initial mission use an opposition class trajectory limits subsequent missions to the MOR option, regardless of its relative performance.

Is MOR the institutional architecture of choice because of the perceived need for an initial opposition trajectory, or because MOR is the best way to support long term Martian exploration?

If all missions after the first will use conjunction class trajectories, it is appropriate to pose two questions. First, can we quantify the penalty for sending the initial expedition on an opposition trajectory? Second, how does the MOR architecture's ability to support long term exploration compare with other mission options (at least until the advent of more efficient propulsion systems)?

Both questions may be answered by comparing different classes of architectures endowed with similar capabilities.

COMPARISON OF ARCHITECTURE CAPABILITIES

To compare and evaluate the different mission design options and to examine the impact of *in situ* propellant manufacture, we have created a human Mars mission simulation model. Inputs to this model include module masses for ascender (cab mass), and transfer modules. Also included are the surface payload, and mass fraction of each modular component, the propulsion system specific impulse, and the velocity increments required for each impulsive maneuver. The model regresses from the mass of the crewed transfer module injected into Earth return trajectory from Mars (TEI), to determine the initial mass required in low Earth orbit. To simplify the analysis, we assume aerocapture at Mars and that no parachutes are used for landing. Initial mass in low Earth orbit (IMLEO) is given by Hill and Peterson [1965]:

$$IMLEO = (M_4 + M_2) \frac{e^{(\Delta v_1/v_1)}}{(1 - S_1 e^{(\Delta v_2/v_2)})} \qquad (1)$$

where:
S_1 is the TMI stage structural mass fraction (including engine).
ΔV_1 is the velocity increment required for TMI.
V_1 is the exhaust velocity of the propulsion system and is given by the product of the acceleration of gravity with the propellant specific impulse (Isp).
M_4 is the mass of the MOR or MFR interplanetary transfer vehicle for TEI module before TEI is accomplished and is given by:

$$M_4 = M_0 \frac{e^{(\Delta v_4/v_4)}}{(1 - (A_4 + S_1)e^{(\Delta v_4/v_4)})} \qquad (2)$$

where:
M_0 is the crew module for the transfer.
ΔV_4, V_4, S_4 are defined as above, but for the TEI component.
A_4 is the aerobrake mass fraction.
If the TMI stage is retained, the values of the structural fractions are modified accordingly.
M_2 is the mass required for the components to be landed on the surface, including the ascender and expedition hardware and is given by:

$$M_2 = (M_3 + M_S)\frac{e^{(\Delta v_2/v_2)}}{(1-(A_2+S_2)e^{(\Delta v_2/v_2)})} \qquad (3)$$

where:

M_S is the expedition surface payload. It is not returned to Earth, but becomes part of the installed surface infrastructure on Mars.

M_3 is either the fully fueled lander or the lander dry weight, depending on whether the in-situ propellant option is exploited.

ΔV_2, V_2, S_2, and A_2 are defined as before, but for these component modules.

M_3 is given by:

$$M_3 = M_a e^{(\Delta v_3/v_3)} \qquad (4)$$

M_a is the ascender dry mass and is given by:

$$M_a = \frac{M_c}{(1-S_3 e^{(\Delta v_3/(Nv_3))})^N}$$

where:

M_c is the Mass of the crew cab on the ascender.
N is the number of ascender stages.

This fundamentally simple analytic tool allows different architectures to be compared, and the relative benefits of specific mission strategies to be assessed, in a normalized way. The results of such a normalized analysis are summarized in the next section.

Using this parameterized model it is possible to compare the capabilities of different classes of missions, representative examples of which were discussed in the previous section. However, some simplifying assumptions are necessary to reduce the large number of parameters that must be defined to describe any one point design. These assumptions must remain consistent among different architectures. As an example, the performance (specific impulse) and the structural mass fraction of the Trans-Mars Injection (TMI) stage should be the same for all categories. The goal is to reduce a mission model to a functional relationship between two parameters. For example, the initial mass in low Earth orbit can be compared to mass delivered to the Martian surface.

The point design examples discussed previously provided the basis for generic Mars mission designs in each category that illustrate simple parametric relationships. The exemplary point-designs for each architecture are useful for providing a model validation for each mission class. These examples are represented as points near or along more general curves that describe encompassing classes of missions. In other words, the point designs are literally points to be found on curves that illustrate each architecture.

To achieve a set of normalized mission profiles it was necessary to alter some of the characteristics of each point design. These characteristics are compared to determine the real advantages of various propellant production schemes within a design and between designs. The impact of *in situ* propellant production (ISPP) is perceived as a gating item when assessing the initial mass required in LEO (IMLEO). Also, the masses of major mission components are systematically varied to illuminate specific mission capabilities or attributes.

All four point designs assume a single, discarded, cryogenic (H_2/O_2) propulsion stage to accomplish trans-Mars injection (TMI) out of Earth orbit. All use aerobrake capture at Mars. For comparison purposes, engine performance, stage, and aerobrake mass fractions were set by the values taken from the NASA MOR study, and applied to all other designs. A propulsive final descent to the Martian surface (descent without parachutes) was assumed, even though this landing mode entails a disproportionate IMLEO penalty for MSR and MFR style missions that require larger landers.

Two types of nuclear stages were also considered for comparison purposes. The disposable nuclear engine is discarded after TMI. MFR and MSR designs employed this type of nuclear stage. The MOR design employed a retained nuclear stage which was kept in Mars orbit to perform the trans-Earth injection assumed for ascent vehicles (TEI). A common crew cab mass of 4 tons was (ascenders) that rendezvous with interplanetary transfer vehicles, whether in orbit or on flyby trajectory.

Mission architectures based on low thrust, electric propulsion systems can be usefully compared to designs using high thrust chemical and nuclear propulsion. However, it must be recognized that electric propulsion represents a significant improvement in propulsive technology, that is only achieved by a major investment and developmental effort.

The mass of mission components launched from the Earth's surface is limited by booster rocket capability. For consistency in the parametric comparison of the designs, we assumed that all missions commenced from low Earth orbit (LEO). (It should be noted that a major advantage touted for the Martin-Marietta MSR point design is the fact that it can be launched directly to Mars from the Earth's surface).

Table 2
MARS MISSION VELOCITY REQUIREMENTS
FOR OPPOSITION AND CONJUNCTION CLASS TRAJECTORIES

	Velocity Increments (m/s)	
	Opposition Class*	Conjunction Class
Trans-Mars Injection	4417	3900
Trans-Earth Injection	3400	1700
Mars Orbital Velocity	5320	5320
Mars Escape Velocity	8400	6700

* Venus Swing-by Assumed

Velocity requirements (shown in Table 2) for the opposition class trajectory were taken from the MOR study. Velocity requirements for the conjunction class trajectory were taken from the MSR and MFR studies. A complete list of architecture specifications is shown in Table 3. Variables include the mass of the interplanetary transfer vehicle (M_o) or crew cab of the MSR ascender (M_c), and the useful mass placed on the Martian surface (M_s). M_s includes habitats, rovers, power generation systems, ISPP equipment, and stockpiles of foodstuffs and spare parts. Also varied was the specific impulse (Isp) of the ascender using optimum values for each propellant option. The primary dependent variable was taken to be the initial mass in low Earth orbit (IMLEO) at mission start.

Table 3
MISSION CHARACTERISTIC SPECIFICATIONS FOR ARCHITECTURE CATEGORIES

Component	MOR	MFR	MSR
TMI Stage: (from Low Earth Orbit)			
Structural Mass Fraction	0.10	0.10	0.10
Cryo-Specific Impulse (I_{sp})	475.0	475.0	475.0
Nuclear I_{sp}	900.0	900.0	900.0
Descent Stage			
Surface Payload	varies	varies	varies
Structural Mass Fraction	0.10	0.10	0.10
Aerobrake Mass Fraction	0.15	0.15	0.15
Chemical I_{sp}	475.0	varies	475.0
Ascender			
Crew Cab Mass	4.0	4.0	varies
Structural Mass Fractio	0.15	0.15	0.15
Chemical I_{sp}	varies	varies	varies
TEI Stage (from Mars orbit or Flyby)			
Transfer Vehicle Mass	varies	varies	N/A
Aerobrake (to Mars Orbit)	0.10	0.10	N/A
Structural Mass Fraction	0.15	N/A	N/A
Cryo I_{sp}	475.0	varies	N/A
Nuclear I_{sp}	900.0	N/A	N/A

Figure 5 and 6 compare the variation in IMLEO with ascender propellant specific impulse, for one or two stage vehicles, for each of the three mission designs. All three designs are assumed to land a 30 ton payload. MOR and MFR have a departure and return transfer vehicle crew module mass of 35 tons. MSR, however, has a transfer crew module mass of 30 tons upon departure and a mass of 15 tons upon return. A MOR case is included with a prefueled (fueled prior to Earth departure) ascender. Pre-fueled ascenders are considered impractical for either MFR or MSR designs, so ISPP should be assumed for these cases.

It is apparent in Figure 5 that a single-stage ascender designed to achieve escape velocity (as with either MFR or MSR missions) is most effective in minimizing IMLEO growth for propellant Isp above 400 seconds. To achieve Mars orbit in a single stage, while minimizing IMLEO, a prefueled ascender propellant should have Isp greater than 375 seconds. However, with ISPP, specific impulse down to 325 seconds will provide ample performance.

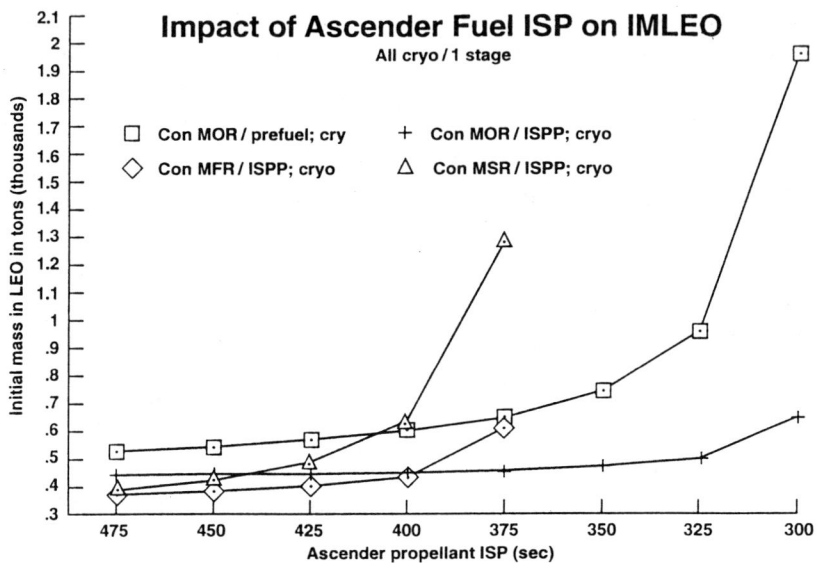

Figure 5   IMLEO dependence on Isp of single stage ascender for MOR (prefueled), MOR (ISPP), MFR, and MSR missions.

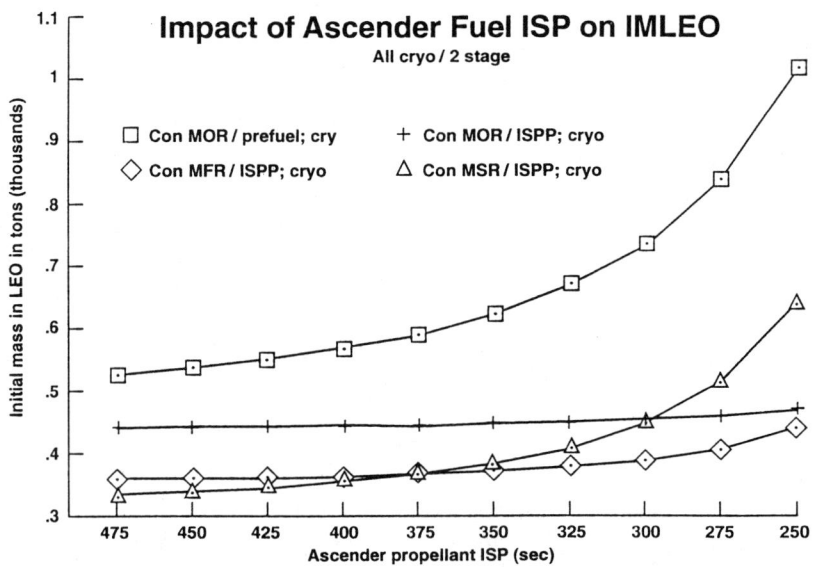

Figure 6   IMLEO dependence on propellant ISP for two stage ascender, prefueled-MOR, ISPP-MOR, MFR, and MSR missions.

Figure 6 shows that IMLEO grows rapidly for two-stage, prefueled MOR ascender propellant Isp below about 300 seconds. On the other hand, ISPP allows the use of propellants with specific impulses below 250 seconds without a major increase in initial mass. Two-stage ascenders for Mars escape should have propellant Isp greater than about 325 seconds to avoid significant IMLEO penalties.

In general, specific impulse lower than 300 seconds penalizes the initial Earth departure mass for all designs but one. The exception is MOR with ISPP. In this case, utilization of CO/O_2 is most beneficial. On the other hand, a two-stage orbital ascender, prefueled with methane, is about as efficient as a single-stage prefueled hydrogen vehicle. If methane ISPP is implemented after the first orbit-to-orbit mission, substantial initial mass savings can be realized for subsequent missions.

Even though the MSR and MFR mission designs allow significant IMLEO reductions, their primary advantages are realized once an expedition reaches Mars. A methane propelled two-stage escape-ascender for either MFR or MSR provides the basis for a capable, single-stage, reusable, orbital-ascender to visit the two Martian moons, Phobos and Deimos, or to provide orbit-surface transfer service if MOR missions supersede early MSR or MFR expeditions. Considering the difficulties in manufacturing or handling either hydrazine or hydrogen, methane seems to be the optimal ascender propellant, even at the expense of initial hydrogen importation. It also provides the performance and evolutionary flexibility wanted in any mission design. We will limit further ISPP comparisons to that option.

We next examine two significant criteria with which to judge the relative merit of the different mission designs. The first is the interplanetary transfer module architecture. The second is the surface capability provided by each mission architectures. We will discuss each in turn.

The mass of the transfer vehicle crew module may be the most straightforward measure of habitability. Greater mass implies larger transport volume per crew member, more compartments, and a capability for artificial gravity. We recognize that coupling crew volume to module mass can be misleading, particularly when comparing dissimilar mission architectures. For example, a MSR crew may spend the greater part of the mission, including the outbound trip and the surface stay, in relatively commodious accommodations, at the expense of enduring the return to Earth at zero-g, in a more cramped ascender cab! Experience on Salyut/Mir has shown that two crewman can live in zero-g aboard a 20 ton space station for periods much longer than the Mars return leg. A crew of three lived comfortably aboard the capacious 35 ton Skylab for three months. Missions may entail very different flight-times and a given mass may become less habitable with increased time in space. Nevertheless, we calculate and compare the IMLEO for different crew module masses for each mission design.

Figure 7 shows the variation of IMLEO with the mass of the crew module of the MOR, MFR and MSR interplanetary transfer vehicles. The comparison includes the MOR case for opposition and conjunction class missions. In this figure, all designs are assumed to employ cryogenic propulsion for TMI. The payload delivered to the Martian surface is fixed at 30 tons. The most obvious feature of this figure is the clear advantage of conjunction class over opposition class missions. It is evident that the MFR mission can accommodate larger crew modules. This is a consequence of avoiding planetary capture, which allows more habitable mass to be launched at TMI per kilogram of propellant expended. Large MFR crew modules are desirable since flight times may exceed three years. The MSR mission habitability appears to be a poor alternative. IMLEO is very sensitive to increases in MSR ascender cab mass so its size is severely restricted.

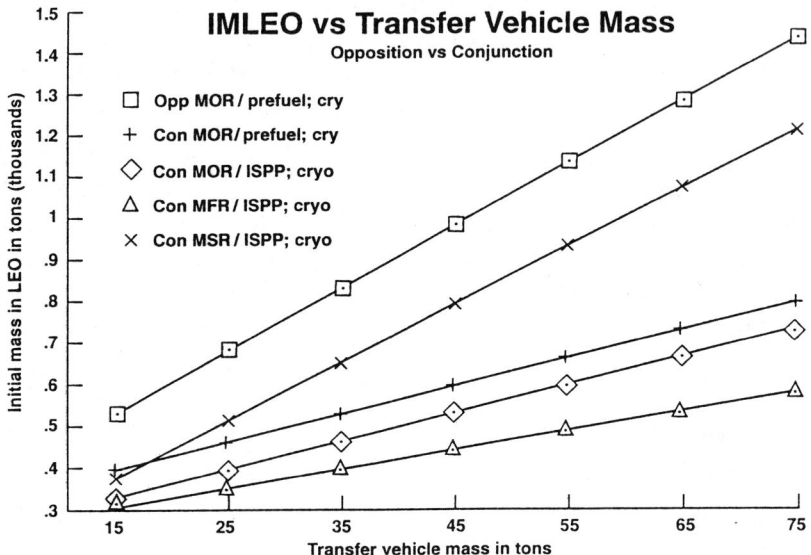

Figure 7 IMLEO dependence on mass of Interplanetary Vehicle Crew Module. Comparison of opposition prefueled-MOR, with conjunction missions using: prefueled-MOR, ISPP-MOR, MFR, and MSR designs, and assuming cryogenic propelled TMI.

Figure 8 shows the significant improvement that can be made by using a thermal nuclear upper stage. All cases represent conjunction class missions. The MFR and MOR case are, once again, seen to be more habitable in flight. Also shown in this figure is the IMLEO for a single stage, reusable, cryogenically propelled spacecraft 'packet' that could operate between Earth orbit and the Mars surface on a regularly scheduled basis. Such a craft could initiate a packet service that would deliver personnel or small payloads to Martian bases with an efficiency rivaling nuclear propulsion. Such service could only be instituted after liquid hydrogen production and handling facilities have been installed on the Martian surface.

A second criteria with which to judge mission architectures relative capability is provided by quantifying the surface capability each delivers to the Martian surface. Mission surface capability can be contrasted by comparing the dependence of IMLEO on the amount of useful cargo delivered to the Martian surface. Cargo delivery is assumed to be accomplished by the same propulsion and landing systems used for crewed spacecraft. If the cargo is delivered separately by solar sail, then the benefits of in-situ propellant production become even more evident.

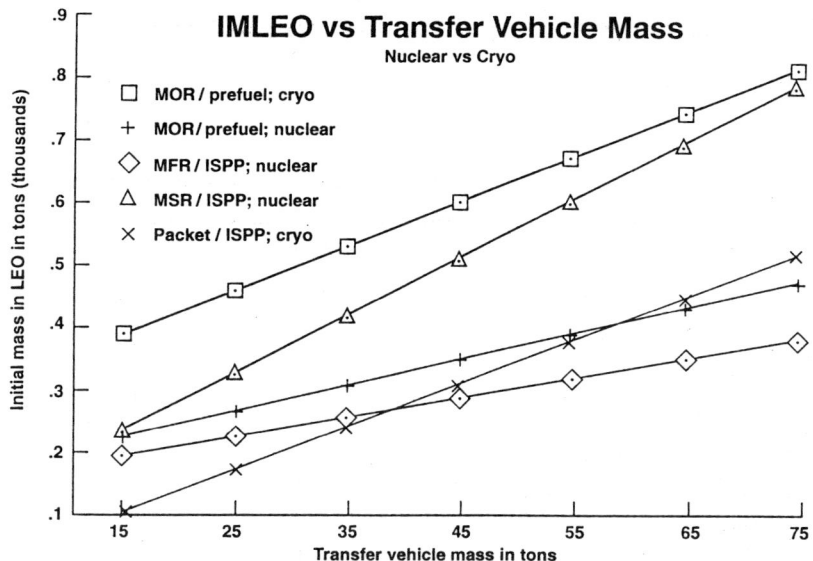

Figure 8 IMLEO dependence on mass of Interplanetary Transfer Vehicle Crew Module. Comparison of conjunction class missions using cryogenic and nuclear primary propulsion, including: cryo-prefueled-MOR, nuclear-prefueled-MOR, nuc-ISPP-MOR, nuc-MFR, and nuc-MSR, and single-stage, cryo-packet.

Comparing useful mass on the surface can be misleading. A MSR mission may land two separate habitats on the surface both of which are suitable for extended occupancy. Even though the ascender crew cab is not meant to be used on the surface, it may service the crew for additional lodging and safe harborage. In the event of a catastrophic failure of the surface habitat, the MSR ascender cab may be only a short walk away. If this occurs, it is conceivable that the crew could survive until a TEI launch window or until resupplied. For this reason, the ascender cab can be included as useful surface cargo for the MSR type mission. In contrast, an MOR mission would require its crew to retreat to an orbiting transfer vehicle where they would remain until the advent of a TEI opportunity, while a MFR exploration team may have to land multiple habitats to ensure crew survival while awaiting the return of the flyby transfer craft.

In Figure 9 we compare the cryogenic opposition class MOR mission with conjunction class missions done with all other architectures. Here, the strength of the conjunction class MSR design is very apparent. More generally, it shows the advantage ISPP missions have in delivering payloads to the surface over those that use pre-fueled landers. With ISPP, surface infrastructure could grow at a rate 50% to 100% faster.

Figure 10 shows the surface capability of MOR and MSR nuclear propelled expeditions versus their respective cryogenic counterparts. A cryogenic MSR expedition is competitive with the nuclear MOR propulsion.

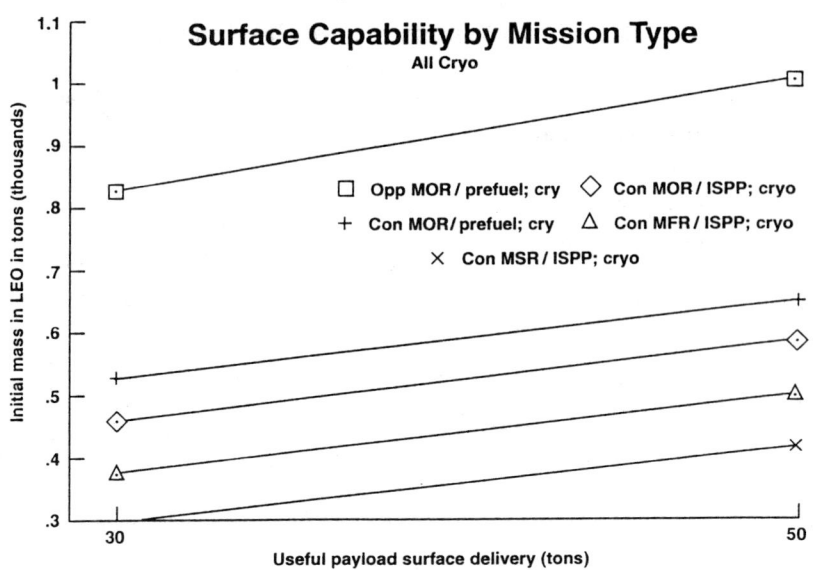

Figure 9 IMLEO dependence on surface payload, comparing opposition prefueled-MOR, with conjunction prefueled-MOR, ISPP-MOR, MFR, and MSR missions. Cryogenic propelled TMI assumed.

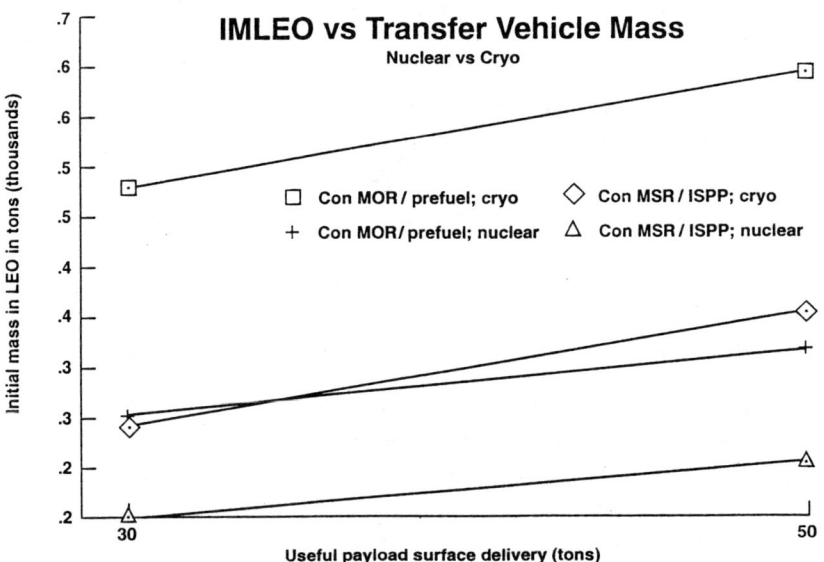

Figure 10 IMLEO dependence on surface payload, comparing conjunction class missions using: cryo-prefueled MOR, nuc-prefueled-MOR, cryo-MSR, and nuc-MSR.

CONCLUSIONS

In summary, it appears that methane ISPP can provide major reductions to IMLEO for any mission design. In the case of an MOR mission, ISPP implementation can occur progressively, with virtually no IMLEO penalty. Methane ISPP can be prototyped by initiating exploration with a two stage methane prefueled ascender. Ascender prefueling can be eventually replaced by ISPP using imported hydrogen. Gradually importation will become unnecessary as locally produced hydrogen becomes available.

For MFR and MSR missions, ISPP is an integrated activity from the outset. Both provide major IMLEO reductions over MOR. However, there are tradeoffs to be made with either design. MFR may trade flight-time for crew habitability in space, while MSR trades crew comfort on the return flight for enhanced surface capability. The superior performance of methane provides major IMLEO savings to both MFR and MSR missions. It also facilitates the evolutionary growth of program capabilities. As in MOR expeditions, methane production can commence with imported hydrogen and evolve to independent surface production of the hydrogen.

As the human presence on Mars expands, methane ISPP should give way to hydrogen production. The human foothold on Mars could then be reasonably sustained solely with cryogenic propulsion systems. If thermal nuclear propulsion becomes a reality, then its capabilities would be greatly magnified with Martian hydrogen. The mass savings achieved by ISPP can convert to greater exploratory capability on the surface or in orbit. Moreover, it may lead the way to regular, cost effective transportation between planets. Mars could then become the much needed "watering-hole" to mitigate the hostility of space.

REFERENCES

Ash, F.L., W.L. Dowler, and G. Varsi, "Feasibility of propellant production on Mars," *Acta Astronautica*, 5, pp. 705-724, 1978.

Baker, D., R. Zubrin, "Mars Direct: Combining near term technologies to achieve a two launch manned Mission," *JBIS*, 43, p. 11, November, 1990.

Clark, B.C., "The Viking results - the case for man on Mars," AAS 78-156, in *The Future United States Space Program*, R.S. Johnson, A. Naumann, Jr., C.W.G. Fulcher, eds., American Astronautical Society, Advances in the Astronautical Sciences, 38-I, pp. 263-278, San Diego, CA, 1979.

Clark, B.C., "Survival and prosperity using regolith resources on Mars," *JBIS*, 42, pp. 161-166, 1989.

Clark, B.C., "A straight arrow approach to the near term human exploration of Mars," Proceedings of *The Case for Mars IV*, Boulder, CO, June 4-9, 1990, T.R. Meyer, ed., American Astronautical Society, Science and Technology Series, San Diego, CA, in press.

Cohen, A., Report on the 90-Day Study on human exploration of the Moon and Mars, prepared by Johnson Space Center, Houston, TX, November, 1989.

French, J.R., "Rocket propellant from Martian resources," *JBIS*, 42, pp. 167-170, 1989.

French, J.R., R.L. Staehle, C.R. Stoker, C. Emmart, and S.B. Welch, "Mission strategy and spacecraft design for a Mars base program," proceedings of *The Case for Mars IV*, held in Boulder, CO June 4-9, 1990, T.R. Meyer, ed., American Astronautical Society, Science and Technology Series, San Diego, CA, in press.

Hill, P.G. and C.R. Peterson, *Mechanics and Thermodynamics of Propulsion*, Addison and Wesley, Reading, MA, 1965.

Lusignan, B., E. Reeves, L. Colin, T. Biford, A. Evitch, Y. Ivshenko, V. Kotin, and E. Narimanov, *The Stanford US-USSR Mars Exploration Initiative: Exectutive Summary*, Stanford Univeristy, School of Engineering, September 1991.

LeCompte, M.A. and J.P. Stets, "Propellant manufacturing on Mars: Issues and implications," *Space Manufacturing: Energy and Materials from Space*, Proceedings of the Tenth Princeton/AAA/SSI Conference, pp. 369-379, May 15-18, 1991.

Meyer, T.R. and C.P. McKay, "The resources of Mars for human settlement," *JBIS*, 42, pp. 147-160, 1987.

Ramohalli, K., W. Dowler, J. French, and R. Ash, "Some aspects of space propulsion with extraterrestrial resources," *J. Spacecraft,* 24, pp. 236-244, 1987.

Ramohalli, K., E. Lawton, and R. Ash, "Recent concepts in missiosn to Mars: Extraterrestrial processes," *J. Propulsion*, 5, pp. 181-187, 1989.

von Braun, W., *The Mars Project*, the Univ. of Illinois Press, Urbana, IL, 1962.

Zubrin, R.M., D.A. Baker, O. Gwynne, "Mars Direct: A simple, robust, and cost effective architecture for the space exploration Initiative," *AIAA 91-0326*, presented at the 29th Aerospace Sciences Meeting, Reno, NV, January 7-10, 1991.

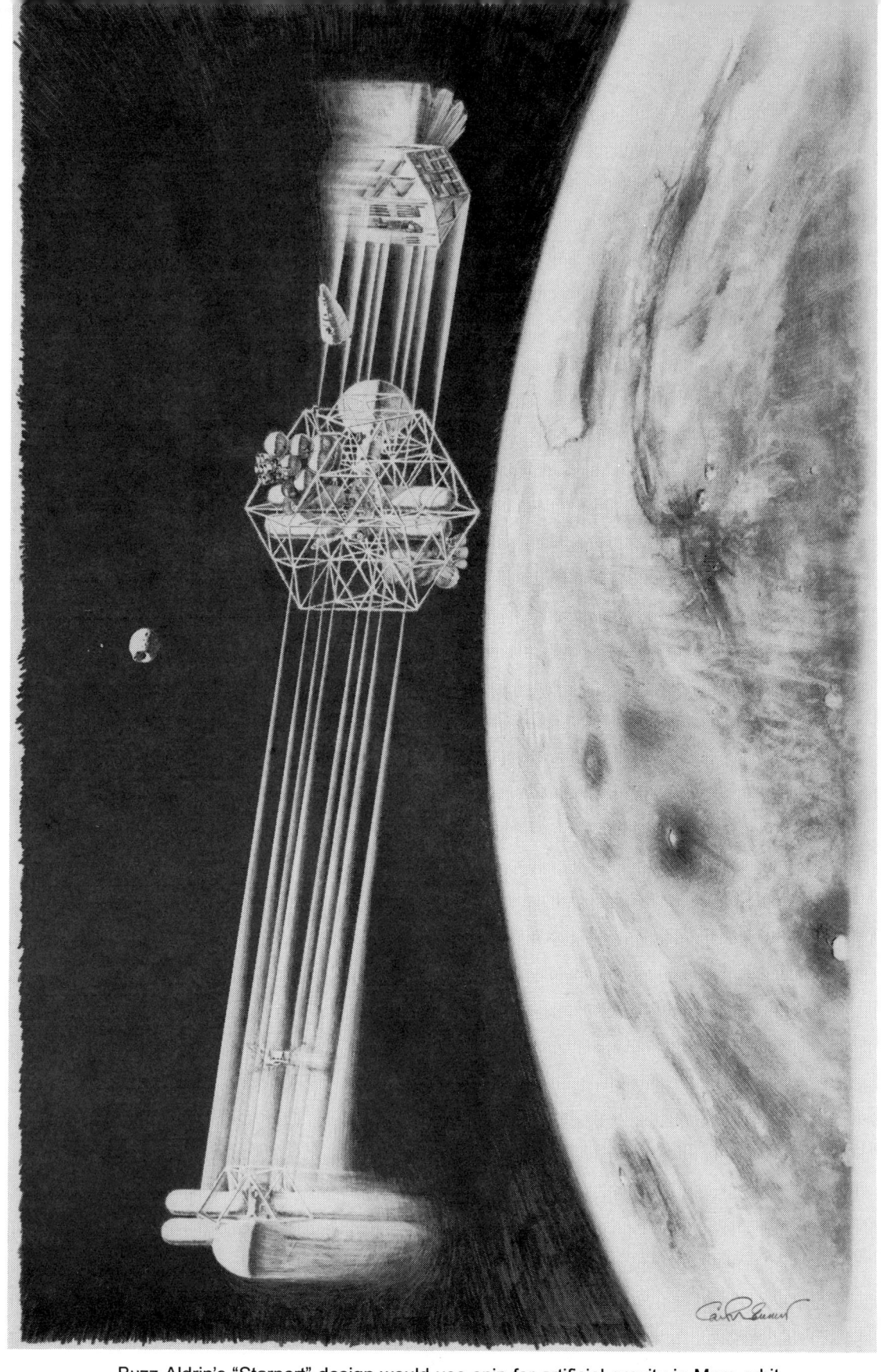

Buzz Aldrin's "Starport" design would use spin for artificial gravity in Mars orbit.

AAS 95-480

Chapter 10

ARTIFICIAL GRAVITY: DESIGN IMPLICATIONS FOR MARS VEHICLES

Lawrence G. Lemke[*]

It is now virtually certain that a human mission to Mars will require active measures to counter the effects of prolonged weightlessness. One approach is to adopt the paradigm of medical intervention in response to an illness, and treat crewmembers individually with pharmaceuticals, physical therapy, and prosthetics. Another approach is to avoid the need for intervention by placing the entire crew in a pseudogravity field. The latter approach has typically been assumed to be very costly in terms of vehicle mass, yet a recent study conducted by the Marshall Space Flight Center has shown that this is not necessarily the case. The specific approach and results of that study are discussed and then extended model of the design algorithm. The resulting model is used to investigate various vehicle design strategies that depart from the groundrules assumed in that study. It is shown that the possibility exists for reducing the mass penalty to about 10% or less of the comparable zero-g vehicle.

INTRODUCTION

The realization that a traveler in a vehicle on a ballistic trajectory in free space would experience a nullification of gravity occurred early in the history of astronautics (e.g., Jules Verne, *Around the Moon*, 1865). Even though the image of space travelers floating around effortlessly in their space ships was used at a very early date to popularize the field of astronautics, schemes for introducing artificial gravity on a large scale were also proposed at equally early dates. Apparently some doubt existed in the minds of the early pioneers as to whether weightlessness would necessarily be the condition of choice for space travel. Accordingly, they sought to answer potential attacks on their nascent field by covering both possibilities.

In 1923, Hermann Oberth described how two individual space vehicles could be connected by a long flexible cable and be spun up to a desired angular velocity by the use of reaction jets thrusting in opposite directions [Oberth, 1972]. This system was proposed for use in the construction of a large Earth orbital station. At the time, access to space by rocket vehicles was not yet a practical reality, the effect of weightlessness on humans was only guessed at, and it is therefore perhaps understandable that Oberth

---

[*] Chief, Advanced Space Projects Branch, NASA Ames Research Center, Moffett Field, California 94087.

did not support his speculation with a quantitative engineering analysis. One may argue whether Oberth's speculation was the intuitive leap of a genius or a simple lucky guess; however, recent studies have shown that it is precisely this concept (and possibly only this concept) that brings artificial gravity within the range of feasibility for inclusion on a human Mars mission.

THE NEED FOR ARTIFICIAL GRAVITY

Sixty-five years after Oberth first proposed his concept, it is now well understood that long-term habitation in a condition of microgravity has cumulative adverse effects on the human organism. Among the more serious of these effects are cardiovascular deconditioning, orthostatic intolerance, muscular atrophy, and skeletal demineralization [Johnson, 1986]. If untreated, these effects call into question the ability of space travelers to return safely and productively to a 1-g or partial-g environment. Some of the medical countermeasures that have been attempted include increased exercise, supplemental dietary calcium, and other pharmaceuticals. To date, none of these countermeasures has been shown to be both effective and free of side-effects for long-duration missions [Space Science Board, 1987].

Depending on the scenario selected, an exploration mission to Mars could involve approximately two years of total trip time, not counting the time spent on the Martian surface. At the end of the trans-Mars trajectory of such journeys the crew may have to endure tens of minutes of deceleration at levels up to 5-g during the aerocapture maneuvers. They may then immediately be required to control rendezvous and docking maneuvers with a cargo vessel previously placed in Mars orbit. Those selected for the surface exploration would undergo further deceleration during the descent, followed by immediate activities to set up the base, preparations for return and, of course, intense exploration of the planet itself, which may last for months. The return trip to Earth would differ primarily in the lack of exploration activities at the end. It is thus easy to imagine loss of life and failure of the mission if one or more of the crew was physically or psychologically incapacitated due to the cumulative effects of microgravity.

Since these problems do not occur in the majority of the normal Earthbound population it is plausible to suppose that habitation in a centrifugal acceleration field that duplicates or approaches Earth-normal gravity will eliminate them altogether. In addition to the human physiological issues, it is also likely that a range of habitability concerns will be simplified the more Earth-like an environment is. For example, liquids will stay in containers, waste and contaminants will not drift freely through the environment, conventional electronic equipment can be convectively cooled, etc. For these reasons, NASA is now actively considering the provision of artificial gravity on long duration planetary missions.

ARTIFICIAL GRAVITY AND HUMAN FACTORS

The only known feasible method for generating so-called artificial gravity is by the use of a centrifuge to impose centripetal acceleration on the contents of the centrifuge.

The major characteristic distinguishing centrifuges intended for long-term human habitation from centrifuges for other purposes is the radius of rotation.

The well-known equation relating desired centripetal acceleration (α), rotation rate (ω), and rotation radius, r, for a centrifuge arm is:

$$\alpha = \omega^2 r \tag{1}$$

where ω is in radians per second. Since there are three design variables and one constraint equation, the design engineer can theoretically choose any two of the variables independently. In actuality, however, human habitability concerns tend to drive the selection of α and ω, thus determining r.

The presumption underlying the interest in artificial gravity space vehicles is that there is some level of artificial gravity, perhaps less than Earth-normal, at which the adverse biological effects of weightlessness can be avoided and the beneficial effects of gravity can be retained. Unfortunately, there is currently no accepted scientific data base that allows a precise determination of the dose-response curve for fractional gravity on humans. The only significant space experience with partial gravity is the time spent by the Apollo astronauts on the Moon, and the duration of those missions ($\approx$ 10 days) was not long enough to observe a measurable effect. In the absence of any evidence to the contrary, the most conservative assumption would be that the crew of a Mars mission must be maintained at 1-g of artificial gravity for both orbital transfer trajectories.

There is also a similar imprecision with respect to the determination of allowable spin rate, although in this case we are not entirely without a data base. In the 1960's a series of experiments was performed at the U.S. Naval School of Aviation Medicine in Pensacola, Florida in which test subjects lived continuously in rotating rooms [Guedry, *et al.*, 1964] for extended periods of time (weeks). At rotation rates of 6 rpm or more virtually every subject initially showed symptoms of debilitating motion sickness. At about 2 rpm and below, very mild symptoms were experienced by only a few subjects—even in individuals highly susceptible to motion sickness.

Eventually, most occupants of the rotating room showed some adaptation (i.e., increased tolerance) to the rotation, even at the higher rates. When adaptation occurred, however, a corresponding period of readaptation was required upon leaving the rotating environment—a fact that could be important for a Mars mission involving artificial gravity, since this period of readaptation could occur during the period of atmospheric entry and rendezvous. In view of this situation, the actual spin rate that will allow long-term human habitation in space can also not currently be known with scientific precision. The value of 2 rpm is being used for planning purposes under the assumption that it is conservative, would exclude only a small fraction of the population from consideration, and would require a negligible readaptation period.

Substituting the nominal values 1-g (9.8 m/s^2) and 2 rpm (0.2094 rad/s) into Eq. (1) shows that the required radius for a human rated centrifuge is about 223 m.

MARS VEHICLES

The 1987 MSFC Study

The design of a Mars mission vehicle configured for artificial gravity was generated during the summer of 1987 as the result of an engineering study directed by the Marshall Space Flight Center with participation from other NASA field centers. The purpose of the study (hereafter referred to as the "MSFC study") was to assess the cost penalty for including artificial gravity in human Mars missions. The approach was to begin with the definition of an existing vehicle for a mission without artificial gravity and determine the mass additions implied by the minimal modifications necessary to include artificial gravity. Because of the brevity of the study, the emphasis was placed on a point design, assuming use of the values, 1g and 2 rpm, for the acceleration level and spin rate, respectively, for the reasons mentioned above.

The starting point for the study was a version of a mission model generated under NASA contract by the Science Applications International Corporation [Science Applications International Cooperation, 1987]. This mission is a so-called split mission option; a robotic cargo vehicle is first placed in Mars orbit using a low energy transfer trajectory and is then followed by a human crew in a piloted vehicle on a faster trajectory. Upon arrival at Mars the crew must rendezvous and dock with the cargo vehicle, refuel the piloted vehicle for the return trip to Earth, and retrieve the supplies for the descent to the Martian surface and exploration.

The results of this study have been presented in detail elsewhere [Schultz, 1987] and are summarized here in Table 1, which emphasizes the differences in component masses for two vehicles of equivalent capability, one with and one without artificial gravity. The mass penalty for artificial gravity on the human vehicle, for the set of assumptions made in the MSFC study is about 20%. Largely as a result of this study, the possible inclusion of artificial gravity on piloted Mars missions is being carried as an option in NASA long-range planning.

Under the design strategy utilized in the study, additional components to implement artificial gravity were required on the piloted vehicle only, thus increasing its mass directly. A heavier piloted vehicle, however, requires more propellant in the boost stage for departure from Earth orbit as well as more propellant held in Mars orbit in the cargo vehicle for the return. Most conventional cost models for space hardware are non-linear with respect to mass, and would therefore predict a total mission cost increment out of proportion to the primary mass increment on the piloted vehicle. Since any previous space projects upon which the cost models could be built are certainly much smaller than a piloted Mars mission, and because the mass additions to the boost stage and cargo vehicle require no additional technology development, the validity of the use of existing models in this case is questionable. Unlike the MSFC study, the parametric analysis which follows will avoid reference to any absolute monetary cost estimates and concern itself simply with estimating the relative mass increase to the piloted vehicle. This is regarded as a reasonable first-order estimate of the relative increase in mission cost.

Table 1
COMPONENT MASSES FOR MARS VEHICLES

Item Description	0-G Vehicle (kg)	1-G Vehicle (kg)
Counterweight End:		
Aerobrake	13,209	14,336
Avionics	0	178
RCS subsystem	0	5,444
Aupport structure	1,287	2,110
Other	21,345	21,345
SUBTOTAL	35,841	43,413
Crew Module End:		
Tether	0	1,966
Deployer	0	2,726
Power Subsystem	2,028	3,840
RCS Subsystem	0	6,219
Support Structure	721	1,812
Other	65,493	65,493
SUBTOTAL	68,242	82,056
SYSTEM TOTAL	104,083	125,469

The following discussion retraces the design assumptions made in the MSFC study in a parametric manner. This results in a simplified mathematical model for the mass penalty associated with including artificial gravity in a human Mars vehicle, calibrated against the MSFC point design vehicle. It thus permits easy analysis of the consequences of altering certain parameter choices in the engineering of such a vehicle.

Mass Additions to the Piloted Vehicle

The general design strategy was to divide the basic elements of the zero-g vehicle into two categories; the crew quarters plus associated life-critical subsystems, and everything else. The crew quarters and associated subsystems (accounting for about 60% of the total mass) were placed at one end of a centrifuge arm (the "Human Module End") and the remaining elements were placed at the other end (the "Counterweight End"). The minimum additional requirements were for a structure to connect the two end bodies and a means of spinning and despinning the vehicle. The resulting vehicle resembles a freely rotating dumbbell with unequal end masses.

A design issue with critically important consequences for the mass penalty associated with this approach is whether the centrifuge arm is of conventional construction (such as a triangulated truss) or not. In general, a centrifuge arm constructed of rigid materials would probably have to be designed to specific stiffness requirements related to the spin frequency in order to avoid the buildup of harmful resonances which could destroy the structure. It has been shown elsewhere [Lemke, 1987] that the mass of a "rigid" centrifuge arm designed to a bending stiffness criterion increases like arm length to the fourth power, whereas the mass of a tension-stiffened tether increases proportional only to length to the first power. Thus, the use of tension-stiffened tethers to connect the two end bodies is very much more mass efficient than any other construction technique for large, freely spinning centrifuges. For this reason, the use of a tether was assumed in the MSFC study and in the model discussed here.

Since the tether connecting the two bodies has no stiffness at zero rotation rate, but the vehicle must start and end at zero rotation rate, the entire spin-up and spin-down operation can not take place at constant radius. It is planned that the two bodies would start hard-docked to each other at essentially zero tether length. A negligible amount of propellant would be used to spin the configuration up to 2 rpm, at which point enough tension would exist in the tethers to allow them to be unfurled to their operating length. The angular velocity would, of course, decrease correspondingly due to conservation of angular momentum. The main spin-up would then proceed from an initial condition of small, but non-zero angular velocity. Spin-down would be the reverse procedure.

The vehicle must be spun up to its operating speed after insertion into the trans-Mars orbit trajectory and despun prior to Mars encounter. This process must be repeated on the return trajectory to Earth, thus implying a minimum of 2 start-stop cycles. In addition, it was decided that EVA could not be performed while the vehicle was spinning; therefore, planning for unpredictable contingencies requiring EVA enroute necessitates including provision for a number of additional starts and stops. The MSFC study assumed two such contingencies, for a total of 4 cycles.

The only two feasible techniques for spinning and despinning a body in free space are mass-expulsion thrusters and angular momentum storage devices. Because of its unusual configuration, the radius of gyration of an artificial gravity Mars vehicle would be one to two orders of magnitude greater than a more conventionally constructed spacecraft of equal mass. Therefore, conventional angular momentum storage devices (momentum wheels) are infeasible for spinning and despinning the vehicle. Very large, high performance momentum wheels can be postulated, but would require an aggressive technology development program and would still be more than an order of magnitude more massive per start/stop cycle than thrusters. For other possible artificial gravity spacecraft that must undergo a large number of start-stop cycles, advanced momentum storage devices may be considered; however, since the total number of start/stop cycles for the MSFC vehicle is 4, the use of mass-expulsion thrusters was assumed. Equal-size, opposite-facing thrusters would be located on each of the two endbodies and would be fired in concert to avoid perturbing the trajectory of the center of mass (CM) of the system.

A number of miscellaneous mass increases also had to be included to account for resizing of the aerobrake, structural, and power subsystems, and for the tether deployer and additional avionics not present on the zero-g vehicle. In reality, the masses of these components would probably vary as some function of the major design parameters of spin rate and g-level. In the MSFC study and in this mathematical model, however, these additions are treated as constants because the functional relationships of the masses to the system parameters were not easily derivable (and are probably discontinuous).

Modeling Approach

As discussed above, the design strategy was to retain all the elements of the zero-g vehicle, include additional elements required to generate artificial gravity, and enlarge some existing components, where necessary. From Table 1, the significant mass differences are:

1. The propellant required for spin-up and spin-down.
2. Additional tanks to store the propellant.
3. The tether to connect the two end bodies.
4. Aerobrake, support structure, and power subsystem enhancements, and the addition of tether deployers and avionics.

Intuitively, items 1, 2, and 3 must themselves be functions of the final total mass of the artificial gravity vehicle, M_{ag}. That is, a more massive vehicle will require more propellant to spin and despin, larger tanks to hold the propellant, and stronger tethers to support the increased weight. It will be shown below that these three items are, in fact, proportional to M_{ag}, with proportionality constants, σ_p, σ_{pt}, and σ_t. On the other hand, as mentioned above, item 4 is treated as a constant. Since the mass of the zero-g vehicle, M_0, is also a constant for the purposes of this model, item 4 can be expressed as proportional to M_0, with proportionality constant, ε.

Mathematically, the design strategy can be summarized as:

$$M_{ag} = M_0 + \sigma M_{ag} + \varepsilon M_0 \qquad (2)$$

where $\sigma \equiv (\sigma_p + \sigma_{pt} + \sigma_t)$.

Collecting terms and rearranging equation (2) gives:

$$M_{ag} = \left[\frac{1+\varepsilon}{1-\sigma}\right] M_0 \qquad (3)$$

For σ and ε small, and ignoring second-order small terms, the quantity in the brackets is approximately $(1 + \sigma + \varepsilon)$. Hence,

$$M_{ag} \approx (1+\Delta) M_0 \qquad (4)$$

where

$$\Delta \equiv [\sigma + \varepsilon] \qquad (5)$$

This gives us a convenient, non-dimensional expression for the fractional mass penalty of a vehicle configured for artificial gravity relative to the equivalent vehicle configured for zero-g.

Propellant Mass

The following is an estimate of σ_p for a system consisting of two bodies of constant mass constrained to rotate around a common CM at constant radius and phase. In order for this model to accurately reflect the actual Mars vehicle, it is necessary that each body employs equal-size thrusters whose thrust vectors are in the plane of rotation and perpendicular to the radius vector, that the mass change due to propellant consumption is a small fraction of the total vehicle mass, and that the mass of the structure interconnecting the bodies is negligible compared to the overall system mass. Simple design choices assure the validity of the first assumption; the derivations below show that the second and third assumptions are also consistent.

Consider the crew compartment, of mass m_1, rotating at radius, r_1, around the CM. If a thruster is supplying a thrust, T_1, to increase the angular velocity, ω, then the equation of motion is:

$$T_1 r_1 = \frac{d}{dt}(J\omega) = m_1 r_1^2 \frac{d}{dt}(\omega) \qquad (6)$$

This leads to:

$$\int_0^{t_f} T_1 dt \approx m_1 r_1 \int_0^{\omega} d\omega \qquad (7)$$

The "$\approx$" is used here because of the non-zero initial angular velocity discussed above. Neglecting the inequality and performing the indicated integration gives:

$$I_{T1} = \omega m_1 r_1 \qquad (8)$$

where I_{T1} is defined as the total impulse exerted on the first body. The conventional definition relating total impulse to specific impulse is:

$$I_T \equiv I_{sp} M g \qquad (9)$$

where M is the total mass of propellant consumed in generating the impulse.

By symmetry, the same amount of propellant is consumed in the spindown, hence, Eqs. (8) and (9) together yield the mass of propellant consumed in one start-stop cycle for the inhabited end body:

$$M_{P_1} = \frac{2m_1 r_1}{gI_{sp}} \tag{10}$$

If we normalize the centrifugal acceleration to the acceleration of gravity, g, through a scale factor, F, then $\omega^2 r_1 = Fg$. Substituting this constraint into Eq. (10) gives:

$$M_{P_1} = \frac{2Fm_1}{\omega I_{sp}} \tag{11}$$

The total propellant consumed is simply twice this amount since the thrusters are equal on both end bodies. Furthermore, it can easily be shown that:

$$m_1 = \frac{M_{ag}}{(1 + m_1/m_2)} \tag{12}$$

Therefore, the total propellant consumed for N start-stop cycles is:

$$M_{P_T} = \left[\frac{NF\Phi}{\omega I_{sp}}\right] M_{ag} \equiv \sigma_p M_{ag} \tag{13}$$

where

$$\Phi \equiv \left[\frac{4}{1 + m_2/m_1}\right] \tag{14}$$

Propellant Tanks

For the type of technology assumed in the MSFC study, the mass of the tank plus insulation was about 9% of the mass of the contained propellant (LH_2 - LOX). For small variations around the nominal case, this relation between propellant mass and tank mass will be taken as a constant. In this case, the tanks are sized to contain one-half the total fuel at any one time, on the assumption that they will be refueled in Mars orbit. Thus,

$$M_{\tan k} = 0.045 M_{P_T} \tag{15}$$

$$\Rightarrow \sigma_{p_t} = 0.045 \sigma_p \tag{16}$$

Tether Mass

The design tension force in the tether (τ) is equal to the suspended habitation volume mass (m_1) times the acceleration level experienced by that mass. i.e.,

$$\tau = \frac{M_{ag} Fg}{(1 + m_2/m_1)} \tag{17}$$

where we have again made use of Eq. (12). Conventional practice is to have the design tension be a small fraction, f, of the tensile strength, S_t, of the tether material. If A is the cross-section area of the tether, then

$$\frac{\tau}{A} = fS_t \tag{18}$$

Combining Eqs. (17) and (18) and solving for A gives:

$$A = \frac{FgM_{ag}}{fS_t(1 + m_2/m_1)} \tag{19}$$

Multiplying by the length of the tether, L, and the tether material density, λ, gives the total mass of the tether, which allows us to solve for σ_t:

$$\sigma_t = \left[\frac{FgL\lambda}{fS_t(1 + m_2/m_1)} \right] \tag{20}$$

It can be shown that L = $(1 + m_1/m_2) * (Fg/\omega^2)$ and that $(1 + m_1/m_2) \div (1 + m_1/m_2)$ = (m_1/m_2). This leads to the final expression for σ_t:

$$\sigma_t = \left[\frac{m_1 \lambda}{m_2 fS_t} \left[\frac{Fg}{\omega} \right]^2 \right] \tag{21}$$

Miscellaneous Mass Additions

From Table 1, the total mass for these items (7,757 kg) represents about 7.5% of M_0. Hence,

$$\varepsilon = 0.075 \tag{22}$$

Results: Substituting Eqs. (14), (16), (21), and (22) into (4) finally allows us to write the parametric expression for the quantity ($\sigma + \varepsilon$):

$$\sigma + \varepsilon = \left[1.045 \left[\frac{NF\Phi}{\omega I_{sp}} \right] + \left[\frac{m_1 \lambda}{m_2 f S_t} \left[\frac{Fg}{\omega} \right] \right]^2 + 0.075 \right] \qquad (23)$$

Numerical Evaluation

The following parameter values may be used in this model:

$m_1/m_2 = 1.897$, $I_{sp} = 460$ sec, $f = 0.115$, $\lambda = 1450$ kg/m^3 (Kevlar), $S_t = 2.758 * 10^9$ N/m^2 (400 Kpsi, Kevlar), $g = 9.807$ m/sec^2

Evaluating the model numerically gives:

$$\sigma + \varepsilon = \left[5.95 * 10^{-3} N \left[\frac{F}{\omega} \right] + 8.34 * 10^{-4} \left[\frac{F}{\omega} \right]^2 + 0.075 \right] \qquad (24)$$

Where N, F, and ω remain as design variables. Recall that in the MSFC study, N = 4, F = 1.0, and ω = 0.2094 rad/s. When these values are used in the model, the expression in the brackets (about 0.207), agrees with the value computed in the MSFC study (0.205) within less than 2%.

VEHICLE OPTIMIZATION

Given this model, it is interesting to consider how well one could do if allowed to depart from the nominal assumptions. We can begin by noticing that both of the variable terms in Eq. (24) vary with the ratio of (F/ω), hence, reducing the design g-level and increasing the allowable spin rate monotonically reduces the mass penalty (as intuition would lead us to believe). However, since the term varying quadratically in this ratio is already the smallest contributor (2%), most of the effect of changing these parameters enters linearly.

As discussed above, the presumed ability of humans to live for prolonged periods of time at a value of F other than 1.0 is conjectural at this time. On the other hand, the premise of long-term habitation of Mars already contains this conjecture within it. Hence, it may be reasonable to assume, for example, that the outbound portion of the trip would be conducted at Mars-normal gravity and the return portion would be conducted at Earth-normal, thus delivering the crew to its destination fully adapted to the local conditions. Under this assumption, the effective value of F would be the average of Earth and Mars gravity, or, about 0.69 g.

The parameter of next largest influence is N, the nominal number of start-stop cycles. For the assumptions made in the MSFC study, the absolute minimum size for N is 2, since the vehicle spin must be started and stopped propulsively on both transfer trajectories. As discussed, additional propellant for one contingency cycle per trajectory was included on the basis of mission safety conservatism, for a total of 4. It is unlikely that a total lack of contingency propellant would ever be acceptable, but it may be possible to dispense with it on the Earth return trajectory. Even though the vehicle would not be able to tolerate two contingencies and still deliver the design level of artificial gravity for the entire mission, it may be acceptable to assume a greater risk of interruption of artificial gravity on the return leg, since recuperation and medical facilities would be awaiting the crew on their return to Earth. Planning for only one contingency start-stop cycle would thus allow an N of 3.

Finally, we should notice Φ, which can be thought of as a "form factor" (Eq. 14). This factor decreases monotonically as more mass is moved from the inhabited end of the vehicle to the uninhabited end. The ground rules for the MSFC study required the use of the same components for the artificial gravity vehicle as for the zero gravity vehicle, with minimum modifications. This requirement indirectly forced most of the mass to be distributed on the habitation end of the system and Φ to take on a value of about 2.62. In general, because of safety considerations, it is desirable to locate most of the subsystems critical to crew survival in proximity to the crew itself. Since most of the subsystems of a mass-constrained vehicle fit this description, it is likely that an artificial gravity vehicle with all the crew located on one end will always result in Φ being significantly greater than 2.0. It is possible, however, to consider the more efficient arrangement where the mass distribution is equal on both ends of the tether (i.e., $\Phi \equiv 2.0$). The most obvious way to achieve this distribution is to have identical vehicles on both ends.

In addition to improving the efficiency of mass distribution, this would also allow a novel option for further savings on spin propellant (called "aerotorquing"). If both endbodies were equipped with aerobrakes, they could be used to dissipate the rotational velocity of the vehicle ($\approx \pm 0.04$ km/s) in addition to the planetary approach velocity already assumed (≈ 4 km/s). This could be accomplished by disconnecting the bodies from each other with the system still rotating and the tether still fully deployed shortly before planetary encounter, thus flinging them apart. This maneuver could be timed such that one body would receive a posigrade Δv and the other a retrograde Δv. A very small amount of propellant would have to be used to stop the residual rotation of each end body prior to atmospheric entry. Effectively, this technique utilizes the planetary atmospheres instead of propellant to dissipate the major portion of the system angular momentum along with the system linear momentum. In this scenario, propellant for one full start-stop cycle with one contingency cycle would give $N \approx 2$.

There may be a mass economy of scale associated with large vehicles not reflected in this model. That is, two small vehicles may weigh more than one large vehicle with twice the per-unit capacity. Hence, the overall empty weight of an artificial gravity vehicle may be greater for a symmetric design than for an unsymmetric design of the same capability. However, since the propellant mass used for spinning and despinning the vehicle is the largest variable the designer has to work with, the option of a symmetric vehicle seems sufficiently attractive that it should probably be studied in greater detail.

An option for further minimization of propellant use that will not be developed here quantitatively lies in the technique of "overdeploying" the tether. This technique consists of starting the spinup maneuver with the endbodies at larger radii than the desired final radii to allow the thrusters to convert their impulses to angular momentum more efficiently. This obviously requires more tether mass while it saves propellant and tank mass. To first order, the mass changes are both proportional to the nominal propellant-plus-tank mass and tether mass, respectively. The optimal condition, therefore, is where the propellant-plus-tank mass equals the tether mass. This argument ignores any mass changes that may have to be imposed on the reel mechanisms in order to move the endbodies dynamically in the artificial gravity field.

The assumption of I_{sp} = 460 sec is consistent with the use of liquid hydrogen and liquid oxygen as propellants. The only foreseeable technology which could yield higher performance in this application is ion-electric propulsion. Unfortunately, the specific thrust level of this technology is so low that the spin-up and spin-down times are prohibitively long. Hence, this parameter is unlikely to be subject to improvement.

Using the most optimistic set of assumptions (F = 0.69, Φ = 2.0, N = 2) gives an estimated mass penalty of about 11%. If future developments validate the ability to use higher spin rates, lower acceleration levels, and the tactic of tether overdeployment, the potential exists for moving the mass penalty below 10%.

SUMMARY & CONCLUSIONS

It is now apparent that human spaceflight of the duration required for human exploration of Mars must have some provisions for counteracting the effects of zero-gravity. The two extreme possibilities are a zero-g vehicle together with some combination of pharmaceuticals, exercise, special garments, and other measures for the inhabitants, or an artificial gravity vehicle. Either approach will have effects on the overall cost-effectiveness of a typical mission that have yet to be accurately compared. Neither approach has yet been experimentally demonstrated to be entirely effective.

If other countermeasures for zero-gravity effects are never discovered or developed, artificial gravity may be an enabling technology for human Mars missions. In less extreme cases, it may be viewed as a mission-enhancing technology, because it makes the voyage conditions more Earth-like. This factor would allow the general use of conventional facilities by the crew during most of the voyage and would possibly allow selection of the crew from a larger population. Perhaps more importantly, it would increase the feasibility of considering the lower-energy but longer duration conjunction class transfer trajectories. Everything else being equal, these missions tend to maximize the time available on the Martian surface for a given launch opportunity and are relatively more robust against missed launch opportunities.

Probably the single most significant result of the MSFC summer study and of this discussion is simply the relative magnitude of the effect of including artificial gravity on human-occupied interplanetary vehicles. *A priori* estimates based either on no engineering analysis at all or on the assumed use of conventional structures typically result in the belief that the mass penalty for including artificial gravity is prohibitive. If the use of tension-stiffened tethers can be assumed, this belief is probably incorrect since the nominal penalty is no greater than approximately 20%, and possibly as low as 10%.

REFERENCES

Oberth, H., "Ways to spaceflight," NASA Technical Translation, TT-F-622, NASA, Washington, DC, January, 1972.

Johnson, P., "Adaptation and readaptation medical concerns of a Mars trip," in *Manned Mars Mission's Working Group Papers,* NASA M002, Vol. 2, pp. 593-600, NASA, Marshall Space Flight Center, Huntsville, AL, June, 1986.

Space Science Board, Commission on Space Biology and Medicine, *A Strategy of Space Biology and Medical Science for the 1980's and 1990's*, pp. 111-116, National Academy Press, Washington, DC, 1987.

Guedry, F., R. Kennedy, C. Harris, A. Graybiel, "Human performance during two weeks in a room rotating at three rpm," *Aerospace Medicine*, pp. 1071-1082, November, 1964.

Science Applications International Corporation, "Piloted sprint missions to Mars," *Report No. SAIC-87/1908*, Schaumberg, IL, November, 1987.

Schultz, D., "A manned Mars artificial gravity vehicle," presented at the PSN/NASA/ESA Second International Conference on Tethers in Space, Venice, Italy, October 4-8, 1987.

Lemke, L., "The use of tethers for an artificial gravity facility," presented at the PSN/NASA/ESA Second International Conference on Tethers in Space, Venice, Italy, October 4-8, 1987.

Chapter 11

NUCLEAR ROCKETS:
HIGH-PERFORMANCE PROPULSION FOR MARS

Clayton W. Watson[*]

A new impetus to human Mars exploration was introduced by President Bush in his Space Exploration Initiative. This has led, in turn, to a renewed interest in high-thrust nuclear thermal rocket propulsion (NTP). The purpose of this paper is to give a brief tutorial introduction to NTP and provide a basic understanding of some of the technical issues in the realization of an operational NTP engine.

Fundamental physical principles are outlined from which a variety of qualitative advantages of NTP over chemical propulsion systems derive, and quantitative performance comparisons are presented for illustrative Mars missions.

Key technologies are described for a representative solid-core heat-exchanger class of engine, based on the extensive development work in the Rover and NERVA nuclear rocket programs (1955 to 1973). The most driving technology, fuel development, is discussed in some detail for these systems. Essential highlights are presented for the 19 full-scale reactor and engine tests performed in these programs. On the basis of these tests, the practicality of graphite-based nuclear rocket engines was established.

Finally, several higher-performance advanced concepts are discussed. These have received considerable attention, but have not, as yet, developed enough credibility to receive large-scale development.

INTRODUCTION

Almost from the inception of the "Nuclear Age," in the last days of World War II, the potential advantage of nuclear energy for propulsion was realized and under study [Serber, 1946; Dixon and Yockey, 1946]. Early attention emphasized Earth-bound applications, such as ICBM or aircraft propulsion, and consideration of basic energetics quickly led to the realization that nuclear energy would be key for exploratory missions into space, and would, perhaps, be essential for more ambitious missions such as inter-

---

[*] Ph.D., Retired, Los Alamos National Laboratory, Los Alamos, New Mexico 87545.

planetary human exploration [Shepherd and Cleaver, 1948-1949; Bussard, 1953; Bussard, 1962].

The advantages of nuclear thermal over chemical propulsion derive from two fundamental features: (1) the enormous energy available per unit mass of fission (or fusion) fuel, compared to chemical-energy sources, and (2) the energy-producing medium in a nuclear system is separate from the thrust-producing propellant, allowing a low molecular-weight propellant such as hydrogen to be used, which greatly increases the propulsive force per unit propellant flow.

The new impetus to human Mars exploration introduced by President Bush in his Space Exploration Initiative (SEI) [Stafford *et al.*, 1991] has led, in turn, to a renewed interest in high-thrust nuclear thermal propulsion (NTP). NTP is not a new nor undeveloped concept [Dewar, 1974]; interest in NTP covers a time span of nearly 50 years, and a great deal of research and development has resulted. The 17-year, ~ 1.5 billion-dollar Rover/NERVA program, for example, proved the feasibility of, and developed full-scale operating versions of fission-driven rocket reactors, with demonstrated performance adequate for SEI Mars missions. A corresponding NTP engine system was also developed, ground-tested, and brought to near-flight status, although the program was canceled before flight-testing was achieved.

Since NTP has a long history and is a technically broad and complex field, only a brief outline can be presented in this paper. The purpose here is to give a brief tutorial introduction to the subject, provide entry points into the literature if further study is desired, and allow a basic understanding of some of the more fundamental technical issues.

PHYSICAL PRINCIPLES

Propulsion Efficiency

Any space maneuver, whether "launch" to some orbit around a gravitational body or transfer from one position in space to another (orbit-to-orbit transfer), is accomplished by imparting an impulse to the maneuvering vehicle to produce a momentum change. The function of a rocket engine is thus to exert a force, $\vec{F}$, for a time, t, on a body of mass, m, to change the velocity, $\vec{v}$, of the body by an amount, $\Delta \vec{v}$. This is accomplished by expending a mass, Δm, of fuel from the maneuvering vehicle. The quantity $\Delta \vec{v}$ thus characterizes the mission requirement, and Δm represents the "cost" in terms of mass expended to achieve $\Delta \vec{v}$.

A rocket engine exerts a force by producing a hot gas (propellant) and exhausting it through an expansion nozzle at a velocity, v_e, with respect to the vehicle. The exhausted propellant, at velocity v_e, produces a force, $F = (dm/dt)v_e$, where dm/dt is the propellant flow rate. The efficiency of the engine is clearly determined by v_e, the force produced per unit flow rate, and this is most frequently defined in terms of the "specific impulse," I_{sp},

$$v_e = g I_{sp}$$

where g is the acceleration of gravity.* (Note that specific impulse has units of velocity ÷ acceleration = seconds.)

In practice, calculation of actual space maneuvers in various gravitational fields is complex; however, basic principles are straightforward. The mass of the vehicle in a given maneuver will be reduced from an initial value, m_0, to a final value, m, after the maneuver, Δv. The mass ratio, m/m_0, is an important measure of the efficiency of the maneuver. For a vehicle in free space (no other force on the vehicle), simple momentum conservation leads to the "rocket equation,"

$$\frac{m}{m_0} = e^{-\Delta v/v_e} = e^{\Delta v/g I_{sp}} \quad (1)$$

which illustrates the importance of I_{sp} in maximizing m, or minimizing m_0, or both.

The key point for NTP is that the propellant exhaust velocity for a rocket engine is related directly to propellant conditions by,

$$I_{sp} \propto \sqrt{\frac{T_c}{M}} \quad (2)$$

where T_c is the propellant total, or "chamber," temperature (before expansion) and M is the propellant molecular weight.

We can now see the attractiveness of NTP. In a chemical rocket, the highest I_{sp} (lowest M) is available from burning H_2 and O_2 to H_2O, with an M of ~ 18. The resulting highest I_{sp} for such an engine is ~ 450 s. Thus, if a nuclear rocket, using H_2 as a propellant, could operate at the same T_c as the H_2/O_2 engine, its I_{sp} would be ~ 1,350 s. In practice, a solid-core nuclear rocket would operate at a somewhat lower T_c, and the actual I_{sp} achieved in nuclear rockets to date is ~ 900 s, still a very impressive enhancement. Figure 1 compares theoretical specific impulses and implied mass ratios for various energy sources based on equations (1) and (2). The incentive for high T_c and low M is clear.

Energy Production and Transfer

The advantage of nuclear energy for space propulsion can be viewed as deriving from the fact that the nuclear system can use a low molecular-weight propellant. An implicit assumption, however, is that a chemical engine cannot similarly transfer chemical energy from a burning fuel to a separate propellant. In practice, this assumption is totally valid because of the enormous difference in energy available per unit mass of fuel—roughly 200 Mev per fission event compared to a few ev per reaction in a chemi-

---

* For example, if exhaust velocity is given in units of m/s, the corresponding acceleration of gravity is 9.8 m/s², and v_e and I_{sp} differ by a factor of 9.8 (~ 10).

cal fuel; i.e., a ratio of ~ 10^6 to 10^7 in energy per unit mass. Nuclear energy is, in this sense, essentially "free" in terms of mass burned, whereas the fuel mass expended for energy production in a chemical engine is so large that the combustion products must also serve as propellant—a separate mass expenditure for propellant cannot be tolerated.

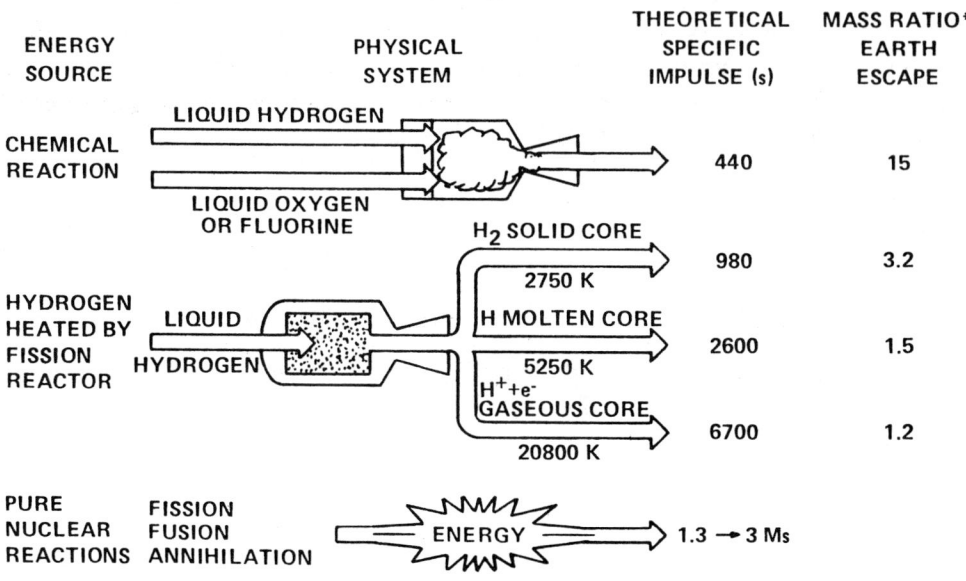

Figure 1 Theoretical specific impulse and implied Earth-escape mass ratio for various propellant-energy sources.

Nuclear energy is not, in fact, literally free in terms of mass expenditure in a space maneuver. The nuclear engine must pay a "fixed" mass penalty in the hardware required to produce and transfer energy to the propellant—the mass of a nuclear reactor, for example—and, this "fixed" mass requires propellant mass to maneuver it as an integral part of the vehicle mass. The net overall advantage of NTP is still large, however, and, as can be seen by examining equations (1) and (2), this advantage increases rapidly as the difficulty (Δv) of the mission increases.

A host of technical problems arise in a nuclear engine in efficiently transferring maximum energy to the propellant, with maximum T_c and minimum reactor mass, as discussed later. The extent to which workable solutions to these requirements can be achieved in a reliable, operational system is what ultimately determines the feasibility and efficiency of NTP.

MISSION PAY-OFFS

A space mission usually has a variety of possible goals, with cost vs. pay-off trade-offs among them that must be evaluated, and many routes by which the mission can be accomplished, with corresponding complex value-functions and trade-offs that must be considered. Thus, there is no well-defined "Mars mission" for which simple "performance" comparisons can be made. The fundamental I_{sp} advantage of NTP over chemical propulsion, however—in conjunction with "reasonable" thrust-to-weight capabilities of demonstrated NTP systems—allows projection of a number of qualitative NTP advantages, plus quantitative comparisons for selected, illustrative cases [Stafford *et al.*, 1991; Bennett *et al.*, 1991; Bussard, 1953; Borowski and Wickenheiser, 1990].

Qualitatively, NTP can: Reduce transit times for long stay-time missions, for the same initial mass in low-Earth orbit (IMLEO) (minimize crew exposure to zero gravity, solar flares, and ambient space radiation, and increase fraction of mission time spent at Mars); Reduce round-trip times for short-stay missions for the same IMLEO; and/or: Reduce IMLEO (propellant mass) for same mission duration, thereby reducing the number of Earth-to-orbit (ETO) launches and/or ETO vehicle lift requirements and mission costs; Allow greater mission design flexibility (allow accomplishment of various missions with a common vehicle design, increase Earth and Mars departure windows, increase propulsion margin for mission variations and aborts).

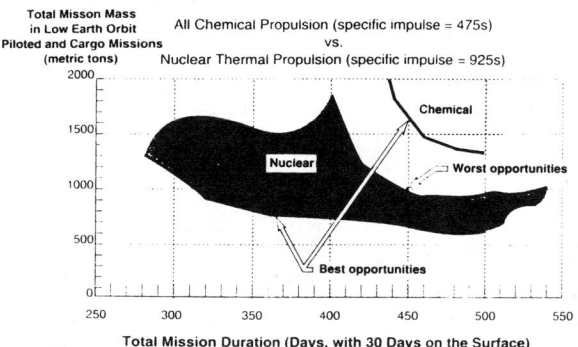

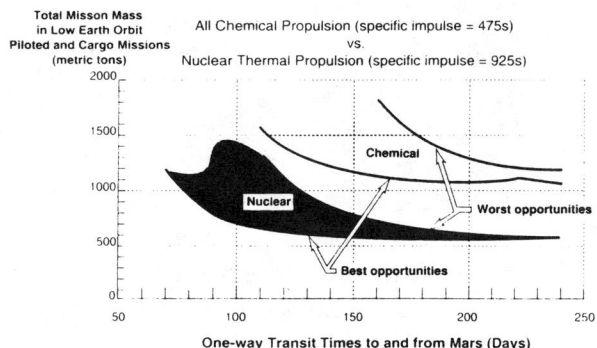

Figure 2 Comparisons of chemical and nuclear thermal (two different assumed specific impulses) propulsion systems for human Mars missions and launch opportunities between 2008 and 2022; long-duration mission = ~ 1000 days with ~ 500 days stay-time; short-duration mission = ~ 500 days with ~ 30 to 100 days stay-time.

Missions to Mars generally fall into one of two categories [Stafford *et al.*, 991]: long-duration missions of ~ 1,000 days with ~ 500 days stay-time on Mars, and short-duration missions of ~ 500 days with 30 to 100 days of Mars stay-time. Illustrative missions of these types are described in [Stafford, *et al.*, 1991], for mission architectures emphasizing trade-offs between two primary concerns: launch costs and crew effects. Launch costs are heavily dependent on IMLEO and argue for lower-Δv mission configurations, correspondingly lower propellant masses and longer transit times. Biomedical and psychological crew effects—prolonged microgravity, space radiation exposure, and confinement times—on the other hand, are strong incentives to reduce transit time. Stafford (1991) evaluated missions for launch opportunities between 2008 and 2022, for chemical (I_{sp} = 475 s) and nuclear (I_{sp} = 925 s) propulsion sources, and for both long- and short-duration missions. Figure 2 shows the resulting comparisons.

Analogous comparisons are shown in Figure 3 for a broader spectrum of potential propulsion options and an illustrative "short-duration" mission [Borowski, 1990]. Although ultimate choices for mission architectures and propulsion systems will depend on a host of issues, many of which are non-technical and most of which are ill-defined and indeterminate at present, the relative trends in Figure 3 are fundamental and likely to persist during the long-term evolution toward the first human Mars exploration.

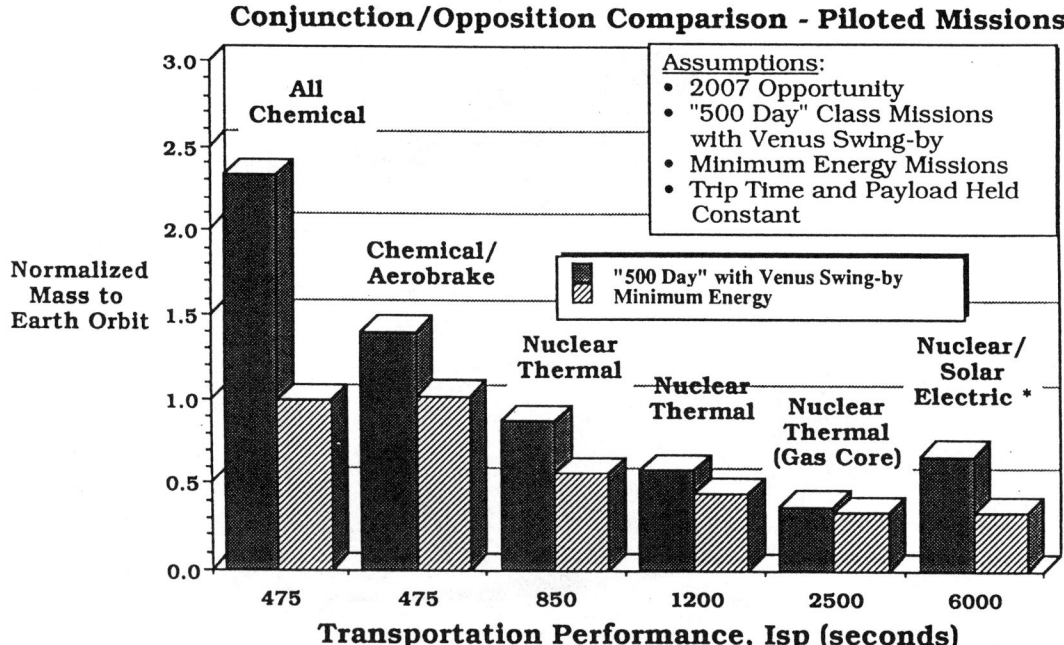

* Electric propulsion low-thrust trajectory trip times not equivalent to impulsive thrust trip times

Figure 3 Mission performance summary—Mars missions.

AN ENGINE CONCEPT

Historically, the primary and most practical approach to NTP has been the solid-core, heat-exchanger nuclear reactor (Figure 4). Liquid hydrogen (LH_2) propellant is pumped through all extra-core components (nozzle, reflector, structures, shield) for cooling, then through the reactor core where it is heated to a temperature determined by the material limits of the core (typically, ~ 2,500 to 3,000 K), and expanded through the nozzle to produce thrust.

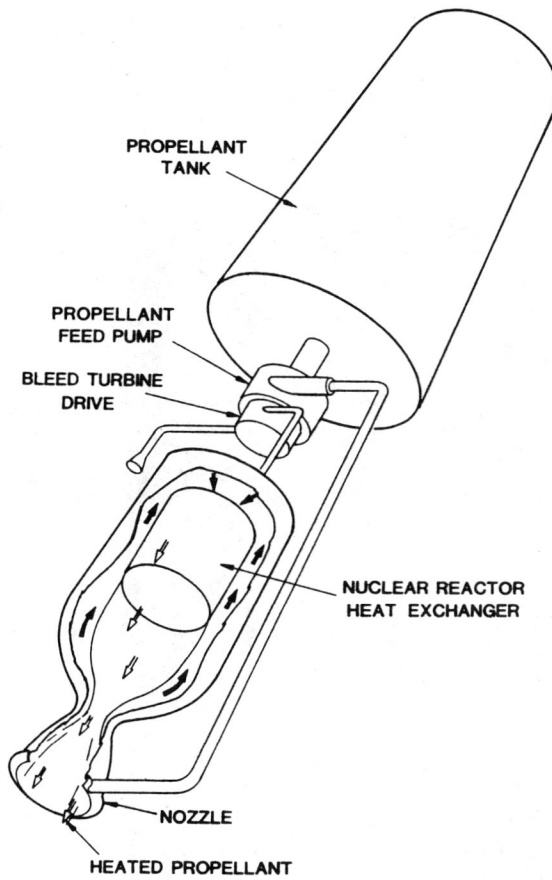

Figure 4 Schematic of a nuclear rocket propulsion motor.

A principal driver in determining the reactor configuration for such a system is the fundamental requirement for nuclear "criticality" in the core [Glasstone, 1955; Weinberg and Wigner, 1958]. A number of important implications and requirements result, generally in an effort to minimize the fixed mass of the reactor—for example, minimizing neutron-absorbing material in the core, providing neutron moderating (slowing-down) core components, highly enriched (uranium) fuel, complex fuel-loading regimes, a neutron-reflector to minimize neutron losses from the core, and minimizing overall core dimensions.

Criticality requirements are only one class of technological problems that must be solved simultaneously, under severe conditions, if the nuclear engine is to achieve the desired high propellant-temperature with low fixed-mass. Neutronics issues, plus associated reactor control and dynamics requirements, must be addressed while simultaneously maximizing power density (heat transfer to the propellant), minimizing overall system mass (materials and structures), and integrating super-light-weight components that must operate reliably at very high performance levels, at temperatures ranging from extremely low (LH_2 at ~ 30 K) to extremely high ($\sim 3{,}000$ K).

Figure 5 shows internal details of a solid-core, heat-exchanger nuclear rocket reactor. The heart of the system is a nuclear-fission reactor core composed primarily of a high-temperature matrix material, preferably a neutron-moderator such as carbon, loaded with uranium fuel. The uranium is highly enriched in U-235 to minimize criticality constraints on the core size and operating regimes.

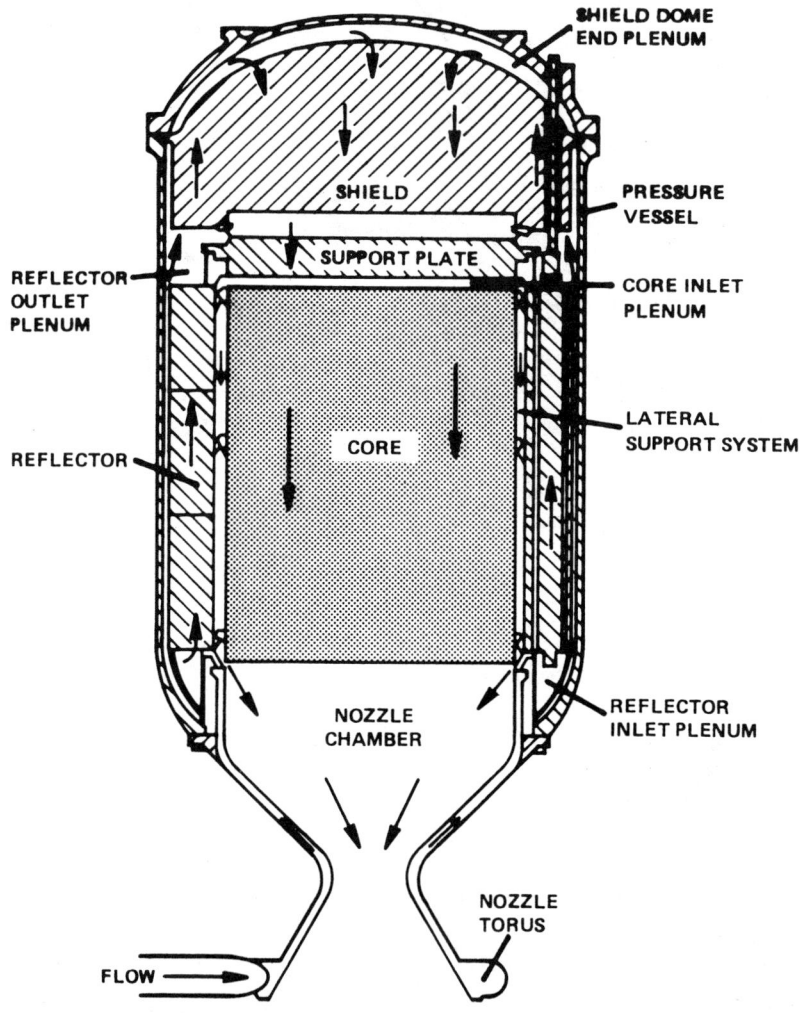

Figure 5 Reactor propellant-flow schematic for a nuclear rocket propulsion motor.

The propellant is carried as LH$_2$ in a slightly-pressurized tank, and, during operation, is fed to the engine by a gas-driven turbopump. A number of functions are performed by the propellant between the LH$_2$ tank and the nozzle outlet, besides ultimately producing the engine thrust. High-pressure fluid from the pump outlet first regeneratively cools the nozzle, then cools the reactor reflector and associated support structures, then the pressure vessel, shield, and core support plate before passing through the reactor core where it is heated to T_c, expanded through the nozzle, and ejected to produce thrust. An intermediate GH$_2$ bleed stream is used to drive the turbopump and then returned to the main flow before entering the core.

Heating of engine components by nuclear radiations emanating from the core is a special problem in NTP engines; the core power and power density are high, the system size and mass are made as small as possible, and resultant neutron and x-ray leakages are thus high. Substantial, detailed, and very careful cooling of all components is required. An internal bulk shield protects the LH$_2$ in the storage tank from excessive boil-off due to radiation heating and also reduces heating in other, external engine components. The shield is also required to reduce radiation doses to crew members for human missions.

The reflector is made of a neutron-moderating material such as Be, and not only enhances criticality of the core but also provides a convenient, low-temperature region for reactor criticality control. Rotating drums in the reflector, with a neutron absorber on part of the drum surface, provide the required neutronic control.

KEY TECHNOLOGIES

Fuel Development

The most important consideration in designing and developing a nuclear engine is the choice of reactor fuel material and configuration. First and foremost, the fuel material must have very high temperature capability, notably, adequate strength above about 2,500 to 3,000 K. Other desirable attributes include low neutron-absorption cross-section, high thermal conductivity, compatibility with a high-temperature uranium compound, reasonable fabricability, compatibility with hot H$_2$, and low mass and molecular weight. Only two classes of materials emerge as possible contenders: refractory metals such as tungsten and its alloys, and carbon-based materials such as graphite and metal carbides.

The metals are all strong neutron absorbers, whereas graphite is not. In addition to having good high-temperature strength (at least in compression), graphite also has high thermal conductivity, is compatible with uranium compounds, has low density, and is a good neutron moderator. It has one major drawback in that it reacts readily with hot H$_2$, and, unless protected with a refractory coating, quickly erodes away. The dominant advantages of graphite materials, however, led to the choice of carbon-based fuel matrix in the Los Alamos Rover program, although considerable effort was also spent on tungsten designs as backup [Bohl, *et. al.*, 1991].

Development, testing, and evaluation of carbon-based fuel elements, especially the performance of protective coatings, was one of the main technology efforts in the Rover nuclear-rocket development program. Overall performance was measured in terms of total run-time, determined ultimately by fuel-element corrosion rates. Good historical summaries exist [Taub, 1975; Kirk and Hanson, 1990b] that outline the myriad difficulties encountered and solved in this very extensive effort, and only a few highlights will be mentioned here. Problems included uranium migration, chemical deterioration in air, dimensional changes, reproducibility, coating destruction, and, most difficult, cracking of coatings due to thermal stress.

This latter problem was most severe in terms of "mid-range" corrosion. The core inlet end has a low corrosion rate because the temperature is low; at the high-temperature outlet end, power-density and thermal gradients are low, so that thermal-stress cracking of the coatings is also low. In between, however, temperature, power-density, and thermal gradients are high, cracks appear in the coatings because of mis-matched coefficients of thermal expansion, and high corrosive mass losses occur through the cracks.

Three fuel materials received the most development at Los Alamos during the Rover program. These are listed below in order of decreasing experience base, but increasing performance potential.

Bead-Loaded Graphite. This fuel consists of a graphite matrix containing 200 μm fuel beads with a 150 μm UC_2 core coated with pyrocarbon to protect the UC_2 from (humid) atmosphere. Surfaces exposed to H_2 were coated with NbC (or ZrC in some later tests), with an overcoating of molybdenum, in some instances, to help seal "mid-range" cracks. Reactor tests showed this fuel to be capable of a T_c of ~ 2,500 K for at least 1 h.

Composite Fuel. This fuel consists of 30 to 35 volume % UC • ZrC dispersed in graphite. The volume % carbide is approximately an optimum trade-off between higher corrosion resistance and reduced thermal stress resistance at higher carbide content. This fuel is capable of a T_c of ~ 2,700 K for (at least) 1 h.

Carbide Fuel. A pure carbide mixture such as UC • ZrC is required to maximize the time-temperature performance of carbon-based fuels, although this material has very poor thermal stress resistance. It is also difficult to fabricate. Nevertheless, by designing the fuel element in pieces, and with considerable additional development, such a fuel might be practical. Some testing of such fuels was accomplished in the latter part of the Rover program, but not enough to establish confidence in this fuel. Estimated T_c performance is ~ 3,000 to 3,200 K (I_{sp} ~ 950 s).

Figure 6 summarizes the fuel-performance experience in the Rover program. These fuel-endurance limits might be extended somewhat with modest additional development effort; however, temperatures in the range 2,500 K to 3,000 K and I_{sp} of ~ 900 to 950 s appear to be the approximate limit of such fuels.

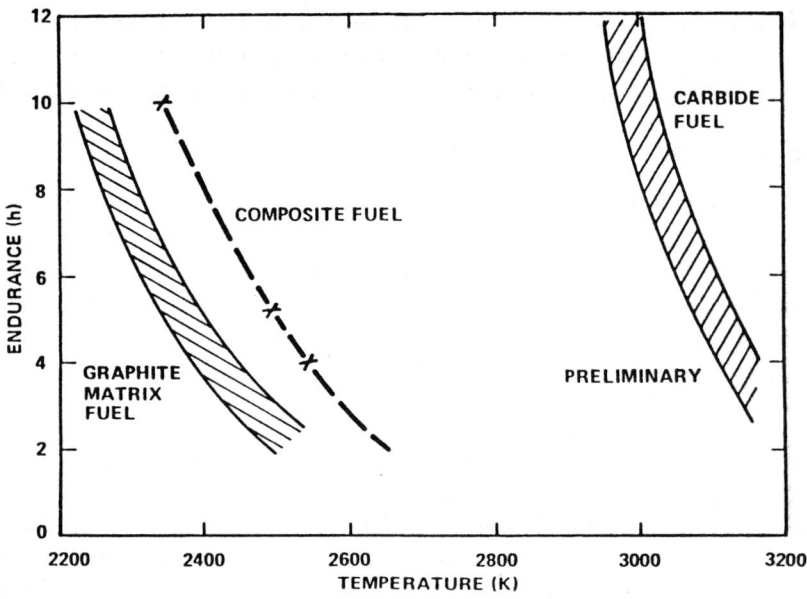

Figure 6 Comparison of projected endurance of several nuclear rocket fuels versus coolant exit temperature.

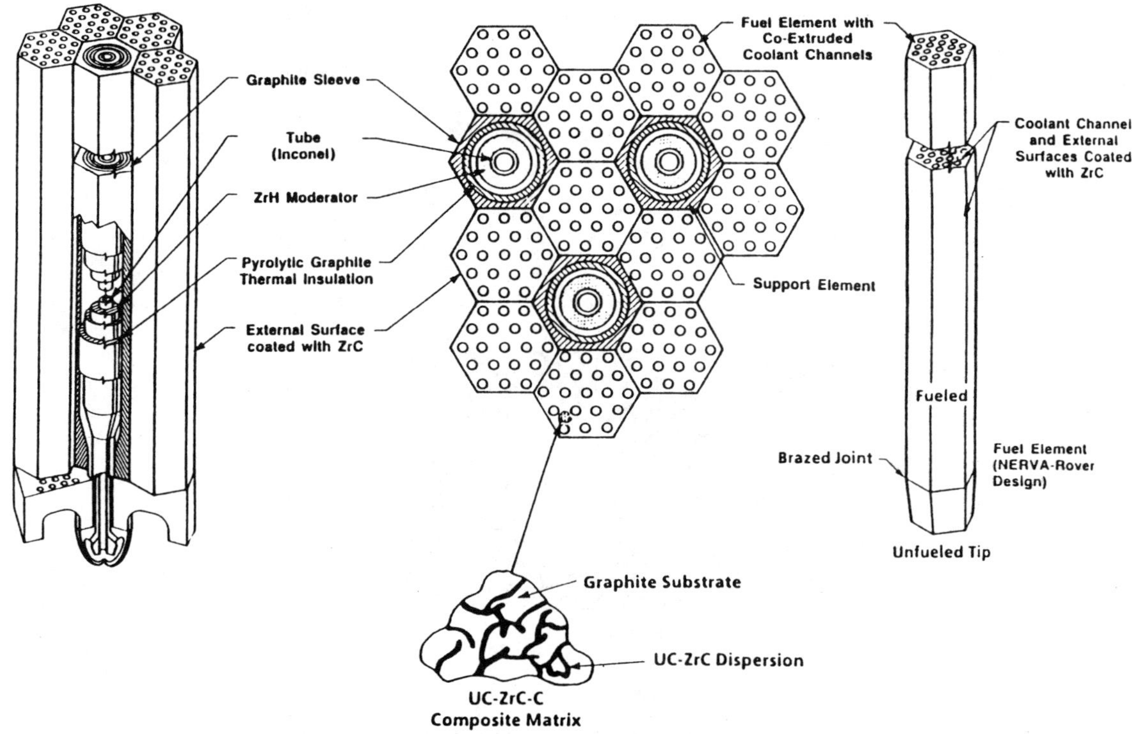

Figure 7 Fuel-element and support-element details for Rover and NERVA nuclear rocket reactor cores. Hexagonal elements are 52-in. long, 0.753 in. across the flats, with nominal coolant channel diameter of 0.100 in.

The basic fuel element concept that evolved is shown in Figure 7—a 52-inch-long, hexagonal, 19-hole, carbon-matrix, U-loaded element that was extruded, fine-machined, and then coated with NbC (or ZrC) on all surfaces to be exposed to hot H_2. The element was 0.753 in across the flats and the nominal coolant channel diameter was 0.100 inch.

Reactor Design

An NTP fission-reactor must function, and be viewed, in several ways simultaneously. It is a device for initiating and sustaining fission chain reactions, a high power-density heat exchanger with internal heat generation, an intense source of nuclear radiation, a mechanical structure with many types of loads under extreme temperature conditions, and a dynamic system that must be monitored and controlled. It is therefore a collection of many components and materials; a few of the more major of these will be outlined here.

The Rover reactors can be used, again, for illustration, with a core made up of solid fuel elements loaded with enriched U (93.15% U-235), as described above. Figure 8 shows a cross-section of such a system. A radial beryllium neutron reflector enhances the criticality of the core, helps flatten the core radial power profile, and, most importantly, houses rotatable neutronic-control drums in an easily managed low-temperature environment. The U fuel-loading is varied radially from element-to-element in the core to flatten the radial power profile to maximize thermal efficiency and T_c. In addition, inlet orifices for each coolant channel match the flow to local power.

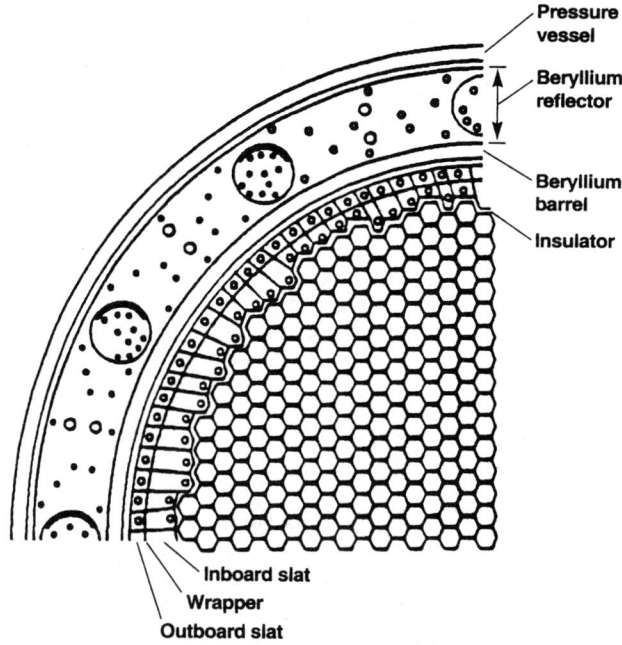

Figure 8 Cross-sectional schematic of a Rover nuclear rocket reactor.

The first requisite for the reactor core is nuclear criticality, at both very low and very high temperatures. Concomitant additional nuclear requirements are adequate control margins over the entire low-to-high temperature range, and controllable dynamic behavior, including rapid start-up and shut-down. Also, detailed radial power profiles must be mapped to establish orficing requirements. Although detailed neutronic calculations are used extensively, the ultimate establishment of the reactor nuclear characteristics is by means of low-power measurements in reactor mock-ups and, finally, in the actual system before full-scale testing.

Another major design requirement is that of supporting the large axial pressure drop across the core during high-power operation. Since graphite has good compressive strength, but poor tensile strength, the fuel elements are supported from the outlet end by means of a support-block/regeneratively-cooled tie-tube assembly (Figure 7). Typically, seven elements are supported in one tie-tube cluster, as shown. The tie-tubes transfer the core loads to an aluminum support plate (Figure 5) at the inlet end of the reactor, thus keeping the core in compression and the support plate in a low-temperature environment.

Other major reactor components include an aluminum pressure vessel, a reflector/core interface lateral-support system, and a shadow shield to protect external engine components and personnel above the core (Figure 5).

Other Technology Issues

A number of auxiliary, but essential, technology areas required new developments in the Rover/NERVA program to accomplish a viable NTP concept that could lead to an operational engine for a human Mars mission. These "engineering" problems generally derived from the extreme conditions under which hardware had to operate—particularly, the temperature extremes, and, in some instances, high ambient nuclear-radiation fields. Pumps, turbines, valves, seals, nozzles, and especially bearings had to operate down to LH_2 temperatures and under high pressures, and undergo many thermal-cycling start-ups and shut-downs. While many of these requirements are now handled in modern LOX/LH_2 rocket engines, the overall complexity and the reliability requirements for a Mars-mission NTP system still provides a challenge.

One unique NTP technology that had to be developed in the Rover program deserves special mention—large-scale LH_2 cryogenic facilities. Prior to the Rover work, LH_2 was essentially a laboratory "curiosity," and quantities as large as a few liters were unusual. At the termination of the program, facilities and operations for storing over one million gallons of LH_2, for handling very large quantities of LH_2 (and GH_2) in complex test-cell facilities, and for supplying LH_2 to reactor tests at several hundred pounds per second for hours were routine—and safe. Very few "accidents," and no serious injuries, occurred throughout some 19 full-scale reactor tests, plus many more auxiliary operations.

THE ROVER AND NERVA PROGRAMS

Overview

The use of nuclear energy for propulsion was under study as early as 1946, when R. Serber of Douglas Aircraft [Serber, 1946] concluded that the most reasonable approach was a "conventional" nuclear reactor heating a low molecular-weight propellant, and, with great prescience, predicted that payload advantages over the best chemical rocket of a factor of two or more were possible, depending "entirely on how well the difficulties of heat transfer and high temperature (material problems) can be solved."

There were many other studies of NTP for rockets, ramjets, aircraft, and space travel, but, it was the potential of a nuclear engine for ICBM propulsion that led initially to the establishment (1955) of a nuclear rocket program, designated the Rover program, at Los Alamos, under the sponsorship of the Air Force and the Atomic Energy Commission (AEC). Rapid improvements in chemical engines, coupled with large decreases in payload (nuclear weapons) size and weight requirements, however, eventually made this application unattractive for such modest propulsion requirements.

Nuclear rocketry had gained momentum, and the emergence of a strong non-military space program in the U.S. (Apollo) turned attention to more ambitious propulsion requirements, and NTP was recognized as being extremely attractive for human interplanetary travel; e.g., to Mars. With the fading military mission and the onset of the U.S. manned space program, NASA replaced the Air Force in a joint NASA/AEC office to manage nuclear rocket programs (1960).

The Los Alamos program had grown rapidly, and a first rudimentary reactor was tested at Jackass Flats, Nevada in 1959. Other tests of improved designs followed, and, by 1961, the program sponsors decided that technology had progressed sufficiently to bring an industrial team on board to develop a flight engine based on the Los Alamos technology. This engine program was designated the Nuclear Engine for Rocket Vehicle Application (NERVA) program, and, in a 1961 competition, Aerojet General was chosen as the NERVA contractor with Westinghouse (Astronuclear) as the reactor subcontractor. Los Alamos was to work closely with the NERVA team to transfer pertinent technologies and continue development and test programs to explore advanced designs.

Los Alamos built and tested 13 reactors before termination of the program in January, 1973, while the NERVA team tested six reactors, two of which were part of engine tests. These tests were merely the visible highlights of a very large and broad research and development effort. Major facilities were built in Nevada for reactor assembly, full power testing, remote disassembly and post-mortem examination, and, at Los Alamos, for fuel-fabrication and electrically-heated testing, post-mortem examination of fuel and other reactor components, critical assemblies, and other component-fabrication and testing facilities. A number of major subcontractors also supported the program; e.g., Rocketdyne, ACF Industries, EG&G, ORNL.

Despite the many technical successes, the NERVA work was stopped in 1971, before engine development could be completed, and the Los Alamos effort was terminated about one year later. Several complex technical and political factors played a role in the termination of these programs, but, in simplest terms, lack of a firm mission was the driving force that stopped the programs before the next major steps into space could be taken.

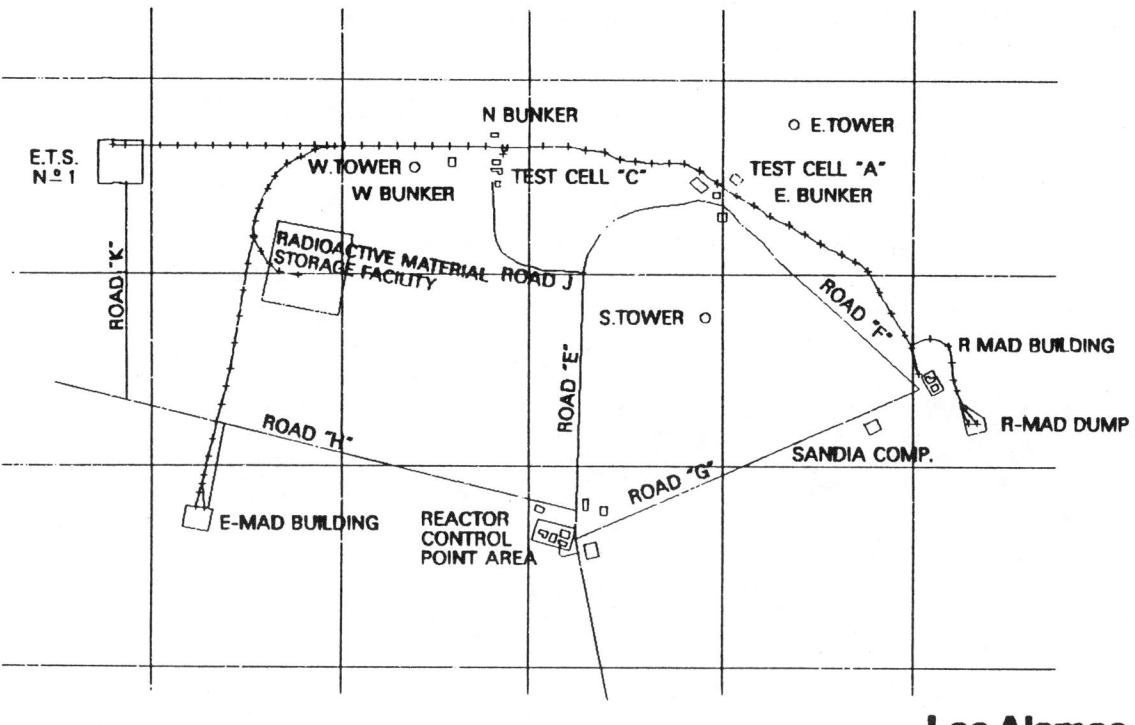

Figure 9   Area layout of the Nuclear Rocket Development Station (NRDS) at Jackass Flats, Nevada, for remote testing of nuclear rocket systems in the Rover and NERVA programs.

Nuclear Rocket Development Station (NRDS)

All Rover and NERVA tests that involved significant nuclear power generation were conducted at the remote NRDS at Jackass Flats, Nevada. Figure 9 shows a layout of the principal NRDS facilities:

Central Control Point (CP) — control complex from which remote operation of the reactor test cells was conducted.

Test Cell A — the original Los Alamos test cell, where the KIWI-A test series, a majority of the NERVA reactor tests, and a variety of "cold" flow, feed system, and component tests were conducted.

Test Cell C — the main Los Alamos test cell, where the Los Alamos reactor power tests (except above) took place.

R-MAD — Reactor Maintenance and Disassembly building; final reactor assembly before test, and remote disassembly and post-mortem examination of "hot" reactors after test.

E-MAD — Engine (NERVA) MAD building.

ETS-1 — Engine Test Stand #1; the (NERVA) engine test cell.

Railroads — for transporting test reactors and engines (both before and after testing) among the various facilities, on specially modified railroad flat cars.

Figure 10 is an aerial view of Test Cell C. Prominent features include dewars for storing $\sim 1.1 \times 10^6$ gallons of LH_2, assorted other tank farms and dewars; e.g., for LN_2, GN_2, and water. Also, in the upper right quadrant in the figure, the rail track to the test cell face (not visible) can be seen, and a removable shed (near the tower) to cover and protect test units on the pad from weather, dust, etc.

Figure 10 Test Cell C at NRDS, Jackass Flats, Nevada, for full-power testing of nuclear rocket reactors.

Figure 11 shows the E-MAD assembly bay with a NERVA "Experimental Engine (XE)" being assembled. (Note: reactors were tested in an up-firing position in Test Cell C, while XE systems were tested in a down-firing configuration into a water effluent scrubbing system, at ETS-1.)

Figure 11 Assembly in the Engine Maintenance and Disassembly (E-MAD) building, NRDS, of the Experimental Engine (XE), before testing.

Summary

Nineteen different reactor systems were tested at power in the Rover and NERVA programs between 1959 and 1972, in seven different series: KIWI-A, KIWI-B, Phoebus (Ph), Peewee (PW), and Nuclear Furnace (NF) in the Rover program (Los Alamos), and NRX and XE in NERVA (Aerojet-Westinghouse).

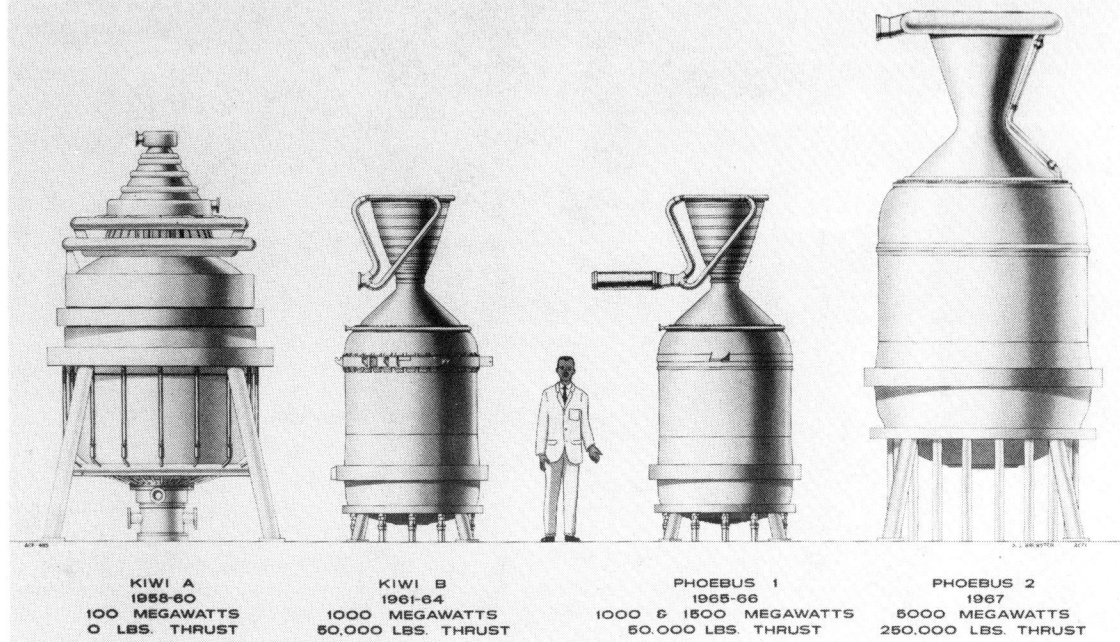

Figure 12 Comparisons of Rover nuclear rocket reactors.

Table 1
SUMMARY OF MAJOR PERFORMANCES ACHIEVED IN ACTUAL ROVER TESTS.*

Power (Phoebus 2)	4100(5000) MW
Thrust (Phoebus 2)	205,000(250,000) lb
Equivalent I_{sp} (Peewee)	848(875 to 900) s
Reactor Specific Mass (Phoebus 2)	2.3 kg/MW
Average T_c (Peewee)	2550 K
Peak Fuel Temperature (Peewee)	2750 K
Average Core Power Density (Peewee)	2340 MW/m^3
Peak Core Power Density (Peewee)	4500 MW/m^3
Total Time at Full Power (NF-1)	109 minutes
Number of Restarts (XE)	28

* Pewee was a small, 500 MW advanced performance reactor designed and tested by Los Alamos for possible use in such missions as an Earth/moon trip, or in orbit-to-orbit transfers; NF-1 was a Los Alamos "Nuclear Furnace" reactor to test advanced fuel elements in full-scale reactor operating environments; XE was the NERVA prototype nuclear rocket engine.

Figure 12 shows comparisons among the various classes of reactors developed and tested in the Rover program; the NERVA reactors were essentially similar to the Phoebus 1 reactor, with potential for upgrading to Phoebus 2 performance levels. Major performance milestones actually achieved, and achievable (in parentheses), are shown in Table 1.

At the end of the program, engines with cumulative full-power operating time in excess of 1 h with specific impulse of ~ 850 s, were achieved; and, technology demonstrations allowed reasonable projections to 10 h engines and I_{sp} of ~ 900 s or greater.

Finally, Table 2 projects potential performance characteristics for the basic 75,000 lb-thrust NERVA engine, progressing from the demonstrated graphite system, through the composite fuel for which successful fuel element tests were accomplished, to the more speculative carbide-fueled system with $I_{sp} \simeq 1{,}040$ s.

Table 2
75 K ENGINE CHARACTERISTICS

Fuel Element	Nozzle Chamber Temp		I_{sp} (sec)	Weights klb	
	K	°R		Reactor	Engine
Graphite	2500	4500	900	10.5	16
Composite	2700	4860	925	12.5	18
Carbide	3100	5580	1040	14.5	20

ALTERNATIVE CONCEPTS

Hydrogen Dissociation

For any interplanetary human space mission, several complex engine performance trade-offs must be considered; and, pay-offs are determined by a variety of interdependent drivers—for example, the desire to minimize travel time for safety reasons, and to maximize "stay" time for mission effectiveness. It is thus desirable for the engine to maximize thrust-to-weight ratio (power density) and I_{sp} (exhaust velocity).

These generally are conflicting goals. High thrust is achievable only through high propellant flow rate, with I_{sp} limited by available propellant energy (T_c), while high I_{sp} is achieved only through some non-thermal-expansion process, such as electrical acceleration of propellant particles, with correspondingly low flow rate (thrust). The pay-off for NTP is in the high-thrust, "medium" I_{sp} regime where substantial payload and minimum travel time are the primary goals, and where improvements in performance can only be achieved by increasing T_c, hopefully without seriously reducing thrust-to-weight.

A potential exception to the above generalizations is the possibility of increasing I_{sp} by reducing propellant molecular weight through the dissociation of the H_2 propellant; either the dissociated hydrogen provides a lower molecular weight propellant, or recombination in the nozzle adds thermal energy to increase T_c, or both. A maximum theoretical specific impulse of $\sqrt{2}$ × hydrogen I_{sp} = 1,200 to 1,300 s would thus be possible with a Rover-type engine.

Figure 13 shows the relevant curves of I_{sp} vs. chamber temperature and pressure. Unfortunately, dissociation is insignificant at realizable T_c for a solid-core engine unless chamber pressure is reduced to a small fraction of the 40 bars required to achieve high power-density in the Rover-NERVA reactors. A lower-pressure system could be designed, but, significant dissociation, with reasonable reactor and nozzle sizes, means lower power and thrust and increased engine operating time. This means, in turn, backing off on reactor temperature, probably below the dissociation range. Overall mission performance could actually be reduced [Kirk and Hanson, 1990b; Kirk and Hanson, 1990a].

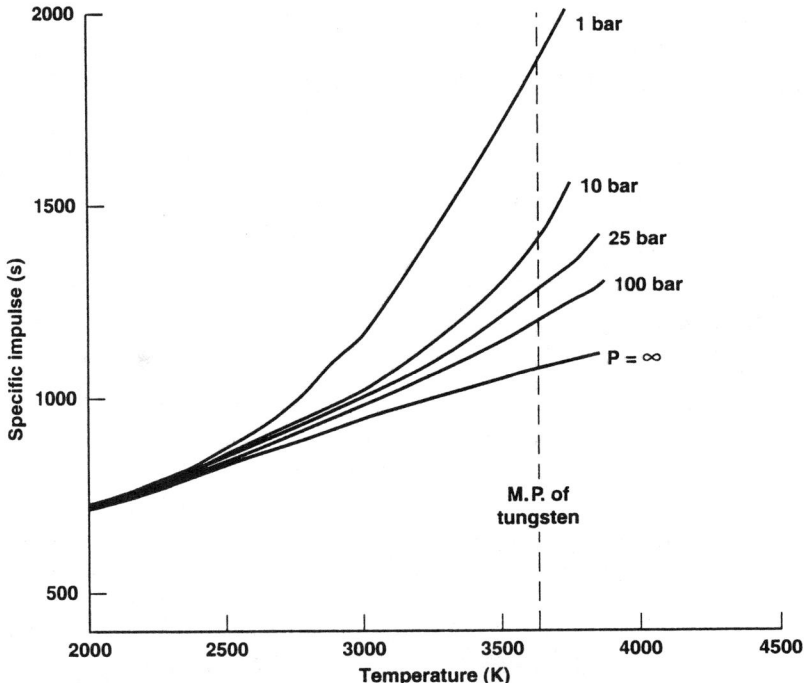

Figure 13   Potential nuclear rocket reactor performance, with hydrogen dissociation, versus chamber temperature and pressure.

Thus, to maintain overall mission performance, the practical limit for a solid-core nuclear engine appears to be ~ 900 to 1,000 s, determined by materials temperature limits on T_c. A wide variety of NTP concepts have been examined, historically, in an attempt to alleviate, or eliminate, these constraints. These approaches generally attempt to relieve the temperature constraints of solid-core systems while still operating in temperature-pressure regimes that maintain thrust-to-weight ratio. Three such concepts are discussed in the following.

Pebble-Bed Reactor (PBR)

The basic PBR concept has received increased attention in recent years because of a desire to improve solid-core power-density, as well as I_{sp}, for launch and cost advantages in a variety of near-Earth DoD missions; e.g., space-tugs and orbit-to-orbit transfer operations [Lenard, 1992]. Some experiments were done on a similar concept in the late 1950s at Los Alamos, and a low-level effort has persisted at Brookhaven National Laboratory (BNL) since that time. A number of recent BNL publications [Powell and Botts, 1983; Botts, *et al.*, 1984; Powell and Horn, 1985], predict very high power-density, rapid start-up, and high exit-gas temperatures to provide significant mission performance increases.

In the PBR concept, small (500 μm) fuel particles are held in a bed between two porous concentric cylinders. Propellant flows radially inward through the bed and then axially inside the inner cylinder to the nozzle chamber. The small particle- and flow-passage dimensions create an extremely good heat transfer geometry, with small ΔT in the particles and between particles and gas. Several of these cylinders are stacked to form the reactor core. Specific impulse of ~ 1,000 s and thrust-to-weight ratio of ~ 30 are projected for a 75,000 lb thrust engine [Lenard, 1992].

A number of technical challenges must be met before the feasibility of the concept can be established and performance predictions realized. These include fuel materials problems and more detailed engineering design work. For example, the extremely large fuel surface area may lead to rapid fuel corrosion rates at high temperatures; and, properly distributing propellant flow to match local fuel-power may pose difficult engineering problems. The "frit" material for the porous cylinders may also pose a severe development problem.

Gas-Core Systems

All solid-core NTP systems are constrained by temperature limitations of the fuel. It is thus attractive to consider a reactor using a very high-temperature gaseous fuel. A number of conceptual designs embodying this idea has been proposed over the years [McLafferty, 1968; Rodgers, *et al.*, 1976; Mensing, 1985], including considerable supporting experimental work on separation techniques; critical assembly experiments were also done at Los Alamos [Barton, *et. al.*, 1977], using gaseous UF_6 for part of the fuel.

The fundamental difficulty in this concept is separating the fissioning plasma from the propellant while still maintaining close thermal coupling to transfer heat efficiently to the propellant. It is impractical (and degrades the I_{sp}) to allow a significant fraction of the fissile fuel to be ejected with the propellant. Proposed separation schemes include centrifugal (vortex) systems, separation with magnetic fields, and separation of fuel and propellant by transparent walls. The latter, termed the "nuclear light bulb," is the most enduring, and the only approach that thus far promises achievement of adequate separation.

Although specific gaseous core reactors have been elaborated for decades in continuing analytical studies, they have never established sufficient credibility to attract substantial development funding.

Orion

The "ultimate" in maximizing thrust and I_{sp} simultaneously in an NTP engine would be achieved by means of the "ultimate" in nuclear-energy production, namely, through nuclear explosions. This is the basis of the Orion concept [Everett and Ulam, 1955]. A series of nuclear explosions produce very high-temperature plasma "propellant" pulses that impinge on an ablatively-cooled "pusher plate." The pusher provides propulsion by means of a large shock-absorber system that transfers reasonable, damped, impulses to the spacecraft.

The concept has received considerable attention [General Dynamics, 1963-1964] and appeared surprisingly practical, at least on paper. Substantial work was done on ablation experiments, on system design, and on "pulse" generation; and a potential for extremely high I_{sp} and thrust was projected. Impressive high-explosive-driven models were also built and operated as demonstrations.

The fundamental Achilles' heel of the concept was the fact that nuclear explosives do not come with small outputs. As a result, the spacecraft was prohibitively large for envisioned missions, even though potential payloads were correspondingly impressive.

A variant on this concept was "Sirius [J. D. Balcomb, *et al.*, 1971]," which assumed relatively small, laser-driven fusion pulses as the energy source. The resulting engine and spacecraft were of a size compatible with human Mars missions, and, of course, projected extremely high thrust and I_{sp}. Unfortunately, no such laser-fusion burns have, as yet, been demonstrated.

SUMMARY AND CONCLUSIONS

On 2 November 1989, President Bush approved a national space policy that affirmed the long-range civil space program goal to "expand human presence and activity beyond Earth orbit into the solar system," with a long-term focus on placing humans on Mars by 2019. This rekindled interest in advanced propulsion concepts such as NTP.

Three basic facts of NTP make it a better-performing option than chemical rockets: (1) nuclear energy comes from a source that can be converted into thermal energy of a separate rocket propellant, (2) chemical combustion is not needed in the propellant, thereby eliminating the need for an oxidizer and allowing use of a low molecular-weight propellant, and (3) nuclear fuel is not limited by chemical heat of combustion, so that many orders-of-magnitude more energy is available than from chemical fuels.

The resulting increase in specific impulse of at least a factor of two over the best chemical rockets means that NTP offers several potential advantages. With the same Initial Mass in Low Earth Orbit (IMLEO) NTP allows: reduced transit times, larger mission "stay" time and/or reduced total mission time, reducing crew exposure to zero-gravity and space-radiation environments; and/or, reduced IMLEO for the mission allows reduced Earth-to-orbit launch requirements and costs, and greater mission design flexibility; e.g., increased departure windows and multi-mission capability with a common vehicle.

NTP has a long history in the U.S., beginning with the first studies after World War II which indicated the benefits and feasibility of nuclear rockets, followed by the Rover and NERVA programs, which demonstrated that nuclear rocket engines could be built and successfully operated for times sufficient for a human mission to Mars. These programs, which were terminated in 1973 because of a lack of post-Apollo missions, left a tremendous technological legacy for future generations to build on for eventual voyages to Mars, and beyond.

The Rover-NERVA program was very successful, technically, with record (and achievable) performances as shown in Table 1. On the basis of results achieved to date, the practicality of graphite-based nuclear rocket reactors and engines has been established; and, technology has been demonstrated to support future space propulsion requirements, using LH_2 as propellant, for thrust requirements ranging from 25,000 to 250,000 lb, with I_{sp} over 850 s, and with full engine-throttle and restart capability. This performance is "commensurate with today's propulsion requirements . . . (and) future NTP technology development for new space exploration initiatives can be directed to incremental performance, reliability, and lifetime improvements [Gunn, 1989]."

REFERENCES

Balcomb, J.D., et al., Nuclear pulsed propulsion study, *Los Alamos Scientific Laboratory report LA-4684-MS,* Los Alamos, NM, May, 1971 (SRD, SRI).

Barton, D.M., et al., "Plasma core reactor experiments," *paper 3D-13, 1977 IEEE International Conference on Plasma Science,* 1977.

Bennett, G.L., et al., "Prelude to the future: a brief history of nuclear thermal propulsion in the United States," *NASA document,* 1991.

Bohl, R.J., et al., Nuclear rocketry review, *Los Alamos National Laboratory report LA-UR-91-20-21,* Los Alamos, NM, 1991.

Borowski, S.K., Mission performance summary, briefing chart, *NASA Lunar & Mars Exploration Program Office,* Mission Development and Operations Office, August, 1990.

Borowski, S.K. and T.J. Wickenheiser, "Nuclear thermal rocket technology and stage options for Lunar/Mars transportation systems," *AIAA Space Programs and Technologies Conference paper, AIAA-90-3787,* Huntsville, AL, Sept., 1990.

Botts, T.E., et al., "A bimodal 200 MW/200 kW reactor for space power," *1st Symposium on Space Nuclear Power Systems,* Albuquerque, NM, Jan., 1984.

Boyer, K. and J. D. Balcomb, "System studies of fusion powered pulsed propulsion systems," *AIAA paper no. 71-636,* AIAA/SAE 7th Propulsion Joint Specialist Conference, Salt Lake City, UT, June, 1971.

Bussard, R.W., Nuclear energy for rocket propulsion, *ORNL Central Files No. 53-6-6,* Oak Ridge National Laboratory, Oak Ridge, TN, 1953.

Bussard, R.W., "Nuclear rocketry—the first bright hopes," *Astronautics,* vol. 7, No. 12, 1962.

Dewar, J.A., Project rover: a study of the nuclear rocket development program, 1953-1963, *unpublished manuscript, U.S. Dept. of Energy,* Washington, DC, 1974.

Dixon, T.F. and H.P. Yockey, A preliminary study of use of nuclear power in rocket missiles, *North American Aviation report NA-46-574,* 1946.

Everett, C.J. and S.M. Ulam, On a method of propulsion of projectiles by means of external nuclear explosions, *Los Alamos Scientific Laboratory report LAMS-1955*, Los Alamos, NM, Aug. 1955(SRD).

General Dynamics, Nuclear pulsed propulsion project (Project Orion), *Technical Summary Report (4 Vols.), General Atomics Division of General Dynamics report No. RTD-TDR-63-3006*, 1963-1964 (SRD).

Glasstone, S., Principles of nuclear reactor engineering, D. Van Nostrand Company, Inc., *Library of Congress card No. 55-8832*, 1955.

Gunn, S.V., "Development of nuclear rocket engine technology," *AIAA paper 89-2386* presented at AIAA/ASME/SAE/ASEE 25th Joint Propulsion Conference at Monterey, CA, July, 1989.

Kirk, W.L., Nuclear rocket propulsion, *unpublished manuscript, Los Alamos National Laboratory, Los Alamos*, NM, March, 1990.

Kirk, W.L. and D.L. Hanson, Comments on low pressure operation of nuclear propulsion reactors, *Los Alamos National Laboratory internal report N-DO-90-24*, Los Alamos, NM, Jan., 1990.

Kirk, W.L. and D.L. Hanson, Practical limits on solid-core nuclear rocket performance, *Los Alamos National Laboratory internal report N-DO-90-84,* Los Alamos, NM, March, 1990.

Koenig, D.R., Experience gained from the space nuclear rocket program (Rover), *Los Alamos National Laboratory, report LA-10062-H*, Los Alamos, NM, May, 1986.

Lenard, R.X., Recent progress in the U.S. Air Force Space Nuclear Thermal Propulsion Program, *U.S. Air Force Phillips Lab unpublished manuscript,* Kirtland AFB, Albuquerque, NM, 1992.

McLafferty, G.H., "Survey of advanced concepts in nuclear propulsion," *Journal of Spacecraft and Rockets*, Vol. 5, No. 10, American Institute of Aeronautics and Astronautics, Oct., 1968.

Mensing, A.E., Gas core nuclear reactors for space propulsion and power, S*pace Nuclear Propulsion Workshop summary report*, Vol. 2, (held at Los Alamos National Laboratory, Dec., 1984), Air Force Space Technology Center, Space Division, Air Force Systems Command, Edwards AFB, CA, June, 1985.

Powell, J.R. and T.E. Botts, "Particulate bed reactors and related concepts," in *Proc. Symp. on Advanced Compact Reactor Systems*, Washington, D.C., Nov, 1982, Energy Engineering Board, National Research Council, Washington, DC, 1983.

Powell, J.R. and F. L. Horn, Direct thrust nuclear propulsion based on particle bed reactors, *Space Nuclear Propulsion Workshop Summary Report*, Vol. 2 (held at Los Alamos National Laboratory, Dec., 1984) Air Force Space Technology Center, Space Division, Air Force Systems Command, Edwards AFB, CA, June, 1985.

Rodgers, R.J., *et al.,* Investigation of applications for high-power, self-critical fissioning uranium plasma reactors, *United Technologies Research Center report R76-912204, also NASA-CR-145048*, Sept., 1976.

Serber, R., The use of atomic power for rockets, *Douglas Aircraft Company*, July, 1946.

Shepherd L.R. and A.V. Cleaver, "The atomic rocket:, Parts I-IV," J*ournal of the British Interplanetary Society,* Vol. 7, No. 5 and 6 and Vol. 8, No. 1 and 2, 1948-1949.

Stafford, T.P., *et al.,* America at the threshold, *Report of the Synthesis Group on America's Space Exploration Initiative*, Superintendent of Documents, U.S. Government Printing Office, Washington, DC, May, 1991.

Taub, J.M., A review of fuel element development for nuclear rocket engines, *Los Alamos National Laboratory report LA-5931*, Los Alamos, New Mexico, June, 1975.

Weinberg, A.M. and E.P. Wigner, the physical theory of neutron chain reactors, *Library of Congress No. 58-8507*, University of Chicago Press, 1958.

A former Soviet design could use two nuclear reactors and ion drive to get to Mars.

AAS 95-482

Chapter 12

NUCLEAR-ELECTRIC PROPULSION FOR HUMAN MARS MISSIONS

Ernst Stuhlinger[*]

A human mission to Mars will require an Earth-to-Mars propulsion system that provides for a short transfer time, sufficient payload capability to assure safety and comfort for the astronauts as well as enough scientific equipment for exploration, redundancy of critical systems, a minimum of orbital assembly work and other in-space maneuvers, fast transit through the Van Allen Belts, and reasonably low mass to be transported from ground to low Earth orbit. An electric propulsion system with ion thrusters and nuclear-thermionic power source could meet these requirements. Technical details of such a system, and of a 6-astronaut mission to Mars with a total mission time of about 400 days, including 30 to 40 days on Mars, will be described. An evolving development program, beginning with small robotic flights to interplanetary targets and finally leading to a human Mars mission, will be outlined.

HUMAN MARS MISSION - A NEW DIMENSION OF SPACE FLIGHT

With a human expedition to Mars, a new era of space exploration will begin. While conventional propulsion systems have served space projects well during the past thirty-five years, including flights to the Moon, it will be appropriate to consider a new class of propulsion systems for human flights to Mars, and also for extended robotic flights to planetary and interplanetary targets. All these missions demand long flight times, large payload fractions, substantial facilities for scientific exploration, high reliability through redundancy, and, particularly on crewed spacecraft, considerable electric power for housekeeping purposes.

A human mission to Mars should not be planned and conducted as an isolated, one-time achievement, but rather as one highlight in an evolving program that uses a new type of propulsion system. It would begin with modest robotic missions to targets in planetary space, progress systematically by modular growth of the propulsion system through more complicated missions, and finally lead to a human Mars mission.

An evolving program of this kind, with a continuing flow of exciting space achievements, would earn and retain the interest and support of the public as well as the government. Such an evolving program with recurrent use of major components, eventu-

---
[*]   3106 Rowe Drive, Huntsville, Alabama 35801.

ally leading to a human Mars mission, has the additional advantage that funding for the program need not be borne only by the Mars mission, but also by a number of other space projects that precede the human flight to Mars.

THE CASE FOR ELECTRIC PROPULSION

In every major rocket project of the past sixty years, the propulsion system was the decisive element whose size, mass, and performance capability determined the detailed design of the system. Concept and details of a human Mars mission, also, must be planned around the propulsion system that will send the astronauts on their Martian round trip.

The propulsion system to be chosen for the Mars mission should have the following features: (1) Its basic technology should be mature, even flight-tested; (2) it should offer a high degree of reliability to assure maximum safety for the astronauts; (3) its efficiency should be sufficiently high to allow a sizable payload, a modest Earth-to-orbit mass, a relatively short travel time, flexibility in the choice of mission parameters, possibility of flight plan changes during flight, liberal accommodations for the astronauts, and ample opportunities for scientific observations; (4) if nuclear power is to be used, testing and operation of the nuclear reactor should not present an undue hazard, or even reasons for concern, on Earth and in space.

Electric propulsion systems, particularly the ion propulsion system, would meet these requirements. Such systems have been under study since the late 1940s. By 1990, more than eighty electric propulsion systems had been working on orbital missions, either for testing purposes, or for orbital control of satellites. Most of these electric propulsion flight projects were carried out by the Soviet Union. Unfortunately, space mission planners in the United States have been reluctant to include electric propulsion systems in their program plans, although space engineers and space scientists have been suggesting and even urging the use of electric thrusters as prime propulsion systems on space missions for more than twenty-five years, as shown, for example, in the regular reports on electric propulsion in American Institute of Aeronautics and Astronautics and other professional journals [Moeckel, 1959; Stuhlinger, 1954 and 1978; Turchi *et al.*, 1992; *Aerospace America*, 1992; Loeb and Popov, 1992].

ELECTRIC SPACE PROPULSION SYSTEMS

Basic Concepts

Electric space propulsion systems are rockets whose exhaust particles are not accelerated by heat energy from chemical or nuclear reactions, but by electric energy; this energy must be provided by an on-board electric power supply. Electric rocket engines can reach far higher exhaust velocities than chemical and nuclear-thermal rocket motors. Among several different kinds of electric rocket systems, which include electro-thermal, arc jet, ion, and plasma jet engines, by far the most appropriate system for a Mars mission will be the electrostatic or ion engine. Its efficiency in transforming electric power into thrust, its ability to obtain the desirable high exhaust velocity, as well as its operational lifetime are substantially higher than those of all other systems.

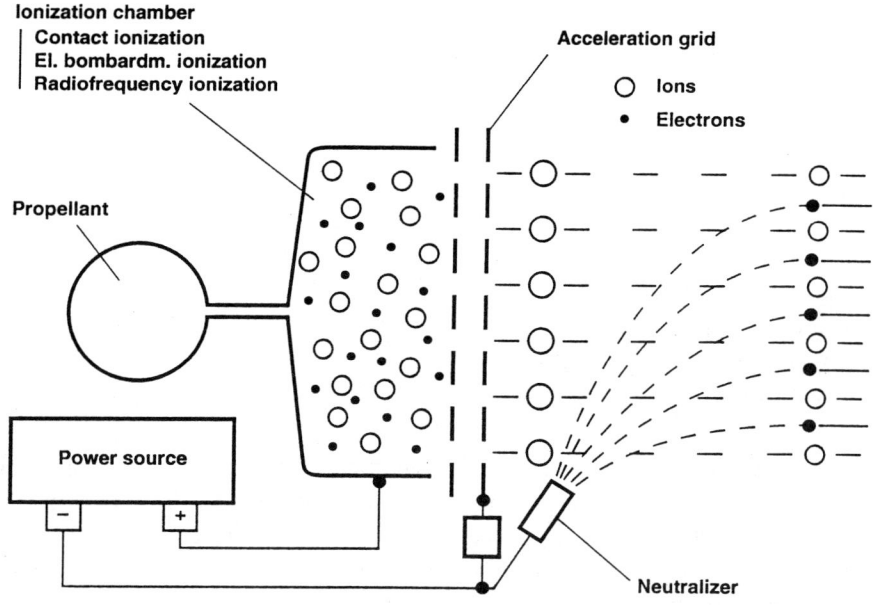

Figure 1   Schematic of ion thruster. The same basic principle holds for surface contact, electron bombardment, and radiofrequency ionization.

Figure 1 shows the principle of an ion thruster. The propellant (either a gas, such as Ar or Xe, or vaporized Hg or Cs) is fed into the ionization chamber where it is ionized either by electron bombardment or by radio frequency fields. The electric field between the grids G1 and G2 accelerates the positive ions to the velocity, v. Electrons are injected into the ion beam after its acceleration to neutralize its positive charge; therefore, the exhaust beam is electrically neutral shortly behind the thruster. The relations between thrust, rate of propellant consumption, beam power, and exhaust velocity are shown by the equations

$$F = \dot{m}v = \frac{2W}{v} \tag{1}$$

where
  F = thrust force
  $\dot{m}$ = propellant consumption
  v = exhaust velocity
  W = beam power.

Note that beam power equals electric power multiplied by electric thruster efficiency, typically between 85 and 95 percent.

While the basic rocket equation

$$u = v \ln \frac{M(f)}{M(e)} \qquad (2)$$

or

$$\frac{M(e)}{M(f)} = e^{-u/v} \qquad (3)$$

where
   u = rocket velocity at burnout
   v = exhaust velocity
   M(f) = rocket mass at start (full)
   M(e) = rocket mass at burnout (empty)

is valid also for electric rockets; the specific properties of electric rockets can be studied in more detail if some further variables are introduced which take into account that the "empty mass" of electric rockets must include the mass of the electric power source; that mass is roughly proportional to the amount of power to be provided to the thrusters. Introducing the "specific power"

$$p = \frac{\text{total electric power}}{\text{mass of power plant}} \qquad (4)$$

the rocket equation (1) can be extended [Stuhlinger 1964] to a form that contains the specific power p of the power plant and also the total time T of propulsion:

$$\frac{M(L)}{M(f)} = e^{-u/v} - \frac{v^2}{2pT}(1 - e^{-u/v}) \qquad (5)$$

where *M(L)* comprises the structural mass and the payload mass.

This equation shows that the payload that can be carried by an electric rocket of a given M(f) at start, a given specific power, *p*, a given terminal velocity, *u,* and for a given propulsion time, *T*, shows a maximum for one particular exhaust velocity, *v*. That optimum exhaust velocity, *v(opt)*, is nearly independent of the desirable terminal velocity. For a higher terminal velocity, the payload mass is lower, but the maximum payload will still be obtained at about the same exhaust velocity (Figure 2).

This behavior can be understood qualitatively: as the exhaust velocity increases from a low value to higher values while the thrust is kept constant, the required electric power—and therefore the mass of the power plant—increases, while the rate of propellant consumption—and therefore the total propellant mass—decreases. At one particular exhaust velocity, the sum of propellant mass and power supply mass will be a minimum, and, because of constant $M(f)$, the payload will be a maximum.

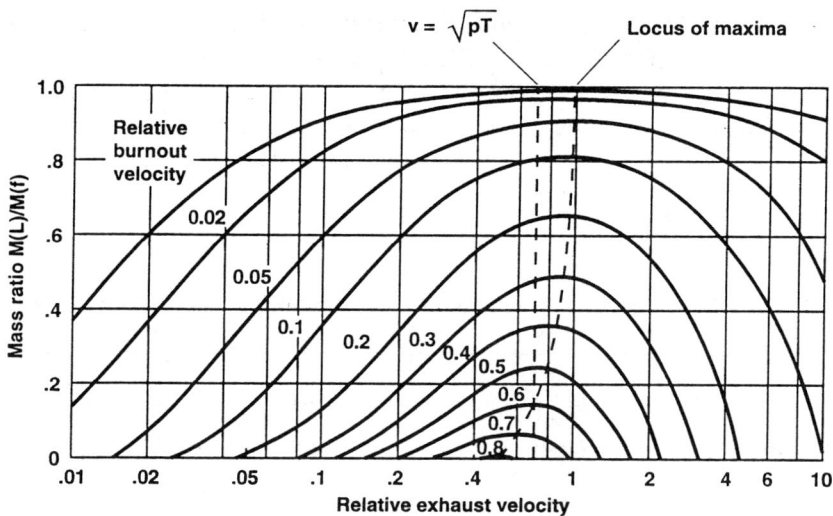

Figure 2 Payload capability of electrically propelled spacecraft as a function of exhaust velocity at burnout and of terminal spacecraft velocity.

Within an accuracy of 5 to 10 percent, this optimum exhaust velocity can be expressed simply as

$$v(opt) = \sqrt{pT} \qquad (6)$$

an equation that is very helpful for a quick assessment of optimum design features of an electric propulsion system for a given mission and a given specific power of the power source.

Table 1 gives an impression of the features of an electrically propelled spaceship suitable for a human Mars mission.

Brief History of Electric Propulsion

Possibilities of electric space propulsion have been mentioned since the end of the last century [Tsiolkovskii, 1951]. Beginning in the late 1940s, systematic studies were made [Stuhlinger, 1954], and about ten years later numerous electric propulsion development projects were started in the United States and a number of other countries. The most successful thrusters, as far as efficiency and systems reliability are concerned, were built from the early 1960s on by Harold Kaufman at the NASA Lewis Research Center (Figure 3) [Kaufman, 1960] and by Horst Loeb at the Giessen University in Germany (Figure 4) [Loeb, 1962]. The Kaufman thruster uses the electron bombardment principle to ionize the propellant; the Loeb thruster applies an electrode-less radio frequency field as ionizing agent. Both types have been analyzed and tested in the laboratory extremely thoroughly; test runs in the thousands of hours were achieved by both types. Kaufman thrusters were built for space applications by the Hughes Research Corporation in California (Figure 5), Loeb thrusters by Messerschmitt-Boelkow-Blohm in Munich in Germany (Figure 6). Overall efficiencies on the order of 80 percent (combined electric and mass efficiencies) were obtained by both thruster types. Kaufman thrusters were successfully tested in space in 1970 in the SERT (Space Electric Rocket Test) program. A very active program for electric propulsion systems of various types has been underway in the Soviet Union for years; more than 80 different thrusters were operated on satellites, either for testing purposes, or for orbit corrections [Zhurin et al., 1983; Loeb and Popov, 1992].

Table 1
DESIGN AND PERFORMANCE FIGURES OF A MANNED MARS SPACESHIP

Initial mass	M (f)	3.2	×	10E5	kg
Propellant mass	M(P)	1.2	×	10E5	kg
Powerplant mass	M(W)	0.7	×	10E5	kg
Structural mass	M(St)	0.3	×	10E5	kg
Payload mass	M(Pl)	1.0	×	10E5	kg
Empty mass	M(e)	2.0	×	10E5	kg
Terminal mass	M(t)	1.6	×	10E5	kg
Total electric power	W(tot)	8	×	10E6	W
Beam power	W(beam)	6.5	×	10E6	W
Specific power	P	1.2	×	10E2	W per kg
Exhaust velocity	v	6	×	10E4	m per sec
Specific impulse	ISP	6.1	×	10E3	seconds
Propellant comsumption	$\dot{m}$	3.8	×	10E-3	kg per sec
'Thrust	F	2.2	×	10E2	N
Propulsion time	T	3.15	×	10E7	sec (1 year)
Acceleration, init.	a(i)	0.7	×	10E-3	m per sec square
Acceleration, termin.	a(e)	1.4	×	10E-3	m per sec square
"Ideal" veloc. increment	Δu (ideal)	2.8	×	10E4	m per sec

Note that empty mass will equal full mass minus propellant. Terminal mass will equal full mass minus propellant, food, water, oxygen, and Mars landing craft, but plus Mars specimens.

Figure 3 Early Kaufman thruster, electron bombardment ionization; NASA Lewis Research Center.

RIT 35

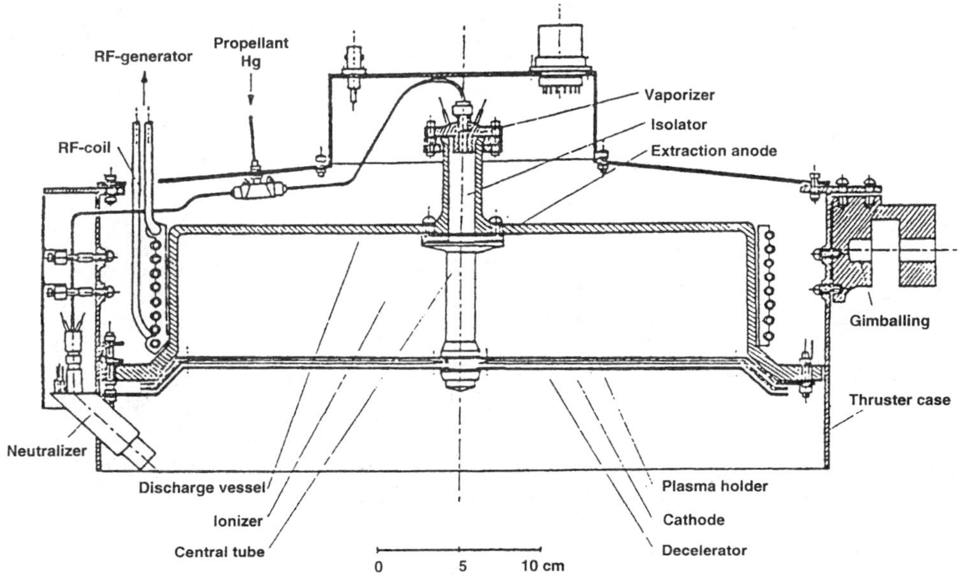

Figure 4a Cross section through a Loeb thruster (RIT 35), showing basic design details.

Figure 4b A modern laboratory test version of a RIT 35 thruster, and also a flight model of the same thruster, designated ESA-XX and built jointly by the University in Giessen (Germany), DASA (Munich, Germany), UKAEA (Culham Laboratories, England), and Fiar (Milan, Italy) are shown in this figure.

Figure 5 Modern Kaufman thruster (Hughes Research Corporation).

Figure 6 Arrangement of six modern Loeb thrusters, type RIT-35. Messerschmitt-Boelkow-Blohm, 1988.

An Evolving Program for Electric Space Propulsion

As soon as the decision is made to develop and use electric thrusters for the propulsion of spacecraft, a systematic program can be started that will evolve from relatively small, one-way missions to targets such as the Earth's magnetotail, asteroids, comets, planetary moons, and planets. These missions will be powered by nuclear-electric and also by solar-electric energy sources. Next, sample return missions will be carried out, still with relatively small spacecraft, powered by larger nuclear-electric energy sources; this mission class will include one-way flights to targets as far away as Pluto. Then, heavier space vehicles will follow for more demanding, still robotic flights to Mars and other planetary targets. Finally, planetary round trip spaceships will be built for human Mars missions.

The two major components of each of these vehicles, the electric thrusters and the electric power supply, will be of a modular nature. Proceeding from a relatively small system propelling an asteroid probe to a large system for a human Mars mission will not require the development of new subsystems, but only an increase of the number of modules. There will be one standard thruster module for small and large space vehicles, and there will be one thermionic generator module to be used in varying numbers on small and large electric power supplies. Also, there will be one type of heavy lift vehicle, to be used with or without attached boosters (depending on the load to be transported into low Earth orbit), and there will be one type of space transfer booster.

In an evolving space vehicle program of this nature, each step builds on the previous step, and also forms the basis for the next step. Each step will have its own funding cycle, and each will lead to technical and scientific achievements in its own right. The total cost will be borne by a number of individual projects. It will not be necessary to commit to the last phase, a human Mars mission, before the previous steps have proven the validity of the concept and the soundness of the technology.

A SIX-PERSON EXPEDITION TO MARS

Mission Plan

Assumptions.

- A heavy lift Earth-to-orbit rocket vehicle will be available with chemical rocket engines. As a two-stager, it can transport 150 tons into a 400 km (250 mile) Earth orbit. With attached booster rockets, its payload capability will be 200 tons (2 boosters) or 250 tons (4 boosters).

- A space transfer booster will be available, propelled by several hydrogen-oxygen engines of the RL-10 type. It will propel each Earth-Mars spaceship after assembly in low Earth orbit at a relatively low acceleration (0.2 g) to near-escape velocity, and into the plane of the Mars orbit. The space transfer booster can also serve as a third stage on the heavy lift vehicle.

- A standard electric (ion) thruster module will be developed on the basis of existing experience and test data. Its size will be optimized for efficiency and reliability.

The same thruster module, clustered at various numbers, will be used for deep space probes, for uncrewed freight vehicles to Mars, and for crewed Mars spaceships.

- A nuclear-electric power source will be developed, based on the thermionic in-core principle, with NaK or Li cooling and electromagnetic pumping of the coolant. The design of the reactor with thermionic elements will be modular. A family of power sources will be built, ranging in electric power from 100 or 200 kWe to about 8,000 kWe (8 MWe). Each of the Mars spaceships will carry two similar reactors, one operating and one spare.

- A standard human habitat will be developed, slightly modified for different applications. It will be needed on orbiting space stations, on the Moon, on Mars, and on Earth-to-Moon and Earth-to-Mars spaceships.

Guidelines.

- First priority should be the safety of the astronauts. Sound design principles, highest quality assurance standards, thorough ground testing procedures, and redundancy of components must be applied throughout the program.

- Accommodations for the astronauts during Earth-Mars transfer should provide for comfort, and also for privacy when desired.

- Flight times from Earth to Mars, and from Mars to Earth, should be as short as possible, not longer than 160 to 180 days (one-way). With a stay time on the Martian surface during this first Mars mission of 30 to 40 days, the total mission time will remain below 400 days.

- Flight time through the Van Allen Belts should be as short as possible, not longer than a few hours.

- The need for high-precision aerobraking maneuvers of returning spaceships near Earth should be avoided. Aerobraking will be needed by the Mars landing vehicles, in combination with parachutes and counter rockets.

- Mission abort capability and the ability to return to the vicinity of the Earth should exist at least during the initial part of the mission.

- Nuclear reactors should not operate at distances from the Earth closer than about 40,000 km. Even after shut-off, a "hot" reactor should not come closer to Earth than about 40,000 km.

Mission Concept. The Mars mission will be carried out by two piloted and one uncrewed Earth-to-Mars spaceships. Each human piloted vehicle will be occupied by three astronauts; two of them will descend to the surface of Mars, while the third remains in the orbiting spaceship. Each human-occupied spaceship will carry one Mars landing craft; the unoccupied spaceship will carry two Mars landing craft. Each landing craft will carry an ascent vehicle. Each of the two landing craft on the uncrewed spaceship will carry a Mars rover. In case of failure, all four astronauts will be able to return from the Martian surface to the spaceships in one ascent vehicle; if necessary, all six astronauts can return from Mars orbit to Earth orbit in one spaceship.

The crewed Mars spaceships will not leave low Earth orbit before the Mars landing craft of the uncrewed Mars spaceship have landed on Mars, and before all their systems have been unloaded, set up, and reported in functional condition by automated procedures.

Mars Spaceship Design

Structure. The Earth-to-Mars spaceship comes in two versions: an uncrewed freight ship and a crewed ship. Both versions use the same propulsion system, power source, radiation cooler, and thrusters; the same basic structure; the same guidance and control system; and the same transfer trajectory. Each crewed ship carries accommodations for the astronauts and one Mars landing craft. It is prepared for a full round trip Earth-Mars-Earth. The uncrewed freight ship carries two landing craft; it will not return to Earth from Mars. Crew and freight spaceships are shown in Figure 7. Table 1 lists estimated component masses.

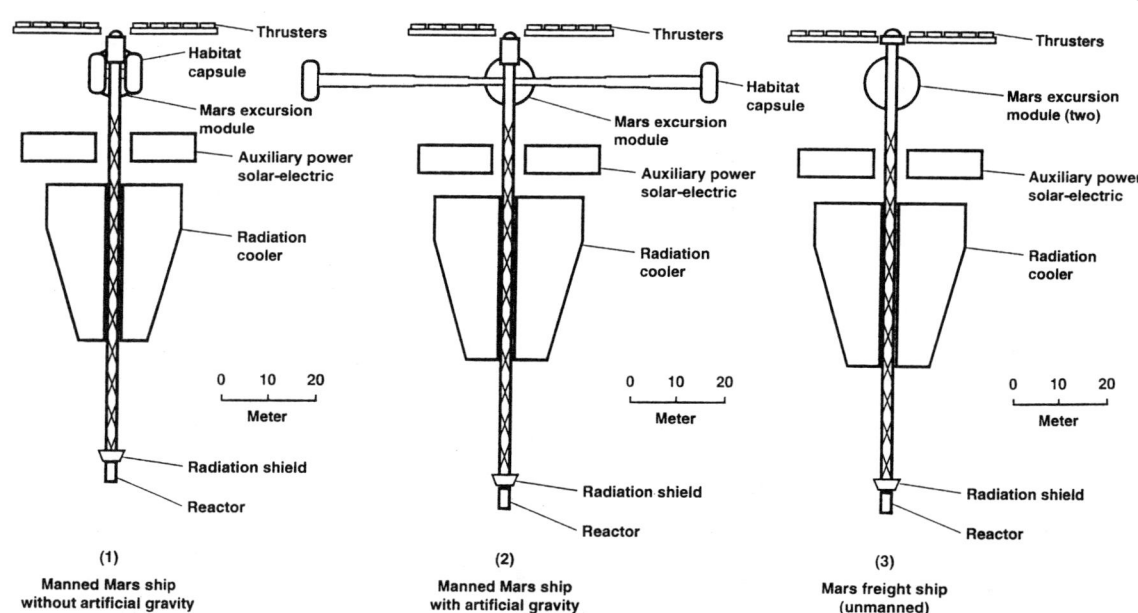

Figure 7 Mars spaceships; (1) crewed ship without artificial gravity, (2) crewed ship with artificial gravity, (3) uncrewed freight ship.

Each crewed ship will have two habitats and one central radiation shelter. They will be connected by tubes big enough to allow passage of an astronaut. Habitats, shelter, and tubes will be pressurized, but each unit has a hatch that can be closed in an emergency (Figure 8). The two habitats will be identical. Each will be equipped with its own life support system, food and water store, guidance and control terminals, and communication system. This complete redundancy will ensure greater safety in case of a failure; it will also provide a degree of privacy for astronauts when desired.

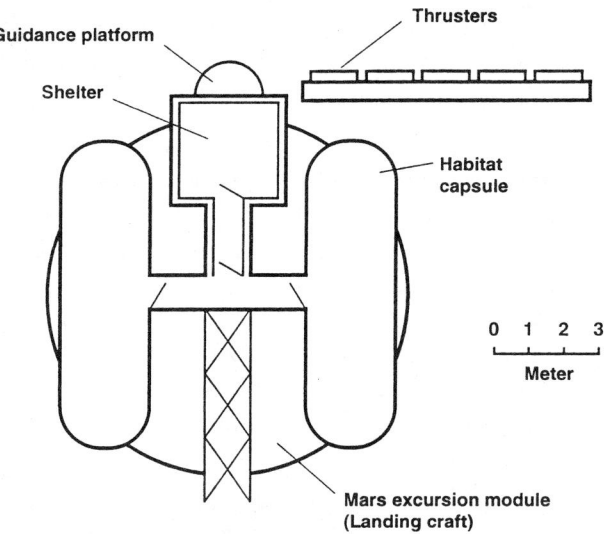

Figure 8 Arrangement of two habitat capsules, radiation shelter, dome for guidance platform, and attached lunar landing craft.

Each spaceship will have several small thrusters of the arc jet or the plasma jet type for attitude control. While these thrusters are less efficient and have shorter operational lifetimes than ion thrusters, they are preferable for attitude control because they have a more compact build and a less elaborate power conditioning system than ion thrusters.

Propellant. Since the early 1960s, developers of electric thrusters considered mercury to be the proper propellant for an electric thruster system because of its high density, high thruster efficiency, small tank volume, and simple feed procedures. Also, the mercury propellant can be used for additional protection around the radiation shelter.

More recently, mercury has come under criticism as a space propellant for several reasons. First, it is a very nasty material to work with because it contaminates test chambers. Second, a tank rupture during Earth-to-orbit transportation would result in severe pollution problems. Third, scientists fear it as a potential contaminant of near-Earth space if thrusters are operated near Earth. Another propellant, argon or xenon, could be used, but with lower thruster efficiency. Also, a gaseous propellant would need a huge tank volume whose surface would have to be protected against puncture by meteoroids. As a shield against space radiation, a gas would be far less useful than mercury.

A human Mars mission concept as described here would largely avoid or circumvent those disadvantages of mercury. Design and performance figures reported here, therefore, are based on mercury as propellant.

Radiation Shelter. A radiation shelter will be needed on Earth-to-Mars spaceships to provide protection for the astronauts while traveling through the Van Allen Belts, and during solar bursts. Stay time within the shelter will always be short (hours), so the protected volume need not be large, nor does it have to offer many commodities. Astronaut shelters of similar design and size will be needed on the surfaces of the Moon and Mars.

Artificial Gravity. The question "to g or not to g" has been debated frequently in connection with long-time space missions. Actually, this problem must be considered from two different viewpoints in connection with a human mission to Mars. First, the question is whether long-time exposure to near zero-g conditions may result in damaging physiological effects that could persist for a considerable time after exposure before they fully disappear. Second, it is mandatory that the astronauts are in full possession of their bodily strength and dexterity immediately upon arrival on the Martian surface. So far, the question whether artificial gravity should be provided during the long transfer from Earth to Mars, and back from Mars to Earth, has not been answered to everybody's satisfaction.

The suggested spaceship design (Figure 7) will lend itself relatively easily to an artificial gravity provision (Figure 9). The tubes connecting the habitats with the central radiation shelter and the exit port will be elongated to a length of 40 m from the center line to the habitat, and the entire spaceship will rotate around its central axis at 1 revolution every 20 sec to provide a radial centrifugal acceleration in the habitats of 0.38 g, equal to the natural gravity on the Martian surface. The guidance platforms (two for redundancy) would be affected by this rotation unless they were placed on a non-rotating mount on the center line of the spaceship, as shown in Figures 8 and 9. They will maintain their inertial orientation with the help of star seekers.

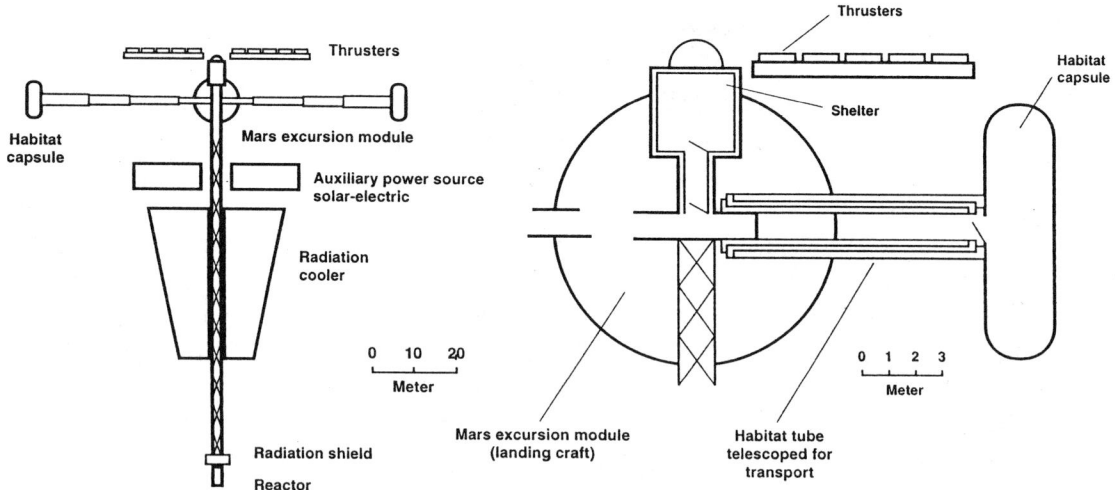

Figure 9 Crewed Mars spaceship with extended, and with withdrawn habitat tubes. Instead of a telescoping tube design, tube sections could also be launched separately and assembled in low Earth orbit.

Orbital Assembly and Checkout

After assembly and checkout on Earth, each spaceship will be disassembled into three portions (Figure 11), each of which will be carried to low Earth orbit by a heavy lift vehicle (Figure 12). The mercury propellant will be transported to low Earth orbit in a number of relatively small, sturdy steel bottles capable of withstanding even severe impacts during a hypothetical failure of the heavy lift vehicle during launch (Figure 10).

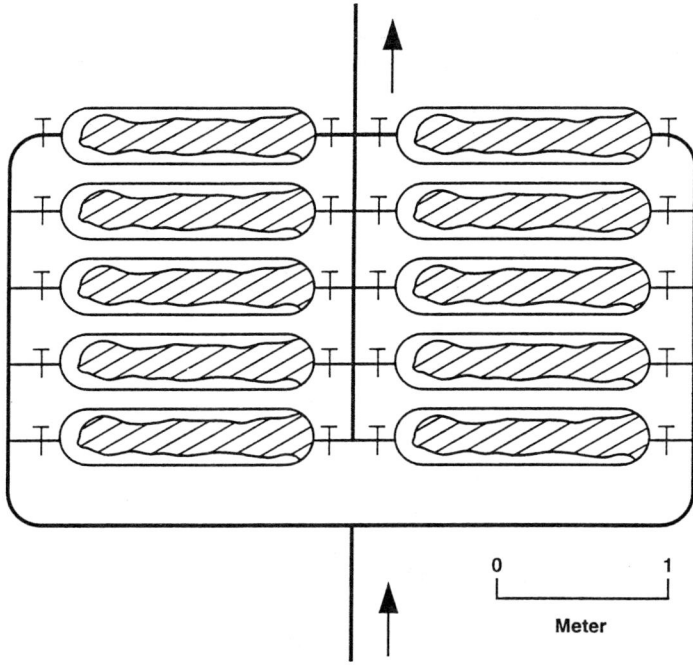

Figure 10 Battery of heavy steel bottles for the transportation of mercury propellant from the ground into the spaceship assembly orbit (low Earth orbit).

Assembly of the spaceships in orbit will only require making mechanical and electric connections between the three portions, connecting four cooling pipes between reactor and radiation cooler, and making a pipe connection between the battery of mercury bottles and the spaceship's mercury tank. Thus, the total amount of "plumbing" and other assembly work will remain modest.

A thorough checkout will follow assembly. The thrusters will be checked out one by one, and with a power source located on a heavy lift vehicle—with argon as propellant in order to avoid mercury contamination of near-Earth space. The reactor, electric connections of the thermionic modules, cooling circuit, pumps, valves, and other components of the power supply system will be checked out while the reactor is working on a very low power level. Working at full power, and testing of the thermionic modules at their nominal temperatures, will be restricted to a very short time to prevent the build-up of a high level of radioactivity in the reactor.

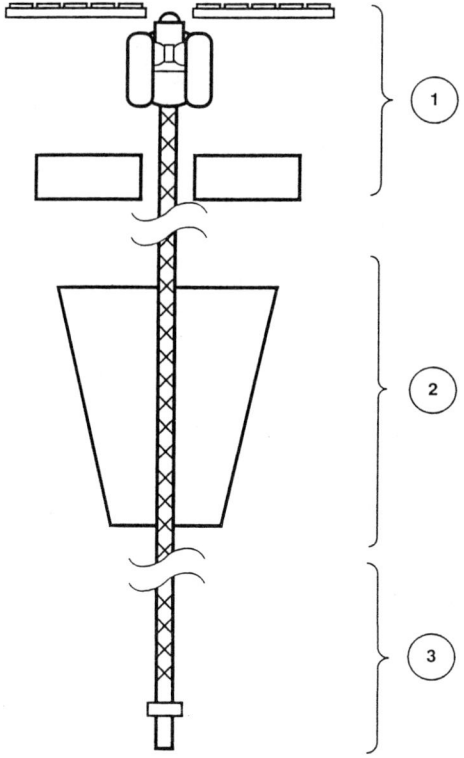

Figure 11 Each spaceship would be launched into low Earth orbit in three separate sections for orbital assembly.

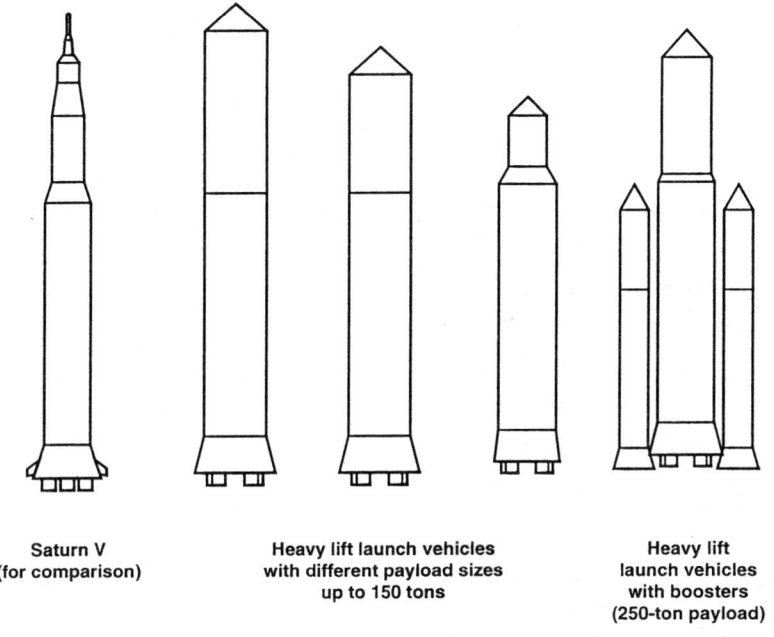

Figure 12 A family of heavy lift launch vehicles for various loads.

Transfer from Earth Orbit to Mars Orbit

Space Transfer Booster. Following the checkout procedures in Earth orbit, each of the three spaceships will be attached to a space transfer booster (Figure 13). Powered by a cluster of hydrogen-oxygen engines of the RL-10 type, a space transfer booster will propel a spaceship to near-escape velocity, and into the Martian orbit plane. The final boosted velocity will be kept below escape velocity so that the spaceship remains in an Earth orbit in case the nuclear-electric propulsion system should fail to begin its operation.

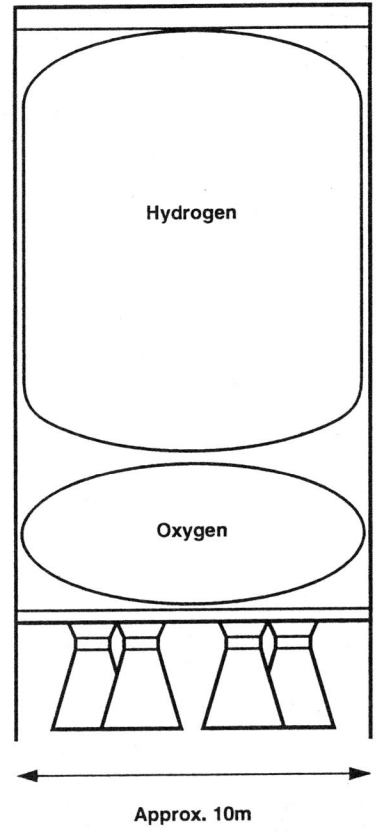

Approx. 10m

Figure 13 Outline of the space transfer booster, needed to accelerate the assembled spaceships from low Earth orbit to near-escape.

Space transfer boosters and their propellants will be transported from Earth to low Earth orbit by heavy lift vehicles; one 150-ton and two 250-ton heavy lift vehicles will be needed for each Mars spaceship (Figure 12).

Earth-Mars Trajectory. After burnout and detachment of the space transfer booster, the spaceship will coast quickly through the Van Allen Belts. When a distance from Earth of about 40,000 km has been reached, the nuclear reactor and the ion thrusters will be started. At that distance, neither the mercury exhaust nor a failed reactor will present a danger to Earth.

Figure 14 shows the approximate trajectory of the spaceship between Earth and Mars. Its velocity, in a reference system in which the Sun is at rest, is shown in Figure 15. In these figures, the trajectory and the velocity of a space vehicle following a Hohmann transfer trajectory (minimum energy transfer) are shown for comparison.

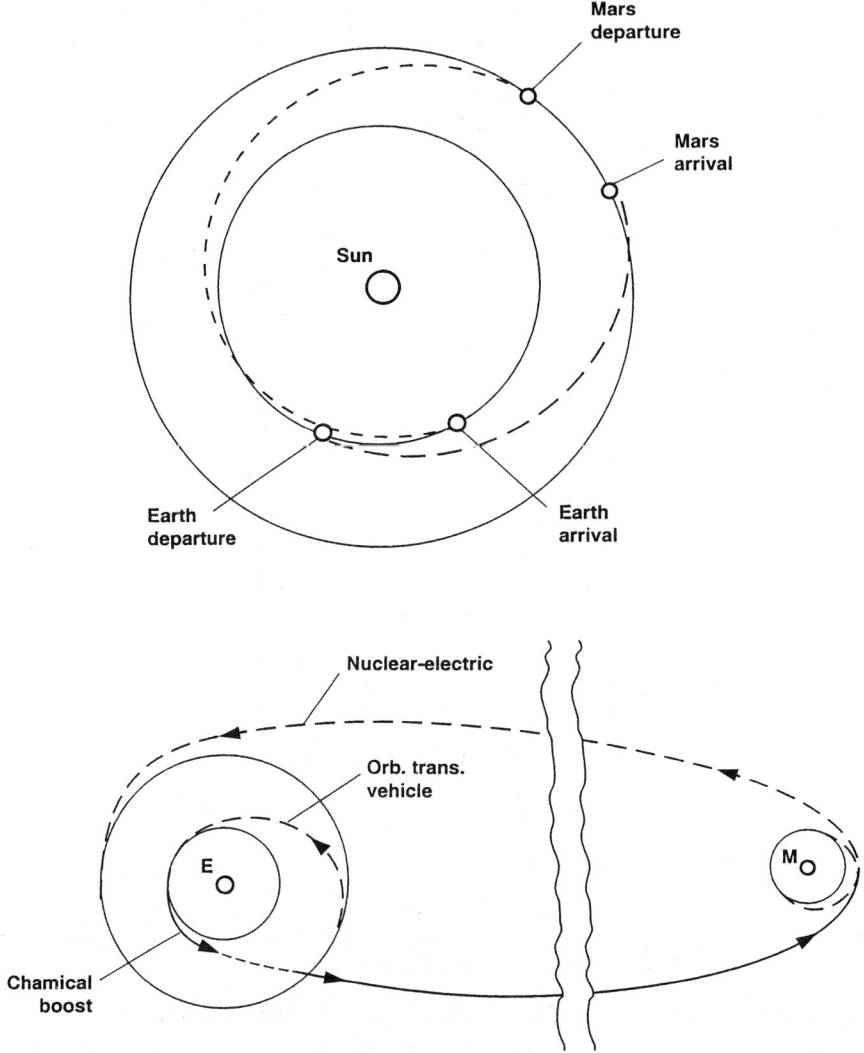

Schematic of manned Mars mission

Figure 14   Spaceship trajectory between Earth and Mars.

During the first part of the transfer trajectory, the spaceship accelerates; later, it reverses its orientation and decelerates. Its average velocity, therefore, is larger than the velocity on a minimum energy transfer trajectory, and its flight time as well as its path are shorter than the Hohmann flight time and path from Earth to Mars.

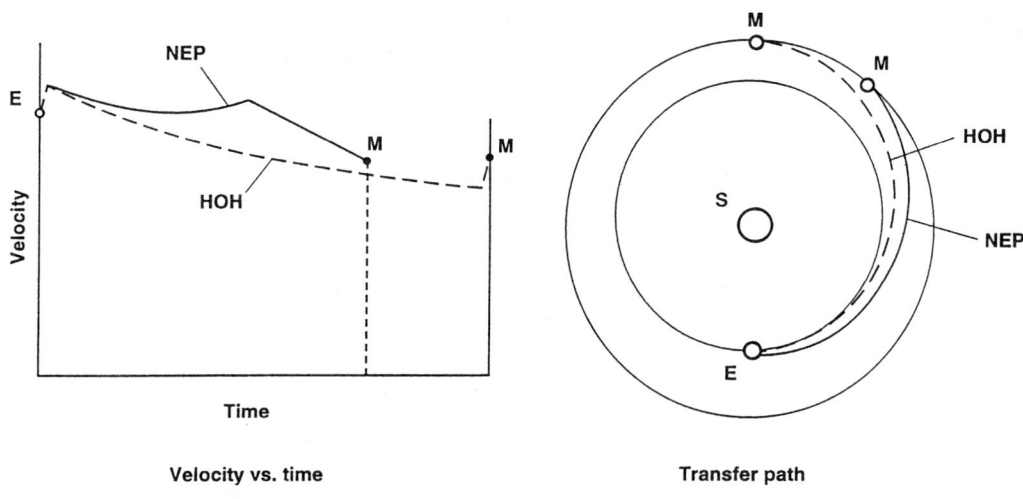

Figure 15 Velocity profile of spaceships between Earth and Mars.

Upon arrival near Mars, the spaceship enters into a high near-circular orbit, and then spirals down to a lower orbital altitude. During its approach toward Mars, and during its spiraling phase, the spaceship changes its orbital plane so that it enters into the planet's equatorial plane. Before the spiraling phase ends in a low Mars orbit, the Mars landing craft (two on the uncrewed freight ship, one on each crewed spaceship) will detach and descend to the surface, thus utilizing much of the required spiraling time of the spaceship for exploration on the surface.

All flight maneuvers of the unoccupied Earth-to-Mars freight ship will be executed either by computer program or by remote control. The two landing craft of the freight ship will land on the equator in close mutual proximity at a point determined earlier during one of the robotic Mars Observer missions. Each landing craft on the freight ship will carry a habitat with a life support system, a radiation shelter, a small roving vehicle, equipment for scientific exploration, communication systems, and an ascent (return-to-orbit) vehicle. Only when the landing has been successful, and when all systems report to be in operable condition, will the two crewed spaceships with the astronauts leave their low Earth orbit for their trip to Mars.

The landing craft on the crewed spaceships will be attached in such a way that the astronauts can access them from the pressurized system through an airlock. Crew capsule, descent stage, ascent stage, and other components of the landing craft may be arranged as shown in Figure 16.

In case of a malfunction on a crewed spaceship during the outbound phase of the mission which forces the astronauts to abandon ship and try to return to Earth on their own (rather than transfer to the other spaceship with a space taxi), the crew capsule, descent stage, and ascent stage can separate as a unit from the landing craft and achieve a return-to-Earth trip as a two-stage rocket system. This rescue mode could be used only during the early part of the Earth-to-Mars transfer because of the limited oxygen, water, food, and electric supply on the crew capsule of the landing craft.

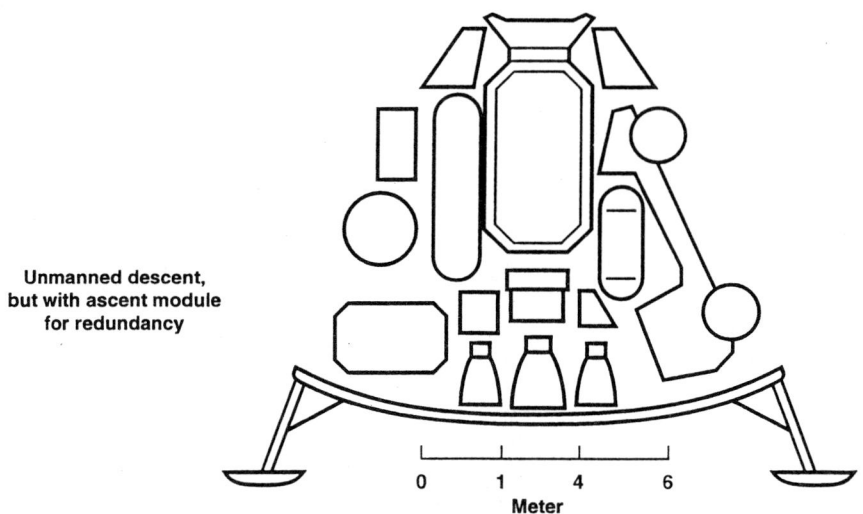

Unmanned descent, but with ascent module for redundancy

Figure 16   Lunar landing craft (Lunar Excursion Module). Uncrewed version without astronauts, but with ascent vehicle and roving vehicle. Crewed version with 2 astronauts and ascent vehicle, but without rover.

Space transfer boost phase, electric propulsion system, trajectory from Earth to Mars, spiraling approach to Mars, descent to the surface, and landing on Mars will be identical for the uncrewed and crewed flights. This commonality not only simplifies design, development, testing, quality assurance, mission planning, mission implementation, and ground control procedures for the overall mission, but also offers a very valuable "real life test" of nearly all flight systems before the astronauts have to embark on their deep space voyage.

The two crewed landing craft will land in close proximity to the uncrewed landing craft. The four astronauts will find two shelters with life support systems (in addition to the shelters and life support systems in their own landing craft), and they will also find two rovers, equipment for scientific exploration, and communication systems to the orbiting spaceships, and directly to Earth. The four landing craft will carry four return-to-orbit vehicles. This multiple redundancy offers a high degree of safety for the astronauts. If not needed in an emergency, nearly all the systems not returning to orbit will remain at the disposal of future astronauts visiting Mars on later missions.

On their return trip to Earth from the Martian orbit, the two spaceships will be propelled only by their electric thrusters. They will be considerably lighter than at the beginning of their mission because one half of the propellant and almost one half of their oxygen, water, and food supply has been consumed, and the landing craft has been dispatched. The return trajectory (Figure 14) begins with an ascending spiral around Mars. As soon as escape from the Martian gravitational field has been achieved, the spaceship decelerates to effect the descent toward the Earth's planetary orbit. Again, an accelerating phase with the thrust vector oriented inward shortens the flight time. Eventually, after several decelerating and accelerating phases, the ship approaches Earth and spirals down into a circular orbit at about 40,000 km altitude, far above the Van Allen Belts, and at a distance where neither the mercury exhaust nor the "hot" reactor will represent any threat to Earth. At that point in space, the ship is met by an orbital transfer vehicle (as used for traffic between low Earth orbit and geosynchronous orbit) that takes the astronauts, their films, samples, etc. quickly through the Van Allen Belts and back to low Earth orbit. From there, descent to Earth takes place in a spaceplane or in a shuttle.

If considered worthwhile, the orbiting spaceships with their nuclear power sources and thrusters could be refurbished and used again on a future mission. Another option would be to use their remaining propulsion capability to propel them back into deep space. As a third possibility, they could simply be left in high orbit where their lifetimes would be extremely long compared to the decay times of the "hot" reactors.

TECHNICAL CHALLENGES

Thermionic converters, recognized more than thirty years ago as potentially attractive electric power sources for space vehicles if used in combination with nuclear reactors, were the subject of considerable theoretical and experimental work at General Dynamics in California, Brown Boveri in Germany, and other companies during the 1950s and 1960s. In spite of encouraging progress, those studies had to be terminated because of lack of funds. More recently, work on thermionic converters was resumed at several places, aiming at nuclear-electric power sources for space use. Such power sources will have the particular advantage that they permit a very compact, well integrated design with no moving parts, except valves in the cooling system. On the other hand, several technical problems still have to be solved before thermionic power supplies will meet the requirement of reliable long-time operation in space. One of these problems is the maintenance of narrow gaps between cathodes and anodes (about 0.1 centimeter), while the cathodes are kept at a temperature on the order of 1,700 K, and the anodes at about 600 to 700 K. Any swelling or warping of the fuel elements, probably UO_2, must be avoided. On the other hand, efficiencies of the source of better than 20 percent, and a power-to-mass ratio of the complete system of about 120 watt per kg may eventually be obtained. A schematic illustration of the thermionic power source is shown in Figure 17.

For a total electric power of 8 MWe, the thermal power output of the nuclear reactor will be about 40 MWth; the highest temperature inside the reactor will be on the order of 1,800 K. These values will be considerably lower than power and temperature of a nuclear reactor for nuclear-thermal propulsion, about 5,000 MWth and 3,000 K. On the other hand, the nuclear-electric power source on a Mars spaceship will have to work continuously for about one year, while the reactor of a nuclear-thermal propulsion system will typically operate four times, for less than an hour each time, during a human mission to Mars. However, even between operating phases the reactor will be a very strong source of radiation.

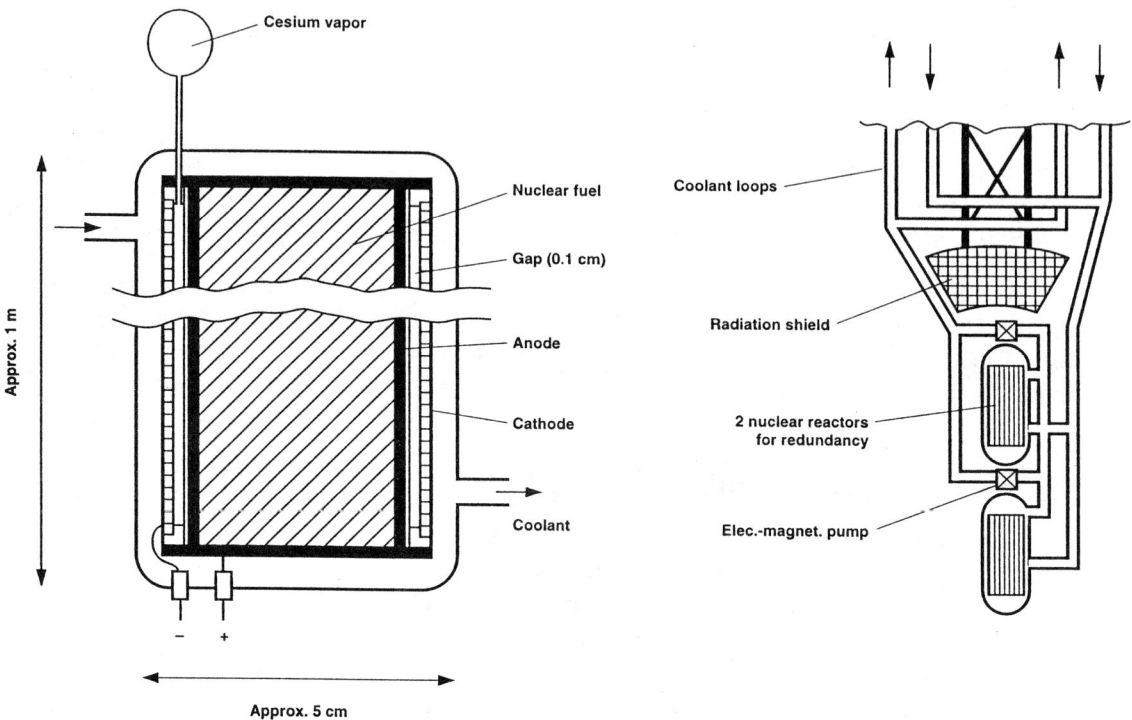

Figure 17 Schematic of nuclear fuel element with thermionic converter, and of nuclear reactor arrangement (one operating reactor, one spare).

Most ion thrusters built and operated in the past have diameters between 5 and 35 cm; they consume electric power between a few hundred and a few thousand watts. Thrusters for crewed Mars spaceships should be larger. Models with diameters of about 1 m and a power consumption on the order of 250 kWe were built and successfully operated at the NASA-Lewis Research Center and also at the Giessen University [Loeb 1992]. For a crewed Mars ship, an optimum thruster size should be determined, large enough to keep the total number of thruster modules manageable, but small enough to assure high efficiency, and to avoid mechanical problems such as warping of the accelerator grids. It is expected that the optimum module size will be a thruster with 0.8 to 1.2 m diameter, and a power consumption of 200 to 300 kWe.

Even if the thrusters used mercury as propellant during the transfer from Earth to Mars, development of the thrusters, as well as long-time test runs on the ground, and functional testing after assembly in orbit, could be made with a gas, such as argon. While fully valid for functional testing, tests with argon would avoid the use of mercury in laboratories, on test stands, and in low Earth orbit.

Commonality of Components

Planners of a spaceflight system, as well as planners of earthbound systems such as airplanes, ships, automobiles, buildings etc., will endeavor to use components that have been developed, tested, and used for other systems before. If they have to develop a new component, they will try to make it useful also for other systems besides the one under development at that time. This "commonality of components" is an important cost-saving element in any new project; it may even be of decisive influence on the acceptance of a new project.

A human Mars mission as outlined here will benefit substantially from wide commonality and multiple use of components. The most important component of every space mission, its propulsion system, will have the same basic nature for a number of robotic and human missions. The major elements of electric propulsion systems—thrusters, power sources, and power conditioning systems—will have a modular structure; smaller missions will use fewer modules than larger missions. Earth-to-orbit heavy lift vehicles will be used for a number of space missions besides human Mars flights, among them Earth-orbiting satellites, space stations, and robotic and human lunar missions. A similar situation exists for space transfer boosters and for orbital transfer vehicles. Nuclear-electric power sources, with power levels between about 100 kWe and 8 MWe, will be needed in the future on geosynchronous satellites, on lunar bases, and on Mars bases; whether they will be acceptable on space stations in low Earth orbits is not known at the present time. Life support systems, some of them including radiation shelters, will be needed on space stations, on lunar bases, on Mars bases, and on long-distance roving vehicles on the Moon and on Mars. Roving vehicles, for short and for long distances, robotic (teleoperated) and human piloted, will be needed on the Moon and on Mars.

Figure 18 shows the commonality relations between systems (components) needed on a human Mars mission as outlined here, or otherwise existing, and space missions of the present and the future. The left-hand column lists systems (components), while the right-hand column lists space missions presently under consideration. The lines connect systems with those missions on which they will be needed.

PROGRAM PLAN

Time Schedule

A project as large and complex, and as expensive, as a human mission to Mars need not be started as a complete project at a given time. Most of its components will be developed in steps, and a number of these steps will lead to achievements in space even before a definite commitment to a Mars mission must be made. In fact, decisions for many of the details of a human Mars mission can be made only after previous steps have been successfully accomplished.

As a first step toward a human Mars mission, development work on a nuclearthermionic power source should begin. Its first goal should be a 100 to 200 kWe space power source, with the final goal of a 8 MWe power source. This component will require the longest lead time among all mission components.

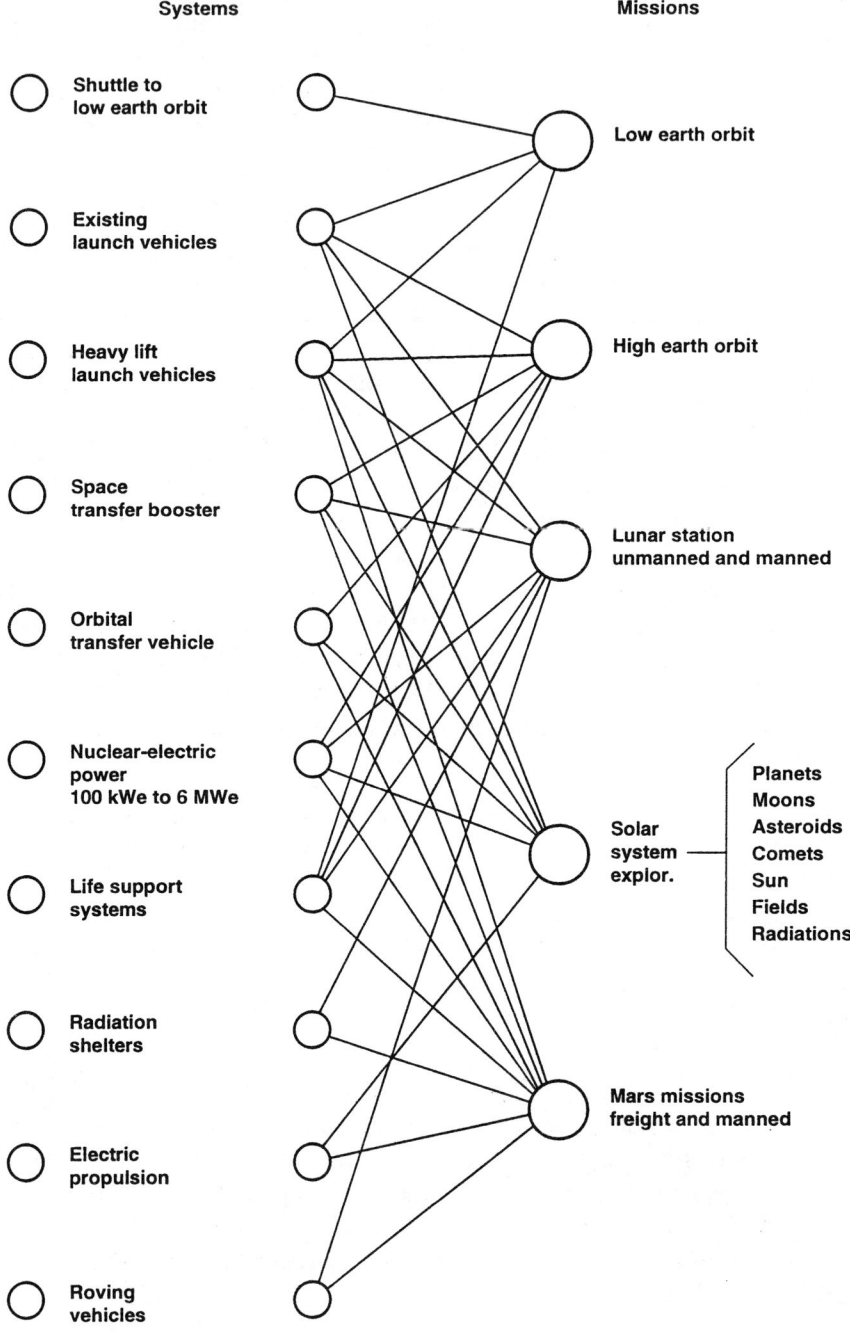

Figure 18 Commonality of systems on a variety of space missions.

Other components, such as optimized ion thrusters, habitats with life support systems, Mars landing craft with ascent vehicles, radiation shelters, roving vehicles, and others will require less development time; their initiation can be staggered in time over a number of years.

The year 2019, about a quarter of a century in the future, happens to be a particularly favorable one for Earth-Mars missions. If that year were targeted for a human Mars mission, the twenty-five intervening years would offer enough time for an evolving program as indicated here. Considering the space related achievements of past twenty-five year periods, for example the time between 1954 and 1979, the twenty-five years from now till 2019 would appear comfortably adequate for an evolving program climaxing in a human Mars mission. However, systematic planning for such a program should begin soon; also, technical development work on the longest-lead-time component, the nuclear-thermionic power source, should begin without much delay (Figure 19).

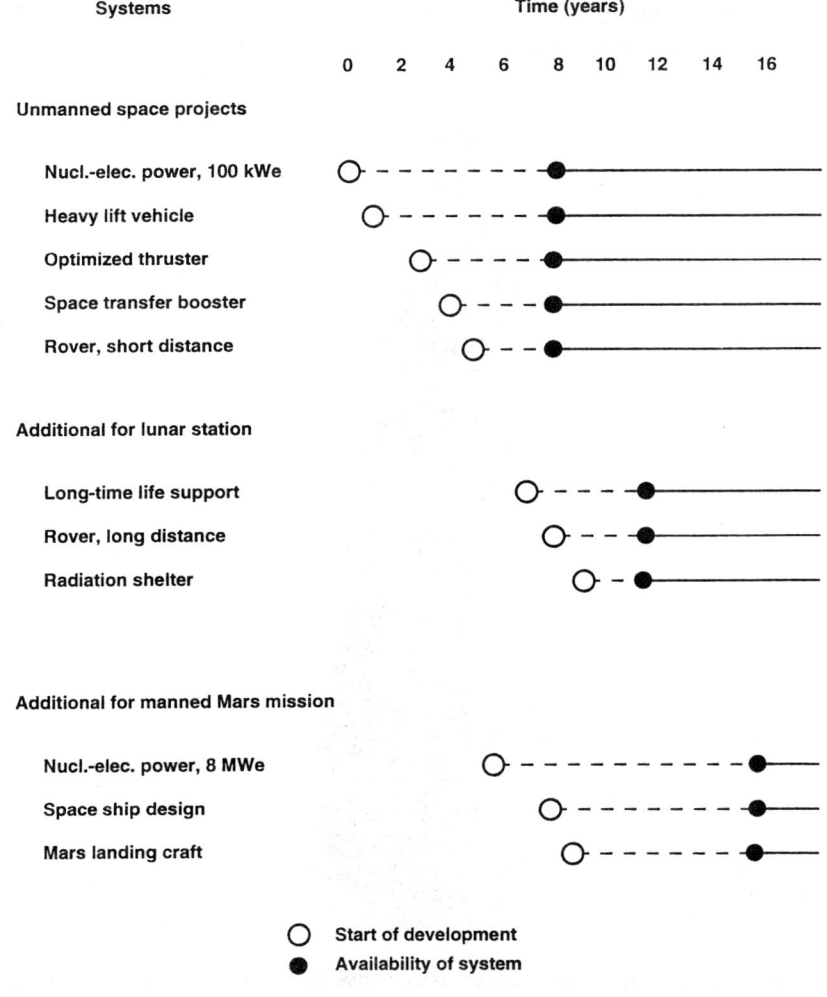

Figure 19 Time schedule for the development and mission availability of various space missions.

International Cooperation

Fruitful cooperation between the United States and a number of European countries has been underway for years. With the dissolution of the Soviet Union, broad-based cooperation between the United States and former member states of the U.S.S.R. will be much easier to implement in the future than this was possible in the past. Several very successful Russian space projects of the past would have an immediate bearing on the program suggested here; among these projects are the heavy lift vehicle Energia; the space station Mir; the space-tested nuclear-thermionic power source Topaz; and more than 80 orbital flights of electric thrusters, among them ion thrusters, plasma thrusters, Hall effect thrusters, and arc jet thrusters [Loeb and Popov, 1992]. In these projects, Russian space engineers and scientists have gained considerable knowledge and experience which would be most valuable for an evolving program as suggested here. Without a national space program of their own, Russian colleagues would certainly be anxious to join American, European, Japanese and perhaps other colleagues in a joint, well managed program as broad and diversified as proposed here. Figure 20 shows the flight version of a Russian nuclearthermionic space power source "Topaz."

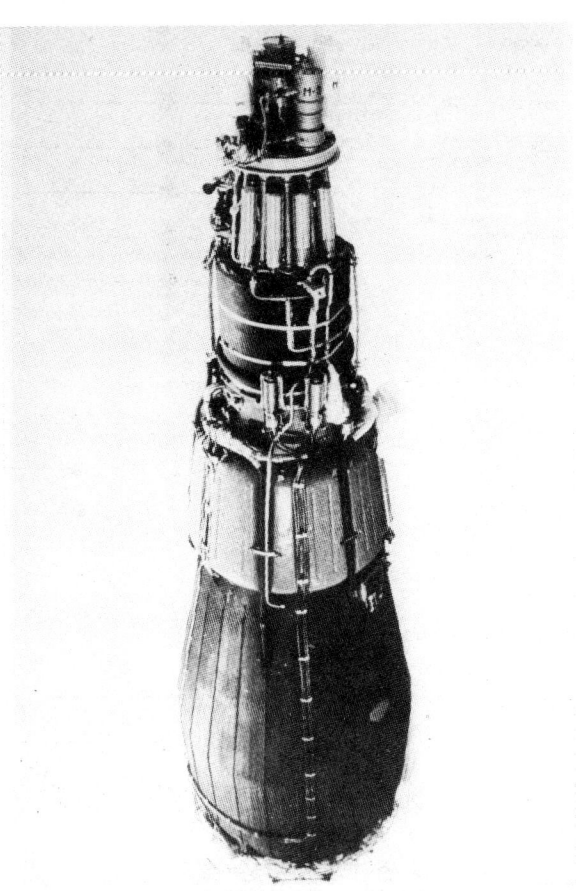

Figure 20 Flight version of Russian nuclear-thermionic space power source "Topaz", 32 volt, 6kWe, launched in 1987 (0.5 year mission) and in 1988 (1 year mission). Y.M. Gryasnov, Red Star Enterprises, Moscow.

Figure 21 gives an impression of the number of launch rockets, space transfer boosters, and Earth-to-Mars spaceships that would be needed for a six-person, one-year expedition to Mars as proposed here.

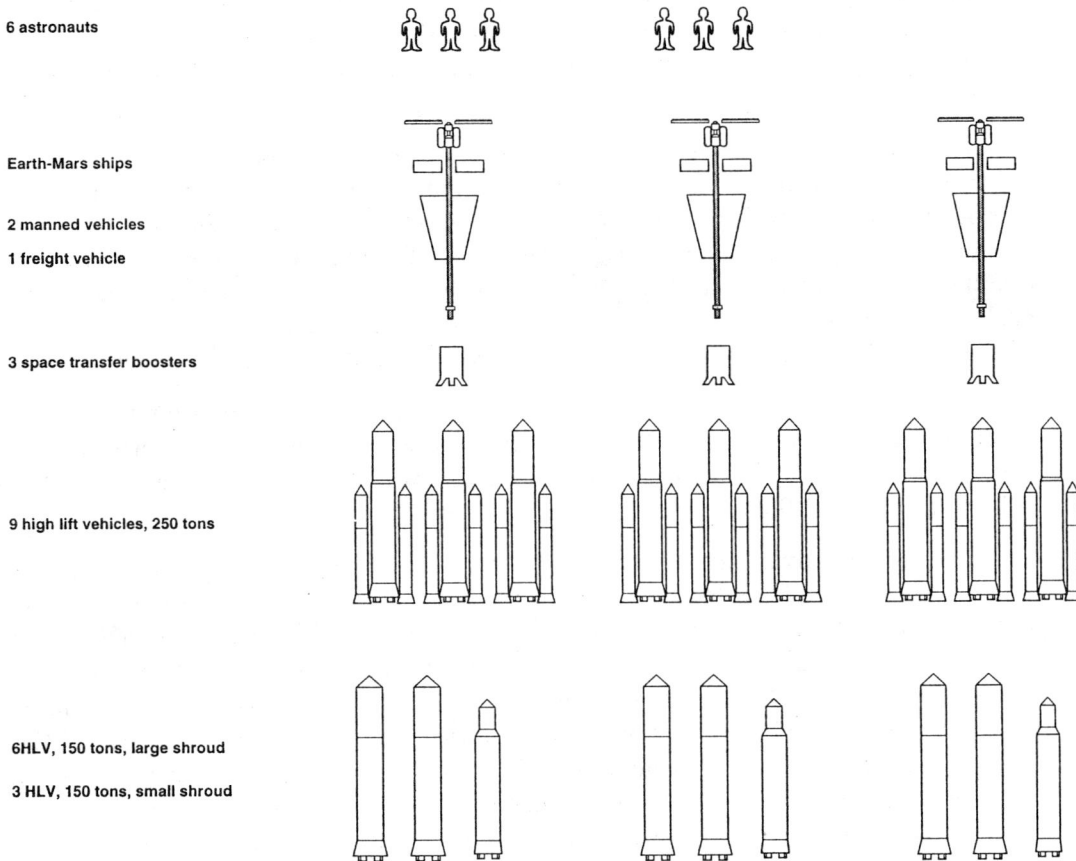

Figure 21 Launch rockets, space transfer boosters, and Earth-to-Mars spaceships needed for a six-person expedition to Mars.

CONCLUSION

In addition to chemical and nuclear-thermal rockets, a third space propulsion system, the electric rocket, has been under study, development, and testing during the past forty-five years. It has reached a state of maturity that definitely warrants its consideration for deep space missions, from relatively small spacecraft, powered by solar-electric power sources, up to and including human expeditions to Mars, powered by nuclear-electric power supplies.

From a flight mechanics standpoint, electric space propulsion systems differ principally from chemical and nuclear-thermal rockets. They provide low thrust forces over long periods of time. These specific impulses can be chosen at will within a wide range. For any particular mission, the specific impulse will be selected by the designer on the basis of an optimization procedure. Depending on mission parameters and on the performance rating of the nuclear-electric power source, an electric propulsion system will have a specific impulse between about 3,500 and 7,000 seconds, equivalent to an exhaust velocity between 35 and 70 km per second. For a Mars mission as presented here, the specific impulse should be about 6,000 seconds. The total impulse generated by an electric propulsion system on a typical human Mars mission, expressed as the total "ideal" velocity increment of the spaceship, will be on the order of 28,000 meters per second.

Electric power for the propulsion system on a flight mission to Mars will be provided by a nuclear-electric power source developing about 8 MWe and operating at a thermal power level of about 40 MWth and a temperature of about 1,800 K. The nuclear reactor will be operating continuously during the entire mission; its power level will be throttled down during the stay in Martian orbit. It should be noted that the reactor of a nuclear-thermal rocket will be operating only for short thrusting periods; during these periods, it will generate on the order of 5,000 MWth at a temperature of about 3,000 K.

The nuclear-electric space system will avoid problems of environmental hazards on Earth by starting its nuclear reactor only after the spaceship has reached a distance from Earth of about 40,000 km. After the reactor has become "hot," it will never come closer to Earth than about 40,000 km.

The specific advantages of nuclear-electric propulsion systems for human expeditions to Mars include the following: Total mission time, with 30 to 40 days on Mars, will be below 400 days. A fleet of three spaceships will be able to carry a total payload to Mars of more than 300 tons. If desired, artificial gravity during the transfer can be provided relatively easily. Since the propulsion system is continuously in operation, trajectory corrections and modifications can be carried out at any time. The ratio of total take-off mass from Earth to payload mass on Mars is substantially greater for electrically propelled spacecraft than for spacecraft propelled by chemical or nuclear thermal systems. Power efficiency and mass efficiency of existing thrusters (ion thrusters) are close to the theoretical limit; work on the propulsion system for a Mars project could begin without further substantial research and development efforts. In contrast, the megawatt nuclear-electric power source will require development work; it will be the pacing element in a human Mars project with nuclear-electric propulsion. The major components of a nuclear electric propulsion system for a human Mars mission will be modular; the same basic modules will be used in propulsion systems for smaller, robotic deep space vehicles that will precede a human Mars mission.

REFERENCES

Aerospace America, "Electric propulsion" (Review article by the American Institute of Aeronautics and Astronautics Technical Committee on Electric Propulsion), 42pp., December, 1992.

Kaufmann, H.R., and P.D. Reader, "Experimental performance of ion rockets employing electron bombardment ion sources," *American Rocket Society*, 1374-60, 1960.

Loeb, H.W., "Ein elektrostatisches raketentriebwerk mit Hochfrequenzionenquelle," *Astronautica Acta*, 8, 1, 49, 1962.

Loeb, H.W., "Large scale radio-frequency ion thrusters for manned Mars missions," *43rd Congress of the International Astronautical Federation*, IAF-92-0619, September, 1992.

Loeb, H.W. and G.A. Popov, Proceedings of the Russian-German Conference on Electric Propulsion Engines and their Technical Applications, Rauischholzhausen, Germany, March 1992; published by H.W. Loeb, *First Institute of Physics*, Justus Liebig University, Heinrich-Buff-Ring, D-6300 Giessen, Germany, 1992.

Moeckel, W. E., "Propulsion methods in astronautics," *Intern. Ser. Aeron. Sci. and Space Flight*, Vol. 2, Pergamon Press, New York, 1959.

Stuhlinger, E., "Possibilities of electric space ship propulsion," *Ber. 5th Internat. Astronaut. Kongr.*, Innsbruck, 81pp., Springer Verlag, Vienna, 1954.

Stuhlinger, E., *Ion Propulsion for Space Flight*, McGraw-Hill Book Company, New York, Toronto, London, 1964.

Stuhlinger, E., "Electric propulsion ready for space Missions," *AIAA Astronautics and Aeronautics*, April, 1978.

Turchi, Peter J., F.M. Curran, J.C. Andrews, J.R. Beattie, and J. Gilland, "Electric propulsion: the future is now," *Aerospace America*, 38pp., July, 1992.

Tsiolkovskii, K.E., *Collected Works*, Izd. Akad. Nauk, SSSR Moscow, Tompervyi, Vol. 1, Aerodynamics, 266pp., 1951; Vol. 2, Reactive Flying Devices, 456pp., 1954.

Zhurin, V.V., A.A. Porotnikov, and S.B. Borisova, *AIAA paper 83-1397*, June, 1983.

Section III

Living in Space: The Human Element

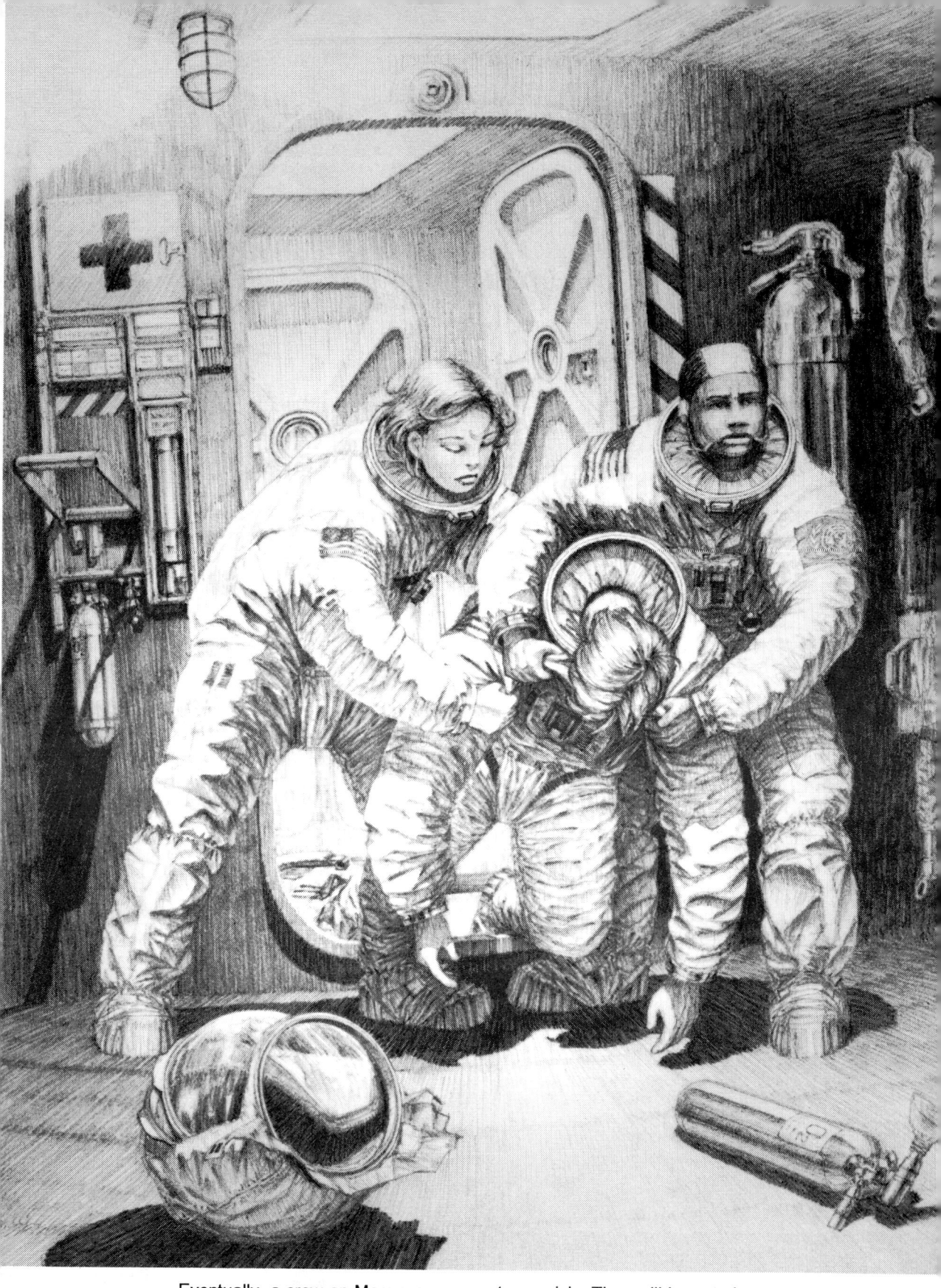
Eventually, a crew on Mars may encounter a crisis. They will have to be prepared.

Chapter 13

BIOMEDICAL ISSUES IN THE EXPLORATION OF MARS

Rosalind A. Grymes, Charles E. Wade and Joan Vernikos[*]

This section introduces the biomedical requirements for human Mars exploration missions. Supporting humans on another planet will be a defining challenge for science and technology. Of central concern are environmental factors and countermeasures as the crew is exposed to new environments, both in transit and at the Martian surface. Mars researchers will venture far beyond the home planet, and care must be taken to provide portable environments that support life and encourage well-being. Physiological changes are expected as a result of the microgravity and radiation elements of space. Countermeasures encompass both pro-active interventions and reactive treatments. The crew will be far from external assistance in an inhospitable environment. They will also, for the first time since the lunar landings, be in an "un-earthly" environment, undertaking an unpredictable adventure. Our continuing efforts to explore new territories are fundamental to our character. At the same time we reach inward, seeking to understand the smallest workings of our inner environments through the life sciences. Earth-like conditions have been central to the evolution of all life on Earth, and have been inseparable from the scientific study of life. Extraterrestrial environments will focus our introspection as never before. We must understand ourselves as a pre-requisite to discovering our place among the planets.

INTRODUCTION

The scientific agenda for a Mars exploration mission will require the mobility, intelligence, and creativity of a human crew. The Viking Lander, an earlier robotic mission, provided exciting new glimpses of the surface. Unmanned exploration tools, augmented by intelligent software control systems, will undoubtedly be used again in preliminary survey missions. While limited in scope, these are essential first steps to human visits. Our most advanced tools can elicit important information, but we cannot experience or interact with the Martian environment without providing a human presence. This presence requires a healthy crew performing at peak efficiency both during transit and on arrival at the planetary surface. The safe delivery, maintenance and return of these valuable researchers is a top priority.

---

[*] Life Science Division, NASA Ames Research Center, Moffett Field, California 94035-1000.

Our collective urge towards exploration is due, in part, to our desire to test the limits of our capabilities. As humans transition between environments, they demonstrate tremendously adaptable physiology and behavior. Humans can adapt to a hot environment by increasing their sweat rate, thus providing both evaporative cooling and a small example of adaptive physiology. Such physiological responses are initiated by a change in the environment which disturbs the normal workings of the body. Behavioral adaptations, in the selection of clothing, housing styles, and preferred foods, are other approaches that increase the productivity of an individual in a non-optimal environment. In addition to the flexibility selected through evolution, we have the ability to impact our environment through the application of reason and the use of tools. Adverse environmental factors can often be negated by technological advances. These triumphs of machines over environment are not limited to the space program. In the example above, of uncomfortable heat, many can elect to turn on an air conditioner rather than endure the heat. Technology obviates the need for physiological or behavioral compensation and adaptation.

The human body regulates many functions within tightly defined limits. Once again, let us use the example of temperature regulation. Normal body temperature is 37°C. If an individual's temperature is elevated, sweating will occur to lower body temperature. If body temperature is decreased, shivering will increase body temperature. Though simplified, this is an example of the "homeostatic" mechanisms which allow the human body to maintain an internal equilibrium. In a mission to Mars, crew members will be exposed continually to the radical changes of space and extraterrestrial environments. On the voyage between Mars and Earth, and on Mars' surface, many factors will impact and threaten the body's homeostasis. Even the supportive habitats of the spacecraft and Martian dwelling represent unusual environments, constrained as they will be by the extreme amount of recycling that must take place within a very small habitable volume. These environmental challenges will induce compensatory physiological changes. Extreme changes in regulatory processes will result in functional impairment of crew. The reversibility of these responses is a critical concern in assuring the exploration team's safe return.

The biomedical support requirements of a Mars exploration mission are defined by **Environmental Factors** and **Countermeasures**. The first is protection from inhospitable elements, and the provision of a portable human-friendly habitat. The second involves augmenting and assisting the adaptive compensatory changes that are induced by microgravity and other elements of the space environment.

ENVIRONMENTAL FACTORS

Life Support Systems

A number of environmental issues can affect the health and safety of an individual during space exploration. A first order example is the internal environment of the spacecraft and space suits. These must be maintained within specific limits, and redundant systems for dependable provision of atmospheric conditions (gas, temperature, humidity), food and water are essential. If a failure of this system occurs, countermeasures will be insufficient and medical care inadequate to prevent the catastrophic loss of the crew. Other significant factors related to the spacecraft and surface habitat are air quality (e.g. particles, toxins) and quantity (fresh/stale exchange rates), light (intensity, quality, duration), aspects of confinement (restricted volume, privacy, forced interaction with other crew members), restricted access to communication or other links with Earth, diversity of esthetic and recreational activities, and stress. Responses to non-gravitational environmental factors could exacerbate the effects of microgravity. For example, cumulative sleep loss or sleep disturbances could make immunosuppression and orthostatic intolerance worse. Current stress theory holds that coping mechanisms are most vulnerable when the external environment is unpredictable and uncontrollable. Space explorers encounter unpredictability as the rule, rather than as the exception. The most reliably stress-provoking situation is one involving actual or perceived inescapability. One can hardly conceive of a condition that fulfills these criteria better than a mission to Mars.

Many of the requirements for a sustainable human-supporting ecosystem are understood from our experience on Earth. The two dominating elements of the space environment are radiation and microgravity. Unlike the example of the air conditioner in heat adjustment, technology is currently unable to affect gravity or radiation. Thus, there are two important distinctions between the air conditioner example and the approaches that can be employed in supporting space explorers. First, while the air conditioner uses physico-chemical principles to modify its surroundings, space life support systems must use knowledge of human physiology to protect and support human adaptive responses. This is an important difference. We are unable to directly influence the gravity and radiation factors of space. Only by understanding the normal workings and homeostatic mechanisms of the human body can we predict and prepare for the changes encountered in space, and on Mars. Some amelioration will be achieved with hardware solutions based on biomedical research. Secondly, failure of the air conditioner will result in discomfort. Failure of life support systems will be fatal.

Radiation Hazards

The radiation hazard posed by solar energetic particles (SEP) and the protons and highly charged energetic particles (HZE) of the galactic cosmic rays (GCR) is a primary concern in evaluating the appropriateness of environmental life support and biomedical countermeasures for Mars exploration teams. The thin atmosphere of Mars offers little protection from either source, and Mars, like Earth's Moon, has no magnetic field to trap charged particles. The approach to SEP radiation is focused on early warning and the provision of "safe havens." Continuous improvement in solar flare monitoring and prediction will contribute to the success of Mars missions. SEP events are episodic and of limited duration, and small shelters designed for temporary occupation can satisfy shielding requirements. However, SEP radiation presents an intense, acute hazard due to the intrinsic high energy spectra. Regolith and terrain features can be used for habitat shielding while exploration teams are stationed on the planetary surface. The human crew (and any experimental or life-support biological organisms) must also be protected during transit between Earth and Mars, both from the external radiation danger and from secondary radiation generated inside the spacecraft.

While SEP bursts occur and dissipate in hours or days, GCR's are present continuously. Shielding from high energy transfer, heavy ion GCR's is described in the chapter by Townsend and Wilson. The many uncertainties associated with GCR quantitation introduce variables and errors in our risk assessments. Lower energy GCR's can be modulated by solar activity, while the measurement precision of high energy particle numbers may be only ± 25% [Space Radiation Health Program Plan, 1991]. Estimates of radiation transport efficiencies through vehicle/habitat materials depend on the nuclear physics of HZE particles and their cross-sectional interactions with various building constituents. These relationships are under study, but conservative estimates should assume errors on the order of 30% [Space Radiation Health Program Plan, 1991]. When interacting measurements are combined, their inherent uncertainties combine. As is often the case, the reliability of the product is most affected by the least precise observation. The accuracy of risk prediction for exploration missions suffers from the combination of cumulative errors in individual risk estimates.

At the biological level, the highly charged, energetic particles (of high atomic numbers) of the GCR have a high rate of energy deposition as they traverse and are absorbed by tissue. Energy deposition rates are most reproducibly ascertained in cell or tissue equivalents—materials approximating the chemical composition, size, and density of a cell or tissue. However, these measurements fail to predict the complexity which characterizes interactions within living systems. Absorbed radiations, as they cross the cell, will have the potential to disturb ongoing enzymatic processes and to leave "memories" of their passage to perturb later intracellular events. In living organisms, physiologic functions are finely balanced. Alterations in one venue will inevitably be felt in many others.

While the possibility exists for acute exposures to life-threatening radiation, particularly from unshielded SEP radiation, the effects of long-term moderate to low-level HZE exposures are the greatest mission-related radiation issue. Some of the effects of GCR radiation are expected from our knowledge of radiobiology, while others may be less predictable. In particular, a significant incidence of cataract lesions is predicted [Worgul et al., 1987; Worgul, 1988; Worgul et al., 1989]. Mutations will occur, and some will have the capacity to progress, in time, to frank tumors. Other mutations may be physiologically "silent," may be removed by the body's immune surveillance systems, or may be localized to the germ cells and revealed only in subsequent generations. Mutations are discussed in more detail below. The less expected consequences of GCR exposure in space are due to the combination of the radiation environment with the cellular and physiologic alterations subsequent to extended microgravity interludes. These may include compromised wound healing and depressed immunologic competence [Cox and Lett, 1989]. Proliferation and differentiation of cells to restore the integrity of damaged tissues from the microvasculature to larger organs requires a tightly linked series of transitions. To the extent that gravity plays a role in defining the architecture of intracellular structures and the nature of extracellular interactions, these activities are at risk in microgravity. Incomplete or inappropriate healing reveals tissue components that are not normally exposed to the cell-associated and circulating immune systems. Auto-immune diseases are characterized by the generation of activated immune cells that are stimulated by normal tissue constituents perceived in abnormal settings. In sum, both the known and unknown biomedical risks related to radiation exposures in space are critical issues in the provision of environmental support facilities for the mission to Mars.

Mutations and/or other insults to the genetic material are among the most critical non-acute biomedical effects of radiation. The genetic information is stored in molecules of DNA, long polymers of deoxyribonucleotides. Joined by phosphodiester bridges branching from deoxyribose (a sugar), the substituents of DNA are differentiated by the identity of their associated "bases"—adenine, guanine (A, G; the purines), cytosine, and thymidine (C, T; the pyrimidines). The entire hereditary information content is contained in the precise sequence of A, G, C, and T found in the polymer. The contiguous sequence of these bases also exactly specifies the sequence identity of a complementary strand of DNA, and the two strands form a stable helical duplex. At each DNA replication event, the helix is unwound in portions, and once again the sequence of each strand exactly determines that of the newly synthesized DNA complement. The precision and integrity of the DNA sequence is the key to life. Some radiation effects may insult the DNA indirectly, impacting the efficiency and/or accuracy of DNA repair enzymes or the production of DNA precursors. Direct DNA damage caused by radiation may be: (1) survivable (little or modest damage), (2) reparable or bypassed (either process may be accompanied by mutation), or (3) lethal.

Mutation events (substitution of any "correct" base with any other base), have several possible outcomes. Cancers are the result of carcinogenic mutations, while benign tumors can follow non-malignant mutations. Some mutations may be silent (causing no observable effects) if the protein described by the mutated DNA remains functional. Mutations of the germ cells (those providing the genetic information for the next generation) can result in sterility, birth defects, or increased cancer incidence/cancer risk through subsequent generations. Determination of the complete sequence of human DNA, and that of several other reference species, has been undertaken as part of the International Genome Project. Known sequences from many organisms, from bacteria to man, are "banked" in computer libraries. Any newly determined sequence can be checked against its normal counterpart for variation. When differences are found, they must be carefully evaluated to sort the normal variables of polymorphism and genetic drift (the expected differences between individuals) from specific induced mutations. Mutations in several genes are known to have far-reaching impacts on cancer risk. These include single point mutations (changes of a single base in the sequence) activating silent "onco"genes (onco = cancer), point mutations which release genes encoding growth-regulating proteins from normal controls, and point or deletion (removal of a stretch of DNA sequence) mutations that alter the protein structure so that important conformational interactions fail to occur (such as repression of gene sequences by DNA-binding proteins). Tumor suppressor proteins are critical in modulating growth related intracellular activities. Mutations inactivating these genes, or deletion events eliminating them, greatly increase the risk of abnormal, unregulated growth (cancer) in affected cells.

Radiation science has defined DNA damage resulting from radiation exposures as a significant and critical cause of biological aftereffects. Thus, space radiation biology will naturally entail the study of DNA mutations and mutation rates. Experimentation in the space environment, however, imposes some important limits on the number and nature of test samples/subjects. Timely access (or, indeed, even the possibility of access) for analytic observation is often problematic. The orbits needed to access appropriate GCR environments may necessitate long waiting times for sample recovery, or non-recoverability. Cell proliferation and associated DNA replication will rapidly alter the ratio of mutated/unaffected cells in any population, as selective pressure for or against a mutation is exerted. Some consequences of low-dose or intermittent radiation exposures appear only in longer time frames. These late changes make measurements of mutation frequencies particularly difficult. Some of these challenges may be met using recombinant DNA methodologies. Tagging, tracing, and directed search for specific DNA sequences can begin to expose many of these questions to productive study.

To quantify risk, whether the variable is mutagenic risk or life-span cancer risk, relationships between the radiation dose and its relative biological effectiveness (RBE) must be determined. Terrestrial radiation biology has created sophisticated models of the relative risks associated with ionizing radiations. The unique domain of space radiation biology has not yet been able to access sufficient data to produce equivalent risk projections. Even in the case of ionizing radiation, extrapolation from the mutation frequencies of laboratory cultured cell populations can only predict trends in the observed responses of whole organisms. Further, the laboratory animals most conveniently studied will have only limited predictive value for human risk assessments. Due in part to our very long life-span, and in part to our intricately inter-related physiological pathways, human risk is not rigorously predicted by experimental measurements carried out in animals. In anticipation of Mars exploration missions, optimal evaluation of the risks associated with radiation exposures requires experiments of two types:

1. Animal experiments and animal-derived cultured cell experiments, designed to interface with the existing database on radiation effects of ionizing radiation;

2. Protocols using human cultured cells (including mixed populations and small organ cultures) to more closely approximate the micro-environment relevant to human response.

The radiation environment of space provides a unique window through which we can extend our perception of the interaction between radiation and living systems. Study of GCR and SEP will further our understanding of the range of effects associated with radiations of varying energy, quality, and RBE. At the same time, the practical challenges of supporting human exploration of Mars absolutely requires this information.

Gravity and Microgravity

Alterations in gravity experienced during the mission to Mars will be a primary influence on crew health. Gravity is a constant force to which all living systems on Earth are exposed, and have been continuously during their evolution. At the organismal level, this force influences a variety of factors, such as morphology and movement. This recapitulates the effects of gravity on intracellular structures and intercellular architectures. Exploration missions will expose the crew to altered gravities. The trip profile will include a transient increase at launch, followed by the weightlessness (microgravity) of flight during the voyage, additional perceived g-forces during deceleration and landing, and exposure to 0.387 g on the surface. The entire sequence will be reversed on the return trip, culminating in exposure to 1 g again on Earth.

We know that exposure to a weightless environment on the journey to and from Mars, and to reduced gravity on the Martian surface, will produce changes in human physiology. The alteration of homeostasis will affect health and performance. Transit between Mars and Earth will require longer exposures to microgravity than have been evaluated to date. In fact, very little data is available to directly assess the effects of extended spaceflight. A few Soviet cosmonauts, on the Mir space station, have remained in microgravity for 6, 9 or 12 months [Gazenko et al., 1988; Vorobyov et al., 1983]. In predicting the risks associated with a Mars exploration mission, we must extrapolate from our observations of shorter flights. Even so, because each mission is different, not just in length but also in profile, much of the flight data consists of observations of each astronaut/cosmonaut singly. Compiling data from a number of crew members is complicated by natural individual variations. Nevertheless, a pattern of changes has emerged. NASA's most useful database remains that prepared from study of the Skylab mission cohort completed almost 20 years ago [Johnson and Dietlein, 1977]. Thus, while flight research results provide the most unarguable information, they are limited in scope and sample size.

Current knowledge about the physiology of gravity effects is derived from both flight and ground experiments. It is the ground research which provides the most accessible testbed for hypotheses. The two facets of space biology experimentation, ground-based and flight research, are interdependent. Ground-based studies have greater opportunity for evaluating many variables cheaply and repeatably. But the key variable in space biology is gravity, relentless and unvarying on the surface of the Earth. Ingenious tilt bed models have been developed that mimic many aspects of the physiological changes predicted to occur in fractional gravity fields. This ground research, and our experiences on the Moon, are the closest we will come to Mars gravity (0.387 g) effects until we land on Martian soil. Questions surrounding the effects of transitions from one gravity field to another remain unresolved. Past astronauts and cosmonauts have recovered from microgravity alterations following their return. They have, however, required assistance and rehabilitation therapy despite in-flight countermeasures and pre-flight conditioning. While Mars' gravity will be less demanding in terms of force than Earth's, it's harsh and relatively unpredictable environment will be a demanding challenge. On Earth, returning explorers have arrived into the care of medical support professionals. On Mars, we do not expect equivalent personnel to be provided.

The hallmarks of the transition from the 1 g environment to that of microgravity are the physiological responses to four major factors:

1. Hydrostatic pressure changes in the vascular and non-vascular body fluids,

2. The removal of stress and strain on body support structures,

3. The absence of signal to gravity receptors of the nervous system, and

4. Changes in the immune systems, as cellular physiology changes in response to microgravity exposure.

Some short-term "adaptive" changes; post flight orthostatic intolerance, musculoskeletal weakness, and early space-sickness; have been described extensively over the years [Nicogossian and Parker, 1982]. Much less is known about the mechanisms of long-term adaptation to microgravity. These are discussed in the following sections.

The Cardiovascular System. Reduced gravity results in deconditioning of the cardiovascular system in humans. This is initiated by a redistribution of fluids in the body. On Earth, in an upright posture, fluid pools in the lower extremities (legs and feet). This fluid moves into the thorax (chest) and head in microgravity, regardless of posture. The volume moved may be as much as 1 liter [Greenleaf, 1989]. Transfer of the bulk of this fluid volume to a central body location (chest instead of legs) activates compensatory cardiovascular, neural and endocrine reflexes. In this adaptive response, overall fluid volume is decreased by increasing the excretion of water and salts by the kidneys. Within twenty-four hours, a new equilibrium is attained, resulting in a blood volume reduction of as much as ten percent.

Over time, reduced blood volume and the lack of cyclic variations due to changes in posture cause a decreased sensitivity to blood pressure differentials in the cardiovascular system [Convertino et al., 1989]. For example, for a fixed decrease in blood pressure at 1 g, the heart rate will increase a defined amount to facilitate the delivery of blood to the tissues. After a period of weightlessness, the heart rate response to the same decrease in blood pressure is muted. This is an example of decompensation of the cardiovascular system, and can lead to a number of problems—including fainting and unconsciousness upon return to a nominal (1 g) gravity environment. The absence of variations in blood flow and the decrease in blood volume may also initiate or exacerbate changes in cardiac muscle morphology. Rats flown for fourteen days in space exhibited a reduction in the cross sectional area of ventricle cardiac fibers of twenty-two percent [Goldstein et al., 1992]. The cross-sectional area measurement relates to the strength of a muscle, thus the hearts of the animals flown in space were weakened. Such changes may impair the heart's ability to increase blood flow in the presence of an increased demand, leading to the decompensation described above.

Orthostatic intolerance is an example of cardiovascular decompensation. Under normal conditions most individuals experience little or no difficulty in moving from a seated to a standing position. However after a period of weightlessness, fainting may occur. Ground-based studies of bed rest subjects have demonstrated that this response can be isolated from the changes in blood volume mentioned above. As a result of this reaction, cosmonauts are moved very slowly from the seated/lying position to standing following very long duration spaceflights. Fainting may be the result of blood pooling in the lower limbs, with inadequate sensing of the subsequent decrease in blood pressure due to flight decompensation. When a decrease in blood pressure is not appropriately recognized within the body, normal compensatory mechanisms (increase in heart rate) are not initiated. Blood flow to the brain is reduced, resulting in loss of consciousness.

Alterations in the cardiovascular system will greatly impact acute medical care. For instance, injuries associated with hemorrhage (bleeding) will be of greater consequence because of reduced initial blood volume. Without rapid intervention, death may ensue. A decrease in blood volume and cardiac function will also impair physical performance. Decrease in blood volume has been associated with reduction in maximal physical work performance and intolerance to increased applied g-forces [Greenleaf, 1990]. Loss of work capacity may hinder the construction of habitats or the duration of surface exploration forays. There are additional unknowns. For example, decrease in g-tolerance could affect the margins of safety incorporated into acceleration/deceleration profiles. Overall, alterations in the cardiovascular system will have diverse impacts on the design of a mission to Mars.

The Musculo-Skeletal System. Gravity is the single greatest influence on the shape, size and strength of bones. The skeletal system has evolved in response to the load placed on the body by gravity, since we moved from aquatic to terrestrial habitats. Biomechanical forces from the contraction of attached muscles apply flexural stress and strain to the bones, in addition to the compressional forces experienced in resisting the pull of gravity. Since the activity and strength of muscles is reduced in a weightless environment, there is a decrease in the stress/strain load imparted to the bone. This is combined with an absolute absence of the resistive load experienced on Earth due to gravity.

Bone tissue is in a constant state of change, continuously being renewed and replaced. During spaceflight this process appears to be perturbed. In flight, there is a net loss of bone minerals from the weight-bearing bones, thus decreasing bone density. This finding correlates directly with lessened bone strength and stiffness in flight animals [Morey-Holton and Arnaud, 1991]. On missions of up to one hundred and seventy days, the density of selected bones was decreased five to nine percent. This loss appears to be progressive, as the percent decrease is a function of flight duration [Morey-Holton and Arnaud, 1991]. Previous work on bone density has focused on the weight bearing long bones of the leg. Recent work in humans, repeated and confirmed in animals, suggests redistribution of bone minerals at the whole body level [Arnaud *et al.*, 1989; Roer and Dillaman, 1990]. Regional bone densitometry demonstrates an increase in the density of the skull and a decrease in the density of the bones of the foot. Furthermore, flight animals show a redistribution of minerals along individual bones [Mechanic *et al.*, 1990]. The distal ends (foot-ward) of the long bones show a decrease in mineral density, while the proximal ends (head-ward) show an increase. Localized bone mineral losses increase the risk of fracture. Breaks, particularly those that involve the lower extremities, are a hazard during landing and surface exploration. The healing of bone fractures appears to be delayed in a weightless environment. In addition, there is no data describing the restoration of bone integrity in altered gravity. Thus the increased potential for bone injury and compromised recovery are cause for concern on a Mars mission.

Alterations in calcium metabolism, the primary mineral of bone, are also of concern. In microgravity, there is a net loss of calcium in some individuals through increases in fecal and urinary excretion of this mineral. Bone is the body's primary reservoir of calcium, and the loss appears to be from this source, as discussed above. The increased urinary excretion of calcium may contribute to the formation of kidney stones—a painful and debilitating problem. The Russians have reported an incident of kidney stones during spaceflight. It remains unclear whether an underlying condition contributed to this isolated occurrence. Nevertheless, the possibility of medical problems associated with decreases in total body calcium in microgravity exists.

During daily activity, the musculature is exercised and conditioned by moving the mass of the body. In microgravity, this workload is greatly reduced, leading to "muscle disuse atrophy"—a lessening of size, weight, and strength. The atrophy of the muscles is not uniform. The muscle groups which are normally associated with the maintenance of posture, those which are constantly resisting gravity, show the greatest degree of atrophy. The decrease in muscle strength and mass is accompanied by intracellular biochemical alterations, by morphological changes in the neuromuscular junction, and by reduced responsiveness of the muscle to electrical, nervous, or hormonal stimulation.

During Skylab missions of up to eighty-four days, leg muscle mass was decreased about ten percent, a change accompanied by a reduction in leg strength on the order of fifteen to twenty percent [Thornton and Rummel, 1977]. Arm strength was also decreased, although to a lesser extent (five to ten percent). The greater decrease in the strength of the legs reflects their role in a 1 g environment, in the movement of body mass and the support of upright posture. In addition to a general decrease in muscle function, there is also a shift in muscle type from fast twitch muscle to the slow twitch variety. The fast and slow phenotypes are defined by the intracellular expression of particular contractile proteins by muscle cells. In development, fetal mammals possess exclusively the slow twitch muscle type. Fast twitch muscle gradually increases in proportion during maturation. Spaceflight, then, recapitulates early development at the cellular level. Is this a result of the similarities between weightlessness and the aquatic environment of early development (both evolutionary development and embryonic/fetal development)? Space life science research may address this suggestive question.

Muscle function decrease may be due in part to alterations in the relationship between the neural inputs and the muscle architecture. Studies in animals have shown changes in the motor endplates of muscle tissues in response to alterations in gravity exposure. These changes were accompanied by a lack of appropriate coordination of gross muscle movement and by muscle atrophy [D'Amelio and Daunton, 1992]. Alterations in fine muscle control, in contrast to changes in the large muscles associated with posture, are probably due to attenuation of neural inputs. Decreases in muscle strength and control could impact both biomedical and safety issues during a mission to Mars. The strength reduction would decrease an individual's ability to react to an emergency, to perform construction tasks or to effectively ambulate in the 0.387 g environment. Loss of fine motor control could hamper simple tasks such as matching a nut and a bolt or operating a joy stick to pilot a vehicle.

Nervous System. Both acute and chronic effects of gravity alterations are evident in the nervous system. Proprioceptive inputs change acutely, as demonstrated by the tendency to overshoot the intended movement of an unweighted limb. This is corrected with relative rapidity through perceptive adaptation, as visual and tactile sensory inputs are re-aligned. As mentioned above, the Mars expedition team will transit between various gravity fields. Temporal duration of motor and proprioceptive adaptation could significantly affect initial performance.

The sense of body orientation is altered as central nervous system inputs from otolith organs, which perceive relative head position, differ in novel gravity environments. Animal experiments show that the morphology of these organs is changed after only four to nine days exposure to altered gravity [Daunton et al., 1991]. The numbers of information carrying synaptic connections are increased in flight animals [Ross, personal communication] and decreased following exposure to 2 g on a centrifuge. The mechanism(s) which directs these changes is not known. Motor coordination, righting reflexes and stability during movement may all contribute to adaptation when experiencing novel gravity fields [Daunton et al., 1991].

Changes in sensory and motor inputs may result in the "motion" sickness of spaceflight. This is often referred to as space-sickness. This syndrome has affected roughly half of the individuals who have flown, generally resolving within three to five days [Reschke, 1990]. While the impact of this response may be minimal on health and safety in the initial days of the journey to Mars, it is not known whether a similar response will be noted upon landing and on the return flight. The nervous system is responsible for the integration of all physiological inputs. Microgravity-related changes in this arena may thus include alterations in psychological well-being, circadian (day/night rhythms), sleep quality, and other sensory perceptions. Decremental changes will adversely impact mission productivity and the safety of the crew.

The Immune System. During the Apollo period, astronauts experienced a significant incidence of bacterial and viral infections during flight until the "Flight Crew Health Stabilization Program," consisting of physical examinations and pre-flight quarantine, was initiated [Hawkins and Zeigelschmid, 1975]. In-flight head colds and gastrointestinal symptoms stimulated interest in the response of the immune system to spaceflight. Experimental results showed that immune cell responses to a variety of stimuli are decreased by over thirty-five percent post-flight [Taylor and Janney, 1992]. The decrease in cell-mediated immunity is accompanied by a decrease in the total number of lymphocytes (white blood cells) and atrophy of lymph organs [Sonnenfeld and Taylor, 1991]. The production of interleukins and interferon, essential to the body's defense against infection, also decreases following microgravity exposure. Interleukins and interferon are important modulators of interactions between lymphocyte subpopulations, as well as regulators of T-lymphocyte production.

Numerous studies have demonstrated immune suppression following spaceflight. In-flight analytic procedures are limited, however, and post-flight analyses do not exclude the possibility that the observed immune suppression is caused not by microgravity, but by the stress of reentry and return to 1 g. In support of the thesis that microgravity is the cause of spaceflight immunosuppression is the observation that the Apollo 12 commander experienced contact dermatitis from an electrolyte paste used in a biosensor apparatus only in-flight [Hawkins and Zeigelschmid, 1975]. Post-flight analysis failed to identify any constituent particular to the batch used on the mission, when compared to those used in pre-flight simulations or those produced in parallel with the flight batch. Contact dermatitis and respiratory allergies to substances encountered on the surface of Mars could pose new problems in the context of systemic immunosuppression. There are a number of factors which could contribute to the suppression of the immune system. Animal and cell culture studies suggest that microgravity can be perceived at the cellular and subcellular levels, quite independent of the physiological and psychological stresses of spaceflight. Interference with the normally regulated functioning of the interactive cellular populations of the immune system is a likely cause of the observed suppression. At present, the influence of radiation exposure and other factors of spaceflight on this system have not been defined.

Prior to a long duration mission, exhaustive procedures will be used to sanitize the spacecraft. However, a variety of organisms will inevitably gain entrance, either as independent travelers or as passengers brought aboard by the crew. While the recycled and reconditioned air and water will be filtered and scrubbed, small closed-loop life support systems are finely balanced within narrow tolerances. Contaminations can escalate rapidly. Normal bacterial flora from one crew member, recycled repeatedly and continuously exposed to radiation, may become pathogenic for any crew member. In combination with this expected increase in infectious challenge, the decreased resistance of immune suppression will be a serious concern. On a long duration mission to Mars, even mildly debilitating conditions (upper respiratory tract infections) are undesirable. Much less manageable are acute outbreaks of communicable diseases with significant morbidities. Dysentery, for example, could have a major negative impact on the success of such a mission. The role of the immune system in wound healing must also be considered. Any open wound, from a scrape to a deep puncture, perforates the body's first line of defense, the skin, with its associated resident immune cells. Preliminary animal studies indicate that healing does not follow a normal pattern in microgravity. Taken together, the accumulated evidence suggests that the four factors discussed above are likely to have serious impacts on the Mars mission. To recapitulate, they are spread of infection, resistance to infection, immunosuppression, and compromised wound healing. Confinement alone seems to result in reduced immune system function, the combined effects of confinement, microgravity and radiation exposure are unknown.

COUNTERMEASURES TO PHYSIOLOGICAL ALTERATIONS DUE TO GRAVITY

A large number of countermeasure procedures have been proposed, developed and tested in both the U.S. space program and those of other space-faring countries. Evaluation of their effectiveness has been confounded by the same factors that make assessment of the effects of microgravity on man difficult: normal variability between individuals, too few systematic observations using identical protocols, variability in mission length and profile, and interaction between individual countermeasures. Countermeasures for the effects of microgravity fall into three main types:

1. Those that prevent adaptive responses to microgravity and therefore must be prescribed throughout a mission.

2. Those that restore or correct a deficit that is only of concern during a transition to a new g field (e.g. return to Earth, landing on Mars) or activity (EVA) and would therefore be administered just before such a transition.

3. Those that depend on pre-flight selection to identify (on a physiological basis) candidates who would be most adaptable to the space environment (e.g. non-fainters, slow bone turnover) or most at risk.

This last approach has not been used extensively (except with respect to motion sickness) but should not be arbitrarily dismissed in planning a mission to Mars.

The menu of current countermeasure approaches has been extensively reviewed. It includes:

- Exercise (of every conceivable variety)
- Mechanical (lower body negative pressure, gravity suit)
- Dietary (fluid loading, mineral supplements)
- Pharmacological (drugs for space sickness, orthostatic intolerance)
- Psychological (biofeedback)
- Special training (preflight-adaptation trainer)
- Artificial g (intermittent or continuously applied).

As a last resort, the concept of designing systems that provide 1 g artificial gravity to crews has been discussed since the inception of the space program. The technical feasibility was tested briefly during Gemini IV. Artificial gravity has been adopted by the science fiction genre, but the diverse capabilities of such energy sources as the dilithium crystals used on the U.S.S. *Enterprise* are as yet unavailable to NASA. In practical terms, the provision of continuous 1 g using tethering or similar engineering concepts is difficult and costly. Deconditioning would inevitably result from the necessary rotation stoppages necessary for Mars approach and landing. This might affect performance in Mars' 0.387 g field in much the same way as residence in a vehicle kept continuously at zero g. Tolerance to the higher g forces of landing could also be affected.

Is it necessary to provide living systems with continuous 1 g to maintain normal health? The answer is unknown, and can only be tested in space. Only in that environment are study subjects removed from Earth's G vector. What is the minimum daily requirement (MDR) for exposure to gravity? We, and other mammals, spend considerable amounts of time in sleep and inactivity. Physiologists analyzing the effects of weightlessness routinely use bedrest, horizontal or with head down tilt, to simulate microgravity effects. The value of periodic exposure to 1 g (or greater), with and without activity, needs to be assessed. If MDR's can be developed that are acceptable to the crew and to mission health specialists, then the more complicated and costly approaches of artificial gravity (continuous or intermittent) can be de-emphasized. Periodic g exposure studies can be conducted using ground based microgravity simulation. Fractional g studies can also be conducted on the ground—using bedrest on tilt tables. The adequacy of gravity experienced on the Martian surface, with and without additional countermeasures, can be evaluated. To mirror the exploration mission, fractional gravity should be studied following an initial period of "weightless" deconditioning.

CONCLUDING REMARKS

In the final analysis, the Mars exploration mission is an adventuresome expedition, and a risky one. Some degree of health deficit will be unavoidable and should be acceptable. Attention to biomedical issues will not prevent the mission, it will facilitate the success of the mission. Gravitational Biology and Space Radiation Biology are scientific disciplines with their futures in the stars and their roots in closely allied Earth-based fields. While there are many gaps in our understanding of the adaptation of human physiology in response to spaceflight, the unknowns are a natural by-product of the youth of both areas. Particularly in the fundamentals of cellular perception of microgravity, and cellular reaction (both immediate and latent) to particulate GCR radiations, there are many questions that beg for our attention.

REFERENCES

Arnaud, S.B., M.R. Powell, R.T. Whalen, and J. Vernikos-Danellis, "Bone mineral redistribution during head down tilt bed rest," *Am. Soc. Gravit. Space Biol.*, Bull. 2, p. 54, 1989.

Convertino, V.A., D.R. Doerr, E.L. Dwain, J.M. Fritsch, and J. Vernikos-Danellis, "Head-down bed rest impairs vagal baroreflex responses and provokes orthostatic hypotension," *J. Appl. Physiol*, 68, pp. 1458-1464, 1989.

Cox, A.B., and J.T. Lett, "The quantification of wound healing as a method to assess late radiation damage in primate skin exposed to high-energy protons," *Adv. in Space Res.*, 9, pp. 125-130, 1989.

D'Amelio, F., and N.G. Daunton, "Effects of space flight in the adductor longus muscle of rats flown in the Soviet biosatellite COSMOS 2044: A study employing neural cell adhesion molecule (N-CAM) immunocytochemistry and conventional morphological techniques (light and electron microscopy)," *J. Neuropath. and Experimental Neurology*, 51, pp. 155-160, 1992.

Daunton, N.G., M.D. Ross, R.A. Fox, M.L. Corcoran, L.K. Cutler, and L.C. Wu, "Effects of chronic hyper-gravity on the righting reflex and vestibular end organ morphology in the rat," *Neurosci. Abst.*, 13, p. 316, 1991.

Gazenko, O.G., E.B. Schulzhenko, A.I. Grigoriev, O. Yu. Atkov, and A.D. Egarov, "Review of basic medical results of the Salyut-F, Soyuz-T 8 month manned flight," *Acta Astronatica*, 17, pp. 155-160, 1988.

Goldstein, M.A., J.P. Schroeter, R.J. Edwards, and I.A. Popova, "Cardiac morphology after conditions of microgravity," *J. Appl. Physiol.*, 73, pp. 94S-100S, 1992.

Greenleaf, J.E., "Hormonal regulation of fluid and electrolytes during prolonged bedrest: implications for microgravity," in *Hormonal regulation of fluid and electrolytes environmental effects*, edited by J. R. Claybaugh and C. E. Wade, pp. 215-232, Plenum Press, New York, 1989.

Greenleaf, J.E., "Perspectives in Exercise Science and Sports medicine," in *Fluid Hemostasis during exercise*, edited by C. V. Gisolfi and D. R. Lamp, pp. 309-346, Benchmark Press, Inc., Carmel, 1990.

Hawkins, W.R., and J.F. Zeigelschmid, "Clinical aspects of crew health," in *Biomedical results of Apollo*, edited by R. S. Johnston, L. F. Dietlein and C. A. Berry, pp. 43-81, US Government Printing Office, Washington, DC, 1975.

Johnson, R.S., and L.F. Dietlein (Eds.), *Biomedical results from Skylab*, 491pp., Scientific and Technical Information Office, NASA, Washington, D.C., 1977.

Mechanic, G.L., S.B. Arnaud, A. Boyce, T.G. Bormage, P. Buckendahl, J.C. Elliott, E.P. Katz, and G.N. Durnova, "Regional distribution of mineral and matrix in the femurs of rats flown on Cosmos 1887 biosatellite," *FASEB Journal*, 4, pp. 34-40, 1990.

Morey-Holton, E., and S.B. Arnaud, "Skeletal responses to spaceflight," in *Advances in Space biology and Medicine*, edited by S.L. Bonting, pp. 37-69, JAI Press, Inc., Columbia, 1991.

Nicogossian, A.E., and J.F. Parker (Eds.), *Space Physiology and Medicine*, US Government Printing Office, Washington, D.C., 1982.

Reschke, M.F., "Statistical prediction of space motion sickness," in *Space and Motion Sickness*, edited by G. H. Cramation, pp. 263-316, CRC Press, Boca Raton, FL, 1990.

Roer, R.D., and R.M. Dillaman, "Bone growth and calcium balance during simulated weightlessness in the rat," *J. Appl. Physiol*, 68, pp. 13-20, 1990.

Sonnenfeld, G., and G.R. Taylor, "Effects of microgravity on the immune system," *SAE Tech Paper Series*, 911515, pp. 1-6, 1991.

Taylor, G.R., and R.P. Janney, "In vivo testing confirms a blunting of the human cell mediated immune mechanism during space flight," *J. Leukocyte Biology*, 51, pp. 129-132, 1992.

Thornton, W.E., and J.A. Rummel, "Muscular deconditioning and its prevention," in *Biomedical results of Skylab*, edited by R. S. Johnston and L. F. Dietlein, pp. 191-197, US Government Printing Office, Washington, DC, 1977.

Vorobyov, E.I., O.G. Gazenko, A.M. Genin, and A.D. Egarov, "Medical results of Salyut-6 manned space flights," *Aviat. Space Environ. Med. Supp.*, 1, pp. S31-S40, 1983.

Worgul, B.V., "Accelerated heavy particles and the lens: V. Theoretical basis of cataract enhancement by dose fractionation," *Opthalmic Res.*, 20, pp. 143-148, 1988.

Worgul, B.V., C. Medvedovsky, P. Powers-Risius, and E.L. Alpen, "Cataractogenesis and cytopathological changes in mouse lenses exposed to 600 MeV/amu iron ions," *Invest. Opthamol. and Visual Science*, 28, p. 282, 1987.

Worgul, B.V., G.R. Merriam, C. Medvedovsky, and D.J. Brenner, "Accelerated heavy particles and the lens. III. Cataract enhancement by dose fractionation," *Radiation Res.*, 118, pp. 93-110, 1989.

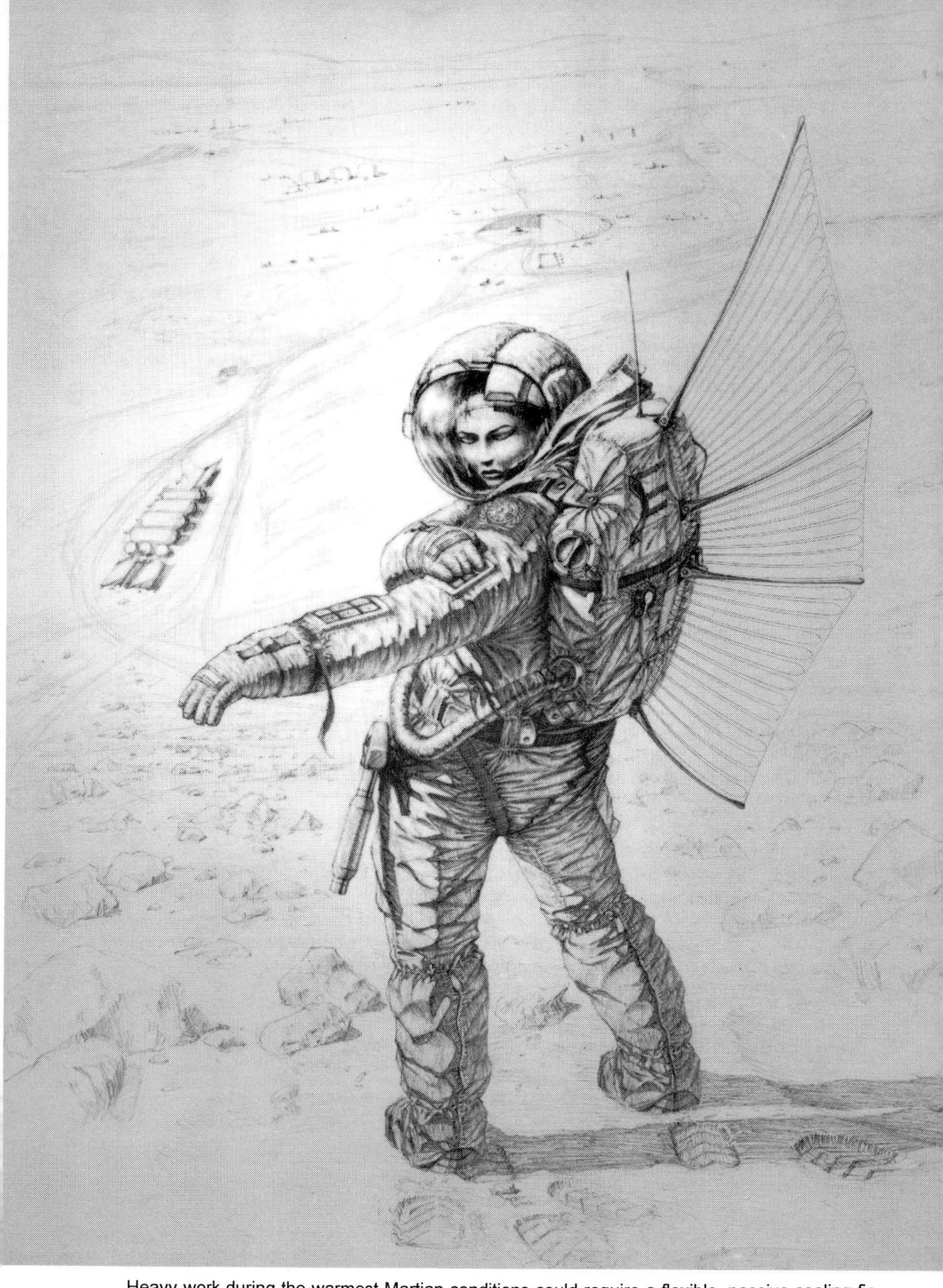
Heavy work during the warmest Martian conditions could require a flexible, passive cooling fin.

Chapter 14

THE HUMAN SIDE OF MARSFLIGHT: A REVIEW OF HUMAN FACTORS ISSUES[*]

Mary M. Connors[†] and Albert A. Harrison[‡]

This chapter is concerned with the psychological and social aspects of Marsflight: how the Mars astronauts are likely to function during flight, how they will relate to one another and to people on Earth, and how they are likely to be affected by their voyage. Through selection and training, spacecraft design, and the structuring of situations and tasks, mission managers can directly affect the crew's compatibility, motivation and interpersonal relations, and indirectly promote public support for subsequent space missions. The conduct of innovative and rigorous behavioral research over the next few decades will contribute to Mars mission safety and success.

INTRODUCTION

Adaptability is one of the most prominent of human qualities. Over the years, people have survived in both the Arctic and the Sahara, and *homo sapiens* has endured countless episodes of famine, disease, and war. In addition, when it comes to exploration we humans pride ourselves on being dauntless and intrepid, willing to endure whatever privation is demanded in order to break new ground, open new territories, or reach new heights. And certainly, the people who are likely to be selected for the first mission to Mars will be especially resourceful, resilient, and tough. So, the first question is: Do we really need to concern ourselves with the psychological, social and cultural aspects of spaceflight?

Our unhesitating answer is "yes." A good understanding of behavioral variables is desirable for any manned space mission, but is especially important for an extended mission such as a Marsflight. Although humans are quite adaptable, this adaptability varies from person to person and in all cases is subject to limits. If the demands of flight become so severe that one or more of the astronauts, or the team as a whole, fails to perform satisfactorily, the entire mission could be endangered. A careful analysis and understanding of the task demands and the psychological and social dimensions of

---

[*] The views expressed in this paper are those of the authors and do not necessarily reflect the views of the National Aeronautics and Space Administration or any other governmental agency.

[†] Aerospace Human Factors Research Division, NASA Ames Research Center, Moffett Field, California 94035.

[‡] Department of Psychology, University of California, Davis California 95616.

Marsflight can help prepare crewmembers for the work to be accomplished and keep flight stress within tolerable limits. Moreover, we have an obligation to space travelers, and to ourselves as sponsoring societies, to do as much as we can to provide them with an environment where they can be fully productive, where they can live happily, and where they can prosper and grow. In this chapter we will review major behavioral issues facing Mars crews; assess how, through a focused planning effort, we can begin to address these issues; and examine how the new alliance between Russian and American space programs can help accelerate progress in this area. Finally, we will offer an overall assessment of the role human behavioral research should play in preparing for exploration missions to Mars.

BEHAVIORAL ISSUES FOR MARS CREWS

An engineer working on a propulsion system for a Mars mission might begin by asking "What aggregate tonnage will have to be moved from where to where over what period of time in order to fulfill mission requirements?" A space life scientist might ask "What amounts of oxygen, water and food are required to support the first manned Mars mission?" The answers to these questions would constitute goals or criteria to be met. Similarly, it is useful for psychologists to ask, "What should we look for in terms of human qualities, organizational arrangements, work performance and behavior in order to ensure a successful mission to Mars?" Only if we have an understanding of the behavioral criteria to be met can we begin to look for methods to satisfy them.

Thus far, there has been only limited opportunity for psychologists and other human factors experts to study behavior in space. However, there have been a substantial number of careful studies of people in spaceflight-analogous environments. These latter environments share with spaceflight the environmental conditions of social isolation, physical confinement, hardship, danger, and other attributes [Harrison *et al.*, 1989, 1991; Stuster, 1986; Vinograd, 1974]. Space analogs include Arctic and Antarctic bases, nuclear submarines, off-shore drilling rigs, remote manufacturing and military sites, fallout shelters, supertankers, undersea research vessels, flight simulators, manned balloons, and experimental manned ecosystems such as Biosphere II. The body of literature on spaceflight-analogous environments is now sufficiently well-established that certain tentative conclusions can be drawn. Based on experience in isolated and confined settings [Gunderson, 1974; Kanas, 1990; Taylor, 1987] and our understanding of what it takes to succeed on a long-duration spaceflight [Connors *et al.*, 1986; Rose *et al.*, 1993] the behavioral criteria we propose for the Mars astronauts include sustained competent performance, emotional stability, constructive interpersonal relations, and positive cultural impact.

Sustained Competent Performance

Mars voyagers will have to be high performers in the sense that each member of the crew must be able and willing to carry out a wide range of duties. Even tasks that are not all that hard to perform under Earth conditions can be surprisingly difficult and time consuming during spaceflight [Harrison and Connors, 1990]. Crewmembers must be among the most competent individuals in their technical specialties, and must be able to perform well despite such hindrances as shifting reference points (if there is no artificial gravity and they "float" around in space) and cumbersome protective gear (in the case of extravehicular activities and surface exploration). Crewmembers must be able to resist the tendencies in confinement to drift into repetitive, routinized modes of thought; to become increasingly suggestible; and to become so engrossed in contemplation that external events fail to register [Barabasz, 1991]. Additionally, crewmembers must be able to perform despite dangerous or threatening conditions.

A continuing high level of motivation is another requirement for sustained effective performance. Since motivation declines during prolonged periods of isolation and confinement [Gunderson, 1974; Kanas, 1990; Taylor, 1987], a precondition for a successful manned Mars mission will be to create, as far as possible, environmental conditions that tend to promote and sustain motivation and involvement. A further requirement will be to identify "self starters" i.e., those individuals who are able to remain challenged and involved for very long periods of time.

Emotional Stability

Marsfarers should be stable in the sense that they think clearly and rationally; display emotional reactions that are appropriate and beneficial in the situation; generally feel good about themselves, their colleagues, and the mission; and refrain from acting in ways that their colleagues find troublesome. Upon return to Earth, the Marsfarers should be able to re-assimilate into their home community with a minimum of disruption. Finally, the overall impact of the experience on their future lives should be positive.

Although isolation and confinement do not always produce adverse psychological effects (indeed, in many cases consequences are innocuous or beneficial [Harrison & Summit, 1991; Suedfeld, 1991a, 1991b]), nonetheless, the prolonged and multiple stresses of Marsflight could pose a serious challenge to some people's emotional stability [Kanas, 1990; Kanas and Fedderson, 1971; Santy, 1983, 1987, 1990; Santy *et al.,* 1991]. Such stressors as being cut-off from family and friends, being able to interact with only a limited number of individuals, cramped and crowded living conditions, a lack of everyday comforts and amenities, unrelenting pressure to perform, extreme danger, and the threat of the unknown could overwhelm individual or group resources. How likely is this to happen? Although results are far from consistent, several researchers of spaceflight-analogous conditions have found that prolonged isolation and confinement leads to psychosomatic complaints, sleep disturbances, and emotional changes that often take the form of mild depression [Gunderson, 1974; Kanas, 1990; Levine, 1991; Palinkas, 1990; Taylor, 1987]. Reports of serious emotional problems under conditions of isolation and confinement are rare, but this may reflect a reluctance to report them rather than a relatively low incidence [Lugg, 1991]. Preparation for extended missions requires: (1) techniques for supporting healthy individuals in abnormally stressful environments; and (2) individualized plans intended to improve the psychological well-being of specific crewmembers and their families [Santy, 1990].

Constructive Interpersonal Relations

Negative interpersonal attitudes and social tensions are commonly-reported consequences of isolation and confinement [Kanas, 1990; Palinkas, 1990; Penwell, 1990]. Social tensions within an isolated and confined group are more likely to result in social withdrawal than in open aggression, since there appears to be tacit recognition that open conflict is unacceptable under these conditions. However, irritability and anger are often directed (or perhaps, redirected) towards outsiders.

Signs of interpersonal conflict are not necessarily all bad. For example, theorists from a Freudian or psychoanalytic tradition might argue that social conflict has a cathartic effect: that is, occasional displays of irritability and hostility may prevent bigger social problems from developing in the same way that releasing small amounts of steam from a boiler can prevent a major "blow up" later on. Theorists from other traditions might come to a similar conclusion based on the belief that social conflict is a necessary precondition for social change, and that social change is necessary for group survival in a changing or unpredictable environment. Thus, somebody may need to "rock the boat" before a group will look critically at its decisions and consider new possibilities. Our requirement for Mars astronauts is that they create, not a tension-free minisociety, but rather one in which neither internal conflicts nor disputes with outsiders take on the characteristics of a "grudge match" or another destructive form.

Positive Cultural Impact

Mars astronauts will be expected to act in ways that are considered acceptable and appropriate by the public [Harrison, 1986]. This does not necessarily mean that the Mars astronauts must be national exemplars, as were the astronauts and cosmonauts of the 1960s in their day. It does mean that their behavior should not be inconsistent with the folkways, mores, customs, and norms of the sponsoring society. Astronauts' behavior influences public confidence in and support for the manned U.S. space program. The image projected by astronauts and cosmonauts has been an implicit concern of the American and Soviet space programs since the earliest days; this is unlikely to change in the future.

PLANNING GUIDELINES

What can we do to promote the kinds of behaviors required for a truly successful manned Mars mission—high competence, sustained motivation, constructive interpersonal relationships, and a good image? We can *select* people who are already likely to behave in ways that are desirable under the conditions of interest, and/or we can *train* people by putting them through programs that help them develop the needed interests and proficiencies. Still another option, *environmental design*, involves creating physical and social environments supportive of mission goals. Finally, we can *structure* activities and tasks in such a way that behavior is channeled along desirable routes.

Selection

The goal of selection is to identify people who are highly likely to meet mission requirements. There are many ideas about the kinds of people expected to do well under exploration flight conditions, but many of these ideas reflect hunch and intuition rather than scientific fact [Palinkas, 1990]. Here, we will consider two general questions. First, what kinds of skills and personality characteristics make some people more attractive candidates for Marsflight than others? Second, how are different crewmembers' characteristics likely to interplay to determine the performance and morale of the overall crew?

Individual Characteristics. Crewmembers must possess two broad sets of qualifications or skills: *technical skills*, which pertain to scientific and technical tasks, and *interpersonal skills*, which pertain to maintaining positive or constructive human relations within the crew [Helmreich *et al.*, 1980; Jones and Annes, 1983; Rose *et al.*, 1993]. While we know more about the technical than the interpersonal requirements of Marsflight, past studies of people in spaceflight-analogous and laboratory environments do provide some helpful leads. First, crewmembers will need to get along well with members of both sexes and with representatives of different cultures [Penwell, 1990]. Prejudicial or stereotypical views will be unacceptable. Additionally, crewmembers will have to be tolerant of culturally-based and other preferences, aversions, and differences in lifestyles.

Research in Antarctica and undersea research environments suggests that experience and breadth of perspective confer an advantage, perhaps giving an edge to candidates who are in their thirties or beyond rather than to those who are in their twenties [Radloff and Helmreich, 1968; Smith and Haythorn, 1970]. Attractiveness, in terms of appearance, cleanliness, and general demeanor is also an important consideration [Rawls *et al.*, 1969]. In general, the greater the number of socially desirable traits associated with a given individual, the more likely he or she is to be seen as fit for duty in space, Antarctica, or under the seas. After listing the qualities prescribed for saturation divers (intellectual brilliance, courage, perseverance, social adaptability, and humility), Biersner [1984] concluded that those who are most suited to work at watery depths should be able to walk on the surface as well.

A particularly intriguing possibility is that people who can combine seemingly opposing traits make good candidates for spaceflight. For instance, in many Western societies, men are expected to be autonomous, independent, somewhat dominating and aggressive, and emotionally inhibited. These are instrumental or task-oriented activities. Women are expected to be warm and nurturing and to openly display their feelings. These are referred to as expressive activities. According to Helmreich and his associates, these activities are orthogonal rather than mutually exclusive, and we might try to select people who are able to demonstrate both sets of skills [Helmreich *et al.*, 1980]. In a study that directly assessed the short-term (two-day) influence of the leader's personality on crew performance, Chidester *et al.* [1990] found that flight crews where the captain had a strong task orientation performed about as well as flight crews headed by captains with both task and interpersonal skills, and better than those where the captain had a more socio-emotive orientation. Since its inception, NASA has selected individuals with very high task orientation, and the Chidester *et al.* study gives support to this practice. However, it does not answer the question of whether the absence of interactive skills leads to performance decrements where they are expected to occur, i.e., over the long term. People who are too heavily oriented towards accomplishment could hurt other people's feelings and increase tensions which are already high due to stresses imposed by spaceflight. Alternately, "androgynous" people, i.e. those who combine the best of the "masculine" with the best of the "feminine," may be in a particularly good position to help the crew reach mission objectives while at the same time promoting cordial interpersonal relations.

Another reason for choosing "androgynous" individuals was pointed out by Jeanne Williams, who served in Antarctica [Williams, 1987]. She noted that it has historically fallen to women in confinement to provide emotional support and nurturing for the entire crew. This additional, undocumented duty can be very draining for women, especially in a mostly male group. Like other responsibilities, nurturing needs to be shared among all crewmembers. In a related vein, Williams also noted that feedback on performance must be different in confinement than in the usual work situation. Many managers tend to emphasize negative feedback, that is, to attempt to adjust subordinate behavior through criticism of deficient performance. Under conditions of isolation and confinement, there are relatively few sources for support and praise, so it becomes essential for managers to encourage and support their subordinates, as well as to correct them. Recent research on air crews suggests another reason for positive, supportive interpersonal styles. Gruff, cool, or remote leadership styles tend to discourage subordinates from volunteering information or following up to be sure that the information is understood [Foushee, 1984; Nicholas and Foushee, 1990]. Such information could be critical for effective command decisions.

A duality of traits may be useful also when considering the tendencies to be outwardly or inwardly directed. One might ask: Who is likely to make a better space traveler—introverts, who are shy and withdrawn, or extroverts, who are sociable and outgoing? On the one hand, introverts should be able to weather separation from family and friends and also respect the needs of other astronauts to be left alone. On the other hand, extroverts can provide social stimulation and help satisfy other astronauts' affiliative needs. In fact, the ideal astronaut may be the person who can be either withdrawn or outgoing depending on the demands of the situation [Connors *et al.*, 1985].

Technical and interpersonal factors are not necessarily independent. For example, a person who is unskilled or incompetent may also be disliked because he or she poses an unacceptable risk to the rest of the crew. In the search for technically competent and highly motivated people, we should try to find individuals who work hard because they find their work intrinsically satisfying or else because they have an interest in improving their proficiency and performance. However, we should avoid those individuals who are "competitive" in that their hard work is motivated by a desire to do better and to look better than other people [Helmreich *et al.*, 1980]. The rationale is that selecting people who find work intrinsically satisfying and who are interested in improving their performance, but who are not interested in outshining others, will promote a high level of achievement without increasing rivalries and interpersonal tensions.

Group Characteristics. There are certain communication strategies that help group members to coordinate their activities and work effectively together. Strategies such as using established patterns of initiation/response [Chidester *et al.*, 1990; Kanki *et al.*, 1991] and employing explicit language to develop shared understanding of problems, goals, plans, etc. [Orasanu and Fischer, 1992] are applicable to groups of varying composition. However, the interplay of different people's personal characteristics also influences how well people function together. An individual's traits find full expression only in the context of other people's attributes. The goal of this kind of analysis is to seek people whose personal qualities intermesh to form a cohesive, efficient, and effective team.

One commonly-reported finding is that people who have similar social, moral, and ethical values get along better than do people who have differing values. Homogeneity or heterogeneity of crews have also be posited to be important factors in the selection of effective training or leadership strategies [Kanki, 1990; Kanki and Gregorich, 1992]. Byrne *et al.* [1971] have shown homogeneity of outlook to be a sufficiently robust factor that the proportion of shared values can be used as a measure of affiliative tendencies. The Mars mission itself will be an important shared value or point of mutual interest for all participants. Commonalties might be further increased by selecting individual crewmembers in such a way that each pair of crewmembers has many shared interests, even though the interests shared vary from subgroup to subgroup. For example, crewmembers A and C may share an interest in chess, B and C may have a common religion, and C and A may come from the same home town. This general pattern could be reiterated within all possible pairs of crewmembers, with the shared interests taking different forms among different pairs. We suspect that shared interests with other crewmembers can strengthen a relationship and provide a buffer against the deleterious impact of minor disputes and squabbles. We do not recommend that all of the different crewmembers have identical values and interests: a diversity of viewpoints and perspectives can be useful when it comes to generating solutions to problems, and also for combating boredom.

Different people's needs may fit together in ways that affect interpersonal relations [Penwell, 1990]. People have compatible needs when their needs are such that the satisfaction of one person's needs results in the satisfaction of the other person's needs. For example, a more dominant person may enjoy the companionship of a less dominant person, but two dominant people may frustrate and antagonize one another. Studies of small groups isolated in laboratory settings for relatively brief periods have shown that in comparison to people with incompatible needs, people with compatible needs adapt better to one another and to the conditions of isolation and confinement [Haythorn, 1973; Smith and Haythorn, 1972]. The broader lesson is that the members of space crews and other groups bear a certain resemblance to the pieces in a jigsaw puzzle: what is important is not so much the shape and color of each individual piece, but the way the different pieces combine with one another to form a coherent and meaningful pattern.

In our search for compatibility, we need to be aware of the complexities of the time factor. Despite concerns about mission duration, time could be an ally as well as an adversary. Whereas the passage of time allows the accumulation of frustrations and hostilities, it also makes it possible to meet challenges and to achieve successes. These later experiences have a tempering effect, helping to turn a loosely-knit aggregate of individuals into a tightly-knit crew.

Mission duration may also influence the relative importance of particular traits. Personality characteristics that may be unimportant early in a flight may be useful as the flight progresses. For example, the ability to provide other people with emotional support may not be important during a brief mission or at the beginning of a prolonged mission, but it may be very important somewhere around the two-thirds mark of a prolonged mission, a point in time where mood and morale are likely to be especially low [Bechtel and Berning, 1991]. Additionally, the kinds of characteristics that make a person initially attractive may make that person unattractive later on or vice versa. There are likely to be important on-going processes that are not all that well understood but that are important for the crew's well-being. "Wearing well," we suspect, is discovered only after countless episodes of "give and take," as each person makes allowances for the other, and takes advantage of the latitude offered in return [Connors *et al.*, 1985]. Because such adjustive processes are fluid and dynamic, we expect that they will be very difficult to study scientifically. Yet they may be critical for the long-term success of isolated and confined groups.

Multinational Crews. There are strong justifications for drawing crewmembers from a multitude of nations. First, it will be useful to have many different nations contribute to the costs of the expedition. It is unlikely that many would be inclined to do so, unless they see some chance of representation within the crew. Second, extending beyond national boundaries enlarges the pool and potentially increases the quality of the top candidates, making it possible to select truly "world class" crewmembers. Moreover, a multinational mission would provide a common purpose (or "superordinate goal") which could help overcome international differences and promote cooperation and collegiality.

At the same time, a multinational expedition raises certain problems. Transit vehicles, surface habitats, gear, and both formal and informal activities must meet the needs of people of differing physiques and cultural tastes. Any hint of prejudice or differential treatment must be eliminated, and it may be necessary to discourage cliques formed along national or ethnic lines. Potential language barriers will have to be overcome. Proficiency levels that are adequate under normal conditions could prove inadequate under conditions of diminished speech intelligibility or during an emergency [Kelly and Kanas, 1992]. In the interests of safety and efficiency, a high level of fluency in a common language will be required. A high level of fluency will also be important for shared humor [Connors *et al.*, 1985], and humor will undoubtedly play a central role in maintaining crew morale. Issues associated with a multinational expedition are not insurmountable, but each will require careful attention to resolve.

Selection Procedures. Presumably, recruitment for the Mars mission will begin with a widely-disseminated announcement of opportunity to participate. If past experience is an indicator, hundreds of individuals could respond. Most will be eliminated on the basis of inadequate education, training or experience, while still others will be eliminated during initial screening interviews. A small, select group will undergo very close scrutiny.

During the early years of the space program, psychological and psychiatric screening of astronaut candidates was extensive. Astronaut candidates completed banks of psychological tests and underwent lengthy interviews for signs of maladjustment or pathology. In retrospect, not all of this time was necessary since the individuals these tests were designed to identify did not reach the interview stage [Jones and Annes, 1983]. Over the years, 30 hours of testing and psychiatric interviews was replaced by two hours of unstructured interviews. Clinical techniques were used to identify those people who were more likely than others to develop emotional or behavioral problems under the stress of future spaceflight [Jones and Annes, 1983]. Currently, efforts are underway to increase the objectivity of the process by devising more structured or standardized psychiatric interviews and by supplementing these interviews with a limited number of useful and validated tests [Santy *et al.*, 1991]. In addition, attention has shifted beyond disqualifying or "selecting out" candidates with psychological liabilities to finding or "selecting in" candidates who will bring exceptional psychological strengths to the mission [Jones and Annes, 1983; Rose *et al.*, 1993].

Ultimately, a three-phase selection approach may prove useful. During the first cut, a large enough group of people could be chosen to fill three to five complete Mars crews. Individual members would then be selected on the basis of their technical and interpersonal skills. In the second phase, these individuals would be formed into crews based on the likelihood that they will prove compatible with one another. The performance of these crews would be carefully monitored as they undergo training in all phases of the Mars mission. Finally, the most effective and trouble-free crew is chosen for the pioneering voyage. The purpose of this approach would be to have a well-formed, coherent, and smoothly-functioning crew in place prior to the moment of liftoff.

Necessarily there will be many candidates who are highly qualified and who have undergone extensive training and yet who are not selected for the initial Mars mission. It will be important to find good uses for these valuable human resources. Some may be promising candidates for other types of mission (e.g., orbital or lunar) and others may be developed for subsequent journeys to Mars. Still other non-chosen candidates may perform outstanding service at remote work sites on Earth, or serve as mission control personnel. The point is that these candidates' personal investments and the investments made in them should not go to waste [Clark, 1989].

Training

Whereas selection strategies are based on identifying people who already possess certain desired characteristics, training strategies involve helping people acquire the requisite skills and interests. The two strategies are complementary, in that one can select the most promising candidates and then have them undergo experiences that further bring them "up to speed." The same themes introduced in the discussion of selection continue in the discussion of training: crewmembers will require training in the psychological and interpersonal as well as technical aspects of Marsflight, and the primary goal of the training is to create an efficient and effective group [Nicholas and Foushee, 1990; Penwell, 1990].

Training for extended missions such as the Mars mission will differ from training for short-duration missions in a number of important ways [Clearwater and Harrison, 1990]. First, because of the sheer length of the time involved, exploration mission simulations will need to be partitioned into many parts, with mission training built primarily around critical activities such as launchings, landings, and so forth. This advanced training will have to impart general as well as particular skills that will serve the astronauts well in a number of different situations. Second, on prolonged missions, facility maintenance becomes important, with the result that the astronauts will have to develop strong skills in such areas as habitat maintenance and equipment repair. Third, Mars mission crews must be trained to protect against time-related effects, such as boredom, inattention, or the mindlessness that may follow from performing a task repetitively. Mars explorers must also be made aware of certain dangerous transition points, such as when a night shift replaces a day shift, where more than the usual number of errors are likely to occur. Finally, on-board training in non-critical areas (and practice in critical areas) will speed mission preparation, reduce the lag between acquiring and exercising skills, and provide something to do during potentially boring intervals.

One training goal is to prepare the crew to perform well under stress. Centuries of experience preparing warriors for combat shows that subjecting recruits to tough, dangerous, and even life-threatening conditions builds self-confidence, reduces fear, and increases the ability to keep calm and alert and to take constructive action under stress. Such training also forges strong interpersonal bonds within groups of trainees. Ranger, Special Forces, and Marine Reconnaissance training, which involves such elements as parachute and survival training, produces team members who are highly committed to each other's well-being, even to the point of making extraordinary sacrifices for one another. Advanced conditioning can be supplemented by training in direct techniques for stress control [Levine, 1991] and mental health maintenance [Santy, 1990].

Training in interpersonal skills is likely to involve three components [Nicholas and Foushee, 1990]. The first phase involves heightening awareness of interpersonal dynamics through lectures and classroom discussions. The second phase is practice, analysis, and feedback. This includes completing psychological tests that reveal interpersonal styles, analyzing one's own role in small group interaction, and so forth. The third phase is strengthening or reinforcing the new skills through repetition. In effect, critical skills must be overlearned; one repetition is not enough. Among the goals of such training are helping leaders develop positive, supportive styles that encourage subordinate participation, and establishing communication patterns to assure that verbal messages are received and fully understood [Foushee, 1984; Nicholas and Foushee, 1990].

As much as possible, training should involve authentic situations and tasks and should include both high quality simulators and authentic spaceflight-analogous or actual spaceflight environments. Antarctica is frequently mentioned as a preferred site to conduct research on human behavior under spaceflight-analogous conditions, and as a training ground for space [Harrison et al., 1989, 1991; Palinkas, 1990]. Since a Mars mission is not likely until some years after the advent of the space station, at least some training may take place on the space station itself [Santy, 1990]. This will provide valuable experience in learning to perform tricky or difficult tasks in cramped areas under conditions of weightlessness.

Ground control is an integral part of the overall Mars team. It will be essential for the flight and ground crews to establish and to set in place effective communication policies that can support the full mission. Flight crews and ground crews must share a basic idea of the goals they are trying to accomplish; a "mental model" of what they are trying to communicate; and a concept of issues of authority, i.e., who has decision-making power for certain circumstances. Beyond that, they must be able to maintain good relations. Ground controllers (and to some extent, astronauts' families) should receive training in support of effective personal relations [Penwell, 1990; Santy, 1990]. It is particularly important in the course of Earth-ground communication that ground controllers be able to relate to each astronaut as a unique person with a distinctive set of preferences, aversions, and interests. Similarly, astronauts need to understand their ground controllers. To the extent possible, the people who will actually serve as ground controllers should assume this role in Marsflight simulations.

Environmental Design

Architectural and other environmental interventions provide another means for promoting criterion behavior. The Marscraft and Marsbase facilities must be designed in such a way as to accommodate the needs of human users, and this theme must be carried through in the choice of supplies, the development of equipment, and the overlay of procedures. The primitive conditions that were acceptable for short duration spaceflights will not prove acceptable for long term sojourns aboard the space station, on the Moon, or on Mars, and it is crucial to get beyond thinking that minimal environments will suffice and that meeting basic life support needs fulfills habitability requirements [Clearwater, 1988; Clearwater and Harrison, 1990].

Interior Space Requirements. According to the best available estimates, required habitable volume per person increases as a function of mission duration, but reaches asymptote at approximately six months [NASA, 1987]. These estimates suggest that 5 m^3 per person provides a tolerable amount of interior space and that 17 m^3 per person offers an optimal amount of interior space given crew sizes between two and twelve and assuming a mission duration of over six months. By habitable interior space, we mean usable space, not space taken by life support and scientific equipment or used for storage.

Many Marsflight reference configurations call for a spacecraft capable of accommodating crews of both sexes and drawn from several different nations. This means that the craft must accommodate people of differing sizes and physiques. If we are to use the space station as a guideline, activity envelopes (the amount of space provided for a person to perform a particular task and the arrangement of work materials within this space) must accommodate all but the largest (top 5%) and smallest (bottom 5%) individuals. Especially in the case of international crews, this range may be extensive.

Noise Control. Spaceflight environments tend to be noisy environments, and we look for ways to keep noise under control. Sound levels in the shuttle have been in the 55-60 dB range. When sound levels exceed this range, performance on complex or intellectual tasks suffers, particularly when exposure continues for a prolonged period of time [Loeb, 1986]. In a survey of 33 shuttle astronauts, Willshire and Leatherwood found that more than half of those who responded noted that shuttle sound levels interfered with speech, disturbed sleep, made it difficult to relax, and in general proved to be a source of annoyance. Seventy-four percent of the respondents felt that there should be reduced background noise on shuttle, and 93% felt that there should be lower noise within the space station [Willshire and Leatherwood, 1985]. A later survey of astronauts and cosmonauts confirms that ambient noise threatens both intracrew and external relations [Kelly and Kanas, 1992, 1993].

Sound control is a good example of a process that is complicated by the requirements of spaceflight. There are two main techniques for reducing sound. One is to use walls or other barriers. Unfortunately, this would involve adding weight to the spacecraft. The other technique is to dampen the sound by transmitting it outside of the space habitat. Dampening is unworkable, at least in spaceflight transit, because spaceflight environments do not have external atmospheres that can absorb the sound, nor are they anchored to a major land mass, such as Earth. Other possible options include material dampening (in which sound energy is lost in the molecules of materials) and friction (conversion of sound into heat), neither of which offer practical advantage. Some limited sound control may be possible through the use of negative sound generators, devices which transmit sound waves that are essentially mirror images of environmental sound waves and which thereby cancel out environmental sounds. However, maximum sound reduction will depend on curtailing it at the point of generation.

Odor Control. Odor control in space is tightly coupled with sanitation. In space, sanitary facilities are likely to be meager, in part because only a limited amount of water can be carried aboard and in part because weightlessness tends to complicate showering, handwashing, and other bathroom rituals. To some extent, we can habituate or adapt to accumulating odors; however, at some point, we can expect dissatisfaction and reduced work performance to occur. Improved hygienic and ventilation systems will be high level requirements for a Mars mission.

Microgravity. In the absence of artificial gravity during transit, microgravity could provide interior designers with a degree of freedom that is not present on Earth. For example, under normal conditions, the height of chests of draws should not exceed the easy reach of users. Under conditions of microgravity, height is not a problem, because the user can "float" to the necessary level. Indeed, the top drawer can be mounted so that it is virtually flush with the ceiling, provided that it is inserted "upside down." In microgravity, there will be no greater tendency for materials to fall from this "inverted" drawer than from any other drawer. Or, any surface—wall or ceiling—could be used as a "floor." Work stations could be located on top of each other, but at 180-degree orientations, so that from the perspective of a person in either one of the work stations, the other work station seems to be "upside down." It thus becomes tempting to consider different ways of capitalizing on weightlessness to make maximum use of interior space.

The potential problem with such designs is that they reduce the extent to which the environment provides an understandable, coherent frame of reference, and hence the ease and speed with which inhabitants can orient themselves to the environment and objects within it. Whereas some people are not bothered by the lack of a true vertical, others experience some disorientation. Disorientation is disruptive to performance and particularly so under emergency or other conditions where fast, error-free behavior is vital. If designers make use of the full 360-degree multi-plane environment of microgravity, ways must be found to ensure high comprehensibility.

Microgravity has many medical consequences. As a result, artificial gravity is often proposed either as a continuous condition or else in dosages sufficient to offset physiological losses. However, the use of artificial gravity could complicate design, as well as behavior and performance considerations, particularly if its use is variable or sporadic.

Privacy. Privacy, in space as elsewhere, is an important consideration [Connors *et al.*, 1985; Harrison *et al.*, 1988]. Because privacy is a psychological rather than a physical need, it is often thought of as a luxury that can be easily set aside. Although privacy can be sacrificed under certain conditions or for brief periods, over the long term, privacy is important to the maintenance of one's sense of personal identity and psychological health [Margulis, 1977].

Privacy exists to the extent that people are able to exert control over the amount and nature of the contact they have with others and the revelations about themselves that are made to others. Being able to restrict or manage social contact serves several important functions. First, it promotes the high degree of concentration that is necessary to perform scientific and technical tasks. Second, being physically or mentally "away" from others provides the "down time" necessary for rest and recuperation. Third, separation helps people manage the images they project to others, and hence the relationships they have with others. For instance, being able to achieve some degree of physical or psychological distance from others decreases the chance that socially devalued behaviors (such as weeping or anger directed towards the other person) could contribute to interpersonal difficulties. Finally, restricting social contact is a precondition for allowing two or three people to interact without interference from other members of the crew. On a Mars mission, space travelers are likely to have need for private conversations with one another, and individuals will want to communicate privately with others on Earth [Connors, 1991a; Kelly and Kanas, 1993].

Mars-bound astronauts will require several kinds of privacy protections—one to apply within the confines of the missions and the other to exert control over interactions with the outside world. For spacecraft privacy, Mars-bound astronauts will require separate sleeping quarters. To the extent possible, individual "bedrooms" should be large enough to allow the occupant to read, watch television, write, and engage in other solitary activities. Additionally, there should be areas where one or two crewmembers could hold conversations apart from other members of the crew. Whereas, admittedly, such private areas could encourage the formation of cliques or subgroups, we feel that the lack of a place to hold private conversations poses the greater risk. For example, if the flight commander needs to "dress down" an individual crewmember, this should not occur in front of others. Private spaces should be relatively sound-protected, at least to the extent that conversations in adjacent "rooms" should not be distracting and the content should not be intelligible. One advantage of having artwork and personal entertainment devices (recorders, disks, reading materials, etc.) is that they provide crewmembers an opportunity to mentally tune one another out.

Communication with the ground will be via electronic media. Electronic media have the capability to ensure that communications are private, but in practice these protections are often overlooked. Electronic privacy violations become particularly worrisome when communication is digitized and when computers and databases intervene. Because of the nature of the venture and the great interest in their activities, Mars travellers may not be able to have as much privacy in their exchanges with the ground as they might otherwise prefer. However, they will need to have some opportunity for private exchanges. They will also need to be able to trust that whatever they assume to be private is, in fact, held private. The critical issues here are not technological; they are policy and procedural. Decisions will have to be made about what conversations and other communications are to be private; the crew will need to be clear on what is and is not included in each category; and mechanisms and protocols will need to be put in place to ensure that the arrangements struck are observed, and are not allowed to dissipate over time.

Interior Design and Aesthetics. Despite volumetric, weight, and materials limitations, Marsflight environments need not be sterile or ugly. Viewports, including windows and cupolas, are essential not only for allowing astronauts to orient themselves to the outside world, but also for preventing people from feeling cramped [Haines, 1991]. Artwork can be chosen for its appeal and its ability to maintain interest over time [Clearwater and Coss, 1991]. Artwork that creates an impression of depth may be of particular value in spaceflight [Clearwater and Coss, 1991]. Because posters are relatively light weight, a number of posters can be carried aboard and changed at different points through the mission. Or, if artwork is presented by means of a liquid crystal or other display, many thousands of works can be easily carried on one laser disk.

Finally, advances in "virtual reality" and similar visualization techniques offer new opportunities. Virtual reality employs computers, three dimensional color displays, stereophonic sound, and other technological means to create convincing sensory impressions of other places, times, or realities. During Marsflight, a virtual walk through a village or forested area while using a treadmill might have great psychological value and increase motivation to exercise. The visual scene and sound effects could be geared to the movement of the treadmill, and the operator offered choices or alternatives as he or she "approaches" divergent walkways or paths.

Structure

The way the mission is put together or structured will also have important behavioral consequences. In the social sciences, structure refers to social influence effects and social rules that help pattern people's relationships to one another and to the work that is to be performed. Structure is reflected in the distribution of authority and tasks.

Command Structure. Command structure is a central issue. Traditionally, manned spaceflight has involved military or quasi-military command structures. Such structures take the form of a tightly-defined hierarchy in which individuals at one level have the authority to direct the activities of individuals at lower levels. There are many questions regarding the optimality of this approach given different crews, scenarios, and mission goals [Nicholas and Foushee, 1990; Penwell, 1990]. Alternatives to hierarchical or autocratic decision making procedures (under which the commander reaches a decision on his or her own) are a range of consultative procedures (under which the commander solicits advice and opinions and then decides) as well as democratic decision-making procedures (under which crewmembers reach decisions through voting or through discussion to consensus). The desirability of one particular form of command structure over another depends on many factors such as size, group composition, etc. In addition, contemporary thought in social and organizational psychology holds that the most appropriate decision-making or managerial strategy is contingent upon such variables as the capabilities and attitudes of the people involved, time constraints, and the extent to which group acceptance of the decision is important [Vroom and Yetton, 1976].

Clearly, it is necessary to have an unambiguous chain of command to ensure that a qualified person will step in if another higher-ranking person is incapacitated. Beyond that, we endorse a mixture of three types of decision-making. The commander would make autocratic decisions when instantaneous action is required. (An example might be a decision to abort a particular maneuver in the interests of safety, such as cutting short extra-vehicular activity based on a malfunction of a life support unit.) The commander would solicit opinion and advice and then reach a consultative decision in the case of mission-related issues not accompanied by strong time pressures. (For example, a change of protocol might be made in order to pursue an unexpected scientific "target of opportunity.") Finally, democratic voting or consensus-seeking would be used when the issue revolves around life within the habitat. (This approach might be utilized when establishing rules for the use of recreational equipment). In all cases, there should be excellent communication between the leader and the crewmembers, and the latter must never hesitate to draw critical information to the leader's attention. Decision-making processes on board will have to be well established beforehand because the time delay involved in Mars-Earth communications may make it difficult or impossible to consult with experts on Earth when rapid decisions are necessary [Connors, 1991b].

Role Structure and Loading. In addition to hierarchy, social structure is reflected in the division of labor or the distribution of tasks. Roles will have to be defined in such a way that, in combination, all of the activities will be performed that are required for the mission. Since there will not be as many people as roles needed, each astronaut will have to perform multiple roles. For example, it may be impossible to include a biologist, a physician, a dentist, and a psychologist. Instead, it may be necessary to find a person who can perform all of these roles. Also, it will be necessary to select people who can be cross trained and provide one another with "backup" in case a primary worker is incapacitated. Aside from the logistical demands of the flight, it may be psychologically desirable to assign each crewmember a range of tasks. Giving each person a variety of duties (some of which do not require a high level of expertise) will allow flexibility in the allocation and rotation of jobs, so that each person receives his or her share of desirable and undesirable tasks.

Role loading—the amount of work associated with each role—will be a major concern. On the one hand, there should not be more work to do than the astronauts can comfortably perform. Because of the high cost associated with launching a person into space, it is very tempting to assign each astronaut very large amounts of work. The problem is that workloads that tax the capacity of individuals tend to result in inefficiencies, mistakes, and, over an extended period of time, can lead to severe illness [Miller, 1978]. In the past, astronauts and cosmonauts have sometimes been over programmed, the most famous incident involving Skylab IV [Weick, 1977]. In this case, heavy work demands contributed to conflict between the crew and mission control, and led to a work stoppage. On the other hand, underloading must also be avoided. Members of the first Mars crew are likely to have strong work needs and want to keep themselves occupied performing meaningful work. The long cruise phase of a Mars flight may be of particular concern in this regard, since there may not be all that much work to do.

Fortunately, there are some useful guidelines for defining appropriate workloads. First, it should be kept in mind that tasks that take a given time to perform under normal Earth conditions may take considerably longer to perform in space. Second, the level of work required can vary substantially throughout the course of the mission. Inflight training and other useful activities will have to be identified to provide space travellers with constructive work during the cruise phase. Third, when people who are to perform the roles have an active say in defining them, a well-balanced workload results [Weick, 1977]. This balance is less likely to be achieved when planners define roles that other people are to carry out.

Coordinating Social and Technical Systems. There are two characteristics of future space missions that will affect social organization and interpersonal relations. First, compared to today's space missions, tomorrow's will involve large numbers of extremely sophisticated automated systems [Connors, 1989; Statler and Connors, 1991]. Unlike Earth-based activities where automation can be introduced essentially as an enhancing or incremental feature, it is unlikely that a Mars mission could even be attempted without a large and significant role being assumed by automated systems. This raises to a critical level the need to address questions about the interactions between people and machines and between human and artificial intelligence.

While both humans and automated systems are often credited with being "intelligent," the nature of that intelligence differs in significant ways. Humans possess extreme adaptability and often perform well in new situations. However, the sustained performance of humans tends to be uneven-excellent at some times, only fair at others. Automated systems can perform some tasks, for instance complex computations, in a tiny fraction of the time a human would require (if the human is capable of performing them at all.) However, automated systems are largely unadaptable, will persist in an activity regardless of its consequences, and are easily confused by new or unplanned events. Automated systems also tend to perform either at a very high level of accuracy, or else to fail in dramatic ways.

The complementarity of humans and automation suggests that, at some future time, a system of automated design and human training could be developed that could outperform our present capabilities to a degree that would effectively comprise a new kind of intelligence [Connors, 1989]. For the present, our task is to understand the capabilities of each system and to organize their respective activities. To do this we must answer questions such as: What work is best performed by humans? What work is best performed by machines? And, how can we effectively integrate the two?

A second characteristic of future space missions involves the necessity of coordinating the activities of a number of individuals at various locations. In the case of the Mars mission, for example, this could well include mission control on Earth, a satellite orbiting Mars, a ground base, and at least one field party. A number of issues impact the design of systems and processes to support such multi-person, decentralized activities. These include (a) decision-making strategies, (b) selection of communication media, and (c) communication delays.

There can be substantial difficulties in coordinating information in order to reach decisions, even if individuals are in the same room [Fischhoff and Johnson, 1990]. Decision-making activities of separated teams are still more complex [Tsitsiklis and Athans, 1985]. For Mars spaceflight, information will be distributed both among co-located individuals and across vast distances, presenting highly complex decision-making challenges. As understanding of decision-making activity in Earth-based distributed teams accumulates and matures, there will be a need to quickly incorporate this understanding into the system design and procedures that support a Mars exploration mission.

The selection of the medium through which the distributed team members communicate (i.e., computer-, audio-, video-based) has long been recognized as a potential influence on both the dynamics of the group and the outcome of the activity [e.g. Johansen, Vallee and Spangler, 1979; Refer, 1991]. For narrowly-drawn, information-transference tasks, the differences may be minor; for tasks with a highly affective component, the differences may be substantial. In general, a video-based medium that provides kinesic, paralinguistic, and linguistic cues leads to a higher sense of "presence" and a fuller or more social environment for exchange than does a narrower bandwidth (computer or audio) medium [Connors et al., 1985]. On a Mars mission, selection of media is likely to be driven primarily by energy or power considerations. Understanding the task/media relationships and the manner in which particular pairings can influence outcomes should also play a significant role in planning for communication needs.

Communication delays result from physical limitations in transmission speed over distance. Although rarely a problem on Earth, the upper boundary of transmission rate described by the speed of light can pose a problem for space-relevant distances. As distance increases, communication between separated parties gradually changes from interactive, to turn-taking, to sequential one-way communication, and finally to quasi-independent messages [Connors, 1991a]. It follows, then, that large distances change the reliance space crews can put on the ground and increase the need for intelligent, on-board systems.

Both human/automation requirements and the kinds of communication systems needed to coordinate activities at different locations are salient concerns that require careful attention. They also may be enabling technologies, that is, technologies that must be in place before the first mission to Mars is attempted [Connors, 1991b].

Social Norms. Finally, an important structural consideration is social norms, that is, shared understandings as to what constitutes unacceptable, optional, and mandatory views and behaviors. Most social norm issues should resolve themselves during training. As crewmembers work together and associate socially over a long period, relationships will begin to mature, hopefully resulting by the time of flight in interactions marked by understanding, loyalty and trust. However, no one can fully anticipate all of the issues that will assume importance on a two- or three-year mission. Disagreements should be aired during preflight training, and processes for reaching agreement should be refined as astronauts coalesce into crews. Crews will need to identify how problematic issues are to be dealt with during flight. Each crew should be free to set up its own mechanisms, but each crew should demonstrate in advance that an approach has been adopted.

Special relationships between or among particular crewmembers also should be addressed prior to flight. Whereas attention is generally drawn to the problems associated with acrimonious or distant relationships, special friendships could also pose problems. Those who develop particularly close relationships could undermine the overall unity and cohesiveness of the crew. For instance, negative consequences could result if a fallout should occur between close friends, or if other crew members perceived themselves to be disadvantaged due to special loyalties or treatments. Consciously limiting personal friendships for the benefit of the larger group has many precedents. One example is found in the brotherhood and sisterhood of religious life. Although here, as in all human settings, friendships develop, cloistered orders tend to guard against special attachments, dependencies, cliques, and gossip groups. Other examples are to be found in Israeli kibbutzim. Potential problems that could result from either special friendships or aloofness towards certain crewmembers are probably best resolved simply by making sure that crewmembers understand the consequences that could follow in this restricted environment.

Ultimately, relationships between and among crew members will be defined by the crew itself, and it is their judgment that will lead to the evolution of norms governing such relationships. Once crews have established their own group norms, peer pressure can be counted on to enforce these norms.

AMERICAN AND RUSSIAN SPACE PROGRAMS

The point has frequently been made that behavioral scientists have played a larger role in the space program of the former Soviet Union than in the American space program. For over thirty years, Russian behavioral scientists have been integral members of the research and support team of the Soviet space program [Garshnek, 1989; Kanas, 1990; Oberg and Oberg, 1986] and their area of activity has been broader than that of their American counterparts. While American scientists have tended to focus on issues of selection, training and ergonomics, Soviet scientists have been concerned also with maintaining human performance and providing for the emotional and social welfare of crews.

In a discussion of social and personality psychology's role in the U.S. space program, Connors [1986] posited several reasons for the differences in behavioral science emphases between the Russian and American programs. One factor related to basic programmatic differences. These differences included longer flight durations and higher relative budgets for the Soviet program as compared with the American program. Longer duration demonstrates the need for behavioral science intervention; higher budgets present the opportunity to address that need. A second factor involved the cultural values of the sponsoring societies. Cultural differences included a Soviet communal world view (tending to emphasize the needs of the group and the benefits of cooperation), as compared with the more individualistic American world view (emphasizing personal initiative and competition). Of course, both programs recognized that cooperation does not work without individual initiative, nor individual initiative without cooperation. However, given the cultural backgrounds from which the U.S. and the Russian programs developed, it is not surprising that the Russian program emphasized early on such issues as how spacecrews relate to one another and how they can work cooperatively together.

Acknowledgment of historical differences between space programs should not lead to the conclusion that the future will mirror the past. On the contrary, the more demanding requirements of future flights should lead to greater concern for a wide range of behavioral issues in all space programs. And, as we are now witnessing, greater cooperation is developing among spacefaring nations. In this more open environment it is possible for both the Russian and the American space programs, as well as the space programs of other nations, to gain from the experience and research of others. For instance, agreements are now being put in place that will allow Russian cosmonauts and American astronauts to participate in each other's missions; to test out each other's equipment; to develop standards for interoperability; and, potentially, to co-develop the next generation of equipment, facilities, and even space platforms. At the same time, opportunities are developing for scientists from various nations to work together, to utilize each other's research facilities and to co-sponsor both simulated and in-flight experiments. This new spirit of cooperation will increase the understanding of what has already been learned concerning human behavior and performance in space, and accelerate the development of forward-looking, exploration-relevant behavioral science research.

BEHAVIORAL ISSUES IN PERSPECTIVE

We have three general statements to make regarding the role of human behavioral research in the manned Mars mission. *First, behavioral issues are important*: motivation and will-power cannot be relied on to suppress anxieties or social conflicts that would otherwise occur. Although psychological and social factors have caused only minor problems in space, these factors can, and have, posed problems for other isolated and confined groups. (For a historical perspective of problems of isolated and confined groups, see Vinograd [1972]; for more recent experiences in the Antarctic, see Palinkas [1990].) Furthermore, as already noted, the extent of such problems may have been underreported, since their acknowledgment can reflect poorly on the group's leader or on the group's sponsors [Harrison, 1986]. Dismissing psychological factors as unimportant not only ignores existing evidence, but overlooks the fact that no one has yet been subjected to the combination of conditions associated with a Mars mission.

Second, psychological and social issues will not resolve themselves. Some people acknowledge the importance of psychological and social variables, but add that by the time a Mars mission is launched, these variables will be well under control. The argument is that as a result of natural growth and development, we will know all that needs to be known about personality, group dynamics, and so forth. Consequently, we need not devote much attention to such factors right now. Unfortunately, this position ignores the fact that, again and again, advances in technology have outpaced advances in our understanding of the technology's human users. As a result of this lag, problems in such areas as vehicle and habitat design and environmental support are likely to be solved long before we can feel confident that psychological factors are under control. If, when the first Mars mission departs many years from now, psychological and interpersonal factors are well understood, this understanding will not have come about as a consequence of cultural evolution but will reflect, instead, active and purposeful research efforts. A strong faith in a better future should not draw us away from an active role in bringing this future into being.

Third, behavioral factors are best regarded not as an area of singular concern, but as one of many integral parts in planning for a Mars mission. On occasion, the argument is made that isolation, confinement and other rigors associated with spaceflight are extreme stressors conducive to psychological and social pathology. Any implication that the stressors of spaceflight will lead inexorably to dire consequences is not supported by the evidence. On the contrary, persons often perform remarkably well even under severe stress. However, we would argue that humans are capable of making errors and misjudgements under even the best circumstances, and that the conditions of spaceflight constitute a factor that adds significantly to the potential for problems. In order to place psychological and social considerations in a true perspective, we should neither deny their potential importance nor dwell upon hypothetical scenarios where psychological factors trigger disaster. Instead, we need to recognize their real significance and encourage serious and sensible research programs intended to better understand and manage them.

CONCLUSION

The future historian may look back upon the Mars mission as the high point in a primitive area of exploration—that of the solar system. As far as we know, aside from Earth, Mars is the only solar system planet that is potentially suitable for human settlement. Temperature extremes, toxic atmospheres, and formidable gravitation make the other planets unappealing. As was the case following our first steps on the Moon, the Mars base may be followed by a lengthy hiatus for purposes of planning and mounting new kinds of missions—interstellar missions—which will involve voyages of hundreds of people taking place over many generations [Finney and Jones, 1984]. We must make sure that the first step, the flight to Mars, is a sure step and that it provides the foundation from which further exploration is possible.

REFERENCES

Barabasz, A.F., "Effects of isolation on states of consciousness," in *From Antarctica to Outer Space: Life in Isolation and Confinement*, A.A. Harrison, Y.A. Clearwater, C.P. McKay, eds., pp. 201-209, Springer Verlag, New York, 1991.

Bechtel, R.B. and A. Berning, "The third quarter phenomenon," in *From Antarctica to Outer Space: Life in Isolation and Confinement*, A.A. Harrison, Y.A. Clearwater, C.P. McKay, eds., pp. 261-266, Springer Verlag, New York, 1991.

Biersner, R.J., "Psychological evaluation and selection of divers," in *Physician's Guide to Diving Medicine*, C.W. Schilling, C.B. Carlson, eds., Plenum, New York, 1984.

Byrne, D., C. Gouaux, W. Griffitt, J. Lamberth, N. Murakawa, M. Prasad, A. Prasad, and M. Ramirez, "The ubiquitous relationship: Attitude similarity and attraction: A cross-cultural study," *Human Relations*, 24, 5, pp. 201-207, 1971.

Chidester, T.R., B.G. Kanki, H.C. Foushee, C.L. Dickinson, and S.V. Bowles, *Personality factors in flight operations:I. Leader characteristics and crew performance in full mission air transport simulation*, NASA TM, NASA Ames Research Center, Moffett Field, CA, 1990.

Clark, B., "Crew selection for a Mars exploration mission," AAS 87-192, in *The Case for Mars III: General Interest and Overview*, C.R. Stoker, ed., American Astronautical Society, Science and Technology Series, 74, pp. 193-204, Univelt, Inc., San Diego, CA, 1989.

Clearwater, Y.A., "Space station habitability research," *Acta Astronautica*, 17, pp. 217-222, 1988.

Clearwater, Y.A., and R.G. Coss, "Functional esthetics to enhance well-being in isolated and confined settings," in *From Antarctica to Outer Space: Life in Isolation and Confinement*, A.A. Harrison, Y.A. Clearwater, and C.P. McKay, eds., pp. 331-348, Springer Verlag, New York, 1991.

Clearwater, Y.A. and A.A. Harrison, "Crew support for an initial Mars expedition," *J. Brit. Interplan. Soci.*, 43, pp. 513-518, 1990.

Connors, Mary M., *Exploring Social/Personality Psychology's Role in the U.S. Space Program*, Discussant at the 94th Annual Convention of the American Psychological Association, Washington, D.C., August, 1986.

Connors, M.M., "Crew System dynamics: Combining human and automated systems" (SAE 891530), paper presented at the Intersociety Conference on Environmental Systems, San Diego, July 24-26, 1989.

Connors, M.M., "Communications issues of space exploration," in *From Antarctica to Outer Space: Life in Isolation and Confinement*, A.A. Harrison, Y.A. Clearwater, and C.P. McKay, eds., pp. 267-280, Springer Verlag, New York, 1991a.

Connors, M.M., "The Role of Human Factors in Missions of Exploration," Paper presented as part of the Space Exploration Initiative Life Science Session at the 21st International Conference on Environmental Systems, San Francisco, CA., July 15-18, 1991b.

Connors, M.M., A.A. Harrison, and F.R. Akins, *Living Aloft: Human Requirements for Extended Spaceflight* (NASA SP-483), US Government Printing Office, Washington, DC, 1985.

Connors, M.M., A.A. Harrison, and F.R. Akins, "Psychology and the resurgent space program," *Am.. Psychol.*, 41, pp. 906-913, 1986.

Finney, B.R., and E. Jones, *Interstellar Migration and the Human Experience*, University of California Press, Berkeley, 1984.

Fischhoff, B. and S. Johnson, The Possibility of Distributed Decision Making, Appendix A in National Research Council (Ed.), *Distributed Decisions*, National Academy Press, Washington, DC, 1991.

Foushee, H.C., "Dyads and triads at 35,000 feet: Factors affecting group process and aircrew performance," *Am. Psychol.*, 40, pp. 885-893, 1984.

Garshnek, V., "Soviet space flight: The human element," *Aviat., Space, Environ. Med.*, 60, pp. 695-705, 1989.

Gunderson, E.K.E., "Psychological studies in Antarctica," in *Human Adaptability to Antarctic Conditions*, E. K. E. Gunderson, ed., pp. 115-131, American Geophysical Union, Washington, 1974.

Haines, R., "Windows: Their importance and function in confining environments," in *From Antarctica to Outer Space: Life in Isolation and Confinement*, A.A. Harrison, Y.A. Clearwater, and C.P. McKay, eds., pp. 349-358, Springer Verlag, New York, 1991.

Harrison, A.A., "On resistance to the involvement of personality, social, and organizational psychologists in the U.S. space program," *J. Soc. Beh. and Pers.*, 1, pp. 315-324, 1986.

Harrison, A.A., "Beyond Earthnocentrism: Anthropology on the high frontier," *Space Power*, 7, pp. 345-352, 1988.

Harrison, A.A., B.S. Caldwell, N.J. Struthers and Y.A. Clearwater, *Incorporation of Privacy Elements in Space Station Design*, (Final Report NASA NAG-2-431), Department of Psychology, University of California, Davis, 1988.

Harrison, A.A., Y.A. Clearwater, and C.P. McKay, "The human experience in Antarctica: Applications to life in space," *Beh. Sci.*, 34, pp. 253-271, 1989.

Harrison, A.A., Y.A. Clearwater and C.P. McKay, eds., *From Antarctica to Outer Space: Life in Isolation and Confinement*, Springer Verlag, New York, 1991.

Harrison, A.A., and M.M. Connors, "Human factors in spacecraft design," *J. Spacecraft and Rockets*, 27, pp. 478-481, 1990.

Harrison, A.A., and J. Summit, "How third force psychologists might view humans in space," *Space Power*, 10, pp. 185-203, 1990.

Haythorn, W., "The miniworld of isolation: laboratory studies," in *Man in Isolation and Confinement*, J.E. Rasmussen, ed., pp. 219-241, Aldine, Chicago, 1973.

Helmreich, R.L., J.A. Wilhelm, and T.E. Runge, "Psychological considerations in future space missions," in *Human Factors in Outer Space Production*, S. Cheston and D. Winter, eds., pp. 1-23., Westview Press, Boulder, 1980.

Johansen, R., J. Vallee, and K. Spangler, *Electronic Meetings: Technical Alternatives and Social Choices*, Addison-Wesley, 1979.

Jones, D.R., and C.A. Annes, "The evolution and present status of mental health standards for selection of USAF candidates for space missions," *Aviat., Space, Environ. Med.*, 54, pp. 730-734, 1983.

Kanas, N., "Psychological, psychiatric and interpersonal aspects of long-duration space missions," *J. Spacecraft and Rockets,* 27, pp. 457-463, 1990.

Kanas, N., and W. Fedderson, *Psychosocial Factors Affecting Actual and Simulated Space Missions* (NASA TM X-58067), Brooks Air Force Base, TX, 1971.

Kanki, B.G., "Teamwork in high risk environments analogous to space" (AIAA-90-3764), in *Proceedings of Space Programs and Technologies '90*, September, 1990.

Kanki, B.G. and S.E. Gregorich, "Team dynamics in isolated, confined environments: Saturation divers and high altitude climbers," in *Proceedings of the AIAA Space Programs and Technologies Conference,* Huntsville, Alabama, 1992.

Kanki, B.G., M.T. Palmer, and E. Veinott, "Communication variations related to leader personality," in *Proceedings of the Sixth International Symposium on Aviation Psychology,* Ohio State University, Columbus, 1991.

Kelly, A.D., and N. Kanas, "Crewmember communication in space: A survey of astronauts and cosmonauts," *Aviat., Space, Environ. Med.,* 63, pp. 721-726, 1992.

Kelly, A.D., and N. Kanas, "Communication between space crews and ground personnel: A survey of astronauts and cosmonauts," *Aviat., Space, Environ. Med.,* 64, pp. 795-800, 1993.

Levine, A.R., "Psychological effects of long duration space missions and stress amelioration techniques," in *From Antarctica to Outer Space: Life in Isolation and Confinement,* A.A. Harrison, Y.A. Clearwater, and C.P. McKay, eds., pp. 305-316, Springer Verlag, New York, 1991.

Loeb, M., *Noise and Human Efficiency,* John Wiley & Sons, New York, 1986.

Lugg, D.J., "Current international human factors research," in *From Antarctica to Outer Space: Life in Isolation and Confinement,* A.A. Harrison, Y.A. Clearwater, and C.P. McKay, eds., pp. 31-42, Springer Verlag, New York, 1991.

Margulis, S.T., "Conceptions of privacy: Current status and next steps," *J. of Social Issues*, 33, pp. 3, 5-51, Summer, 1977.

Miller, J.G., *Living Systems Theory,* McGraw-Hill, New York, 1978.

NASA, *Man-Systems Integration Standards ST-3000,* National Aeronautics and Space Administration, Washington, D.C., 1987.

Nicholas, J.M., and H.C. Foushee, "Organization, selection and training of crews for extended spaceflight: Findings from analogues and implications," *J. Spacecraft and Rockets,* 27, pp. 451-456, 1990.

Oberg, J.E., and A.R. Oberg, *Pioneering Space,* McGraw-Hill, New York, 1986.

Orasanu, J. and U. Fischer, "Team cognition in the cockpit: Linguistic control of shared problem solving," in *Proceedings of the 14th Annual Conference of the Cognitive Science Society,* Erlbaum Associates, Hillsdale, NJ, 1992.

Palinkas, L.A., "Psychosocial effects of adjustment in Antarctica: Lessons for long-duration spaceflight," *J. Spacecraft and Rockets,* 27, pp. 471-477, 1990.

Penwell L.W., "Problems of intergroup behavior in human spaceflight operations," *J. Spacecraft and Rockets,* 27, pp. 464-470, 1990.

Radloff, R., and R. Helmreich, *Groups under Stress: Psychological Research in Sealab II,* Appleton-Century-Crofts, New York, 1968.

Rawls, R., A.E. Hopper, and D.J. Rawls, *Variables Thought to Determine Personal Space: An Opinion Sample,* Institute of Behavioral Research, Texas Christian University, 1969.

Reder, Stephen and R.G. Schwab, "The communicative economy of the workgroup: Multi-channel genres of communication," in *Proceedings of the Conference on Computer Supported Cooperative Work,* (Portland, 26-28 September, 1988), Elsevier Sci. Pub. Ltd., 1991.

Rose, R.M., R.A. Helmreich, L. Fogg, and T.J. McFadden, "Assessments of astronaut effectiveness," *Aviat., Space, Environ. Med.,* 64, pp. 789-794, 1993.

Santy, P.A., "The journey out and in: Psychiatry and space exploration," *Amer. J.. Psychiat.,* 140, pp. 519-527, 1983.

Santy, P.A., "Psychiatric support for a health maintenance facility (HMF) on Space Station Freedom," *Aviat., Space, Environ. Med,* 58, pp. 1219-1224, 1987.

Santy, P.A., "Psychological health maintenance on Space Station Freedom," *J. Spacecraft and Rockets*, 27, pp. 482-485, 1990.

Santy, P.A., A.W. Holland, and D.M. Faulk, "Psychiatric diagnoses in a group of astronaut applicants," *Aviat., Space., Environ. Med.,* 62, pp. 969-973., 1991.

Smith, S., and W.W. Haythorn, "Effects of compatibility, crowding, group size, and leadership seniority on stress, anxiety, hostility, and annoyance in isolated groups," *Journal of Pers. Soc. Psychol.,* 22, pp. 67-97, 1972.

Statler, I., and M.M. Connors, "Issues on Combining Human and Non-Human Intelligence," Paper presented at SOAR Conference, Albuquerque, NM, 1990.

Stuster, J.W., *Space Station Habitability Recommendations Based on a Systematic Comparison of Analogous Conditions* (NASA CR-E3943), Anacapa Sciences, Santa Barbara, 1986.

Suedfeld, P., "Groups in isolation and confinement: Environments and experiences," in *From Antarctica to Outer Space: Life in Isolation and Confinement*, A.A. Harrison, Y.A. Clearwater, and C.P. McKay, eds., pp. 135-146, Springer Verlag, New York, 1991a.

Suedfeld, P. (Ed.) "Polar psychology," *Environment and Behavior*, 23, 653-808, 1991b.

Taylor, A.J.W., *Antarctic Psychology* (DSIR Bull. No. 24), Science Information Publishing Center, Wellington, NZ, 1987.

Tsitsiklis, J. and M. Athans, *On the Complexity of Distributed Decision Problems*, Laboratory for Information and Decision Systems, M.I.T., Cambridge, MA, 1985.

Vinograd, S., *Studies of Social Group Dynamics Under Isolated Conditions: Objective Summary of the Literature as it Relates to Potential Problems of Long Duration Spaceflight* (CR-2496), Sciences Communication Division, George Washington University, Saint Louis, 1974.

Vroom, V., and P.W. Yetton, *Leadership and Decision Making,* University of Pittsburgh Press, Pittsburgh, 1976.

Weick, K., "Organization design: Organizations as self-designing systems," *Organization Dynamics*, pp. 31-46, Autumn, 1977.

Williams, J., "Life in Antarctica," presentation to the conference on Human Experience in Antarctica: Applications to Life, Sunnyvale, CA, August, 1987.

Willshire, K.F., and J.D. Leatherwood, *Astronaut Survey of Shuttle Vibroacoustic Environment* (PIR No. SD-6), Acoustics Division, Structures Directorate, NASA Langley Research Center, Hampton, VA, 1985.

Three piloted spacecraft embark along the ecliptic over a summer morning in South America.

AAS 95-485

Chapter 15

FROM THE GREAT VOYAGES OF EXPLORATION TO MISSIONS TO MARS

Ben Finney[*]

Experiences between scientists and naval personnel who sailed together around the world on the great voyages of scientific exploration of the late 18th and early 19th centuries suggest that relations between scientists and those charged with operating the spacecraft on multi-year Mars missions could be problematic. Although the first oceanic exploration voyages carried sizable complements of civilian scientists, conflict between them and naval personnel led to a reduction in numbers of scientists carried on later voyages, with a consequent impact on scientific research. Although much of the conflict can be attributed to the great hazards of navigation during that era, and the need for naval personnel to have unfettered control of their vessels, judging from the continued tensions between scientists and ship's personnel on modern oceanographic vessels, sub-cultural differences were also basic. Scientists and operational personnel working together for long periods in the confines of ships of the sea or of space bring different goals, attitudes and work habits with them. Recognizing these differences, and working out ways to accommodate them, may be crucial to the success of the long missions to Mars.

INTRODUCTION

As currently being discussed, missions to Mars and return will require humans to be away from Earth, hurtling through space in transit and exploring Mars itself, for periods significantly longer than previously experienced by cosmonauts and astronauts. The endurance record in space is currently held by cosmonauts Vladimir Titov and Musa Manarov, who spent 366 days orbiting Earth aboard *Mir*, and most other spaceflights have been far briefer. Yet, those who discuss human Mars missions routinely talk of multi-year ventures. For example, a recent study from NASA's Lunar and Mars Exploration Office of such mission profiles classes them into 500-day and 1,000-day missions [Lineberry and Soldner, 1990]. As missions in the 500-day class allot only short, 30-60 day, stay times on Mars, it seems likely that pressures to maximize stay times for science purposes will favor missions of the 1,000-day class for they provide for stay times in the order of 400 to 600 days.

---

[*] Department of Anthropology, University of Hawaii, Honolulu, Hawaii 96822.

Mission times of upwards of three years raise serious concerns about hardware reliability, operations and costs, as well as our ability to keep people alive and well in the cramped confines of spacecraft and on the surface of Mars. In addition, as has become increasingly obvious from the Soviet experience aboard *Mir*, as well as from a few hints dating back at least to the 84-day Skylab mission, that maintaining good relations among the crew members must be a priority for the success of long missions [Finney, 1987]. Most discussions of crew relationships have focused on psychological compatibility and small group dynamics, although recently issues of multi-cultural and both-sex crews have been addressed. At the risk of being accused of "borrowing trouble," in this paper I would like to raise another human issue which might well turn out to be crucial on long Mars missions, the relationship between operational and scientific personnel.

Reaching Mars, landing on the planet and then successfully returning from there to Earth will severely test the skills of those who operate the spacecraft involved. Even with a high degree of automation, operational personnel would have to have an encyclopedic knowledge of propulsion, navigation, life support and other systems involved, and have the skills needed to intervene when problems arise. Similarly, those charged with carrying out geological, biological and other research on Mars would have to be highly qualified and exhaustively trained to take maximum opportunity offered (at such a tremendous cost!) by a stay on such an intriguing planet as Mars. By a great stretch of the imagination, it might be possible to envisage recruiting and training extraordinary individuals capable of carrying out both the demanding operational tasks and the daunting scientific ones. More likely, however, the crew will be composed of both operational personnel (pilot astronauts/cosmonauts, engineers, etc.) and scientists of various disciplines, which raises questions about how these two groups of people will relate to one another on such a hazardous and demanding mission as a multi-year expedition to Mars and return.

Problems between operational personnel and scientists on multi-year research voyages are not just the stuff of science fiction. They have happened before—on missions around our own planet, though over the ocean not through space. I refer to those globe-girdling expeditions during which wooden sailing vessels spent years at sea sailing around the world mapping islands and coasts, collecting specimens, surveying geodesically and geologically and probing the ocean itself and its creatures. Take, for example the first voyage into the Pacific made by Captain Cook aboard *H.M.S. Endeavour*, or the voyage of *H.M.S. Challenger*, the first ship dedicated solely to oceanography to sail around the world. On these and other such voyages made in 18th and 19th centuries groups of officers and men charged with the running of the vessels, and scientists charged with carrying out research, sailed together on expeditions lasting as long as four years, at times cooperating and at times clashing over issues of command, mission priority and day to day operations. In so doing, they learned by sometimes bitter experience how difficult it can be for people of differing backgrounds, training and goals to work together in the cramped confines of a rolling ship, and during forays ashore, always with the knowledge that an unseen reef or other hazards at sea or on shore could mean that some or all aboard would never see their home country again.

NASA has recognized the analogy between the exploration of the sea and space by naming all its orbiters after famous sailing ships used for exploration and research: the *Endeavour* and the ill-fated *Challenger* after the aforementioned vessels; the *Columbia* after the ship that explored the mouth of the Columbia River (and gave the river its name); the *Discovery* after the ship Hendrik Hudson used to explore Hudson's Bay; and the *Atlantis* after the schooner operated by the Woods Hole Oceanographic Institution. In this paper, I would like to explore this analogy further by examining the relations between those officers and men who commanded and sailed the ships of the so-called Age of Scientific Discovery, and those geologists, botanists, biologists, linguists and others who conducted the research at sea and on the islands and coasts encountered along the way in order to provide an empirical basis for thinking about how to handle problems in human relations that could hamper the conduct and accomplishments of missions to Mars, and some day beyond.*

THE AGE OF SCIENTIFIC DISCOVERY

If the first circumnavigation of our globe were to be any indication of things to come in space, people would recoil in horror at the prospects of multi-year missions to Mars or elsewhere in our solar system. In 1519 Magellan sailed from Spain with five ships carrying 236 officers and men. Three years later there returned to Seville only one of those ships, with just 18 men aboard. Disease, mutiny, hostile islanders, belligerent fellow Europeans, and disasters at sea had taken their awful toll. Even the visionary Magellan did not return; he was killed on a beach in the Philippines while meddling in local politics.

But this pioneering voyage belongs to the first Age of Discovery when nascent European sea powers sailed the seas not for scientific purposes, but to find new ways to reach Asia and gain the spices and other rich products to be had there. It is the better organized, safer voyages of the second age of discovery, a period beginning in the late 18th century that historians have labeled the Age of Scientific Discovery, that are of interest to us for they were focused primarily on exploration and research. [Goetzmann, 1986]. In fact, in terms of the mobilization of government resources and the expenditures of large sums for scientific ends, and in terms of intense national competition between exploring nations, this age foreshadows the space race of the 1960s [Dunmore, 1969]. The maritime powers of the day invested considerable sums to send their ships on great, globe-girdling voyages in which, not unlike later space ventures, the goals of scientific discovery were intertwined with those of national prestige and aggrandizement.

The focus of this competition was on the Pacific, then a largely uncharted expanse initially thought to hold rich islands and perhaps the fabled *Terra Australis*, the Southern Continent considered by prominent geographers of that era as necessary to balance the great land mass of the Northern Hemisphere. The competition began in earnest during the late 1700s with Britain and France sending expedition after expedition into the Pacific and then round the world on exploring missions that lasted up to three years, and

---

* This chapter expands upon Finney [1990].

sometimes longer. Spain, then well into its decline as a maritime power, halfheartedly joined in by sending a few expeditions. Russia entered the field in the early 1800s with a series of well carried out voyages, while the United States, the last great power to join the competition, did not send out an expedition until the late 1830s.

The British

The most famous scientific voyage of this era was that undertaken by Captain James Cook aboard the bark *Endeavour*. In 1768 Britain sent the *Endeavour* to the newly discovered South Sea island of Tahiti. Although the expedition was to search for the Southern Continent as it traversed the South Pacific, the primary goal of the voyage was astronomical: to observe from Tahiti the transit of Venus across the face of the Sun and thereby provide data needed to calculate the distance between Earth and Sun. This project, which was part of an international scientific effort to measure exactly the distance between the Earth and the Sun so that it could be used as the basic unit of astronomical measure, was promoted by the Royal Society, funded by the Crown, and assigned to the Royal Navy for execution.

Who should command such an expedition—a scientist or a man of the sea—was an immediate issue. Seventy years earlier, in 1698, the great Edmond Halley, the astronomer who had first proposed using the transit of Venus for measuring the Earth-Sun distance, had been commissioned by the Admiralty to command an expedition which has been called "the first sea journey undertaken for a purely scientific object" [Thrower, 1981]. Halley's assignment, which he himself had promoted and for which he had secured Royal Society backing, was to study magnetic variation throughout the Atlantic. Scientifically, the voyage was a success. Halley's chart of magnetic variation was immediately acclaimed and became the standard for years to come (although his goal of using magnetic variation to determine longitude proved unattainable). However, Halley had to cut the voyage short and return to England earlier than planned because of insubordination among his officers resentful at having a landsman in command. Halley complained that one in particular, a Lieutenant in the Royal Navy, had a habit of countermanding his orders when his back was turned. The officer was reprimanded, and Halley was sent out on two more, shorter expeditions. But, evidently the idea of having a landsman command a naval vessel, even one so skilled in the theory and practice of navigation as Halley, did not sit well with the Lords of the Admiralty for when it came time to choose a commander for the *Endeavour*, they would not entertain the idea that a civilian scientist might command.

The Astronomer Royal, Nevil Maskelyn, was asked for his recommendation as to who should command the expedition to observe the transit of Venus. He proposed Alexander Dalrymple, a passionate advocate of oceanic exploration who was later to become the first hydrographer of the Royal Navy. The Royal Society seconded Dalrymple's nomination, and forwarded his name to the Admiralty as their choice to lead the expedition. Dalrymple, however, was not a professional seamen and had not served in the Royal Navy. Whatever experience at sea he might have had, the Lords of the Admiralty were not about to appoint him to command a vessel. Such an appointment, they wrote to the Royal Society, would be "totally repugnant to the rules of the Navy." The First Lord of the Admiralty amplified this stand, vowing that "he would suffer his right hand to be cut off rather than sign such another commission as had gone to Halley in the *Paramour Pink* in 1698—a civilian in command of a naval vessel on a scientific voyage, whose difficulties with his officers had been painful" [Beaglehole, 1974].

There could be no question. A professional seaman, an officer in the Royal Navy, must command—not a scientist. But who? The Admiralty made an inspired choice: a young officer named James Cook who had risen from able seaman to master of his own vessel. Cook, who was already renowned for his navigating and surveying skills, was promoted to lieutenant and given command of the *Endeavour*. A complement of scientists and their assistants sailed with him: astronomer Charles Green who earlier had served as assistant to the astronomer royal; Joseph Banks, a young and wealthy botanist who financed his own way and that of his assistants, including Daniel Solander, a pupil of Linnaeus considered to be the ablest botanist in Britain; and two artists assigned to draw islands, geological formations, scenes and plants and animals discovered.

Cook's task was to sail the *Endeavour* to Tahiti so that the transit of Venus could be observed, and then to reconnoiter the Pacific for other islands and to search for the hypothesized Southern Continent, and, in general, to facilitate the observations and collecting of the scientific staff. The job of the scientists and their assistants, civilians all, was to make the astronomical observations, and to describe and illustrate the plants and animals they found, as well as the island societies encountered. This division of labor between those who sailed the ship and the scientists was seen as the obvious, rational formula to be followed.

It seemed to work well during that great voyage. Cook sailed the *Endeavour* to Tahiti without incident. The transit of Venus was observed, and the data were later delivered to the astronomical community in England. A search westwards across the South Pacific in latitudes revealed no Southern Continent between 30° and 40°S latitude, though it brought the *Endeavour* to New Zealand which the English were then able to reconnoiter. Next they explored the east coast of Australia, then sailed through the Torres Straits for Batavia, and from there to England via Cape Town to complete the three-year voyage.

According to Cook's biographer, during this magnificent circumnavigation the great navigator "very efficiently brought the scientists to their material" [Beaglehole, 1974]. As a result, in addition to making major astronomical and geographical observations, a wealth of specimens, drawings and experience with the South Sea islanders was gained. Furthermore, except for an ill-fated stay in pestilential Batavia, the voyage was a healthy one. The scientists and seamen apparently worked well together, particularly for Cook and Banks, who had emerged as the chief scientist aboard. For example, in addition to carrying out his botanical duties, the linguistically talented and culturally sensitive Banks was most useful in developing good relations with their Tahitian hosts—so essential if the English were to be allowed to conduct their astronomical observations unmolested. The formula of having sailor command, and scientists do the research seemed to have worked well.

But, this synergistic partnership between Cook and Banks unraveled once they reached England. The seeds of discord were, if not initially sown, well fertilized when the press proclaimed the voyage to be a triumph of Banks, not Cook. So lionized by press, public, and royalty was Banks that it was immediately assumed that he would lead a second expedition to the South Seas to extend his researches. The attention received led Banks to recruit an oversized scientific entourage, and to require that an extra deck be added to the expedition's ship, the *Resolution*, to house his people and their scientific gear. Cook, who had been chosen to lead this second expedition, at first acquiesced. But, when during sea trials the *Resolution* proved impossibly top-heavy, he ordered the extra deck cut off. Banks objected, but the Admiralty backed their man. Banks quit the expedition in disgust. Although later Banks, in his capacity of President of the Royal Society, was to be an influential voice in promoting subsequent British expeditions into the Pacific, he was never again to sail on one of them.

The departure of Banks did not, however, spell an immediate end to the formula of having naval officers command scientific expeditions and civilian scientists carry out the scientific work. This second Cook expedition, which had its primary purpose to search in the high latitudes of the South Pacific for the Southern Continent, sailed with a number of scientists: two professional astronomers, and, as replacements for Banks, the naturalist Johann Reinhold Forster and his son Georg. During the voyage Cook was able to lay to rest forever the idea of a continent in the temperate latitudes of the South Pacific (though he correctly guessed that a land mass was to be found at higher latitudes), and also to survey many islands, and—thanks to one of the first practical chronometers and the aid of the astronomers on board—to fix their exact positions. In addition, during the voyage the Forsters were able to continue the botanical work started by Banks, and to undertake other natural historical and anthropological research. However, the relationship between Cook and Johann Forster was not, apparently, a happy one. We do not know exactly what happened. However, Cook's biographer considers the "dogmatic, humourless, suspicious, censorious, contentious, demanding" naturalist made life extremely difficult for Cook and others during the voyage [Beaglehole, 1974]. Forster, for his part, gives vent in his journal to his own frustrations as to what he considered unwarranted interference by Cook and others in his scientific projects [Hoare, 1982].

Once back in England, Cook was promoted and pensioned off with half pay. He was soon tempted back to sea again, however, when command of a third grand expedition to the Pacific was offered. This time the central object was the search for the Northwest Passage, a seaway linking the North Atlantic and the North Pacific so that England could trade directly with Asia by a northerly route, with the charge of making scientific observations along the way. On this voyage, however, the division of labor between science and seamanship that had prevailed on the previous two voyages was drastically modified. Only three civilians sailed with Cook: an artist; an astronomical observer sent by the Board of Longitude, and a botanical collector. They were technicians, not independent scientists. Conspicuously absent were any prominent scientists of the stature of Banks or the elder Forster. Instead, Cook was to utilize the services of his officers to carry out many of the scientific functions on the voyage: Lt. James King who because of his study of astronomy at Oxford and Paris could make precise astronomical observations; a young William Bligh who was an expert draftsman and surveyor; and the three surgeons who had talents in natural history, writing and drawing.

We really do not know whether Cook, the Admiralty, or both were behind this virtual exclusion of civilian scientists and their substitution by naval officers. However, when Lt. King called on Captain Cook before the voyage to pay his respects and to express his regret that no scientific person was going on the voyage as before, King was answered with a tirade in which Cook reportedly said "Curse all the scientists and all science into the bargain!" [Beaglehole, 1974].

After his experience with the exhilarating, but overly ambitious Banks, and then the cantankerous Forster, Cook apparently wanted nothing more to do with prominent scientists. He had a job to do, to search for the northwest passage, and he apparently wanted no one on board who would question his methods or carp at him to go here and there for purposes of making scientific observations or of collecting unique specimens. If there was anything scientific to be done, he would rely on his officers and the malleable technicians on board. For the English naval establishment, the honeymoon between scientists and seamen was over and a separation, if not a divorce, was in order.

The French

The French followed a similar progression as their British rivals. They too started sending civilian scientists out with their Pacific expeditions, and they too ended up virtually banning civilians and having naval officers with scientific training and inclination make the scientific observations and do the scientific collecting.

The first French voyage into the Pacific with great scientific pretensions sailed in 1766 under the command of Bougainville. Bougainville's expedition was less serious in a strict scientific sense, however, than Cook's near contemporaneous first voyage. Although Bougainville did bring along the distinguished naturalist Commerson, who had been nominated for the post by the great Buffon, the motivation for the voyage was more geopolitical than scientific, and the voyage had little impact other than to fuel French ambitions in the Pacific and to reinforce romantic ideas of natural man, and accompanying critiques of the ancien regime—largely based on an inaccurate portrayal of Tahitian society penned by an enraptured but uncritical Commerson.

With subsequent voyages, French scientific exploration came to fore, large complements of scientists were carried, and an effort was made to dovetail the work of scientific and naval personnel. But French naval enthusiasm for science was not to last. Friction between scientists and seamen grew as the complement of scientists expanded. When Baudin sailed in 1800 to survey the then largely unknown Australian continent, he carried no less than 22 scientists and their assistants on board. The voyage was most difficult. Baudin was incapacitated from consumption which finally killed him on the way back to France, food ran short, and scientists and seaman kept getting in one another's way. The officers and sailors were especially resentful of the mass of scientific gear carried, the amount of food consumed by the "useless" scientists, the stink of their rotting shell specimens, and above all the time "wasted" in collecting natural history specimens ashore. When, for example, a young French officer met the English navigator Matthew Flinders, who had beat the French to the south Australian coast, he is reported to have said to Flinders: "Captain, if we had not been delayed so long picking up shells and collecting butterflies in Van Dieman's Land (Tasmania) you would not have discovered the South Coast before us" [Dunmore, 1969].

Thereafter no French expedition was to be so burdened by scientists. In fact, in 1817 Louis de Freycinet sailed for the Pacific with not a single civilian scientist on board. Instead he had his young lieutenants handle the telescopes and make oceanographic and geophysical measurements, and had the two surgeons and the pharmacist double as botanists and zoologists. Freycinet had learned from personal experience on the Baudin expedition that "a group of scientists is likely to be a source of trouble, and he preferred to have scientific work carried out by naval officers, whom he could easily control" [Dunmore, 1969].

The French naval officers, like their British rivals, were apparently glad to be rid of civilian scientists cluttering their decks. For scientific tasks, the employment of "naval officers with some scientific training, including a knowledge of natural history, became the rule." In comparison to civilian scientists, these naval men were considered "far more valuable because they would obey orders in a way no passenger, no guest, however bound by agreements or a sense of duty, ever would" [Dunmore, 1969].

The Russians

The Russians did not begin their grand exploratory voyages into the Pacific until 1803 when Krusenstern sailed from the Baltic port of Kronstadt for the North Pacific. Over the next three decades some 30 ships were to sail from there to the Pacific, many making the circumnavigation after completing their tasks. Although most of the Russian voyages were linked in one way or another with the needs of Russian outposts on both sides of the Bering Straits, scientific goals were proclaimed for many of their voyages, and the ships commonly carried a scientist or two and often an artist as well. I do not know the literature well enough to say if the trend of starting out with grand scientific projects and then of later downgrading science and scientists was operative in the Russian case. However, from the journal of the naturalist Chamisso on the Kotzebue expedition (1815-1818) it is apparent that Russian seamen and scientists were not immune from conflict. Like Forster, Chamisso [1856] complains of the restraints placed upon him by a captain who he says did not understand the needs of science.

The Americans

Although the Americans may have been late on the scene, once aroused, their scientific ambitions were anything but modest. Americans did not start thinking seriously about sending a major exploring expedition into the Pacific until more than a half-century after those first voyages by Bougainville and Cook. Coming so late upon the scene, they apparently wanted to play catch-up with a vengeance.

Through a grand scientific voyage the United States, said expedition promoter Jeremiah Reynolds before the House of Representatives on April 3, 1836, could "wipe out, at one glorious effort, the taunting imputation so long cast upon the American character" that we are but unlettered receivers of knowledge from Europe, and "throw back on Europe, with interest and gratitude, the rays of light we have received from her" [Stanton, 1975]. Reynolds then told the assembled congressmen that the object of the expedition would be "to collect, preserve, and arrange every thing valuable in the whole range of natural history . . . and accurately to describe that which cannot be preserved; to secure whatever may be hoped for in natural philosophy; to examine vegetation, from the hundred mosses of the rocks, throughout all the classes of shrub, flower, and tree, up to the monarch of the forest; to study man in his physical and mental powers, in his manners, habits, and disposition, and social and political relations; and above all, in the philosophy of his language; to examine the phenomena of winds and tides, of heat and cold, of light and darkness . . . ."

To those amazed listeners who were asking themselves how all this might be done, Reynolds calmly offered the formula already discredited in British and French naval circles: "By an enlightened body of naval officers, joining harmoniously with a corps of scientific men, imbued with the love of science." Chosen wisely, the officers and the "lights of science" would, "like stars in the milky-way, shed a lustre on each other, and all on their country" [Stanton, 1975].

Many influential people drank all this in. Congress voted funds for a naval expedition carrying civilian scientists. And members of the budding American scientific establishment lined up for the promised positions as botanists, geologists, astronomers, philologists, zoologists and so on in what Reynolds had promised would be a "scientific faculty, complete in all departments" [Stanton, 1975]. But when, after much wrangling and many delays, Charles Wilkes, a young naval officer with scientific ambitions, was chosen to command the expedition, he radically pruned the ranks of the scientists. Wilkes may have been acquainted with the history of British and French scientific expeditions, or perhaps it was just his seamen's reasoning that made him follow the European model. Whatever the case, he decided that all work pertaining to astronomy, surveying, hydrology, geology, geodesy, magnetism, meteorology and physics in general would be handled exclusively by his naval officers.

Wilkes also tried to cut down the size of the rest of the scientific complement, but did not totally get his way [Stanton, 1975]. Notable among those civilians who were finally allowed to remain on board were Horatio Hale, a young Harvard graduate who escaped the axe only through political intervention, and whose linguistic and anthropological researches conducted on the expedition were to bring him worldwide fame, and James Dwight Dana, a young geologist who during the voyage made outstanding contributions to the knowledge about volcanism from his study of the Pacific islands.

The expedition, known as the "United States Exploring Expedition" sailed from Norfolk in 1838, and returned in triumph to New York four years later. On board the three main ships were twelve civilians: naturalists and their assistants, plus the geologist and linguist mentioned above. Wilkes proved to be a stern disciplinarian, and after the voyage was court marshaled and reprimanded for his treatment of the crew. Although no official charges were lodged by the scientists, many complained privately that Wilkes had hindered their scientific work, in particular by severely limiting their time ashore. For example, during the voyage the geologist Dana had written from Valparaiso to the Harvard botanist Asa Gray to congratulate him and other scientists cut from the expedition by Wilkes for their "narrow escape from Naval servitude" [Stanton, 1975].

The Wilkes expedition was the last experiment for many years to come in joining large numbers of civilian scientists with naval personnel on a grand voyage of exploration. After its return, and especially given the rancorous disputes that followed between the Navy, various other government agencies and the scientific establishment over the care and housing of the enormous scientific collections brought back and the publication of the research results, there was no attempt to follow so difficult a model. During the following several decades, most of what science was done at sea was carried out by naval officers, although like their British and French counterparts, American naval vessels might carry a naturalist or two when sent on an exploring mission.

Too Much, too Soon?

The model, born of the enthusiasm of the Enlightenment, of having groups of civilian scientists and naval personnel work cooperatively at doing science and exploration at sea was, judging from the experience reviewed, prematurely applied.

Of course, if we take the scientists' point of view, it is easy to sympathize with, for example, Forster, who, when frustrated by restrictions placed upon his research work ashore, rails against "people who know nothing of Sciences & hate them, never care whether they are enlarged & knowledge increases or not" [Hoare, 1982], or with Commerson, his French colleague on the Bougainville expedition, when he describes his ship as "that hellish den where hatred, insubordination, bad faith, brigandage, cruelty, and all sorts of disorders reign" [Dunmore, 1969]. After all, if the voyages were really for science, it would seem logical that every effort should have been made to facilitate research in every way.

Nonetheless, a good case can be made that the naval officers charged with sailing their ships half-way around the world to explore the globe's largest ocean, and then with getting back safely home, had good reasons for restricting science and scientists. In an era when ocean voyaging was so hazardous, when a sudden storm or an unseen reef could wreck a ship leaving absolutely no prospect for rescue, and when scurvy and other illnesses regularly took their toll, these seamen had their hands full just keeping their ships intact and their crews alive. To carry out the minimal geographical explorations required by their orders was a big enough job in itself. All the rest—the collecting, the sketching, the inquiring into native customs—was frosting on the cake to be done only if these activities did not compromise the safety and schedule of the expedition.

In the end, some scientists realized this. For example, however frustrated he may have felt while at sea on the U.S. Exploring Expedition, the geologist Dana had to admit that Wilkes respected science, and that all the scientists were brought home safely, and with loads of botanical and zoological specimens, drawings, rock samples, vocabularies and descriptions of Indian tribes and island societies, and the like. Furthermore, Dana admitted that he doubted that with any other commander would the scientists "have fared better, or lived together more harmoniously" [Stanton, 1975].

But to ban civilian scientists altogether from the voyages was certainly too drastic a solution to this problem. Some naval officers, most notably Cook, may have been able to make important scientific observations, but their activities were necessarily limited by their all important operational duties, as well as their lack of training. An argument can even be made that learned natural scientists could actually enhance the safety of a voyage through their judicious advice to naval officers on how to relate to the islanders along the way. The great Cook did not live to see England again after leaving on the third voyage. After finishing his survey of the Northwest Coast, and finding no passage to the Atlantic, Cook returned to Hawaii, an archipelago he had discovered on his way north. While anchored at Kealakekua Bay, a ship's boat was stolen. An enraged Cook went ashore to take the high chief hostage to ensure the return of the craft. An angry crowd gathered, and when Cook tried to take the chief back to the ship the Hawaiians killed the navigator. Forster wrote later that the absence of an "educated, scientific gentleman" on the third voyage had contributed to the tragedy. In Forster's opinion, Banks and Solander on the first voyage, then Forster himself and his son on the second voyage, had been civilizing influences on Cook; their presence had prevented Cook from acting too rashly with the islanders. As, however, there were no such "educated, scientific gentlemen" on the third voyage, there was no one to restrain the great navigator from making that fatal error of judgment [Hoare, 1976, Beaglehole, 1961].

DARWIN AND THE VOYAGE OF *H.M.S. BEAGLE*

While the great expeditions with large groups of scientists that ushered in the era of scientific exploration may have become models to avoid as the problems they engendered came to fore, not all naval ships sent out on expeditions banned scientists. A civilian naturalist or two were often invited to sail, and their research interests were facilitated, although always subordinated to the safety and needs of the ship and its primary mission. Darwin's voyage on *H.M.S. Beagle* bears witness to how scientifically fruitful such a naval expedition carrying a single scientist can be, for during the nearly five years he was at sea and roaming exotic shores and islands Darwin collected data and gained insights which were to eventually result in his epochal work, *The Origin of Species*. Furthermore, it is apparent that it was a happy voyage from the point of view of naval-scientific relations.

Late in 1831 *H.M.S. Beagle* sailed from Plymouth bound for South America on a voyage that took the brig back and forth along the Atlantic and Pacific coasts of South America and finally across the Pacific and around the world. Unlike the voyages made by Cook and his contemporaries, this was not voyage bent on discovering unknown lands, but rather an expedition focused upon surveying the coasts and harbors of South America for British naval and commercial use. Nonetheless, the Captain, Robert FitzRoy, invited a young naturalist, Charles Darwin, along on the voyage, probably more for companionship than any great scientific ambitions, however.

Darwin and FitzRoy apparently got along famously, enjoying one another's company throughout the long voyage. Darwin's junior status and unprepossessing manner were probably important here. When they sailed, he was hardly a full-fledged scientist. Rather, Darwin was a 22-year old student with some training in botany and geology and other sciences who had yet to decide on a career. Furthermore, even though FitzRoy was only four years older, Darwin readily accepted his authority and decisions, however much he might disagree with them. Judging from Darwin's letters and journal, this was not always easy, for FitzRoy was a taskmaster at a time when naval captains had immense authority. As Darwin puts it in his diary, "the difficulty of living on good terms with a Captain of a Man of War is much increased by its being almost mutinous to answer him as one would answer anyone else; and by the awe in which he is held, or was held in my time, by all on board" [Barlow, 1933].

FitzRoy too was an unusual individual. A meticulous and tireless surveyor, FitzRoy had an inquiring mind and was a lover of scientific inquiry who later was to gain recognition as one of the founders of meteorology. In fact, the two respected one another's competencies and enjoyed discussing their respective discoveries, ideas and theories, particularly over meals.

Nonetheless, however ideal they were as shipmates and intellectual companions on the long voyage, their friendship was to become greatly strained in later years, ironically because FitzRoy could not accept Darwin's work on evolution, work that he himself had played such an important role in fostering by enabling Darwin to make the very observations that were to lead to his revolutionary theory. Darwin sadly writes about how FitzRoy became "very indignant with me for having published so unorthodox a book (for he became very religious)" [Barlow, 1933].

OCEANOGRAPHIC RESEARCH

Once oceanic voyaging became less hazardous and more routine, once steam displaced or at least supplemented sail, once accurate charts and navigation instruments made it possible to navigate with precision, and once preserved foods and the ability to easily obtain provisions from ports along the way banished hunger and scurvy, then naval officers could afford to relax a little. The idea of taking groups of scientists on extended cruises once again surfaced, for the learned gentlemen could now be accommodated much more easily than before. For example, in 1872, a hundred years after Cook, the Royal Navy's *Challenger*, a three masted, square-rigged wooden vessel with a steam engine for maneuvering along the coast and in harbors, sailed on a round-the-world oceanographic cruise that marked a new era of doing science at sea. This modern and well-equipped vessel carried six marine scientists—complete with their dredging and sampling gear, and laboratories as well. The ship, her officers and her crew were totally dedicated to making research possible on this pioneering oceanographic voyage.

If we are to believe the head of the scientific staff, Charles Wyville Thompson, "the somewhat critical experiment of associating a party of civilians, holding to a certain extent an independent position, with the naval staff of a man—of-war, has for once been successful. Captain Nares and Captain Thompson both fully recognized that the expedition was intended for scientific purposes . . . . All the naval officers, without exception, assisted the civilians in every way in their power, and in the most friendly spirit . . . ." [quoted in Linklater, 1972]. That, of course, is the official version; a search through the diaries of the scientists and officers in question might reveal some cracks in this facade. For example, however smoothly officer-scientist relations may have proceeded smoothly by official reckoning, there are hints in the journal kept by an assistant purser of conflict between the "scientifics," as he called them, and the below-decks crew [Rehbock, 1990].

It is tempting to assume that, given the subsequent growth of oceanographic research wherein especially designed ships carry oceanographers and other scientists on research cruises throughout our planet's oceans, the good relations and research facilitation idealized by the *Challenger*'s chief scientist would become routine, and that any problems between ordinary seamen and scientists such as those hinted at by the *Challenger*'s purser would have been worked out during the more than a century of subsequent experience with oceanographic cruises. However, casual inquiries among my oceanographic colleagues, and a study of the relations between scientists and crewmen carried out by anthropologist H. Russell Bernard, indicate that even routine oceanographic cruises are far from problem-free.

Currently, in the United States oceanographic research is primarily conducted by oceanographic institutes and universities which send research vessels on extended cruises that may last for several months. The research vessels are essentially floating laboratories for the conduct of research in physics, chemistry, geology and biology at sea. Although the officers and crew are civilians, most have served in the Navy and all are professional seamen. The scientific party, which is usually equal in number to the officers and crew, is made up of a chief scientist, typically the principal investigator whose grant funds the cruise or a major portion thereof, plus other scientists, graduate students and technicians.

While resident at the Scripps Institute of Oceanography during 1973, Bernard was able to go to sea on a research cruise to observe behavior at sea first hand, and to interview both scientists and seamen at sea and ashore. He concludes from his research that tension between the two groups is inevitable because they essentially form two separate sub-cultures with different values and goals [Bernard and Killworth, 1973, 1974]. For example, the scientists, who typically only join the vessel for short periods of time, are totally interested in gathering data in relation to highly focused research projects, while the seaman, for whom the vessel is a home for many months at a time, are primarily interested in the smooth operation of the vessel and in maximizing their time ashore. Class differences between highly educated scientists, and, in particular, those ordinary seamen with a high school education only, further demarcate these two subcultures.

Bernard cites numerous examples of conflict between, on the one hand, data-hungry scientists who cannot understand why crewmen won't do everything, including giving up time in port, to help them wring every last bit of information out of the sea, and, on the other hand, seamen who do not want to be disturbed in their routine by outsiders who don't seem to understand how a ship should be operated and who insist on the priority of their research. While in a dispute, aggrieved scientists may rightly argue that the whole purpose of a cruise is to undertake research, if they press their case too hard their projects may be sabotaged. Bernard cites the case of a dispute over a refrigerator, originally designated to house biological specimens, in which the crew had been storing their liquid refreshments. When their supplies were evicted, and the refrigerator was locked, the crewmen countered by breaking open the refrigerator and dumping all the biological specimens into the sea, thereby ruining weeks of costly work.

DISCUSSION

From the perspective of the history of the grand voyages of exploration of the 18th and 19th centuries, what has occurred during the first stages of piloted space flight begins to look more natural, if not inevitable. During the first decade of the American piloted space flight, when methods of getting into space and orbiting Earth had to be worked out, pilot astronauts drawn from the military test pilot fraternity were overwhelmingly dominant. When scientists were recruited for the Apollo flights to the Moon, the operational requirements of these extraordinary voyages, and perhaps also the natural reluctance of established pilot astronauts to see their monopoly on space flight broken, put them at the end of the line for flight opportunities. When budget cuts caused cancellation of the last flights, this meant that only one scientist astronaut ever reached the Moon.

During the shuttle era there appears to have been a conscious effort to fuse pilot astronauts and mission specialist astronauts (i.e., scientist astronauts) into a tightly knit team. But, mission specialists are generally not, to use the jargon of organized science, the "Principal Investigators" on the experiments they carry out in orbit. Those scientists who conceive the experiments, and have the primary responsibility for analyzing the data brought back from orbit and writing up the results, stay on the ground. This seems to have at least partially shifted the spacecraft operations-scientific research interface from the arena of interpersonal relations within the orbiter to communications between space and ground. Furthermore, judging from conversations with mission specialists proud of their astronaut training and qualifications, it seems that on shuttle flights the role of the outsider, with all that implies for misunderstanding and potential conflict, has been thrust upon the payload specialists, who are, in effect, visiting scientists/technicians who receive less operational training and do not really belong to the elite astronaut corps.

Modern oceanographic research experience provides one model of how scientists could be accommodated on long missions to Mars. However, the contemporary oceanographic situation also indicates that if space research were to be routinized to the extent that ocean research now is, we would still have to reckon with sub-cultural differences and resultant tensions between scientists and those pilots and other crew members whose job it will be to enable researchers to carry out their tasks in space. Oceanic solutions to these problems may not, however, be directly transferable to space. For example, whereas Bernard proposes that one mechanism for controlling conflict between scientists and oceanographic ship personnel is the maintenance of "sub-cultural privacy" wherein the latter attempt to shut out scientists from their own world aboard ship, it would be most difficult to assure any such sub-cultural privacy in a cramped spacecraft. Nonetheless, even if specific oceanic solutions may not be applicable to space, the extensive human experience gained in doing science at sea should not be forgotten. To paraphrase Santayana, those who forget the mistakes of the past may be compelled to relive them in the future.

REFERENCES

Barlow, N., *Charles Darwin's Diary of the Voyage of the H.M.S. "Beagle"*, pp. xxvi-xxviii, Cambridge University Press, Cambridge, 1933.

Beaglehole, J.C., *The Voyage of the Resolution and Adventure 1772-1775*, Vol 2, pp. xlvi-xlvii, Cambridge University Press, Cambridge, 1961.

Beaglehole, J.C., *The Voyage of the Endeavour 1768-1771*, Cambridge University Press, Cambridge, 1968.

Beaglehole, J.C., *The Life of Captain James Cook*, pp. 125, 699, 502, Haykluyt Society, London, 1974.

Bernard, H.R. and P. Killworth, "On the social structure of an ocean-going research vessel," *Social Science Research*, 2: pp. 145-184, 1973.

Bernard, H.R. and P. Killworth, *Scientists at Sea: A Case Study in Communications at Sea.* Report BK-103-74, Code 452, Contract N00014-73-4-0417-0001, Office of Naval Research, 1974.

Dunmore, J., *French Explorers in the Pacific*, Vol. 2, pp. 29-65, 384-389, Clarendon Press, Oxford, 1969.

Finney, B., "Anthropology and the humanization of space," *Acta Astronautica*, 15, pp. 189-194, 1987.

Finney, B., Scientists and seamen, *The Human Experience in Antarctica and Its Application to Life in Space,* A.A. Harrison, Y.A. Clearwater, and C.P. McKay, eds., pp. 89-101, Springer Verlag, New York, 1990.

Goetzmann, W.H., *New Lands, New Men*, Viking, New York, 1986.

Hoare, M.E., *The Tactless Philosopher*, Hawthorn, Melbourne, Australia, 1976.

Hoare, M.E., *The Resolution Journal of Johann Reinhold Forster 1772-1775*, Volume 3, 551pp., Haykluyt Society, London, 1982.

Lineberry, E.C. and J.K. Soldner, "Mission profiles for human mars missions," paper presented at *American Institute of Aeronautics and Astronautics* Space Programs and Technologies Conference, Huntsville, Alabama, September 25-28, 1990.

Linklater, E., *The Voyage of the Challenger*, 274pp., John Murray, London, 1972.

Rehbock, P. F., *At Sea with the 'Scientifics': The Challenger Letters of Joseph Matkin*, unpublished manuscript in the possession of the author, 1990.

Stanton, W., *The Great United States Exploring Expedition of 1838-1842*, pp. 32, 41, 63, 137, University of California Press, Berkeley, 1975.

Thrower, N.J.W., *The Three Voyages of Edmond Halley in the Paramore 1698-1701*, 16pp., Haykluyt Society, London, 1981.

von Chamisso, A., *Reise um die Welt mit der Romanzoffischen Entdeckungs Expedition in den Jahren 1815-18*, Weidmann, Berlin, 1856.

AAS 95-486

Chapter 16

THE INTERPLANETARY RADIATION ENVIRONMENT AND METHODS TO SHIELD FROM IT

L. W. Townsend and J. W. Wilson[*]

An overview of the space radiation environment relevant to a human Mars mission is described. General methods for incorporating the nuclear and electromagnetic interactions of these radiations into radiation transport models which mathematically describe the propagation of the radiation fields through spacecraft/habitat structures are presented. Estimates of radiation exposures for human Mars missions are given. A discussion of additional research needed to reduce exposure estimate uncertainties is presented.

INTRODUCTION

A major source of concern for mission planners and spacecraft designers involved in planning human missions to Mars is cumulative exposures of crews to penetrating space radiation. Within the past several years, it has become obvious that much attention must be given to mitigating the harmful effects of these radiations by careful design of spacecraft and surface habitats for a lunar or Martian outpost.

The three primary sources of energetic space radiations, as shown in Figure 1, are (1) galactic cosmic rays (GCR), (2) solar particles emitted in flares, storms, and the solar wind, and (3) trapped particles (electrons and protons) in the Van Allen belts. Galactic cosmic rays are high-energy particles which originate outside our solar system but mainly from within our galaxy. They appear to arrive at a point in space from all directions with equal intensity (i.e., they are isotropic). The energy variations and ranges of intensity of all these radiations are depicted in Figure 2. Because of the temporal and spatial variations of these radiations, the actual environment encountered by a spacecraft during a Mars mission depends upon such factors as spacecraft location and time spent within the Earth's magnetic field, mission trajectory, mission duration, solar modulation effects on the GCR radiation levels, and solar activity levels. In low-Earth orbit (LEO), the effects of the Earth's magnetic field on the radiation environment directly depend upon the inclination and altitude of the orbit. For low-altitude (<500 km), low-inclination (near equatorial) missions such as Space Station *Freedom*, the main radiations encountered are trapped protons and electrons from the Van Allen belts and some very

---

[*] High Energy Science Branch, Space Systems Division, NASA Langley Research Center, Hampton, Virginia 23665-5225.

high energy GCR particles. In geosynchronous orbits (GEO), the outer Van Allen belt electrons, solar particle events (SPE), and to a lesser extent, GCR particles, are the main contributors to the radiation environment. For missions beyond the Earth's magnetosphere, such as a human Mars mission, the major radiation considerations are energetic GCR and SPE particles. For GCR particles, the major concern is chronic exposure to the high-energy, heavy ion component (HZE particles). They alone account for most of the GCR exposure [Townsend 1990a] and are more effective than electrons and high-energy protons in causing damage to biological tissue. Galactic cosmic ray particles alone can provide career-limiting exposures to crews on long-duration missions. Extremely large SPEs, consisting of large numbers of energetic protons emitted by the Sun, are rare events which can be potentially lethal to astronauts in thinly shielded spacecraft. For any space mission, radiation exposures from *all* sources must be considered and integrated together for purposes of estimating exposure levels and shielding requirements. An excellent reference which describes space radiation interactions and shielding in great technical detail was recently published [Wilson 1991b]. The reader is urged to search that work for detailed discussions of all techniques and issues discussed in the present paper.

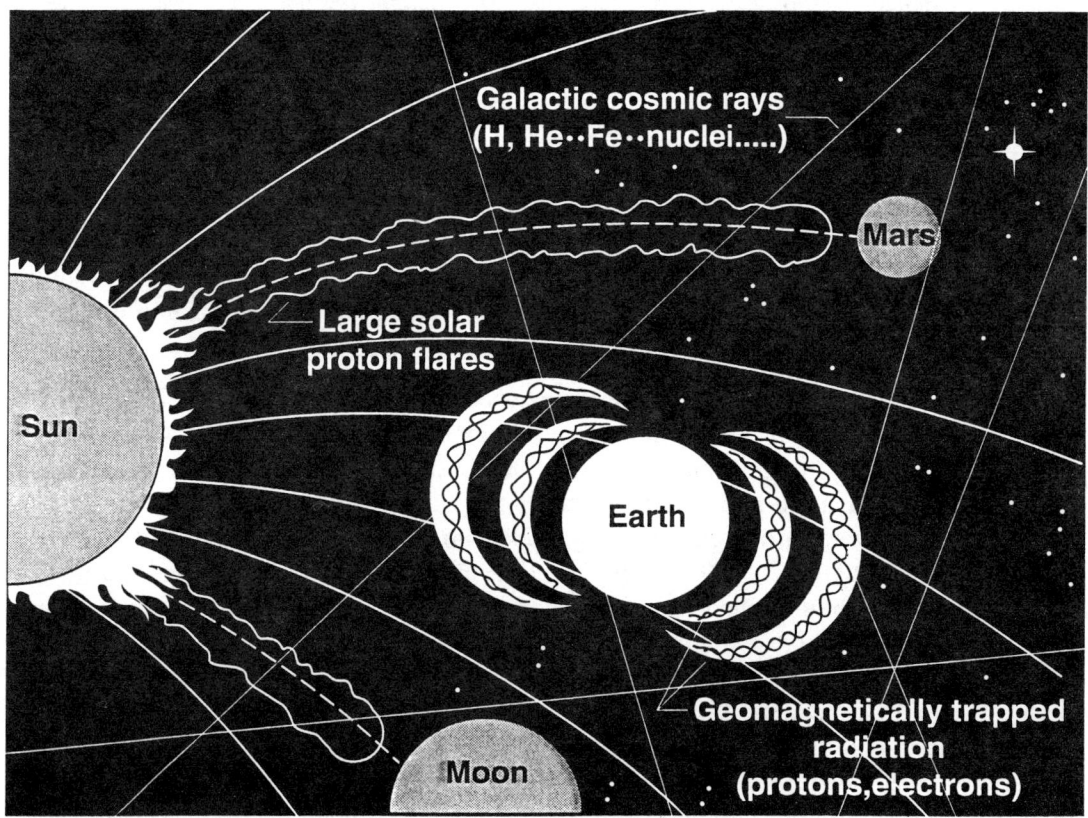

Figure 1 Sources of space radiation include the geomagnetically trapped radiation of the Van Allen belts, large solar flares emanating from the Sun, and galactic cosmic rays which are ever-present in interplanetary space.

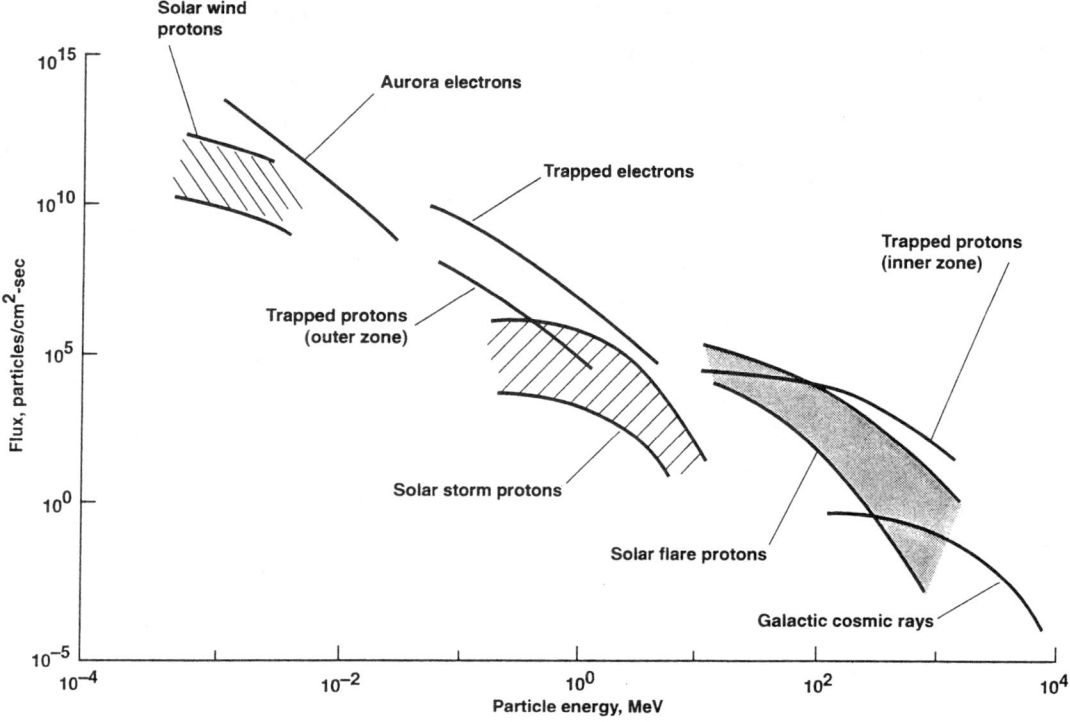

Figure 2 Space radiation environment. The flux distributions are displayed for each type of particle, its source or location, and range of energies. Only particles with energies above 1 MeV are important for radiation protection considerations.

In this paper, an overview of the space radiation environment relevant to a human Mars mission will be described. Then, general methods for incorporating the complex nuclear and electromagnetic interactions of SPE and GCR into radiation transport models which accurately describe the propagation of radiation fields through spacecraft/habitat structures will be outlined. Next, important aspects of radiation biology issues and risk assessment methods will be discussed. The space radiation transport computer codes, BRYNTRN (BaRYoN TRaNsport) and HZETRN (HZE TRaNsport), developed at Langley Research Center, will be briefly described and used to investigate spacecraft shielding strategies. Estimates of radiation exposures for human Mars missions (transit and surface habitat) will be presented. Finally, additional research needs to reduce exposure estimate uncertainties will be presented.

SPACE RADIATION ENVIRONMENT

Evaluations of spacecraft and surface habitat shielding requirements, and estimates of the radiation exposures to be accrued by the crews inhabiting them, cannot be made without a knowledge of space radiation particle types, energies, and fluences (number of particles/unit area) which may be encountered on a mission. Rather than using actual measured data sets, which are often incomplete or imprecise, environment models obtained from considerations of all available data are typically developed for use in actual calculations. In this section, the various types of environmental space radiations and their models will be briefly described.

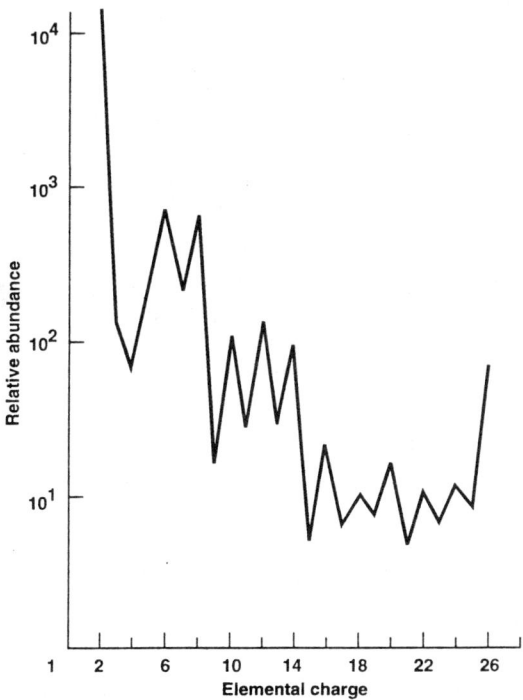

Figure 3 Elemental abundances of galactic cosmic rays by charge number. The abundances are shown relative to silicon (Z = 14) which has been assigned the value of 100.

Galactic Cosmic Rays

The GCR environment consists of approximately 98 percent protons and heavy ions and approximately 2 percent electrons and positrons. Because of their reduced intensities, GCR electrons and positrons are relatively unimportant to human exposure considerations and will be ignored in the remainder of this discussion. Although the relative abundances of protons, helium, and heavier ions vary slightly with energy, approximately 88 percent of these particles are protons, nearly 10 percent are helium isotopes (mainly alpha particles, which are ^{4}He nuclei), and the remaining 1-2 percent are heavy ions (charge number greater than two). Elemental abundances (normalized to silicon = 100) for GCR ions from helium (charge number, Z = 2) to iron (Z = 26) are displayed in Figure 3. The data were taken from Simpson [1983]. Because of their minuscule abundances, ions heavier than iron are unimportant for astronaut exposure consideration and therefore can be safely ignored. GCR ions are not monoenergetic but possess broad energy dependencies, such as displayed in Figure 4. They are also isotropic, arise mainly from sources outside the solar system (possibly supernovae explosions), and possess intensities which are significantly modulated by solar activity levels. During periods of maximum solar activity (called solar maximum), GCR intensities are reduced because the incoming ions are deflected away by the interplanetary magnetic fields carried out from the Sun by the increased volume of plasma in the solar wind which streams out into the solar system. During periods of minimum solar activity (called solar minimum), GCR intensities are the largest observed for that particular solar cycle because the incoming ions are less affected by the reduced interplanetary magnetic field strengths. Occurring over an approximate 11-year cycle, this modulation effect can change ion fluxes by more than a factor of 2 for ion kinetic energies between 10 MeV/nucleon and 10 GeV/nucleon. There are also intensity variations between solar cycles. For example, the 1977 solar minimum period had larger GCR particle fluxes (number of particles/unit area/time interval) than the 1965 minimum. These ions make only a minor contribution to total radiation exposure during near-equatorial LEO missions (such as Space Station *Freedom*) because the dipole-like nature of the Earth's magnetic field deflects the lower-energy ions resulting in substantially reduced GCR fluxes for energies below several GeV/nucleon. For human interplanetary missions, such as to the planet Mars, galactic cosmic rays will likely be the major source of radiation exposure to the crews.

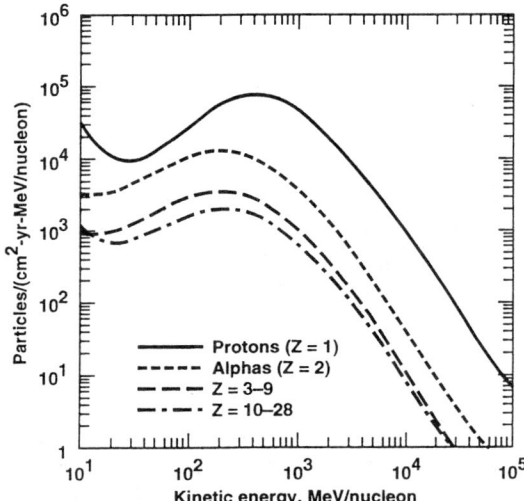

Figure 4 Galactic cosmic ray ion spectra for solar minimum conditions according to the model of Adams [1986]. Curves are shown for protons (Z = 1), alpha particles (Z = 2), the cumulative flux of lithium through fluorine ions (Z = 3-9), and the cumulative flux of neon through nickel ions (Z = 10-28).

For use in radiation exposure and shielding calculations, Adams and collaborators [1986] at the Naval Research Laboratory (NRL) have formulated a GCR environment model which provides energy spectra for all elements from hydrogen (Z = 1) through uranium (Z = 92). The ion fluxes displayed in Figure 4 were obtained from this model. Later in this paper, we will use it to estimate crew exposures and shielding requirements for a Mars mission. Recently, Badhwar and O'Neill [1991] have developed an improved GCR environment model which permits evaluation of exposures for specific time periods (e.g., 1977 solar minimum) using the actual measured fluxes/fluences from that period. It is likely that the Badhwar and O'Neill model will eventually replace the NRL model as the model of choice for future studies.

Solar Particle Events

A major concern for radiation protection purposes is the random occurrence of an extremely large solar particle event (SPE) involving large emissions of charged particles (protons, alphas, and heavier ions). Recent measurements of alphas and heavier ions in an SPE, however, were generally limited to particle energies of less than 20 MeV/nucleon [Shea 1990]. Therefore, we will consider only SPE proton fluences when estimating crew exposures and radiation shielding requirements.

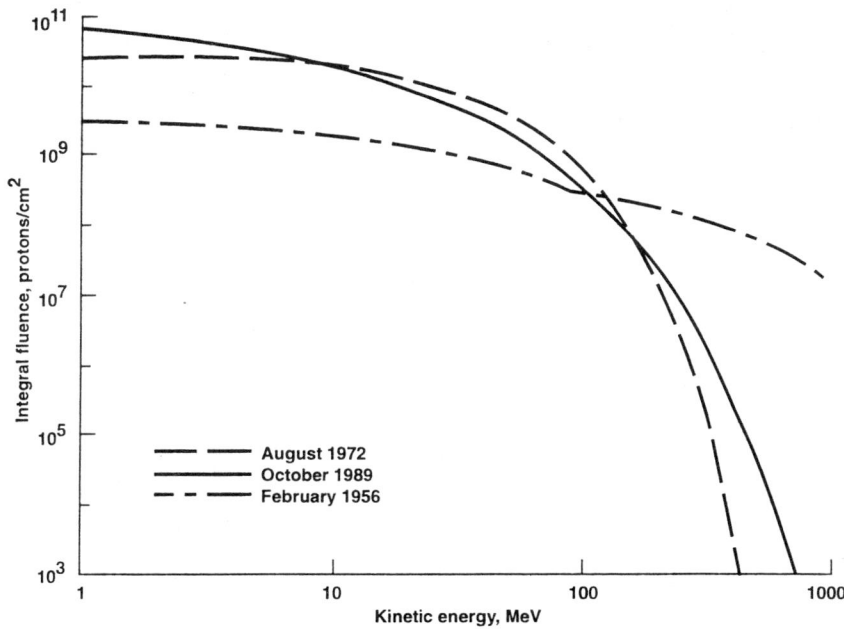

Figure 5 Integral fluence spectra for three of the most intense solar particle events ever observed. Displayed are the number of total protons/cm^2 (integral fluence) in the event having energies greater than the values on the abscissa.

Because of the complexities of solar processes, not all solar flares produce solar particle events. In addition, proton fluences associated with a particular SPE are not always large enough or energetic enough to pose a significant health hazard to the crew. The largest events, which can have total fluences exceeding 10^{10} protons/cm^2 with energies greater than 10 MeV, happen infrequently and sporadically. Typically occurring near periods of maximum solar activity, but not necessarily at the time of solar maximum, events large enough to pose a significant hazard to crews may occur up to several times during a particular solar cycle, or may not occur at all. Unfortunately, the ability to accurately forecast the occurrence of a large SPE for weeks, months or years in advance is virtually nil. Present technology will provide reasonable SPE forecasts for periods of time up to approximately 24 hours in advance [Heckman 1988]. Figure 5 displays integral fluence spectra for three of the largest and most penetrating events ever recorded [Wilson 1976, King 1974, Townsend 1991]. The February 1956 event had the hardest (most energetic) spectrum and was therefore the most penetrating. Fortunately, its intensity was much less than either of the other two events. For analysis purposes, the February 1956 SPE is typically used for estimating exposures and shield requirements where substantial quantities of shield/structural material, such as for habitats on the Martian surface, are available. For thinly shielded spacecraft, the August 1972 or October 1989 SPE spectra are often used as inputs into shielding and exposure calculations because they have the largest fluences and give the most limiting crew exposures [Townsend 1991]. Differential and integral fluence spectra for SPEs are usually provided by the Space Environment Laboratory of NOAA (National Oceanic and Atmospheric Administration) from measurements obtained by instruments onboard Interplanetary Monitoring Platform (IMP) and Geostationary Operational Environmental Satellites (GOES).

Van Allen Belts

The Earth's magnetic field is roughly similar to that of the dipole field associated with a simple bar magnet. The field axis is tilted away from the Earth's rotation axis by 11° in a plane passing through 70.9°W longitude. The field center is further displaced from the Earth's center by 489 km toward a point in the northwest Pacific Ocean (20.4°N, 147.3°E). Energetic protons and electrons, and some low-energy, heavy ions (ignored hereafter because they are unimportant for human exposure considerations) are trapped by this magnetic field and produce what are known as the Van Allen radiation belts, named for their discoverer, James Van Allen. The trapped protons and electrons gyrate about and oscillate along the magnetic field lines, bouncing back and forth between mirror points in the northern and southern hemispheres. In addition, there is an azimuthal drift about the Earth (westward for protons, eastward for electrons). Figure 6 displays the distributions of these trapped protons and electrons. Low energy integral fluences are depicted on the right-hand side and higher-energy fluences on the left. The stable trapping zone occurs at the lowest altitudes (less than 4 Earth radii) near the equatorial region and is known as the "inner" zone. Particle populations are probably supplied mainly by the decay of neutrons produced by particles colliding with the atmosphere. The "outer" zone is marked by short particle lifetimes because of radial drift and is kept populated by intense injection of particles through the magnetic tail region. The "inner" belt is primarily dominated by energetic protons and the "outer" belt by electrons. Because of the magnetic field axis tilt and displacement, there is a region in the south Atlantic, near the Brazilian coast, called the South Atlantic Anomaly (SAA), where the inner Van Allen radiation belts come very close to the Earth's surface. Typical low-altitude, low-inclination (near equatorial) orbits, such as this are usually flown by the Space Shuttle (and planned for Space Station *Freedom*), are relatively free of radiation except for passages by the spacecraft through the SAA (typically six passes per day). GCR and SPE radiations are usually not a problem because the Earth's magnetic field largely deflects them away. The only caveat to this generalization is the possibility of increased SPE exposures in LEO if geomagnetic storm conditions significantly alter the Earth's magnetic field (see Wilson [1990a] for further details). For high-inclination (nearly polar) orbits, SPE and GCR fluences will be a significant component of the incident radiation environment. For human missions to Mars, the trapped radiation environment will provide radiation exposures which are likely to be a minor component of the total dose received by the crew [Nealy 1991], if rapid transits through the belts are made. The currently used models of the trapped environment, provided by NASA Goddard Space Flight Center, are designated as AP-8 for protons and AE-8 for electrons [Gaffey 1990].

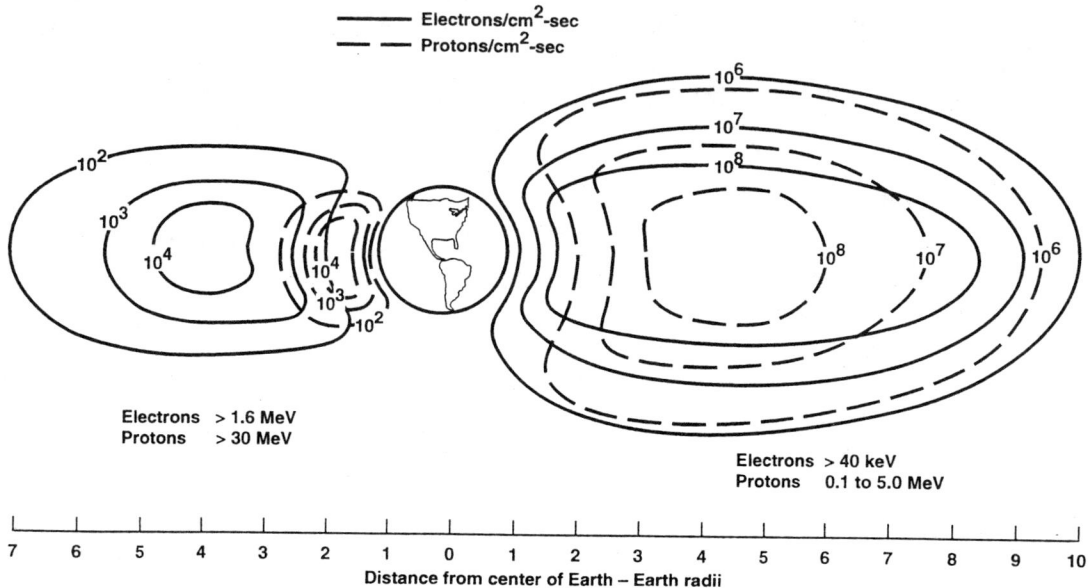

Figure 6 Near-Earth trapped radiation environment [Parker, 1973]. Low energy integral fluxes are shown on the right while the higher energy integral fluxes are shown on the left.

Artificial Sources

Artificial sources of radiation which may result in crew exposures include nuclear power sources (propulsion and power), such as nuclear reactors and radioisotope thermal generators (RTG), as well as radioactive sources for medical procedures and diagnoses. As appropriate for a particular mission scenario, these radiation sources must be considered when assessing shield requirements and crew exposure levels. Especially important is the potential impact of radiations from an operating nuclear reactor on spacecraft shielding and crew exposure assessments.

RADIATION INTERACTIONS AND TRANSPORT

As high-energy space radiations traverse bulk matter, such as a habitat or spacecraft, their radiation fields change composition through interactions with the materials in their paths. As a result of these interactions, the internal radiation environment within the habitat/spacecraft can differ appreciably from the incident external environment. These alterations in the incident radiation environment depend upon the thickness, geometry, and material composition of the vehicle. They are described by transport models which relate the transmitted fluxes (as a function of spatial location, kinetic energy, and directions of particle motion) to the incident fluxes.

Physical Interactions

The main interaction processes involved in the transport of these radiation fields through bulk matter are (1) ionization energy losses through collisions with atomic electrons, (2) nuclear elastic and inelastic collisions, and (3) nuclear breakup (fragmentation) and electromagnetic (EM) dissociation interactions. Nuclear breakup and EM dissociation are particularly important because fragmentations result in the production of reaction products which alter the elemental and isotopic composition of the transported radiation fields. The radioactive decay contributions of the transported fragments are ignored because their decay times are typically much longer than the time required for the radioactive fragment to exit the spacecraft or undergo a subsequent nuclear collision. Radioactive decay of vehicle internal components and structures should be considered, however, in any final crew exposure determination.

Stopping Power. In passing through a material, an ion loses a substantial fraction of its energy to atomic excitation of the material. Therefore, the dominant term in any shielding calculation is energy loss through ionization; that is, collisions between the incoming charged particle (proton, electron, or heavy ion) and the orbital electrons of the atoms of the shielding material. The interaction is through Coulomb scattering. The cross section is inversely proportional to the square of the energy transfer which is usually quite small. For electrons bound in atomic orbitals, another possibility involves excitation energy transfer where the orbital electrons transition to less-tightly bound orbits, but the atoms are not ionized. Again, the cross section varies as the inverse of the square of the energy transfer. A third process that can be important (especially for incident electrons) is the Coulomb interaction with the atomic nucleus, which results in multiple-scattering effects. For a typical ion, there are $\sim 10^6$ atomic collisions/cm of travel in ordinary matter each involving energy transfers which are small compared with the ion kinetic energy. Therefore, energy loss by atomic collision is usually expressed in a continuous slowing down approximation. These three processes are schematically depicted in Figure 7. Details of methods used to incorporate these atomic collision processes into the transport codes are given elsewhere [Wilson *et al.*, 1991a]. Typical uncertainties range from several percent (energies above several MeV/nucleon) to approximately 40 percent for some heavy ions at low energies (below several MeV/nucleon).

Nuclear Absorption. The flux of fluence of particles propagating through shielding is also reduced by collisions where the nuclear fields of the colliding nuclei interact. Unlike atomic collisions, the mean-free path for nuclear collisions is large (on the order of a centimeter), and the energy transfer can be large in comparison to the total ion kinetic energy. Often, however, the range of the particle is comparable to its nuclear collision mean-free path so that cosmic rays typically undergo one or more nuclear collisions before coming to rest. The nuclear collision usually alters (loss of mass and charge) both the incident ion and the struck nucleus. This attenuation or removal of flux by nuclear collision is usually expressed as a macroscopic absorption cross section (the inverse of the mean-free path) given by

$$\sigma_i = \sum_j \rho_j \sigma_{abs}(i,j) \qquad (1)$$

where the r_j are the elemental constituent-number densities for the target, and the s_{abs} are the microscopic nuclear absorption cross sections for the nuclear collision pair. These are often obtained from parameterizations [Townsend 1986, Wilson 1988a] or from formal nuclear scattering theory [Townsend 1985a]. The absorption cross sections used in equation (1) must be accurate because the transported radiation flux in a propagation calculation depends exponentially upon these cross sections. Therefore, relatively small errors are compounded as the radiation field passes through the shield material. Typical cross section uncertainties are on the order of a few percent at energies greater than 25 MeV/nucleon [Townsend 1985a].

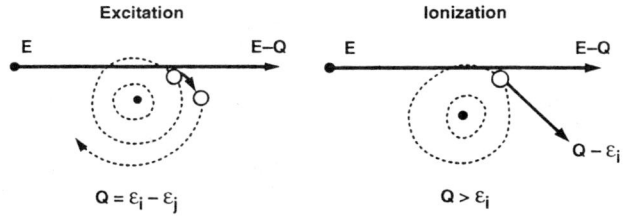

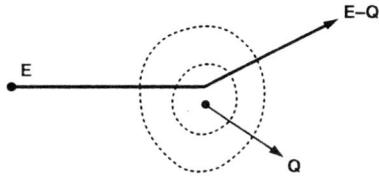

Figure 7 Diagrams of atomic interactions. The interactions with atomic orbital electrons include excitation, where the orbital electron is excited to a higher level in the atom, and ionization, where the orbital electron is completely knocked out of the atom. There is also a Coulomb interaction with the nucleus of the atom whereby the incoming particle scatters and the atomic nucleus recoils. The symbols in the figure are incident particle energy (E), final particle energy (E—Q), energy transfer (Q) by the collision, and the e's represent energies for the various atomic orbits.

Nuclear Fragmentation. Although nuclear collisions attenuate the propagating ion fluxes, they also serve as the sources of new energetic particles. The process of transferring kinetic energy into secondary particle production occurs through several different mechanisms, including direct knockout of nuclear constituents and resonant excitation followed by particle emission. At high energies, secondary particle production is dominated by fragmentation processes whereby the incident projectile nucleus and/or the target nucleus breaks up into lighter mass constituents. Both nuclear and coulomb fields contribute to these breakups.

Accurate fragmentation cross sections are necessary because they directly yield the fluence spectra of secondary and other subsequent-generation reaction products present in the propagating radiation fields. As will become apparent in the later sections of this paper, fragmentation products of the incident GCR environment become the dominant source of crew exposure for moderate thicknesses of spacecraft shielding. Unfortunately, it is totally impractical to measure all the necessary cross sections because transport calculations require values for all incident GCR species (mainly hydrogen through iron) colliding with any target material (spacecraft structure/shielding, an astronaut's body) over several decades (factors of 10) of incident particle energies. Thus, literally thousands of fragmentation cross sections must be known in order to carry out shielding and exposure studies.

At present there is no suitably accurate theory for predicting nuclear fragmentation (breakup) cross sections for all collision pairs at all energies of interest. Detailed quantum-mechanical formulations based upon optical potential considerations are being developed [Townsend 1985b]. These methods are too complex to use in a transport calculation. As an alternative, a semiempirical abrasion-ablation model has been developed [Wilson 1987a]. Although computationally simpler and faster, the model conserves fragment mass and charge, is physically realistic, has only a single adjustable parameter, and agrees reasonably well with most available experimental data. Typical uncertainties and differences between cross section data sets, taken by different groups, for the same collision pairs at the same or nearly the same incident energies may be as large as a factor of 2.

Electromagnetic Dissociation. Because the dissociation of projectile and target nuclei by their interacting Coulomb fields may be important for some heavier nuclei at high energies, the electromagnetic dissociation cross-section contributions, s_{em}, must be added to the nuclear fragmentation cross section, s_{nuc}, to yield the total fragmentation cross section. Methods for estimating these have been developed and parameterized for use with this fragmentation model [Norbury 1988] and are being incorporated into the GCR transport codes.

Other Nuclear Interaction Processes. In addition to the interactions previously discussed, space radiations will induce other nuclear collision processes which must be included where appropriate in a complete radiation transport calculation. Among these interactions are nuclear elastic scattering, quasielastic scattering, nucleon nonelastic scattering (where the target nucleus remains in an excited state whose decay to the ground state is ignored), and target fragmentation events. The latter are particularly important for the nucleonic component of the propagating radiation fields. For example, at 10 A GeV (10 GeV/nucleon), target fragments contribute nearly 60 percent of the total dose equivalent resulting from incident protons, over 40 percent resulting from incident alpha particles, and only 0.04 percent resulting from incident iron ions. Detailed methods for incorporating all these interactions into a shielding/exposure calculation are described elsewhere [Wilson 1989, Wilson 1990b].

High-Energy Heavy Ion (HZE) Transport Theory

The propagation of these radiation fields is described by a Boltzmann equation, which can be derived from considerations of mass and energy conservation. Its solutions give values of particle fluxes and energies everywhere within and exiting the boundaries of the target medium. For GCR particles, the typically large ion kinetic energies allow one to neglect changes in particle direction because of collisions. This is called a straightahead approximation. The one-dimensional Boltzmann equation is then written [Wilson 1983, 1984]

$$\left[\frac{\partial}{\partial x} + \sigma_i(E) - \frac{\partial}{\partial E} S_i(E)\right] \phi_i(x, E) = \sum_i \sigma_{ij}(E) \phi_j(x, E) \qquad (2)$$

The terms on the left-hand side of equation (2) represent particle flux losses because of translation ($\partial/\partial x$), nuclear absorption processes (s_i) and energy loss from atomic/nuclear collisions (S_i). The term on the right-hand side represents flux gains resulting from fragmentation processes. In equation (2), f_i is the flux of type i ions at position x with motion along the x axis and energy E in units of A MeV (or MeV/nucleon), $s_i(E)$ is the corresponding macroscopic nuclear absorption cross section in units of cm^{-1}, $S_i(E)$ is the change in E per unit distance (e.g., the stopping power per unit projectile mass), and $s_{ij}(E)$ is the fragmentation cross section, in units of cm^{-1}, for producing ion i from a collision by ion j. Methods of evaluating these cross sections and stopping powers were discussed in the previous section. Solution methods for equation (2) are usually optimized for the specific transport problem being investigated. For accelerator-beam studies, solutions can be obtained using analytic methods in a perturbation expansion [Wilson 1984]

$$\phi_i(x, E) = \sum_{n=0}^{\infty} \phi_i^{(n)}(x, E) \qquad (3)$$

where n = 0 denotes the primary beam, n = 1 denotes secondaries produced by fragmenting of the primaries, n = 2 denotes tertiaries produced by fragmenting of the secondaries, etc. For the GCR transport problem, the incident GCR particles possess broad energy spectra; therefore, a different solution technique is required for computational efficiency [Wilson 1987b]. This is accomplished by using the first two terms of the perturbation expansion in an iterative procedure which effectively sums subsequent generations of reaction products to all orders. Taking the initial point on the boundary then allows propagation to any arbitrary depth within the interior. The computational algorithm has been verified to within 2-percent accuracy by comparison with an analytical benchmark [Wilson 1988b].

Nucleon Transport Theory

To characterize the radiation fields produced by primary and secondary nucleons (neutrons and protons) from incident solar particle events and galactic cosmic rays, a deterministic coupled neutron-proton transport code, BRYNTRN, has been developed [Wilson 1989]. Because of the high energies involved, the straightahead approximation is again introduced into the Boltzmann equation to yield

$$\left[\frac{\partial}{\partial x} - \frac{\partial}{\partial E} S(E) + \sigma_p(E)\right] \phi_p(x, E) = \sum_j \int_E^\infty f_{pj}(E, E') \phi_j(x, E') dE' \tag{4}$$

for protons. For neutrons, there is no stopping power contribution S(E) so that it becomes

$$\left[\frac{\partial}{\partial x} + \sigma_n(E)\right] \phi_n(x, E) = \sum_j \int_E^\infty f_{nj}(E, E') \phi_j(x, E') dE' \tag{5}$$

where the f_{ij} are differential cross sections for elastic and nonelastic collision processes, and the remaining symbols are defined following equation (2). Solutions to equations (4) and (5) have been extensively described elsewhere [Wilson 1989, 1991b]. These computational algorithms were verified to within 1-percent accuracy by direct comparison with an analytical benchmark solution for a continuous space proton input spectrum [Wilson 1988c].

RADIATION HEALTH AND RISK ASSESSMENT

At the cellular level, biological damage produced by energetic space radiations include cell killing, cell mutation, and neoplastic transformations. The latter may result in the formation of neoplasms or new and abnormal growths, such as tumors. Current knowledge of the biological effects of space radiations largely results from fundamental radiation biophysics and radiation therapy research. The space radiation environment is unique, however, because its exposures are almost entirely the result of particulate radiations. Although the effects of these radiations have been studied using systems at various levels of biological organization, there is a paucity of information concerning SPE/GCR proton and heavy ion irradiation effects in humans. Therefore, extrapolations of experimental animal data to humans are often made [NCRP 1989]. Because there is a substantial quantity of radiobiology research literature available, we will refer the interested reader to several excellent reviews which are available [NCRP 1989, NAS/NRC 1967, and Tobias 1974].

In this section, dosimetric quantities will be defined and health effects from interplanetary space radiations discussed. The issue of what radiation limits/guidelines apply to a human Mars mission will also be addressed.

Dosimetric Quantities and Units

Using the transport methods described earlier, the transmitted fluxes/fluences of radiation at a depth x with energy E can be calculated. Note that it is particle flux/fluence which is transported. Dose and dose equivalent are not transported—they are derived quantities calculated from the transported fluxes/fluences.

After the particle fluxes for each ion species (proton or HZE) have been determined, the dose (energy absorbed per gram of target material) is computed from

$$D_i(x, > E) = A_i \int_E^\infty S_i(E') \phi_i(x, E') \, dE' \tag{6}$$

The symbol $D_i(x, > E)$ refers to the dose at depth x resulting from all of the ith ionic species whose kinetic energies are greater than E. The S. I. unit of absorbed dose, the gray (Gy), is defined such that 1 Gy is equal to the absorption of 1 joule of energy per kilogram of material. An earlier unit, the rad, continues to be widely used. For conversion purposes, 1 Gy equals 100 rad. Because different target materials have different stopping powers, $S_i(E¢)$, the absorbed dose computed from equation (6) is a specific quantity which differs for different target materials. In this paper, all doses will be computed in water, a material whose radiation absorption properties closely resemble body tissue.

It is well known that different radiation types cause different amounts of biological damage per unit absorbed dose. HZE particles and low-energy protons deposit high rates of energy per unit track length and are more effective than electrons and high-energy protons which lose less energy per unit track length. The rate of energy loss per unit track length is called LET (linear energy transfer). Hence, radiations which lose high rates of energy are called high-LET radiations (e.g., HZE particles and low-energy protons). Electrons and high-energy protons are examples of low-LET radiations. The quantity that expresses the biological effect of interest for all types of radiations on a common scale is called "dose equivalent." Hence, for risk assessment purposes, the dose equivalent is obtained from

$$H_i(x, > E) = A_j \int_E^\infty Q_i(E') S_i(E') \phi_i(x, E') \, dE' \tag{7}$$

where Q denotes the quality factor which relates absorbed dose to risk for a particular radiation type. These are linear-energy-transfer (LET)-dependent quantities obtained from ICRP-26 [1977]. The S. I. unit of dose equivalent is the Sievert (Sv). An earlier unit, the rem (1 rem = 0.01 Sv), is also commonly used.

Equations (6) and (7) are used to compute absorbed dose and dose equivalent contributions for the propagating ionic species (protons and HZE particles) in the transport calculations. For target fragments and nuclear recoils produced by propagating protons, the absorbed dose and dose equivalent are given by

$$D_p^*(x, >E) = \sum_j \int_E^\infty \overline{E}_j{}' \, \sigma_{jp}(E') \phi_p(x, E') \, dE' \tag{8}$$

and

$$H_p^*(x, >E) = \sum_j \int_E^\infty \overline{E}_j{}' \, Q_j(E') \sigma_{jp}(E') \phi_p(E') \, dE' \tag{9}$$

where $\overline{E}_j{}'$ is the average recoil energy of the jth fragment. For propagating neutrons, all of the dose and dose-equivalent results from target nuclear recoils and fragments and are computed using equations identical in form to (8) and (9). Dose and dose equivalent contributions from target fragments produced by propagating HZE particles are obtained from similar methods [Wilson 1990b].

Health Effects

Health effects of space radiations fall into two general categories: (1) early or acute effects and (2) late or delayed effects.

Early/Acute Effects. Early effects occur within hours, days, or weeks following a high dose exposure of the whole body. They are clinically significant only when doses greater than 1-2 Gy are received in short time periods, such as minutes to hours [NCRP 1989]. It is likely that only a large SPE will produce such exposure levels and then only outside the Earth's magnetosphere. GCR exposure rates are too low to cause acute effects. Lower doses can, however, produce significant cell or organ injury, even if not clinically dangerous. Table 1, taken from NCRP Report No. 98, [NCRP 1989] lists health effects as a function of exposure level. The main concerns are proliferating cells of renewing tissues and organs, such as bone marrow, gonadal tissues, intestinal epithelium and the skin. As indicated in the table, acute whole body doses in the range 2 to 4 Gy can be lethal, usually with 20 to 40 days after exposure. Another significant concern for crews is vomiting, which could be lethal for someone in a helmeted space suit. Other acute effects which may occur include erythema (skin reddening) and epilation (hair loss) at about 6 Gy and moist desquamation (skin blistering) at about 18 Gy.

Late/Delayed Effects. Late effects usually manifest themselves months or years after exposure. These include tissue damage, cancer induction, lens opacifications, fertility changes, and developmental abnormalities in offspring. Late effects are classified as stochastic or nonstochastic. Stochastic effects are those in which the probability of producing the effect, but not the severity of the effect, increases with increasing dose. Examples are neoplasms and genetic damage. Nonstochastic late effects are those in which the severity of the effect increases with increasing dose, and there are threshold doses below which no effect is clinically detected. Examples are opacification of the ocular lens (cataracts) and graying of the hair. The principle concern for late effects of radiation is cancer induction. Current career limits for Space Station *Freedom*, for example, are based upon a lifetime excess cancer risk of 3 percent for crew exposures to radiations in LEO. An excellent overview of radiation effects can be found in NCRP Report No. 98 [NCRP 1989].

Table 1
RADIATION EXPOSURE LEVELS NEEDED TO INDUCE ACUTE HEALTH EFFECTS
(From NCRP Report No. 98)

Effect	Dose (Gy)
Blood count changes (population)	0.15 - 0.25
Blood count changes (individual)	0.5
Vomiting threshold	1.0
Mortality threshold	1.5
LD_{50}, minimal support care	3.2 - 3.6
LD_{50}, supportive medical treatment	4.8 - 5.4
LD_{50}, bone marrow/blood stem cell transplant	11.0

LD_{50} = dose of radiation which kills, within a specified time period, 50 percent of the individuals in a population.

Radiation Quality and Limits. Recall that the quantity which relates absorbed dose to risk is the quality factor (Q). The appropriate values of Q to use, obtained from ICRP Report No. 26 [ICRP 1977], are plotted in Figure 8 as a function of LET. The relatively large values of Q associated with high-LET radiations (above 100 keV/micron) are derived from studies of carcinogenesis and mutagenesis and do not apply to acute or early effects or to late non-cancer organ effects. Therefore, the appropriate radiation quantity to use to estimate acute exposure effects, such as might result from a large SPE, is absorbed dose (Gy) and not dose equivalent (Sv).

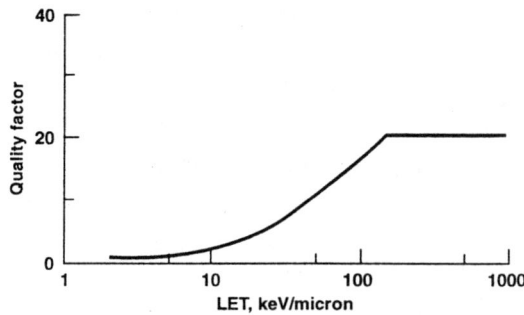

Figure 8 Quality factor as a function of linear energy transfer (LET) according to ICRP Report No. 26 [1977]. The quality factor is used to relate radiation risk to the absorbed dose.

In all future situations, radiation protection must be planned to ensure that astronaut exposures are kept As Low As Reasonably Achievable, the ALARA principle. At present no radiation limits exist for a human mission to Mars. The limits recommended by the NCRP in Report No. 98 pertain to LEO missions, such as SSF. No specific limits for human exploratory missions were recommended. The SSF limits, shown in Table 2, are useful as guidelines for discussion/comparison purposes but should not be used as actual design limits, especially since NCRP Scientific Committee 75 has been reconstituted for the purpose of replacing NCRP Report 98 with a new document possibly containing revised SSF limits. (One of us, L. W. T. is a member of this committee). The 30 days and annual limits in Table 2 were established in order to avoid acute or nonstochastic effects in critical organs. As mentioned previously, the career limit is primarily based upon limiting excess cancer risk.

Table 2
NCRP REPORT 98 DOSE EQUIVALENT LIMITS
APPLICABLE TO LOW EARTH ORBIT MISSIONS

	BFO (cSv)	Eye (cSv)	Skin (cSv)
Career	100 - 400*	400	600
Annual	50	200	300
30 Days	25	100	150

*depends on age and gender

SPACE RADIATION TRANSPORT CODES

BRYNTRN

For transport of nucleons (neutrons and protons) through arbitrary target materials, a deterministic nucleon (baryon) transport code (BRYNTRN) has been developed. Detailed descriptions of the computational procedures and input interaction data bases have been published [Wilson 1988a, 1989, 1991b]. The computational algorithm has been verified to within one-percent accuracy by direct comparison with an analytical benchmark solution [Wilson 1988c] for a continuous space proton input spectrum. The BRYNTRN code is also computationally efficient. An analysis of a large SPE for shield thicknesses ranging up to 100 g/cm^2 in areal density typically can be completed in about 30 minutes of running time on a VAX 11-785 computer. Comparisons of dose estimates from BRYNTRN with Monte Carlo computer codes are in impressive agreement [Shinn 1990a]. The current version of the code will accept continuous spectral distributions from GCR/SPE protons and trapped proton spectra as input. There are no limitations in the transport code regarding the number of layers of target material or their thicknesses. The maximum number of constituents in each layer is usually limited to five but only to limit the sizes of the dimensional arrays in the code. The code can be easily modified to handle more than five constituent materials in a particular layer, if required.

The calculations utilize LET-dependent values of Q from ICRP-26 [ICRP 1977]. Options in the computer code permit the use of recently proposed revisions [ICRU 1986, Sinclair 1990] to Q if desired. The code output includes particle fluxes/fluences, dose, and dose equivalent from propagating neutrons and protons, their target nuclear fragments, and their target nuclear recoils. All dosimetric quantities are given as a function of distance into the target material. Integral LET spectra are also provided. The main shortcomings of this code are that the present versions do not include mesons or electromagnetic cascade effects. In addition, target-decay contributions (beta and gamma rays) are not yet included.

HZETRN

For transport of galactic cosmic rays (nucleons and HZE particles) and their reaction products through arbitrary target materials, a deterministic transport code was developed by coupling the HZE component of an earlier GCR transport code [Wilson 1987b] with the nucleon transport code BRYNTRN. This coupling produced a single, complete, computationally efficient computer code for use in GCR shielding and exposure studies. Detailed descriptions of the computational procedures and input interaction data bases are published elsewhere [Townsend 1992a, Wilson 1991a]. The computational algorithm used has been verified to within 1.5-percent accuracy by direct comparison with an analytical benchmark solution [Wilson 1988b]. The current version of the code uses the incident GCR spectrum obtained from the Naval Research Laboratory CREME model [Adams 1986]. If desired, combined GCR and SPE spectra can be analyzed using the code [Townsend 1990b]. As was the case for BRYNTRN, there are no limitations in HZETRN regarding the number of layers of target material on their thicknesses. The maximum number of different constituents in each layer is presently limited to five but can be easily modified to accommodate more than five, if necessary. The calculations utilize LET-dependent values of Q [ICRP 1977] with options to use recently proposed revisions [ICRU 1986, Sinclair 1990] if desired. The code output includes flux/dose/dose-equivalent contributions from propagating neutrons, protons, alpha particles, high energy, heavy ions, and their subsequent generation reaction products. LET distributions are also calculated. All dosimetric quantities can be listed as a function of LET or distance into the target material.

The main shortcomings of HZETRN are as follows:

1. Except for tissue targets, mass number 2 and 3 fragment contributions are neglected.

2. All secondary particles produced by HZE interactions are assumed to be produced with a velocity equal to that of the incident particle.

3. Meson and electromagnetic cascade contributions are ignored.

4. Contributions from target nuclear decays are ignored.

These shortcomings are not conservative and therefore result in an as yet unknown underestimation of the exposure.

SPACECRAFT SHIELDING STRATEGIES

Providing adequate radiation protection measures for a human mission to Mars is unlike any previous experience in the history of the human space program. For the first time, crews will be exposed to biologically significant fluences of high-LET galactic heavy ions, for which no risk estimates are directly available from previous human exposure. In addition, typical Mars scenarios assume mission durations of 1-3 years. Such lengthy missions, beyond the protection of the Earth's intrinsic magnetic field, could result in a greater than 10-percent probability of occurrence of a very large, mission-threatening SPE [Feynman 1988]. Unlike LEO missions, where return to the Earth's surface can be accomplished quickly, crews on a Mars mission may be months or years away from any possible return to Earth. If the spacecraft is en route to or from Mars, the only available shielding is what has been carried onboard. Adding additional shielding, if needed, is not an option; unlike the situation on the Martian surface where regolith may be available for use. Therefore, inadequate shielding may present a potential health hazard to the crew; however, being too conservative may result in large weight penalties for unnecessary shielding—thereby adversely impacting the mission in a different manner. Two variables which directly impact crew exposures and vehicle shield design are shield-material composition and vehicle geometry. The goal is to optimize both in order to maximize crew protection while concomitantly minimizing shield total mass. Strategies for doing both will be discussed. Finally, the possible use of electromagnetic shielding as an alternative to bulk material shielding will be considered.

Material Composition

Utilizing the space radiation transport methods and codes described in the previous sections, estimates of dose and dose equivalent for energetic radiations impinging upon any conceivable shield material can be made. The usual procedure is to transport the incident SPE or GCR spectrum through the desired slab thickness of shield material followed by an additional slab thickness of water to simulate human soft tissue. Skin dose/dose equivalent is usually approximated by a 0-cm thickness of water. Blood-forming organ (BFO) dose/dose equivalent is approximated by a 5-cm thickness of water. Later in this paper, we will present some organ results using an actual computerized man model. Therefore, to avoid confusion, we will reserve the use of the terms skin, eye, and bone marrow (or BFO) for actual organ-exposure estimates and use the terms "0-cm" and "5-cm" to refer to calculations using the water-thickness approximation.

It is customary in radiation-protection analyses to use areal density (mass per unit area) rather than actual thickness as a shielding "thickness" unit, since different shield materials with the same areal density will have the same total mass, even though their actual thicknesses differ. In terms of actual density r and thickness t, the areal density is obtained from the expression

$$\text{areal density} = rt \tag{10}$$

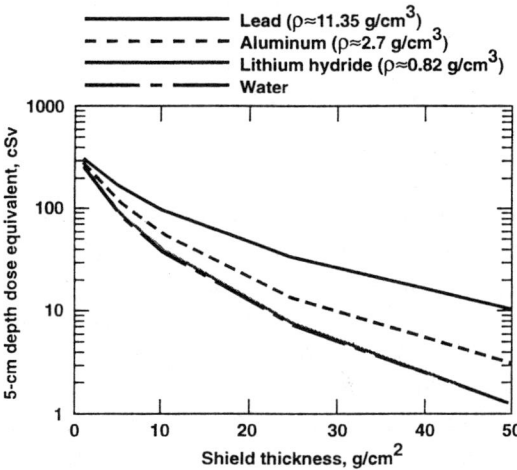

Figure 9 5-cm depth dose equivalent versus shield thickness for the sum of the six 1989 solar particle event fluences. Results for four different candidate shield materials are displayed.

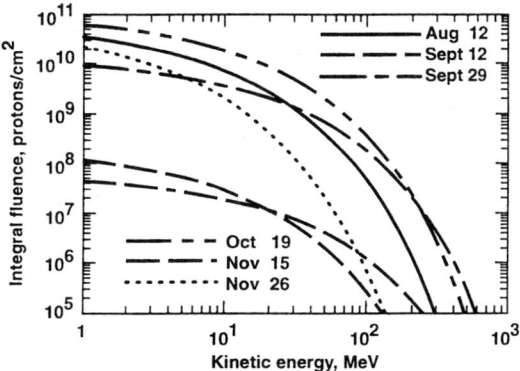

Figure 10 Integral fluence spectra for the six large solar particle events of 1989. The curves are based upon data from the GOES-7 satellite [Sauer, 1990].

Figure 9 displays 5-cm depth-dose-equivalent results for several different candidate shield materials for the combined spectra of the six large SPEs, which occurred in the Fall of 1989 [Simonsen 1991a]. The individual integral fluence spectra for these six SPEs are displayed in Figure 10. The data were obtained from measurements on the GOES-7 satellite [Sauer 1990]. From Figure 9, we note that water (H_2O) and lithium-hydride (LiH) are nearly identical in shielding efficacy. Either is a substantially better SPE shield material than metals, such as aluminum (Al) or lead (Pb). At approximately 15 g/cm^2 shield thickness, the 5-cm depth dose equivalent is 0.25 Sv (25 rem) for LiH and H_2O, 0.38 Sv (38 rem) for Al, and nearly 0.80 Sv (80 rem) for Pb. Reducing the exposures behind Al and Pb to 0.25 Sv would require nearly 19 g/cm^2 of Al or 32 g/cm^2 of Pb. The main reason that low, atomic-weight materials, such as H_2O and LiH, are more effective than Al or Pb is because the propagating particles are mainly nucleons, which lose a larger fraction of their kinetic energies when colliding with target nuclei of equal or nearly equal masses. Hydrogen target nuclei do not break up and contain no neutrons; therefore, target fragment and recoil contributions to the dose/dose equivalent are minimal. These results clearly suggest that hydrogenous compounds of low atomic weights are the materials of choice for maximizing SPE protection while minimizing total shield mass.

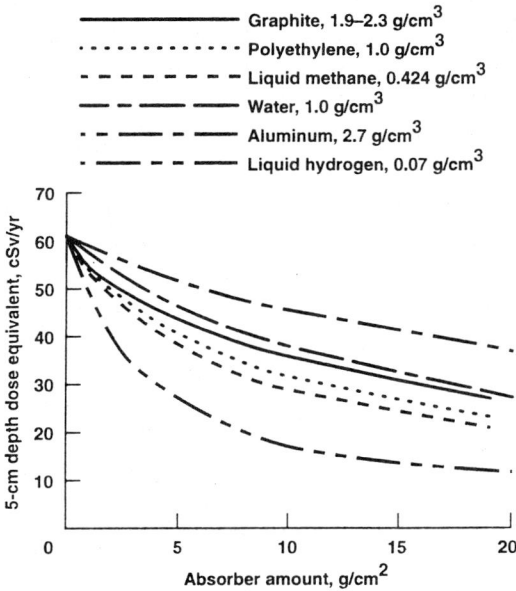

Figure 11 Solar minimum galactic cosmic ray exposure as a function of shield material type and thickness. Note that aluminum is the least effective shield material of the six materials displayed.

For galactic cosmic rays at solar minimum, the annual 5-cm depth-dose equivalent is plotted in Figure 11 as a function of thickness for various candidate shield materials. The obvious material of choice is liquid hydrogen which is used in chemical propulsion. The least desirable material of those considered, from a total shield-mass consideration, is aluminum. Polyethylene (CH_2) is the most effective solid. Comparing CH_2 with graphite (C) clearly indicates the improvement in shielding performance resulting from the presence of hydrogen in the target material. These findings suggest that the use of low atomic-weight materials, with a significant hydrogen component, should be the material of choice, from a GCR protection perspective, for spacecraft structure and shielding. Vehicle design should also maximize the use of intrinsic shielding, such as that provided by liquid hydrogen fuel.

Geometry Effects

Aside from selecting materials to optimize space radiation shielding, a proper understanding of geometry effects is also important. The results presented in Figures 9 and 11 in the previous section are calculations for a monodirectional beam of particles normally incident on a planar layer of shield material, i.e., a "slab" calculation. For a one-dimensional transport calculation using the straight-ahead approximation, the flux/dose/

dose equivalent at a particular depth in the "slab" is equivalent to the flux/dose/dose equivalent at the center of a spherical shell of the same thickness in an isotropic radiation field. This situation is depicted in Figure 12. Within the interior of the spherical shell, the dose/dose equivalent is largest at the center. This is easily visualized by considering that the wall thickness is uniformly constant in every radial direction from the center of the spherical shell. For any other interior point, the *minimum* thickness is the shell thickness along the radii passing jointly through the center and the point under consideration. All other directions from the point under consideration pass through the shell with a path-length which is greater than the radial thickness. Therefore, the dose/dose equivalent will usually be lower because the average thickness of the shielding distribution is greater than the radial shell thickness. Similar results hold for other geometric shapes. For a cylindrical volume, the maximum dose/dose equivalent point will be on the cylinder central axis. For a cubical volume, it will be at the center of the cube. Knowing the dose/dose equivalent distribution within a three-dimensional object allows the designer to arrange internal components so as to minimize crew or radiation-sensitive equipment exposures.

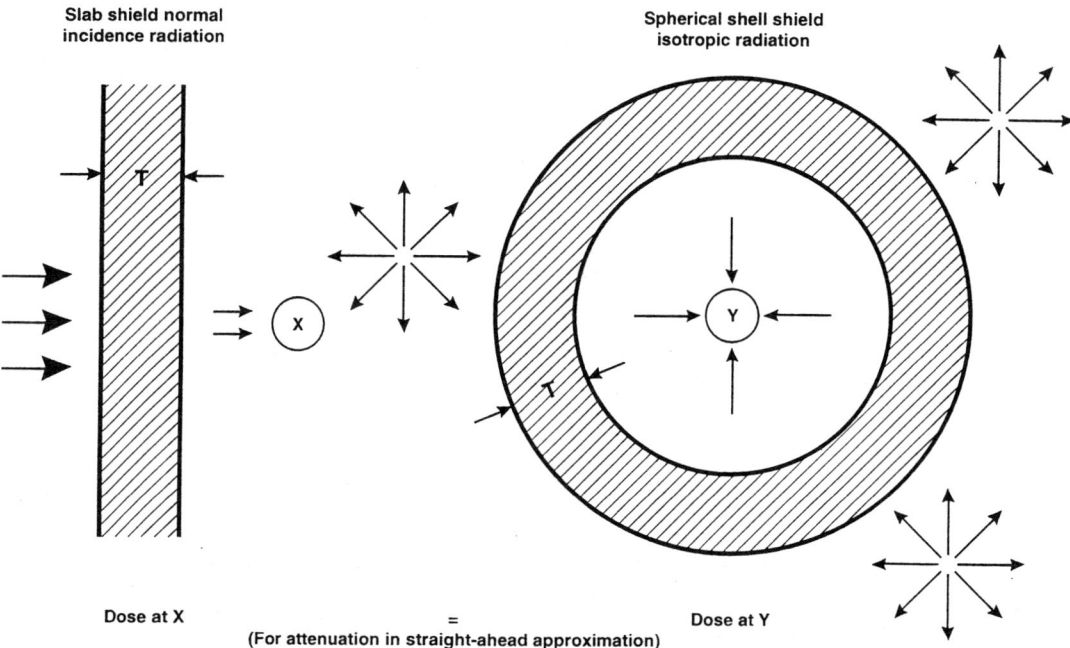

Figure 12   The calculation equivalence of slab and spherical-shell shields. The dose at point X, behind a slab of thickness T, resulting from normally-incident radiation is equal to the dose at the center of a spherical-shell (point Y) of thickness T, which is imbedded in an isotropic radiation field (e.g. galactic cosmic rays). Because the dose at the center of the sphere is the maximum value for any point inside the sphere, and spherical shells are the optimum spacecraft configuration to minimize shield mass, this equivalence is especially useful for evaluating realistic mission scenarios with simple calculations.

Besides internal dose/dose equivalent distribution differences, geometry also directly affects total shield mass through the choice of vehicle shape. From solid geometry, it is well known that a sphere encloses a specified volume within the minimal possible surface area. Any other geometric solid enclosing the same volume will have a larger surface area. For example, a sphere enclosing a 1,000 m^3 volume has a radius of 6.2 meters. Its total surface area ($S = 4\pi R^2$) is 483.6 m^2. A cube enclosing a 1,000 m^3 volume has a length/width/height of 10 meters, and a total surface area ($S = 6 L^2$) of 600 m^2, a 24-percent increase over that of the sphere. Because the total shield mass is the product of the areal density and the shielded volume surface area, a spherical shield configuration will have a significantly smaller mass than a cubical configuration for the same areal density (thickness) enclosing the same total volume.

Next we consider a simple example to illustrate the potential differences which may occur if an "average" shield thickness rather than the actual shield thickness distribution is used to estimate exposures/shielding requirements. Imagine two identical spherical volumes (labeled A and B). Enclose sphere A with a uniform aluminum shield distribution having an areal density of 10.3 g/cm^2. Enclose sphere B with 0.5 g/cm^2 aluminum over two thirds of its surface and with 30 g/cm^2 aluminum over the remaining one third (an "average" of 10.3 g/cm^2). Using the transport methods described previously, the solar minimum GCR dose equivalent at the center of sphere A is estimated to be 0.57 Sv/yr. For sphere B, it is 0.70 Sv/yr, nearly 23 percent larger than for sphere A. Clearly, shield distribution, as well as shield thickness and configuration, is also important.

Human Geometry Effects. Typical assessments of crew risk and related shielding requirements are made using the assumptions that skin and blood-forming organ (BFO) exposures are well approximated by equivalent tissue spheres of radii 0-cm and 5-cm, respectively [Nealy 1991, Townsend 1990b, Simonsen 1991a]. Although generally conservative [Shinn 1990b], such approximations may result in substantial weight penalties when used to estimate spacecraft shielding requirements. Therefore, more precise evaluations of shielding requirements and body organ risk assessments must include body self-shielding factors obtained from models of the actual human geometry.

A computerized anatomical man (CAM) model, developed by Kase [Kase 1970] and refined by Billings and Yucker [1973], has been used for various studies [Townsend 1991, for example]. The model depicts a 50th percentile U.S. Air Force man weighing 73.5 kg and standing 175.5 cm in height. The body organ shield distributions for the skin and bone marrow are displayed in Figure 13. These are obtained by using a 512-ray distribution over the 4π solid angle surrounding a particular organ site. For the skin and bone marrow the distributions are averages over 33 different sites each in the body. For the ocular lens of the eye (not shown in Figure 13), which is a localized organ, only one site is used. For actual organ-exposure calculations, the incident SPE or GCR spectra are transported through the appropriate thickness of shield material, and then through an additional 100 g/cm^2 of water. The dose/dose equivalent at a specific organ site is then obtained by folding the dose/dose equivalent versus water depth with the appropriate organ-depth distribution curve. Sample results for average bone marrow dose equivalent resulting from the GCR spectrum at solar minimum are displayed in Figure 14. Note that the predicted organ exposures are smaller than the predicted 5-cm equivalent sphere exposures. Clearly, the latter are conservative if used to represent BFO exposures. Predictions of organ doses and dose equivalents, as a function of aluminum shield thickness, are made elsewhere [Townsend 1991a].

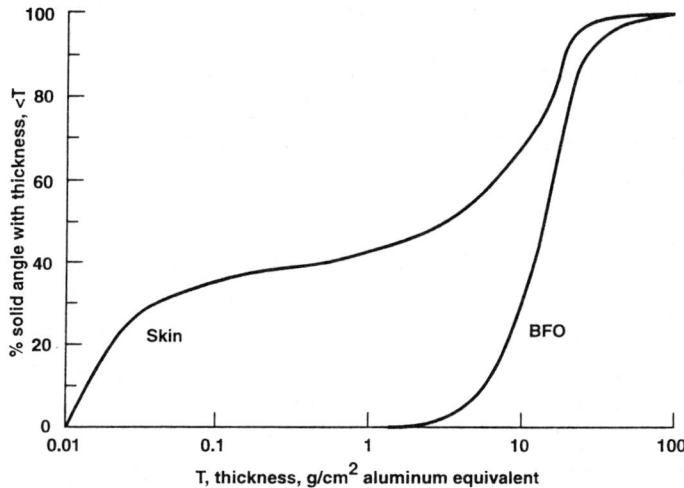

Figure 13 Computerized anatomical man (CAM) body-self-shielding distributions for the skin, and blood-forming organs (BFO) for a 50th percentile Air Force male.

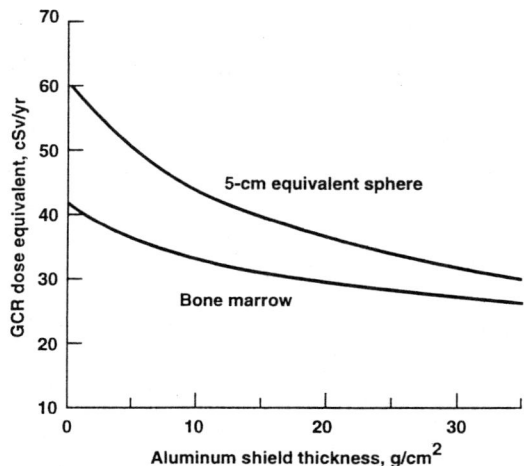

Figure 14 Solar minimum GCR doses equivalent versus aluminum shield thickness. Curves are shown for the bone marrow (33 sites average) and the 5-cm water equivalent sphere approximation. Note that the equivalent sphere approximation gives a conservative overestimate for all shield thicknesses.

Electromagnetic Shielding

Nearly 30 years ago, it was recognized that electromagnetic (EM) fields might offer substantial shield mass reductions over bulk passive shields for protecting space crews from harmful space radiations. Proposed electromagnetic shielding concepts included the use of electrostatic fields [Vogler 1964], plasmas [Levy 1966], and static magnetic fields, which are confined to a finite region of space by design [Brown 1962,

Bernert 1965], or unconfined [Paluszek 1978]. The focus of each of these methods was to shield against space protons and electrons. Except for the unconfined magnetic field configuration, none of the analyses considered GCR particles. For the unconfined field, magnetic shielding appeared to offer substantial weight savings over bulk passive shields for large colonies but not for smaller vehicles likely to be used for exploratory missions [Paluszek 1978].

Of the remaining EM shielding methods, the most plausible are the confined magnetic field configuration and the use of electrostatic fields. Studies of both, as potential GCR shield candidates, have been made. For the confined magnetic field configuration, proposed as a candidate for a crewed Mars transportation vehicle, the study [Townsend 1990b] showed the effectiveness of the concept for SPE protection and its total lack of effectiveness for GCR protection. Although confined magnetic fields (in this case 4 Tesla in magnitude) can be used to deflect most SPE protons, these fields are virtually transparent to the more energetic GCR particles. The same study then clearly demonstrated that through proper design configuration and selection of shield materials, lower exposures could be obtained from passive bulk shielding for the same total mass as that needed to generate and support the confined magnetic field configuration. A study of the possible use of electrostatic fields as a GCR shield was made in 1984 [Townsend 1984]. Because of the large voltages required (over an order of magnitude larger than current state-of-the-art) and vacuum breakdown considerations, the concept was deemed to be useless as a GCR shield candidate. In summary, EM shielding appears to be viable for SPE protection but virtually useless for GCR protection. Therefore, its further consideration should be limited to applications where GCR protection is not required, such as for a short excursion lunar rover.

MARS MISSION RADIATION EXPOSURES

Radiation exposures likely to accrue during a human mission to Mars will be highly dependent upon the specific mission scenario. For example, consideration must be given to radiation exposure accumulated at a LEO node, such as Space Station *Freedom*. The exposure accumulated during the transit through the Van Allen belts will depend upon the transit time and trajectory chosen. The major sources of crew exposures during the transit to and transit from Mars will be from GCR and SPE particles. If a nuclear power/propulsion system is used, then exposures from it will also be accrued during the transit phase of the mission. Finally, exposures on the Martian surface must be considered.

In this section estimated exposures for the transit to/from Mars, and for Martian surface operations, will be made. An excellent overview of anticipated exposures for the various phases of lunar and Mars missions has been recently published [Simonsen 1991a]. We will summarize many of its relevant findings in this work.

Transit To and From Mars

Within the past several years, there has been a flurry of activities involving analyses of anticipated radiation exposures for interplanetary/Mars missions [e.g., Nealy

1990a, 1990b, 1991; Simonsen 1991a; Townsend 1989a, 1989b, 1990a, 1990b, 1990c, 1991, 1992a, 1992b; Letaw 1988]. We will consider: (a) GCR exposures during solar minimum and maximum periods; (b) SPE exposures in free space; (c) combined GCR and SPE exposures over a solar cycle; and (d) a nuclear powered sprint mission to Mars.

GCR Exposures. Galactic cosmic ray dose and dose equivalent estimates as a function of aluminum shield thickness are displayed in Figure 15. Curves for both solar minimum and solar maximum periods are shown. Also displayed are recent measurements made during the last solar cycle (cycle XXI) by instruments onboard Soviet Prognoz satellites [Kovalev 1989]. The agreement between measurements and calculation at 1 g/cm^2 aluminum shielding is excellent for solar minimum. At solar maximum the calculations significantly underestimate the data, possibly indicating that the GCR environmental model fluxes are too small. The data shown at 9 g/cm^2 aluminum thickness are measurements whose shielding thicknesses are uncertain because of a lack of knowledge of the actual spacecraft mass distribution surrounding the detectors.

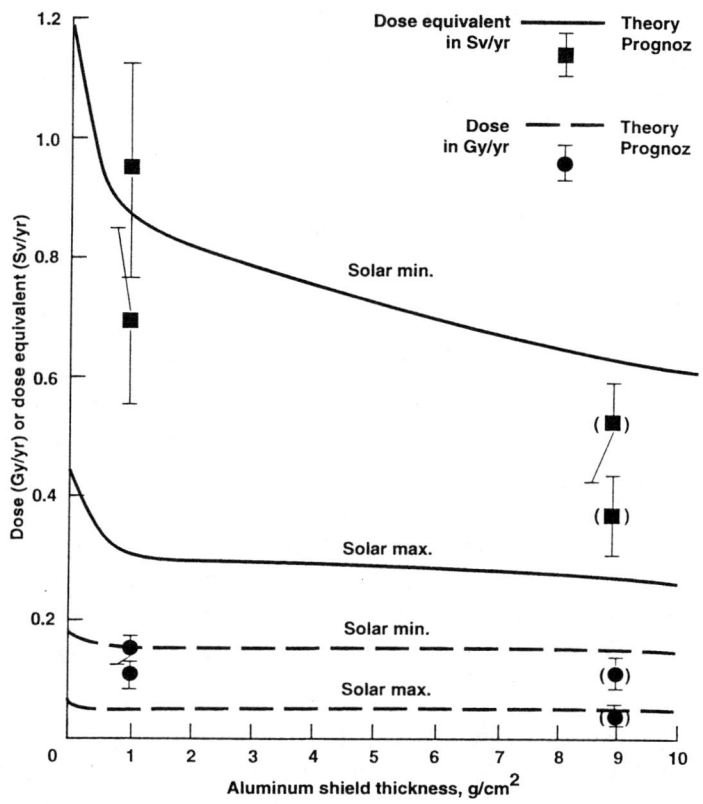

Figure 15 GCR exposure behind an aluminum shield compared with spacecraft measurements. The curves are theoretical estimates using the HZETRN space radiation transport computer code. The symbols are actual measurements from the Soviet Prognoz series of satellites. The parentheses around the symbols at 9 g/cm^2 shield thickness indicate that the actual depths of the interior measurements are uncertain.

From Figure 15, we note that the estimated dose equivalent during solar minimum is nearly twice the value during solar maximum for any displayed shield thickness. Because the maximum GCR exposure will occur near solar minimum activity periods, we will limit ourselves, for present discussion purposes, to that part of the solar cycle. Table 3 lists bone marrow and 5-cm depth dose equivalents, for the solar minimum GCR spectra obtained from the NRL CREME model [Adams 1986], as a function of aluminum shield thickness. Note that the 5-cm equivalent sphere approximation is conservative for all shield thicknesses considered. A nominal spacecraft thickness of 5 g/cm^2 aluminum yields 50 cSv/yr for the 5-cm equivalent sphere approximation and an estimated average bone marrow exposure of nearly 37 cSv/yr. An analysis of the components of this estimated exposure to the bone marrow indicates that 62 percent comes directly from incident GCR ions which have penetrated the spacecraft and astronaut's body self-shielding, 33 percent comes from projectile fragmentation products of the incident ions, and 5 percent results from target fragments produced in the blood-forming organs. As discussed earlier, aluminum is not the most effective shield material. Choosing polyethylene or water would result in thinner and lighter shield masses for the same exposure (~ 3 g/cm^2 for 50 cSv/yr) or lower exposures for the same total mass as aluminum.

Table 3
ANNUAL SOLAR MINIMUM GALACTIC COSMIC RAY DOSE EQUIVALENT IN UNITS OF CSV AS A FUNCTION OF ALUMINUM SHIELD THICKNESS

Shield Thickness (g/cm^2)	Bone Marrow	5-cm Equivalent Sphere
0	41.5	60.2
0.5	40.9	58.9
1.0	40.3	57.7
1.5	39.8	56.6
3.0	38.4	53.5
5.0	36.7	50.0
10.0	33.5	43.5
30.0	27.3	32.1

SPE Exposures. For most previous analyses, the large solar particle events of August 1972, November 1960, and February 1956 were used to estimate potential exposures and shield requirements in deep space [Nealy 1990b; Townsend 1989a, 1989b, 1990b, 1990c, for example]. Solar cycle XXI ('1975-1986) was relatively quiet with no SPEs of these magnitudes recorded. Solar cycle XXII, however, has been much more active. In the fall of 1989, a series of six large SPEs occurred over the period from mid-August to mid-December. The largest of these began on October 19 and lasted until early November. Its integral fluence, displayed in Figure 5, was comparable to that of August 1972—generally considered to be the standard to which all other SPEs are compared. The August 1972 SPE, however, yields larger estimates of dose and dose equivalent to the skin, eye, and bone marrow for aluminum shield thicknesses up to 15 g/cm². At 20 g/cm² aluminum thickness, the organ exposures are nearly identical at approximately 15 cSv to the skin and ocular lens and approximately 5 cSv to the bone marrow [Townsend 1991].

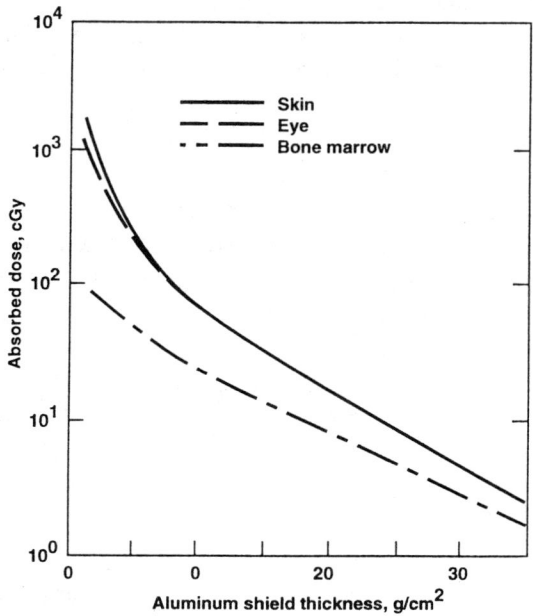

Figure 16 Organ doses versus aluminum shield thickness for the cumulative fluence for the August through December 1989 solar particle events.

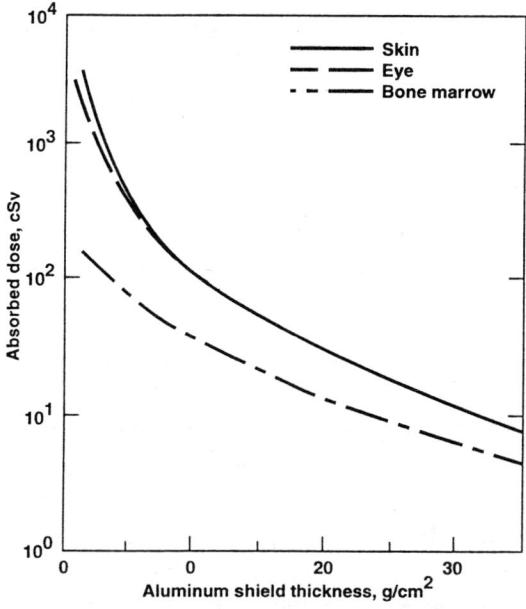

Figure 17 Organ dose equivalents versus aluminum shield thickness for the cumulative fluence for the August through December 1989 solar particle events.

Figures 16 and 17 display the cumulative organ doses and dose equivalents for the sum of the 1989 flares shown in Figure 10. Taken together, these six SPEs give an estimated 2.2 Gy/3.5 Sv to the skin, 2.0 Gy/3.0 Sv to the ocular lens, and 0.47 Gy/0.68 Sv to the bone marrow behind 5 g/cm^2 of aluminum shielding. For this same shielding, the August 1972 event yields 2.7 Gy/4.2 Sv to the skin, 2.4 Gy/3.7 Sv to the eye, and 0.43 Gy/0.64 Sv to the bone marrow [Townsend 1991]. All these exposures exceed the LEO guidelines of NCRP Report 98. At least 10 g/cm^2 of aluminum shielding is needed to prevent exceeding those guidelines. Rather than shielding an entire spacecraft to at least 10 g/cm^2 of aluminum, the crew can be provided with such wall thicknesses in a storm shelter for use during a large SPE.

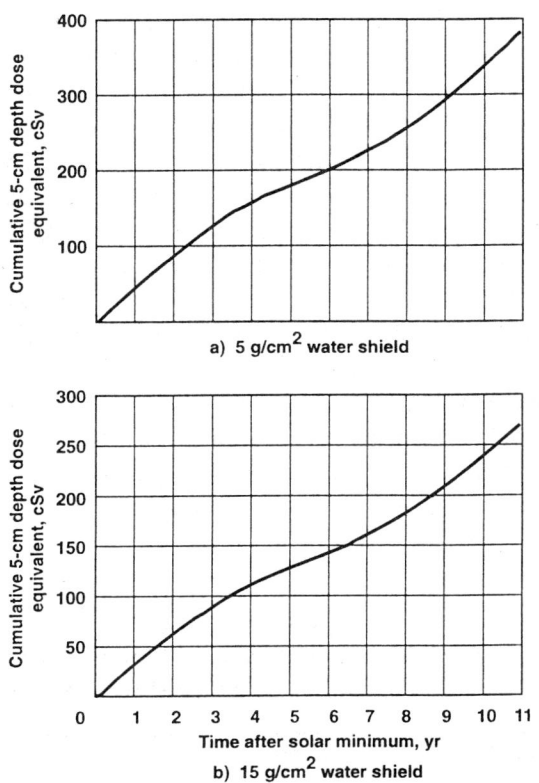

Figure 18 Cumulative total incurred 5-cm depth dose equivalent throughout the 11 year solar cycle as a function of time after the solar minimum. Results are displayed for two different water shield thicknesses.

Solar-Cycle Analysis. Nealy and collaborators [Nealy 1990b] have performed an analysis of dose and dose equivalent variation behind various thicknesses of water shielding as a function of mission length and time in cycle XXI after solar minimum. The study used the proton fluence spectra for the 55 largest solar flares of that solar cycle and the GCR environmental model of NRL [Adams 1986] as inputs into the BRYNTRN and HZETRN transport codes. It does not include any large SPEs of the August 1972 variety. Figure 18 displays cumulative 5-cm depth-dose equivalent results as a function of time after solar minimum behind 5 g/cm^2 and 15 g/cm^2 of water and shielding. This type of analysis can be used to estimate 5-cm depth-dose equivalent for long-duration missions that begin and end at arbitrary times during the solar cycle. For example, the total dose equivalent for a 3-year mission beginning at solar minimum is predicted to be 1.25 Sv for 5 g/cm^2 of water shielding. However, if the 3-year mission begins 4 years after solar minimum, then the total dose equivalent is estimated to be approximately 0.65 Sv. This apparent dose equivalent reduction, although valid for cycle XXI (1975-1986), would be greatly modified in the current cycle (XXII) by the presence of the large SPEs occurring in the fall of 1989. These would double the 0.65 Sv estimates to nearly 1.3 Sv, thereby negating any apparent advantage in choosing a transit period during solar maximum rather than during solar minimum.

Nuclear-Powered Sprint Mission. Another attractive means of reducing Mars mission exposure estimates is to reduce mission duration by using a nuclear-thermal-rocket (NTR) for a high-speed transit. Such an analysis has recently been carried out [Nealy 1991] for a 500-day mission occurring near solar maximum (to minimize GCR exposures) using a proposed NTR-powered crewed vehicle concept. The results are summarized in Table 4. Details of the analysis will not be presented here. Instead, we briefly describe the major study assumptions. First, the crew module is assumed to be shielded by hydrogenous or other low-density materials (such as H$_2$O, foodstuffs, etc.) to an areal density of approximately 25 g/cm^2 shielding. Second, the total solar flare contribution is given by the cumulative results from the fall 1989 SPEs. Finally, the GCR dose equivalent assumes 16 hours/day spent in the thickly shielded crew module and 8 hours/day spent elsewhere in a thinly shielded (2 g/cm^2) part of the vehicle. The total estimated exposure of 48 cSv makes this scenario attractive and certainly worthy of additional detailed analyses which further assess the effects of environmental uncertainties, computational method uncertainties, and various assumptions regarding specific spacecraft designs and scenarios.

Table 4
5-CM DEPTH DOSE EQUIVALENT ESTIMATES BY SOURCE FOR A HYPOTHETICAL 500-DAY MARS SPRINT MISSION

Nuclear Thermal Rocket (NTR) Firings	< 1.1 cSv
Transit Through Trapped Belts	~ 1.5 cSv
30-Day Surface Stay (GCR)	< 1.0 cSv
Solar Flare Contribution	7.7 cSv
GCR Contribution	36.7 cSv
Total	48.0 cSv

Transit Results Summary. From these analyses, it is recommended that at least 5 g/cm^2 of aluminum shielding be provided throughout the transit. The estimated GCR exposures to the bone marrow are then expected to be in the range of 20-40 cSv/yr, depending upon the phase of the solar cycle. During large SPEs, such as occurred in August 1972 or during fall 1989, at least 10 g/cm^2 of aluminum shielding should be easily accessible in the form of a storm "shelter." The resultant BFO dose equivalents from combined SPE/GCR sources (including a large August 1972 type SPE) should be approximately 1.5 Sv for a 3-year mission total transit period. Because of improved shielding efficacy, water or some other hydrogenous, low-atomic-weight material is preferred over aluminum as a shield material. A nuclear-powered sprint mission may result in a reduced exposure level. One such scenario of a 500-day sprint mission yielded a 5-cm depth-dose equivalent total estimate of 48 cSv.

On the Martian Surface

Except for an earlier study by Letaw [1988], most of the recent analyses of potential space radiation exposures during Martian surface operations have been published by Simonsen and collaborators [Simonsen 1990a, 1990b, 1990c, 1990d, 1991a, 1991b]. These results will be briefly summarized in this section.

Because of the presence of a thin, carbon-dioxide (CO_2) atmosphere, the radiation environment on the Martian surface is expected to be less severe than on the lunar surface. In addition, the planet's surface will provide complete shielding for one-half of the 4 p total solid angle surrounding any dose point on the surface. Unlike the Earth, Mars has no intrinsic magnetic field strong enough to deflect ions in the incoming GCR or SPE spectra. Therefore, estimating doses/dose equivalents for crews on the surface of Mars involves accounting for the shielding provided by the planet itself, its atmosphere, and any vehicle (rover) or habitat structural shielding. To date no studies using actual human geometries on the surface are available; therefore, the results quoted herein are for 0-cm or 5-cm depth-dose equivalent in water behind the specified shielding.

Atmosphere Shielding. For analysis purposes, a pure CO_2 atmosphere is assumed. The Committee on Space Research (COSPAR) has developed a high-density model and a low-density model of the atmosphere. The vertical thickness of the atmosphere is 16 g/cm^2 for the low-density model and 22 g/cm^2 for the high-density model. The surface doses/dose equivalent at various altitudes in the atmosphere are computed for CO_2 absorber amounts along slant paths in the atmosphere ranging from the overhead to the horizon (zenith angles from 0° to 90°) in 5° increments. Total exposures are then determined by integrating all doses/dose equivalent for each particular slant range over solid angle at the target point. Typical 5-cm depth-dose-equivalent results for a solar minimum GCR environment and the indicated SPEs, taken from Simonsen [1990a], are displayed in Table 5 as a function of altitude in the atmosphere. No vehicle/habitat shielding is assumed.

Martian Regolith Shielding Analysis. For habitats on the planet's surface, shielding provided by local resources, such as Martian regolith, may be used to reduce excessive launch weight requirements from Earth. In one recent study [Simonsen 1991b], estimates were made of dose and dose equivalent on the surface of Mars after transport and attenuation of the incident space radiations through the Martian atmosphere and through additional shielding provided by Martian regolith. The incident radiations analyzed were the solar minimum GCR spectra and the February 1956 SPE. The former was chosen because it should yield the largest GCR exposure for any time during a particular solar cycle. The latter was chosen because the February 1956 SPE had the hardest (most energetic) proton spectrum and should therefore yield the largest dose/dose equivalent for moderate total shield thicknesses ($\sim$ 25-50 g/cm^2). For transport through the atmosphere, the COSPAR low density model (16 g/cm^2 CO_2 vertical thickness) is assumed (a conservative assumption, since the high-density atmosphere is thicker). The spherical geometry of the atmosphere was considered by accounting for the increased thickness with increasing zenith angles. On the surface (0-km altitude), the atmosphere provides 16 g/cm^2 of CO_2 protection. This increases to nearly 60 g/cm^2 CO_2 at 75°. For a surface elevation of 8 km, the CO_2 atmosphere thickness values are 7.5 g/cm^2 at 0° and nearly 28 g/cm^2 at 75°. From Table 5, we note that the annual 5-cm depth dose equivalent received from the combined GCR and February 1956 SPE spectra, at the surface (0 km) is 21.8 cSv. At 8-km altitude, it is 29.4 cSv.

Table 5
5-CM DEPTH DOSE EQUIVALENTS (cSv) FOR
THE COSPAR REFERENCE MARS ATMOSPHERE MODEL

Source	Atmosphere Model	Dose Equivalent at Altitude (km)			
		0	4	8	12
Galactic Cosmic ray	High-density	10.5	12.0	13.7	15.6
	Low-density	11.9	13.8	15.8	18.0
Solar Particle Event 8/72	High-density	2.2	4.8	9.5	17.4
	Low-density	4.6	9.9	18.5	30.3
Solar Particle Event 2/56	High-density	8.5	10.0	11.7	13.4
	Low-density	9.9	11.8	13.6	15.3

For transport through additional shielding provided by Martian regolith, a regolith model must be assumed. In this case, the model, based upon the chemistry of the Viking Lander site, assumes that the normalized weight percentages are 58.2 percent SiO_2, 23.7 percent Fe_2O_3, 10.8 percent MgO and 7.3 percent CaO. The density at the site varies from 1.0 g/cm^3 to 1.8 g/cm^3. The analysis procedure for the regolith shield calculations was to transport the flux spectra at 0 km and 8-km altitudes in the atmosphere through the added regolith shielding.

One early conception of a Martian habitat assumed a cylindrical module (Space Station *Freedom* derived) that is 8.2 m in length and 4.45 m in diameter. The module is positioned lengthwise on the Martian surface and covered with regolith as shown in Figure 19 (a). Calculations were performed for various regolith thicknesses. Figure 20 displays the maximum 5-cm depth dose equivalent in the central cross-section plane of the shielded module as a function of regolith shield thickness. Note that the regolith provides little added protection over that already provided by the CO_2 atmosphere. Most of the exposure reduction due to the added regolith occurs in the first 20 g/cm^2 of regolith thickness. Very little reduction in 5-cm depth dose equivalent is achieved for regolith thicknesses greater than 20 g/cm^2. For 20 g/cm^2 of regolith shielding, the annual GCR contribution to the 5-cm depth dose equivalent is 10 cSv/yr at 0 km and 11.2 cSv/yr at 8 km. The contribution from the SPE (February 1956 spectrum) is 6.3 cSv at 0 km.

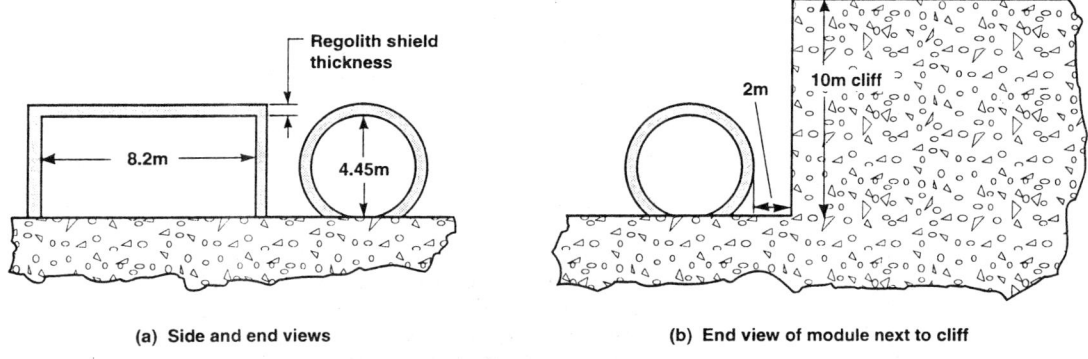

Figure 19 Configurations for cylindrical habitat modules with regolith shielding on the Martian surface [Simonsen 1991a].

Additional reductions in accrued exposures might be attained by utilizing shielding provided by the natural terrain on the Martian surface. In their recent study, Simonsen and collaborators [Simonsen 1991a, 1991b] assumed that the habitat was located 2 m away from a 10-m-high cliff, as depicted in Figure 19 (b). From Figure 20, we note that the cliff further reduces the 5-cm depth dose equivalent at 0-km altitude by 2-3 cSv/yr for the GCR, and by 1-1.5 cSv for the February 1956 SPE. Note also that the shielding provided by the cliff and atmosphere alone results in a 5-cm depth dose equivalent which is lower than the dose equivalent accrued with 20 g/cm^2 of regolith shielding and no cliff.

For radiation protection on the Martian surface, mission planners must decide if the slight reductions in crew exposures warrant the expenditures of time and energy to bury the habitat in regolith. A better alternative appears to be the use of natural terrain features found on the Martian surface, especially for short-duration surface missions of 30-90 days. For longer stay times of a year or more, regolith shielding will provide additional crew protection and should be seriously considered.

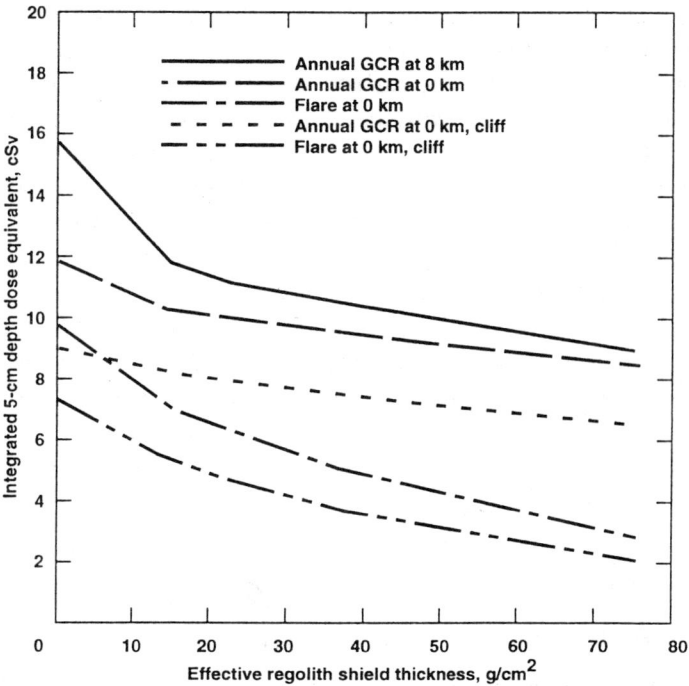

Figure 20 Maximum dose equivalent in the central cross-sectional plane of a habitat module on Mars as a function of the regolith shield thickness [Simonsen 1991a] using the configurations in Figure 19. Curves are shown for (1) the annual GCR contribution at the average surface elevation (0 km) and at an elevation of 8 km, (2) the dose equivalent due to a large solar flare at 0 km elevation, and (3) the annual GCR dose equivalent and the contribution from a large flare at 0 km when the module is located next to a cliff.

ADDITIONAL RESEARCH NEEDS

The previous discussion of methods and results hardly addressed the uncertainties in the computational methods and models and their possible impacts on the estimated exposures and shielding requirements. In this section, a brief assessment of the estimated uncertainties will be provided. Then, additional research needs in the areas of transport code development and their nuclear reaction input data will be described. Finally, a brief description of space-based validation will be presented.

Assessing Uncertainties

Estimates of interplanetary crew radiation exposures from galactic cosmic rays and solar particle events are fraught with various uncertainties. Among the most significant are:

1. Uncertainties in the GCR radiation environment measurements and their models.

2. Uncertainties in the atomic and nuclear interaction models and data.

3. Approximation used in radiation transport methods and their computer codes.

4. Uncertainties in spacecraft shielding distribution.

5. Uncertainties in SPE fluence characteristics and probability of occurrence.

6. Uncertainties in risk assessment from space radiations, especially high-LET GCR particles.

Uncertainties in the GCR environmental model, CREME, vary with energy. A reasonable uncertainty bound is ~ 50 percent [Adams 1986]. The main source of uncertainty appears to be the effect of solar modulation on the GCR spectra for times between solar maximum and solar minimum.

The main source of uncertainty in the atomic and nuclear interaction models and data appears to be in the nuclear fragmentation (breakup) cross sections. The experimental data are very few in number, and variations between data sets may be as large as 50 percent or more. The resultant uncertainty in estimated dose/dose equivalent is a function of shield thickness and increases with increasing thickness. No actual analyses of the magnitude of this uncertainty are currently available for any spacecraft shield material. Recommendations for further research in this area are addressed in the next subsection.

None of the current generation of space radiation transport codes (e.g., BRYNTRN or HZETRN) have been validated in careful laboratory experiments. Conversely, none of the currently used laboratory high-energy, heavy ion transport codes can be readily adopted for space radiation calculations. Future research needs in this area will be addressed later in this section.

The effects on dose equivalent estimates of uncertainties in spacecraft shielding distributions were demonstrated in an earlier section. With the advent of CAD (computer-aided design) techniques, accurate models of spacecraft shielding distributions should be readily available to design engineers for exposure analysis purposes.

The forecasting of large SPEs will be vitally important for human Mars missions. Presently, the NOAA Space Environment Laboratory uses the intensities of x-rays and radio emissions from the Sun to estimate the likelihood of an SPE and its peak proton flux. For 24-hour predictions during the recently completed solar cycle XXI, only 10 percent of the events which occurred were not predicted. The false alarm rate, however, was about 50 percent [Heckman 1988]. Accurate forecasting on events weeks, months, or years in advance does not exist; therefore, on-board instruments for energetic proton detection from an SPE must be available to enable the crew to quickly repair to the SPE storm shelter when necessary.

Finally, uncertainties in risk assessment from high-LET space radiations may be larger than all other sources combined [NCRP 1989]. A detailed set of research recommendations was delineated in NCRP Report 98 [NCRP 1989]. The interested reader is referred to that document for more details.

An assessment of the overall impact of uncertainties in 5-cm depth-dose equivalent, due to GCR particles, on shield mass requirements has been made by Townsend and collaborators [Townsend 1990c]. There it was found, for water shielding, that an exposure uncertainty of 10 percent required a 43-percent increase in water shield thickness. For a 20-percent exposure uncertainty, a 129-percent increase in shield thickness was needed. For a 50-percent exposure uncertainty, a 1,000-percent increase in shield thickness was required. The need to better understand, quantify, and reduce uncertainties is clear from these results.

Nuclear Reaction Data

As previously discussed, the main source of uncertainty in the input nuclear and atomic data lies in the nuclear fragmentation cross section data base. The experimental data base is very sparse—probably much less than 1 percent of the necessary cross sections have been measured. The underlying physics of the fragmentation process is reasonably well understood from a qualitative perspective. Quantitative cross section estimates, however, may be inaccurate by as much as 30-50 percent for heavy mass fragments and by over a factor of 2 for protons, neutrons, and the light ions (alphas, lithium, etc.). In addition, the dependence on incident ion kinetic energy of the fragmentation cross sections is poorly known because of a paucity of measurements. Recently, a combined NASA Langley Research Center/Lawrence Berkeley Laboratory research program has been established to systematically measure fragmentation cross sections for a variety of incident projectile ions, at GCR energies, colliding with targets of interest for spacecraft shielding and also representative of human tissue. All fragment elemental cross sections will be measured for each incident energy and ion collision pair. Neutron production cross sections and some isotopic cross sections for various elemental fragments will also be obtained. The experimental data will be used to develop newer, quantitatively accurate nuclear fragmentation models for use in the transport calculations—thereby significantly reducing the present uncertainties in the estimated exposures and shield requirements.

Transport Code Development and Validation

Current-generation GCR transport codes are still in the early stages of development and validation. Analytic benchmarks [Wilson 1988b] serve to substantiate the accuracy of the particular solution algorithm, but do not address questions of the one-dimensional transport equation as a valid approximation to the real three-dimensional transport problem. None of the current GCR transport codes have been validated by comparison with laboratory beam experiments—nor can they be because of the methodologies used in them. Conversely, existing laboratory HZE transport codes cannot be easily adapted for space radiation shield design calculations because of the prohibitively large computational requirements entailed in their use. What is needed is a single transport code which will readily accept either monoenergetic beams of a single ion species (laboratory beams) or broad-energy beams of several ions (space radiation) as input into the same code. The validity and accuracy of the space radiation code can then be established directly by careful laboratory beam experiments.

Recently, an analytic solution for high-energy, heavy ion transport was found in terms of a Green's function whose series solution is rapidly convergent for most practical applications. This technique can be applied with equal success to laboratory beams and galactic cosmic rays, thereby permitting directly laboratory validation of the resultant space shielding code [Wilson 1990c]. A complete GCR transport code, using this formalism, is presently in the initial stages of development.

Laboratory validation of HZE transport codes using neon beams and water columns of variable thicknesses has been underway for some time [Wilson 1984, Schimmerling 1989, Shavers 1990]. These validation methods are now being extended to include other projectile ions and target materials in a collaborative effort involving NASA Langley Research Center/Lawrence Berkeley Laboratory personnel.

Space-Based Validation

Validation of GCR radiation exposure predictions and spacecraft shielding requirements in space is difficult at best because of an inability to accurately measure and adequately control all the experimental variables. Uncertainties in the input spectra, especially if modified by the Earth's magnetic field, combined with uncertainties in spacecraft mass distribution, detector response characteristics, and sensitivities all serve to make analyses of differences between prediction and measurement difficult at best. Space-based validation may be useful as a supplement to laboratory validation as it is not practical to measure exposures for all GCR species and energies in the laboratory. Also, there is no single ion which is an accurate representation of the complete GCR spectrum; therefore, no single laboratory beam experiment can duplicate the effects of the complete GCR environment. Instead, space validation is useful as a "final" check of transport codes and input data after their uncertainties have been quantified and their methodologies validated by carefully controlled laboratory experiments.

One possible space validation concept previously under consideration for GCR exposure experiments was a reusable, robotic satellite called LifeSat. It was designed by NASA to carry biological specimens to regions of space beyond near-Earth orbit, where cosmic rays are hazardous to human exploration. Plans were for four flights beginning in the mid- to late-1990s. Supporting the biological measurements would be physical measurements of the number, energy, and charge of particles contributing to the specimen exposures inside the spacecraft. The principal questions LifeSat was intended to answer were:

1. How valid are radiobiological models developed in ground-based studies for predicting radiation effects in space?
2. How well are the shielding properties of materials in space predicted by data from ground-based studies?
3. How do other space-related bioeffects combine with radiation effects?

Unfortunately, congressional approval was not obtained for this mission.

CONCLUDING REMARKS

Accurately estimating astronaut radiation exposures and shielding requirements for a manned Mars mission is a complex endeavor involving many diverse research areas over a broad range of scientific and engineering disciplines. The present calculational methods and codes should be considered as interim tools. The exposure results and shielding requirements obtained from them are useful but not definitive. Significant uncertainties exist, which are being quantified and reduced through careful, systematic research. Concepts and ideas can and should be explored, however, to enable man to extend his presence further into space. The next decade of research and planning will provide the design tools needed to reliably and accurately estimate radiation protection requirements for solar system exploration.

REFERENCES

Adams, J.H., Jr., Cosmic ray effects on microelectronics, Part IV, *Naval Research Laboratory Memorandum Report 5901* (revised), Washington, D.C., December, 1986.

Badhwar, G.D. and P.M. O'Neill, "An improved model of galactic cosmic radiation for space exploration missions," paper presented at 22nd International Cosmic Ray Conference, Dublin, Ireland, August, 1991.

Bernert, R.E. and F.J. J. Stekly, "Magnetic radiation shielding using superconducting coils," in *Second Symposium on Protection Against Radiation in Space*, NASA Special Publication No. SP-71, pp. 199-209, 1965.

Billings, M.P. and W.R. Yucker, Summary final report, The computerized anatomical man model, McDonnell Douglas Astronautics Company Report No. MDC G4655, 1973.

Brown, G.V., "Magnetic radiation shielding," in *Proceedings of the International Conference on High Magnetic Fields*, MIT Press and J. Wiley & Sons, New York, pp. 370-378, 1962.

Feynman, J., T.P. Armstrong, L. Dao-Gibner, and S. Silverman, "A new proton fluence model for E > 10 MeV," in *Interplanetary Particle Environment* Conference Proceedings, NASA Jet Propulsion Laboratory Publication No. 88-28, J. Feynman and S. Gabriel, eds., pp. 58-71, Pasadena, April 1988.

Gaffey, J.D., Jr. and E. Bilitza, D., "The NSSDC trapped radiation model facility," American Institute of Aeronautics and Astronautics Paper No. *AIAA-90-0176*, 1990.

Heckman, G.R., "Solar proton event forecasts," in *Interplanetary Particle Environment* Conference Proceedings, NASA Jet Propulsion Laboratory Publication No. 88-28, J. Feynman and S. Gabriel, eds., pp. 91-100, Pasadena, April, 1988.

International Commission on Radiological Protection (ICRP), Recommendations of the commission, *ICRP Publication 26*, Pergamon Press, New York, 1977.

International Commission on Radiation Units and Measurements (ICRU), The quality factor in radiation protection, *ICRU Report 40*, April, 1986.

Kase, P.G., Computerized anatomical model man, Air Force Weapons Laboratory Technical Report No. AFWL-TR-69-161,

King, J.H., "Solar proton fluences for 1977-1983 space missions," *Journal of Spacecraft & Rockets*, 11, pp. 401-408, 1974.

Kovalev, E.E., I.A. Muratova, and V.M. Petrov, "Studies of the radiation environment aboard Prognoz satellite," *Nuclear Tracks and Radiation Measurements*, 14, pp. 45-48, 1989.

Letaw, J.R., R. Silberberg, and C.H. Tsao, "Galactic cosmic ray doses to astronauts outside the magnetosphere," in *Terrestrial Space Radiation and Its Biological Effects*, NATO ASI Series A, Vol. 154, P. D. McCormack, C. E. Swenberg, and H. Bucker, eds., pp. 663-673, Plenum Press, New York, 1988.

Levy, R.H. and G.S. James, G.S., "Plasma radiation shielding for deep-space vehicles," *Space/Aeronautics,* 45, pp. 106-120, 1966.

National Academy of Sciences/National Research Council (NAS/NRC), *Radiobiological factors in manned spaceflight*, W. H. Langham, ed., National Academy Press, Washington, D. C., 1967.

National Council on Radiation Protection and Measurements (NCRP), Guidance on radiation received in space activities, *NCRP Report No. 98*, July, 1989.

Nealy, J.E., L.C. Simonsen, H.H. Sauer, J.W. Wilson, and L.W. Townsend, Space radiation dose analysis for the solar flare of August 1989, *NASA Technical Memorandum No. TM-4229*, December, 1990a.

Nealy, J.E., L.C. Simonsen, L.W. Townsend, and J.W. Wilson, "Deep-space radiation exposure analysis for solar cycle XXI (1975-1986)," 20th Annual International Conference on Environmental Systems (ICES), Williamsburg, VA, July 1990, SAE Paper No. 901347, 1990.

Nealy, J.E., L.C. Simonsen, J.W. Wilson, L.W. Townsend, G.D. Qualls, B.G. Schnitzler, and M.M. Gates, "Radiation exposure and dose estimates for a nuclear-powered manned Mars sprint mission," in *Proceedings of Eighth Symposium on Space Nuclear Power Systems*, Part 2, M.S. El-Genk, and M.D. Hoover, eds., pp. 531-536 (CONF-910116), 1991.

Norbury, J.W., F.A. Cucinotta, L.W. Townsend, and F.F. Badavi, "Parameterized cross sections for coulomb dissociation in heavy ion collisions," *Nuclear Instruments & Methods*, B 31, pp. 535-537, 1988.

Paluszek, M.A., "Magnetic radiation shielding for permanent space habitats," in *The Industrialization of Space*, R.A. Van Patten, P. Siegler, E.V.B. Stearns, eds., (Proceedings of the Twenty-Third AAS Annual Meeting), American Astronautical Society, Advances in the Astronautical Sciences, 36-I, pp. 545-574, San Diego, CA, 1978.

Parker, J.F., Jr. and V.R. West, eds., Bioastronautics data book, Second ed., NASA Special Publication No. *SP-3006*, 1973.

Sauer, H.H., R.D. Zwickl, and M.J. Ness, Summary data for the solar energetic particle events of August through December, 1989, *Space Environment Laboratory Report (Unnumbered), NOAA*, February, 1990.

Schimmerling, W., J. Miller, M. Wong, M. Rapkin, J. Howard, H.G. Spieler, and B.V. Jarret, "The fragmentation of 670 A MeV Neon-20 as a function of depth in water. I. Experiment," *Radiation Research*, 120, pp. 36-71, 1989.

Shavers, M.R., S.B. Curtis, J. Miller, and W. Schimmerling, "The fragmentation of 670 A MeV Neon-20 as a function of depth in water. II. One-generation transport theory," *Radiation Research*, 124, pp. 117-130, 1990.

Shea, M.A. and D.F. Smart, "A summary of major solar proton events," *Solar Physics,* 127, pp. 297-320, 1990.

Shinn, J.L., J.W. Wilson, J.E. Nealy, and F.A. Cucinotta,, "Comparison of dose estimates using the buildup-factor method and a baryon transport code (BRYNTRN) with Monte Carlo results," *NASA Technical Paper No. TP-3021*, October, 1990a.

Shinn, J.L., J.W. Wilson, and J.E. Nealy, Reliability of equivalent sphere model in blood-forming organ dose estimation, *NASA Technical Memorandum No. TM-4178*, April, 1990b.

Simonsen, L.C., J.E. Nealy, L.W. Townsend, and J.W. Wilson, "Space radiation dose estimates on the surface of Mars," *Journal of Spacecraft & Rockets*, 27, pp. 353-354, 1990a.

Simonsen, L.C., J.E. Nealy, L.W. Townsend, and J.W. Wilson, "Radiation exposure for manned Mars surface missions," *NASA Technical Paper No. TP-2979*, March, 1990b.

Simonsen, L.C., J.E. Nealy, L.W. Townsend, and J.W. Wilson, "Ionizing radiation environment at the Mars surface," in *Engineering, Construction and Operations in Space II, Proceedings of Space '90*, S.W. Johnson and J.T. Wetzel, eds., pp. 748-758, 1990c.

Simonsen, L.C., J.E. Nealy, L.W. Townsend, and J.W. Wilson, "Space radiation shielding strategies for Martian habitats," 20th Annual International Conference on Environmental Systems (ICES), Williamsburg, VA, July, 1990, *SAE Paper No. 901346*, 1990d.

Simonsen, L.C. and J.E. Nealy, "Radiation protection for human missions to the Moon and Mars," *NASA Technical Paper No. TP-3079*, February, 1991a.

Simonsen, L.C., J.E. Nealy, L.W. Townsend, and J.W. Wilson, "Martian regolith as space radiation shielding," *Journal of Spacecraft & Rockets*, 28, pp. 7-8, 1991b.

Simpson, J.A., "Elemental and isotopic composition of the galactic cosmic rays," in *Annual Reviews of Nuclear and Particle Science*, 33, J. D. Jackson, ed., Annual Reviews, Inc., Palo Alto, pp. 323-381, 1983.

Sinclair, W.K., "Recent developments in estimates of cancer risk from ionizing radiation," 20th Annual International Conference on Environmental Systems (ICES), Williamsburg, VA, July 1990, *SAE Paper No. 901344*, 1990.

Tobias, C.A. and P. Todd, *Space Radiation Biology And Related Topics*, Academic Press, New York, 1974.

Townsend, L.W., Galactic heavy ion shielding using electrostatic fields, *NASA Technical Memorandum No. TM-86265*, September, 1984.

Townsend, L.W. and J.W. Wilson, Tables of nuclear cross sections for galactic cosmic rays: absorption cross sections, *NASA Reference Publication No. RP-1134*, May, 1985a.

Townsend, L.W., J.W. Wilson, and J.W. Norbury, "A simplified optical model description of heavy ion fragmentation," *Canadian Journal of Physics*, 63, pp. 135-138, 1985b.

Townsend, L.W. and J.W. Wilson, "Energy-dependent parameterization of heavy ion absorption cross sections," *Radiation Research*, 106, pp. 283-287, 1986.

Townsend, L.W., J.E. Nealy, J.W. Wilson, and W. Atwell, "Large solar flare radiation shielding requirements for manned interplanetary missions," *Journal of Spacecraft & Rockets*, 26, pp. 126-128, 1989a.

Townsend, L.W., J.W. Wilson, and J.E. Nealy, "Space radiation shielding strategies and requirements for deep-space missions," 19th Annual International Conference on Environmental Systems (ICES) San Diego, CA, July 1989, *SAE Paper No. 891433*, 1989b.

Townsend, L.W. and J.W. Wilson, "Interaction of space radiation with matter," 41st Congress of the International Astronautical Federation, Dresden, FRG, October 1990, *Paper No. IAF/IAA-90-543*, 1990a.

Townsend, L.W., J.W. Wilson, J.L. Shinn, J.E. Nealy, and L.C. Simonsen, "Radiation protection effectiveness of a proposed magnetic shielding concept for manned Mars missions," 20th Annual International Conference on Environmental Systems (ICES), Williamsburg, VA, July 1990, *SAE Paper No. 901343*, 1990b.

Townsend, L.W., J.E. Nealy, J.W. Wilson, and L.C. Simonsen, Estimates of galactic cosmic ray shielding requirements during solar minimum, *NASA Technical Memorandum No. TM-4167*, February 1990c.

Townsend, L.W., J.L. Shinn, and J.W. Wilson, "Interplanetary crew exposure estimates for the August 1972 and October 1989 solar particle events," *Radiation Research*, 126, pp. 108-110, 1991.

Townsend, L.W., J.W. Wilson, F.A. Cucinotta, and J.L. Shinn, "Galacticosmic ray transport methods and radiation quality issues," *Nuclear Tracks and Radiation Measurements*, 20, pp. 65-72, 1992a.

Townsend, L.W., J.W. Wilson, J.L. Shinn, and S.B. Curtis, "Human exposure to large solar particle events in space," Committee on Space Research (COSPAR) XXVIII Plenary Meeting, The Hague, The Netherlands, July 1990, *Advances in Space Research*, 12, pp. (2) 339-(2) 348, 1992b.

Vogler, F.H., "Analysis of an electrostatic shield for space vehicles," *AIAA Journal*, 2, pp. 872-878, 1964.

Wilson, J.W. and F.M. Denn, Preliminary analysis of the implications of natural radiations on geostationary operations, *NASA Technical Note No. TN D-8290*, September, 1976.

Wilson, J.W., "Heavy ion transport in the straightahead approximation," *NASA Technical Paper No. TP-2178*, June 1983.

Wilson, J.W., L.W. Townsend, H.B. Bidasaria, W. Schimmerling, M. Wong, and J. Howard, "^{20}Ne depth-dose relations in water," *Health Physics*, 46, pp. 1101-1111, 1984.

Wilson, J.W., L.W. Townsend, and F.F. Badavi, "A semiempirical nuclear fragmentation model," *Nuclear Instruments & Methods B*, 18, pp. 225-231, 1987.

Wilson, J.W., L.W. Townsend, and F.F. Badavi, "Galactic HZE propagation through the Earth's atmosphere," *Radiation Research*, 109, pp. 173-183, 1987.

Wilson, J.W., L.W. Townsend, W.W. Buck, S.Y. Chun, B.S. Hong, and S.L. Lamkin, Nucleon-nucleus interaction data base: Total nuclear and absorption cross sections, *NASA Technical Memorandum No. TM-4053*, August, 1988a.

Wilson, J.W. and L.W. Townsend, "A benchmark for galactic cosmic ray transport codes," *Radiation Research*, 114, pp. 201-207, 1988b.

Wilson, J.W., L.W. Townsend, B.D. Ganapol, S.Y. Chun, B.S. Hong, and W.W. Buck, "Charged particle transport in one dimension," *Nuclear Science & Engineering*, 99, pp. 285-287, 1988c.

Wilson, J.W., L.W. Townsend, J.E. Nealy, S.Y. Chun, B.S. Hong, W.W. Buck, S.L. Lamkin, B.D. Ganapol, F. Khan, and F.A. Cucinotta, "BRYNTRN: a baryon transport model," *NASA Technical Paper No. TP-2887*, March, 1989.

Wilson, J.W., J.E. Nealy, W. Atwell, F.A. Cucinotta, J.L. Shinn, and L.W. Townsend, "Improved model for solar cosmic ray exposure in manned Earth orbital flights," *NASA Technical Paper No. TP-2987*, June 1990a.

Wilson, J.W., J.L. Shinn, and L.W. Townsend, "Nuclear reaction effects in conventional risk assessment for energetic ion exposure," *Health Physics*, 58, pp. 749-752, 1990b.

Wilson, J.W., L.W. Townsend, S.L. Lamkin, and B.D. Ganapol, "A closed-form solution to HZE propagation," *Radiation Research*, 122, pp. 223-228, 1990c.

Wilson, J.W., S.Y. Chun, F.F. Badavi, L.W. Townsend, and S.L. Lamkin, "HZETRN: A heavy ion/nucleon transport code for space radiations," *NASA Technical Paper No. TP-3146*, December 1991a.

Wilson, J.W., L.W. Townsend, W. Schimmerling, J.E. Nealy, G.S. Khandelwal, F.A. Cucinotta, L.C. Simonsen, F. Khan, J.L. Shinn, and J.W. Norbury, Transport methods and interactions for space radiations, *NASA Reference Publication No. RP-1257*, December 1991b.

Section IV

Being There: Living and Working on Mars

Fresh vegetables are cut inside an inflated greenhouse while blanket-carrying cables are checked outside.

Chapter 17

MOVING IN ON MARS:
THE HITCHIKERS' GUIDE TO MARTIAN LIFE SUPPORT

Penelope J. Boston[*]

Life support encompasses all elements necessary to produce the total living *environment* for Mars inhabitants and the integration of those elements. This living environment will be present to varying degrees in the primary habitations, in remote safe-haven facilities, and in mobile vehicles for use in long distance exploration of the planet. Specific mission details will dictate the scope of life support planning for any given scenario.

Life support planners face myriad choices of structure type, lighting for living, working, and plant growth, protection from ultra-violet and ionizing radiation, provision of breathable air and consumable water, methods of food production and processing, and control of microbial and chemical contamination.

The integration of all these components into one flexible, reliable system is at once the most critical and the most challenging aspect of life support research. Computer modeling as a predictive tool, real-world whole system experiments, and development of reliable automation and robotics are essential areas of investigation which must reach a high degree of development before large, ecologically-based, managed life support systems can be confidently employed on the surface of Mars.

INTRODUCTION

It's a lazy Friday afternoon. You are planning a weekend hike into the mountains. You make a list of items to take along; extra clothes, perhaps some cold-weather gear or mosquito repellent, freeze-dried food, water or iodine tablets to make local streams drinkable, a topo map of the area, and a few plastic bags to store trash. If you are too fastidious to allow your oral bacteria to triumph unchecked for a day or two you might also take a toothbrush. The critical test of your gear is "Can I carry this in my pack and not collapse under the load or fall over backwards from the weight?" A simple test. If, on the other hand, you are making arrangements to scale the North Face, your equipment will include a great number of additional items, including oxygen, the means to

---

[*] Complex Systems Research, 7079 Redwing Place, Niwot Colorado 80503.

keep your extremities from freezing, and Sherpas or pack animals to help you carry this elaborate assemblage of equipment to some base camp. The critical test of your gear for this trip is "Will this stuff keep me alive and functioning long enough to reach my goal and hopefully return safely to my starting point." However, if, you are planning on moving to the Yukon to set up permanent housekeeping and mine for gold, your list of goods will grow very much longer, including as many amenities and comforts of civilization as your pocketbook, imagination, and transport facilities can handle. You may arrange for planes to drop additional supplies to you at regular intervals. When you arrive at your new Yukon home, you will situate yourself near reliable sources of water, collect local food materials, and learn to use whatever you can find at hand to increase your comfort and the reliability of essential supplies. You've come to stay and you mean business.

The spectrum of possible human life support scenarios for Martian missions is analogous to our Friday afternoon choices. How we approach the topic of human sustenance on such missions depends on our overall intent. Do we intend to just visit on a one or several time basis—an interplanetary backpacking trip? Do we intend to stay for a few years worth of missions and then abandon the planet—an expedition? Or do we intend to go, to stay, and make human life on Mars a reality for our posterity—that is, truly move in? The decisions we make and the research we do in the near future will determine which of these options are ultimately open to us if we choose to take them.

In the broad sense of the term, life support involves *everything* required to keep the crewmembers alive, functional, and sane. Obviously, this covers a tremendous range of needs including medical and psychiatric care, recreational facilities, adequate living space, good air to breathe, pure water and nutritious and palatable food to eat, removal and possible recycling of waste material, and control of possible pathogenic microorganisms and contaminating chemicals. The medical, psychiatric, and exercise requirements are a field by themselves and are addressed elsewhere in this volume (see chapter by Grymes *et al.*). The design of living space and recreational needs constitute the areas of ergonomics and aesthetics and are discussed in Harrison and Connors [this volume]. In this paper I shall concentrate on the input/output needs of human explorers on planetary missions, namely, food, water, breathable air, waste disposal and recycling, contamination control, and integration of all these factors into one unified system.

The underlying primary goal of all life support planning must be to create a low stress environment in which explorers can work, relax, and feel safe from the outside environment. It is a task of daunting complexity to accomplish this under novel and difficult circumstances.

TYPES OF MISSIONS

Effects of Mission Types on Life Support Planning

The specific technology and skills required for Mars exploration will depend upon the specific type of missions we envision. Perhaps more importantly, the entire philosophy and psychology of the design teams, engineers, scientists and crew will depend upon mission intent.

In many ways, the problem of living on Mars is simpler than the problem of living in space. Mars has a surface (14.4 x 10^7 km^2), a gravitational field (0.34 g), and a day/night cycle similar to Earth.* It offers the sensory perception of up from down, and essentially unlimited possibilities for expansion of initial facilities. It provides gatherable and extractable resources and the opportunities for increasing local "wealth" based on these resources.† Easier, cheaper methods of shielding against ionizing radiation are available than for spacecraft (e.g. mainly, heaps of dirt piled on top of buried habitations). In addition, even the tenuous Martian atmosphere provides adequate shielding from cosmic radiation for many applications.

Continuous human presence‡ on Mars is easier to plan for in some respects than quick trips to the planet. Anyone who has ever packed for a scientific field expedition knows the many pitfalls of the latter! ("Where is the wrench that fits *that* piece of equipment? What do you mean, you thought I brought it!") The goal of permanent habitation offers the possibilities of gradual bootstrapping, gradual accumulation of the material wealth of Mars resources, development of an indigenous manufacturing capability, space for storage of surplus supplies and reservoirs, ability to expand, and the familiarity of a landscape. Ultimately, it may offer the opportunity for the natural development of an entire Martian culture responsible for its own maintenance and welfare.

Developing the technology to provide reliable life support for the most ambitious missions will be a significant contributor towards a decision to actually mount such missions. With our current state of life support expertise, we are severely restricted in the options we can presently consider. We *must* develop more sophisticated life support methods now so that this area will not constitute a mental block in the minds of mission planners of the near future. If we plan our activities around quickie, Apollo-style missions, we are essentially backpacking and life support research now should be oriented towards this limited kind of mission. If this is our choice, we must make sure that the short-term scientific goals are met at the expense of devoting crew energies to life support activities. If we anticipate a few well-planned, limited duration missions (the expedition), we still go as visitors. More attention will be devoted to life support but it will still be a lower priority with considerable reliance on expendable components and materials.

If we expect our reach for Mars to culminate in a continuing human presence, then we must approach the development of life support systems in a sequential, step-by-step fashion, each phase of planning and each early mission building on the life support

---

* The Martian day, a sol, is 24 hours 39.5 minutes long [Smith and West, 1983].

† Water is the most obvious precious material which could serve as a medium of exchange for a Mars base. Clark and Pettit [1989] have also suggested hydrogen peroxide as a medium of exchange because it can serve as a storage form of oxygen, water, and energy. In addition, it can be used as a monopropellant, antifreeze, bleach, disinfectant, chemical reactant, and explosive.

‡ The possible types of human habitation on Mars range from a single continually inhabited research base, to a series of bases, to a full scale human colony. I prefer to avoid the distinction since the latter can be seen as an organic outgrowth of the Mars base. I will refer to all possibilities for permanent human inhabitance as "facilities."

capabilities of the last. We will go as residents, nest builders, for the long haul. If we choose this third road, then early missions can be devoted largely to erecting the structures and establishing the infrastructure and protocols necessary to sustain an increasingly sophisticated and reliable, self-sufficient life support system. Table 1 shows the comparison of various mission scenario trade-offs which enable us to determine the cross-over point at which self-sufficiency becomes advantageous.

Table 1
SELF-SUFFICIENCY TRADE-OFFS

Study	O_2 loop	H_2O Loop	Food
Spurlock et al., 1975	30 days	6 months	3–14 years
Henson & Henson, 1977	28 days	8 months	4 years

In this paper, I will consider primarily the third, continuous habitation option. My own preference predisposes me to consider Mars as a permanent human goal. Not only is it the next logical step in the advance of human civilization, but it is the only way to truly study the science of another world, an entire planet's worth of mysteries.

Venues for Life Support Systems

The chronologically first life support systems encountered during Mars missions will be those aboard spacecraft which transport humans to the planet. As transit times are shortened by improved propulsion systems, the strain on space travelers will decline and so will the demands on spacecraft life support systems. The constraints on mass and volume and the problem of zero-gravity or perhaps artificially provided gravity (by means of rotating craft) severely limit many of the options for life support systems aboard transit vehicles. Of course, considerable recycling of air and water will still be essential. However, extensive *in situ* biological food production, and ecologically-based systems seem both less essential and less appealing than for a Mars or lunar habitats. Kliss and MacElroy [1989] have designed a system for providing fresh salad and vegetable materials for spacecraft or space station use. This level of limited dietary augmentation by *in situ* food production is probably nutritionally and psychologically beneficial.

Life support devices and procedures currently in use aboard the Space Shuttles, the Russian Mir station, and under development for Space Station will be the most suitable for application to the design of interplanetary vehicles. We will presumably continue to amass this kind of experience and knowledge which will be directly applicable to Mars transit vehicles. For recent reviews of these topics see Waligora *et al.*, 1990 and Boston, 1989a and b.

Various types of structures for a Mars facility must have life support provisions. These range from the primary habitats and work areas, to remote installations serving as temporary shelter for field parties, to life support aboard surface vehicles, and surface spacesuits.

STRUCTURES, LIGHT, AND BEAUTY

Craters, caves or other stable geological depressions are an obvious choice as pre-excavated sites for Martian structures. Pre-fabricated or built *in situ*, interior structures would be protected from micrometeorites, dust storms, and various types of radiation.

Horz [1985] has suggested the use of lava tubes on the Moon as a pre-excavated, pre-existing enclosure in which to build lunar base components. Such tubes which are well-known on Earth may also exist on Mars in volcanic regions and be available for use. They could also be used unmodified as garages or storage areas for equipment protected from radiation, micrometeorites, and the ubiquitous Martian dust. The interior structures could be relatively flimsy since they could be suspended from the tube ceiling, and will require minimal weight-bearing components. The thermal regime inside a tube would be more constant than the surface of Mars.

One suggestion for habitat construction on site involves the use of ceramic materials in an interesting blend of new high technology components and ancient principles developed for the buildings of ancient Persia [Khalili, 1986]. Using fused, glazed adobe material possibly lined with other impervious material to create an air-tight barrier, Khalili suggests flowing material into already existing craters, lava tubes or other partially excavated structures on the Moon. Such ideas could also be applied to Mars.

Melting of indigenous rock and soil by heated penetrators or boring devices to produce melted glass casings or linings for holes has been explored in some detail by Rowley and Neudeker [1985]. They conclude that the techniques they consider are relatively insensitive to rock types and conditions. If true, such construction methods would be highly versatile for use on Mars. Barrel-vault brick structures built from indigenous Martian materials, sealed, and covered with regolith are another idea suggested by Mackenzie [1989]. Mortar and glass produced from Mars materials and scrap from discarded landers and other machinery can also be employed as building material. This author calculates that 7 m of regolith above a brick barrel vault structure would be adequate to contain a 1 atmosphere internal pressure. Boyd *et al.* [1989] have suggested the use of indigenous Martian clays, salts and water to create duracrete and composite building materials for Mars construction. These investigators have experimentally produced materials from Mars analog materials to prove the feasibility of their ideas.

Even while providing protection from the rigors of the external environment, it is neither practical nor desirable to make our Mars explorers into troglodytes.* Many activities will have to be performed on the surface both near and distant from the base. In addition, the view of the external vista may prove to be important to psychological health.

Simple, light-weight inflatable structures have been suggested as surface plant growth modules [Boston, 1981] and as surface work areas [McKay *et al.*, 1986] called "airshells." Surface structures could provide direct access to visible light for habitation and plant growth. Advantages include more usable volume per mass of structure, ease of construction, and rapid deployment. Such structures could perhaps be for temporary use

---

* Prehistoric cave dwellers.

only. The primary disadvantage is exposure to radiation. Ultra-violet radiation can be combated by the use of uv-opaque, uv-resistant materials. In addition, uv-cured plastics and other materials could be easily employed using the native ultraviolet flux. Roberts [1989] has suggested that inflatable structures could also be covered with dirt as shielding from ionizing radiation. His calculations show that an inflatable structure on Mars with an internal pressure of 101.4 KPa can support a covering of 1.6 g cm^{-3} density dirt to a depth of 17 m. This is well in excess of a depth required for shielding. He points out that abrasion from dirt or rock particles must be taken into consideration and a resistant layer should be employed between inflatable structure and over-lying material. Possible collapse upon puncture by micrometeorites or other punctures can be prevented by the use of internal supports placed after inflation [Roberts, 1989] or by an external rigid structure constructed after inflation and attached to the inflatable module.

Possibly the best solution is a hybrid of the various structural possibilities tailored for each specific use. This is the historical pattern of human habitation and the common sense way to proceed. Some structures could be envisioned which would have surface-exposed and buried areas. Such multi-part structures would have aesthetic as well as practical value. Estimates of required space for human living quarters varies with the assumed duration in the space, the number of people in the group, the nature of work and daily operations and the psychological make-up of individuals [Connors *et al.*, 1985]. A range of space estimates is presented in Table 2.

Table 2
LIVING SPACE REQUIREMENTS

Study	Volume	# People	Per person
Breeze, 1961	17 m^3 (600 ft^3)	1	17 m^3 (600 ft^3)
Fraser et al., 1968	7-11.8 m^3 (250-415 ft^3)	1	7-11.8 m^3 (250-415 ft^3)
Nelson (this volume) Biosphere II			
Actual volume Habitat	10,677 m^3	8	1335 m^3
Usable volume	153,908 m^3	8	19,239 m^3
Spacious house (8000 ft^2)	1722 m^3	4	430 m^3
Typical ranch house (2000 ft^2)	437 m^3	4	110 m^3
Apartment (800 ft^2)*	174 m^3	2	87 m^3
Small bedroom (8' × 10')*	17.2 m^3	1	17.2 m^3

* A height of 2.4 m is assumed.

Lighting and power requirements depend intimately upon exact details of base design. Use of natural light with ultra-violet wavelengths filtered out would be convenient when possible, however, seasonality and periodic dust storms are drawbacks. Solar collectors and photovoltaics have been suggested, however, disadvantages of this type of system involve deposition of dust on collector surfaces and prolonged obscuration (up to 50% reduction) of light by dust storms [Geels et al., 1989]. This level of illumination is similar to a cloudy day on Earth.* Nuclear power has been suggested [e.g. French, 1985; Isenberg and Heller, 1989].

The SP-100 reactor designed under the auspices of NASA is considered a good model for the sort of small reactors thought appropriate for a Mars habitat. For power comparisons see Table 3. Artificial lighting for plant growing areas is very energy costly although exact sizing varies greatly with diurnal cycles chosen, and preferred intensities of the species being grown. For plant growth lighting requirements see Table 4.

Table 3
SUGGESTED POWER SOURCES

Study	Power	Mass	Advantages	Disadvantages
French, 1985 SP-100 unit thermoelectric	100 kWe	3000 kg	Operates day and night. Unaffected by dust storms or seasonal changes.	7–10 year service life presents nuclear disposal problem.
Isenberg and Heller, 1989 SP type unit	300 kWe	7200–7500 kg	Operates day and night. Unaffected by dust storms or seasonal changes.	7–10 year service life presents nuclear disposal problem.
Galecki and Patterson, 1987	3 MWe	30,000 kg	Operates day and night. Unaffected by dust storms or seasonal changes.	7–10 year service life presents nuclear disposal problem.
Geels et al., 1989; solar incident at surface with photovoltaics	180 W m^{-2}	Not stated	No radioactive disposal problems.	Interference from dust storms and seasonal changes.
Angelo & Buden, 1985 RTG[1] RDCG[2] Reactors[3] Reactors[4]	Up to 500 We 0.5 kWe–10 kWe 10 kWe–1000 kWe and 1–10 MWe units 10–100 MWe			

[1] (Radioisotope thermoelectric generator)
[2] (Radioisotope dynamic conversion generator)
[3] Heat pipe, solid core or thermionic.
[4] Solid core, pellet bed, fluidized bed gaseous core.

---

* On Earth, a moderately cloudy day provides about 1/16 of the illumination of a sunny day.

Table 4
ENERGY REQUIREMENTS FOR LIFE SUPPORT ORGANISMS

	Photosynthetic efficiency		Typical field efficiency	Solar	Artificial lighting	
	Theoretical	Practical			RTG	Photovotaics
Schneider, 1973	10–12% (higher plants)	5–6% (higher plants)	—	—	—	—
Calvin, 1974	—	—	0.02–0.10% (higher plants)	—	—	—
Stokes, et al., 1981	—	—	—	5% (algae)	1.0% (algae)	0.15% (algae)
				—	24% (H_2 bacteria)	3.7% (H_2 bacteria)
Cullingford & Schwartzkopf, 1990	300–600 µmol m^{-2} s^{-1} (minimum photosynthetic light)			0.5–1 kW per m^2 growing area		
Biosphere II Nelson (this volume)	—	—	—	1.1 kW/m^2 # (intensive agriculture area)		0.11 kW/m^2 provided by fossil fuel for entire area

*RTG – Radioactive thermal generator
#2403 kW for 2232 m^2
**1500 kW for 13,766 m^2

PROTECTION FROM RADIATION

The surface of Mars is bombarded by a variety of natural radiation. The surface is subject to a strong ultra-violet radiation flux, 6-7 x 10^3 ergs cm^{-2} in the 2,000-3,000 angstrom range [Boston, 1985]. Ultra-violet wavelengths shorter than this are filtered out by the thin carbon dioxide atmosphere.

There are several types of damaging ionizing radiation. Galactic radiation comes from explosive events outside our solar system. It is composed of very high energy particles, HZE (high z energy where z is an element), protons, alpha particles, and nuclei of heavy metals. The fluxes of these destructive high energy particles is quite low [Harding, 1989]. Protons account for about 87%, alpha particles account for 12%, and heavy particles ranging from lithium to tin account for an additional 1% [Nicogossian and Nachtwey 1989]. Solar radiation especially from flare events, can have high energies and high fluxes (10^6 to 10^8 electron volt proton fluxes) according to Harding [1989]. High energy protons can possess energies of 10-500 MeV [Olson and McCarson, 1974].

Ultra-violet radiation causes breakage and incorrect repairing of the chemistry of the DNA molecules within cells (known as dimerization of thymine). Ionizing radiation forms ion pairs in cells, which subsequently react with cell water to form free radicals. Hydroxyls and hydrogen peroxides do great violence to the cell's chemistry resulting in serious biological damage [McCormack and Nachtwey, 1989]. Gamma-rays and x-rays ionize atoms within cells. High energy protons interact with atomic electrons producing an array of secondary particles. These include neutrons, pions, heavy particles, lower energy secondary protons, and gamma rays all of which wreak varying degrees of cellular damage as they pass through the cell. Neutrons exchange energy when colliding with hydrogen nuclei. Unfortunately, all biological organisms are very rich in hydrogen and are vulnerable to neutron damage [Lundquist, 1979].

Protection from ultra-violet is a relatively simple matter consisting of simply avoiding exposure of organisms and hardware to surface fluxes. Very thin layers of material are adequate shields against this kind of radiation. Ultra-violet opaque materials are currently available to provide resistant building materials. The effect of the strong flux will be to reduce the functional lifetime of structures and materials exposed to them.

Protection from various ionizing radiations is a much more complex matter. However, on the Mars surface use of indigenous regolith and rocks is probably the most practical solution, at least for habitats. Protection in spacecraft is considerably more difficult because of mass constraints. Harding [1989] claims that 2 g cm^{-2} is a sufficient density of radiation shielding for average exposure levels in space except for the largest solar flares. Letaw and Clearwater [1986] cite 20 g cm^{-2} as adequate for long term human exposure. On the high end, 30-40 g cm^{-2} density shielding is recommended against very high energy HZE particles by McCormack and Nachtwey [1989]. The idea to use existing caves and lava tubes discussed above in the section on structures provides more than sufficient radiation protection for habitats. For surface activities, other alternatives must be found. Rovers can be shielded with the same materials used for spacecraft shielding. For protection from solar flares, "storm shelters" of higher density shielding can be incorporated into the vehicles themselves or perhaps be located along established routes.

The problem of radiation protection from solar storms while space suits are in use on the unprotected surface must be solved more satisfactorily than at present if inhabitants are to spend prolonged periods on the surface. New high technology alternatives to large amounts of mass and lead for radiation shielding are being developed. Yaffe and Mawdsley [1991] have developed a new type of radiation shielding for protection from medical x-rays. They mix barium and tungsten powder with the vinyl or rubber matrix of medical radiation garments. The resulting product weighs 20% to 30% less than similar lead garments while providing better coverage in the range of salient energies than just lead alone. Gold particles embedded in a foam matrix is also being developed as an ultra lightweight radiation shielding for medical applications. Whether such advanced solutions will be able to reduce the weight of shielded space suits to an acceptable level is debatable. At least, we will be dealing with only 0.34 g on the Martian surface.

AIR AND WATER

Mars has an atmosphere. This is the single most important feature of the planet for creating and maintaining a life support system. From this atmosphere, we can extract water, and we can make a breathable gas mixture for humans and life support plants, microbes, and animals. Various estimates of air and water requirements are listed in Table 5.

Table 5a
AIR AND WATER (PER PERSON PER DAY)

Study	Air (O_2)	Drinking (H_2O)	Washing (H_2O)
Harding, 1989	0.9 kg (2.0 lb)	2.5 kg (5.5 lb)	1.5 kg (3.3 lb)
Rummel & Volk, 1987	—	4.6 kg (10.1 lb)	18 kg (39.6 lb)
Quattrone, 1984	0.8 kg (1.8 lb)	4.2 kg (9.3 lb)	1.1 kg (2.5 lb)
Gustan & Vinopal, 1982	0.84 kg (1.8 lb)	3.1 kg (6.8 lb)	—
Modell & Spurlock, 1980	"4 times body weight/yr" (0.01 x body weight/day)	"8 times body weight/yr" (0.02 x body weight/day)	—
Sharpe, 1969	0.9 kg (2.0 lb)	3.6 kg (7.9 lb)	5.4 kg (11.9 lb)
Alman et al., 1968	—	2.3 kg (5.0 lb)	—

Table 5b
TOTAL CONSUMABLES (PER PERSON PER DAY)

Study	Mass	Type of material
Salisbury, 1990	24 kg	Air, food, water, and stored waste
	19 kg	Packing for above consumables
	43 kg	Total for above
Rummel, 1990	6.5 kg	Air, food and water
	13 kg	Wash water
	19.5 kg	Total for above

Air

Various methods for mining the Martian atmosphere for breathable gasses have been assessed [Meyer and McKay, this volume]. Of course, once the breathing mixture is created the bulk of it will be continuously recycled with newly extracted gasses being added to make up for leakage and air-lock losses. A hybrid system using both higher plants and physical/chemical systems for air recycling has been designed [Lee and Brown, 1989]. These investigators control temperature, humidity level, and partial pressures of oxygen and carbon dioxide in a system with a Plant Habitat, a Crew Habitat and separate Electrochemical O_2 and CO_2 separators. Plant Habitat moisture is condensed out and returned to the plants as water or as humid air. As the Crew Habitat CO_2 levels rise, this air is sent to the CO_2 separator which also separates oxygen. The mixture of CO_2 and O_2 is then sent to the Electrochemical O_2 separator where the oxygen is stripped out and returned to the Crew Habitat. The CO_2 is sent to the Plant Habitat. In total, this appears to be a fairly simple and sensible design. It incorporates both living and non-living components of the system to perform the required air regeneration functions over the wide range of conditions to be experienced in a Mars habitation. Many schemes for production of oxygen from biological sources have been advanced since early in the space program. Many such schemes have relied upon algae for this purpose [e.g. Smernoff *et al.*, 1986; Eley and Meyers, 1964; Myers, 1954]. Averner and coworkers [1985] have reported upon a system containing a mouse and algae with a closed-loop gas cycle. The Soviets are said to have produced enough oxygen for one person using the algal species *Chlorella* in 8 m^2 area [Salisbury, 1990].

Choice of the inert portion of the breathing mixture is an unresolved issue as yet. On Earth, nitrogen comprises 79% of our air. On Mars, nitrogen may be too precious a resource to squander on this inert use. Argon gas is present in small quantities (1.6%) in the Mars air and can be extracted for use relatively easily [Meyer and McKay, this volume]. The long-term effects of breathing argon or an argon/nitrogen mixture are unclear but it has been suggested that breathing argon or helium can reduce dangers of compression sickness as inhabitants go from interior habitat pressures to typical surface spacesuit or rover vehicle pressures [Harding, 1989].

Water

Mars possesses several known and several suspected sources of water. The polar caps have water ice in the perennial ices and the layered deposits [Squyres, 1985]. The Martian atmosphere contains small but extractable quantities of water, 0.135% by volume [Meyer and McKay, 1984]. Geological features on the surface of the planet attributable to flowing water have led to suggestions for significant permafrost layers present poleward of about 30° latitude north and south. Water bound geochemically to silicates may also be minable. A much better understanding of water sources on Mars is critical information which must be gathered before good design choices can be made [Squyres, 1985].

Once the water is obtained, the technologies of water purification and recycling are highly developed and readily available [for good reviews see, Hermann and Wydeven, 1990 and Highsmith *et al.*, 1990]. Substantial quantities of water are required for plant growth and other agricultural uses. This water can be recycled within the agricultural sphere, however gradual contamination by metals or other difficult to remove compounds may limit its indefinite recycling potential. This may be rectified by use of reverse osmosis and ion exchange technologies or filtration through a microbial/soil bed. In addition, there will be unavoidable losses from the habitats and any expansion of the facilities will necessitate the continued capability for water extraction.

FOOD

Food Requirements

Unlike primitive human groups of the past, or the initial colonization efforts of Europeans in the New World, the "primitive" Martian inhabitants will not have access to the hunter-gatherer mode of gleaning a natural environment for food. They must immediately become agrarians. Growing food is a basic human activity of great antiquity. Not only does it nourish us physically, but those who engage in it also draw psychological sustenance from it. Gardening as a hobby attracts 90 to 100 million people in the United States alone. Food production for a Mars base or colony will include the growth of plants, the growth of microbes, and chemical and microbiological processing of food.

Table 6a
NUTRITIONAL REQUIREMENTS—TOTAL MASS AND ENERGY (PER PERSON PER DAY)

Study	Total food	Energy
Nicogossian, 1989 (after Sauer & Rapp, 1981, 1983)	—	3000 kcal (recommended) 2793 kcal (typical for Apollo missions)
Harding, 1989	1.5 kg (dry)	—
70 kg man at rest	—	2000 kcal
" office work	—	2400 kcal
" heavy activity (e.g. mining)	—	4500 kcal
Rummel & Volk, 1987	0.85 kg	—
Busta et al., 1986	0.5 kg (fiber not incl.)	—
Quattrone, 1984	2.5 kg	—
Gustan & Vinopal, 1982	0.6 kg (dry)	2800 kcal
Modell & Spurlock, 1980	"3 times body weight per yr" (0.008 x body weight/day)	—
Sharpe, 1969	0.6 kg	—
Alman et al., 1968	—	2737 kcal
Latham, 1965		
55 kg man	—	2200–3000 kcal
47 kg woman	—	1800–2500 kcal
pregnant "	—	2200–2900 kcal
lactating "	—	2700–3400 kcal

Various estimates of human nutritional needs are available (see Table 6). Our lack of experience on other planetary bodies with different environmental conditions means that we will have to guess at the actual nutritional requirements of Mars inhabitants. We must interpolate between terrestrial data and that gathered for astronauts and cosmonauts in Earth orbit. In addition, individual variations between people can cause differences in the requirements for a given nutrient to vary by factors between 3 and 10 [Dubos, 1979].

Table 6b
NUTRITIONAL REQUIREMENTS—SPECIFIC NUTRIENTS

Study		Protein	Fat	Carbs	Crude fiber or ash	Ca	P	Na	K
Nicogossian, 1989 Apollo	(Recmd)	56 g	—	—	—	800 mg	800 mg	3450 mg	2737 mg
	(Actual)	102.4 g	99.8 g	372.1 g	21.1 g (ash)	1168 mg	1646 mg	5101 mg	2728 mg
						Fe	Mg	Zn	Cl
	(Recmd)					18 mg	350 mg	15 mg	—
	(Actual)					13 mg	249.1 mg	—	11.2 mg (as NaCl)
Rambant & Johnson, 1989		80–160 g	Max 150 g 1% fat as linoleic acid	350–400 g	5–10 g (fiber)	—	—	—	—
Harding, 1989 70 kg man at rest		70 g	65 g	370 g	—	—	—	—	—
Busta et al., 1986		70 g	80 g	350 g	—	—	—	—	—
Alman et al., 1968		105 g	53 g	497 g	—	—	—	—	—
Latham, 1965									
55 kg man		60–70 g	—	—	—	—	—	—	—
47 kg woman		55–65 g	—	—	—	—	—	—	—
pregnant woman		85 g	—	—	—	—	—	—	—
lactating woman		95 g	—	—	—	—	—	—	—

The secondary effects of spaceflight on human physiology undoubtedly affect nutrition. While it takes less energy to locomote in lower gravity, the reduction of body mass and atrophied muscle mass may account for an increase in the need for carbohydrates (14%) and decrease in intake of fats (by 21%) experienced by Skylab astronauts [Michel *et al.*, 1974]. Skylab astronauts lost body fat even though caloric requirements should have been amply met [Rambaut and Johnson, 1989]. A decrease in metabolic functioning as reflected in an increase in the thyroid hormone thyroxin was reported for Skylab crewmembers [Sheinfield *et al.*, 1974]. Also, a heavy physical workload can essentially double the caloric requirements of a sedentary lifestyle (see Table 6).

Higher Plants

Because of our huge cultural experience and daily reliance on higher plants, great emphasis has been placed on their role in the prospective Mars diet.

Intensive horticulture of wheat in closed environments has received over a decade of attention at the labs of Frank Salisbury and colleagues [Salisbury, 1990]. They have tested over a thousand different wheat strains and are breeding ultradwarf (less than 40 cm high), very high-yielding varieties themselves. They have experimentally manipulated all the environmental variables including light intensity and duration, temperature, planting density, and nutrients. Their yields have reached 60 g m^{-2} day^{-1} compared to the world's field grown wheat record of 12-14 g m^{-2} day^{-1}. Yields this high require high light intensities. The power necessary to provide this light will undoubtedly be the limiting factor controlling actual yields. In addition, many other crops cannot be grown with the high density efficiency of wheat. When these other realistic considerations are included Salisbury [1990] has roughly estimated 50 m^2/person as a reasonable size for space farms [Salisbury, 1990]. Table 7 presents various estimates for the necessary size of food growing areas. As can be seen from these figures, requirements per person range from very optimistic estimates based on maximum yields of high-density monoculture [Gitel'son, 1977] to the actual size of the Biosphere II experiment [Nelson and Dempster, this volume]. The latter provides a more realistic picture of the yields which could be expected under the actual exigencies of daily life in a closed habitat. All of the other estimates in Table 7 compare favorably to the field-grown, conventional agriculture derived estimate of MacElroy and Averner [1978]. The 1,800 Kcal day^{-1} diet of the Biospherians may be marginal or insufficient for Mars Base inhabitants, a point to bear in mind when scaling life support systems.

Table 7
PLANT GROWTH AREA

Study	Area	# People	Per person
Nelson (this volume)	2232 m^2 (24,000 ft^2)	8	279 m^2 (3000 ft^2)
Cullingford & Schwartzkopf, 1990	20–30 m^2	1	20–30 m^2
Bugbee & Salisbury, 1989; Salisbury, 1990	13 m^2 (minimum) 5000 m^2	1 100	13 m^2 (minimum) 50 m^2 (ample)
Root zone depth	10–20 cm (minimum) 100 cm (ample)	—	—
Oleson & Olson, 1986	56.9 m^2	1	56.9 m^2
Vasilyev, 1984 Bios 3	60 m^2	2 (for food) 4–5 (for oxygen)	30 m^2
Hoff et al., 1982	24 m^2	1	24 m^2
Gitel'son, 1977	14 m^2	1	14 m^2
MacElroy & Averner, 1978. Based on conservative field yields as basis for comparison.	820 m^2	1	820 m^2

Other crops which are being studied by NASA in confined environment monocultures include potatoes, soybeans, and lettuce and other oil-producing crops [Tibbitts, 1989; Tibbits and Alford, 1982]. The Soviets have studied wheat, chufa, various vegetables, potatoes, carrots, radishes, tomatoes, cucumbers, kohlrabi, dill, and peas [Ivanov and Zubaraeva, 1985]. In a study by Hoff and coworkers [1982], it was concluded that a minimal complete diet required 10 main crops. These include soybean, peanut, wheat, rice, potato, carrot, chard, cabbage, lettuce and tomato.

Cullingford and Schwartzkopf [1989] have suggested growing jojoba to provide lubricating oils for life support machinery. Guyaule could be grown for its rubber. Varieties of crops not commonly grown in the West should be assessed for extra-terrestrial use. For example, mizuna (also known as Japanese mustard spinach) is growing very densely in a cramped tub in the author's garden in the cement-like subsoil which passes for garden loam in Colorado. It has suffered outrageous abuse, regularly wilting to rag-like, prostrate strings only to be immediately revived upon application of water. It remains sweet, tender and not bitter even when flowering and going to seed unlike conventional spinach which becomes inedible when bolting. Many such indestructible yet delectable plants exist in the garden repertoire of other cultures or perhaps growing wild as yet undiscovered by horticulture.

A large-scale, ongoing project at the NASA Kennedy Space Center is known as the Kennedy Breadboard Project. Begun in 1985, an old vacuum test chamber was converted into a sealable module for testing the hydroponic growth of higher plants in highly controlled, closed situations [Salisbury, 1990; Prince and Knott, 1989; Averner, 1989; Boston, 1988]. The growing area available is 54 m^3 and high pressure sodium lamps are used for illumination. Wheat, soybeans, potato, lettuce, and rice are plant choices for this experimental system.

Choices for rooting higher plants include ordinary dirt, inert solid media like rockwool or vermiculite, or suspension through an inert surface like Styrofoam. If the latter options are chosen, liquid nutrient solution must provide all plant growth requirements. This growing method is called hydroponics. Bugbee and Salisbury [1989] use a 1:1:1 mixture of peat, perlite, and vermiculite in their hydroponic wheat-growing system with seeds sewn into rockwool and later transplanted.

Another method which has been employed to deliver nutrients to roots is known as "aeroponics." This involves suspension of the plants in air with the exposed roots sprayed continuously with an aqueous nutrient solution. This can allow for good use of vertical space, but some deleterious effects on plant stocks have been detected with time. Of course, seedstock need not be produced in this way. Such a method could be reserved for very high density production activities.

Dirt

In an analysis of Martian regolith as a base for creating agricultural soil, Banin [1989] tentatively concludes that with initial aqueous removal of excess sulfates and chlorides, and addition of some macro and micro nutrients it may provide a reasonable growth medium for plants. He points out that smectite clays can act as a buffer retaining the pH in the plant-compatible range, however, fine porosity may lead to compaction and poor water availability to roots. In their recent exhaustive survey of the known and inferred properties of Martian regolith, Stoker et al. [1993] point out that the material is planet-wide and apparently highly consistent in texture and chemistry. This would enable the same soil amelioration and amendment techniques to be employed in habitats placed anywhere.

Methods for the transformation of lunar regolith into soil for agriculture have received relatively concentrated recent attention [see Ming and Henninger, 1989]. The physical and chemical properties have been assessed based on existing samples of lunar materials [Ming, 1989; Helmke and Corey, 1989; Whitney, 1989]. Potential toxicity is discussed by Hossner and Allen [1989].

The role of microbes in soil-building from regolith has been discussed vis-à-vis the Moon [e.g. Alexander *et al.*, 1989]. The presence of clays in Martian soil will be a boon in providing microhabitats for microbes and retaining plant nutrients [Stotzky, 1989]. Microbes are critical in mobilizing essential elements for plant nutrition [Ehrlich, 1989].

Microbes as Food

Single cell protein (SCP) production by microbes has been studied intensively by industry and has frequently taken the attention of life support investigators [Kihlberg, 1972; Moo-Young, 1976; Litchfield, 1977]. While direct consumption of microbes is not usually palatable or very digestible, with subsequent processing, high quality food items can be produced. The upper limit of direct consumption of whole algae or yeast seems to range between 100-200 g day^{-1} person^{-1} [Stokes *et al.*, 1981]. Microbial cells seem essentially unaffected by microgravity or reduced gravity which renders them a good choice for Mars and in-flight food production [Klein, 1981].

Protein content of most microbes varies between 7% and 12% of dry matter which is higher than most other common foods. They can be continuously produced in high densities in closed fermentation or growth chambers with minimal water requirements compared to higher plants. They can be treated in many ways more like chemicals than like living organisms which lends itself to automation and rapid processing. Waste disposal problems are small and many organisms can be fed on plant, human, animal and industrial wastes. Photosynthetic organisms require only carbon dioxide, water and light as do higher plants, and they photosynthesize in a non-oxygen atmosphere.

Another advantage of microorganisms is the ease of storage of inactive cultures [Boston, 1984]. Backup cultures and a library of various strains can be maintained in the freeze-dried state (lyophilized) for indefinite periods of time. This could serve as an insurance policy against system failure or uncontrollable contamination of production cultures by undesirable microbial competitors.

Another set of talents that microbes possess which will be invaluable to Martian inhabitants is their ability to produce a range of chemicals for industrial use, plastic-like materials, and fuels. This area is beyond the scope of this paper but an interesting review of the subject can be found in Zeikus, 1980.

Microbial Processing

Microbial processing of inedible or low-quality plant materials is of great antiquity. For thousands of years foods have been fermented to produce shoyu, tofu, miso, tempeh, bread, beer, wine, yogurt and cheese [Hesseltine and Want, 1967]. In addition, a substantial industry has arisen to produce flavorings, enzymes, vitamins, and amino acids as food additives [Kihlberg, 1972].

Many plant or microbe-produced foods gain in palatability and nutritional value by further microbial processing. These "engineered foods" are both traditional types mentioned above and high technology products like soy meat substitutes [Henry, 1979]. Textured vegetable protein has been successfully engineered to taste and appear like a variety of products from meat and fish to nuts, and mushrooms.

Chemical Processing

The chemical food processing industry is so immense in our own society that a discussion apropos Mars seems almost pointless. The endless arrays of new junk, convenience, and novelty foods which parade across our grocery store shelves are testimony to the excesses of creativity which can be employed when sufficient profit motive is provided. Mars life support planners can tap into this rich storehouse of technology when the basics of food production choices have been worked out.

Aquaculture and Animals

Organisms which are adapted to live in a buoyant medium like water are probably a good choice for the low gravity environment of Mars. Fish like the apocryphal *Tilapia* (which grow rapidly in unbelievably dense concentrations) and various crustaceans have been considered for life support systems. One terrestrial aquaculture system which actually operated on a commercial basis used human waste as fertilizer, and algae, shrimp and fish to balance nutrient and gas levels [Costa-Pierce, 1984]. Aquaculture has also been extensively studied by the Biosphere II researchers [Nelson and Dempster, this volume].

Food and Culture

The sociological aspects of food include its function as a centerpoint of human social activity. It is hard to imagine a social gathering without the provision of food or at least coffee and donuts. To our species, food connotes safety, belonging, helpfulness and acceptance. Bluth [1982] has suggested that space explorers share at least one meal a day to contribute to social cohesiveness building. The down side of food consumption is a tendency to excessive emphasis on it in an impoverished, boring, or confined environment [Connors *et al.*, 1985]. This is far more likely to be a problem aboard the spacecraft that delivers the Mars inhabitants than it is on the Martian surface. However, one can imagine junk food binges during prolonged and isolating global dust storms.

RECYCLING AND DISPOSAL OF WASTE

Waste materials include inedible portions of food plants, microbes and animals, human metabolic wastes and faeces, wastes from biomass production of chemicals, fuels, building materials, and from various industrial and maintenance processes conducted in the habitations (see Table 8). Using Draconian methods of recycling, much of the organic material can be completely oxidized back to carbon dioxide, water and ash. However, this wastes much of the energy stored in the complex bonds of these molecules. A more energy wise approach is to use as much waste material as possible directly as secondary input to other processes. However, recycling tradeoffs depend upon the energy and labor costs involved. For example, recycling 99% of some material may cost more per unit than recycling a more modest percentage using some significantly less costly process.

Table 8
WASTE PRODUCTION (PER PERSON PER DAY)

Study	CO_2	Resp. H_2O + Evap.	Urine	Feces	Wash water	Plant by-product	Waste heat	Other
Harding, 1989	0.8 kg (1.8 lb)	—	2.5 kg (5.5 lb)	0.2 kg (0.44 lb)	1.5 kg (3.3 lb)	—	—	8 g dead epithel.
Light work in temp. climate		0.9 L	1.5 L	Fecal H_2O 0.1 L				0.5 L (Flatus)
			2.5 L Total H_2O loss					
in desert			10.0 L Total H_2O loss					
Busta et al., 1986	—	—	—	—	—	1.75 kg (3.85 lb)	—	—
Sharpe, 1969	1.0 kg (2.2 lb)	2.5 kg (5.5 lb)	1.5 kg (3.3 lb)	0.16 kg (0.35 kg)	—	—	12,660 kJ	—
Gustan & Vinopal, 1982	1.0 kg (2.2 lb)	1.8 kg (4.0 lb)	1.5 kg (3.3 lb)	0.2 kg (0.44 lb)	—	—	2800 kcal	—
Quattrone, 1984	1.0 kg (2.2 lb)	1.8 kg (4.0 lb)	2.0 kg (4.4 lb)	0.13 kg (0.30 lb)	1.2 kg (2.5 lb)	—	—	—
Rummel & Volk, 1987	1.1 kg (2.4 lb)	2.1 kg (4.5 lb)	3.0 kg (6.6 lb)	0.16 kg (0.35 lb)	18 kg (40 lb)	—	—	—

Typical agricultural residues are approximately 10-30% lignin, 25-50% hemicelluloses,* and 25-40% cellulose. These materials can find alternative uses. For example, Volk and Rummel [1989] have suggested an array of uses for waste cellulose from plants to make furniture, interior wall covering, fiber for clothing and other textiles, myriad paper products, cellophane and other cellulose-ester plastics [Milgrom, 1987].

---

* The term "hemicellulose" refers to a variety of pentose polymers.

Cellulose, lignins and other plant products are so useful that it has led Volk and Rummel [1989] to suggest that these plant materials be produced purposely rather than just as a by-product of Martian agriculture. They suggest that this could form the basis of a growth industry on Mars independent from Earth as the population increases and the need for increased habitat accelerates. The abundance of CO_2 in the Martian atmosphere as a relatively easily acquired commodity [Meyer and McKay, this volume] makes more permanent uses of fixed carbon feasible. Cellulose also ties up water (0.5 g per 1 g of cellulose) and the cost of its extraction must be figured into the cost of permanent cellulose uses [Volk and Rummel, 1989]. A beneficial by-product is the 1.2 kg of O_2 produced per 1 kg of cellulose produced.

Busta and colleagues [1986] have explored uses for non-edible residues from wheat in controlled growth systems. They estimate 1.75 kg per person per day could be produced by such systems. Options they consider include meat animal feed, direct microbial conversion to either mushrooms, or fermented biomass as animal feed, or directly as cell-free protein for processing into human foods. Breakdown product processes considered include enzymatic or acid hydrolysis to sugars. These sugars could be either consumed directly or fed to fungi and yeasts to produce fermented foods or the production of enzymes for use in industrial processes. Conversion of lignocelluloses into microbially fermentable sugars by enzymatic or physical/chemical methods are commonly suggested [Petersen and Baresi, 1990]. Useful species of bacteria and fungi can use these sugars as growth substrate. Alternatively, anaerobic digestion of lignocelluloses can produce useful substances like methane. It has been recently demonstrated that pretreatment of cellulose with microwave radiation increased conversion to glucose by enzymatic hydrolysis from 37% in untreated material to 92% efficiency [George and Cullingford, 1990].

Electrooxidation of organic materials in wastewater is advantageous because no expendable chemicals are used in the process. A system for purification of water from urine and waste water has been designed and experimentally demonstrated [Hitchens *et al.*, 1990].

Recycling of plant growth water and nutrient solutions is simpler than the provision of drinking water. However, gradual concentration of metals in recycled plant nutrient solutions can be a problem. Methods of removal such as organic chelation or ion exchange can be used to ameliorate this problem.

Scrap metal and other debris from defunct machinery and building materials should be readily reusable if care is taken in original choice of materials.

Mining wastes from processes like permafrost extraction could become an unsightly and hazardous nuisance. The only practical remedy is to backfill the pits created by these activities [Briggs and Sacco, 1985]

Nuclear waste may become a disposal issue in the future if power supplies like the SP-100 nuclear reactor being developed by NASA come into common use [French, 1985; Isenberg and Heller, 1989].

Currently designed units of this sort are estimated to have a 7 to 10 year lifetime. As we know from the terrestrial experience, safe permanent disposal of nuclear wastes is a highly controversial subject. No doubt, if sufficient time elapses with humans living on Mars, the inhabitants will face some of the same difficulties that we on Earth are facing with this issue.

CONTAMINATION CONTROL

Microbial Pathogens on Humans

There are three major kinds of possible microbial contamination within the life support system: specific human pathogens, opportunistically pathogenic organisms which are harmless at low levels but can be hazardous when growing out of control, and recently mutated pathogenic strains which arise from harmless strains. Control of the first category is perhaps the easiest. High levels of cleanliness and decontamination procedures carried out on hardware, living materials, and people prior to launch, isolation of crewmembers until development time of pathogens has passed prior to launch, and maintenance of a clean environment after launch should be able to successfully control these strains of organisms.

The case of opportunistic pathogens is more complicated and difficult. These organisms can be normal human gut or surface bacteria which may proliferate given the right environmental conditions. Further, when stressed, humans are more prone to the development of disease states from ordinarily benign microflora. Normal plant bacteria and fungi can achieve pathogenicity on humans especially in humid states or where the physiology or nutrition of the people has been compromised. Broad spectrum fungicides and anti-biotics must be available to treat these conditions on a case-by-case basis. The best policy is to minimize the opportunities for these opportunists by keeping the environment within relatively low humidity levels and preventing the buildup of organisms within water supply and air filtration devices. An ounce of prevention is worth a pound of anti-biotics. Experience with legionnaire's disease over the last several decades here on Earth is an object lesson in the importance of keeping air and water filtration equipment decontaminated [Stout and Yu, 1985].

The third microbial contamination problem, newly arisen pathogens, is the most difficult to assess. The development of new strains of microbes in the new environment of a Mars habitation is *inevitable*. Steps to limit the varieties of microbes of all types which accompany the Martian inhabitants will not prevent the development of new strains arising from those that do arrive on Mars. The possibility of sending totally germ-free (axenic) people and other organisms is non-existent. Producing such microbeless people and organisms is impossible in a practicable way. To produce axenic organisms for laboratory experiments requires that young be aseptically removed from the egg or the parent animal and raised in completely sterile environments. Clearly, this is not feasible (for humans at least). For plants, sterilizing the seed material or the surfaces of generative plant parts does not prevent further colonization when the plant material begins to grow. Perhaps more importantly, all higher organisms have *evolved* in conjunction with accompanying microbes and depend upon these normal residents for many services. Repelling invasion by potentially pathogenic alien microbes which are not a part of our normal flora is one of the essential services that they perform [Roth and James, 1988]. We depend upon our gut bacteria to detoxify certain normal breakdown products of digestion and to synthesize some vitamins and growth factors which we cannot do ourselves.

Pathogens which are common both to us and to other animals is a major argument against inclusion of animal-derived foods at least in early habitations on Mars. Common sense would suggest that the more closely related we are to a given species the more likely we are to share pathogens. This is generally true. Therefore, in particular, it is wise to avoid mammal and bird species. Animals more removed from us evolutionarily like fish and crustaceans would be better choices as a source of animal food. Even insects like bees may be beneficial or even essential because of their pollination services to crop plants and the provision of honey. One can imagine honey becoming a sought-after delicacy and status symbol at a Mars habitation.

Pathogens on Plants and Microbes on Life Support Equipment

If growing conditions for higher plants include high-density monoculture and high humidity, plant pathogens could run rampant and devastate the food production. Some human bacteria can also become plant pathogens under the correct circumstances [Starr, 1979]. Disease resistant plant strains must be used when possible. Many of these are available for all commonly cultivated food crops.

Control of algal gunk on water recycling and agricultural equipment may partly be a matter of striking a compromise between high light intensities for plant growth and minimizing out-of-control growth of algae. Humidity or seeping water greatly stimulates surficial algal growth as anyone with a home aquarium can attest. Surface materials which repel microbes or which prevent adhesion could help prevent colonization of unwanted levels or strains from fouling equipment and water handling devices. A study by Bloom and colleagues [1990] suggests that several surface types, surface alterations, and stirring can help reduce levels of colonization in water systems.

Bacteria and viruses in water are always a potential human health problem but can be controlled by ozone, heated iodine or chlorine [Herrmann and Wydeven, 1990]. Several pathogenic bacteria in the mycoplasma and ureaplasma groups survive well in water at high temperatures [Kundsin and Perkins, 1990]. This could be devastating in a water supply system and suggests that water purification should be implemented using *multiple* methodologies.

Compartmentalization of growing chambers to allow for emergency isolation of contaminated areas is essential. Such compartmentalization is a good idea, in general, to allow for the tailoring of specific growth conditions for specific crops.

Chemical Contamination

The primary defense against chemical contamination of the environment is to introduce as little as possible in the first place. For example, pesticides, biocides and fungicides, complex organic solvents and substances, drugs, anti-biotics, plastics and materials which outgas or degrade rapidly, and sources of potentially toxic heavy metals must be avoided or minimized. Substances given off by plants, animals and humans may be unavoidable and have to be dealt with. Respiratory by-products like carbon dioxide, carbon monoxide and methane are relatively straightforward to deal with and several technologies already exist to scrub them from the air [e.g. McCray *et al.*, 1989; Stinson and Montz, 1988]. Catalytic oxidation, condensers, and adsorbents which can be reversibly regenerated are all available for use. A hybrid system using higher plants and chemical/physical processing has been designed by Lee and Brown [1989].

More complex substances like ethylene or terpenes given off by plants into the air require additional removal techniques. One comprehensive air filtration scheme which is desirable because of its versatility is the use of microbes to take up the metabolic by-products and other air and water chemicals and use them as substrates to support their own growth. Bacteria in soil beds also degrade organic compounds like methane, CO, SO_2 and H_2SO_4 [Cullingford and Schwartzkopf, 1990; Bohn and Bohn, 1986]. In addition, if these microbes are themselves useful products to the life support system, the benefit is multiplied. A version of this idea using soil bacteria in a soil matrix is being employed in the Biosphere II experiments underway in Arizona [Nelson and Dempster, this volume; Frye and Hodges, 1990]. This patented system relies on the simple concept of running "used" air through the soil/microbe system with the microbial uptake of the contaminants and the resulting purification of the air. Another method of air decontamination using bacteria involves use of a liquid reactor [Binot and Paul, 1989]. Air is bubbled through a liquid culture of bacteria. The bacteria themselves are adhered to a solid surface.

Non-metabolic sources of toxins in current spacecraft include spills and leaks of various devices and storage areas [Coleman, 1989] and particulates like lint, dust, shed epithelium and hair (8 g day^{-1} person^{-1}, Harding, 1989], and outgassing and combustion products from items like wire insulation, gaskets, epoxies and the like. Compounds given off by burning or over-heating and fusing of hardware components have included highly toxic items like hydrogen cyanide and benzonitrile [Coleman, 1989]. Current NASA guidelines for Space Shuttle include SMAC limits (spacecraft maximum allowable concentration) for each compound which could be detected aboard the vehicles [Coleman, 1989]. Such guidelines will continue to be developed as our experience with confined spaces increases. Currently, individual air samples are collected and analyzed after the shuttle returns to Earth. This will not be adequate for Mars habitation. Real-time monitoring of chemical contaminants in air and water must be part of any life support system. The prolonged nature of exposure to substances in the habitat environment could lead to chronic exposure disorders if levels are not monitored and controlled continuously. In addition, automated monitoring and control will considerably reduce the labor burden on Martian crewmembers.

A system for cleaning the air after a spacecraft fire has recently been designed using a series of disposable particulate removal filter cells and toxic gas removal beds [Sribnik *et al.*, 1990]. Such a system allows cleaning of the air in emergency situations without venting and replacing the entire habitat atmosphere. The unit takes less than a cubic foot of space, weighs less than 24 kg (52 lbs) and uses only 50 watts of power. It can clean the combustion products of 10 kg (22 lbs) of combustible materials.

Odor

One other significant element of chemical contamination is the presence of odors. The psychologically preferred state according to Connors *et al.* [1985] is that of no odor. Some of the response to odor is visual and perceptual as well as olfactory. In one study [Helmreich *et al.*, 1979], odors were considered more offensive when perceived to be coming from visible animal cages. Clearly, attention must be given to separating humans from potentially offensive or strong odors both visually and chemically.

Pleasing or evocative odors may also have a place in a Mars habitat as they have from time immemorial on Earth. Because the olfactory system is so evolutionarily ancient, it is also intimately associated with our emotional system. This allows scent to be used to affect mood and to some extent behavior. Judicious use of scents may be helpful to Mars explorers very far away from friends, family, and traditions of their own cultures. Commercial and marketing use of artificially produced scent has reached an incredibly high degree of sophistication and maturity.

Noise

Excessive noise in the environment can produce effects ranging from minor annoyance to physical harm. Adverse impacts on human performance under noisy conditions include a narrowing of attention particularly with complex tasks, higher error rates on tasks, lowered morale, fatigue, distraction, sleep disturbance, heightened blood pressure, greater levels of aggression and reduced levels of altruistic behavior [Connors *et al.*, 1985; National Research Council, 1977]. As a general guideline for design, the overall indoor noise level should not exceed 45 dB [Connors *et al.*, 1985]. Of course, specific work areas can exceed this, if necessary, perhaps with the provision of individual hearing protection devices. Special, very low-noise, relaxation facilities for individuals to use intermittently would be beneficial. For example, on Earth visits to parks and hiking in the wilderness serve this purpose. In a mature space habitat, park-like environments could be provided.

Pleasant sound including music also has a place in the Mars environment. People often find the sounds of gently trickling water pleasant. Other environmental sounds generally associated with nature including bird sound, ocean waves, rain and the like are considered by many to be relaxing. The beneficial effects of music on human behavior is legendary. "Music hath charms to soothe the savage breast" [Congreve, 1697]. Fried and Berkowitz [1979] have documented this in a study which found that helping behaviors were encouraged by soothing music.

INTEGRATION AND OPTIMIZATION OF SYSTEMS

Now we come to perhaps the most important of all facets of the life support problem, integration of all the subcomponents to provide a unified whole. The system is useless unless the components interact successfully and do not destructively interfere with each other under certain circumstances.

Effect of Competing Life Support Philosophies on System Integration

Virtually everyone is agreed that at least some degree of self-sufficiency is necessary for Mars life support systems. One essential component of self-sufficiency is a relatively closed system. The less leakage and wastage that occurs, the less energy and time will be expended on replacing material. This imperative has led to the idea that closed ecologically based systems would be a useful way in which to design self-sufficient life support environments.

Two distinct philosophies of how to achieve this self-sufficiency have emerged. These two approaches to the study and development of closed ecologically-based life support systems conform to the two primary philosophies of how science should be carried out. The reductionist approach, the classical modus operandi of science, seeks to build a closed life support system by adding single components at a time, gradually increasing the complexity of the whole system, while understanding the individual effects of each addition on that system. The other approach is more observational in nature and takes its philosophical roots both from the natural philosophy of the last century and from holistic methods of study which are coming back into vogue. This philosophy asserts that the whole is greater than the sum of the parts. Therefore the study of individual components cannot fully elucidate the behavior of the whole system. The *system* must be considered as the entity for analysis, not the components.

Both of these philosophies have implications for system integration and each has its proponents. The most pragmatic current approach is to sidestep the philosophical debate and begin at both ends of the problem. When we understand as much as we can about both levels of operation, i.e., the individual components and behavior of whole model systems, then we may be able to construct an actual life support system with some degree of confidence.

The more reductionistic approach has traditionally relied on primarily non-living, engineered solutions for system management. The more holistic approach in its most extreme form relies primarily on biological controls on materials exchange and recycling. We will undoubtedly end up with systems which combine both of these features for the greatest reliability. I view this as the "Managed Ecosystem" path to successful life support.

Automation and Robotics

Even in natural ecosystems, life support (i.e., gathering, hunting, farming) is very labor intensive. No matter what the eventual design of Mars life support systems, automation of many life support duties will be imperative if the limited human labor available in a Mars base is to be used optimally. Tasks which could be profitably automated include monitoring and adjustment of air composition, physical parameters like light and temperature, some planting and seedling care, some transplanting tasks, plant sampling, harvesting of uniform growth habit plants, some stages of food processing, e.g. milling wheat [MacElroy et al., 1985], production of microbial biomass and food products, agricultural waste recycling, other waste recycling, and water recycling.

Automation of as many agricultural functions as possible could reduce the necessity of crew entry into areas which may have environments specifically tailored for plants and not humans. Such atmospheres could be very high humidity and contain high levels of carbon dioxide which could be difficult for humans to tolerate [MacElroy et al., 1985.] It could also limit exposure to possible plant-associated microbes which could become opportunistic pathogens.

Automatic sensing for monitor and control of the atmosphere is a critical function. The technologies to provide this are currently developing. For example, a new class of electrochemical gas sensors has been recently developed which have fast response time, long operating life, high tolerance to changing temperature and voltage, and high sensitivity to many gasses [Venkatasetty, 1990]. Water quality monitoring for use aboard Space Station is under development [Niu et al., 1990]. One of the design criteria for this system is "automated operation with minimal crew involvement." An automated thermal control system for Space Station and future planetary missions is similarly in the design phase [Troy, 1990].

Computer Simulations

Simulations of whole CELSS* operation are invaluable tools for working out the dynamics and failure modes of such complicated systems. One recent effort is able to compare various mission scenarios, using different starting parameters and a series of both planned and unscheduled events to investigate the robustness and flexibility of the system in question [Cullingford, 1990]. One important feature of these simulations is the time delay sometimes present between an event and its effects. These authors attribute this feature to the intricate interdependencies of components of CELSS systems. With computer simulations, long expanses of time can be compressed greatly and allow for the emergence of such effects in simulations designed to last for months and years of computer-simulated time. Such predictive power could be invaluable in alerting investigators to potential problems in real trials of actual life support systems undergoing testing.

Test Case - Bios 3

Soviet scientists have kept 2 and 3 person crews for periods of between 4 and 6 months in a closed ecological system known as Bios 3 during the years from 1972 through 1984 [Vasil'yev, 1984; Salisbury, 1990]. This system had a 63 m^2 growing area and another habitation section. All crops were grown hydroponically including wheat, chufa and a variety of garden vegetables. An initial endowment of food at the beginning of the experiments was supplemented with grown items which accounted for 75-80% of the total food consumption during the experiments. Some of the air filtration was accomplished by passage through the growing chambers. Organic contaminants were re-

---

* This acronym has been used to mean "Closed Ecological Life Support Systems," "Controlled Environmental Life Support Systems" and other permutations.

moved by thermal catalytic filters. Water condensate was collected, purified and recycled both for drinking and other purposes thus closing the water cycle. The Bios 3 experiments were a significant landmark in demonstrating the viability of the CELSS concept.

Test Case - Antarctic Planetary Testbed Proposal

An ambitious plan to construct a test facility in the Antarctic to study human and machine response to isolated, hostile environments was put forward by a group at the University of Houston [SICSA, 1988]. Their plan relies on international public and private participation and funding to construct the rather elaborate facilities they envision. After construction carried out by a crew of twelve people, scientific and technological research will be conducted. Initial plans call for overwintering in the facility by six to eight of the crewmembers. Amongst the items on the list of accomplishments they propose to achieve, the SICSA group envisions testing of partially closed life support systems including reclamation of waste and recycling systems. Sampling of soils and material processing under challenging conditions and the testing of automated and robotic systems are other key test items on the agenda. Obviously, many other relevant features of the isolated Antarctic winter environment are apropos to Mars. Long periods of external darkness (in Antarctica, seasonal and on Mars due to months-long dust storms), limited communication with the outside world, and the generally inhospitable external conditions will simulate very well the psychological and social strains on crewmembers. The main goal of the group in putting this proposal forward is to "Serve as tangible expression of commitment to future planetary initiatives." [Anderson *et al.*, 1990; SICSA, 1988]. Relevance to Mars exploration of the Antarctic experience is particularly poignant in the area of human sociology and adaptation. For a discussion of this issue see Palinkas [1989], Harrison *et al.* [1989] and Harrison and Connors [this volume].

Test Case - Biosphere II

The minimum size of an ecological life support system which can be largely self-sustaining is presently unknown. The Biosphere II project in Oracle, Arizona is one privately funded, closed ecological life support system project attempting to answer this question (amongst others). Their underlying philosophy enunciated by Nelson [1990, and Nelson and Dempster, this volume], centers around the notion of redundancy. This experimental closed ecosystem relies on multiple organisms and multiple possible food chain pathways. These process pathways will perform the vital ecosystem functions. This is a form of insurance, since if one species dies out the other redundant components take over the ecological relationships of the extinct species. The Biosphere II facility has a very elaborate system of environmental monitors and controls. The performance of this experimental system is providing unmatched insights into the complexities of large life support system control.

EXISTING TECHNOLOGY VS. NEW DEVELOPMENT

There are many areas of technology presently being developed for Earth-bound purposes which could be modified for use in space and extra-terrestrial environments.

An exploration of these current frontiers of development is both useful and heartening to would-be space adventurers. At least we can be assured that we may have numerous basic tools at our disposal when we begin to seriously develop life support systems for real long-duration human missions.

Areas in which development has been progressing include current work in the area of high yield food production, current work in the area of biological waste recycling, advances in the area of biological contamination control, and significantly, the broad field of genetic engineering may provide organisms specifically tailored for various life support purposes.

One area of supreme importance which is not being explored by non-space interests is a significant improvement in the reliability and robustness of all kinds of engineered components for life support systems. Typical failure rates and the accessibility of spare parts for terrestrial applications is impractical and unsafe for Mars missions. Conventional industry standards in commercial engineering are inadequate for Mars missions. Even performance of engineered components within NASA and the space community at large have been disappointing . . . one need only to contemplate the Hubble Space Telescope difficulties to realize this. Possibly the development of von Neumann machines will be necessary to assure self-repairability of engineered devices.

CONCLUSIONS

Our ambitions for long-duration human missions are more constrained by the current state of life support technology than by any other area of technology. This is partly because life support involves the creation and maintenance of the *entire environment* of the Mars inhabitants. This is a massive task. The sophistication and reliability of life support systems will dictate the time and distance that human missions will be able to go in the future. One of the most critical problems to emerge from consideration of life support technology is that of developing adequate control and integration capabilities to manage very complex systems. Although such systems will be very complex, they will probably still be below the minimum size necessary for a meaningful degree of self-regulation. Hence, they must be engineered and at least partly artificially controlled.

In addition to solving the purely technological life support problems, life support systems must also pass a much more rigorous test than we ask of most engineered systems on Earth. They must be usable conveniently, hospitably, and (above all) reliably by actual human beings engaged in a potentially hazardous, physically and mentally demanding endeavor. Ideally, Martian explorers would have to pay as little attention to their life support as we in the affluent First World pay to our air conditioners, refrigerators and microwave ovens. Mars crews must ultimately be able to rely on and take for granted their life support machinery and techniques as much as possible. This degree of nonchalance will not be achievable on the first mission or even the first few missions but will be a product of gradually increasing familiarity and facility in handling the day-to-day needs of humans in the Martian environment. Indeed, in our current spacefaring, we have not achieved the necessary degree of reliability with *any* systems to date. One only needs to consider the repeated delays and glitches in Space Shuttle launch attempts to realize that reliability is something we have not yet achieved in the space program.

The development of sophisticated life support methods to sustain humans on space missions is its own reward. However, that is not to say that it has no other utility for humans on Earth. The sustenance of human life and the understanding of closed systems are vitally important to us as a species on a planet beleaguered by our excessive numbers and disproportionate impact. In many parts of the Earth, we already exceed the carrying capacity of the environment. The antiquated and ill-considered notion that we will somehow alleviate the crowding of the Earth by shipping excess populations into space is clearly ridiculous to anyone who understands Malthusian dynamics and the biology of an expansionist, generalized species like ourselves. More importantly and realistically, our Earth-based ecological problems may become more tractable in light of the knowledge required to develop closed ecosystems for space use.

The advance of human civilization and knowledge seems to progress by accident and serendipity rather than by design. Indeed, designed communities have a very poor track record. Perhaps this is why all utopian communities have met with failure in the end. At the risk of being an apologist for space, I submit that the value of the intense scrutiny that space-motivated research will bring to bear on the solution to problems of human sustenance and well-being will exceed attempts to advance our knowledge of these areas directly. I say this not because we *cannot* do it otherwise, but because history has shown that we *will not* do it otherwise. We seem to need a specific and pragmatic goal in order to justify basic scientific investigations. The relative budgets of pure versus applied investigation in our society tells the story.

A permanent human presence on Mars for scientific exploration is a worthy goal. In addition to the value of inhabiting Mars itself, it will also be the testing ground for any further expansion of humans into alien environments. If we can make it on Mars, we can make it on other planets of similar inhospitable but potentially exploitable character elsewhere in our solar system or perhaps circling distant suns. Just as organisms on Earth have had to acquire new skills and physical characteristics to adapt to new niches, so too will humans acquire new skills and tools in learning to adapt to the space and planetary environments. By doing so, we will undoubtedly change our culture, our view of ourselves, and perhaps some of us may alter our physiology. Isolation of populations from each other and the experience of different resulting selection pressures are the classical means of creating new species from old. In biological terms, we will speciate, that is, gradually evolve from one species into many. This is the way of life on our planet, and perhaps the way of life everywhere in the universe.

REFERENCES

Alexander, D.B., D.A. Zuberer, D.H. Hubbell, "Microbiological considerations for lunar-derived soils," in D.W. Ming and D.L. Henninger, eds., *Lunar Base Agriculture: Soils for Plant Growth,* Amer. Soc. Agronomy, Inc., Madison, WI, p. 255, 1989.

Alman, P.L., D.S. Dittner, *Metabolism,* Fed. Amer. Soc. Experim. Biol., Bethesda, MD, 1968.

Anderson, D.T., C.P. McKay, R.A. Wharton, J.D. Rummel, "An Antarctic research outpost as a model for planetary exploration," *J. Brit. Interplanet. Soc,* 43, pp. 499-504, 1990.

Angelo, J.A., Jr., D. Buden, "Power requirements for the conquest of Mars," AAS 84-177, in C.P. McKay, ed., *The Case for Mars II,* American Astronautical Soc., Science and Technology Series, 62, pp. 497-516, San Diego, CA, 1985.

Averner, M.M., "Controlled ecological life support system," in D.W. Ming and D.L. Henninger, eds., *Lunar Base Agriculture: Soils for Plant Growth,* Amer. Soc. Agronomy, Inc., Madison, WI, 255pp., 1989.

Averner, M.M., B. Moore, I. Bartholemew, R. Wharton, "Atmosphere behavior in gas-closed mouse-algal systems: An experimental and modeling study," in R.D. MacElroy, D.T. Smernoff, and H.P. Klein, eds., *Controlled Ecological Life Support System,* NASA Conference Pub. 2378, NASA, Moffett Field, CA, pp. 39-46, 1985.

Banin, A., "Mars soil - A sterile regolith or a medium for plant growth," AAS 87-215, in C.R. Stoker, ed., *The Case for Mars III: Strategies for Exploration -General Interest and Overview,* American Astronautical Soc., Science and Technology Series, 74, pp. 559-571, San Diego, CA, 1989.

Binot, R.A., P.G. Paul, "BAF - An advanced ecological concept for air quality control," *Proceedings of 19th Intersociety Conference on Environmental Systems,* San Diego, CA, 1989.

Bloom, W., S. Pope, J.C. Richardson, A.T. Mikell, Jr., "Bacterial selectivity in the colonization of surface materials from groundwater and purified water systems," in A.W. Carlson and J.E. Swider, eds., *Space Station Environmental/Thermal Control and Life Support Systems,* SP-829, SAE, Warrendale, PA, pp. 101-111, 1990.

Bluth, B.J., "Staying sane in space," *Mech. Eng,* 4, pp. 24-29, 1982.

Bohn, H.L., R.K. Bohn, "Soil bed scrubbing of fugitive gas releases," *J. Environ. Sci. Health,* A21, pp. 561-569, 1986.

Bohn, H.L., R.K. Bohn, "Soil beds weed out air pollutants," *Chem. Eng.,* April 25, 1988.

Boston, P.J., "Low-pressure greenhouses and plants for a manned research station on Mars," *J. British Interplanetary Soc.,* 34, pp. 189-192, 1981.

Boston, P.J., "Critical life sciences issues for a Mars Base," AAS 84-167, in C.P. McKay, ed., *The Case for Mars II,* American Astronautical Soc., Science and Technology Series, 62, pp. 287-332, San Diego, CA, 1985.

Boston, P.J., "Mars mission life support," AAS 86-177, in D. Reiber, ed., *The NASA Mars Conference,* American Astronautical Soc., Science and Technology Series, 71, pp. 487-507, San Diego, CA, 1988.

Boston, P.J., "Mir space station," F.N. Magill, ed., *Magill's Survey of Science: Space Exploration Series,* pp. 1025-1031, Salem Press, Inc., Pasadena, CA, 1989a.

Boston, P.J., "Space Shuttle living conditions," F.N. Magill, ed., *Magill's Survey of Science: Space Exploration Series,* pp. 1634-1640, Salem Press, Inc., Pasadena, CA, 1989b.

Boyd, R.C., P.S. Thompson, B.C. Clark, "Duricrete and composites construction on Mars," AAS 87-213, in C.R. Stoker, ed., *The Case for Mars III: Strategies for Exploration - General Interest and Overview,* American Astronautical Soc., Science and Technology Series, 74, pp. 539-550, San Diego, CA, 1989.

Breeze, R.K., Space vehicle environmental control requirement based on equipment and physiological criteria, ASD-TR-61-161, Wright-Patterson AFB, Ohio, November, 1961.

Briggs, R., A. Sacco, Jr., "Environmental considerations and waste planning on the lunar surface," in W.W. Mendell, ed., *Lunar Bases and Space Activities of the 21st Century,* Lunar and Planetary Institute, Houston, TX, pp. 423-430, 1985.

Bugbee, B.G., F.B. Salisbury, "Controlled environment crop production: Hydroponic vs. lunar regolith," in D.W. Ming and D.L. Henninger, eds., *Lunar Base Agriculture: Soils for Plant Growth,* Amer. Soc. Agronomy, Inc., Madison, WI, 255pp., 1989.

Busta, F.F., D.R. Heldman, M. Karel, G. Kempler, D.L. Kaplan, D.R. Beem, P.L. Russell, Conversion of inedible wheat biomass to edible products for space missions, American Inst. Biological Sciences, Washington, D.C., 1986.

Calvin, M., "Solar energy by photosynthesis," *Science,* 184, pp. 375-381, 1974.

Clark, B.C., D.R. Pettit, "The hydrogen peroxide economy on Mars," AAS 87-214, in C.R. Stoker, ed., *The Case for Mars III: Strategies for Exploration - General Interest and Overview,* American Astronautical Soc., Science and Technology Series, 74, pp. 551-557, San Diego, CA, 1989.

Coleman, M.E., "Toxic hazards in space operations," in A.E. Nicogossian, C.L. Huntoon, and S.L. Pool, eds., *Space Physiology and Medicine,* 2nd edition, Lea and Febiger, pp. 315-327, Philadelphia and London, 1989.

Congreve, W., *The Mourning Bride: A Tragedy,* printed for Jacob Tonson, London, 1697.

Connors, M., A.A. Harrison, F.R. Akins, *Living Aloft: Human Requirements for Extended Spaceflight,* NASA, Washington, D.C., 1985.

Costa-Pierce, B.A., "Intensive food production systems for a lunar base station," in M.B. Duke, chairman, *Lunar Bases and Space Activities in the 21st Century Abstracts,* Solar System Exploration Div., NASA, Houston, TX, 1984.

Cullingford, H.S., W.P. Bennett, W.A. Holley, J.G. Carnes, P.S. Jones, "CELSS simulations for a lunar outpost," in A.F. Behrend and R.P. Reysa, eds., *Advanced Environmental/Thermal Control and Life Support Systems,* SP-831, pp. 223-232, SAE, Warrendale, PA, 1990.

Cullingford, H.S., S.H. Schwartzkopf, "Conceptual design for a lunar-base CELSS," in A.F. Behrend and R.P. Reysa, eds., *Advanced Environmental/Thermal Control and Life Support Systems,* SP-831, pp. 247-254, SAE, Warrendale, PA, 1990.

Dubos, R., "The intellectual basis of nutritional science and practice," in M. Chou and D.P. Harmon, Jr., eds., *Critical Food Issues of the Eighties,* Pergamon Press, New York, 1979.

Ehrlich, H.L., "Role of microbes to condition lunar regolith for plant cultivation," in D.W. Ming and D.L. Henninger, eds., *Lunar Base Agriculture: Soils for Plant Growth,* Amer. Soc. Agronomy, Inc., Madison, WI, 255pp, 1989.

Eley, J.H., J. Myers, "Study of a photosynthetic gas exchanger: A quantitative repetition of the Priestly experiment," *Texas J. Science,* 16, pp. 296-333, 1964.

Fraser, T.M., "Confinement and free volume requirements," *Space Life Sciences,* vol. 1, pp. 428-466, 1968.

French, J.R., "Nuclear powerplants for lunar bases," in W.W. Mendell, ed., *Lunar Bases and Space Activities of the 21st Century,* Lunar and Planetary Institute, Houston, TX, pp. 99-106, 1985.

Fried, L., L. Berkowitz, "Music hath charms . . . and can influence helpfulness," *J. Appl. Soc. Psychol,* 9, pp. 199-208, 1979.

Frye, R., C. Hodges, "Soil bed reactor work of the environmental research laboratory of the University of Arizona in support of the research and development of Biosphere 2 in biological life support technologies," M. Nelson and G. Soffen, eds., *NASA CP-3094,* NASA and Synergetic Press, Oracle, AZ, 1990.

Galecki, D.L., M.J. Patterson, "Nuclear powered Mars cargo transport missions utilizing advanced ion propulsion," *AIAA Paper 87-1903,* Also, NASA TM-100109, NASA, Washington, D.C., 1987.

Geels, S., J.B. Miller, B.C. Clark, "Feasibility of using solar power on Mars: Effects of dust storms on incident solar radiation," AAS 87-266, in C.R. Stoker, ed., *The Case for Mars III: Strategies for Exploration - Technical,* American Astronautical Soc., Science and Technology Series, 75, pp. 505-516, San Diego, CA, 1989.

George, C.E., H.S. Cullingford, "Microwave irradiation of cellulose and enzymatic hydrolysis of waste paper for long space missions," in A.F. Behrend and R.P. Reysa, eds., *Advanced Environmental/Thermal Control and Life Support Systems,* SP-831, pp. 211-215, SAE, Warrendale, PA, 1990.

Gitel'son, I.I., "Problems of creating biotechnical systems of human life support," *NASA Tech. Translation TTF 17533,* NASA, Washington, DC, 1977.

Grymes, R., C.E. Wade, J. Vernikos, "Biomedical issues in the exploration of Mars," This volume, Chapter 13.

Gustan, E., T. Vinopal, "Controlled ecological life support system: A transportation analysis," *NASA-CR-166420,* NASA-Ames Research Center, Moffett Field, CA, 1982.

Harding, R., *Survival in Space: Medical Problems of Manned Spaceflight,* Routledge, London and New York, 1989.

Harrison, A.A., M. Connors, "The human side of Mars flight: A review of human factors issues," This volume, Chapter 14.

Harrison, A.A., Y.A. Clearwater, C.P. McKay, "The human experience in Antarctica: Applications to life in space," *Behavioral Science,* 34, pp. 253-271, 1989.

Helmke, P.A., R.B. Corey, "Physical and chemical considerations for the development of lunar-derived soils," in D.W. Ming and D.L. Henninger, eds., *Lunar Base Agriculture: Soils for Plant Growth,* Amer. Soc. Agronomy, Inc., p. 255, Madison, WI, 1989.

Helmreich, R., J. Wilhelm, T.A. Tanner, J.E. Sieber, S. Burgenbach, *A Critical Review of the Life Sciences Project Management at Ames Research Center for the Spacelab Mission Development Test III,* NASA TP-1364, 1979.

Henry, W., "Future of engineered foods," in M. Chou and D.P. Harmon, Jr., eds., *Critical Food Issues of the Eighties,* Pergamon Press, New York, 1979.

Henson, H.K., C.M. Henson, "Closed ecosystems of high agricultural yield," *Space Manufacturing Facilities, (Space Colonies),* pp. 105-114, American Institute of Aeronautics and Astronautics, New York, 1977.

Herrmann, C.C., T. Wydeven, "Physical/chemical closed-loop water recycling for long-duration missions," in A.F. Behrend and R.P. Reysa, eds., *Advanced Environmental/Thermal Control and Life Support Systems,* SP-831, pp. 233-246, SAE, Warrendale, PA, 1990.

Hesseltine, C.W., H.L. Want, *Biotechnol. Bioeng,* 9, pp. 275-288, 1967.

Highsmith, A., B.M. Kaylor, C.J. Reed, E.W. Ades, "Evaluation of water treatment systems producing reagent grade water," in A.W. Carlson and J.E. Swider, eds., *Space Station Environmental/Thermal Control and Life Support Systems,* SP-829, pp. 113-117, SAE, Warrendale, PA, 1990.

Hitchens, G.D., O.J. Murphy, L. Kaba, C.E. Verostko, "Electrooxidation of organics in waste water," in A.F. Behrend and R.P. Reysa, eds., *Advanced Environmental/Thermal Control and Life Support Systems,* SP-831, pp. 179-188, SAE, Warrendale, PA, 1990.

Hoff, J.E., J.M. Howe, C.A. Mitchell, "Nutritional and cultural aspects of plant species selection for a controlled ecological life support system," *NASA Contractor Report #166324,* Ames Research Center, Moffett Field, CA, 1982.

Horz, F., "Lava tubes: Potential shelters for habitats," in W.W. Mendell, ed., *Lunar Bases and Space Activities of the 21st Century,* pp. 405-412, Lunar and Planetary Institute, Houston, TX, 1985.

Hossner, L.R., E.R. Allen, "Nutrient availability and element toxicity in lunar-derived soils," in D.W. Ming and D.L. Henninger, eds., *Lunar Base Agriculture: Soils for Plant Growth,* p. 255, Amer. Soc. Agronomy, Inc., Madison, WI, 1989.

Isenberg, L., J.A. Heller, "The SP-100 space reactor as a power source for Mars exploration missions," AAS 87-224, in C.R. Stoker, ed., *The Case for Mars III: Strategies for Exploration - General Interest and Overview,* American Astronautical Soc., Science and Technology Series, 74, pp. 681-695, San Diego, CA, 1989.

Ivanov, B. and O. Zubareva, "Bios 3," *Soviet Life,* April, 1985, pp. 22-25, 1985.

Khalili, N., "Magma, ceramic and fused adobe structures generated in situ," in W.W. Mendell, ed., *Lunar Bases and Space Activities of the 21st Century,* pp. 399-403, Lunar and Planetary Institute, Houston, TX, Also reprinted in the author's own book, *Ceramic Houses: How to Build Your Own,* 1986, pp. 203-208, Harper and Row, San Francisco, CA, 1985.

Kharatyan, S.G., "Microbes as food for humans," in M.P. Starr, ed., *Annual Reviews of Microbiology,* 32, pp. 301-327, Ann. Revs., Inc., Palo Alto, CA, 1978.

Kihlberg, R., "The microbe as a source of food," in C.E. Clifton, ed., *Annual Reviews of Microbiology,* 26, pp. 427-466, 1972.

Klein, H.P., "U.S. biological experiments in space," *Acta Astronautica,* 8, pp. 927-938, 1981.

Kliss, M., R.D. MacElroy, "Salad machine: A vegetable production unit for long duration space missions," in A.W. Carlson and J.E. Swider, eds., *Space Station Environmental/Thermal Control and Life Support Systems,* SP-829, pp. 81-88, SAE, Warrendale, PA, 1990.

Knott, W.M., R.P. Prince, "CELSS Breadboard project at the Kennedy Space Center," in D.W. Ming and D.L. Henninger, eds., *Lunar Base Agriculture: Soils for Plant Growth,* Amer. Soc. Agronomy, Inc., Madison, pp. 155-163, WI, 1989.

Kundsin, R.B., R.E. Perkins, "Survival of mycoplasmas and ureaplasmas in water and at elevated temperatures," in A.W. Carlson and J.E. Swider, eds., *Space Station Environmental/Thermal Control and Life Support Systems,* SP-829, pp. 97-100, SAE, Warrendale, PA, 1990.

Latham, M.C., *Human Nutrition in Tropical Africa,* FAO-UN, Rome, p. 243, 1965.

Lee, M.G., M.F. Brown, "Hybrid air revitalization system for a closed ecosystem," in A.F. Behrend and R.P. Reysa, eds., *Advanced Environmental/Thermal Control and Life Support Systems,* SP-831, pp. 171-177, SAE, Warrendale, PA, 1989.

Letaw, J.R., S. Clearwater, *Radiation Requirements on Long-duration Space Missions,* Severn Communications Corp, 1986.

Levine, J.S., "Solar radiation incident on Mars and the outer planets," *Icarus,* 31, pp. 136-145, 1977.

MacElroy, R.D., M.M. Averner, "Space ecosynthesis: An approach to the design of closed ecosystems for use in space," *NASA Tech. Mem. 78491,* NASA, Washington, DC, 1978.

MacElroy, R.D., H.P. Klein, M.M. Averner, "The evolution of CELSS for lunar bases," in W.W. Mendell, ed., *Lunar Bases and Space Activities of the 21st Century,* pp. 623-633, Lunar and Planetary Institute, Houston, TX, 1986.

Mackenzie, B.A., "Building Mars habitats using local materials," AAS 87-216, in C.R. Stoker, ed., *The Case for Mars III: Strategies for Exploration - General Interest and Overview,* American Astronautical Soc., Science and Technology Series, 74, pp. 575-586, San Diego, CA, 1989.

McCormack, P.D., D.S. Nachtwey, "Radiation exposure issues," in A.E. Nicogossian, C.L. Huntoon, and S.L. Pool, eds., *Space Physiology and Medicine,* 2nd edition, pp. 328-348, Lea and Febiger, Philadelphia and London, 1989.

McCray, S.B., R.W. Wytcherley, D.T. Friesen, R.J. Ray, "Preliminary evaluation of a membrane-based system for removing CO_2 from air," in A.F. Behrend and R.P. Reysa, eds., *Advanced Environmental/Thermal Control and Life Support Systems,* SP-831, pp.159-166, SAE, Warrendale, PA, 1989.

McKay, C.P., T.R. Meyer, P.J. Boston, "Mars research base infrastructure," in S.M. Welch, and C.R. Stoker, eds., *The Case for Mars: Concept Development for a Mars Research Station,* pp. 95-108, NASA, Jet Propulsion Laboratory, Pasadena, CA, 1986.

Meyer, T.R., C.P. McKay, "Concepts for using Martian resources," This volume, Chapter 19.

Meyer, T.R., C.P. McKay, "The atmosphere of Mars - Resources for the exploration and settlement of Mars," AAS 81-244, in P.J. Boston, ed., *The Case for Mars,* American Astronautical Soc., Science and Technology Series, 57, pp. 209-230, San Diego, CA, 1984.

Michel, E.L., J.A. Rummel, C.F. Sawin, M.C. Buderer, J.D. Lem, in R.S. Johnston and L.F. Dietlein, eds., *Proceedings of the Skylab Life Sciences Symposium,* NASA TM X-58154, p.723, NASA, Houston, TX, 1974.

Milgrom, L., "Lignin: cornucopia of chemicals," *New Scientist,* 8 Oct., 40, 1987.

Ming, D.W., D.L. Henninger, *Lunar Base Agriculture: Soils for Plant Growth,* Amer. Soc. Agronomy, Inc., p. 255, Madison, WI, 1989.

Ming, D.W., "Manufactured soils for plant growth at a lunar base," in D.W. Ming and D.L. Henninger, eds., *Lunar Base Agriculture: Soils for Plant Growth,* Amer. Soc. Agronomy, Inc., pp. 93-105, Madison, WI, 1989.

Modell, M., J. Spurlock, "Closed-ecology life support systems (CELSS) for long-duration, manned missions," American Society of Mechanical Engineering, 79 ENAs-27, 1979.

Myers, J., "Basic remarks on the use of plants as biological gas exchangers in a closed system," *Aviation Medicine,* 1954.

National Research Council, *Noise Abatement: Policy Alternatives for Transportation,* Rep. ISBN-0-309-02648-2, Analytical Studies for the US Environmental Protection Agency, Washington, D.C., 1977.

Nelson, M., "The biotechnology of space biospheres," in M. Asashima and G. Malacinski, eds., *Fundamentals of Space Biology,* pp. 185-200, Springer-Verlag, Berlin, 1990.

Nelson, M. and W.F. Dempster, "Living in space: Results from Biosphere 2's initial closure, an early testbed for closed ecological systems on Mars," This volume, Chapter 18.

Nicogossian, A.E., "Countermeasures to deconditioning," in A.E. Nicogossian, C.L. Huntoon, and S.L. Pool, eds., *Space Physiology and Medicine,* 2nd edition, pp. 294-311, Lea and Febiger, Philadelphia and London, 1989.

Nicogossian, A.E., D.S. Nachtway, "Orbital flight," in A.E. Nicogossian, C.L. Huntoon, and S.L. Pool, eds., *Space Physiology and Medicine,* 2nd edition, pp. 47-58, Lea and Febiger, Philadelphia and London, 1989.

Niu, W., D. Burchfield, G. Snyder, K. Conklin, "Development of a water quality monitor for Space Station Freedom life support system," in A.W. Carlson and J.E. Swider, eds., *Space Station Environmental/Thermal Control and Life Support Systems,* SP-829, pp. 127-136, SAE, Warrendale, PA, 1990.

Oleson, M., R.L. Olson, "Controlled ecological life support systems (CELSS) conceptual design option study," *NASA Contractor Rep. 177421,* NASA, Moffett Field, CA, 1986.

Palinkas, L.A., "Antarctica as a model for the human exploration of Mars," AAS 87-194, in C.R. Stoker, ed., *The Case for Mars III: Strategies for Exploration -General Interest and Overview,* American Astronautical Soc., Science and Technology Series, 74, pp. 215-228, San Diego, CA, 1989.

Petersen, G.R., L. Baresi, "The conversion of lignocellulosics to fermentable sugars: A survey of current research and applications to CELSS," in A.F. Behrend and R.P. Reysa, eds., *Advanced Environmental/Thermal Control and Life Support Systems,* SP-831, pp. 89-100, SAE, Warrendale, PA, 1990.

Quattrone, P.D., "Extended mission life support systems," AAS 81-237, in P.J. Boston, ed., *The Case for Mars,* American Astronautical Soc., Science and Technology Series, 57, pp. 131-162, San Diego, CA, 1984.

Rambaut, P.C., P.C. Johnson, "Nutrition," in A.E. Nicogossian, C.L. Huntoon, and S.L. Pool, eds., *Space Physiology and Medicine,* 2nd edition, pp. 202-213, Lea and Febiger, Philadelphia and London, 1989.

Roberts, M., "The use of inflatable habitation on the Moon and Mars," AAS 87-217, in C.R. Stoker, ed., *The Case for Mars III: Strategies for Exploration - General Interest and Overview,* American Astronautical Soc., Science and Technology Series, 74, pp. 587-593, San Diego, CA, 1989.

Roth, R.R., W.D. James, "Microbial ecology of the skin," in L.N. Ornston, A. Balows, and P. Baumann, eds., *Annual Review of Microbiology,* 42, pp. 441-464, Ann. Revs., Inc., Palo Alto, CA, 1988.

Rowley, J.C., J.W. Neudecker, "In situ rock melting applied to lunar base construction and for exploration drilling and coring on the Moon," in W.W. Mendell, ed., *Lunar Bases and Space Activities of the 21st Century,* pp. 399-403, Lunar and Planetary Institute, Houston, TX, 1985.

Rummel, J., T. Volk, "A modular BLSS simulation model," in R.D. MacElroy and D. Smernoff, eds, *Controlled Ecological Life Support Systems,* COSPAR Advances in Space Research, 7, 4, pp. 59-67, Pergamon Press, 1987.

Rummel, J.D., "Planetary protection and back contamination control for a Mars rover sample return mission," AAS 87-197, in C.R. Stoker, ed., *The Case for Mars III: Strategies for Exploration - General Interest and Overview,* American Astronautical Soc., Science and Technology Series, 74, pp. 259-263, San Diego, CA, 1989.

Rummel, J.D., "Long term life support for space exploration," in A.F. Behrend and R.P. Reysa, eds., *Advanced Environmental/Thermal Control and Life Support Systems,* SP-831, pp. 67-73, SAE, Warrendale, PA, 1990.

Salisbury, F.B., B. Bugbee, "Plant productivity in controlled environments," *Hort. Sci,* 23, pp. 293-299, 1988.

Salisbury, F.B., "Controlled environment life support systems (CELSS): A prerequisite for long-term space studies," in M. Asashima and G. Malacinski, eds., *Fundamentals of Space Biology,* pp. 171-183, Springer-Verlag, Berlin, 1990.

Sauer, R.L., R.M. Rapp, "Food and nutrition," in S.L. Pool, P.C. Johnson, Jr., and J.M. Mason, eds., *Shuttle OFT Medical Report: Summary of Medical Results from STS-1, STS-2, STS-3, and STS-4,* NASA TM-58252, pp. 53-62, NASA, Washington, D.C., 1983.

Sauer, R.L., R.M. Rapp, "Food and nutrition," in S.L. Pool, P.C. Johnson, Jr., and J.M. Mason, eds., *STS-1 Medical Report,* NASA TM-58240, NASA, Washington, D.C., 1981.

Schneider, R.T., "Efficiency of photosynthesis as a solar energy converter," *Energy Conversion,* 13, pp. 77-85, 1973.

Sharpe, M.R., *Living in Space,* p. 192, Doubleday, Garden City, NJ, 1969.

Sheinfeld, M., C.S. Leach, P.C. Johnson, "Plasma thyroxine changes of the Apollo crewmen," *Aviat. Space Environ. Med,* 46, pp. 47-49, 1975.

SICSA, Internal report, College of Architecture, Univ. of Houston, Houston, TX, 1988.

Smernoff, D.T., R.A. Wharton, M.M. Averner, "Operation of an experimental algal gas exchanger for use in a CELSS," in R.D. MacElroy and D.T. Smernoff, eds., *Controlled Ecological Life Support System,* NASA Conference Pub. 2480, pp. 15-25, NASA, Moffett Field, CA, 1987.

Spurlock, J., M. Modell, "Technology requirements and planning criteria for closed life support systems for manned space systems," *Final Report, Task B, Contract NO. NASw-2981,* NASA Off. Life Sci., Washington, D.C., 1978.

Spurlock, J., M. Modell, D. Putnam, L. Ross, J. Pecoraro, "Evaluation and comparison of alternative designs for water and solid waste processing systems for spacecraft," Soc. Automotive Engin., Inc., Warrendale, PA, *Final Report NASA NASw-2349,* 1975.

Squyres, S.W., "The problem of water on Mars," in W.W. Mendell, ed., *Lunar Bases and Space Activities of the 21st Century,* pp. 817-823, Lunar and Planetary Institute, Houston, TX, 1986.

Sribnik, F., P.J. Birbana, J.J. Faszcza, T.A. Nalette, "Smoke and contaminant removal system for Space Station," in A.W. Carlson and J.E. Swider, eds., *Space Station Environmental/Thermal Control and Life Support Systems,* SP-829, pp. 41-49, SAE, Warrendale, PA, 1990.

Starr, M.P., "Human-borne bacteria as pathogens on plants and humans," *Ann. Intern. Med.,* 90, pp. 708-710, 1979.

Stinson, R.G., M.E. Montz, "Space Station EVA Test Bed overview," *SAE Technical Paper Series, Paper No. 881060,* Society of Automotive Engineers, Warrendale, PA, 1988.

Stoker, C.R., J.L. Gooding, T. Roush, A. Banin, D. Burt, B.C. Clark, G. Flynn, O. Gwynne, "The physical and chemical properties and resource potential of Martian surface soils," *Near Earth Resources,* in press, 1993.

Stokes, B.O., G.R. Petersen, W.W. Schubert, W.A. Mueller, "Unconventional processes for food regeneration in space: An overview," *ASME Technical Paper,* 81-ENAS-35, American Society of Mechanical Engineers, New York, NY, 1981.

Stotzky, G., "Microorganisms and the growth of higher plants in lunar-derived soils," in D.W. Ming and D.L. Henninger, eds., *Lunar Base Agriculture: Soils for Plant Growth,* p. 255, Amer. Soc. Agronomy, Inc., Madison, WI, 1989.

Stout, J.E., V.L. Yu, "Ecology of *Legionella pneumophila* within water distribution systems," *Appl. Environm. Microb,* 49, pp. 221-228, 1985.

Tibbits, T.W., "Plant considerations for lunar base agriculture," in D.W. Ming and D.L. Henninger, eds., *Lunar Base Agriculture: Soils for Plant Growth,* p. 255, Amer. Soc. Agronomy, Inc., Madison, WI, 1989.

Tibbits, T.W., D.K. Alford, "Controlled ecological life support system use of higher plants," *NASA Conference Pub.* 2231, NASA, Moffett Field, CA, 1982.

Troy, L.K., "Thermal resource allocation for Space Station Freedom and future planetary missions," in A.W. Carlson and J.E. Swider, eds., *Space Station Environmental/Thermal Control and Life Support Systems,* SP-829, pp. 173-183, SAE, Warrendale, PA, 1990.

Vasil'yev, V., "Five-month experiment in isolation habitat with biological life support," *Trud,* 84, 4, Summarized in *Soviet News Abstract Publication,* Battelle Laboratory, Columbus, OH, 1984.

Venkatasetty, H.V., "Electrochemical amperometric gas sensors for environmental monitoring and control," in A.F. Behrend and R.P. Reysa, eds., *Advanced Environmental/Thermal Control and Life Support Systems,* SP-831, pp. 167-169, SAE, Warrendale, PA, 1990.

Volk, T., J.D. Rummel, "The case for cellulose production on Mars," AAS 87-232, in C.R. Stoker, ed., *The Case for Mars III: Strategies for Exploration -Technical,* American Astronautical Soc., Science and Technology Series, 75, pp. 87-94, San Diego, CA, 1989.

Waligora, J.M., R.L. Sauer, J.H. Bredt, "Spacecraft life support systems," in A.E. Nicogossian, C.L. Huntoon, and S.L. Pool, eds., *Space Physiology and Medicine,* 2nd edition, pp. 104-120, Lea and Febiger, Philadelphia and London, 1989.

Whitney, G., "Geochemistry of soils for lunar base agriculture: Future research needs," in D.W. Ming and D.L. Henninger, eds., *Lunar Base Agriculture: Soils for Plant Growth,* Amer. Soc. Agronomy, Inc., Madison, WI, 1989.

Yaffe, M.J., G.E. Mawdsley, "X-ray protection materials," *Health Physics,* 1991.

Zeikus, J.G., "Chemical and fuel production by anaerobic bacteria," in M.P. Starr, ed., *Annual Reviews of Microbiology,* 34, pp. 423-464, Ann. Revs., Inc., Palo Alto, CA, 1980.

AAS 95-488

Chapter 18

LIVING IN SPACE: RESULTS FROM BIOSPHERE 2'S INITIAL CLOSURE, AN EARLY TESTBED FOR CLOSED ECOLOGICAL SYSTEMS ON MARS[*]

Mark Nelson[†] and William F. Dempster[‡]

On September 26, 1991 a crew of eight people passed through the airlock beginning an experimental habitation of Biosphere 2, a closed ecological system built in the Arizona desert north of Tucson. Two years later they emerged—somewhat thinner but against considerable odds in overall good health and with a viable life support system. The project marked the first long duration habitation by humans in a closed environmental system—and not surprisingly the first two years of its operation included a multitude of problems, including several which were unanticipated. The initial two year closure in Biosphere 2 revealed the sharp fluctuations in atmospheric cycling that will be expected in small closed systems because of its concentration of living biomass and small air volumes. CO_2 during the two years ranged from under 1,000 ppm to over 4,000 ppm. Oxygen was depleted from the atmosphere by reactions with organic C in the system's soils, dropping from an initial 20.9% (ambient) to around 14% after 16 months of closure, when additional oxygen was injected to sustain the crew. Food production supplied over 80% of the eight person crew's nutritional needs and was strongly influenced by seasonally fluctuating light levels and unexpected insect problems. Lowered caloric intake with a nutritionally dense diet produced sharp drops in blood cholesterol and other health improvements previously seen in laboratory trials of similar diets. The created ecosystems grew rapidly, with large increases in biomass evidenced in tree canopy development. Some ecosystem changes, particularly in the desert, were observed as the biomes developed. Wastewater treatment and water recycling was accomplished during the two year closure. The lessons from the Biosphere 2 experiments may prove valuable in preparation for the challenges of utilizing Martian resources and creating at first limited weight and volume life support systems and eventually permanent habitation modules on Mars.

---

[*] During the design, construction and two year closure of Biosphere 2, Mark Nelson was Director of Environmental and Space Applications and William Dempster was Director of Systems Engineering for Space Biospheres Ventures. Mark Nelson was a member of the eight person biospherian crew, 1991-1993.
[†] Institute of Ecotechnics, 24 Old Gloucester St., London WC1 3AL, England.
[‡] EcoFrontiers, Rt. 2 Box 271, Santa Fe, New Mexico 87505.

INTRODUCTION

Biosphere 2 is the first testbed created for complex ecosystem bioregenerative life support on a long-term basis (50-100 years) to determine its viability and dynamics over time. The Biosphere 2 facility is essentially materially closed (with an annual air leakage rate under 10%), energetically open to electricity and sunlight, and covers some 3.15 acres in its airtight footprint, including over seven million cubic feet of volume. The name Biosphere 2 was chosen to emphasize that the Earth's biosphere (Biosphere 1) is the only biosphere known to science. Biosphere 2's structure includes a human living and work area, agricultural zone including waste recycling and potable water system, five areas modeled on natural ecosystems: rainforest, savannah, desert, marsh and ocean, and via air ducts is connected to two variable volume chambers ("lungs") permitting expansion/contraction of the internal atmosphere without incurring leakage.

Research and development for Biosphere 2 has included work on a number of technologies for potential space application as components of smaller systems including soil beds for air purification, aquatic plant waste water recyclers, non-polluting analytic and monitoring labs, multi-level cybernetic systems for system operation and analysis and sustainable soil-based agricultural systems. These were tested in the Biosphere 2 Test Module, a 17,000 cubic foot facility, which has advanced the field by closing the loop for the first time in air and water purification and in recycling of human metabolic waste products.

While the concept of the biosphere is scarcely a hundred years old, understanding the workings of our global biosphere is a much more recent scientific endeavor. We are just on the threshold of coming to a proper appreciation for the complex, adaptive and evolutionary life system that has enabled life on Earth to flourish for at least 3.8 billion years. Recent scientific findings have changed our notion of this biosphere from being simply the fortunate beneficiary of favorable planetary and geological conditions to being a more active shaper of Earth's environment. The Biosphere 2 project is the first attempt to create a man-made biosphere where similar processes as occur in our global environment may be studied but on a scale which permits experimentation and detailed analysis. Started at the end of 1984, research and development for the project preliminary to the first two year closure experiment spanned some seven years during which time component technologies and ecological research was conducted in the Biospheric Research and Development Center at the project site in Oracle, Arizona. The construction of the Biosphere 2 facility itself was a four year endeavor from its groundbreaking in 1987 until its initial closure experiment was begun in September 1991. The motivations behind its creation were many, including creating a new type of laboratory for studying biospheric processes such as biogeochemical cycles, the viability and interaction of small analog ecosystems to those found on the Earth, and as a testbed for developing new environmentally beneficial systems (e.g. biological means of wastewater regeneration and air purification, sustainable non-polluting agricultural systems, and laboratory techniques which minimize the use of toxic chemicals).

In this chapter we will focus on its potential value as a baseline for studying the dynamics of life support systems that may be used for long-term habitation in space.

While much simpler life support systems, evolving from purely physico-chemical ones, will be required in the early phases of space habitation, it seems inevitable that to provide an evolutionary basis for such expansion into space, we will require the added ecological stability and potential that more complex life systems, biospheres, will offer.

LIFE SUPPORT REQUIREMENTS - THE DRIVER

Calculations of the quantities of critical variables (air, water, food) needed for human life support are essential for understanding at which point bioregenerative systems for spacecraft and space stations will become competitive with the approach currently used, physico-chemical systems supported by resupply of water and food from Earth. These quantities of water, air and food also underline the importance of developing bioregenerative life support systems. The quantities are simply too large to consider for long-term resupply from Earth, and such a long supply-line poses significant safety hazards. Finally, developing the ability to recycle and utilize local resources in creating bioregenerative systems, is essential to achieving long-duration habitation and the eventual expansion of the human population on Mars.

Rummel and Volk developed computer modeling and simulation of bioregenerative life support systems using estimates of daily human requirements [1987]. These estimates for metabolic needs may in fact be low, since their study based the diet on nutritional needs being met solely by wheat ignoring nutritional complexities and the need for variety. These calculations (given in grams/person/day) estimate food inputs at 855 g, drinking/food preparation water at 4,577 g, water in food, 128.3 g, wash/flush water at 18,000 g and oxygen (for food metabolism) at 804.6 g. The development of more efficient technologies for water utilization and reclamation in space may, of course, reduce the quantities required for uses other than metabolism.

Metabolic by-products of each human in space are at present a problem, but can become significant resources for bioregenerative life support systems. For example, waste products may provide valuable organic material to help amend Mars soils to support crops. These outputs they estimated as: Water: water in urine, feces 3,025.5 g/person/day, metabolic water (vapor) 406.0, perspiration water (vapor) 1,680.0, wash/flush water 18,000.0, Solids: feces, urine, sweat solids 161.4, CO_2: from metabolized wheat 1,092.3.

From similar projections to these, Modell and Spurlock have estimated that "in the course of a year, the average person consumes three times his body weight in food, four times his weight in oxygen, and eight times his weight in drinking water. Over the course of a lifetime, these materials amount to over one thousand times an adult's weight" [1980].

The implications of these calculations are clear: extended, not to speak of permanent, human presence in space makes necessary "closing the loop" in the regeneration of air, food and water involved in human life support. Little wonder the Soviet space program took as its goal: "We must grow our own apples on Mars!"

ECOLOGICALLY-BASED SYSTEMS: THE HUMAN FACTORS

In addition to the above necessity for developing bioregenerative systems for space, the design team at Biosphere 2 sought to create systems that would meet additional requirements of making an enjoyable, safe, reliable and satisfying environment for its inhabitants. While crews have survived in submarines, cramped space stations and underground isolation chambers for periods of months to years, these types of sterile and mechanical environments are hardly conceivable as permanent habitations for people. There are also significant concerns about the long-term reliability and stability of such systems if they are to be used in space, outside the safety net of the Earth's biosphere.

We are just beginning to unravel the functions that our biosphere performs in biogeochemical cycles, air and water purification, creating free energy and increasing ecosystem organization, maintaining vital life parameters etc. The Biosphere 2 project attempted to change our paradigm about what life and permanent habitation in space will be like. All life that we know, exists in a biosphere, which is essentially its life support system. To expand life including humans in space on a permanent basis, we must begin to consider that we will be ultimately building biospheric units there. The problem (and research opportunity!) is that there are many unknowns about how to miniaturize and operate a biospheric system. This was the fundamental challenge of the Biosphere 2 experiment.

THE HISTORICAL SETTING

The Biosphere 2 project built on the experience gained from experiments conducted over the past three decades in bioregenerative life support. Especially in the U.S. and Russia such research dates back to the beginning of the space age. Initial work concentrated on two species systems, predominantly using *Chlorella vulgaris*, a fast-growing green algae, as a "partner" for humans. Both Russian and American researchers were able to engineer systems that provided air and water recycling using small algae tanks. The Russians at the Institute of Biomedical Problems, Moscow, conducted 15 and 30 day experiments with people in small chambers [Shepelev, 1972]. However, the goal of making the Chlorella also suffice for food supply was never realized, as ingestion of more than 30-50 grams per day caused a variety of gastro-intestinal health problems in man.

The next major step was the inclusion of higher plants as food sources. This step was first taken at the Institute of Biophysics in Krasnoyarsk, Siberia. After previous work with algae systems in Bios-1 and 2, their 315 cubic meter apparatus powered by artificial lights, Bios-3, was the locale for a series of experiments from 1972-1984. In it, crews of two and three people lived for periods as long as six months. Inside the air was nearly completely regenerated, although catalytic burners were needed to handle trace gas buildups, the water was purified by plant evapotranspiration, and the hydroponic cropping area of about 400 square feet grew up to eleven grain, vegetable, oilseed and root crops. These crops met about half of their nutritional requirements. Human wastes, except for some of the urine, were not processed inside the facility, but were "exported," and some food, including dried meat for additional protein, was "imported" [Terskov et al., 1979].

The volunteer crews included doctors, engineers and agronomists. They had access to phones, TV and newspapers which were delivered through their airlock. They harvested their wheat and other crops, processed them and baked bread and cooked meals in their kitchen. When carbon dioxide levels dropped to about 300 ppm (a bit lower than normal atmospheric concentrations of 350 ppm), they oxidized some of the straw from the wheat crops, pushing CO_2 levels up to about 1,400 ppm to maximize plant photosynthesis. CO_2 poses no particular human health problem at levels below 10,000 ppm, although optimal levels for plants are not yet well characterized. Crew members' health was intensively monitored, and although some simplification of their intestinal microbiota occurred due to the limited diversity of their life environment [Lebedev and Petrov, 1971], they maintained good health during their stay inside Bios-3 [Terskov et al., 1979].

The Bios line of research is a landmark in the field of closed ecological systems. It moved the field a long way towards fulfilling the goal of space rocketry pioneer Tsiolkovsky's "closed ecocycles" in space greenhouses by creating the first systems where man is not simply a mass-exchange unit, but an active participant and manager of his life support system [Terskov et al., 1979, Gitelson, 1989].

The NASA efforts, directed through the CELSS (Controlled Ecological Life Support Systems) program since 1978, have supported research in a variety of universities and at the NASA Ames Research Center, NASA Johnson Space Center and NASA Kennedy Space Center to conduct basic research and engineering applications of the various components necessary for life support. Much of this work has focused on high-yield systems of biomass and food production. Here as well as with other closed ecological system research, there is potential for important spin-off benefits "in technology applicable to partially closed, high intensity food production systems useful on Earth and to basic discoveries in plant science that might allow advances in food production technology within ongoing, long-term crop improvement programs" as an early NASA study noted [Modell and Spurlock, 1980].

In 1986 the Breadboard Project, NASA's most ambitious higher plant-based CELSS program, was begun at Kennedy Space Center. The goal of their Biomass Production Chamber is to demonstrate the production of food for human life support, water recycling and atmospheric gas control. While experiments of this type had been previously performed, the Breadboard project is considerably scaled up from previous laboratory-sized research. Support laboratories are investigating associated questions of waste recycling, food preparation and overall data management. Human closure experiments are presently scheduled for later in the 1990s [Knott, 1990; Averner, 1990].

RESEARCH AND DEVELOPMENT IN PREPARING FOR BIOSPHERE 2

The program of research and development that was required to prepare for Biosphere 2 included four years of operation in the Biosphere 2 Test Module with closed systems demonstrating that bioregenerative life support systems can close the loop in water, air and food as well as recycling waste products. Development and testing of automated real-time analytic systems for monitoring air and water quality in closed ecological systems was conducted. This was part of a multi-level cybernetic system developed for managing systems and processing data from some 2,000 data points and 800 sensors in Biosphere 2. Laboratory and Test Module research has demonstrated the air regeneration capabilities of "soil bed reactors" which forces air through the agricultural soil, thus exposing the airstream to the soil microbes which can metabolize potentially toxic trace gases. Biosphere 2 research has broken new ground in the development of this technology by the integration of this air regeneration by the soil bed reactors with crop production in agricultural soils. Preceding the two year closure experiment were four years of operation in Biosphere 2 Test Module with closed systems, from 1987-1990, including experiments with humans lasting up to 21 days [Alling *et al.*, 1993; Nelson *et al.*, 1992a; Alling *et al.*, 1990]

This research phase also included the design and engineering of variable volume chambers ("lungs") as a solution to pressure differential problems [Dempster, 1994], developing a training program for crew members to handle complex biological and technical systems, selection and development of non-polluting technologies compatible with operation inside a closed ecological system for analytical and biomedical laboratories and food production/processing. A marsh aquatic plant system for recycling of human wastes and household water was developed. Breakthroughs in air-tight sealing technologies were also needed to approach the degree of closure aimed for in Biosphere 2 [Dempster, 1994; Nelson *et al.*, 1992a; Nelson, 1990.]

Biosphere 2 Test Module: A Total Systems Laboratory

One of the lessons from the Biosphere 2 endeavor is that complete bioregenerative systems are feasible, however much they may be improved and made more volume/mass efficient in future developments. Just as the Apollo program in the 1960s accelerated development timelines by instituting "all-up systems testing" rather than exhaustive component by component analysis, SBV moved to develop needed innovative technologies and include them in complete bioregenerative systems testing. To accomplish this acceleration of development, the Biosphere 2 Test Module was designed and constructed in 1985 as a precursor to Biosphere 2 and as a testbed for individual and integrated systems components.

The Biosphere 2 Test Module (Figure 1) is a 17,000 cubic foot materially-closed ecological facility, the largest such facility in the world prior to the completion of Biosphere 2. It was designed to test both the engineering and structure planned for the much larger Biosphere 2, and life system interactions in conditions of a closed ecological system. In operating the testbed, there have been progressive approximations towards a successful integrated system. For example, experiments have tested two sealing methods, several generations of analytic/sensor systems and the first application of the variable volume chamber concept.

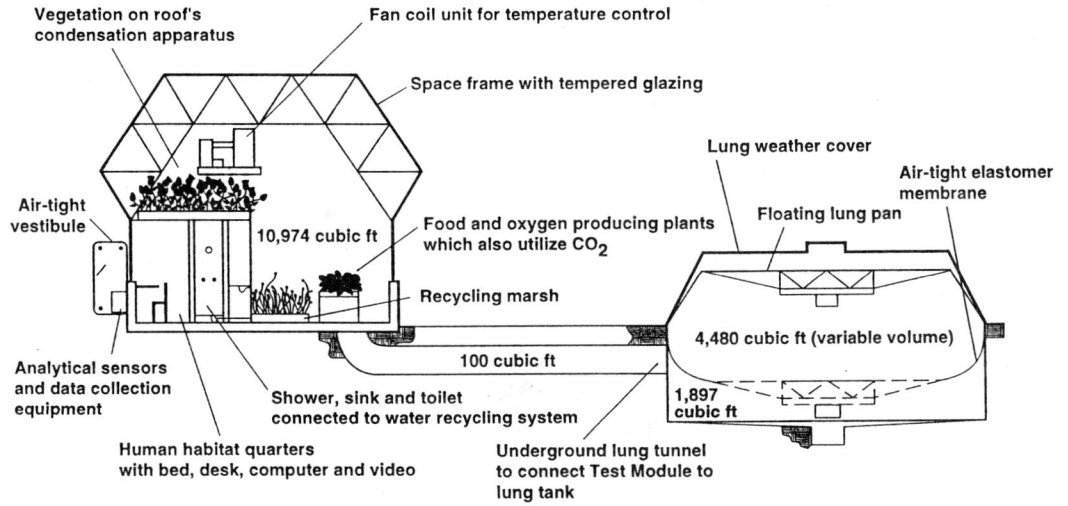

Figure 1 Section view of the Biospheric Research and Development Test Module with lung at Biosphere 2 site, Arizona. This is a schematic of how the Test Module was configured for human habitation experiments, 1988-1990, preceding the closure of Biosphere 2. Its main structure is approximately 7 m on a side by 8 m tall, with a variable volume of approximately 480 cubic meters (Nelson et al., 1993).

The Biosphere 2 Test Module is sealed underground with a stainless steel liner and is connected via an air duct with a variable volume chamber ("lung"). This lung allows the atmosphere to expand and contract as do the lungs of Biosphere 2 which are described in the following section. The Test Module achieved a leakrate of 24% per year [Nelson et al., 1992a].

The Biosphere 2 Test Module was the first bioregenerative facility to achieve air purification through biological means (vs. catalytic burners), water cycling, and human waste and domestic graywater waste recycling. Food was grown to supply nutrition during the short-term human closures, but the limited growing area was not adequate to support long periods of human habitation. Over 60 person-days were logged in experiments, including a three-week closure in November 1989 [Alling et al., 1993, Nelson et al., 1992a, Alling et al., 1990].

INNOVATIVE BIOREGENERATIVE TECHNOLOGIES

Some of the innovative technologies that SBV developed, tested in the Test Module, and are utilized in Biosphere 2, address crucial issues for bioregenerative systems:

Food Production - Intensive, Non-Polluting and Sustainable

SBV and its prime consultant for the agricultural section, Environmental Research Laboratory of the University of Arizona, began with trials of hydroponic and aeroponic cropping techniques. A variety of reasons underlay the subsequent switch to soil-based agriculture. One, of course, is that hydroponics depends on a supply of chemical nutrients that cannot be produced from within the system. In addition, soils in a closed system can play a significant role in air purification, either through the soil bed reactor technology (discussed below) or simply through passive diffusion through the soil facilitating the diverse metabolic capabilities of soil microbial populations. Another advantage of using soil is that it simplifies creation of waste recycling systems for animal and human wastes and inedible portions of crops. Traditional and energetically low intensity technologies like compost or utilizing plant/microbe systems for wastewater regeneration become available options [Nelson et al., 1994, Nelson et al., 1993b].

Composting and marsh wastewater systems (see below) are far less energy-consumptive than alternatives like wet oxidation or incineration. Marsh wastewater and compost systems operate by time-tested biological mechanisms. The biomass produced in the marsh wastewater system in turn can be composted to produce high quality topsoil for replenishing the nutrients that crops removed from the soil. An important requirement for the agriculture of the bioregenerative system is that it must be sustainable as well as highly productive. There are numerous historical examples of sustainable soil agriculture but none thus far of a hydroponic system that can persist without "complex outside additions in the form of fertilizers and pesticides" [Glenn and Frye, 1990].

Another requirement of an agricultural system in a materially closed system is that it must be virtually pollution-free. Even in the nearly seven million cubic foot volume of a facility like Biosphere 2, water, soil and air buffers are so small, and cycling times so rapid, that there is no way of introducing pesticides and herbicides without serious health hazards. The water cycle is a few weeks and CO_2 in air has a residence time of about 4 days in Biosphere 2 [Nelson et al., 1993a]. Therefore, a variety of biological and cultural methods of pest and disease control (known as Integrated Pest Management) must be utilized for the agricultural crops.

These disease and pest control techniques include: selection of resistant crops, using many small plots and rotating crops, switching between several cultivars (varieties) of the major crops, maintenance of "beneficial insect" populations (ladybugs, praying mantis, parasitic wasps etc.) to control pest insects, intercropping and manual control. In addition "safe sprays" such as soap, light oil or *Bacillus thuringensis* may be employed [Nelson et al., 1993b, Leigh et al., 1987; Glenn et al., 1990].

For the first time in a closed ecological life support system, a complete nutritional diet was the goal and domestic animals were included. The diet for the eight crew members of Biosphere 2 included milk (from African pygmy goats) eggs (from the system's domestic chickens), meat (from the goats, chickens and Ossabaw feral pygmy pigs), and fish (from Tilapia grown in the rice/azolla water fern paddies). In addition, a wide range of vegetables, grains, starches and fruit are grown (Figure 2). Biosphere 2 maintains semi-tropical temperatures in the agriculture area (60-85 deg. F.) permitting both temperate and tropical varieties to be grown.

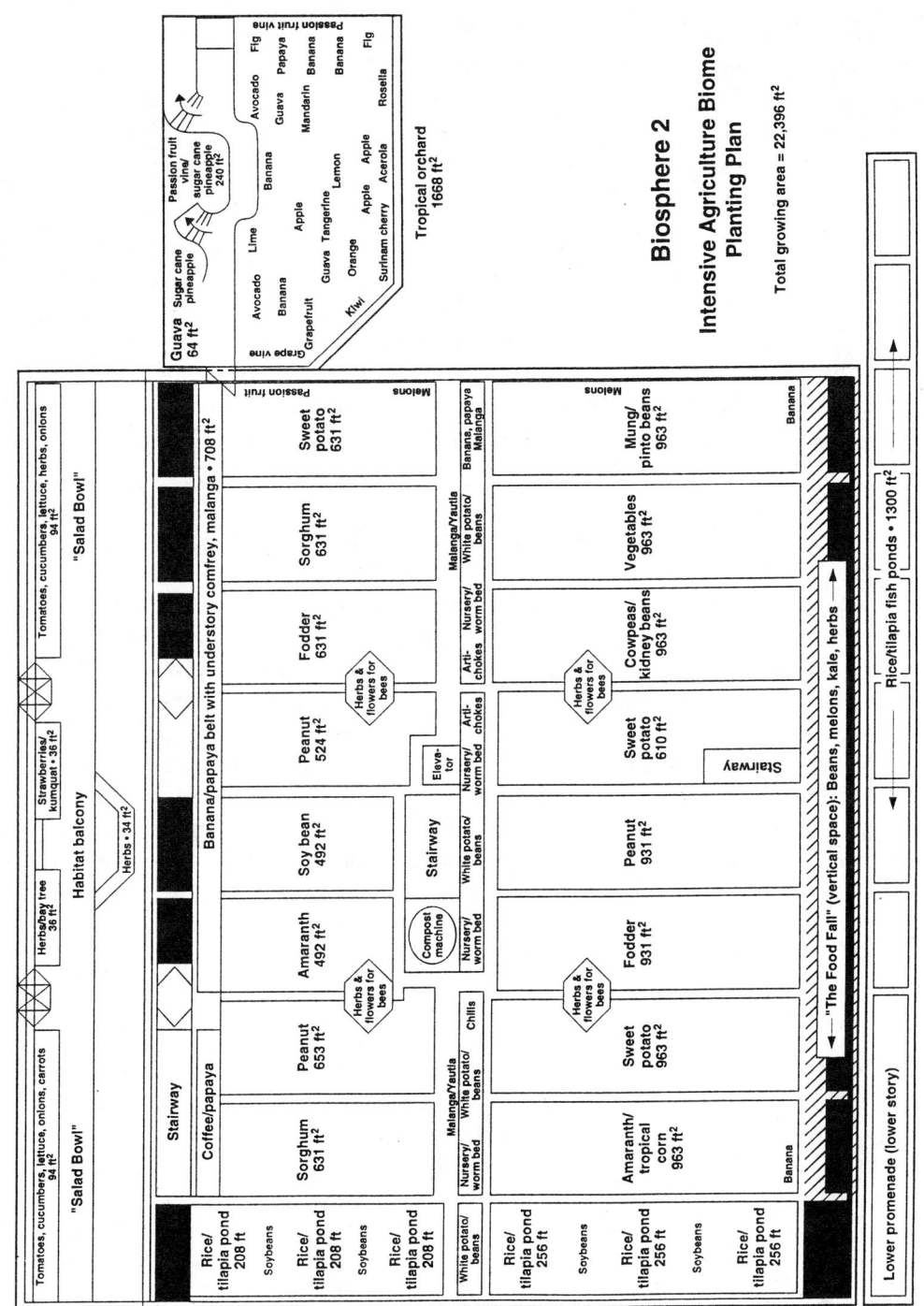

Figure 2 Schematic illustrating the polycultural planting plan in the Biosphere 2 agriculture areas. Apart from perennial trees and rice paddies, most planting beds were rotated in crop two to three times per year. Staggered planting/harvesting assisted in labor scheduling and in assuring a steady supply of food [Nelson, Silverstone and Poynter, 1993].

In all, a total of over 86 crops (including herbs) were utilized during the first closure experiment. Though the diet includes some animal products, fat is in short supply and peanuts as a source of vegetable fat is an important crop. A computer program keeps track of nutrient intake and helps plan forward planting of crops to ensure a balanced diet [Allen, 1991].

The entire agricultural area must produce the fodder crops necessary for the animal food as well as direct human food crops. The reliance on ambient sunlight reduced by 50-55% in passing through the glazed envelope also limits area productivity and might differ in a space application where advantage may be taken of enhanced artificial light techniques to boost yields. Biosphere 2 marked the first time domestic animals have been utilized and that such a variety of crops will be harvested and processed in a closed system. This has required the use of a variety of processing equipment to lower labor requirements [Silverstone and Nelson, *in press*].

Air Purification - Soil Bed Reactors

The addition of soil also opens the way to potential solution to one of the most vexing of problems—the maintenance of air quality. The great diversity of outgassing products from anthropogenic, biogenic and technogenic sources combined with the small volumes and rapid cycling times of atmospheric components in closed systems create a significant hazard for toxic gas buildups. In Apollo, Skylab and Space Shuttle cabins, for example, there were 300-400 gases identified, and significant concerns about unanticipated reactions between such outgassing products [Nicogossian and Parker, 1982; Rippstein and Schneider, 1977]. These air contamination concerns occured in spite of the significant flushing of the air volume through the carbon dioxide removal system, and other measures such as "exclusion of material, equipment isolation, absorption using charcoal, or absorption of soluble substances on the condensate in humidity-control devices. The results of numerous studies performed in anticipation of a Space Station indicate that these methods would be inadequate for longer missions, larger crews and the anticipated greater variety of equipment" [Hord, 1985]. The conventional solutions to this problem include filtering methods using charcoal or catalytic oxidation which will require substantial energy costs and/or expendable parts, such as filters.

In addition to natural soil and vegetative interactions with the air, Biosphere 2 was designed so that the soil of the agricultural area can also function as a "soil bed reactor" when blowers are turned on to force air upwards through its soils. This technology, originally developed in Europe for control of industrial odors, passes the air volume through an active soil to expose it to the metabolic action of the microbial populations there. The diversity and high numbers of microbes are capable of metabolizing an extraordinary range of trace gases that could pose a toxicity problem [Bohn and Bohn, 1986]. In preparation for Biosphere 2, research was conducted at the Environmental Research Laboratory and in the Biosphere 2 Test Module to determine the efficacy of soil bed reactors in removing specific contaminants and those generated in a complex bioregenerative facility. This research demonstrated the ability of SBRs to scrub trace gases and maintain air quality while simultaneously ensuring good aeration and cropping productivity in a working soil [Frye and Hodges, 1990; Nelson *et al.*, 1992a]. The entire air volume of Biosphere 2 can be pumped in less than a day through its soil bed reactor. In addition, the significant volumes of soil and diversity of soil types incorporated within Biosphere 2—along with its abundant vegetation—means that even without use of the soil bed reactor, there will be large biological interactions with the atmosphere.

Waste Recycling Systems

SBV worked with the consultation of B. C. Wolverton, now retired from NASA Stennis Space Center, on the development of waste recycling systems which utilize aquatic plants and their associated microbes to purify water streams containing human and animal wastes and domestic graywater from the human residences and kitchen/laundry. These "constructed marsh" systems utilize the wastes to produce an abundance of plant growth valuable for animal fodder and compost material. The advantage of marsh systems is that they have low maintenance and energy requirements and produce valuable byproducts. As Schwartzkopf and Cullingford note: "Many previous CELSS concepts have incorporated high energy methods of waste degradation such as wet oxidation or super critical wet oxidation. In the process, all of the energy stored in the chemical bonds of the waste materials is lost. By using either bioregenerative technologies or appropriate physicochemical technologies . . . some of the chemical bond energy can be provided to the system by converting wastes into low complexity materials which can be used as foodstocks for bacteria, algae or higher plants" [1990].

Some plants grown in the marsh waste treatment system can be used for animal fodder and grow rapidly in the nutrient rich waters. After leaving the marsh waste treatment system, the water is added to the irrigation supply of the agricultural crops which thus benefit from any remaining nutrients. A similar marsh wastewater system is employed for any chemical effluent that may occur from internal workshops and laboratories, taking advantage of the fact that aquatic plants will concentrate heavy metals, thus isolating them from soil and water contamination. A final advantage of this wastewater system is that the high rate of growth and transpiration of aquatic plants make them valuable sources of potable water through condensation of water vapor [Wolverton. 1986; Wolverton and MacDonald, 1979].

Analytic and Monitoring Systems

To operate Biosphere 2 it was necessary to develop unique capabilities for the analytic and monitoring systems. They had to be highly automated so they are operable with a minimum of human time, produce data about environmental parameters in real-time so that corrective steps might be taken, sufficiently flexible to be able to deal with the wide variety of potential analyses of concern, minimize reliance on consumable supplies and produce little or no pollution since they will operate in a closed ecological system. These requirements, of course, are directly analogous to the needs an operating Mars base will have of its monitoring systems.

To achieve these objectives, SBV developed a series of automated sensing systems which monitors continuously eleven key trace gases from each of the biomes of Biosphere 2 and key water variables of nitrite and nitrate concentrations. A computerized information processing system records data from over 800 sensors distributed throughout Biosphere 2 as well as providing remote control of equipment and data base inquiry and analysis [MacCallum *et al.*, 1990].

All the consumables required by the analytical lab, with the exception of several high-quality chemicals needed in small quantities are produced inside Biosphere 2. These include pure air, pure water, liquid nitrogen, gaseous nitrogen, hydrogen and oxygen. Glove boxes and scrubber systems contain and neutralize solvent or acid fumes [MacCallum *et al.*, 1990, Van Thillo *et al.*, *in press*].

Cybernetic Systems

To assist the crew in operating Biosphere, many of the control and management functions are automated using an artificial intelligence system termed the "Biosphere 2 Nerve System." It is this system (developed by SBV and a team from Hewlett-Packard) that is programmed to carry out the complex and routine control of the infrastructure of electromechanical devices, airhandlers, pumps and valves, that maintain overall environmental parameters [van Thillo *et al.*, *in press*]. It includes environmental sensing and response (sensors and actuators); local data acquisition and control; system supervisory (by biome) monitoring and control; overall Biosphere 2 monitoring, optimization, information analysis, reporting and historical archiving; and telecommunications networks.

Software for these systems include programs specifically designed for life support system use. These are carbon dioxide modeling and real-time monitoring, thermodynamic modeling, simulation and real-time control, global monitoring of overall system status, nutrition/diet planning and crop production scheduling.

BIOSPHERE 2 - THE FIRST MAN-MADE BIOSPHERIC LIFE SUPPORT SYSTEM

Unlike previous closed ecological systems of the CELSS and Bios-3 types which included essentially only one type of ecosystem—an agricultural one—in addition to its human habitat, Biosphere 2 is designed to be a "biospheric" system. That is, it includes several distinct ecosystems (analogous to Earth's biomes which contain characteristic climate, soil and flora/fauna) with the aim of determining whether such a system might enable long-term life support. Biosphere 2 was designed to support (and be operated by) a human crew of eight. It is also virtually materially-closed to exchanges with the outside atmosphere and underlying geology. Biosphere 2 is energetically-open; that is, it receives energy inputs both from incident solar radiation plus electrical power, heating and cooling from an energy plant outside its airtight structure. Biosphere 2 is also informationally-open; it receives information from outside scientists, technicians and engineers, as well as being connected via computer telecommunications, telephone, video, radio and TV. It also sends out a wide variety of communications and data.

Biosphere 2's 3.15 acre footprint and seven million cubic foot volume [Table 1] is sealed above ground with a laminated glass mounted on a spaceframe and below ground by a stainless steel liner [Allen and Nelson, 1988; Augustine, 1987]. Its atmospheric leak rate is documented at less than ten percent per year [Dempster, 1994]. Two variable volume chambers ("lungs") connected to the main structures by underground air ducts allow for expansion and contraction without leakage and without dangerous pressure differences between inside and outside that could break the envelope of Biosphere 2. Each lung consists of a large cylindrical air tank (vertical axis) sealed on top by a flexible impermeable membrane which rises and falls as Biosphere 2's atmosphere expands and contracts. The variable volume of the two lungs combined is about 1,500,000 cubic feet. Each lung is further enclosed by a dome within which the air pressure can be controlled within a critical range by use of fans. This patented arrangement permits the air pressure within Biosphere 2 to be maintained anywhere from modestly positive to very slightly negative relative to external barometric pressure. This gives the powerful capability to limit as well as to determine the leak rate [Dempster, 1994] (See further discussion of leak rate later in this chapter).

Table 1
AREAS AND VOLUMES OF BIOSPHERE 2

Areas	Square feet	Square meters	Square acres	Hectare
Glass surface	170,000	15,794	3.90	1.58
Footprints				
Intensive agriculture	24,020	2,232	0.55	0.22
Habitat	11,592	1,077	0.27	0.11
Rainforest	20,449	1,900	0.47	0.19
Savannah/ocean	27,500	2,555	0.63	0.26
Desert	14,641	1,360	0.34	0.14
West lung (airtight portion)	19,607	1,822	0.45	0.18
South lung (airtight portion)	19,607	1,822	0.45	0.18
TOTAL AIRTIGHT FOOTPRINT	137,416	12,766	3.15	1.28
Energy center	30,000	2,787	0.69	0.28
West lung (weathercover dome)	25,447	2,364	0.58	0.24
South lung (weathercover dome)	25,447	2,364	0.58	0.24
Ocean water surface area	7,345	682	0.17	0.07
Marsh surface area	4,303	400	0.10	0.04

Volumes	Cubic feet	Cubic meters		
Intensive agriculture	1,336,012	37,832		
Habitat	377,055	10,677		
Rainforest	1,225,053	34,690		
Savannah/ocean	1,718,672	48,668		
Desert	778,399	22,042		
Lungs (at maximum)	1,770,546	50,137		
TOTAL	7,205,737	204,045		
Soil, water, structure, biomass	671,635	19,019		
Air	6,534,102	185,026		
Ocean water (1,000,000 gallons, approx)	133,690	3,786		
Fresh water (200,000 gallons, approx)	26,738	757		

The life systems of Biosphere 2 are housed in two wings, which are connected in water/air circulation but which have insect screens to prevent flying insects from moving from one to the other. Like the global biosphere, Biosphere 2 is composed of various "biomes," with differing soils, climate regime and vegetation. Five areas patterned on tropical wilderness areas are housed in the eastern wing, which is some 540 feet long and from 100-140 feet wide. Biosphere 2 had some 3,000 species of plant and animal at initial closure. This "species-packing" strategy was designed so that should losses occur, there is a good chance that other species might fill required food-web econiches.

The biomes of Biosphere 2 are: a tropical rainforest with neotropical, predominantly Amazonian species; savannah with species from Australia, South America and Africa; a coastal fog desert area patterned on Baja California; an estuarine marsh ecosystem collected in the Everglades of Florida; and a tropical oceanic system with coral reef collected off Yucatan and in the Caribbean, shallow lagoon and beach.

The western wing includes two man-made biomes: an agricultural area (including rice/fish paddies, fodder plants, tropical orchard, chickens/goats/pigs and vegetable/grain systems) for growing a complete diet for the eight-person crew plus recycling wastewater and inedible biomass; the human habitat, with living and working areas for the crew.

Table 2
ENERGY OF BIOSPHERE 2

Electrical peak demands	Kilowatts		
Biosphere 2 airtight enclosure	1500		
Energy center	1500		
TOTAL	3000		
Generating capacity*	3750		
(*Fossil fueled generators)			
Cooling peak demands	Refrigeration (tons)	Kilojoule / hour	
Intensive agriculture	900	11,395,000	
Wilderness	1900	24,056,000	
TOTAL	2800	35,451,000	
Heating peak demands	BTU / hour	Kilojoule / hour	
Intensive agriculture	3,400,000	3,587,000	
Wilderness	7,000,000	7,385,000	
TOTAL	10,400,000	10,972,000	
Solar energy entering glass*	BTU / hour	Kilojoule / hour	Kilowatts
Intensive agriculture	8,200,000	8,651,000	2,403
Wilderness	17,500,000	18,462,000	5,129
TOTAL	25,700,000	27,113,000	7,532
(*Peak)			

Energetically, the life systems are powered by ambient sunlight, and technical systems including those required for thermal control are powered by external co-generating natural gas electrical generators [Table 2]. Heated, chilled or evaporatively-cooled water as needed for thermal control passes through Biosphere 2 in closed loop piping into air-handler units where heat exchange occurs. Evaporative cooling water towers outside

Biosphere 2 dissipate rejected heat. The airhandlers provide airflow of such heated/cooled air over the system's ecosystems, providing thermal regulation. Temperature parameters have been set in accordance with the normal tolerances of the various biomes, and range from winter lows of about 15 deg C to summer highs in the desert and savannah of around 38 deg C. The agriculture and human habitat areas have the tightest temperature controls to provide better growing conditions for crops and comfort for the crew [Silverstone and Nelson, in press; Nelson et al., 1993b; Nelson, 1990].

BIOSPHERE 2 PERFORMANCE DURING THE TWO YEAR CLOSURE

Food Production and Agriculture

The agricultural cropping area of 2,000 sq m produced about 80% of the nutritional requirements of the eight biospherians during the two year closure. This was lower than expected, and utilizing three month's food supply which had been grown in Biosphere 2 before closure and some seedstocks resulted in average daily caloric intake over the two years of about 2,200 calories, 73 g of protein and 32 g of fat per person (Table 3). Caloric intake was somewhat lower during the first year, and that factor along with adaptation to the diet, resulted in most of the 10-20% weight loss experienced by crew members (Tables 4 and 5). The combination of restricted caloric intake with high nutritional density produced a large decline in blood cholesterol levels and other healthy physiological adaptations as had been observed in previous laboratory studies of such a dietary regime [Walford et al., 1992]. Modest gains or a virtually leveling off of weight marked the final year of the closure experiment [Silverstone and Nelson, in press].

Light was one of the obvious limiting factors for the Biosphere 2 agriculture. Comparison of crops grown in differing seasons and light conditions during the two year closure reveals that yield was strongly correlated with incident light [Silverstone and Nelson, in press]. This was anticipated—the site in Arizona has a daylength which varies from 9.5 hours of daylight at the winter solstice to 14.5 hours at the summer solstice (correlating with some 25-65 Einsteins/sq m of incident light on the outside on cloudless days) and light attenuation by glass and structural shading lowers incident light by 50-60%. However, what was unanticipated was that the two years of closure would be ones with strong El Niño southern oscillation conditions producing an unusual number of stormfronts especially during the winter months. This proved to be important not only in lowering crop yields, but on the effort to limit seasonal increase of carbon dioxide. One of the major agricultural improvements effected in the transition period after Mission One was the installation of artificial lights to supplement solar input, especially during the winter months. In addition, selection and experimentation with crops more adapted to lower light and other Biosphere 2 environmental factors should improve agricultural production in future closure experiments.

Insect and disease problems were caused by two types of mite, mealy bug, aphid, powdery mildew, root knot nematode and cockroaches. The most serious of these proved to be the broad mite, which proved resistant to a number of control techniques. Its damage caused a shift from white potatoes to sweet potatoes among starch crops, and to increased reliance on lablab bean since both these crops proved highly resistant to attack. Crops of importance included grains: rice, wheat, sorghum; starches: sweet potato, taro; bean: lablab; oil seed: peanuts; vegetables: beet, chili, tomatoes, squash, greens and lettuce; fruit: banana, papaya. The goats were the outstanding producers of the domestic animals [Silverstone and Nelson, *in press*].

Table 3
FOOD PRODUCTION IN BIOSPHERE 2 DURING THE TWO YEAR CLOSURE EXPERIMENT, 1991-1993 (Silverstone and Nelson, *in press*).

Biosphere 2 intensive agriculture production for 8 biospherians
September 1991 to September 1993

Crop	Yields Kg/sq. meter Year 1	Yields Kg/sq. meter Year 2	Total 2 yr. yield Kg	g/ person /day	Protein/ person /day	Fat/ person /day	Kcal/ person /day
Grains							
Rice	0.29	0.20	276.47	46.76	3.52	0.85	167.82
Sorghum	0.24	0.17	189.83	32.10	4.24	0.64	106.63
Wheat	0.22	0.14	191.87	32.45	4.29	0.65	107.78
Starchy veg.							
White potato	0.63	1.42	240.41	40.66	0.86	0.48	31.22
Sweet potato	2.25	1.95	2765.13	467.64	6.68	1.34	494.36
Malanga	n/a	n/a	101.61	17.18	0.41	0.03	18.04
Yam	n/a	n/a	19.50	3.30	0.07	0.00	4.36
High fat legumes							
Peanut	0.10	0.15	147.42	24.93	6.59	12.20	146.03
Soy bean	0.15	0.34	21.41	3.62	1.32	0.66	14.56
Low fat legumes							
Lab lab bean	n/a	n/a	143.79	24.32	3.88	0.33	84.33
Pea	0.05		14.52	2.45	0.61	0.02	8.42
Pinto bean		0.02	27.49	4.65	1.15	0.04	15.94
Subtotal			4111.95	695.41	32.47	17.20	1183.55
Vegetables							
Beans green	1.22	1.03	24.95	4.22	0.08	0.00	1.36
Beet greens	2.64	0.63	432.28	73.11	1.33	0.37	15.38
Beet roots	2.83	2.20	760.23	128.57	1.88	0.18	56.94
Bell pepper			65.73	11.12	0.10	0.05	2.67
Carrots	2.25	4.39	224.76	38.01	0.34	0.07	16.07
Chilli			125.19	21.17	0.42	0.04	8.47
Cabbage	3.42	2.78	152.82	25.84	0.30	0.04	5.91
Cucumber			48.13	8.14	0.04	0.01	0.04
Eggplant			244.94	41.42	0.44	0.03	11.10
Kale	8.30		11.29	1.91	0.06	0.01	0.93
Lettuce	5.37	2.44	150.59	25.47	0.24	0.04	3.37
Onion			139.23	23.55	0.28	0.06	7.99
Pak choi	3.91	3.47	45.29	7.66	0.11	0.01	0.98
Snow pea	0.29		0.91	0.15	0.01	0.00	0.12
Squash seed			10.43	1.76	0.51	0.82	9.70
Summer squash	4.88	4.39	512.56	86.68	0.93	0.19	16.72
Swiss chard	12.69	14.65	141.52	23.93	0.43	0.05	3.93
S. pot. greens			64.41	10.89	0.31	0.03	2.72
Tomato			352.90	59.68	0.53	0.11	11.72
Winter squash	4.15	4.05	342.47	57.96	1.05	0.10	36.82
Subtotal			3850.63	651.22	9.40	2.20	213.57
Fruit							
Apple			0.57	0.10	0.00	0.00	0.06
Banana			2170.92	367.15	2.36	10.49	220.29
Fig			54.34	9.19	0.07	0.03	6.70
Guava			53.50	9.05	0.06	0.04	3.64
Kumquats			4.17	0.71	0.00	0.00	0.42
Lemon			10.12	1.71	0.01	0.00	0.34
Lime			3.63	0.61	0.00	0.00	0.16
Orange			5.56	0.96	0.01	0.00	0.33
Papaya			1215.64	205.59	0.81	0.15	52.87
Subtotal			3518.53		3.31	10.71	284.79
Animal products							
Goat milk			841.84	142.37	4.58	5.59	99.05
Goat meat			16.96	2.87	1.02	0.48	7.54
Pork			58.74	9.93	1.70	2.06	26.11
Fish			10.21	1.73	0.32	0.07	2.03
Eggs 257			14.29	2.42	0.29	0.27	3.86
Chicken meat			8.07	1.37	0.25	0.20	2.92
Subtotal					8.17	8.68	141.50
Total Produced					53.35	38.80	1823.41

Table 4
AVERAGE PROTEIN, CALORIES AND FATS CONSUMED PER PERSON PER DAY OVER THE 24 MONTHS OF CLOSURE (Silverstone and Nelson, *in press*).

Month	Protein (g)	Kcal	Fats (g)
Oct 91	67	1789	28
Nov 91	62	1925	21
Dec 91	74	2183	27
Jan 92	62	1948	23
Feb 92	74	2247	37
Mar 92	74	2206	32
Apr 92	72	2092	29
May 92	71	2043	28
Jun 92	72	2038	30
Jul 92	62	2227	32
Aug 92	66	2307	28
Sep 92	72	2491	34
Oct 92	68	2307	37
Nov 92	70	2304	34
Dec 92	88	2345	30
Jan 93	82	2225	30
Feb 93	74	2247	37
Mar 93	99	2282	34
Apr 93	79	2204	29
May 93	69	2190	30
Jun 93	72	2337	40
Jul 93	67	2252	32
Aug 93	75	2397	39
Sep 93	72	2609	38
Overall average	73	2216	32

Table 5
WEIGHTS OF THE EIGHT BIOSPHERIANS OVER THE TWO YEAR CLOSURE PERIOD (Silverstone and Nelson, *in press*).

(Body weight in Kg)

Crew Member	Sep 91	Mar 92	Sep 92	Mar 93	Sep 93
Male 1	67.2	60.5	56.8	61	59.2
Male 2	68.2	59.1	54.5	58	57
Male 3	67.3	55.5	54	58	56
Male 4	94.5	73.6	67.5	71	72.5
Female 1	75	64.5	66	69	70.5
Female 2	55.9	50.9	49.5	52	52.2
Female 3	59.1	52.3	51	53.5	53.9
Female 4	52.7	42.8	44	48	46.8

Crew ingenuity resulted in the utilization of virtually every conceivable spot within the agricultural area and in tight intercropping where light permitted. Biospherians became highly attuned to helping Biosphere 2's plants capture all available "sunfall" entering the facility, both in the agriculture and wilderness areas. Some 200 sq m of additional food producing planters in the agriculture area were created in this fashion, contributing hundreds of kilograms of "extra" food during the second year of closure [Alling and Nelson, 1993].

> The diet and cooking became an extremely important part of everyday life in the Biosphere. The standard of the cooking on a particular day would have an effect on the general spirits of the crew . . . variety was extremely important. A new taste or a new dish became a real treat and every effort was made to enhance cuisine so as to avoid the monotony of the same foods. Much of the variety was provided by the various fruits, vegetables, herbs and spices and the different milk products such as cheese and yoghurt . . . much of the crew's social life became centered around food. Holidays were celebrated with huge feasts [Silverstone and Nelson, *in press*].

Ecosystem Development and Changes

Even before closure, since Biosphere 2's biomes were installed about a year earlier, there had been rapid growth in the various ecosystems. Biomass increased some 50% in the rainforest between initial measurement in the fall of 1990 and July of 1991 when many of the trees were resurveyed [Petersen *et al.*, 1992]. This rapid development of tree canopies continued throughout the two year closure in rainforest, savannah and marsh biomes. Many *Leuceana glauca* trees, for example, designed to be "early successional" trees in the rainforest to prevent sun damage to more light sensitive species, grew to be 12-16 m tall, and were cut down during the first transition to facilitate growth of later successional trees. The "gingerbelt" which forms the outer perimeter of the rainforest, and the "bamboo belt" which shields the rainforest from potential salt drift from the ocean, also showed prolific growth [Nelson *et al.*, 1993a].

A detailed resurvey of plants was conducted in all the biomes during the transition period following Mission One and before a second crew began a closure experiment (Mission Two) in March 1994. Those data are still being analyzed for publication. But it was apparent during the course of the two year closure that one of the most striking ecosystem developments was a shift in the desert biome from the cactus/succulent vegetation dominance originally envisioned to a system more dominated by shrubs, annuals and, in areas, grasses—thus more resembling a coastal succulent scrub ecosystem. This shift may have been occasioned by the strategy of keeping the desert watered longer and thus active to assist in carbon dioxide management during the low-light months it is normally active. Other factors may have been responsible as well such as water condensate from the glazing especially in winter and by relative humidity remaining higher in the system than would normally be the case in coastal, fog deserts [Nelson *et al.*, 1993].

Species losses in the coral reef were fewer than anticipated and evidence of coral reproduction was discovered during the transition surveys. Whiteband disease affected some brain corals during the two year closure. Manipulation of the ocean's pH was done with addition of buffering chemicals in response to the high levels of atmospheric carbon dioxide in the facility [Nelson et al., 1993]. Nutrient removal to maintain the low levels of nitrates and nitrites found in coral reefs was accomplished by banks of algae scrubbers under artificial lights which treated waters from both ocean and marsh. To further lower nutrient levels, skimmers operated by air bubblers were constructed by the crew during the two years [Alling and Nelson, 1993].

There was a noticeable decline in flying insects during the two year closure and loss of two bird species. Galago ("bushbaby", a prosimian) conception and birth occurred within Biosphere 2 during the closure [Nelson et al., 1993a]. Explosion of some populations including a species of ant and cockroach has stimulated further research and potential control methods [Silverstone and Nelson, in press].

Management required in the ecosystems has varied. The explosive growth of a few plants—notably several vines—morning glory and passionfruit in rainforest and savannah; C4 grasses in savannah and desert; macro-algae in the ocean; necessitated human intervention to prevent loss of other species through shading. In other cases, e.g. the thornscrub ecotone between savannah and desert, canopy development has tended to reduce understory invaders. Biosphere 2 was originally species-packed, so some loss of species and emergence of hierarchies of dominance was expected to naturally develop. In some cases, biospherians intervened, acting as deliberate "keystone predators" in such small synthetic ecosystems. In the ocean, lobsters and trigger fish were culled by biospherians to prevent excessive predation. Several ecosystems were managed to assist in the control of carbon dioxide. For example, savannah grasses and rainforest gingerbelt were pruned and the cut biomass dry-stored to slow decomposition, while the rapid regrowth of the vegetation assisted in sequestering carbon dioxide during low light seasons. The regulation of active/dormant seasons in savannah, thornscrub and desert could be manipulated in the interests of atmospheric management as well [Alling and Nelson, 1993; Nelson and Alling, 1993].

Carbon Dioxide Dynamics

Fluxes of biogeochemical elements can be rapid in small, closed ecological systems because of the high concentrations of biotic elements, and small buffer capacities. It is useful in discussing carbon dioxide dynamics in Biosphere 2, to understand that even in a facility with an atmospheric volume of some six million cubic feet (170,000 cu m), that a concentration of 1,500 ppm in its atmosphere is only equal to about 100 kg of carbon [Nelson et al., 1993a]. In addition, the ratios of distribution of organic carbon are quite different in Biosphere 2. Unlike the Earth with a 1:1 ratio of carbon in plant biomass to atmospheric carbon, Biosphere 2's ratio is on the order of 100:1. When we compare soil organic carbon to carbon contained in the atmosphere, the Earth's ratio is 2:1 while Biosphere 2 has a ratio three orders of magnitude greater [Nelson et al., 1994; Earth estimates from Bolin and Cook, 1983].

Another factor accounting for Biosphere 2's carbon dioxide fluctuations is that the entire vegetated area is photosynthetically active during the daytime, while at night plant and soil respiration is dominant. This results in day/night fluctuations of up to 500-600 ppm for sunny days. In addition, Biosphere 2's carbon dioxide record during closure shows the strong effect of seasonal light variations. For example, June 1992 had an average carbon dioxide concentration of around 1,050 ppm, while December 1991 had an average atmospheric concentration of about 2,450 (Figures 3 and 4) [Nelson *et al.*, 1994; Dempster, 1993].

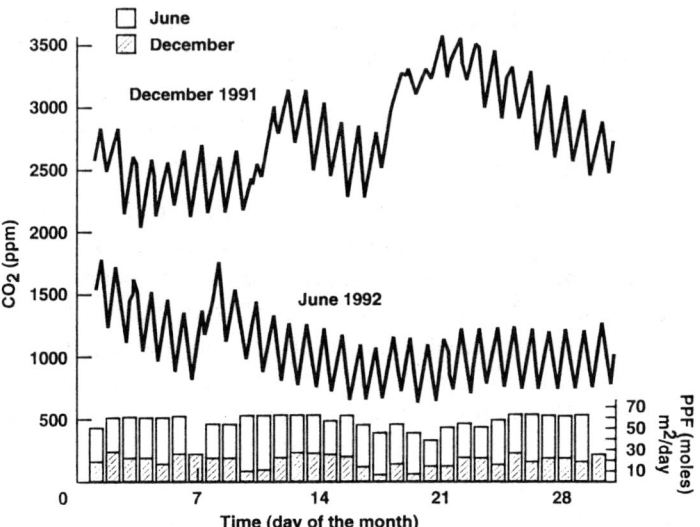

Figure 3 Atmospheric CO_2 dynamics within Biosphere 2 during December 1991 and June 1992. The overlapped bar values given for photosynthetic photon flux (PPF) are for total daily incident sunlight at the project site; internal light levels vary depending on location in the facility but average 40-50% of ambient sunlight. Cloudy days have a large impact on CO_2 dynamics and normal day/night variations are large, because photosynthesis dominates during the daylight hours drawing CO_2 levels sharply down, and soil and plant respiration at night lead to large rises (Nelson *et al.*, 1993).

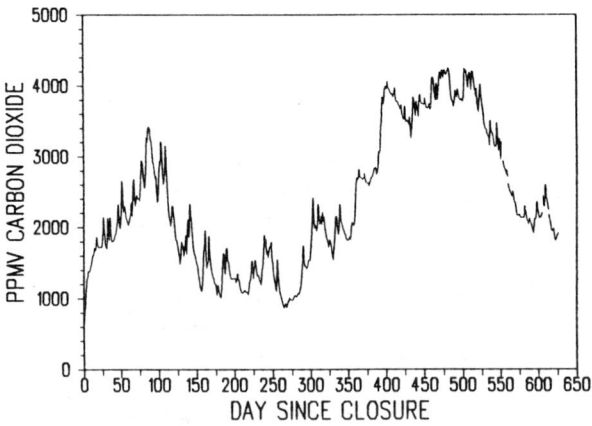

Figure 4 Graph of Biosphere 2 carbon dioxide levels, 26 September 1991 to 13 June 1993 [Dempster, 1993, reprinted with permission from SAE Technical Paper No. 932290 © 1993 Society of Automotive Engineers, Inc.].

To assist in the management of CO_2 a physico-chemical precipitator with capacity to lower CO_2 levels by 100 ppm/day was used. The chemical sequence of reactions is reversible, as the $CaCO_3$ formed could be heated to release CO_2. Other measures taken by the crew to increase photosynthesis and decrease respiration included pruning to stimulate regrowth and dry storage of cut biomass, lowering night time temperatures, cessation of composting and minimizing soil disturbances during the winter [Nelson *et al.*, 1994].

Labor Requirements

Analysis of the crew time spent in various tasks reveals that agriculture and food-related jobs was the largest component requiring about 45% of total work hours (Figure 5). This included food processing, care and feeding of domestic animals and cooking. There was a slight reduction during the second year as the agricultural soil became easier to till, and more efficient means of accomplishing tasks were developed [Silverstone and Nelson, *in press*].

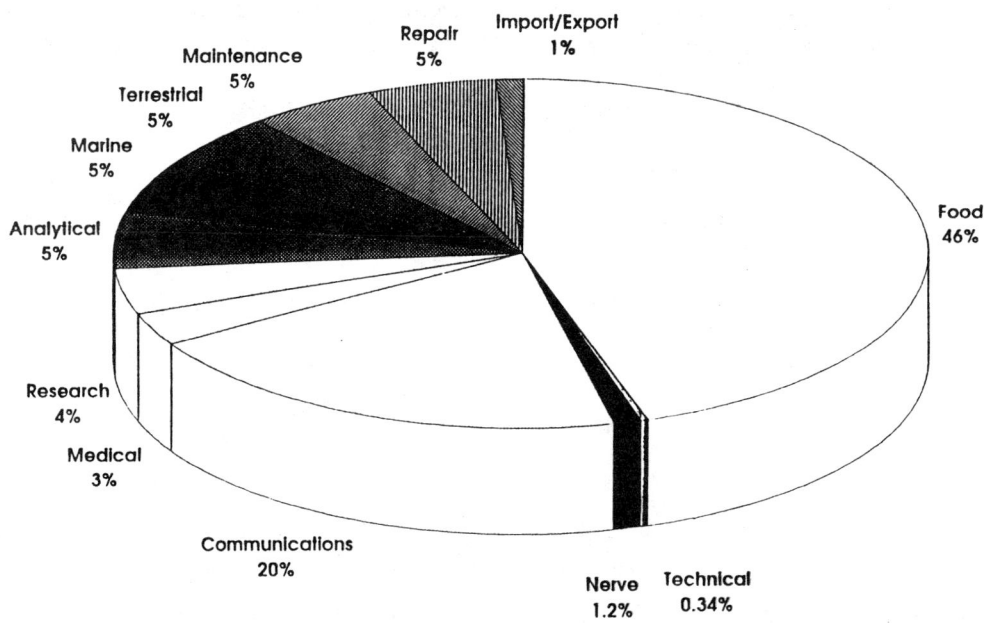

Figure 5 Biosphere 2 crew time spent on various tasks from March 1992 - February 1993 [Van Thillo *et al., in press*].

On average, two thirds of crew time was spent on operations and one third on research and communications as opposed to the 50/50 split originally envisioned. A decision was made halfway through the two year closure to export some of the laboratory equipment originally inside the system, and to permit scheduled exports/imports of scientific samples and instrumentation. This was done with the goals of reducing crew time required and to increase the amount of research that could be accomplished. Each crew member worked an average of 66 hours per week. The cooking duties were taken in turn, with each member doing three meals (one day) every eight days. This required on average eight hours of time. Sundays were days off for the crew, as were holidays observed on the outside—but every day food preparation, domestic animal tending and system checks were required [Van Thillo et al., in press; Alling and Nelson, 1993].

Material Closure and Determination of Leakrate

It is crucial to recognize when a closed ecological system is sufficiently closed to demonstrate its capabilities for fully recycling all materials or, conversely, the progressive increase or decrease of some. Biosphere 2 achieved a major step towards complete closure since its atmospheric leak rate is so small that material balances can be confirmed within narrow limits. The atmospheric leak rate has been measured by two independent methods at approximately 10 percent per year or less [Dempster, 1994] (see Figure 6). It should be explicitly noted in this context that an "atmospheric leak rate of X% per time period" means that over the given time period, X% of the atmosphere is replaced by foreign matter (ambient air) outside the enclosure, while (100-X%) of the atmosphere is matter originally contained in the system regardless of whether or not it has undergone transformations by chemical reactions. Thus when we speak of leakage we do not mean simply inward or outward leakage, but net exchange.

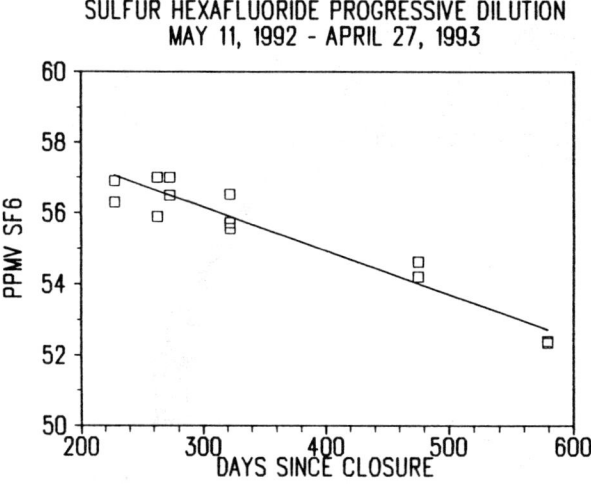

Figure 6 Sulfur hexafluoride progressive dilution May 11, 1992 - April 27, 1993. Determinations like this one with sulfur hexafluoride and helium establish that Biosphere 2's leak rate was less than ten percent during the two year closure, 1991-1993 [Dempster, 1993, reprinted with permission from SAE Technical Paper No. 932290 © 1993 Society of Automotive Engineers, Inc.).

One of the methods employed to determine leakage of Biosphere 2 was to spike the atmosphere with sulfur hexafluoride and with helium as trace gases and to measure their progressive dilution over time. A year's observation of these trace gases confirmed the 10 percent per year estimate [Dempster, 1993]. These two gases were selected, in part, as representatives of extreme ends of the spectrum of molecular weights; sulfur hexafluoride (mol. wt. 146) being heavy and helium (mol. wt. 4) being light. This selection is diagnostic of leak rate differences due to bulk leakage through flaws in the envelope vs. permeation through material or molecular diffusion through microscopic pathways. Since the progressive dilution of both gases was similar, it rules out the possibility that permeation or molecular diffusion are major contributors to the observed leak rate [Dempster, 1994].

The second method involved a test period of constant measured overpressure of about 150 Pa relative to the outside ambient air pressure. This caused outward leakage as observed by the rate of volume decrease of the two lungs, which in turn led to an estimate of the total cross-section of open leak pathways in the entire facility. Subsequently, the inside-outside pressure differential was controlled to vary within approximate limits of +/− 8 Pa which results in a calculable average rate of inward and outward exchange with a net volume change of zero. [A full discussion of this technique as well as the trace gas method is found in Dempster, 1994]. Both methods yielded initial estimates of about 5% per year exchange rate for Biosphere 2 with the higher estimate of 10% per year being attributable to excursions of the pressure control beyond the +/− 8 Pa range.

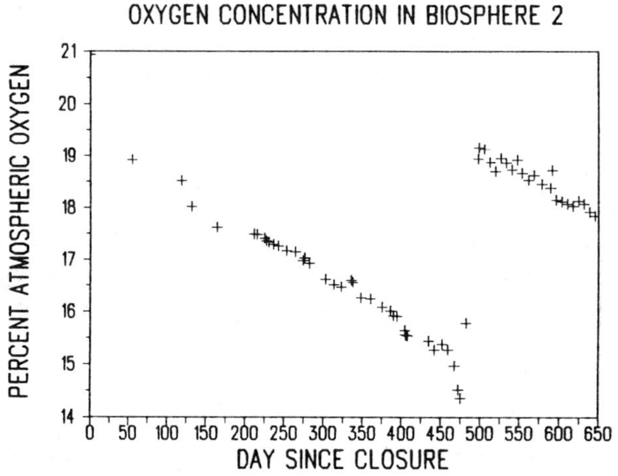

Figure 7 Oxygen concentrations in the Biosphere 2 atmosphere during the two year closure. The jump from approximately 14% to 19% shows the insertion of oxygen in January 1993, sixteen months after the commencement of the two year closure [Dempster, 1993, reprinted with permission from SAE Technical Paper No. 932290 © 1993 Society of Automotive Engineers, Inc.).

Oxygen Loss and Carbon Dioxide

After the initial closure in September 1991, there has been an imbalance between respiration and photosynthesis resulting in a decline in oxygen in Biosphere 2 (Figure 7). At first this appeared to be a mystery because atmospheric oxygen was observed to decline but the corresponding amount of CO_2 that would be expected from the respiration reaction did not appear in the atmosphere. The chemical equation of the reaction is

$$O_2 + CH_2O \longrightarrow CO_2 + H_2O$$

which shows that for each mole of oxygen lost there should be a mole of carbon dioxide produced and *vice versa* for the reverse reaction, photosynthesis. Thus, the sum of O_2 and CO_2 in the atmosphere should be constant. If the sum declines, then either O_2 or CO_2 can be regarded as "missing."

As previously described, the internal equipment of Biosphere 2 included a chemical scrubber to capture CO_2, but its operation only accounted for sequestering of about 1.6% of atmospheric carbon dioxide between September 1991 - June 1993, while about 12% of atmospheric oxygen disappeared during the same period. This left unaccounted more than a 10% loss of atmospheric oxygen.

A detailed investigation [Severinghaus *et al.*, 1994] employing isotopic analysis for carbon-12: carbon-13 ratios in soils, biomass, atmosphere, CO_2 scrubber product and the structural concrete of Biosphere 2 revealed that the interior structural concrete of Biosphere 2 had sequestered substantial quantities of CO_2 that roughly accounted for the "missing" amount by the reaction:

$$CO_2 + Ca(OH)_2 \longrightarrow CaCO_3 + H_2O$$

Other possible sinks for the missing oxygen or carbon dioxide include formation of caliche ($CaCO_3$ precipitate) in the soils, oxidation of reduced forms of nitrogen, sulfur, iron, or hydrogen. However, if such processes are occurring they appear to have played a secondary role to the amounts which are confirmed to have been taken up by the concrete.

The Biosphere 2 agricultural soil and top soil in the wilderness areas are highly enriched with organic matter which promotes rapid microbial respiration and evolution of CO_2, while the glazed structural envelope reduces the light available for plant growth, decreasing photosynthesis and slowing oxygen production. Future closed system experiments may strike a closer balance by using less rich soils and by using supplementary artificial lights.

Humans as Participants in Closed Ecological Systems

A somewhat subtle but important result of the two year closure experiment relates to the human dimension of living in a small biospheric system. As mentioned, the design of Biosphere 2 was motivated in part by the recognition that creating a place of beauty and high diversity of niche was important as the system is not only functional life support but effectively "the world" for the crew for the time that it is inhabited. Each of the eight biospherians of Mission One reported a heightening of awareness of their connection to this world. It is so small that every action is seen to have an impact—for better or worse—on its functioning. There are no "anonymous" actions—the feedback loops are virtually instantaneous. Nor can one mistake that an action in one part of the system will not have consequences elsewhere [Nelson and Alling, 1993].

In a paper written while still experiencing the reality of life inside Biosphere 2, two of its crew expressed it thus:

Our personal experience during the past nineteen months within this closed system has been extremely satisfying. Living as an integral component in our small world, both responsible for maintaining it and benefiting from its support, has been as rewarding as it has been challenging. It has changed our perspectives on the role of humans in all closed systems, whether they be artificial systems like Biosphere 2 or natural closed systems like Biosphere 1, our Earth's biosphere. We participate in a partnership with our biosphere to enhance its well-being by using our own resources, as well as by calling on an extensive network of scientists and engineers on the outside and employing technologies designed to assist in creating desired environmental conditions. There is a new harmony in this effort because our daily experience confirms the fact that we rely on the life systems for survival, and at the same time, the ecological systems depend on our efforts to maximize production and sustain overall health. In a small closed ecological system the equation 'our biosphere's health equals our health' becomes dramatically evident [Nelson and Alling, 1993].

STEPS TO MARS

The National Commission on Space, chaired by Thomas Paine, in its far-reaching vision of the next fifty years in space noted:

> A biosphere is an enclosed ecological system. It is a complex, evolving system within which flora and fauna support and maintain themselves and renew their species, consuming energy in the process. A biosphere is not necessarily stable; it may require intelligent tending to maintain species at the desired levels. Earth supports a biosphere; up to now we know of no other examples. To explore and settle the inner Solar System, we must develop biospheres of smaller size, and learn how to build and maintain them [National Commission on Space, 1986].

Some of the design of Biosphere 2 has been geared to optimize its value as a research tool for the study of ecosystem and biospheric functioning of Earth's planetary biosphere. Certainly, accommodating factors including the radiation environment, ambient atmospheric pressure and suitability of *in situ* materials for structure makes it unlikely that a biospheric system on Mars will look like Biosphere 2. However, the experience that has and will come from the operations of this first ground-based prototype of a permanent complex life and technical infrastructure yields valuable insights and data about the performance and stability of such systems.

Many of the innovative bioregenerative technologies that the Biosphere 2 project has developed may find application in the initial and near-term life support systems for early Mars exploration and settlement. Totally closed and recycling systems using bioregenerative technologies will probably evolve from and replace physico-chemical life support systems and partially bioregenerative ones. While the drawback of bioregenerative systems lies in their mass-requirements; the bulk of these are elements like water, soil, air that can be obtained from Martian resources. This will require the development of extraction techniques and bringing initial equipment to Mars.

Biosphere 2 used a soil-based system for the ecological functions that soil microbes play, and the ready completion of recycling steps. However, there will certainly be a place for hydroponic or aeroponic systems for food production in Mars habitations, especially in early stages of development. Banin has indicated that "on the basis of existing knowledge it is cautiously suggested that from the physical and chemical viewpoints, the Martian soil may constitute an appropriate medium for plant growth" [Banin, 1989; Banin *et al.*, 1988]. To supplement the plant nutrients already present in the Mars soil will require amendments with organic material and microbial inoculations. To a large extent this may be accomplished by the composting of flight and base crews' waste products. Then the types of systems pioneered by Biosphere 2: soil bed reactors, marsh wastewater systems and sustainable intensive agriculture may be constructed virtually entirely from local Martian resources.

The training of successive crews of "biospherians" in Biosphere 2 and in future testbeds in the operation of tightly integrated ecological and technical systems may also be very relevant to the adaptability that will be required of the Martian pioneers. They will make the transition to living—not simply exploring—in space.

REFERENCES

Allen, J., *Biosphere 2: The Human Experiment*, Penguin Books, N.Y., 1991.

Allen, J. and M. Nelson, *Space Biospheres*, 2nd Edition, London/Tucson, Synergetic Press, 1988.

Alling, A. and M. Nelson, *Life Under Glass: The Inside Story of Biosphere 2*, Biosphere Press, Oracle, AZ, 1993.

Alling, A., M. Nelson, L. Leigh, R. Frye, T. MacCallum, and N. Alvarez-Romo, "A summary of the Biosphere 2 test module research," appendix chapter in *Microcosms and Mesocosms in Scientific Research* by R. Beyers and H.T. Odum, Springer-Verlag, New York, 1993.

Alling, A., L. Leigh, T. MacCallum, and N. Alvarez, "Biosphere 2 test module experimentation program," in *Biological Life Support Systems*, M. Nelson and G.A. Soffen, eds., Synergetic Press, Tucson, 1990 and National Technical Information Service publication, NASACP-3094, pp. 23-32, 1990.

Augustine, M., "Biosphere 2: The closed ecology project," AAS 86-119, in *The Human Quest in Space*, G.L. Burdett and G.A. Soffen, eds., American Astronautical Society, Science and Technology Series, 65, pp. 243-254, San Diego, CA, 1987.

Averner, M., "The NASA CELSS program," in *Biological Life Support Systems*, M. Nelson, and G. Soffen, eds., pp. 45-46, Synergetic Press, Tucson, 1990.

Banin, A., "Mars Soil - A sterile regolith or a medium for plant growth?," AAS 87-265, in *The Case for Mars III*, C.R. Stoker, ed., American Astronautical Society, Science and Technology Series, 74, pp. 559-571, San Diego, CA, 1989.

Banin, A., G.C. Carle, S. Chang, L. Coyne, J. Orenberg, and T. Scattergood, "Laboratory investigations of Mars: Chemical spectroscopic characteristics of a suite of clays as Mars soil analogs," in *Origins of Life and The Evolution of the Biosphere*, Kluwer Academic Publishers, 18, pp. 239-256, 1988.

Bohn, H.L. and R.K. Bohn, "Soil bed scrubbing of fugitive gas releases," *J. Environ. Sci. Health,* A21, pp. 561-569, 1986.

Bolin, B. and R.B. Cook, eds., *The Major Biogeochemical Cycles and their Interactions*, John Wiley & Sons, N.Y., 1983.

Dempster, W.F., "Methods for measurement and control of leakage in CELSS and their application and performance in the Biosphere 2 facility," *Advances in Space Research*, 14, 11, pp. 331-335, 1994.

Dempster, W.F., "Biosphere 2: System dynamics and observations during the initial two year closure trial," *SAE Technical Paper Series, no. 932290*, presented at the 23rd International Conference on Environmental Systems, Colorado Springs, CO, July 12-15, 1993.

Dempster, W.F., "Effect of air leaks on ecosystem processes in closed ecological systems," paper presented to the ASHRAE Tri-County Chapter meeting Montebello, CA, December 3, 1991.

Dempster, W.F., "Biosphere 2: Technical overview of a manned closed ecological system," *SAE*, 19th Intersociety Conference on Environmental Systems, San Diego, CA, July, 1989.

Frye, R. and C. Hodges, "Soil bed reactor work of the environmental research laboratory of the University of Arizona in support of the research and development of Biosphere 2," in *Biological Life Support Technologies*, M. Nelson and G. Soffen, eds., NASA CP-3094, pp. 33-40, 1990 and Synergetic Press, Oracle, AZ, 1990.

Gitelson, J., "Methods for creating biological life support systems for man in space," in *Space Manufacturing 7: Space Resources to Improve Life on Earth*, 9th Princeton/AIAA/SSI Conference, B. Faughnan, and G. Maryniak, eds., pp. 237-239, AIAA, Washington D.C., 1989.

Glenn, E.P. and R. Frye, "Soil bed reactors as endogenous control systems of CELSS," in Workshop on Artificial Ecological Systems, Proceedings of Meeting in Marseilles, France, sponsored by DARA and CNES, pp. 41-58, October, 1990.

Glenn, E., C. Clement, P. Brannon, and L. Leigh, "Sustainable food production for a complete diet," *Hort Science*, 25, 12, pp. 1507-1512, December, 1990.

Hord, R.M., *Handbook of Space Technology: Status and Projecbons*, 242pp., CRC Press, Boca Raton, FL, 1985.

Knott, W., "The CELSS breadboard project: plant production," in *Biological Life Support Systems*, M. Nelson, G. Soffen, eds., pp. 47-52, Synergetic Press, Tucson, 1990.

Lebedev, K.A. and R.V. Petrov, "Immunological problems of closed environments and gnotobiology," *JPRS 54331*, National Technical Information Service, Springfield, VA, 1971.

Leigh, L., K. Fitzsimmons, M. Norem, and D. Stumpf, "An introduction to the intensive agricultural biome of Biosphere 2," in *Space Manufacturing 6*, B. Faughnan and G. Maryniak, eds., American Institute of Aeronautics and Astronautics, Washington D.C., 1987.

MacCallum, T., M. Nelson, J. Allen, L. Leigh, A. Alling, and N. Alvarez-Romo, "The Biosphere 2 project: Applications for space exploration and Mars settlement," (to be published in *The Case for Mars IV*, Science and Technology Series, T.R. Meyer, ed., American Astronautical Society, San Diego, CA), paper presented at the Case for Mars IV Conference, Boulder, CO, June 4-8, 1990.

Modell, M. and J. Spurlock, "Rationale for evaluating a closed food chain for space habitats," in *Human Factors of Outer Space Production*, Cheston and Winter, eds., AAS Selected Symposium, 50, 1980.

National Commission on Space, *Pioneering the Space Frontier*, Bantam Books, New York, 1986.

Nelson, M., W.F. Dempster, N. Alvarez-Romo, and T. MacCallum, "Atmospheric dynamics and bioregenerative technologies in a soil-based ecological life support system: Initial results from Biosphere 2," in *Advances in Space Research*, 14, 11, 1994.

Nelson, M., T. Burgess, A. Alling, N. Alvarez-Romo, W. Dempster, R. Walford, and J. Allen, "Using a closed ecological system to study Earth's biosphere: Initial results from Biosphere 2," *BioScience*, 43, 4, pp. 225-236, 1993a.

Nelson, M., S. Silverstone, and J. Poynter, "Biosphere 2 agriculture: Testbed for intensive, sustainable, non-polluting farming systems," *Outlook on Agriculture*, 13, 3, pp. 167-174, CAB International, Oxford, U.K., 1993b.

Nelson, M. and A. Alling, "Biosphere 2 and its lessons for long-duration space habitats," in *Space Manufacturing 9: The High Frontier - Accession, Development and Utilization*, 11th Princeton/AIAA/SSI Conference, Faughnan, B., ed., pp. 280-287, AIAA, Washington D.C., 1993.

Nelson, M., L. Leigh, A. Alling, *et al.*, "Biosphere 2 test module: a ground-based sunlight-driven prototype of a closed ecological system," in *Advances in Space Research XXIV (4): Natural and Artificial Ecosystems*, MacElroy, R., Averner, M. *et al.*, eds., pp. 151-156, Pergamon Press, 1992a.

Nelson, M., J.P. Allen, and W.F. Dempster, "Biosphere 2: a prototype project for a permanent and evolving life system for a Mars base," in *Advances in Space Research XXIV (4): Natural and Artificial Ecosystems*, R. MacElroy, M. Averner, *et al.*, eds., pp. 211-217, Pergamon Press, 1992b.

Nelson, M., "The biotechnology of space biospheres," in *Fundamentals of Space Biology*, G. Malacinski, ed., pp. 185-200, Japan Scientific Press, Springer Verlag, 1990.

Nicogossian, A. and J.F. Parker, *Space Physiology and Medicine*, NASA SP-447, pp. 285-292, Washington D.C., U.S. Government Printing Office, 1982.

Petersen, J., A. Haberstock, T. Siccama, K. Vogt, D. Vogt, and B. Tusting, "The making of Biosphere 2: frontiers in synthetic ecology," *Restoration and Management Notes*, 10, pp. 158-168, 1992.

Rippstein, W.J. and H.J. Schneider, "Toxicological aspects of the Skylab program," in *Biomedical results from Skylab*, R.S. Johnston and L.F. Dietlin, eds., NASA SP-377, Washington D.C., U.S. Government Printing Office, 1977.

Rummel, J. and T. Volk, "A modular BLSS simulation model," in Controlled Ecological Life Support Systems, R.D. McElroy and D.T. Smernoff, eds., COSPAR, *Advances in Space Research*, 7, 4, pp. 59-67, Pergamon Press, 1987.

Schwartzkopf, S. and H. Cullingford, "Conceptual design for a lunar-Base CELSS, in engineering, construction and operations," in *Space II, Proceedings of Space 90*, S. Johnson and J. Wetzel, eds., American Society of Civil Engineers, 1990.

Severinghaus, J.P., W.S. Broecker, W.F. Dempster, T. MacCallum, and M. Wahlen, "Oxygen loss in Biosphere 2," *EOS, Transactions of the American Geophysical Union*, 75, 3, pp. 33, 35-37, January 18, 1994.

Shepelev, Y. Y., "Biological life support systems," in *Foundations of Space Biology and Medicine*, Academy of Sciences USSR/NASA joint publication, Moscow/Washington, D.C., 1972.

Silverstone, S. and M. Nelson, "Food production and nutrition in Biosphere 2: results from the first mission," Sept. 1991 to Sept. 1993, presented at the COSPAR Scientific Assembly, Hamburg, July 1994, *Advances in Space Science* (*in press*).

Terskov, I.A., J.I. Gitelson, B.G. Kovrov, *et al.*, *Closed System: Man-Higher Plants (Four Month Experiment)* translation of Nauka Press, Siberian Branch, Novocibirsk, NASA-TM-76452, 1979.

Van Thillo, M., W.F. Dempster, N. Alvarez-Romo, G.Hudman, T. MacCallum, S. Silverstone, M. Nelson, and A. Alling, "From theory to practice: The initial two year closure of Biosphere 2," (to be published in *The Case for Mars V*, Science and Technology Series, P.J. Boston, ed., American Astronautical Society, San Diego, CA), paper presented at the Case for Mars V Conference, Boulder, CO, May 26-29, 1993.

Walford, R.L., S.B. Harris, and M.W. Gunion, "The calorically restricted low-far nutrient-dense diet in Biosphere 2 significantly lowers blood glucose, total leukocyte count, cholesterol, and blood pressure in humans," *Proc. Natl. Acad. Sci.,* 89, 11, pp. 533-537, 1992.

Wolverton, B.C. and R.C. McDonald, "The water hyacinth: from prolific pest to potential provider," *Ambio*, 8, 1, 1979.

Wolverton, B.C., "Aquatic plants and wastewater treatment (an overview)," chapter for Proceedings of a Conference on Research and Applications of Aquatic Plants for Water Treatment and Resource Recovery, Orlando, Florida, 1986.

Martian resources are tapped for living and fuel and stored in reusable tank farms.

Chapter 19

USING THE RESOURCES OF MARS FOR HUMAN SETTLEMENT

Thomas R. Meyer[*] and Christopher P. McKay[†]

A Mars exploration program that employs *in situ* resource utilization (ISRU) technology to produce bulk consumables such as air, water and fuel from Martian resources has the potential to be more robust, and economical, than programs that depend on bringing bulk consumables from Earth.

The key life-support compounds O_2, buffer gas for breathable air (Ar/N_2), and H_2O are available on Mars and can be extracted from the atmosphere. Water may also be available from soil water and ground ice. The soil can be used as radiation shielding and could provide many useful industrial and construction materials. The soil will also serve as a plant growth medium. Compounds with high chemical energy, such as rocket fuels, can be manufactured *in situ* on Mars. Solar power, and possibly wind power, are available and practical on Mars.

Preliminary engineering studies indicate that autonomous processes can be designed to extract and stockpile Martian consumables. The utilization of O_2, obtained from Martian atmospheric CO_2 will probably be practical even on initial human missions.

The ability to utilize these materials in support of a human exploration effort greatly expands the potential scope of human activities on Mars. ISRU is also the foundation for future expansion, self-sufficiency, and eventual habitation of Mars.

INTRODUCTION

Mars is unique in that it has both an environment that does not preclude humans and all of the resources necessary to support life in some accessible form on the surface. Thus, Mars is the best candidate in the solar system, outside of Earth, on which to establish self-sufficient human settlements.

A strategy for the exploration of Mars that utilizes resources from the Martian environment to support the exploration effort offers the potential for wider exploratory

---

*   Boulder Center for Science and Policy, P.O. Box 4877, Boulder, Colorado 80306.
†   Space Science Division, NASA Ames Research Center, Moffett Field, California 94035.

capabilities, longer stay times, lower transportation costs and increased safety over systems which are fully closed and/or must bring consumables from Earth. It also provides the foundation for a permanent self-sufficient scientific research outpost which could be the prototype Martian colony.

For human missions the production of oxygen is one of the most important applications of ISRU. The extraction of oxygen from atmospheric carbon dioxide may be practical from the very first mission. From the Mars atmosphere alone it is possible to produce all of the essential life support consumables, oxygen, nitrogen/argon buffer gas, and water (see Figure 1). Fuels such as carbon monoxide, methane and methanol as well as the necessary oxygen can also easily be made from the Martian atmosphere [Ash, 1978, Stancotti, 1979, Meyer, 1981, Hepp *et al.*, 1991, McMillen *et al.*, 1996]. Other vital consumables that can be made from the atmosphere and soil include fertilizers, construction materials, and many other industrial compounds [Meyer, 1981, Meyer and McKay, 1984]. Studies of possible methods for utilizing Martian resources have been conducted by Ash, 1978, Clark, 1984, 1985, 1989, Meyer, 1981, Meyer and McKay, 1984, 1989, Ramohalli, 1992, Hepp *et al.*, 1991, McKay *et al.*, 1993, and Stoker *et al.*, 1993.

Figure 1 Artists concept of a gas extractor at a Mars base (painting by Carter Emmart).

In this chapter we discuss the rationale and potential impact of using Martian resources on the exploration program, summarize our knowledge of useful Martian resources, estimate materials requirements for various mission elements, and we conclude with a review of resource utilization technologies that have been proposed for use on Mars.

IMPACT OF ISRU ON MARS EXPLORATION

The basic rationale for *in situ* resource utilization in the near term is that it leverages exploration capabilities and the potential for greater scientific returns. In the long run, ISRU is the central enabling technology for future expansion, self-sufficiency, and eventual habitation of Mars.

The use of Martian resources will have a significant impact on many aspects of the Mars exploration program. ISRU will influence the overall mission architecture, particularly the interplanetary and surface transportation elements. It will affect the design and operation of many of the major Mars base elements including habitats, greenhouses, life support systems, consumable gas generation and reservoirs, supply caches, power systems, fuels and energy production and storage, and construction systems. ISRU will also influence the types of research objectives and field activities that are feasible.

Because of the inherent remoteness of the actual Martian feedstocks and the possibility of unknown factors in the Martian environment, ISRU systems can not be fully tested on Earth. Therefore one of the early activities on Mars will be to deploy, test and integrate these systems, gradually making them operational as their reliability is proven. As implementation progresses, the incorporation of ISRU technology will increasingly alter the character of the program, the physical infrastructure on Mars, and the support requirements.

The ready availability of locally produced supplies of air and water could reduce the complexity of Mars surface habitat designs and their life support systems. With ISRU, the need for tight closure, total recycling and complex toxicogenic filtering of the air supply could be relaxed, thus allowing the use of simpler semi-closed life support systems where losses could be continuously made up from freshly produced air supplies. Toxic buildups could be managed through controlled leakage of the cabin air into the outside environment. Similarly, although it is very scarce on Mars, extreme measures for complete recovery of water could also be relaxed. Regardless of the design, habitats that must rely on fixed reservoirs of consumables will tend to be more complex, costly, and short lived, and will operate at a higher risk than those that employ ISRU. The primary tradeoff will be between the design features of the habitats and those of the ISRU systems which themselves may serve multiple needs. Thus, ISRU has the potential to be an enabling technology that can make long duration missions and permanent bases practical on Mars.

Food production on Mars could benefit extensively from the use of Martian resources, but except perhaps for small scale experimentation, the production of food will most likely commence at a later stage. Greenhouse atmospheres could be provided simply by compressing the raw Mars atmosphere which is 95 percent carbon dioxide, to a few hundred millibars. Supplements of oxygen and nitrogen extracted from the Martian atmosphere could also be added. Alternatively, greenhouses supplied with breathable air may be integrated with crew habitats as has been demonstrated in Biosphere 2. In either case, the initial charge and makeup gases needed by the greenhouses could be provided by ISRU systems. Moreover, the leakage in large structures is more difficult to minimize and thus ISRU systems may be the only practical way to maintain them. The Martian soil could be conditioned to serve as a plant growth medium and fertilizers could be produced from other constituents of the Mars atmosphere. The greenhouses themselves may utilize naturally protective land formations and their roofs may be partially covered with Mars soil to protect sensitive crops from radiation. Like other bulk consumables, the quantities of water required for Martian greenhouses would almost certainly be preferable to obtain locally rather than be transported from Earth. As exploration activity on Mars grows and the need for greenhouses increases, ISRU technology will facilitate this expansion and could provide the necessary supplements to make up for losses due to leakage and the export of food and other agricultural products for use outside of the greenhouses.

One of the largest potential impacts that *in situ* resource utilization can have on a Mars exploration program is on the transportation elements. Although the extent of its applicability depends on the chosen mission scenarios and technology to be implemented, chemical fuels production could be utilized by surface mobility vehicles, ascent/decent vehicles and interplanetary spacecraft. Chemical fuels can also be utilized for heating habitats, laboratories, and field stations and for powering heavy equipment such as drilling, mining, resource processing and construction machinery. Although both fuels and oxidants can be produced by processing the Martian atmosphere, more abundant sources of hydrogen either from Mars soil water or imported from Earth may be desirable.

ISRU has many other important applications that can be implemented as the need and overall supporting infrastructure on Mars grows. These include the manufacture of construction materials including insulation, plaster, cement, glass, ceramic and metals, fertilizers, energy storage compounds, explosives, reagents and process industry feedstocks.

Thus the development of resource utilization technology should become an on-going component of a Mars exploration program and will enable it to evolve toward self-sufficiency. Furthermore, the initial assessment and utilization of resources to provide life support at the first Mars base will provide the basis for determining if Mars has the inventories of water, nitrogen and carbon dioxide necessary for the fabrication of a habitable state for the entire planet—a possible long-term goal of human settlement [McKay *et al.*, 1991].

CRITERIA FOR EVALUATING RESOURCE UTILIZATION TECHNOLOGIES

In a systems level design, tradeoffs must be made among various approaches to the production of resources. In making these tradeoffs numerical comparisons or figures of merit are useful [Ramohalli *et al.*, 1992]. Two particularly important figures of merit for resource utilization are the mass gain factor, Γ, and the production energy, ε [Table 1]. The mass gain factor is the ratio of the mass of the consumable to be produced over the life cycle divided by the mass of the equipment and proportion of the power supply needed on Mars to produce it. The mass gain factor expresses the most critical reason for using Martian resources, namely to avoid the requirement of transporting the resources from Earth. Clearly, the mass gain must be large to make the risk and investments in *in situ* production worthwhile and it should be of order 100 or 1,000.

Table 1
KEY FIGURES OF MERIT IN RESOURCE UTILIZATION

Figure of Merit	Symbol	Definition	Desired Range
Mass Gain Factor	Γ	Total Mass Produced / System Mass	100 – 1000
Production Energy	ϵ	Energy to Produce 1 kg of Product	< 10 kW-hr/kg

The production energy represents the cost to produce resources on Mars. All methods for resource extraction considered have large energy requirements and since power will most probably be the limiting factor in the development of a Mars base, this figure of merit is particularly important to an overall systems design. Based on rough estimates of the resources demand [Meyer and McKay, 1984, McKay *et al.*, 1993] and the power available [Haberle *et al.*, 1993] we suggest that $\varepsilon > 10$ kW-hr/kg.

The most desirable applications for ISRU are those that have the largest requirements for bulk consumables. ISRU is also favored for missions that are longer in duration or wider in scope, where the ISRU facilities have a long lifetime, and where the same ISRU systems can be used by successive missions. There may, however, be other reasons for deploying ISRU technologies such as for crew safety or to enhance selected exploration capabilities that would outweigh unfavorable mass gain and production energy criteria.

MARS RESOURCES FOR EXPLORATION AND SETTLEMENT

A definitive survey of resources available on Mars is a prerequisite for exploration and settlement. Not only must the properties of the planet itself be assessed, but the atmosphere, possible availability of water resources, and possible sources of energy must be found.

Physical Properties of Mars

Mars is located 1.52 AU from the Sun, and its surface receives about half of the sunlight that is incident on the surface of Earth. The mass of Mars is about one tenth that of the Earth and the surface gravity is 0.38 of the Earth value. Because it lacks an appreciable magnetic field or a thick atmosphere, the surface of Mars is exposed to much higher levels of cosmic radiation, solar flare protons, and solar ultraviolet light than the Earth [Letaw *et al.*, 1987]. The length of the day is nearly identical to Earth but the seasons are about twice as long. The mean temperature on Mars is 215°K, but the temperature varies considerably with latitude and season. Because Mars also lacks an appreciable greenhouse effect, the temperature also varies considerably between night and day. The mean temperatures at the equator are about -60°C with daily maximums as high as +15°C and nightly lows of about −100°C [Kieffer *et al.*, 1977]. The physical properties of Mars are summarized in Table 2 [McKay, 1983].

Table 2
PHYSICAL PROPERTIES OF MARS

Property	Mars	Earth
Mass	6.418×10^{23} kg	5.983×10^{24} kg
Equatorial Radius	3397.2 km	6378.4 km
Surface Gravity	3.73 m s^{-2}	9.80 m s^{-2}
Escape Velocity	5.02 km s^{-1}	11.18 km s^{-1}
Radius of Stationary Orbit	20,430 km	42,240 km
Surface Temperature Range	130 to 300 K	
Mean Temperature	215 K	288 K
Surface Pressure	6 to 15 mb	1 bar
Albedo	0.25	0.3
Solar Day (1 "sol")	24h 39m 35.238s	24h
Sidereal Year	686.97964 days, 668.59903 sols	365.26 days
Obliquity of axis	25 deg	23.5 deg
Eccentricity	0.0934	0.0167
Mean Distance to Sun	1.52 AU, 2.28×10^8 km	1 AU 1.49×10^8 km

It appears from the Viking images that Mars never experienced large-scale horizontal lithospheric plate motions (plate tectonics), the most important producer of mineralization on Earth. However, Cordell [1985] has proposed an analog of Mars with Africa which has many examples of mineralization not related to plate tectonic processes. Most of the important features of Mars, crustal swells, rifts, volcanism, water, and stationary plates, are also displayed in Africa and are associated with mineral deposits. The major ore deposits of Africa are correlated with rifts and up welling of the mantle due to stationary geothermal hot spots. Mars may be geophysically similar in this respect. Additionally, deposits associated with Martian volcanism and impact craters are also likely [Cordell, 1985]. There are no direct measurements of the composition of major geological units on Mars; this will have to await future measurements, such as those planned for Mars Surveyer.

The Viking experiments only sampled the upper 30 cm of the surface, and the two sites were remarkably similar despite a separation of 25° in latitude and 180° of longitude. The material sampled at these sites is almost certainly a windblown mantle of dust that is probably uniformly spread over most of the planet. The depth of this æolian mantle is unknown.

The Atmosphere of Mars

The composition of the Mars atmosphere is predominantly carbon dioxide together with small, but potentially useful percentages of nitrogen, argon and water vapor. The values for the atmospheric gases on Mars listed in Table 3 are in terms of the volume fraction of the atmospheric components, except for the entry for dust which is the "clear" sky value expressed as a mass fraction. From these gases alone it is possible to prepare breathable air, water, rocket propellant, fertilizer, and other useful compounds and feedstocks.

Table 3
COMPOSITION OF MARS (Owen et al., 1977) AND EARTH AIR (Verniani, 1966)

Gas	Concentration	
	Mars	Earth
CO_2	95.3 %	0.03 %
N_2	2.7	78.08
Ar	1.6	0.93
O_2	0.13	20.9
H_2O	0.03*	~ 2*
CO	0.07	0.12 ppm
Ne	2.5 ppm	18 ppm
Kr	0.3 ppm	1.1 ppm
Xe	0.08 ppm	0.087 ppm
O_3	0.03 ppm*	40 ppb
Other		<7 ppm
Dust	~ 10 ppm/m*	—

* variable

The total pressure of the Martian atmosphere averages about 8 hPa* and varies from 6 to 10 hPa with season as the major constituent, CO_2, condenses to form the seasonal polar caps.

Atmospheric Dust

The dust loading in the atmosphere, m, is related to the optical opacity of the dust, τ, in the visible range, by the following equation derived by Pollack et al., [1979]

$$m = 5 \times 10^{-4} \tau \ g \ m^{-2}. \tag{1}$$

The value of dust loading in Table 3 is that associated with "clear" sky conditions between dust storms ($\tau \approx 0.5$). During a global dust storm the dust loading is typically 10^{-3} g cm^{-2}. The total column mass of atmosphere on Mars is about 20 g cm^{-2}. The e-folding time constant for the settling of the atmospheric dust is about 60 sols [Zurek, 1982]. The composition of the atmospheric dust is presumably similar to the small grained soil material analyzed at the Viking Lander sites.

Dust storms tend to originate in the southern hemisphere at times near perihelion and southern summer solstice. A database of occurrence of global storms is shown in Table 4 [Zurek, 1982]. High winds are associated with the lofting of dust during a major storm, but these winds occur only in the region of the initial storm; the dust is carried globally but intense surface winds do not accompany it. This is clear from the Viking

---

* 1 hPa = 10^2 Pascals = 1 mbar.

Lander results, which showed maximum wind speeds of about 10 m s⁻¹ even though two global dust storms occurred during the mission.

Table 4
DATE, L_s AND LOCATION OF MAJOR MARTIAN DUST STORMS FROM ZUREK (1982)

Date	L_s	Initial Location
1909 (Aug)	—	—
1911 (Nov)	—	—
1922	192	—
1924 (Oct)	—	—
1924 (Dec)	237	Isidis Planitia
1939	—	Utopia?
1941 (Nov)	—	South of Isidis
1943	310	Isidis
1956	250	Hellespontus
1958	310	Isidis
1971 (Jul)	213	Hellespontus
1971 (Sep)	260	Hellespontus
1973	300	Solis Planum, Hellespontus
1977 (Feb)	205	Thaumasia Fossae
1977 (Jun)	275	—
1979	225?	—

Sources of Water on Mars

There is extensive geological evidence that, at one time, Mars had a lot of water freely flowing on its surface [Carr, 1981, 1987]. This evidence includes the valley networks, which are dendritic runoff systems some of which suggest that there was rain on Mars. The valley networks appear to be quite old, about 3.8 Gyr.

Surface conditions today cannot support liquid water, therefore the runoff channels are evidence that the conditions on Mars have changed considerably over time. The outflow channels, apparently caused by rapid large scale flows of water, indicate that Mars' inventory of water is large.

There is no direct measurement of the amount of water on Mars and estimates vary considerably. Table 5 compares the amount of CO_2, N_2 and H_2O in Mars' atmosphere today, the amount that Mars should have if it formed from the same material that the Earth did (Earth scaling), and the range of scientific estimates of Mars volatile inventory [McKay and Stoker, 1989]. Of the values listed in Table 5 perhaps the most interesting are those based upon geological evidence for fluvial erosion; Carr [1986] has estimated that Mars had an initial endowment of water that would correspond to a layer over 1 km thick covering the planet. If Mars had this much water, then models of planetary formation suggest that it also had large amounts of CO_2 and N_2, much larger than observed in the present atmosphere.

It is probable that Mars still retains most of its volatiles. Although there has been non-negligible loss to space, the current loss rate of volatiles is insufficient to deplete these initial estimated reservoirs. If the current rate of atmospheric escape has remained constant for water (as 2H and O) at 6×10^7 cm^{-2} s^{-1} [Liu and Donahue, 1976, McElroy et al., 1977] and N_2 at 5.6×10^5 cm^{-2} s^{-1} [Fox and Dalgarno, 1983]; the total loss over the past 4.5 billion years would be 2.5 m of water and 1.4 hPa of N_2. While escape rates, particularly of N_2 may have been higher in the past (as much as 30 hPa of N_2 is estimated to have escaped, [McElroy et al., 1977], the current view is that the bulk of the Martian water is tied up in subsurface reservoirs, probably as permafrost [Squyres and Carr, 1986].

Table 5
ESTIMATES OF THE AMOUNT OF KEY COMPOUNDS ON MARS

	CO_2 (hPa)	N_2 (hPa)	H_2O (meters)
Amount in Mars' present atmosphere	~ 10	0.2	~ 7×10^{-6}
Original endowment, scaling based on Earth	27,000	300	1,200
Original endowment, range of current estimates	200 – 20,000	2 – 300	6 – 1,000

In terms of accessibility for utilization by human explorers, the sources of water on Mars may be divided into four categories: (1) the atmosphere, (2) the polar caps, (3) a deep subsurface permafrost or aquifer layer, and (4) the soil. All of these sources have been suggested as resources that could be utilized to support a human presence on Mars.

Atmospheric Water. Throughout most of the year, the Martian atmosphere contains as much water as it can hold with respect to nighttime temperatures. In fact, water frost was occasionally observed at Viking Lander 2 site at 48° latitude [Hart and Jakosky, 1986]. However, at the low Martian temperatures, even saturated conditions imply a very small amount of total water. The column amount of water determined from the Viking Orbiter as a function of latitude over the Martian year shows values ranging from 1 to 90 precipitable microns [Jakosky and Farmer, 1982]—the equivalent thickness of the layer if the water condensed into liquid. These values are 10,000 times smaller than the typical value on the Earth.

The largest amount of atmospheric moisture on Mars is found in the northern latitudes above 60°, and at low elevations in both hemispheres. Northern hemisphere spring is the wettest season. The annual average water vapor in the Martian atmosphere as a function of latitude and longitude [Jakosky and Farmer, 1982] shows a strong correlation with surface topography with lower elevations having much more water. The lowest place in the southern hemisphere, Hellas Basin (−45°S, 290°W) has the highest annual averaged water content in the southern hemisphere and at its peak has more than twice the atmospheric water than any other place on the planet [Jakosky and Farmer, 1982].

Lower elevations on Mars tend to be wetter both in the northern and southern hemisphere. The bottom of Hellas Basin is more than 4 km below the planetary reference level and thus, the pressure there is about 44% higher than the planetary average—reaching as high as 15 hPa.

Although small compared to Earth [see Table 5], the total amount of water in the Martian atmosphere, about 1.3 km^3 (~ 360 billion gallons), it is huge compared to the requirements of a human research base. For comparison, this amount of water could supply the needs of McMurdo Station, the main U.S. research base in Antarctica for 33,000 years—assuming peak summer usage with 1,000 people on station and no recycling [McKay, 1985]. Clearly, there is enough Martian atmospheric water to supply the needs of a research base.

The biggest obstacle to extraction of water from the Martian atmosphere is its low concentration. Table 6, adapted from Meyer and McKay [1984] shows the fraction of water at saturation in the Martian atmosphere over a range of frost points (the temperature at which the atmosphere is saturated). Pollack *et al.* [1977] determined a frost point temperature at Viking 2 of 195°K by monitoring atmospheric fog formation at night through its effect on the view of the Martian moon Phobos. The frost point was determined to be 191°K at Viking 1 and 196°K at Viking 2 by early morning plateaus in the daily temperature curves [Hess *et al.*, 1977].

Table 6
MAXIMUM WATER CONCENTRATION IN THE MARTIAN ATMOSPHERE VS. FROST POINT TEMPERATURE

Frost Temperature (°C)	Atmospheric Water	Partial Pressure of Water
−40°	1.6 %	129 μb
−50°	0.49	39.4
−60°	0.135	10.8
−70°	323 ppm	2.59
−80°	66.7 ppm	0.53
−90°	11.7 ppm	0.09

These frost points correspond to a partial pressure of water vapor over ice of between 0.04 and 0.09 Pa. At the time of these measurements the total vertical column abundances of water vapor [Farmer *et al.*, 1977] were equivalent to a layer of liquid 6 to 18 μm thick at the Viking sites. If this column amount of water was distributed with a scale height equal to that of the atmosphere it would represent a surface pressure of 0.02 to 0.07 Pa of water. For a nominal partial pressure of water of 0.1 Pa, in order to extract 1 kg of water the amount of Martian air that would need to be processed is 19,500 kilograms or 10^6 m^3—assuming 100% extraction efficiency.

Extraction of atmospheric water, even for long-term or large-scale utilization, would not present problems of depletion of a local resource since the atmospheric water would be resupplied by exchange with the polar caps and regolith. One clear advantage of the use of atmospheric water during the exploration phase is that this source is certainly the best characterized at the present time and will be even better understood if the Mars Surveyor mission is successful.

Polar Caps. The Martian polar caps are composed of two components: the CO_2 seasonal cap that forms in winter and the permanent cap, composed predominately of water ice, that persists throughout the year [Clifford, 1987]. The permanent caps are the potential sources of water.

The permanent north polar cap is very roughly circular and centered about the pole. It has a diameter of about 1,000 km, extending down to 80°N, with occasional ice deposits as far south as 75°N. The permanent south polar cap is much smaller, only about 350 km in diameter and is centered at 86°S, 30°W. The thickness of the polar ice deposits is uncertain. Estimates based upon the topology of surface features on the caps suggest that the northern cap has a thickness of 4-6 km at its center near the north pole. Similarly the south polar deposits are believed to reach a maximum thickness of 1-2 km. The wintertime temperatures at the polar caps tend to be close to the CO_2 frost points, ~ 150°K, and can reach upwards of 200°K in the summer.

Deep Subsurface Water. There is compelling evidence that Mars has a permafrost that is rich in water. Geomorphological and topographical features such as craters with rims that look like mud slides suggest the fluid-like flow of soil material and are thought to be the result of significant ice present in the soil. Such features tend to be found poleward of 40° in both hemispheres [Squyres and Carr, 1986], indicating permafrost in these locations.

The geomorphology can be used to crudely estimate the fraction of ice in these soils, and values of 5-10% have been suggested as possible. It was previously thought that this near-surface water-rich permafrost does not occur near the equatorial regions. However, recent modeling work by Paige [1992] suggests that, for possible ranges of surface conditions, ground ice could be stable at the equator. It is important to note that, unlike the strongly bound water of hydration observed in the soils at the Viking Lander sites, this water would be water ice. This ground ice does not necessarily exist right at the surface (note that it was not found within the top few centimeters at the northern Viking 2 site at 48°N) but would exist at depths below the limit of summertime defrosting—a few meters. At depths of a kilometer or more, temperatures reach levels high enough that melting would occur. It is possible that liquid water reservoirs exist at these depths. If the water contained salts, the resulting solution would be liquid at colder temperatures, resulting in liquid existing closer to the surface.

Soil Water. Data from the Viking landers shows that water was released from the soil upon heating to high temperatures [Biemann *et al.*, 1977]. The samples were obtained from the near surface and not kept cold before analysis—thus, any easily-bound or adsorbed water may have been lost. Upon onset of heating, most samples lost little H_2O at 200°C, but all evolved significant and approximately equal amounts at temperatures of 350°C and 500°C. The total amount of water released at 350°C was ~ 0.3% by weight and further heating to 500°C resulted in a net release of about 1%. Because these data are from the Viking GCMS (Gas Chromatograph Mass Spectrometer), which was not designed to detect water, these results are highly uncertain [Biemann et al., 1977] but the soil water may be an available resource and needs further evaluation. Nevertheless, the soil water content is believed to be higher than the GCMS results and may be as much as 15%. Stoker et al. [1993] suggest that salts could contain up to 10% water of hydration and clays, if present, could also contain large quantities of interlayer absorbed water.

The soil at the two Viking landing sites had very similar elemental composition which could suggest that a uniform mantle of æolian dust covers most of the Martian surface. Likewise, the similarity of the water analysis at the two Viking sites would suggest that similar amounts of water would be contained in the dust in other locations. The high temperatures required to release this water indicates that it is fairly tightly bound and it is therefore probably not affected by the range of temperatures occurring on the surface of Mars. Thus, it is probable that windblown soil anywhere on the planet will have similar properties [see, e.g. Stoker et al., 1993].

The possibility of fairly stable brine solutions at or just below the Martian surface has been suggested [Zisk and Mouginis-Mark, 1980] as a possible explanation of strong Earth-based radar reflectivities that are seasonally variable and indicate unusual smoothness near Solis Lacus and Noachis-Hellespontus. These same areas were suggested by Huguenin et al. [1979] to be possible sites of near-surface water activity. The stability of near-surface brine solutions has been discussed by Zent and Fanale [1986], who conclude that certain brines may exist for periods of 10^7 years under current conditions.

The Soil of Mars

The soil of Mars is the potential source of many useful resources. Its most important and immediate use may be to provide mass for radiation shielding. In addition, it could be a substrate for plant growth, and a source of water, building materials, and a variety of other useful compounds [Clark, 1979]. Direct measurement of the properties of the Martian soil is available only from the two Viking Lander sites. Based upon photogeological analysis of Viking orbiter imaging the likely distribution of many geological units has been identified. Units of volcanic (presumably basaltic), sedimentary, and windblown origin have been identified [Carr, 1981].

__Composition of Surface Material.__ Our understanding of the soil composition and structure on Mars is based primarily on the direct measurement of the elemental composition and soil properties at the two Viking landing sites. In addition, analysis of the SNC meteorites (thought to have originated on Mars) and ground based spectral observations provide information for current models of the Martian soil [Stoker et al., 1993]. Based upon photogeological analysis of orbiter imaging the likely distribution of many geological units has been identified. Units of volcanic (presumably basaltic), sedimentary, and windblown origin have been identified [Carr, 1981].

The elemental abundance of the loose debris at the Viking landing site was determined by the Viking X-Ray Fluorescence Experiment for elements with atomic number

12 (Mg) and above. Thus, there was no direct detection of such important elements as H, C, N, or O. The results are shown in Table 7 [Clark et al., 1977]. Most abundant were silicon (21% by mass) and iron (13%), both assumed to be in the form of oxides SiO_2 and Fe_2O_3, respectively. Magnesium, aluminum, sulfur, calcium, and titanium were also detected at the one to a few percent level. Sulfur is about 10 to 100 times more abundant than in terrestrial rocks and soils. Phosphorus, while not listed because its signal is masked by S and Si, is nevertheless thought to be present [Toulmin et al., 1976] especially since it is found in the SNC meteorites at concentrations of about 0.3% by weight [Stoker et al., 1993]. Thus, all the major elements of life (C, H, N, O, P, S) are present on the surface of Mars, and most of the trace elements required, such as Fe, Mg, Al, have been directly detected as well. Stoker et al. [1993] do suggest that potassium (0.08% by weight on Mars compared with ~ 0.83% on Earth) may be in short supply.

Table 7
ELEMENTAL COMPOSITION OF THE VIKING 1 LANDER SITE

Element	Percent by Mass
Mg	5.0 ± 2.5
Al	3.0 ± 0.9
Si	20.9 ± 2.5
S	3.1 ± 0.5
Cl	0.7 ± 0.3
K	< 0.25
Ca	4.0 ± 0.8
Ti	0.5 ± 0.2
Fe	12.7 ± 2.0
L*	50.1 ± 4.3
X**	8.4 ± 7.8
Rb	< 30 ppm
Sr	60 ± 30 ppm
Y	70 ± 30 ppm
Zr	< 30 ppm

* L is the sum of all elements not directly determined.
** If the detected elements are all present as their common oxides (Cl excepted) then X is the sum of components not directly detected, including H_2O, NaO, CO_2, and NO_x.

The mineralogical state of the elements in Table 7 was not directly detected, but based upon chemical considerations, the mineralogy of most of the elements has been inferred [Toulmin et al., 1977, Clark et al., 1982]. These results are shown in Table 8. The fine-grained debris at the Viking landing sites probably consists of palagonite [Singer, 1982], a weathered basalt (weathering product of mafic volcanic glass) or iron-rich clays such as montmorillonite or nontronite [Clark et al., 1977, Banin and Margulis, 1983], with minor amounts of kieserite ($KSO_4 \cdot H_2O$), calcite ($CaCO_3$), and rutile (TiO_2) [Stoker et al., 1993].

Table 8
INFERRED MINERALOGY OF THE VIKING 1 LANDER SITE

Compound	Percent by Mass
SiO_2	44.7
Al_2O_3	5.7
Fe_2O_3	18.2
MgO	8.3
CaO	5.6
K_2O	< 0.3
TiO_2	0.9
SO_3	7.7
Cl	0.7
SUM*	91.8

* Assuming K_2O is counted as O.

Significant quantities of adsorbed volatiles are contained in the Martian soil. Trace amounts of O_2 (70-790 nanomoles cm^{-3}) and other gases were released upon humidification [Oyama and Berdahl, 1977, 1979]. The soil is an unlikely source of O_2 since to supply 1 kg of oxygen by humidification would require over 40 m^{-3}, about 60 metric tons of soil (assuming 790 nmoles O_2 cm^{-3}). In addition, when heating to 500°C, soil samples released from 50 to 500 ppm CO_2 [Biemann et al., 1977]. This is too dilute to be a useful resource given the CO_2 in the atmosphere.

The release of O_2 upon wetting and the absence of organic material in the Martian soil have led to the suggestion that the Martian top regolith layer contains highly oxidizing compounds such as H_2O_2 and other peroxides and superoxides [Klein, 1978, 1979]. It is hypothesized that these could be produced by ultraviolet radiation (190-300 nm) reaching the Martian surface since there is no absorbing ozone layer in the atmosphere. If all of the oxygen (up to 790 nanomoles cm^{-3}) released from the soil was from the decomposition of H_2O_2, then the concentration of this oxidant in the soil would be ~ 30 ppm. The pH of the soil is not known.

Theories of the formation of Mars and its evolution suggest that there may be huge deposits of carbonate materials (as yet undetected). These carbonates may be locally concentrated in ancient dry lake basins and valleys on the Martian surface [McKay and Nedell, 1988]. Based on the Viking elemental analyses [Toulmin et al., 1977, Clark et al., 1982], the most likely carbonates to occur on Mars are $CaCO_3$ (calcite), $MgCO_3$ (magnesite), $CaMg(CO_3)_2$ (dolomite), $FeCO_3$ (siderite), and possibly $MnCO_3$ (rhodochrosite). Based upon estimates of the initial endowment of N_2 on Mars [McKay and Stoker, 1989] there may also be large quantities of nitrates associated with the putative carbonate deposits. In addition to volatiles bound into the soil in mineral phases (such as carbonates and nitrates), there may be significant amounts of CO_2 and H_2O adsorbed in a porous regolith layer. There may be as much as 4,000 kg m^{-2} adsorbed CO_2, which is equivalent to a surface pressure of over 100 hPa [Fanale and Cannon, 1979].

Physical Properties of Surface Material. The mechanical operations of the Viking Lander and remote observations from the orbiters have been used to determine the thermal and mechanical properties of the soil. These results have been applied to the design issues related to rover vehicles on Mars, in particular the tractability of the surface materials. Tables 9 and 10 list some of the properties of the soil. The soils of Mars are rich in salts. Clark and Van Hart [1981] suggest that the likely assemblages are dominated by $(Mg,Na)SO_4$, $NaCl$, and $(Mg,Ca)CO_3$. There appears to be a global crust of salt enriched materials on the surface [Jakosky and Christensen, 1986].

Table 9
THERMAL AND DIELECTIC PROPERTIES OF THE MARTIAN SOIL

Average Visible Albedo[a]	0.25, Range: 0.09 - 0.43
Surface Thermal Inertia[a] ($\sqrt{\kappa\rho C}$)	$1.6 - 11 \times 10^{-3}$ cal cm^{-2} s$^{-1/2}$ K^{-1}
Specific Heat Capacity[b], C	0.145 cal g^{-1} K^{-1}
Thermal Conductivity[b] (κ)	2×10^{-4} cal cm^{-1} s^{-1} K^{-1}
Thermal Emissivity[b]	~ 1
Dielectric Constant	2.3 - 3.5

[a]From Kieffer et al. (1977)
[b]Not measured, these are nominal values used by Kieffer (1977)

Table 10
MECHANICAL PROPERTIES
(page 4521, Moore et al., 1977)

Property	VL-1, Sandy	VL-1, Rocky	VL-2
Bulk Density, (ρ) kg m^{-3}			
Soil	1000–1600	1200–1600	1100–1480
Rock		2900	2600
Particle Size %			
> 2 cm	0	25	20
Clods and fines	100	75	80
Cohesion of soil, N m^{-2}	10–100	10–10^4	10–1000
Cohesion of rock, N m^{-2}	—	> 10^4	> 10^4
Angle of internal friction, deg	30–45	30–45	30–45
Penetration resistance, N m^{-3}	3×10^5	6×10^6	6×10^6
Adhesion, N m^{-2}	1–100	—	—
Coefficient of sliding friction	0.3–0.5	0.3–0.5	—

Energy Sources on Mars

Many of the energy sources available on Earth do not appear to be available on Mars. Geological information on geothermal energy sources on Mars is incomplete, but current estimates suggest that the geothermal heat flux is about 35 mW m^{-2} [Toksöz and Hsui, 1978] (compared to an average value on Earth of $\sim$ 80 mW m^{-2}). Due to the absence of plate tectonic activity there do not appear to be any localized concentrations of geothermal flux. Mars does not have an oxidizing atmosphere and is not thought to have vast reserves of buried hydrocarbons, such as oil or coal, that would burn in such an atmosphere. Thus native energy resources may be limited to two types only: solar power and wind power.

Wind Power on Mars. Wind power on Mars is possible since Mars does have an atmosphere. The feasibility of wind power on Mars has been analyzed by Haslach [1989]. The winds measured by the Viking Lander averaged about 5 m s^{-1}, with peak speeds up to 10 m s^{-1} (22 mph). From the data at these non-optimal sites (for wind power) and model results of the circulation of Mars' atmosphere, Haslach suggests that a well chosen site may have an average wind of about 14 m s^{-1}. The power, P, that can be extracted from the wind, per unit area, is given by

$$P = 1/2 \, \rho \, v^3 \qquad (2)$$

where ρ is the density of the Martian air and v is the wind speed. Although the density of the Martian air is over 100 times less than the air at sea level on Earth, Haslach's design could provide about 30 kW with a area of about 200 m^2.

Solar Power on Mars. Mars is further from the Sun than the Earth and thus receives less solar radiation. The values for the solar constant at the top of the Martian atmosphere averaged over the Martian orbit are listed in Table 11. The average value is 43% less than the corresponding value at the Earth's orbit. However the atmosphere and clouds on the Earth reflect about 30% of the incident sunlight while the transmission of the Martian atmosphere during clear conditions (minimal dust loading) is about 80% averaged over the day. Thus the amount of sunlight that reaches the surface of Mars is very close to 50% of the terrestrial value.

Table 11
THE SOLAR CONSTANT AT MARS ORBIT

Orbital Mean	590.0 W m^{-2} (Earth = 1371 W m^{-2})
Perihelion	718.0
Aphelion	493.0

Available Solar Energy. Since Mars' orbit has a high eccentricity, there is considerable difference in the solar radiation incident on the planet at perihelion and at aphelion. Perihelion occurs about 46 days before southern summer solstice. Furthermore the amount of sunlight that a given place on the planet receives varies seasonally due to the duration of sunlit hours and the solar zenith angle of the Sun at noon. The annual insolation for Mars is given in Figure 2 which shows the maximum, minimum and mean daily averaged insolation as a function of latitude [Haberle *et al.*, 1993].

The solar energy incident at the top of the Martian atmosphere is attenuated as it passes through the atmosphere. Radiation with wavelengths less than about 190 nm is absorbed by the atmospheric CO_2. The dust in the Martian atmosphere absorbs and scatters a significant fraction of the sunlight throughout the visible and near infrared region of the spectrum. Since the dust has a relatively high single scattering albedo, much of the light that is removed from the direct sunlight by dust will still reach the surface as scattered light. Thus it is important to include the scattered light in any computation of the total light incident on the surface of Mars.

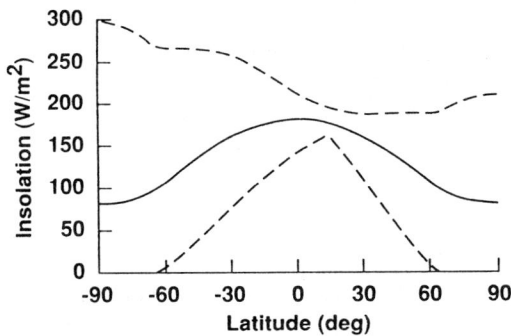

Figure 2 Mars annual insolation. At each latitude, the daily averaged insolation varies during the course of the year. Maximum (top dashed), minimum (bottom dashed), and mean (middle solid) daily averaged insolation are illustrated in this figure [from Haberle *et al.*, 1993].

The visible optical depth measured at the Viking landing sites during relatively clear periods was between 0.2 and 0.5 [Colburn *et al.*, 1989]. Haberle *et al.* [1993] suggest that because of the limited sample there may be even greater global variability than was recorded and that optical depths of 0.1 to 1 should be considered typical for the background dust haze. According to Hunt [1979], during global dust storms, some of which last for months, the optical depths can range from 1 to 6. Occasionally even higher values can be expected from local storms that typically last a few days [Haberle *et al.*, 1993].

Using irradiance data developed by Pollack *et al.* [1990], Haberle *et al.* [1993] have calculated that at normal incidence, for optical depths of 5, almost 200 W m^2 are still available. Even with the Sun only 30° above the horizon, 50 W m^2 are available. This is contrary to suggestions that dust storms would render solar-powered systems ineffective (especially if an all-sky, scattered light collector is used).

CONSUMABLES REQUIREMENTS ON MARS

Water, oxygen, and buffer gas are the critical life support requirements. Local sources of these consumables could provide inputs to closed life support systems, ameliorate the requirements and effects of total closure, make up for supplies lost due to consumption, internal absorption, leakage, and crew egress and ingress. They would also allow more extensive field operations and the use of systems where recycling of expended supplies is impractical.

To initiate food production in greenhouses on Mars, significant amounts of water and fertilizers will likely be necessary. Water and nutrients will be needed to make up for leakage and losses due to harvested food being taken away. The air supply will need to be adjusted according to plant requirements or to human requirements if the greenhouse is to be coupled with crew habitats.

One of the largest bulk consumables needed by Martian expeditions is rocket propellant for ascent vehicles. Propellants and fuels for rockets, surface vehicles, heating and electrical power generation can be made from Martian resources.

In this section we will discuss consumables requirements for life support, greenhouses, and transportation, and present a general requirements model to supply these and other components of a mature Mars base. For a broader treatment of life support requirements issues see Boston [Chapter 17, this volume].

Requirements for Life Support

All humans share a common need for food, oxygen and potable water. These requirements will vary among individuals and with the situation. Although we lack experience with how the amounts will vary on other planetary bodies with different environmental conditions, we have a fairly good understanding of the kinds of life support consumables needed and can be confident that these will not change. Therefore we can begin plans for how to provide these materials.

A representative list of life support inputs and outputs for a typical human being appears in Table 12 [MacElroy et al., 1992].* These needs for water depend sensitively on the design of the habitat, the technology incorporated in such items as clothing, dishes and toilets, as well as the level of fastidiousness of the crew. Obviously, these demands are highly uncertain; MacElroy et al., [1992] estimated the demand for washing water to be 20 kg per person-day. The life support requirements listed in Table 12 are similar to those used in a previous analysis by Meyer and McKay [1984] except that they included wash water (at 1.2 kg/person-day) in their analysis [see Figures 1 and 2 of Meyer and McKay, 1984].

Table 12
AVERAGE REQUIREMENT FOR HUMAN LIFE SUPPORT
(Adapted from MacElroy et al., 1992)

Inputs (kg/person-day)		Outputs (kg/person-day)	
Oxygen	0.83	Carbon Dioxide	1.0
Food		Solid Waste	
Dry weight	0.668	Dry weight	0.1
Water content	0.552	Water content	0.2
Water Intake		Liquid Wastes	
Drinking Water	1.85	Urine	1.5
Water added to Food	0.72	Perspiration	1.82
Total Water	3.122		3.52
Total Mass	4.62		4.62

---

*   A more detailed review of human life support requirements is given by Boston, this volume.

In addition to the consumables listed in Table 12, breathable air must contain an inert buffer gas to prevent oxygen toxicity and spontaneous combustion hazards. On Earth, of course, the buffer gas is N_2. CO_2 becomes toxic to the blood buffer system at levels of about 10 hPa [Parker and West, 1973]. The only candidates for buffer gas on Mars appear to be N_2 and Ar, the only inert gases with any appreciable concentration in the atmosphere (see Table 3). Nitrogen and Argon are both narcotic (well known to deep sea divers) but only above certain levels, ~ 300 kPa for Ar and ~ 400 kPa for N_2 [Parker and West, 1973]. These are well above the partial pressures they would have within the Mars habitat, 100 kPa (Earth normal) or less. Although the buffer gas is not consumed directly it will have to be resupplied against loss by leakage and airlock operations. Estimating this requirement is difficult but based upon early space station designs Meyer and McKay [1984] suggested a value of 0.56 kg/person-day.

It is important to note that humans are a net source of water as well as CO_2. Fundamentally this is a result of the metabolism of food to provide energy. This is best illustrated by considering the oxidation of organic material (represented elementally by CH_2O):

$$CH_2O + O_2 \longrightarrow CO_2 + H_2O + 524 kJ/mol \tag{3}$$

By comparing the water item totals in Table 12 it is apparent that the average human represents a net source of 0.4 kg/person-day. Thus, given efficient washing methods and recycling of wash and human waste water it is entirely possible that a human habitat supplied with food would not have any requirement for water whatsoever. Working with the figures in Table 12 we can determine the level of recycling of human waste water (3.32 kg/person-day) required to supply the potable water demands (2.57 kg/person-day). If only 77% of the human waste water were returned as potable water, then the human life support requirements for water *per se*, would be zero. Even if the food is completely dehydrated (water demand now 3.122 kg/person-day) the potable water requirements can still be met by recycling, now at a level of 94%.

In the initial phase of human exploration of Mars it is likely that food will be transported from Earth and that levels of recycling in the range of 90-95% can be achievable [Meyer and McKay, 1984, MacElroy *et al.*, 1992]. Thus, additional water will probably not be required as a life support consumable during this phase—although supplies of water may be needed for uses other than life support.

Of course, the above analysis depends on the fact that water is being transported to the habitat in the form of wet or dry organic matter—food. If food is produced locally on Mars then the water that is locked up in the chemical structure of the food must be provided to the plants. When a greenhouse is included in the mass budget of the life support system then water is required to make up for inevitable losses and non-recyclable waste material. Meyer and McKay [1984] have considered, in some detail, a possible life support system for a Mars habitat with and without a greenhouse included. They estimate that with a greenhouse the requirement for water is about 0.23 kg/person-day to make up for leakage and non-recyclable waste.

Requirements for Plant Growth and Greenhouses

After the initial phase of exploration, food production is likely to begin. Here we consider the consumables requirements for the long term operation of a greenhouse that may be part of a Mars base or settlement. There are many questions that have not yet been answered regarding how best to grow food on the surface of Mars, but we might begin by considering the relevant similarities and differences with Earth.

Mars has a rotation rate similar to that of the Earth and we assume that its surface gravity (0.38 g) would be adequate for long term biological adaptation. Since Mars is 1.52 times further from the Sun than is Earth, the amount of sunlight incident on the planet is only 43% of the terrestrial value. Even during dust storms the amount of sunlight incident on the Martian surface [Meyer and McKay, 1989, Haberle *et al.,* 1993] is greatly in excess of the minimum light levels needed for photosynthesis, so light will not be a limiting factor, *per se*.

However, a major uncertainty is whether crops can grow in greenhouses on the surface exposed to the harsh radiation environment. The thin atmosphere on Mars (0.2 meters equivalent thickness in the zenith compared to 10 meters on the Earth) and its negligible magnetic field provide little protection against cosmic radiation and solar flare protons. Humans must be protected from this radiation, particularly during times of strong solar flares, but six meters of Mars dirt covering the habitat section of the base would provide shielding equivalent to the column mass of the Earth's atmosphere. However, growing food requires large areas that may be impractical to cover with dirt. Furthermore, if the greenhouses are buried, light could become limiting and may have to be provided by artificial lighting or perhaps piped in with mirrors or fiber optics.

From an engineering point of view, surface greenhouses seem more desirable than those that are buried; they are simpler to build and maintain, can operate on natural lighting, and they can readily be expanded to accommodate base development. Engineering issues with surface greenhouses include making the shell strong enough to withstand the pressure differential, leak repair methods, availability of gas reserves in case of accidental air loss and compartmentalization to reduce risk, and thermal maintenance of the greenhouse.

The feasibility of surface greenhouses may depend on the development of crop strains that are tolerant of the Martian radiation environment. Radiation events that destroy the crops may be rare (perhaps once per solar cycle, 11 yr) and it may not be impractical to accommodate an occasional severe loss of crops given an ample reserve. Obviously, this method can not be applied to radiation shielding requirements for humans or for plants used as a source of future seeds [Boston, 1985]. For surface greenhouses relying on natural light, strong transparent plastics or other materials must be developed that can withstand the rigorous thermal regime and radiation—particularly UV—on Mars. These are important areas for further work if a Mars base is to become self-sustaining. No design studies of Martian greenhouses that include all the relevant aspects have been published. Figure 3 is an artist's conception of Martian greenhouses attached to a habitat that has been constructed from crew landing spacecraft.

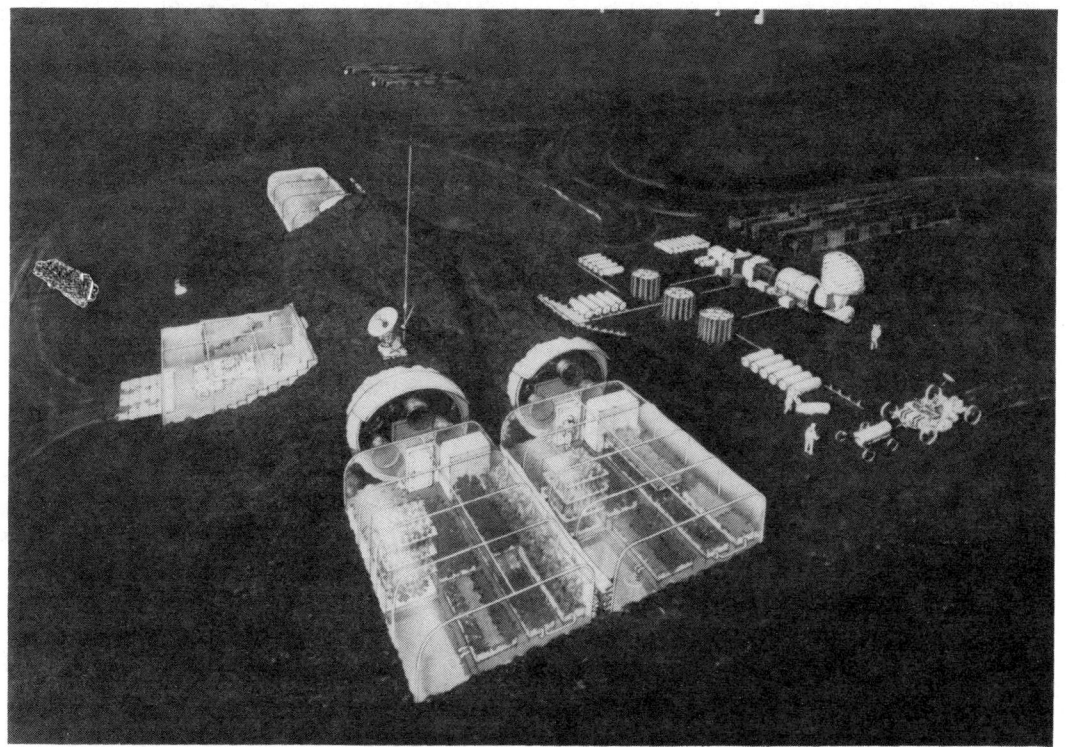

Figure 3 Artists concept of greenhouses attached to a Martian habitat (painting by Carter Emmart). The feasibility of surface greenhouses may depend on the development of crop strains that are tolerant of the Martian radiation environment. Strong transparent plastics or other materials must be developed that can withstand the rigorous thermal regime and radiation, particularly UV, on Mars.

Regardless of how the greenhouses are built, the plants will require CO_2, nutrient fertilizer, and water. The CO_2 can be obtained directly from the Martian air, only slightly modified since most plants show improved growth under high CO_2 concentrations. The CO_2 level on Earth today (~ 0.35 hPa) is low enough that certain plants (particularly C-3 plants) are often limited in their ability to obtain CO_2 for photosynthesis and about 0.15 hPa is a practical lower limit for CO_2 [McKay et al., 1991]. There is no clear upper limit on CO_2 for plants [Rogers et al., 1983] although concentrations over 1 hPa can have adverse effects on certain species [Hicklenton and Jolliffe, 1980, Huber et al., 1984]. Seckback et al. [1970] reported on photosynthetic algae that thrive under pure CO_2. While some O_2 is required by most plants for aerobic mitochondrial respiration, most higher plants actually prefer oxygen levels well below the current value. The requirements vary depending on the plant but, in general, net primary production increases as O_2 is reduced from the ambient level of 210 hPa until a point is reached where the O_2 concentration is low enough (~ 20 hPa) to cause metabolic complications [Salisbury and Ross, 1985]. It may be possible to adapt plants to accept even lower O_2 levels (~ 1 hPa) because the mitochondrial enzyme that requires O_2 has such a strong affinity for O_2 that it can function with O_2 levels of 0.1 hPa [Salisbury and Ross,

1985]. Nitrogen is required for all organisms and the amount of atmospheric nitrogen must be high enough to provide for biological nitrogen fixation. Bacteria can fix nitrogen at levels of 10 hPa and less [Klingler *et al.*, 1989].

As discussed above, food production in a greenhouse requires that water be provided as a life support resource, whereas without a greenhouse water recycling is probably sufficient to provide life support water given the net production of water by metabolism.

All the inorganic nutrients necessary for plant growth can probably be obtained from the Martian soil [Stoker *et al.*, 1993], and the waste products from the habitat can supply the organic fertilizer. Excess waste by-products of plant production and from the crew can be recycled by aquatic and microbial subsystems, decomposed by soil-bed reactors or disposed of outside of the greenhouse.

Using the soil as the plant growth medium may be more advantageous than hydroponics on Mars since the soil is ubiquitous and may provide a simpler, more reliable method. Without samples of the Martian soil it is difficult to determine how much initial soil preparation will be necessary, however, it is possible that it will need to be washed of salts, oxides, and toxins before being adjusted for pH and fertilized.

The average temperature within the greenhouse must be warm enough for liquid water and we assume that it must be greater than 10°C. Plants and anaerobic microbes can tolerate total pressures much lower than Earth normal (~ 1,000 hPa). In addition to a few hPa of N_2 and O_2 and 0.15 hPa of CO_2, the atmosphere need only support the vapor pressure of water at greenhouse temperatures (6.1 hPa at 0°C). Thus, pressures as low as 10 hPa may be permissible for an atmosphere suitable for plants and anaerobic microorganisms. Preliminary work on low pressure greenhouses [Boston, 1981, Schwartzkopf and Mancinelli, 1991, Daunicht and Brinkjans, 1992, Andre and Massimino, 1992] supports this possibility but more long term studies with diverse plants are needed. The important advantage of operating surface greenhouses at low pressure is that this reduces the structural strength required to maintain the pressure differential against the extremely low ambient pressure. The disadvantage is the necessity for humans to work in a spacesuit and the inability to couple common systems with the high pressure habitat and to benefit from organic process of oxygen production, carbon dioxide reduction, and toxicogenic control afforded by certain plant species.

In order to assure the stability of environmental conditions, Martian greenhouses will almost certainly require supervisory monitoring and control systems, mechanical subsystems for materials handling and processing, and significant expenditures of energy. If greenhouses are to be coupled to human habitats, even greater mechanical support measures and expenditures of energy are likely to be required since humans generally have narrower environmental tolerances than plants. The most desirable greenhouse would be an analog of Earth, a materially closed, bioregenerative, self-stabilizing biosphere driven entirely by solar energy, with no requirement for active monitoring and mechanical subsystems, but this goal is probably not realizable. A key reason for this is the inherently small scale of practical-sized greenhouses in comparison to the scale of the Earth.

One consequence of their small size is that greenhouses are unable to support natural solar driven weather systems that could provide thermal regulation, humidity control, and air and water transport. Some indication of the scale required to support actual weather systems can be gotten from considering the Vertical Assembly Building at NASA Kennedy Space Center. The VAB is the world's largest hollow building measuring 218 m x 179 m x 160 m high. It is known to have a micro-climate because it occasionally produces precipitation in the form of a thin mist, but its weather phenomena pale in comparison to Earth where cloud ceilings are typically measured in kilometers and the lateral dimensions of many storm systems are also of that same order. Thus in Martian greenhouses, the functions performed by weather phenomena will, for the most part, need to be performed by mechanical subsystems.

Another important consequence of their small size is the reduced ability of greenhouses to buffer dynamic mass and thermal energy flows. These arise, for example, from the dynamic gas production activity of the contained biomass, from impacts from farming practices, from external weather factors, and so on. A critical requirement is that the greenhouse system maintain environmental stability over the expected range of perturbations. In a passive system designed to rely on natural ecosystems operating in equilibrium, the size of the material and thermal buffers determines the degree of system stability and overall robustness. With decreasing scale, the size of the buffers in the greenhouse starts to become inadequate to maintain stable system operation within environmental tolerances. An inadequately buffered system becomes increasingly sensitive to perturbation, unstable and prone to crashing. An alternative approach to assuring system stability that will function in lieu of having large passive buffers is to use active monitoring, feedback, and mechanical support systems. For example, when sensors determine that biomass production has become inadequate to consume enough carbon dioxide in a greenhouse/habitat, CO_2 scrubbers might be activated, or if the CO_2 is too low, heating of the reservoir of stored carbonate could return it to the atmosphere. (Although Mars has abundant CO_2 in the atmosphere, the last step would presumably be needed to recover the CaO for future use.)

$$CO_2 + 2NaOH \longrightarrow Na_2CO_3 + H_2O \qquad (4)$$

$$Na_2CO_3 + CaO + H_2O \longrightarrow CaCO_3 + 2NaOH \qquad (5)$$

$$CaCO_3 + heat \longrightarrow CaO + CO_2 \qquad (6)$$

The penalty for use of an active system however usually translates into requirements for sophisticated monitoring systems, mechanical support equipment for materials processing, and perhaps significantly increased energy costs.

The need for this has been borne out by experience with Biosphere 2 in Oracle, Arizona, a materially closed biosphere/habitat designed to support humans, animals and plants and which incorporated seven different ecosystems containing a total of some 3,000 different species.

In Biosphere 2, the world's largest air tight building with a foot print of 12,766 m^2 and a volume of 204,045 m^3, the ratio of the atmospheric mass to biomass is some 350 times smaller than that of Earth [Fogg, 1994]. Thus the contained atmosphere, one of the key buffers in the Biosphere 2 system, is significantly less capable of stabilizing the partial pressures of O_2 and CO_2 produced by the greenhouse biota than is the external Earth environment. The production and consumption of these gases varies considerably with time of day and in response to seasonal biological growth and decay cycles. It is also perturbed by occasional impacts from pests, disease, harvesting and consumption, cultural activity, and accidents. Most importantly, it is a function of external environmental factors such as seasons, clouds, external weather, and Martian dust storms that control the solar energy budget and hence the temperature and light levels inside the greenhouse.

During the first closure of Biosphere 2, the use of mechanical scrubbers to remove excess CO_2 proved necessary. (For detailed discussion of Biosphere 2, see Nelson and Dempster, Chapter 18 in this volume.) In the winter when consumption outstripped production, the partial pressure of CO_2 rose to some ten times that of the outside air. Secondly, a significant and unanticipated sink for oxygen existed which required the importation of large quantities of supplemental oxygen in order to maintain the oxygen partial pressure at levels comparable to Earth-normal and to avoid a life threatening situation [Nelson and Dempster, Chapter 18, this volume].

Investigations suggest that soil microbes were primarily responsible for the oxygen loss, but the anticipated evidence of this, excessive carbon dioxide production, was masked by the absorption of a significant portion of the carbon dioxide into the newly poured concrete structural materials inside Biosphere 2 [Severinghaus *et al.*, 1994]—the carbon dioxide being absorbed in the curing process for Portland cement. The existence of the soil microbes themselves was attributed to the initial system having been biased by an overly rich endowment of organic rich top soil when Biosphere 2 was built.

While these difficulties are unlikely to recur in future biosphere designs, they illustrate the kind of unanticipated perturbations which greenhouses must be designed to withstand. A practical sized greenhouse, because of its scale, is vulnerable on two counts; first because human activity or intervention can easily have system-wide effects, and secondly, because its buffers, for practical reasons, may be proportionately many times smaller then those of the Earth. Thus, while it is advantageous to design greenhouses to rely on natural ecosystems operating in equilibrium, it appears they will almost certainly need to be assisted by active monitoring, feedback and control systems, materials processing subsystems, and significant inputs of energy. These may include mechanical watering and irrigation systems, heating, cooling and humidity control, systems for the production of oxygen, nitrogen and water, and scrubbers for carbon dioxide and trace contaminants. On Mars, added to this will be the energy costs of maintaining the internal pressure at a level at least several hundred millibars above Mars ambient.

With the notable exception of mechanical systems for materials production and pressurization, most of these systems were provided in Biosphere 2. The energy required to operate Biosphere 2 is supplied by an on-site natural gas power station. According to one working model, the energy required to operate Biosphere 2 is ~ 100 kWe/person. This compares with an estimate for a 150-person Mars settlement designed by the Japanese Ohbayashi Corporation of ~ 50 kWe/person [Ishikawa et al., 1990]. On Earth the normal values for the per capita use of primary energy such as available from fossil fuels, averages ~ 6 kW/person and ~ 0.7 kWe/person for the industrialized nations, and ~ 2 kW/person and ~ 0.2 kWe/person world-wide [Hall et al., 1993, Fogg, 1994].

On Mars the possibility of obtaining fresh supplies of resources from the local environment plus the possibility of dumping wastes outside of the greenhouse raises the possibility of using different strategies. Tradeoff studies will be needed to determine the best mix of internal recycling versus material exchange with the outside environment.

Requirements for Transportation

The transportation elements needed for Mars exploration include the spacecraft systems used for travel to and from the planet, plus the surface vehicles required for field research and support activities. Regardless of the exploration program architecture, any program that utilizes chemical fuels will require large quantities of propellant and oxidizers to be available on the Martian surface for use by the ascent spacecraft and surface vehicles. This fuel must either be imported from Earth or produced from Martian raw materials.

An estimate of the amount of fuel needed on the Martian surface is given in the engineering analysis of the propellant requirements for early Mars missions developed by Zubrin et al. [1991]. In his "Mars Direct" plan, two launches of a heavy lift booster from Earth are required for every four-person two-year mission. The first booster launch delivers an unfueled and unmanned Earth Return Vehicle to the Martian surface where it autonomously fills itself with methane/oxygen bipropellant manufactured primarily out of indigenous resources. Once it is fueled, the second booster brings the crew.

In this scenario, a feedstock of 6 tons of hydrogen, plus a small processing plant and a 100 kWe nuclear reactor are brought from Earth. These facilities are used to process carbon dioxide from the Martian atmospheric to produce 107 tons of methane/oxygen propellant. Of the amount produced, 96 tons is required to fuel the Earth Return Vehicle, leaving 11 tons for use for the extensive surface mobility required for Mars exploration during the nominal 2 year mission.

An alternative fuel that may be better suited for use in surface vehicles is methanol. Methanol can be stored as a liquid at nearly all Martian temperatures and pressures [McMillen et al., 1996, Clark et al., 1991]. It can be produced from atmospheric carbon dioxide and Earth-derived hydrogen or from indigenous Martian water. Its combustion is slightly less energetic than that of methane but it does not require bulky cryogenic storage on board the vehicle.

Presumably a nuclear reactor would be used to supply the energy needs of a Mars base, but heating requirements for emergency backup and for use in remote locations could be met by chemical fuels such as methanol which is easily stored.

Requirements Model for a Mars Base

Many different components of a Mars base could utilize locally produced consumables. The largest use for consumables will be for transportation, mainly for ascent vehicles. Surface transportation systems are likely to require on the order of a tenth as much. Consumables will also be needed for crew systems such as habitats, laboratories and greenhouses, and for EVA systems. A small amount may be needed for use with science packages and field science. As the base develops, potentially large quantities of consumables will be needed for mining, resource processing, manufacturing, construction, and for power and energy storage facilities.

Some of the consumables that might be produced include water, oxygen, nitrogen, for life support; chemical fuels and oxidizers for rockets and surface vehicles; fertilizers and greenhouse atmosphere gases; manufacturing and construction materials such as shielding, insulation, cement, plaster, ceramic/glass, metals, and sealants; consumables for machinery such as coolants and lubricants; reagents and feed stocks for materials processing; and explosives, drilling mud, water and leaching materials for mining operations.

Table 13
POTENTIAL RESOURCE USE BY MARS BASE COMPONENTS

Surface Component Consumables Requirements Model					
	Type of Consumable				
Surface Component	H_2O	O_2	N_2	Fuel	Other
Habitat	m	m	m	m	Shielding (l)
Greenhouse	l	m	m	m	CO_2 (l), NH_3 (l), NO_3 (l), Shielding (l)
Laboratory	m	m	m	m	Reagents (s), Shielding (l)
EVA Exposure Suit	s	s	s		
Portable Science	s			s	Reagents (s), Explosives (s)
Ascent Vehicle	s	l	s	l	
Ascent Veh. Servicer	m	m	m	m	Shielding (l)
Rover, pressurized	m	m	m	m	Lubricant (s), Coolant (s)
Rover, unpressurized				s	Lubricant (s), Coolant (s)
Balloon					H_2
Airplane	s	s	s	m	Lubricant (s), Coolant (s)
Heavy Eqmt				s	Lubricant (s), Coolant (s)
Drilling/Mining	m			m	Drill mud (l), Explosives (s), Lube (s)
Materials Processing	l	l	s	m	Reagents (l), Feedstocks (l)
Construction	m				Building materials (m), Sealants (m)
Power Facility	m				Working fluids (s), Coolants (s)
Chem Energy Storage					Reagents (l)

Amount (s, m, l) = small, moderate, and large volume users.

Table 13, shows a first order consumables requirements model for a Mars base. Life support compounds and chemical fuels are the materials most likely to be produced on Mars during the initial exploration phase. The indicated quantities are order of magnitude estimates of the amounts needed, and amounts may be highly scenario dependent.

UTILIZING THE MARTIAN RESOURCES

Preliminary engineering studies have identified candidate processing methods that could be adapted to Mars. Many of these are based upon industrial process technologies already in use on Earth. In this section we will review some of the candidate technologies that have been proposed.

Mining Martian Resources

Materials from both the atmosphere and soil will be needed as feedstocks for consumables production. Some requirements can only be met by using the regolith; for example, the use of the soil for shielding habitats from radiation. On the other hand, nitrogen, the buffer gas needed in a breathing mixture, is, as far as we know, only available from the atmosphere. Still other important compounds such as water can be obtained from both the atmosphere and the regolith, though at significantly different energy and logistic costs.

Obtaining resources from the soil generally involves an initial exploration effort followed by transportation of equipment to the mining site, excavation, and mechanical handling of ore and debris. Furthermore, it may be necessary to place mining equipment at multiple locations to obtain all of the required resources. Nevertheless, the soil provides the only source of metals, building materials, salts, and other solids and therefore the necessary mining and handling equipment will need to be developed. Water can be extracted from the soil, and although the processing of solid material to obtain water may be difficult, the proportion of water in the soil is much higher than in the atmosphere. It is likely that relatively pure ice deposits can be found in higher latitudes. Also, based on recent modeling work by Paige [1992], as much as 5-10% may even be found at equatorial latitudes.

By contrast, atmospheric mining has comparatively few logistic constraints on the emplacement of equipment. This allows scientific and other mission logistic priorities to prevail. Since the atmosphere is a ubiquitous gas, there is no requirement for exploration, mobility of mining equipment, excavation, or handling of regolith debris. Furthermore, atmospheric gas extraction equipment is more amenable to unattended operation. It would be feasible to land robotic atmospheric processing plants in almost any location to accumulate supplies of essential consumables in advance of the arrival of human crews.

The Martian atmosphere is well endowed with useful resources. It could supply water, breathable air, rocket propellant and may other compounds, (see Figure 4). Even ammonia fertilizer can be synthesized from the atmospheric nitrogen. But an important disadvantage is that the energy cost to obtain atmospheric water may prove to be high compared to extracting it from certain soil deposits; consequently any compounds containing hydrogen will be expensive to produce and recycling measures will be desirable. However, during the early stages of exploration, higher energy costs may be more easily accommodated than the complexity of surface mining operations and their associated logistical costs.

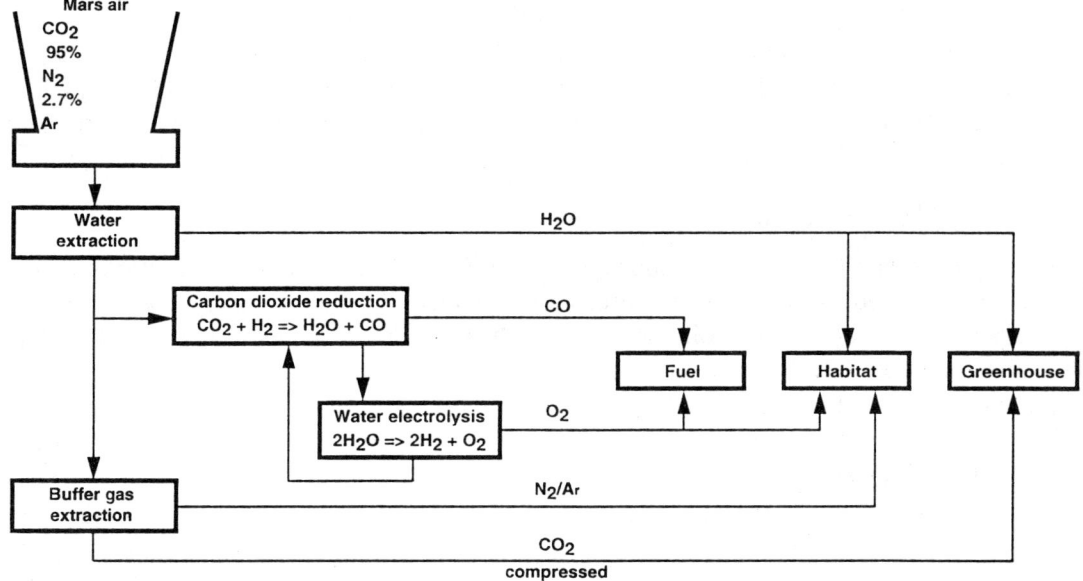

Figure 4 Gas Extractor Schematic Diagram. Martian air is compressed, cooled and processed in multiple stages to produce water, breathable air, and fuel for a Mars base. The water is precipitated out, nitrogen and argon is separated for use as a buffer gas, and carbon dioxide is reduced to produce oxygen and carbon monoxide fuel. Surplus compressed carbon dioxide may also be used to pressurize the greenhouse.

The logistic simplicity of mining the atmosphere makes possible the early implementation of *in situ* production of key consumables. This will enhance the mission and exploratory capability on Mars by providing more abundant supplies of consumables, by reducing the complexity of recycling equipment, and by reducing the cost of shipping bulk consumables from Earth. In the long run, after an infrastructure is established on Mars and the cost and difficulty of surface operations decreases, lower cost consumables derived from the soil will probably replace those mined from the atmosphere except for oxygen and nitrogen.

Resource Extraction and Processing

Many of the processes required to manufacture consumables on Mars can be adapted from conventional terrestrial process chemistry techniques. Tables 14 and 15 show compounds which can be made from the atmosphere and regolith along with representative processes for their manufacture. Some of the materials listed are likely to be produced only by a mature outpost or settlement after a substantial supporting infrastructure is established.

Table 14
MATERIALS WHICH CAN BE MADE USING MARS AIR

Material	Representative Process
H_2O	dehumidification of Mars air
O_2	reduction of CO_2, Sabatier process
N_2/Ar	liquefaction, fractional distillation
CO	reduction of CO_2, Sabatier process
H_2O_2	auto-oxidation, electrolysis
NH_3	electrosynthesis
N_2H_4	Raschig process
HNO_3	Oswald process
N_2O_4	produced from HNO_3
HCOOH	electrochemical reduction of CO_2
CH_4	catalytic hydrogenation of CO

Table 15
MATERIALS WHICH CAN BE PRODUCED FROM MARS SOIL

Material	Representative Process
H_2O	evaporation of indigenous ice/permafrost
H_2O_2	electrolysis of H_2SO_4 soln., vac. evap.
O_2	electrolysis of water
S	from sulfides, sulfates
Fe	from amorphous Fe-oxides, magnetic minerals
Ti	from titanomagnetite, ilmenite
Al	molten electrolysis of oxides
Mg	molten electrolysis of epsomite
Ceramic	from clay, SiO_2, Al_2O_3, H_2O
Glass	from SiO_2, Al_2O_3, MgO, CaO, K_2O
Duricrete	from silicates, salts, iron minerals, CO_2
Cement	from silicates, water
Plaster	from gypsum/calcium sulfate

Life Support Materials

Breathable air requires two components: oxygen and a suitably inert buffer gas. On Mars the most likely candidate buffer gas is a mixture of Ar and N_2, which together make up about 5% of the Martian atmosphere. Oxygen and water can also be obtained from the atmosphere. Table 16 lists the amount of Mars air that must be processed to obtain these materials.

Obtaining breathable air and water from the Martian atmosphere is appealing for many reasons. This resource is ubiquitous and fairly well described with present knowledge. Processing gases and liquids is much simpler and more reliable than handling solids. Furthermore the concentration of CO_2 and Ar/N_2 in the atmosphere is virtually constant with day and season. However, water vapor varies considerably.

Table 16
MASS FLOW RATES AND ENERGY REQUIREMENTS FOR RESOURCE EXTRACTION

Resource	Source	Amount to be processed	Energy required
1 kg Ar/N_2	Atmosphere	1,700 m^3	9.4 kW-hr[a]
			ideal = 0.1 kW-hr
1 kg H_2O	Atmosphere	10^6 m^3 (T_{frost} = 196 K)	103 kW-hr[a]
			ideal = 0.3 kW-hr
	Hydrated soil	100 kg	10 kW-hr[b]
1 kg O_2	Atmospheric CO_2	70 m^3	12 kW-hr[c]
			ideal = 2.5 kW-hr
	Soil oxidant	60 tons (O_2 = 790 nmoles cm^{-3})	?

[a] Meyer and McKay (1984)
[b] Stoker et al. (1983)
[c] see text

It is useful to consider the idealized work to obtain the needed life support resource from the atmospheric reservoirs. The ideal work to obtain buffer gas, water and oxygen from the Martian atmosphere is shown in Table 16.

Buffer Gas. Turning now to realistic engineering designs for obtaining these resources, Meyer and McKay [1984] have done a detailed study of the extraction of buffer gas from the Martian atmosphere and their design is still the best available for providing this important life support resource.

The main problem with providing buffer gas on Mars is in removing the unwanted CO_2. Fortunately CO_2 condenses at a much higher temperature than either Ar or N_2 and so this provides a practical method for separating these gases. The properties of CO_2 dictate the operational pressures and temperatures of the condensation system. In order to liquefy the CO_2, thereby simplifying the separation, it is necessary to compress the Martian air to above the triple point pressure of CO_2, 5.1 atm. The compression is the main energy requirement in the process. At this pressure most of the CO_2 condenses. The remaining CO_2 is then condensed to solid form by cooling via a Joule-Thompson expansion from 5 atm to a working pressure of 2 atm. Meyer and McKay [1984] estimated that the amount of work required to produce 1 kg of buffer gas by this method was 9.4 kW-hr. This is over a factor of 100 larger than the theoretical work because large amounts of Martian air must be compressed to extract the needed buffer gas. Meyer and McKay [1984] also considered the possibility of separating the Ar and N_2 if there was some requirement for this individual gas. This problem is similar to the separation of O_2 and N_2 from liquid air.

Water from the Atmosphere. The same basic compression, cooling and condensation process was used by Meyer and McKay [1984] in their design for extraction of water from the Martian atmosphere. Here the amounts of Martian air that must be processed are truly enormous (see Table 16). However, since water condenses at a higher temperature than CO_2 it is not necessary to liquefy the CO_2. Thus, the bulk of the Martian air that is compressed remains gaseous and, after the trace levels of water are removed by condensation, this compressed gas can be used to power regenerators as it expands back to the ambient pressure. This regenerates some of the energy expended in the initial compressors. However, because of the wide dynamic range of the compression scheme (1,500 to 1) the ratio of the size of the compressor stages varies considerably and the small stages at the high pressure end become a source of inefficiency. The total energy required to produce 1 kg of water was calculated to be 103 kW-hr. Clapp [1985] optimized the process suggested by Meyer and McKay [1984] and obtained a energy requirement of 69 kW-hr/kg water.

In these designs the main problem is the compression of the low pressure Martian air to a high operational systems pressure (5-10 atms). Table 17 [reproduced from Meyer and McKay, 1984] shows one scheme for achieving a 12 atm pressure on Mars. The process has been divided into four stages to improve the overall efficiency and to reduce the pressure rise in each stage to practical levels (~ 6 times). Note that the efficiencies span the range for very large compressors (0.8, Kerrebrock [1977]) to very small compressors (0.2, Ash *et al.*, [1982]).

Table 17
CALCULATION OF WORK TO ISENTROPICALLY COMPRESS MARS AIR FROM 0.008 TO 12 ATMOSPHERES IN 4 CYCLES [from Meyer and McKay, 1984]

Cycle	1	2	3	4
P inlet (atm)	0.008	0.05	0.3	2
P outlet	0.05	0.3	2	12
T inlet (K)	253	253	253	253
T outlet	373	371	378	366
Ideal work (kJ/kg)	106	106	110	101
Assumed efficiency	0.8	0.6	0.4	0.2
Real work	132	177	275	505

Total ideal work = 432 kJ/kg = 0.118 kW-hr/kg of Mars air processed
Total real work = 1089 kJ/kg = 0.303 kW-hr/kg of Mars air processed
Overall efficiency = 0.38

Because the buffer gas and water are minor gases in the Martian atmosphere it is possible that direct compression and cooling of the entire Martian gas mixture—basically the calculation presented by Meyer and McKay [1984]—are not the optimal way to separate these gases. It is possible that separation methods that directly select the target molecules are more efficient. There has been some preliminary work along these lines employing molecular sieves to extract the buffer gas [Jones *et al.*, 1985]. However, there is not a good estimate of the energy required to achieve the separation. For water there may be a variety of ways to preferentially select this molecule from the Martian atmosphere. This could include adsorption by molecular sieve substances, solution with hygroscopic salts such as PO_5, hydration of salts such as gypsum ($CaSO_4$), and possibly even hydration of the natural Martian soil. In each case, the hydrated material would have to be processed (presumably by heating) and the collected water harvested. The now anhydrous material would then be re-exposed to the Martian environment where it would passively collect a new batch of water. This concept is appealing in its simple and direct approach to extracting water but the kinetics and energetics of the system have not yet been sufficiently analyzed to allow a meaningful comparison to the compression/cooling methods.

Water from the Soil. One of the most important tasks of any future Mars base will be to obtain adequate supplies of water. Other than the atmosphere, there are three potential sources of water on the Martian surface: the polar caps, the permafrost, and water of hydration in the soil.

The water of hydration in the soil can be accessed from anywhere on the surface and will yield that water for a smaller energy cost per kilogram of water than will the atmosphere.

One possible method of extracting the water of hydration is to use microwave heating. Microwaves of the appropriate frequency can be used to selectively heat only the dipolar water molecules in the soil. This method of releasing water vapor will be significantly more efficient than doing so by convective heating of the entire bulk soil material. Engineering estimates show that a nominal microwave water extractor for Mars would dry about 100 kg of soil, yielding 1 kg of water in twelve hours using 1.0 kW of power [Gwynne, private communication]. This is more efficient than heating the bulk soil where the energy cost has been estimated by Clark [1989] to be 9 kWe/kg, and it is significantly more efficient than the energy cost of 102 kW-hr/kg obtained by Meyer and McKay [1984] for obtaining water from the atmosphere.

The Viking data indicate that abundant amounts of ice exist year-round at the northern polar cap of Mars. However mining ice from the poles may prove mechanically and logistically impractical. Jones *et al.* [1985] suggest that a small outpost be placed near the northern polar cap. From this location ice-harvesters could go out and collect ice to transport to a main base in equatorial regions. Finally, others speculate that in the future, a pipeline, similar to the Alaska pipeline, will be built to bring water to the lower latitudes of Mars.

If all that is needed is a few kilograms of water each month to offset leakage from a small manned base, then methods of obtaining water from the local environment would be preferable. On the other hand if one is gathering water to use for the production of rocket propellant to return people to Earth, then one will need thousands of kilograms [Baker and Zubrin, 1990]. This level of usage will require industrial scale equipment to obtain and process the water. At moderate volume, transportation costs are likely to exceed those for the actual extraction and processing. However with economies of scale, shipping water from polar sources may prove to be competitive with other methods.

There is evidence that Mars has a permafrost that is rich in water. Patterned ground and topographical features such as craters with rims that look like mud slides suggest the fluid-like flow of soil material, and are thought to be the result of significant ice present in the soil. Such features are found only poleward of 40° in both hemispheres [Squyres and Carr, 1986], indicating permafrost in these locations. The geomorphology can be used to crudely estimate the fraction of ice in these soils, and values of 5-10%

have been suggested as possible. Recent modeling work by Paige [1992] suggests that, for possible ranges of surface conditions, ground ice could be stable at the equator. It is important to note that, unlike the strongly bound water of hydration observed in the soils at the Viking Lander sites, this water would be simply water ice.

Several methods of extracting this water ice permafrost have been proposed. These include borehole heating, gallery heating, and excavation. From the standpoint of minimizing the moving of soil, the simplist method is borehole heating.

Borehole heating could be used in large or small scale operation. The basic implementation is to drill a five to fifteen centimeter diameter hole several meters into the ground. The hole is then heated by one of several methods. One can place a metal rod down the hole fitted with an electrical heating element. Alternatively a heat pipe could be inserted into the ground and heated by a solar concentrator focused at the top to warm the permafrost around the pipe. In any case water vapor would to rise to the surface where it could be captured in a tent.

Oxygen. Providing O_2 on Mars is much more straightforward than obtaining buffer gas or water since O_2 is to be found in the major constituent of the atmosphere, CO_2. Since the initial suggestion of Ash *et al.* [1978] there has been considerable effort directed toward understanding O_2 production from the Martian atmosphere. This work has resulted in a prototype system now operating at the University of Arizona. This system is based on the zirconia membrane separation process in which CO_2 at high temperatures (1,200°C) thermally decomposes to CO and O_2. The O_2 is then preferentially transported across the zirconia membrane [Ash *et al.*, 1978, 1982]. The currently operting unit produces 2 kg of O_2 per day using 288 watts of electrical and 2,000 watts thermal power—about 27 kW-hr kg^{-1} [Ramohalli, 1995]. As the unit is improved, particularly in terms of insolation and packaging, the thermal heat load should drop significantly. Frisbee [1986] estimated the power required for mature O_2 production systems based on this approach to be about 12 kW-hr kg^{-1}.

Agricultural Materials

Current progress is being made to condition lunar regolith into soil suitable for agriculture [Boston, Chapter 17 in this volume, Ming and Henninger, 1989]. Similarly, Banin [1989] and Stoker *et al.* [1993] have suggested that Martian regolith has potential for use as an agricultural medium. It is difficult to know how much initial soil preparation will be necessary before the Martian regolith can be used as a plant growth medium, however it is possible that it will need to be washed of salts, oxides, and toxins before being adjusted for pH and fertilized. Banin [1989] suggests that, with initial aqueous removal of excess sulfates and chlorates and injection of macro and micro nutrients into the soil, a viable growth medium for plants can be produced. Stoker *et al.* [1993] suggest that all inorganic nutrients necessary for plant growth can probably be obtained from the Martian soil and waste products from the habitat could be used to supply organic fertilizer. In any case, for the initial soil preparation on Mars, synthetic fertilizers in addition to other nutrients will probably be required to convert the regolith into a growth medium.

The only currently known source of nitrogen on Mars is in the atmosphere. The mechanical extraction of large quantities of atmospheric nitrogen can be done by compression and cooling to remove the carbon dioxide, followed by fractional distillation to separate out the Argon and other trace gases. Once having nitrogen and a ready source of hydrogen which should also be obtainable on Mars, it is possible to produce ammonia.

One of the most important processes for the production of synthetic ammonia is through the direct reaction of hydrogen and nitrogen. The reaction evolves 46 kJ/mole of ammonia is favored by low temperatures.

$$3H_2 + N_2 \longrightarrow 2NH_3 - 46 \text{ kJ/mol } NH_3 \qquad (7)$$

Once having ammonia, a host of fertilizer products can be manufactured. Using the Oswald process, ammonia together with oxygen can be used to produce nitric acid. From nitric acid and using additional ammonia, sodium carbonate, or calcium carbonate one can produce ammonium, sodium and calcium nitrates. From ammonia and carbon monoxide one can produce urea. The production of ammonium phosphates and ammoniated superphosphates is also straightforward. All of these methods can be adapted for Mars from commercial processes on Earth.

Once the Martian soil has been initially conditioned and chemically fertilized, for the process of soil revitalization, it is highly feasible that organic methods of nitrogenation can be employed to sustain the Martian agriculture.

Nitrogen is an essential component of amino acids, nucleotides and ammonia, all precursors of protein and nucleic acids, which are the blueprint for living organisms. Despite its biological importance, N_2 can only be converted into a useful form by a few species of bacteria. It is therefore necessary to salvage and reuse biologically available nitrogen through a cascade of events known as the nitrogen cycle.

Agricultural practices try to optimize the availability of nitrogen by applying either chemical fertilizer (nitrates, nitrites) or by rotating crops to replenish the nitrogen lost in crop yield. Whereas growth of a crop such as corn depletes soil of nitrogen, leguminous plants such as alfalfa, beans and peas replenish it by fixing atmospheric N_2 into biologically useful forms.

The process of nitrogen fixation is the first step in the nitrogen cycle. It involves the cooperation of the host plant which provides the material that the bacteria lack, and root-nodule bacteria, which supply nitrogen-fixing enzymes. Further nitrification by soil bacteria, in which the bacteria derive energy by oxidizing ammonia, results in the formation of nitrite (NO_2) and nitrate (NO_3). Both plants and bacteria have nitrate reductases that then reduce nitrate back into ammonia [Lehninger, 1982].

Thus, Mars agricultural operations need not be permanently dependent on chemical fertilizers. Given the high cost of operating *in situ* processing facilities, it is highly desirable to use natural processes to revitalize the soil where possible. Nevertheless, operations on Mars are likely to be undergoing constant expansion for a long time to come, so the capability to manufacture chemical fertilizers should become a permanent part of the surface infrastructure.

Chemical Fuels

Our understanding of the environment and climate history of Mars obtained from the Mariner and Viking data suggests that over the past 4 billion years conditions on Mars have not been suitable to support life. Life is the principal mechanism by which stored energy resources such as hydrocarbon deposits and atmospheric oxygen are created on a planet. Planets without life tend to an equilibrium state, where all of their reactions are spent, and where the atmospheric gases and surface materials have reached their lowest energy state.

Thus, we should not expect to find energy resources on Mars such as petroleum, natural gas, or other deposits of organic materials from which to manufacture propellants, and, as our observations show, there will be essentially no free atmospheric oxygen to use as an oxidizer. As a consequence, all fuels that we might wish to manufacture from Martian raw materials will require as much input energy to produce as the resulting fuel will yield upon combustion, less process inefficiencies. In spite of these disadvantages, the cost of shipping propellant from Earth to the surface of Mars is so great that *in situ* propellant production is a highly attractive alternative. In fact, the propellent needed on Mars for Earth-return spacecraft represents the single largest bulk consumable required for human exploration missions.

One way to estimate the savings realizable from producing propellant on Mars is to consider the amount of initial mass in low Earth orbit (IMLEO) required to ship one ton of mass from the surface of Mars to Earth. Using optimistic assumptions including aerobraking at Mars and no allowance for fuel tank mass and engines, Landis [1992] estimates that for every ton of payload returned from the surface of Mars to the vicinity of Earth it requires about 2.6 tons of fuel on Mars. Shipping this fuel to Mars requires an additional 4.3 tons in Low Earth Orbit (LEO). Thus manufacturing the return fuel on Mars would reduce the IMLEO by nearly a factor of seven. Zubrin [private communication], using the same assumptions but including gravity and aerodynamic losses in the assumed return ΔV from Mars surface to trans-Earth injection, obtains a savings factor of nearly 10. Although these values will vary with the type of hardware and mission scenario, the fact is these numbers are only theoretical and do not include tankage (mass of engines, fuel tanks, etc.). For an actual mission they will be significantly higher, but even a savings factor of 7 to 10 represents a huge savings if fuel for the Earth-return vehicle can be manufactured on Mars.

To complete this picture, however, we must also consider the amount of the fuel needed to ship a ton of payload (the return fuel needed on Mars) from the Earth's surface to LEO. The Russian Energia, one of the most capable heavy lift launchers in the world, has a mass at takeoff of 2,400 tons and can deliver a payload of 100 tons to LEO, a ratio of 24 to 1. A Titan 3 has a mass at takeoff of about 680 tons and can deliver 15 tons to LEO, for a ratio of about 45 to 1. Assuming about 90% of the launch mass is fuel and taking an average of the ratios for the two vehicles, a reasonable figure for the number of tons of fuel required to lift one ton of payload from the Earth's surface to LEO is about 30.

So combining these two factors and using Zubrin's number, we see that for every ton of propellant manufactured on Mars, this represents a savings of about 300 tons of fuel at Earth launch. Considering that an Earth-return spacecraft on Mars requires roughly 100 tons of fuel to return a crew of 4 persons to Earth [Zubrin *et al.*, 1991], the ability to produce fuel from indigenous resources on Mars translates into a savings of some 30,000 tons of fuel per return trip. In other words, *in situ* propellant production on Mars makes the difference between a human exploration program that is financially feasible and one that is infeasible.

The most valuable resource on Mars is water. This is true, in part, because of its scarcity but also because of its importance to life, and to a vast array of material applications. As discussed above, water is known to be obtainable from the atmosphere, the soil and the polar caps, but from none of these without an energy cost and, in the case of the latter two, a significant logistical effort.

Using electrolysis and a liquefaction plant it would be possible to produce LH_2/LOX (hydrogen/oxygen) fuel on Mars.

$$2H_2O(l) \longrightarrow 2H_2 + O_2 + 571 \text{ kJ/mol } O_2 \tag{8}$$

But even though LH_2/LOX has the highest specific impulse (I_{sp}) of all the candidate chemical fuels, the relative difficulty of liquefaction, handling and storage of the hydrogen, makes it unattractive, particularly for such a remote application. Moreover, the relative scarcity of hydrogen on Mars motivates the consideration of alternative hydrocarbon fuels. The virtue of hydrocarbon fuels is that they offer, along with a relatively high I_{sp}, a high hydrogen leverage; that is where the ratio of the stoichiometric mass of the hydrocarbon fuel to the mass of the hydrogen is high compared to that in hydrogen/oxygen fuel. Hydrocarbon fuel is particularly desirable for use as rocket propellant where there is no opportunity to recover and recycle the hydrogen from the exhaust as there might be from a rover.

Another abundant and versatile Martian resource is atmospheric carbon dioxide. Carbon dioxide is uniquely important to Mars because it is a resource for manufacturing both fuels and oxidant, and it is ubiquitous. Given a source of hydrogen, carbon dioxide can be reduced by a variety of methods to produce oxygen and carbon monoxide. From carbon monoxide, using Fishcer-Tropsch chemistry, many hydrocarbon fuels and higher hydrocarbons can be synthesized.

From carbon dioxide alone, it is possible to produce carbon monoxide/oxygen fuel which is a viable candidate for powering Martian Earth-return spacecraft, (see Figure 5). This fuel has the advantage that it does not require the use of scarce hydrogen and it can be made directly from easily obtainable atmospheric CO_2.

$$2CO_2 \longrightarrow 2CO + O_2 + 566 \text{ kJ/mol } O_2 \tag{9}$$

While CO/O_2 has relatively poor performance (I_{sp} 259 sec) compared to hydrocarbon fuels, the escape velocity on Mars is sufficiently low that practical spacecraft designs are feasible. The vehicle mass ratio* for a two-stage CO/O_2 powered Earth-return vehicle is 3.39 [French, 1985]. This is very competitive with single-stage spacecraft designed to use hydrocarbon fuels (6.34 for CH_4O_2), but the simplicity of a single stage vehicle is generally preferable even if the mass ratio is somewhat greater. Carbon monoxide/oxygen fuel could also be used in surface vehicles. Both fuel components would need to be tanked as compressed gases since their boiling points are below Mars coldest temperatures.

Figure 5 Artists concept of an Earth Return Vehicle being refueled on Mars using fuel produced from indegenous resources (painting by Carter Emmart). For every ton of material that can be produced on Mars, roughly 300 tons of fuel that would have been needed to transport it from Earth are saved.

Certain requirements on Mars are likely to generate carbon monoxide as a waste product; for example the reduction of CO_2 to produce oxygen for habitats. The carbon monoxide represents valuable stored energy if it can be utilized (283 kJ/mol, depending on the accounting), but to extract useful work requires yet another oxidant. Occasions where oxygen is a waste product are perhaps rare, but one example would be from the electrolysis of water to prepare hydrogen for use in dirigibles or balloons. On Mars hydrogen is probably the only option for use in balloons since sources of helium are normally associated with fossil fuel gas deposits. Hydrogen has the disadvantage that it embrittles some metals but it is otherwise safe to use since Mars does not have an oxygen atmosphere.

---

* Mass ratio is defined as the mass at liftoff divided by the mass remaining when all of the fuel is expended.

Carbon monoxide also has other valuable applications not related to propulsion, for example as a reducing agent for the winning of metals from metal ore oxides as discussed below. These in turn may yield valuable oxygen useful for propulsion.

Three hydrocarbon fuels that have been the focus of study for applications on Mars are methane, methanol, and acetylene [Ash, 1978, Baker and Zubrin, 1990, McMillen *et al.*, 1996, Landis, 1992].

The feasibility of producing methane/oxygen fuel from indigenous Martian materials has been proposed by Ash [1978] and Stancati [1979] and has been studied extensively by others. Baker and Zubrin [1990] have developed a detailed Mars mission architecture called "Mars Direct" based on *in situ* production of methane/oxygen propellant.

Methane can be easily produced using the Sabatier reaction in which hydrogen and atmospheric carbon dioxide are reacted in the presence of nickel cluster catalysts at approximately 400°C. The reaction is exothermic, goes to completion and can be controlled to minimize unwanted by-products.

$$CO_2 + 4H_2 \longrightarrow CH_4 + 2H_2O(g) - 164.8 \text{ kJ/mol } CH_4 \qquad (10)$$

In this process, half of the reacted hydrogen goes to methane and the other half goes to water. By splitting the resulting water by electrolysis, the oxygen produced can be stored and the resulting hydrogen becomes available to be recycled back to the Sabatier reactor. By iterating this cycle, essentially all of the original hydrogen will be shifted into methane. The net result is that all of the hydrogen is fully utilized, however the amount of oxygen produced is only half of the stoichiometric quantity required for combustion of the methane.

$$CH_4 + 2O_2 \longrightarrow CO_2 + 2H_2O(g) - 802 \text{ kJ/mol } CH_4 \qquad (11)$$

Thus it is necessary to produce half again as much oxygen by some means, (or a bit less since methane engines are usually operated slightly rich, CH_4/O_2 1:3.6 [Zubrin *et al.*, 1991]). One option for producing additional oxygen would be to bring additional Earth-derived hydrogen in order to produce additional methane and water.

Yet another option that requires only additional power but no additional hydrogen is to use the same hydrogen twice. This is accomplished by pyrolyzing the methane produced in the process described above, yielding elemental carbon and recovering all of the original hydrogen.

$$CH_4 \longrightarrow C + 2H_2 + 793 \text{ kJ/mol } CH_4 \qquad (12)$$

Now the Sabatier methanation and electrolysis steps as described above are repeated a second time thus restoring the original quantity of methane while producing again as much oxygen [Zubrin *et al.*, 1991]. An advantage of this approach is that none of the steps involve incomplete reactions where hydrogen may be lost, but a disadvantage is that a large quantity of solid carbon waste is produced that must be handled.

Another option is to use the zirconia membrane separation process proposed by Ash [1978] for the reduction of atmospheric carbon dioxide discussed elsewhere in this paper. Perhaps the most revolutionary option would be to use an as yet undemonstrated method described by Hepp *et al.* [1991] for the photochemical reduction of carbon dioxide driven by the solar ultraviolet flux incident on the Martian surface.

In the event that methane/oxygen fuel is produced using indigenous water on Mars, the problem of producing extra oxygen is resolved because it is obtained from the initial electrolysis of the water resource. Figure 6 illustrates the process for the production of methane/oxygen fuel from Martian water and carbon dioxide.

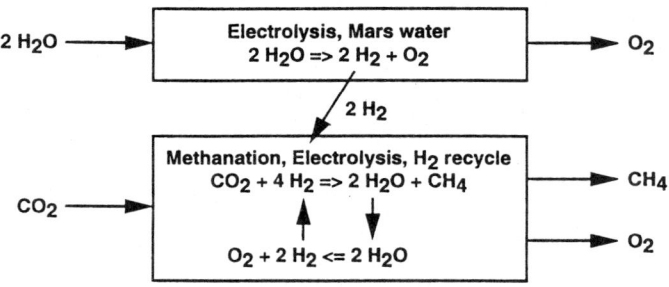

Figure 6 Methane/oxygen synthesis using Mars water and atmospheric carbon dioxide. Indigenous Martian water and atmospheric carbon dioxide can be used to produce methane/oxygen propellant. Since rocket engines operate best when run slightly methane rich, a small amount of surplus oxygen is produced by this process.

It is useful to consider the relative merits of different fuels that might be produced on Mars using Earth-derived hydrogen. The principal motivation for bringing hydrogen from Earth is to avoid the added cost, complexity and uncertainty of obtaining hydrogen from indigenous water on Mars. In the Mars Direct scenario [Zubrin *et al.*, 1991], 6 tons of hydrogen is brought from Earth and reacted with Martian carbon dioxide to produce 107 tons of CH_4/O_2 bipropellant. The mass leverage* provided by this approach is about 18:1.

The mass leverage is an important figure of merit for hydrocarbon fuels. But the physical mass leverage alone does not accurately reflect the true advantage since the various fuels may differ in their I_{sp}. However a better figure of merit results if we compute the normalized leverage where the physical mass ratio is reduced by the ratio of the fuel I_{sp} to that of hydrogen/oxygen fuel I_{sp}. Alternatively, the same information can be expressed as the product of the physical mass leverage and the I_{sp}. Physically this product represents the effective specific impulse of the Earth-derived components when leveraged by the Mars-derived components. It also graphically illustrates the advantage of *in situ* propellant production on Mars. Values for the Effective I_{sp} are presented in Table 18, where Earth-derived hydrogen/oxygen fuel is taken as the baseline.

Whether the concern is the scarcity of hydrogen on Mars or the high cost of shipping consumables from Earth, in either case the *in situ* production of hydrocarbon fuels is desirable. Another kind of tradeoff to be considered when selecting a fuel technology

---

* Mass leverage is defined as the ratio of the mass of the propellant divided by the mass of the Earth-derived components (e.g. hydrogen).

is the complexity of the *in situ* production process required to produce the fuel as well as the oxidant. Even complex processes may become acceptable since with increasing mass leverage it becomes increasingly feasible to utilize the scarce indigenous water on Mars [Hepp et al., 1991].

Table 18
IN SITU PROPELLANT MASS LEVERAGE[a] FOR EARTH-DERIVED HYDROGEN RELATIVE TO TERRESTRIAL H_2/O_2 FUEL

Fuel	Reaction Stoichiometry		I_{sp}[b]	Leverage[c]	Effective I_{sp}[d]
Hydrogen (baseline)	H_2	$+ 1/2\ O_2$ (Earth)	460	1	1
Hydrogen	H_2	$+ 1/2\ O_2$ (Mars)	460	9	4,140
Methane	CH_4	$+ 2\ O_2$	380	20	7,600
Ethane	C_2H_6	$+ 7/2\ O_2$	365	24	8,760
Ethylene	C_2H_4	$+ 3\ O_2$	373	31	11,563
Acetylene	C_2H_2	$+ 5/2\ O_2$	410	53	21,730
Methanol	CH_3OH	$+ 3/2\ O_2$	315	20	6,300
Ethanol	C_2H_5OH	$+ 3\ O_2$	330	24	7,920
Carbon Monoxide	CO	$+ 1/2\ O_2$	259	–	–

[a] After Hepp et al., 1991.
[b] I_{sp} values Zubrin (Private communication)
[c] Physical Mass Leverage = Mass Fuel/Mass Hydrogen
[d] Effective I_{sp} of the Earth-derived H_2 when leveraged by the Mars-derived components (Leverage × I_{sp}).

The production of acetylene is an example that has been studied [Landis, 1992]. Its production involves two steps, the preparation of methane followed by its pyrolysis. Acetylene offers a very high hydrogen leverage [see Table 18], and it has a higher I_{sp} and better storage properties than methane. Its principal drawback is its thermodynamic instability which makes it a potential explosion hazard. Possible solutions include storing it as a liquid and dilution with other gases [Hepp et al., 1991]. Acetylene is produced commercially by the Sachsse process. This involves the partial oxidation of methane which burns to provide energy for the pyrolysis of the remaining methane, thereby producing acetylene. The by-products of the reaction are carbon monoxide and hydrogen.

$$6CH_4 + O_2 \longrightarrow 2C_2H_2 + 2CO + 10H_2 + 341\ kJ/mol\ C_2H_2 \qquad (13)$$

Once an infrastructure is in place on Mars and adequate supplies of resources have been developed, the incremental cost of adding the capability to produce additional consumables will decrease dramatically. For this reason the option of importing consumables from Earth will cease to be a practical consideration and the utility of importing hydrogen and the mass leverage gained theyby will loose its high significance.

There are reasons other than mass leverage and I_{sp} that may dominate the choice of a particular fuel. One possible example is methanol. Methanol is a good candidate for use in Martian surface vehicles because it has superior storage and handling properties and can be used with conventional internal combustion engine technology. It is slightly less energetic than methane but it is particularly attractive because it can be stored as a liquid at Mars ambient pressure and at nearly all Martian temperatures [McMillen et al., 1996, Clark et al., 1991]. Methane by contrast requires cumbersome cryogenic containment on the vehicle.

$$CO + 2H_2 \longrightarrow CH_3OH(l) - 128 \text{ kJ} \qquad (14)$$

$$CO_2 + 3H_2 \longrightarrow CH_3OH(l) + H_2O(g) - 87 \text{ kJ} \qquad (15)$$

Methanol can be produced from carbon monoxide and hydrogen yielding no additional by-products. Alternatively, if it is produced from carbon dioxide and water using electrolysis as illustrated in Figure 7, the resulting methanol and oxygen will be in the stoichiometrically correct proportions for combustion. Methanol can also be used in fuel cells, it has potential applications as a storage medium for hydrogen, and it has other industrial and agricultural uses. Methanol technology is also a potential dual-use technology because methanol is the best candidate for use as the transition fuel from petroleum to natural gas for terrestrial transportation systems [McMillen et al., 1996].

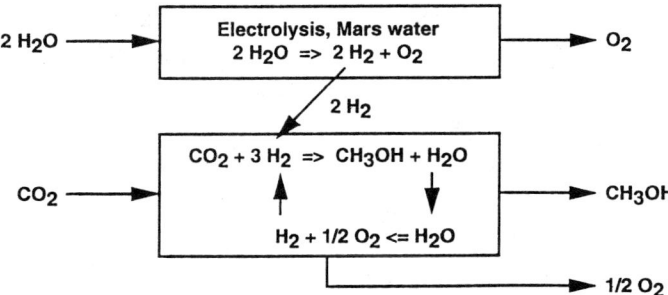

Figure 7 Synthesis of methanol/oxygen fuel using Mars water and atmospheric carbon dioxide. Methanol is a good candidate for use in Martian surface vehicles because it has has superior storage and handling properties and can be used with conventional internal combustion engine technology.

There is currently no ideal companion oxidant for methanol that has equally appealing properties, although hydrogen peroxide has been studied for this application. H_2O_2 readily produces oxygen by catalytic decomposition and it does not require cryogenic storage as does oxygen. However, the melting point of hydrogen peroxide is only $-2°C$, therefore electric or radioisotope heaters would be required to maintain it in a liquid state [Clark, 1991]. Other candidate oxidants that could be produced from Martian resources are nitrogen tetroxide (N_2O_4) and nitric acid (HNO_3).

Construction Materials

Plaster and Cement. For a practical method of utilizing *in situ* materials for the construction of buildings on Mars, a substance called "duricrete" has been proposed. On Earth, analog Martian soils have been synthesized and then thoroughly wetted and allowed to dry [Boyd et al., 1989]. The high salt content produces an extremely strong material somewhat similar to concrete.

Duricretes have been made using combinations of water, 12% $MgSO_4$, 1% NaCi, 85% clay (Pennsylvania nontronite or Wyoming Bentonite), and 2% Fe_2O_3 [Boyd *et al.*, 1989]. Experiments to test the tensile strength of duricrete have shown that strengths comparable to concrete can be achieved. Experiments have also shown that duricrete may crack upon drying, but with the addition of a minimum of fiber or cloth reinforcement, consolidated briquettes and plates of very high tensile strength can be achieved [Stoker *et al.*, 1993].

Ceramic and Glass. It is probable that pure silica sand or abundant feldspar will not be found in the Martian soils. Thus the production of optical-quality glass on Mars may be difficult and expensive. The abundance of iron on Mars may compound this problem since transparent glass must be largely free of iron. However, chemical treatments of the soil or the discovery of local concentrations of silica-rich soil can not be ruled out. In any case, opaque glass could always be made and used as construction material [Stoker *et al.*, 1993].

Lower quality ceramics should be easier to produce, either by baking hydrous materials or by partial fusion of anhydrous materials. Although kaolinite-rich soils are considered to be best for ceramic applications, such soils are probably lacking on Mars. Smectite and palagonite which appear to be more abundant are not known to produce good ceramics because of the expansion properties of smectities. Both are poorly crystaline and contain iron [Stoker *et al.*, 1993].

Metals Fabrication. Because of the availability of water and salts on Mars and the likely prospect that acids, bases and other reagents can be produced from *in situ* resources, it should be feasible to use hydrometallurgical treatments which are similar to those used on Earth for the processing of iron ore [Stoker *et al.*, 1993]. But some differences in the process chemistry for refining Martian minerals is likely because of the lack of free oxygen, abundant organic carbon and hydrocarbons, and because of the need to conserve water, particularly hydrogen. Recycling of reagents (ideally, closed-loop processing) will be necessary due to the high cost of Mars or Earth-derived materials, and in order to minimize the environmental impact of surface operations. The major energy sources will be solar, and perhaps nuclear power. Since there are no known stored energy resources on Mars, energy can not be based on the combustion of organic resources, as on Earth. The major candidate reducing agents for metals production is likely to be hydrogen (H_2), carbon monoxide (CO), and perhaps eventually from biomass.

The chemical and thermodynamic constraints on metal processing are the same on both Earth and Mars, however oxidation of metals exposed to the Martian atmosphere will be minimal. Geochemical abundances of common elements are probably comparable to Earth, although mechanisms for the concentration of desirable elements in ore deposits are more restricted on Mars (although not nearly so much so as on the Moon). Metals such as iron (Fe) are much easier to reduce from their oxides or other compounds than are other common metals such as aluminum (Al) or magnesium (Mg). Relatively simple chemical techniques can be used for their recovery. The production of aluminum and magnesium on Earth is typically done by a relatively sophisticated process (molten salt electrolysis). Given sufficient chlorine and a source of electricity this may also be possible on Mars [Stoker et al., 1993].

Iron oxides can probably be concentrated from ordinary Martian soils or from local ore deposits, such as weathered sulfide deposits (gossans, e.g., Burns [1988]) or from palerplacer deposits (former alluvial concentrations of dense minerals), should these be found to occur. Iron oxides can presumably be concentrated magnetically, as could unoxidized meteorite fragments which may be rich in nickel and available locally. A method of aqueous acid leaching of iron-bearing soils followed by neutralization to iron hydroxides and roasting to iron oxide (Fe_2O_3) could be used for Martian soils lacking a separate or crystalline iron oxide phase [Stoker et al., 1993].

Once an iron oxide concentrate is obtained, it can be reduced to native iron by reduction by either hydrogen gas (H_2 obtained by electrolysis of water) or carbon monoxide gas (CO), a potential by-product of other operations on Mars such as oxygen or fuel production that involves the reduction of atmospheric carbon dioxide.

For H_2 gas, the equation is:
$$Fe_2O_3 + 3H_2(g) \longrightarrow 2Fe + 3H_2O(g) \tag{16}$$

(Note that a similar process has been proposed for the reduction of lunar ilmenite, $FeTiO_3$; hematite is, however, easier to reduce than ilmenite.)

Using CO gas the corresponding equation is:
$$Fe_2O_3 + 3CO(g) \longrightarrow 2Fe + 3CO_2(g) \tag{17}$$

In contrast, the reduction is conventionally done on Earth using organic carbon (e.g. coal):
$$Fe_2O_3 + 3C \longrightarrow 2Fe + 3CO(g) \tag{18}$$

In the pre-industrial era, charcoal or wood was used and Martian-grown biomass might similarly be used for iron reduction.

The manufacture of high-strength steel for tools, machinery, vehicles, containers, and the like, would require further refining involving the introduction of carbon and the addition of alloying metals such as manganese and nickel [Rosenqvist, 1983]. The capability to smelt and process iron would be a valuable asset for a Mars base early on because of the time and cost to transport items from Earth [Stoker et al., 1993].

Copper, like iron, can be easily reduced from its oxides. However, unlike iron, copper will not be universally distributed in Martian soils. It is geochemically a rarer element. It was not detected in the Viking XRF experiment or in SNC meteorites. However, it may be found to occur in soils formed by the weathering of sulfide concentrations at the bases of lava flows [Burns, 1988].

Radiation Shielding. One of the concerns associated with placing humans on Mars is the possible exposure to solar and galactic radiation. Mars does not have an ozone layer capable of filtering solar ultraviolet radiation nor does it have a strong magnetic field (though it would only deflect the less energetic particles). Most importantly, its column mass of atmosphere is some 60 times smaller than that of Earth; thus

Mars receives a significantly high flux of radiation at the surface. The column mass of the Earth's atmosphere is ~ 1 kg/cm^2 which is equivalent to a layer of water ~ 10 m thick. On the other hand, the Martian atmosphere has a column mass of only about 16 g/cm^2 or the equivalent of 0.15 meters of water.

A readily available resource for providing radiation shielding is the Martian regolith. Loose surface soil appears to be widely distributed on the surface and could easily be used to cover the tops of habitats [Stoker *et al.*, 1993]. Figure 8 is an artist's conception of Martian habitats being covered with regolith. For a regolith density of 1.6, it would require about 6 meters of soil to equal the column mass of the Earth's atmosphere. Unlike lunar soil, Martian soils are likely to have better shielding properties since they are likely to contain abundant amounts of lighter elements (such as H, C, N) that are effective for shielding from neutrons produced by primary cosmic rays. [Stoker *et al.*, 1993].

Figure 8 Artists concept of habitats at a Mars base being covered with soil for radiation protection (painting by Carter Emmart). Loose surface soil appears to be widely distributed on the surface and could easily be used to cover the tops of habitats [Stoker *et al.*, 1993].

The main radiation danger during a mission occurs during interplanetary flight and then predominantly during solar flares. The surface of Mars by comparison is relatively safe. The planet itself reduces the radiation level to half of what is normally experienced in space, plus modest additional shielding is provided by the atmosphere. Since early missions are likely to be of limited duration on the Martian surface, and since for most astronauts it may be the only mission of their career, lesser amounts of habitat shielding—one meter of regolith is often proposed—or even none at all may be feasible while still allowing the total dose to remain within the career dose limits set by the National Commission on Radiation Protection for astronauts [NCRP, 1989]. However there remains the problem of shielding during high radiation fluxes from solar flares. For such events a storm shelter of a full six meters of regolith will still be necessary. But since solar flares occur very infrequently (about once per solar cycle, 11 yr), can be monitored and predicted with several minutes warning, and last for only a matter of days, normal working and living quarters may not need to be designed for this requirement.

CONCLUSION

The criteria for implementing resource utilization systems depends on a complex set of factors including assuring mission safety and success, providing a basis for enhanced exploratory capability, establishing self-sufficiency to enable future expansion of the initial base, and a concern for the overall economic feasibility of operations.

A logical progression for the development of materials processing systems on Mars would begin by first establishing production facilities for the primary consumables (fuel, O_2, H_2O, N_2, etc.). These in turn could provide feedstocks necessary to support additional mining, materials processing and recycling facilities.

The economic criteria for selection of local materials to be processed will be based primarily on which materials have the largest mass and transportation costs. Secondary factors include the energy cost of processing, material storage life, complexity and convenience of the processes, transportation weight of the processing equipment, reliability, maintenance requirements and service life, and a consideration for the necessary reagents, supplies and replacement components which must be supplied from Earth.

The simplicity inherent in mining the Martian atmosphere makes feasible the early implementation of *in situ* production of key consumables. This will enhance the mission and exploratory capability on Mars by providing more abundant supplies of consumables, by reducing the complexity of recycling equipment, and by reducing the cost of shipping bulk consumables from Earth.

In the long run, the development of *in situ* resource utilization technology for Mars finds its greatest justification when the vision goes beyond simply maintaining scientific outposts and supporting field operations to that of laying the foundation for future colonies on Mars. Here the main objective becomes the development of the means for future expansion and the capability for self-sufficiency on Mars. The central technical challenge for Martian ISRU will be to develop a process technology that is optimized for the resources available on Mars and its environment. In view of Mars' potential destiny as an abode for future civilizations, it must be highly adapted and reliable, integrated, environmentally sound, and able to operate in perpetuity, ultimately without dependence on Earth.

REFERENCES

Andre, M., and D. Massimino, "Growth of plants at reduced pressures: Experiments in wheat technological advantages and constraints," *Adv. Space Res.*, 12, 5, pp. 97-106, 1992.

Ash, R.L., W.L. Dowler, and G. Varsi, "Feasibility of rocket propellant production on Mars," *Acta Astronautica*, 5, pp. 705-724, 1978.

Ash, R.L., R. Richter, W.L. Dowler, J.A. Hapson, and C.W. Uphoff, "Autonomous oxygen production for a Mars return vehicle," Paper presented at the 33rd Congress of the International Astronautical Federation, Paris (September 1982), IAF publication, 1982.

Baker, D. and R. Zubrin, "Mars Direct: Combining near-term technologyies to achieve a two-launch manned Mars mission," *JBIS*, 453, 11, pp. 519-525, November, 1990.

Banin, A. and L. Margulies, "Simulation of Viking biology experiments suggests smectites not palagonite, as Martian soil analogs," *Nature,* 305, pp. 523-526, 1983.

Banin, A., "Mars soil: a sterile regolith or a medium for plant growth," AAS 87-215, in *The Case for Mars III: Strategies for Exploration*, C.R. Stoker, ed., American Astronautical Society, Science and Technology Series, 74, pp. 559-571, San Diego, CA, 1989.

Biemann, K., *et al.*, "The search of organic substances and inorganic volatile compounds in the surface of Mars," *J. of Geo. Res.*, 82, p. 28, September 30, 1977.

Boston P.J., (This Volume, Chapter 17): "Moving in on Mars: the hitchhikers guide to Martian life support."

Boston, P.J., "Low pressure greenhouses and plants for a manned research station on Mars," *J. Brit. Interplanet. Soc.*, 34, pp. 189-192, 1981.

Boston, P.J., "Critical life support issues for a Mars base," AAS 84-167, in *Case for Mars II*, C.P. McKay, ed., American Astronautical Society, Science and Technology Series, 62, pp. 287-331, San Diego, CA, 1985.

Boyd, R.C., P.S. Thompson, and B.C. Clark, "Duricrete and composites construction on Mars," in *Case for Mars III*, C.R. Stoker, ed., American Astronautical Society, Science and Technology Series, 74, pp. 539-550, San Diego, CA, 1989.

Burns, R.G., "Gossans on Mars," *Proc. Lunar and Planetary Society XVIII*, pp. 713-721, Lunar and Planetary Institute, Houston, TX, 1988.

Carr, M.H., *The Surface of Mars*, Yale University Press, New Haven, CT, 1981.

Carr, M.H., "Mars: A Water-Rich Planet," *Icarus*, 68, pp. 187-216, 1986.

Carr, M.H., "Water on Mars," *Nature*, 326, pp. 30-35, 1987.

Clapp, M., "Water supply for a manned Mars base," AAS 84-181, in *The Case for Mars II*, C.P. McKay, ed., American Astronautical Society, Science and Technology Series, 62, pp. 557-566, San Diego, CA, 1985.

Clark, B.C, A.K. Baird, H.J. Rose, Jr., P. Toulmin, III, R.P. Christan, W.C. Kelliher, A.J. Castro, C.D. Rowe, K. Keil, and G.R. Huss, "The Viking X-Ray fluorescence experiment: Analytical methods and early results," *J. Geophys. Res.*, 1977.

Clark, B.C., "Chemistry of the Martian surface: Resources for the manned exploration of mars," AAS 81-243, in *The Case for Mars*, P.J. Boston, ed., American Astronautical Society, Science and Technology Series, 57, pp. 197-208, San Diego, CA, 1984.

Clark, B.C., "The H-atom resource on Mars," AAS 84-179, in *The Case for Mars II*, C.P. McKay, ed., American Astronautical Society, Science and Technology Series, 62, pp. 527-535, San Diego, CA, 1985.

Clark, B.C., "Survival and prosperity using regolith resources on Mars," *JBIS*, 42, 4, pp. 161-166, April, 1989.

Clark, B.C., "The Viking results — The case for man on Mars," AAS 78-156, in *The Future United States Space Program*, R.S. Johnston, A. Naumann, Jr. and C.W.G. Fulcher, eds., American Astronautical Society, Adv. Astronaut. Sci. Series, 38-I, pp. 263-278, San Diego, CA, 1979.

Clark, B.C., A.K. Baird, R.J. Weldon, D.M. Tsusaki, L. Schnabel, and M.P. Candelaria, "Chemical composition of Martian fines," *J. Geophys. Res.*, 87, pp. 10059-10067, 1982.

Clark, B.C., J.L. Kalkstein, and S. Meyer, "The Case for the methanol powered planetary rover," *IAF-91-447*, 42nd Congress of the Intl Astronautical Federation, Montreal, Canada, 1991.

Clark, B.C., and D.C. Van Hart, "The salts of Mars," *Icarus*, 45, pp. 370-387, 1981.

Clifford, S.M., "Polar basal melting on Mars," *J. Geophys. Res.*, bf 92, pp. 9135-9152, 1987.

Colburn, D., J.B. Pollack, R.M. Haberle, "Diurnal variations in optical depth at Mars," *Icarus*, 79, pp. 159-189, 1989.

Colvin, J., P. Schallhorn, and K. Ramohalli, "Full system engineering design and operation of an oxygen plant," *J. Propulsion Power*, 8, pp. 1103-1108, 1992.

Cordell, B.M., "A preliminary assessment of Martian natural resource potential," AAS 84-185, in *The Case for Mars, II*, C.P. McKay, ed., American Astronautical Society, Science and Technology Series, 62, pp. 627-640, San Diego, CA, 1985.

Daunicht, H.-J. and H.-J. Brinkjans, "Gas exchange and growth of plants under reduced air pressure," *Adv. Space Res*, 12, 5, pp. 107-114, 1992.

Finale, F.P. and W.A. Cannon, "Mars: CO_2 adsorption and capillary condensation on clay -significance for volatile storage and atmospheric history," *J. Geophys. Res.*, 84, pp. 8404-8414, 1979.

Fanale, F.P., "The water and other volatiles of Mars," AAS 86-160, in *The NASA Mars Conference*, D.B. Reiber, ed., American Astronautical Society, Science and Technology Series, 71, pp 157-174, San Diego, CA, 1986.

Farmer, C.B., D.W. Davies, A.L. Holland, D.D. LaPorte, and P.E. Doms, "Mars: Water vapor observations form the Viking orbiters," *J. Geophys. Res.*, 82, pp. 4225-4248, 1977.

Fogg, M.J., *Terraforming: The Engineering of Planetary Habitability*, Chapter 2, SAE International, in press (1994).

Fox, J.L. and A. Dalgarno, "Nitrogen escape from Mars," *J. Geophys. Res.*, 88, pp. 9027-9032, 1983.

French, J.R., "The impact of Martian propellant manufacture on early manned exploration," AAS 84-178, in *The Case for Mars II*, C.P. McKay, ed., American Astronautical Society Science and Technology Series, 62, pp. 519-526, San Diego, CA, 1985.

Frisbee, R.H., "Mass and power estimates for Martian in-situ propellant production systems," *JPL report D-3648* (Pasadena: Jet Propulsion Laboratory), 1986.

Haberle, R.M., C.P. McKay, J.B. Pollack, O.E. Gwynne, D.H. Atkinson, J. Appelbaum, G.A. Landis, R.W. Zurek, and D.J. Flood, "Atmospheric effects on the utility of solar power on Mars," in J. Lewis, M.S. Matthews, and M.L. Guerrieri, *Resources of Near-Earth Space*, pp. 845-885, University of Arizona Press, Tucson, 1993.

Hall, D.O., *et al.*, "Biomass for energy: supply prospects," in *Renewable Energy*, Johansson, T.B., *et al.*, eds., pp. 593-652, Island Press, London, 1993.

Hart, H., and B.M. Jakosky, "Composition and stability of the condensate observed at the Viking Lander 2 site on Mars," *Icarus*, 66, pp. 134-142, 1986.

Haslach, H.W., Jr., "Wind energy: A resource for a human mission to Mars," *J. Brit. Interplant. Soc.*, 1989.

Hepp, A.F., G.A. Landis, and D. Linne, "Material processing with hydrogen and carbon monoxide on Mars," *NASA Technical Memorandum 104405*, May, 1991.

Hess, S.L., R.M. Henry, C.B. Leovy, J.A. Ryan, and J.E. Tillman, "Meteorological results from the surface of Mars: Viking 1 and 2," *J. Geophys. Res.*, 82, pp. 4559-4574, 1977.

Hicklenton, P.R. and P.A. Jolliffe, "Alterations in the physiology of CO_2 exchange in tomato plants grown in CO_2-enriched atmosphere," *Canadian J. Bot.*, 58, pp. 2181-2189, 1980.

Huber, S.C., H.H. Rogers, and F.L. Mowry, "Effects of water stress on photosynthesis and carbon dioxide partitioning in soybean glycine-max cultivar bragg plants grown in the field at different CO_2 levels," *Plant Physiol.*, 76, pp. 244-249, 1984.

Huguenin, R.L., K.J. Miller, and W.S. Harwood, "Frost-weathering on Mars: Experimental evidence for peroxide formation," *J. Mole. Evol.*, 14, pp. 103-132, 1979.

Hunt, G.E., "Thermal infrared properties of the Martian atmosphere. 4. Predictions of the presence of dust and ice cloudsfrom Viking IRTM spectral measurements," *J. Geophys. Res.*, 84, pp. 2865-2874, 1979.

Ishikawa, Y., T. Ohkita, and Y. Amemiya, "Mars habitation 2057: Concept design of a Mars settlement in the year 2057, *Journal of the British Interplanetary Society*, 43, pp. 505-512, 1990.

Jakosky, B.M., P.R. Christensen, "Global duricrust on Mars: Analysis of remote-sensing data," *J. Geophys. Res.*, 91, pp. 3547-3559, 1986.

Jakosky, B.M. and C.B. Farmer, "The seasonal and global behavior of water vapor in the Mars atmosphere: Complete global results of the Viking atmospheric water detector experiment," *J. Geophys. Res.*, 87, pp. 2999-3019, 1982.

Jones, D., C.F. Webb, M.R. LaPointe, H.M. Hart, and A. Larson, "The Retrieval, storage, and recycling of water for a manned base on Mars," AAS-84-180, *The Case for Mars II*, C.P. McKay, ed., American Astronautical Society, Science and Technology Series, 62, pp. 537-556, San Diego, CA, 1985.

Kerrebrock, J.L., *Aircraft Engines And Gas Turbines*, MIT Press, Cambridge, 1977.

Kieffer, H.H., T.Z. Martin, A.R. Peterfreund, B.M. Jakowsky, E.D. Miner, and F.D. Palluconi, "Thermal and albedo mapping of Mars during the Viking primary mission," *J. Geophys. Res.*, 82, 4, pp. 249-4, p. 292, 1977.

Klein, H.P., "The Viking biological experiments on Mars," *Icarus*, 34, pp. 666-674, 1978.

Klein, H.P., "The Viking Mission and the search for life on Mars," *Rev. Geophys. and Space Phys.*, 17, pp. 1655-1662, 1979.

Klingler, J. M., R.L. Mancinelli, and M.R. White, "Biological nitrogen fixation under primordial Martian partial pressures of dinitrogen," *Adv. Space Res.*, 9, 6, pp. 173-176, 1989.

Landis, G.A., and D.L. Linne, "Acetylene fuel from atmospheric CO_2 on Mars," *Journal of Spacecraft and Rockets*, 26, 2, pp. 294-296, 1992.

Lehninger, A.L., *Principles of Biochemistry*, Worth Publishers, Inc., New York, pp. 636-637, 1982.

Letaw, J.R., R. Silberberg, and C.H. Tsao, "Radiation hazards on space missions," *Nature*, 330, pp. 709-710, 1987.

Liu, S., and Donahue, T.M. "The regulation of hydrogen and oxygen escape from Mars," *Icarus*, 28, pp. 231-246, 1976.

McElroy, M.B., T.Y. Kong, and Y.L. Yung, "Photochemistry and evolution of Mars' atmosphere: A Viking perspective," *J. Geophys. Res.*, 82, pp. 4379-4388, 1977.

MacElroy, R.D., M. Kliss, and C. Straight, "Life support systems for Mars transit," *Adv. Space Res.*, 12, 5, pp. 159-166, 1992.

McKay, C.P., "Antarctica: Lessons for a Mars exploration program," AAS 84-156, in *The Case for Mars II*, C.P. McKay, ed., American Astronautical Society Science and Technology Series, 62, pp. 79-87, San Diego, CA, 1985.

McKay, C.P., "Space and planetary environment criteria guidelines for use in space vehicle development," 1982 Revision (Volume 1), NASA Technical Memorandum 82478, 1983.

McKay, C.P., T.R. Meyer, P.J. Boston, M Nelson, T. MacCallum, O. Gwynne, "Utilizing Martian resources for life support," in *Resources of Near-Earth Space*, J. Lewis, M.S. Matthews, and M.L. Guerrieri, eds., pp. 819-843, University of Arizona Press, Tucson, 1993.

McKay, C.P. and S.S. Nedell, "Are there carbonate deposits in Valles Marineris, Mars?," *Icarus,* 73, pp. 142-148, 1988.

McKay, C.P. and C.R. Stoker, "The early environment and its evolution on Mars: Implications for life," *Rev. Geophys. Space Phys.,* 27, pp. 189-214, 1989.

McKay, C.P., O.B. Toon, and J.F. Kasting, "Making Mars habitable," *Nature,* 352, pp. 489-496, 1991.

McMillen, K.R., T. Meyer, and B.C. Clark, "Methanol, a fuel for earth and Mars," in *The Case For Mars V*, P.J. Boston, ed., American Astronautical Society, Science and Technology Series, (in press), San Diego, CA, 1996.

Ming, D.W., "Manufactured soils for plant growth at a lunar base," in *Lunar Base Agriculture: Soils for Plant Growth*, D.W. Ming, and D.L. Henninger, eds., American Society of Agronomy, Inc., Madison, WI, pp. 93-105, 1989.

Meyer, T.R., "Extraction of Martian resources for a manned research station," *J. British Interplanetary Soc.,* 34, pp. 285-288, 1981.

Meyer, T.R. and C.P. McKay, "The atmosphere of Mars - Resources for the exploration and settlement of Mars," AAS 81-244, in *The Case for Mars*, P.J. Boston, ed., American Astronautical Society, Science and Technology Series, 57, pp. 209-232, San Diego, CA, 1984.

Meyer, T.R. and C.P. McKay, "The resources of Mars for human settlement," *J. British Interplanet. Soc.,* 42, pp. 147-160, 1989.

Moore, H.J., R.F. Hutton, R.F. Scott, C.R. Spitzer, and R.W. Shorthill, "Surface materials of the Viking landing sites," *J. Geophys. Res.,* 82, pp. 4497-4523, 1977.

Nachtwey, D.S., "Radiological health risks," *Proceedings of the 19th Intersociety Conference on Environmental Systems*, SAE Technical Paper 891432, San Diego, CA, 24-26 July, 1989.

National Council on Radiation Protection and Measurements (NCRP), "Guidance on Radiation received in Space Activities," *NCRP Report No. 98*, National Council on Radiation Protection and Measurements, Bethesda, Maryland.

Nelson, M. and W.F. Dempster, (This Volume, Chapter 18), "Living in space: Results from biosphere 2's initial closure, an early testbed for closed ecological systems on Mars," 1995.

Owen, T., K. Biemann, D.R. Rushnek, J.E. Biller, D.W. Howarth, and A.L. Lafleur, "The composition of the atmosphere at the surface of Mars," *J. Geophys. Res.,* 82, pp. 4635-4639, 1977.

Oyama, V.I. and B.J. Berdahl, "A model of Martian surface chemistry," *J. Mol. Evol.,* 14, pp. 199-210, 1979.

Oyama, V.I. and B.J. Berdahl, "The Viking gas exchange experiment results from Chryse and Utopia surface samples," *Geophys. Res.,* 82, pp. 4669-4676, 1977.

Paige, D., "The thermal stability of near-surface ground ice on Mars," *Nature,* 356, pp. 43-45, 1992.

Parker, J.F., Jr., V.R. West, eds., *Bioastronautics Data Book*, NASA SP-3006, Washington DC, 1973.

Pollack, J.B., "Climate change on the terrestrial planets," *Icarus,* 37, pp. 479-533, 1979.

Pollack, J.P., D. Colburn, R. Kahn, J. Hunter, W. Van Camp, C.E. Carlston, and M.R. Wolf, "Properties of aerosols in the Martian atmosphere, as inferred from Viking Lander imaging data," *J. Geophys. Res.,* 82, pp. 4479-4496, 1977.

Pollack, J.B., T. Roush, F. Witteborn, J. Bregman, D. Wodden, C.R. Stoker, O.B. Toon, D. Rank, B. Dalton, and R. Freedman, "Thermal emission spectra of Mars (5.4-10.5 µm): Evidence for sulfates, carbonates, and hydrates," *J. Geophys. Res.,* 95, 1990.

Ramohalli, K., "Generating power for Mars exploration," *Mars Underground News*, Fall 1994, 6, 4, 1995.

Ramohalli, K. and K.R. Sridhar, "Extraterrestrial materials processing and related transport phenomena," *J. Propulsion Power*, 8, pp. 687-696, 1992.

Ramohalli, K., T. Kirsh, and B. Preiss, "Figure-of-merit approach to extraterrestrial resource utilization," *J. Propulsion*, 8, *pp.* 240-246, 1992.

Rogers, H.H., J.F. Thomas, and G.E. Bingham, "Response of agronomic and forest species to elevated atmospheric carbon dioxide," *Science,* 220, pp. 428-429, 1983.

Rosenqvist, T., *Principles of Extraction Metallurgy*, 2nd ed., McGraw-Hill, New York, 1983.

Salisbury, F.B. and C.W. Ross, *Plant Physiology*, Wadsworth, Belmont, CA, 1985.

Schwartzkopf, S.H. and R.L. Mancinelli, "Germination and growth of wheat in simulated Martian atmospheres," *Acta Astron.*, 25, pp. 245-247, 1991.

Seckbach, J., F.A. Baker, and P.M. Shugarman, "Algae thrive under pure CO_2," *Nature*, 227, pp. 744-745, 1970.

Severinghaus, J.P., W.S. Broecker, W.F. Dempster, T. MacCallum, and M. Wahlen, "Oxygen Loss in Biosphere 2," *EOS, Transactions*, American Geophysical Union, 75, 3, pp. 33-37, Jan 18, 1994.

Singer, R.B., "Spectral evidence for the mineralogy of high-albedo soils and dust on Mars," *J. Geophys. Res.*, 87, pp. 10159-10168, 1982.

Squyres, S.W. and M.H. Carr, "Geomorphic evidence for the distribution of ground ice on Mars," *Science*, 231, pp. 249-252, 1986.

Stancati, M.L., J.C. Neihoff, and W.C. Wells, "*In situ* propellant production: A new potential for round-trip spacecraft," *AIAA Paper No. 79-0906*, Presented at the AIAA/NASA Conference on Advanced Technology for Future Space Systems, Hampton, VA, May, 1979.

Stoker, C.R., J.L. Gooding, T. Roush, A. Banin, D. Burt, B.C. Clark, G. Flynn, and O. Gwynne, "The physical and chemical properties of and Resource potential of Martian surface soils," in *Resources of Near-Earth Space*, J.S. Lewis, *et al.,* eds., University of Arizona press, Tucson, 1993.

Toksöz, M. N. and A.T. Hsui, "Thermal history and evolution of Mars," *Icarus*, 34, pp. 537-547, 1978.

Toulmin, P., III, A.K. Baird, B.C. Clark, K. Keil, and H.J. Rose, Jr., "Preliminary results from the Viking X-ray fluorescence experiment: The first sample from Chryse Planitia, Mars," *Science*, 194, pp. 81-84, 1976.

Toulmin, P., III, A.K. Baird, B.C. Clark, K. Keil, H.J. Rose, Jr., R.P. Christan, P.H. Evans, and W.C. Kelliher, "Geochemcial and mineralogical interpretations of the Viking inorganic chemical results," *J. Geophys. Res.*, 82, pp. 4625-4634, 1977.

Verniani, F., "The total mass of the Earth's atmosphere," *J. Geophys. Res.*, 71, pp. 385-391, 1966.

Zent, A.P., F.P. Fanale, J.R. Salvail, and S.E. Postawko, "Distribution and state of H_2O in the high-latitude shallow subsurface of Mars," *Icarus*, 67, pp. 19-36, 1986.

Zisk, S.H. and P.J. Mouginis-Mark, "Anomalous region on Mars: Implications for near-surface liquid water," *Nature,* 288, pp. 735-738, 1980.

Zubrin, R.M., D.A. Baker, O. Gwynne, "Mars direct: a simple, robust, and cost effective architecture for the space exploration initiative," *AIAA 91-0326*, 29th Aerospace Science Conf., Reno, NV, Jan, 1991.

Zurek, R.W., "Martian great dust storms: An update," *Icarus*, 50, pp. 288-310, 1982.

A small dune buggy is deployed off a large, less maneuverable, pressurized rover.

Chapter 20

MARS ROVERS

Benton C. Clark[*]

Once humans land on Mars, methods of reaching different areas of the surface will be necessary for the productive exploration of the planet. Cross-country vehicles and also eventually "hoppers," which can bypass intervening distances to reach sites of high priority, will be needed. Over very short ranges, unpressurized transportation will be adequate, but for serious exploration fully pressurized and conditioned vehicles must provide astronauts with "shirt-sleeve" environments and living quarters. Safety requirements will combine with these considerations to place important limits on the acceptable engineering subsystems that can be used to provide these capabilities.

INTRODUCTION

Rovers will be a very high priority item for missions to Mars because they enormously extend the range over which exploration can be accomplished. For robotic missions, such as the Viking Landers, sampling by a robot arm is practical to about 3 meters from a lander and terrain can be imaged only to the local horizon, which in a rugged setting can be as close as 1 to 3 km. Rovers can expand the available sampling area by a factor of at least 30,000 times and the viewing by 100 to 1,000 times for each 100 km of rover travel. An astronaut wearing a heavy spacesuit and portable life support system (PLSS) can hike at most 10 to 15 km across the Martian terrain in a day; half this distance is the farthest they will ever be allowed to venture from base camp when traveling by EVA (extravehicular activity, i.e., the astronaut in his spacesuit). Even with rover transportation of the type used by Apollo, the radius of exploration could not be permitted to exceed the 10-15 km limit. This is because the Apollo rover was unpressurized and the spacesuit's portable life support system (PLSS) could support the astronaut for less than 8 hours. Thus, unless the rover is fully pressurized and has stores of necessities such as food, breathing oxygen, and water, humans would be constrained to very localized exploration on the planetary surface. Lacking this would mitigate many of the scientific and exploration advantages of sending humans to Mars since robotic rovers, though working at much slower and less efficient rates, could, over the course of a year or so, cover all of the territory that the human mission could, and with no risk of life.

---

[*] Planetary Sciences Laboratory (B0560), Martin Marietta, Denver, Colorado 80201.

Pressurized rovers are thus indispensable for the human exploration of Mars. The trip is, otherwise, hardly worth the trouble.

MARS ENVIRONMENT AND ROLE OF THE HUMAN DRIVER

It is perhaps ironic that for future human missions to Mars, there may be no greater need for the experience, reaction skills, and resourcefulness that we normally attribute to test pilots than for driving a vehicle over the unpredictable and hazardous Martian terrain. By way of contrast, the various rocket burns and the spaceship entry-to-landing sequence almost certainly will be computer controlled, with the "pilots" having control of only top level override or abort decisions. Final landing site selection may require some piloting, although more likely a previous rover mission will have certified a specific landing site to which the lander craft navigates by previously emplaced radio transponder beacons for a precision-guided touchdown. But from the current indications, compounded by a lack of certainty in the physical properties of the various materials on the Martian surface, a great deal of skill may be essential for efficient and reliable transportation via roving vehicles.

Relevant Environment Attributes

Mars appears quite unlike the lunar case, with many geological factors contributing to a surface expected to be much more challenging than the Moon. For one, the Viking landing sites were rock-rich. It is well known by hikers and jeepers that loose rocks are one of the most difficult trafficability hazards, from many standpoints. Worse, some of the soils appear to have very poor load-bearing capability, as evidenced by the ease of hoeing by the surface sampler arm and the sinkage of lander footpad number two on Viking Lander-1, Figure 1. The almost 2.5 times higher force of gravity at the surface of Mars compared to the Moon, when added to the fact that lunar regolith has excellent bearing strength because of its coarser soil and other factors related to the lack of an atmosphere, also necessitate more engineering concerns for roving on Mars. Drifts of Martian surface fines could have properties similar to loose mounds of sifted flour. That rocks can be buried within such drifts is verified already by Viking-1. Indeed, boulders apparently can act as wind deflectors to cause deposition of material, especially on the lee side, and eventually burying not only themselves but other rocks in the vicinity, as is evident in Figure 2. The astronaut driver may need to call upon all of his or her skills and abilities to react intelligently in adverse and unforeseen conditions in order to keep from loosing control of the rover. Roll over, downslope sliding, wheel burial, entrapment by obstacles, and impacting hidden rocks are just some of the hazards, as portrayed in Figure 3. Within younger volcanic terrain may be found sharp, clinkery and glassy, frothy rocks (such as in so-called "aa" flows) or thin lava crusts (such as in "pahoehoe" or "elephant-hide" flows, or lava tube roofs), posing additional hazards. Astronauts will need not only driving skills but also the scientific expertise of the geologist, the skill of the mountaineer, and the perceptions of the explorer. Of all the mission jobs that Mars astronauts will undertake, Rover Pilot is undoubtedly one of the most, if not the greatest challenge.

Figure 1 (a) Rocky terrain at Viking Lander site-2.

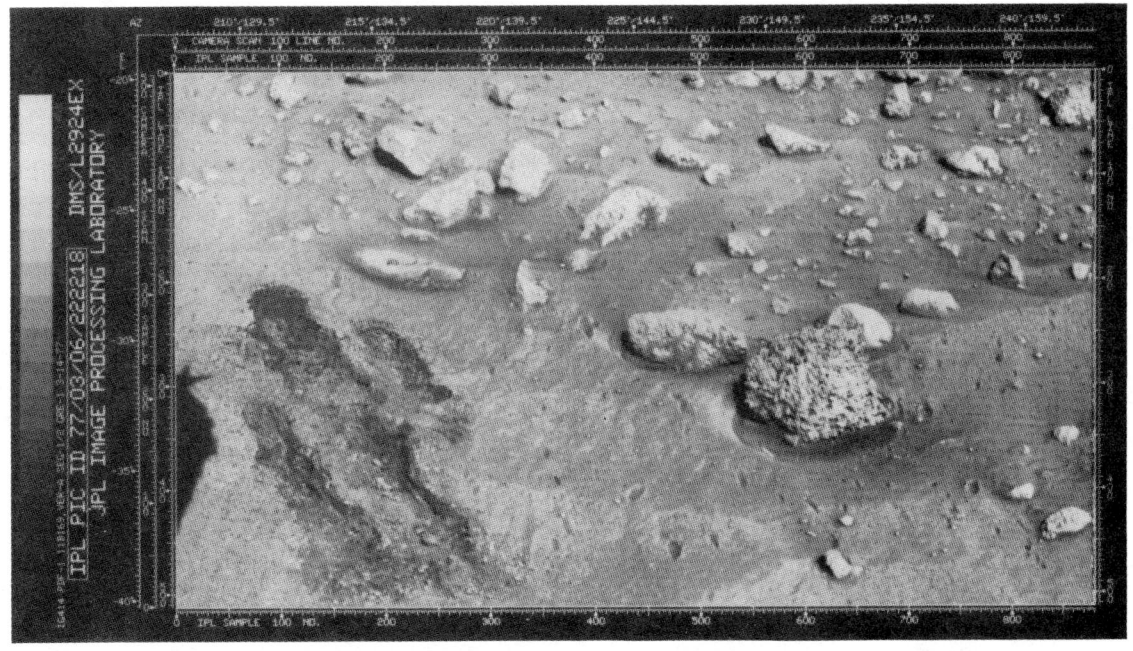

Figure 1 (b) Sample trenches and impact pits in drift material.

Figure 1 (c) Footpad #3 on firm soil.

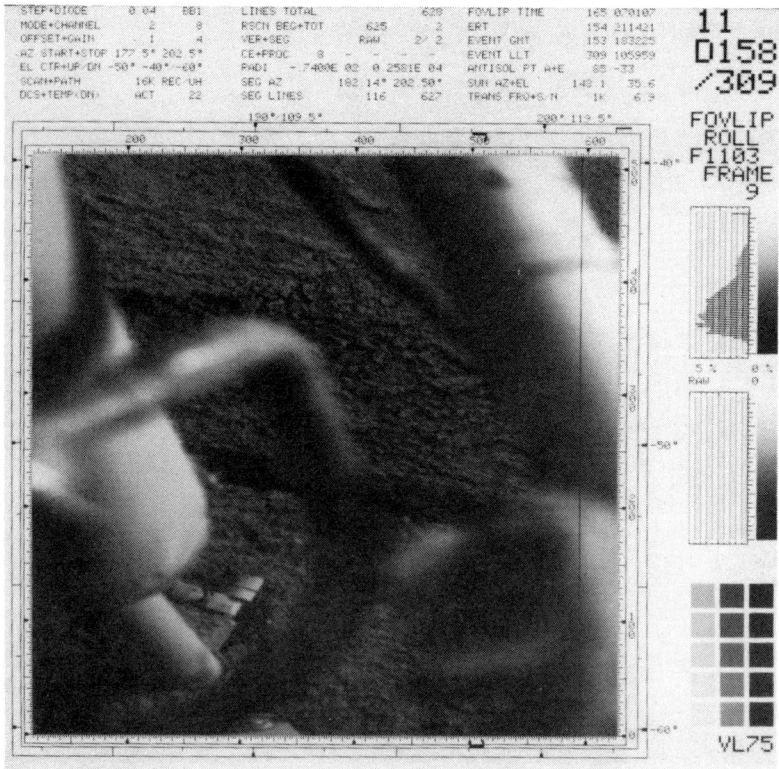

Figure 1 (d) Footpad #2 buried in loose soil (16 cm deep).

Figure 2   Boulders and windblown drift deposits on Mars.

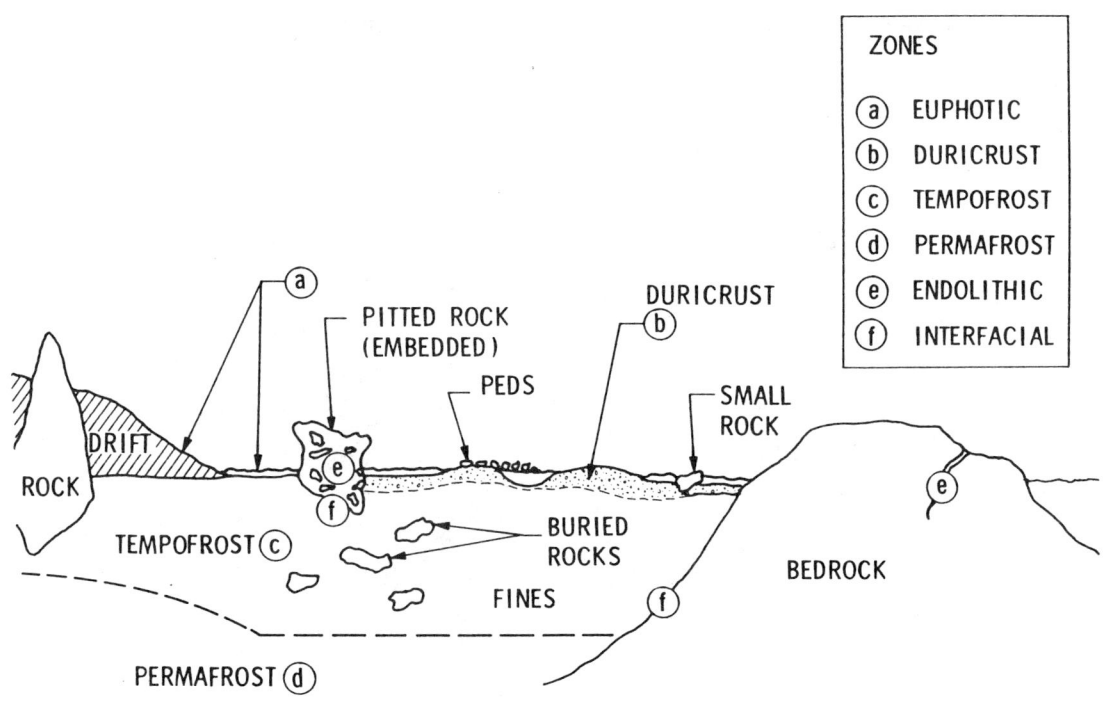

Figure 3   Schematic of terrain types on Mars.

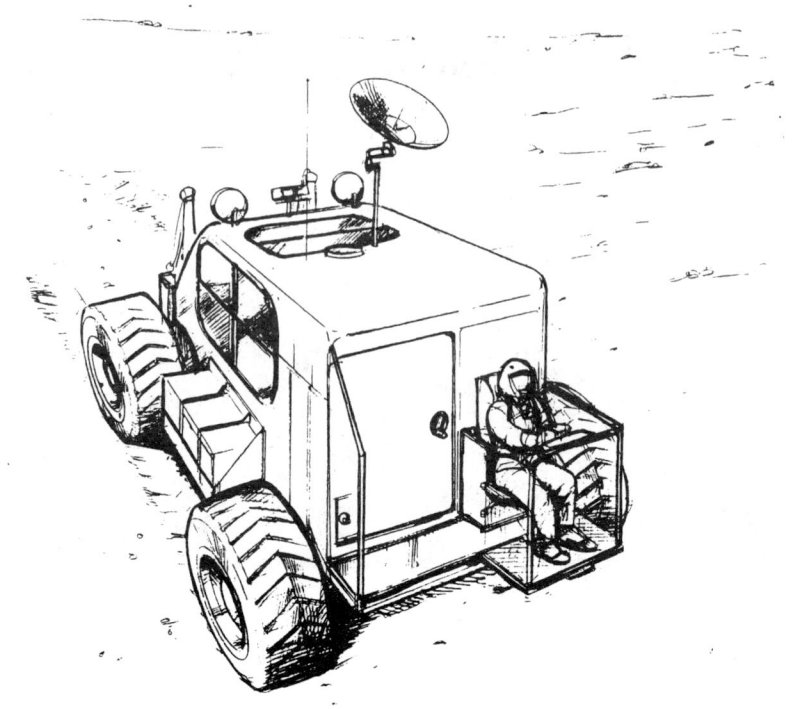

Figure 4 Pressurized rover with external riding seat and airlock.

Figure 5 Astronaut at work outside the rover.

Human Accommodation

Even with astronauts on-board the rover, it would not always be convenient for them to explore if they always must don spacesuits, vent the entire cabin, and perform EVA's. An airlock could be provided, so that one member could egress and ingress without the necessity of the driver suiting up each time. It has been suggested, however, that an external seat be provided so that the EVA explorer could ride outside for much of the day (Figure 4) and simply hop off when in interesting areas (Figure 5). It would be highly desirable to have external analytical tools, rock and soil processing equipment, and storage bins for taking samples back to base. A small sample airlock could be used to pass especially interesting material into the cabin for inspection by the astronaut who was normally driving and for more detailed study inside the shirt-sleeve environment. Providing external robotic arms which can be operated in a telepresence mode by the astronauts will also eliminate the necessity for some of the EVA. These arms could not only be used to sample, but also to pass some specimens inside to the astronauts.

MOBILITY

Mobility Options

There are three fundamental methods of vehicle locomotion usually considered for movement on a planetary surface. The first two are by rolling, using either wheels or moving tracks. The third is by walking, i.e., by successive placement of footpads. Depending on the exact nature of the terrain, any of the three can be the most efficient. For example, on soils of very low bearing strength, a tracked system can significantly outperform the others in achieving distance traveled for minimum energy expenditure. This is because of the relatively limited amount of support area for wheels and footpads and the consequent tendency to burial of the contact elements during motion. On smooth, durable ground the rolling wheel is of course superior both for efficiency and for speed. However, in highly jumbled terrain, such as lava flows with high relief, areas with dense rock populations (e.g., talus slopes), or highly carved sediments the only approach with the possibility of good energy expenditure is the walking mode. The walker is, however, inconsistent with the speed of travel that is needed for efficient exploration by humans. Indeed, humans are such adept walkers themselves, that the preferred *modus operandi* might be to provide for short-range hops of the entire vehicle, interspersed by EVA exploration on foot.

Energy Cost of Mobility

Achieving the highest practical efficiency is certainly one major objective, but the overall system must be also engineered to satisfy trafficability, safety, reliability, and mission requirements, including exploration and/or work tasks. Considering the generalized planetary roving problem, a wheeled system with individually driven and steered wheels could provide the most flexible system. It may also be desirable to provide for electrically powered jacks to move individual wheels up and down, e.g., to level the cabin for safer slope traversal or to clear obstacles.

For a trafficability efficiency model, a mobility power relationship of 1 kWh/km, averaged over likely Mars terrain, may be considered representative of a large pressurized rover. This factor will vary enormously as a function of the details of the topography, the population of obstacles, nature of the surface (especially soil compaction and abundance of loose rocks), weight of the vehicle, design of the wheels, etc. It is assumed here that the driving skills of the operator are very high and that sensor-controlled power train algorithms are computer implemented (e.g., positive traction drive for each wheel, coordinated steering, c.g. management, regenerative braking).

Design Options

Both electrical and mechanical power will be needed to operate the rover. Drive power for wheels, legs, or tracks could be mechanically transmitted from a central power source, but most designers have concluded that direct drive with minimized mechanical linkages using electrical motors would be a preferred approach. Major engineering design issues will result from the harshness of the natural Martian environment: temperatures dipping to more than $-100°C$ at nighttime (freezing of lubricants; brittle phase transitions in many materials); the extremely fine-grains and possible grittiness of dust (causing friction or lock-up of moving parts); and the strong UV and highly oxidizing molecules in the atmosphere (degrading organic materials over time). Critical parts may have to be isolated and conditioned by enclosures (e.g., flexible boots over moving interfaces) and active thermal control (e.g., heaters). Indeed, since the rover is part habitat and part surface transportation system, it presents all of the demands of both. No other mission element will interact so vigorously and intimately with the Martian surface environment. As a system, the rover is surely one of the greatest engineering challenges of the mission.

PRESSURIZED CABIN DESIGN

A pressurized rover provides the advantage that astronauts can work and live in a much more comfortable environment than the very constraining spacesuit. Indeed, operational spacesuits have never been designed for more than 6 to 8 hours of occupancy because of obvious problems with life support, including food preparation, waste products disposal, etc. The pressurized cabin must provide, however, all the functions of a full environmental control and life support system (ECLSS). Cabin volumes are sufficiently small that if the ECLSS were to fail, the astronauts would be incapacitated within a few hours because of oxygen starvation and/or carbon dioxide toxicity. Oxygen may be supplied as boiloff from a tank of cryogenically-stored liquid or by dissociating hydrogen peroxide.

Removal of carbon dioxide can be done chemically or by physical absorption. Assuming that the pressurized rovers do not require more than 2 to 4 weeks of supplies, the method of choice will be to use LiOH canisters, which have only the drawback of irreversibly taking up the CO_2. At the planetary base or for very long-term rover expeditions, a recyclable system such as zeolite adsorbent which can periodically be baked at higher temperature to vent CO_2 back into the atmosphere will be preferred. The power consumption of such a device is quite high compared to the simpler LiOH system.

Humidity control is also a major challenge. Many techniques have been tested and designed for human spaceflight. In most systems flying today, humidity is controlled through condensation onto a heat exchanger cooled to the desired dew point. This requires a source of cold water, which will not be a problem, except that the water will have to be chilled, in turn, through a heat exchanger operating some fluid which does not freeze at Martian ambient temperatures. Drinking water could be taken as a bulk consumable, prestored in tanks. More efficient would be purification of water recovered from the humidity control system and/or from the products of combustion in an engine (see section on Power, below). Food is the remaining key consumables component of the ECLSS. A distinct advantage of the Martian environment is that is it consistently cold. Frozen food is a major possibility, but it must be remembered that a power penalty will be entailed just for thawing and heating the food during preparation. Also, the food would have to be protected from long exposure to the Martian ambient pressure because severe dehydration ("freezer burn") could result.

Because of the drive to keep the rover mass as low as feasible, the habitable volume within a pressurized rover crew cabin will undoubtedly be quite small (see Figures 6a and 6b), crowded with equipment, and not amenable to as much freedom as would otherwise be desired. These missions will be uncomfortable! But the rewards will more than compensate for those whose goal is the exploration of the unknown. The minimum for many-day missions will be to provide for reclining seats or erectable beds. Personal hygiene facilities should also be provided, with a suitable degree of privacy.

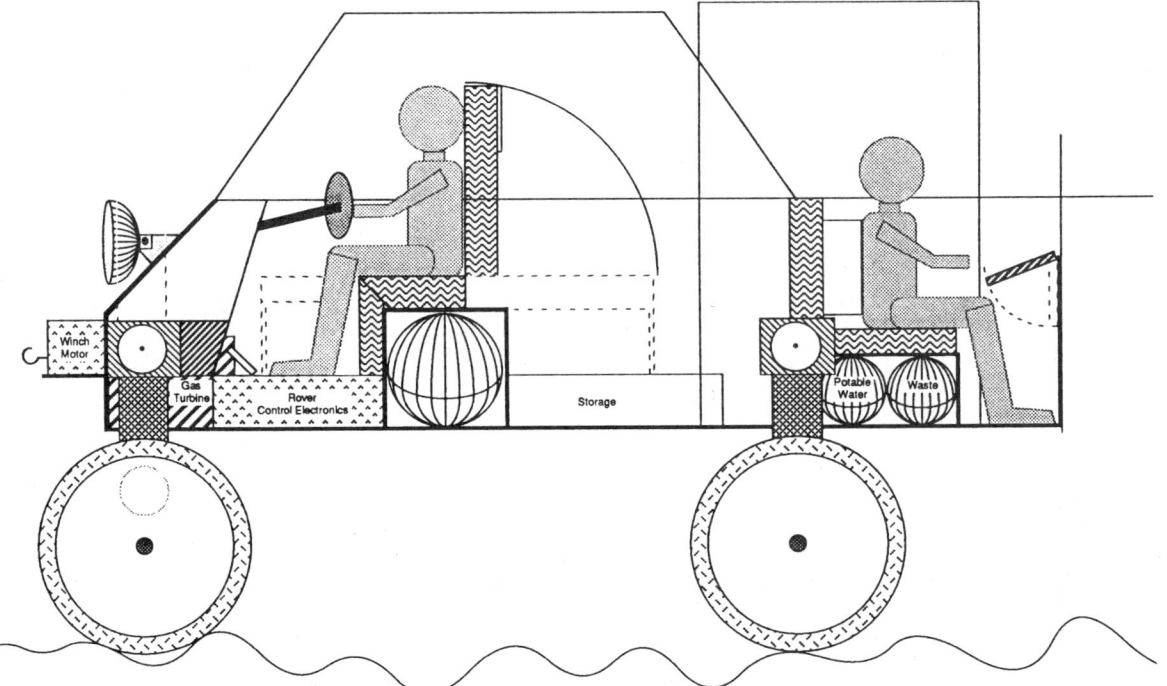

Figure 6a Two-person crew cabin for medium range rover exploration (Side view).

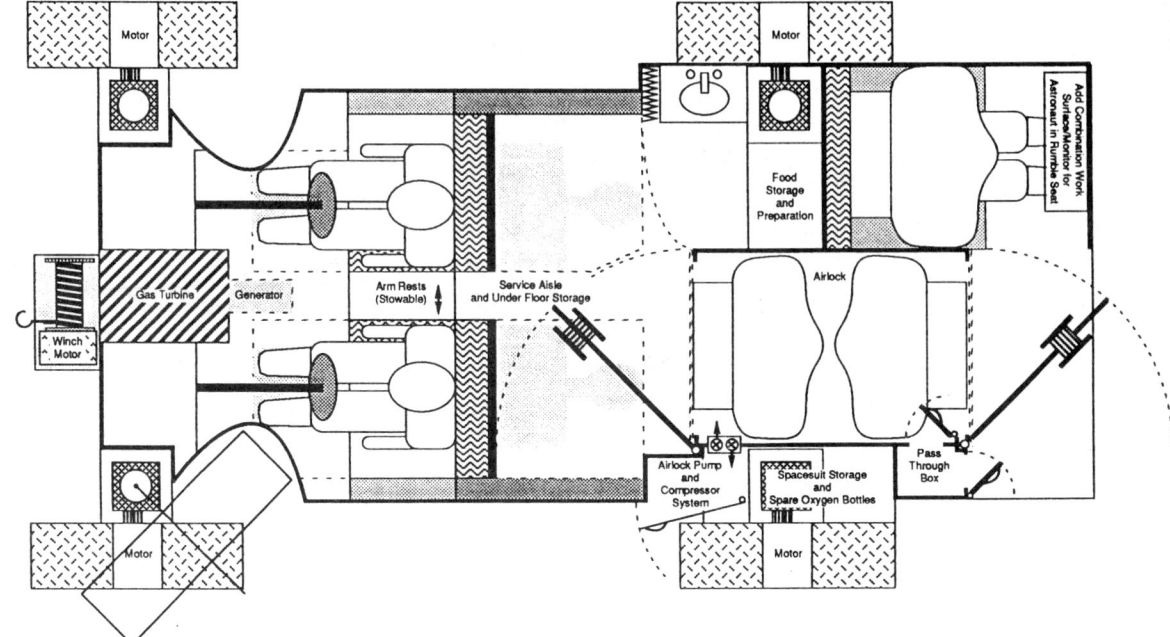

Figure 6b Two-person crew cabin for medium range rover exploration (View from above).

POWER

Perhaps no problem is greater than obtaining the necessary power for achieving mobility over a difficult terrain, especially when that power source must be long-lived, reliable, robust against the environment, and safe. Planetary surface rover operations will be characterized by large variations in power needed, depending upon whether the vehicle is in a major mobility mode or simply stationary. Power consumption for life support, communications, and avionics equipment will also vary on both a routine and unscheduled basis. At times, up to 60 kW$_e$ or more may be required for traversing rugged terrain or achieving high velocity. At other times, the electrical draw could be as low as a few 100's or just 10's of watts to power support electronics. In most scenarios, it will be necessary at certain times to rapidly increase power. Such surges could be supplied by power stored in rechargeable batteries, but significant losses would be incurred.

Nuclear Power Options

Nuclear power sources can satisfy the first three requirements, but not necessarily the latter. For small robotic vehicles, requiring modest power sources of up to a few hundred watts, conversion of the heat energy of radioactive decay into electrical energy is highly suited, as attested by the successful use of plutonium-powered Radioisotope Thermoelectric Generators (RTGs) on virtually all deep space missions, as well as the Viking Landers and lunar surface experiment packages. Although they are extraordinarily low in external emissions, considering the extremely high levels of activity that must be employed, the radiation hazard to humans which spend long times in their vicinity is excessive. As shown in Figure 7, in a two-week period at a distance of two meters from a 1000 W_e RTG, a human would receive more than 0.1 Sv (10 rem) of dose equivalent. This artificial hazard is unacceptable and the high energy gamma ray and neutron emissions can be decreased only insignificantly even by massive shields.

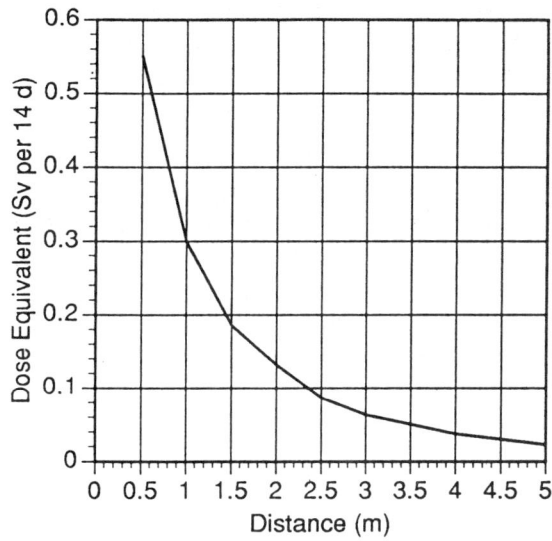

Figure 7 Exposure dose from RTG nuclear energy source.

The situation can be improved by up to a factor of three by increasing conversion efficiency, through replacement of the thermopile with an active system such as a Brayton closed cycle generator. Known as the Dynamic Isotope Power System (DIPS), the only disadvantage is the additional concern for long-life and reliability that machinery entails compared to the passive thermoelectric approach. An additional option is for the power source, whether RTGs or DIPS, to be contained in a follower trailer, as shown in Figure 8. In this concept the radiation exposure is decreased by the inverse-square-of-the-distance effect and perhaps also by a shield made of Martian soil interposed between rover and isotope package. Again, disadvantages are entailed, including serious increase in risk of successful negotiation of complex terrain.

Figure 8 Trailer with nuclear power source.

Solar Power Option

Solar power is, of course, more benign and appealing. However, it is not at all available during dusk or nighttime. Rechargeable batteries for energy storage are useful also for providing peak power when needed. However, considerable energy is lost during each charge-discharge cycle, especially when the batteries must operate over a large temperature range driven by external environmental temperature fluctuations of as much as 100°C from day to night. The Martian atmosphere always contains some dust, a condition aggravated during local or global dust storms. However, even in the worst conditions, scattered solar light is available as a diffuse energy source. Pointing a large solar array for maximum output is thus advantageous only during the clear days. In addition, just the act of roving, sampling, and astronaut EVA activities will raise dust clouds which would contaminate solar cell surfaces. Dust removal may be difficult because of the dry and presumably highly electrostatic conditions that prevail. Finally, the size of the necessary array is incompatible with typical rover dimensions. For example, if an array completely overhung, in umbrella fashion, a typical large pressurized two- or four-person rover, it could be expected to produce only a few hundred watts (diurnal average). This is about ten times less than would be required for sustained living and roving over Mars.

Chemical Energy Option

The indicated method of obtaining power for propulsion is through conversion of stored chemical energy into mechanical energy, perhaps via an electrical intermediate. Primary batteries are untenable except for the short traverses by unpressurized systems. For longer-range travel, fuel cells (FC) or heat engines might be required. Suggested systems have included hydrogen/oxygen, methane/oxygen, methanol/hydrogen peroxide, and other combinations to operate FCs or internal combustion engines.

Table 1
SPECIFIC ENERGY CONTENT OF CHEMICAL REACTANTS (USING GASEOUS OXYGEN) AND ELECTRICAL OUTPUT FROM ENGINE/GENERATOR

Fuel	Heat of combustion (kWh/kg of reactants)	Electrical Output* (kW_eh/kg of reactants)
Hydrogen	3.73	1.26
Methane	2.78	1.05
Methanol†	2.66	1.00
Ethanol	2.41	0.91
Methanol	2.23	0.84
Carbon monoxide	1.78	0.54

* Assumes 50% engine efficiency for alcohols, 45% for methane and hydrogen, and 40% for all others; 94% mechanical-to electrical conversion; and 80% power handling efficiency (storage, conditioning, distribution).
† After thermal dissociation into CO and H_2.

Chemical Energy Density. Hydrogen/oxygen has the highest specific energy and therefore would be the obvious choice for mass-constrained space applications if it were not for the fact that storage and handling create major problems because of the difficulty of storing cryogenic fluids, even in the somewhat cold Martian atmosphere. The boiling points of all other liquids listed in Table 1 are higher than hydrogen by a very significant amount and therefore would be easier to store. Both hydrogen and methane could be stored as pressurized gases. This necessitates heavy tanks in order to withstand the high static pressures and to minimize the possibility of tank rupture with explosive force in the event of an accident. Compared to liquid storage, the quantity of relatively low-density gases does not provide much range for the vehicle (for example, even a very large spherical tank, one-meter in diameter, filled with gaseous methane compressed to 250 bars would contain less than 100 kg of fuel).

Cryogenic Fuels. In the vacuum of space, it is possible to store cryogenic liquids, as has been already demonstrated on many human and robotic spaceflights. However, space storage achieves good thermal isolation (from the relatively warm spacecraft structures and solar thermal radiation) by using multilayer insulation (MLI) blankets, which are effective only in vacuum. On Mars, the atmospheric pressure of 6 millibars is sufficient to totally compromise the effectiveness of MLI because of conduction and convection losses. Furthermore, winds on Mars transport additional heat into the tanks via atmospheric gas which can be as much as 150°C higher in temperature than the cryogens. Considering the Martian average ambient temperature of −55°C, liquid methane would have to be stored at 80 bars pressure (1,100 psia), whereas methanol stores without pressurized tanks being necessary. Both hydrogen and carbon monoxide (CO) storage would be even more challenging because of the extraordinarily high pressures. An answer for the cryogenic fuels would be to utilize vacuum-jacketed tanks with MLI between the double walls, but the tanks will be heavy and penetrations for fluid access will provide heat leaks which must be compensated by allowing boiloff, or overcome by active means such as a refrigerator. The consequences of a loss of vacuum from leakage of Martian atmospheric gas into the tank jacket would be severe and cause a pressure-relief valve to vent fuel. If venting were not successful, explosive conditions could result, as occurred in the Apollo 13 accident. Foam insulation can be used instead of an evacuated MLI jacket, but controlled boiloff of cryogen would be required unless uninterrupted active refrigeration could be guaranteed.

Non-Cryogenic Fuels. Ethanol and methanol are highly suited to liquid storage anywhere on the surface or underground on Mars. Both fuels have an extremely wide liquid range between the melting point and the temperature at which the vapor pressure is one bar (i.e., the "boiling point"), a characteristic resulting from the highly polar nature of the alcohol molecule. They will thus be easily stored as liquids, as long as at least a modicum of thermal insulation is provided to damp out the diurnal extremes. Methanol can be readily dissociated into hydrogen and carbon monoxide, at temperatures as low as 200°C in the presence of suitable catalysts. Either or both can be used as fuels.

Stored hydrogen or the hydrogen from methanol dissociation can also be used in a H_2/O_2 fuel cell to directly obtain electrical energy without need of a combustion engine and generator. The first major application in the space program for fuel cells was to power the Gemini spacecraft, utilizing the solid polymer electrolyte and feed streams of hydrogen and oxygen from cryogenic stores of the liquids. Also known as the proton exchange membrane (PEM), this system has seen continued development and is a viable candidate for future space applications even though not flown during the past two decades. Apollo and the Shuttle have also employed an H/O chemical energy source, but using the alkaline fuel cell with KOH electrolyte. Both types of FC's achieve about 70% efficiency in conversion of chemical energy into electrical output, but the alkaline type is particularly well suited to continuous operation at a more or less constant power level, whereas the PEM type readily accommodates wide-ranging duty cycles and has demonstrated long life. In addition, the KOH system is deleteriously affected by CO_2, a potential problem in the Martian environment.

Oxidizers. Candidate oxidizers for various fuels include oxygen, nitrogen tetroxide (N_2O_4), nitric acid (HNO_3), and hydrogen peroxide (H_2O_2). Pure oxygen could be stored in either the gaseous (GOX) or liquid (LOX) state. As alluded to above, gaseous-state storage tanks are volumetrically large and heavy compared to the mass of gas stored. On the other hand, LOX would be just as difficult to store as would methane or other cryogens, requiring the double-walled, vacuum insulated tanks discussed above. Tankage factors, i.e., the ratio of dry tank mass to propellant mass, can be quite high. Other candidate oxidizers are more easily stored, although all would be susceptible to freezing and would require electrical heaters and/or radioisotope heaters units (RHU) to maintain the liquid state. The H_2O_2 has the advantage that it can easily be catalytically dissociated to release oxygen, water, and heat, three necessities of a life support system.

Fuel/Oxidizer Selection. Overall, the selection of fuels and oxidizers for a chemically-powered rover system will require additional and more detailed engineering study. Many options are available and more than one design solution may be selected if several different rover systems are developed for the advanced exploration of Mars.

OTHER ENGINEERING SYSTEMS

Communications

Communications may be by line-of-site to the base. With the presumably low H_2O content of Martian soil, and the fact that it is undoubtedly hard-frozen ice rather than liquid, ground penetration by radio waves should be excellent and it could be possible to transmit successfully over the horizon. Satellites and/or the spaceborne orbiting return vehicle can also provide relay capability between the rover and the base site.

Navigation

Guidance and navigation will benefit from major recent advances in miniaturized optical-based gyroscopes, as well as satellite triangulation, radio fixes, and strategically placed transponder beacons. As usual, solar and stellar navigation will provide a backup to the other methods. The minuscule Martian magnetic field rules out the use of a traditional compass.

SAFETY

There are two fundamental aspects of safety for human space missions: System Safety (survival of the hardware) and Personnel Safety (survival of the crew). In the case of rover exploration, however, the two are highly intertwined: if the rover itself becomes disabled when exploring, the astronauts are in peril; if the astronauts leave, the rover may only be able to proceed via teleoperation, and then at significantly increased risk.

As alluded to above, it is concern for safety that provides the motivation for the decision that EVA and unpressurized roving is to be allowed only out to certain distances from a major life support system (pressurized habitable volume). This would be quite problematical if a single rover were to explore alone. Thus, it seems likely that the normal mode will be a convoy of two or more rovers, each with the capacity to rescue and transport back to base the additional crewmembers of the other. In such a convoy, one rover might pathfind, with the other rover operated more conservatively to reduce the possibility of becoming mired. In cases where the pathfinder had successfully crossed an area, the emergency return vehicle could follow in its tracks. When the pathfinder encountered trouble, the backup rover could seek an alternative, safer (although possibly more circuitous) route.

Extraction from obstacle traps (e.g., sinkhole, burial in sand, wedging, etc.) should be an important element of contingency planning. Incorporation of winches and augers into the equipment complement is one possible method of recourse. One or both of the sampling arms might have sufficient strength to assist in escape from obstacles, or one rover may abet the other in these scenarios. It may be especially valuable to have an external driving capability, with the astronauts at vantage points outside sending radio signals to command the rover actions, Figure 9.

UNPRESSURIZED ROVERS

Although the case has been made above that pressurized land rovers will be the most indispensable mobility component of the surface exploration infrastructure, there are numerous applications for unpressurized vehicles as well. A two- or more person unpressurized rover could have a somewhat extended range if it had provisions for umbilical connections to replace the spent oxygen tanks, and to provide backup cooling capacity. However, replacing food and water would be somewhat more difficult, and waste management might be most difficult of all. It is therefore unlikely that a multiple-person bare vehicle will be initially needed.

Personal Rover

On the other hand, especially useful would be a very small rover similar in function to the one-person all-terrain vehicles (ATV) that are a popular item for consumer recreation. For safety, the ATV should include a roll cage which completely protects the astronaut against physical injury in case of a major accident and, most importantly, protects the spacesuit against life-threatening rip or rupture of the pressure garment. For maximum robustness against incapacitation of the vehicle, a requirement that it be lightweight enough that the astronaut can physically turn it upright or extract it from a terrain trap, perhaps with the aid of simple lever bars or small winches. It also would be advantageous if the ATV could be passed through an airlock (perhaps with minor disassembly) so that it could be repaired in a shirt-sleeve environment.

Figure 9 External driving and rover escape from obstacles.

Around the base camp, such a vehicle would be used to speed EVA activities which must occur at sites remote from the habitat. It could also provide for moving small masses of equipment from point to point. On a pressurized sortie mission, an ATV could be towed or transported along as an exploration aid. Thus, when one or more astronauts are performing an EVA, they could utilize the smaller vehicle to move into terrain that would be more difficult and hazardous for the larger vehicle to explore. Examples would include areas such as one with high slopes, a boulder field, a small arroyo (dry stream bed), or a crater rim and interior. The small vehicle might also serve a rescue function for return to base or retrieval of a downed astronaut.

Robotic Rovers

There actually is a whole class of rovers, Figure 10, including robotic, unpressurized rovers which will be of great utility even for the human explorers. These rovers could be of any type: roller, tracker, or walker. Their uses range from mundane tasks like transport of heavy loads at the base to intrepid robots climbing talus slopes, rappelling over cliffs to explore layers in the wall rock, or to reach the bottoms of impact craters or volcanic calderas. These robotic rovers could be operated by humans, using joysticks and telecommunications—i.e., telerobotic control. Exploration by human-robot cooperation is one of the most promising avenues for expanding the scope and depth of missions to Mars. For example, is will be a practical impossibility for humans to conduct site explorations over all the planet, at least until widespread colonization of the planet becomes a reality. But telerobotic explorers, being much less demanding in the support equipment and supplies than living beings, could be landed at many different locations around the Martian globe, including sites too hazardous for landing humans. They could be operated from Earth, but the communications delays of 8 to 40 minutes would greatly constrain their progress. Humans at Mars, whether at a lander base camp or in orbit, could teleoperate with communication delays of less than two seconds. Since the rate at which progress can be made in moving over difficult territory is exponential with time delay, an enormous leverage in exploration can be accomplished by having the human operators at Mars. Of special promise for exploration would be the technique of telepresence, whereby the operator need not issue explicit commands, but rather controls the robot rover by his or her head, eye, and limb movements, perhaps supplemented further by voice commands. This will be an extremely important capability because of the high cost (high value) of astronaut time during the mission.

Figure 10 Human and robotic rovers for Mars exploration.

PLANETARY HOPPING

Hoppers are a method of transportation that have not been routinely exploited in the past. On Earth, air transport (aircraft, helicopters, balloons, dirigibles, etc.) is convenient and diverse. But large flat areas which could serve as natural runways are unlikely on Mars. The Martian atmosphere, with its basal pressure only 5% of Earth's, is too thin to support any but the most vigorous of aerodynamic flight. On the other hand, this very sparseness allows rocket hops with little energy-soaking drag. Ballistic hopping by a single initial impulse is hazardous unless the landing site is known in advance and can be confidently achieved. Controlled landing requires active rocket control and eliminates other possible approaches, e.g., of stored elastic energy in a spring, unless they also provide rocket braking (perhaps parachute assisted). Rocketry consumes large quan-

can be confidently achieved. Controlled landing requires active rocket control and eliminates other possible approaches, e.g., of stored elastic energy in a spring, unless they also provide rocket braking (perhaps parachute assisted). Rocketry consumes large quantities of propellants because of the energy costs of nulling downward velocity and hovering in the gravity field of the planet. On-site manufacture of propellants is the obvious answer to these limitations. Some propellants are much more difficult to manufacture than others. It would be especially challenging to accomplish manufacture of easily stored, high performance rocket propellants such as monomethyl hydrazine and nitrogen tetroxide. Liquid hydrogen, methane, and oxygen all can be produced simply from water (and carbon dioxide, in the case of methane), but must then be compressed, liquefied, and stored at cryogenic temperatures. Nuclear thermal propulsion holds the promise of overcoming the propellant manufacture problem by directly purifying carbon dioxide from the Martian atmosphere. Although of lower performance, system design studies have indicated the feasibility of a rocket based upon this concept.

SUMMARY

The capacity to rove about a planetary surface is a fundamental requirement for adequate exploration, especially for a distant planet such as Mars. Many technological approaches can be envisioned and must be applied to provide pressurized, "shirt-sleeve" conditions as the astronauts traverse hundreds of kilometers or more. Via rocket launch they also may hop to new locations from which more conventional roving may be accomplished. There exists a definite role for teleoperated, robotic vehicles on all human exploration missions, to extend man's presence to areas that are of lower priority or too great a hazard for direct human exploration.

ACKNOWLEDGMENTS

The detailed rover designs, including many of the special features (see Figure 6) were accomplished by M. G. Thornton. The sketches of rover applications (Figures 4, 5 and 8) and painting of Figure 9 are by R. S. Murray.

REFERENCES

The following references are recommended for additional information concerning planetary surface transportation.

Clark, B.C., J.L. Kalkstein, S. Meyer, "The case for the methanol powered planetary rover," *42nd Congress of the International Astronautical Federation*, IAF-91-447, Montreal, Canada, 1991.

Clark, B.C., "A day in the life at Mars Base 1," *Jour. Brit. Interplanet. Soc.,* 43, pp. 489-498, 1990.

Griffin, B.N., "A mobile habitat for early lunar exploration," 42nd Cong. of Intl. Astron. Fed., *IAA-91-l628*, October, 1991.

Griffin, B.N., D.L. Thrasher, and B. E. Wallace, "A pressurized rover for early lunar exploration," *AIAA 982-1488*, March, 1992.

Zubrin, R., "Nuclear rockets using indigenous Martian propellants," *25th AIAA/ASME Joint Propulsion Conference*, AIAA 89-2768, Monterey, CA, 1989.

Zubrin, R., "Long range mobility on Mars," *Jour. Brit. Interplanetary Soc.*, May, 1992.

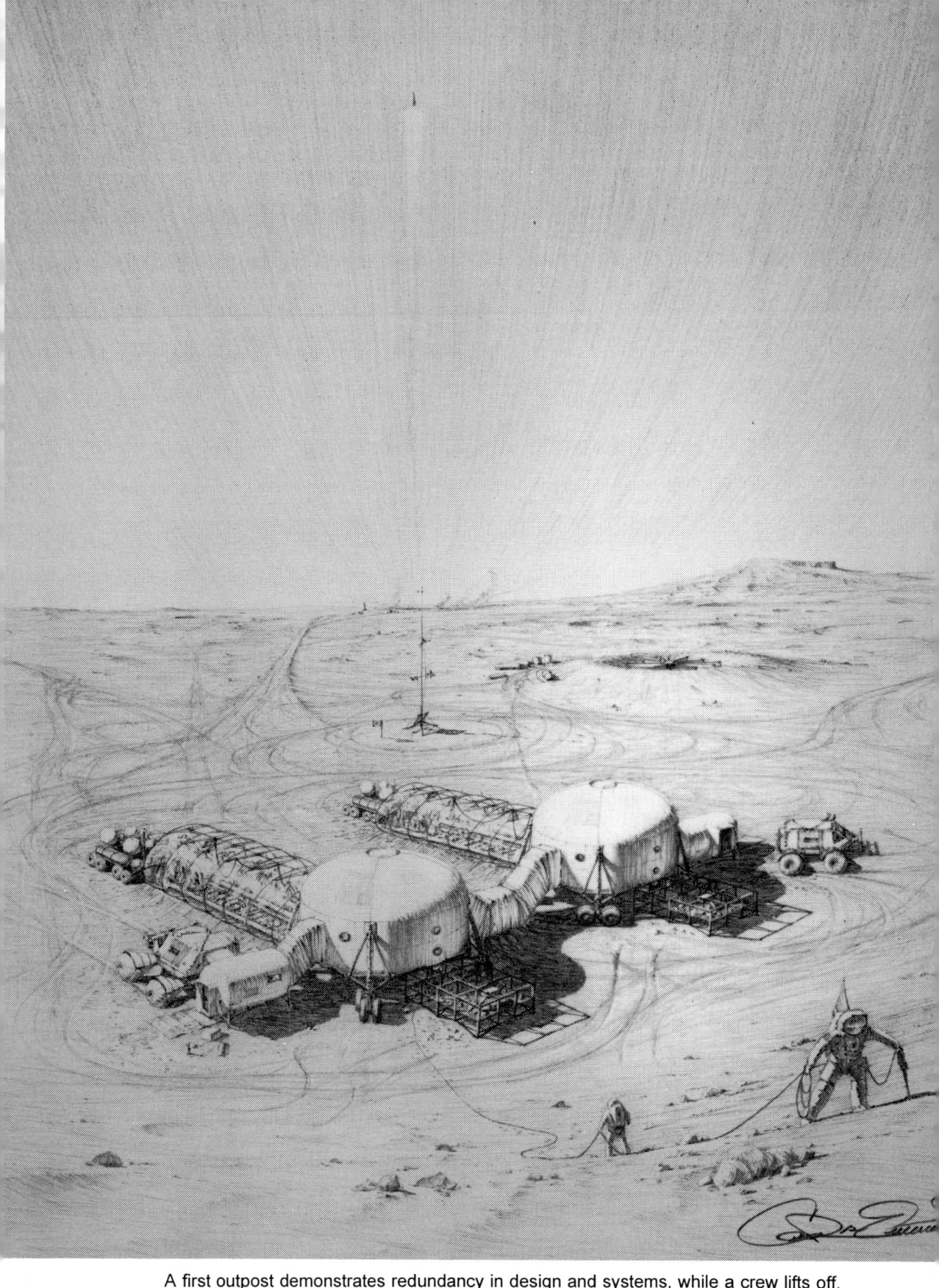

A first outpost demonstrates redundancy in design and systems, while a crew lifts off.

Chapter 21

FIRST MARS OUTPOST HABITATION STRATEGY

Marc M. Cohen[*]

This chapter describes the results of the Mars Surface Mission portion of the 1992-1993 NASA Mars exploration and habitation study. This study proposed two major objectives for Mars exploration: (1) scientific exploration and (2) determining the suitability of Mars as a future home for human habitation. These objectives lead to a problem decomposition that divides a human Mars mission into two portions: "Getting There" and "Being There." This paper addressed the "being there" aspect of the mission. The mission study followed a methodology that posed and answered a sequence of questions:

What are the key issues for Mars exploration?
What evidence do we seek and where should we look for it?
What are the best means to find it?
What support functions will we need on Mars?
What capabilities must we deliver to Mars?

This study framed five components to answer these questions: the objectives and overview, mission design logic, safety philosophy, habitation strategy, and design evaluation. The overview elaborates the two objectives, providing a perspective on how they determine precursor missions, human occupancy of a First Mars Outpost, and the surface mission elements to support habitation.

The mission design logic explicates the assumptions, constraints, and implications of the problem decomposition. These assumptions include: a first mission departure date in 2008, six crew members, 600 day surface stay time, plentiful energy, and decoupling of the habitat from the trans-Mars vehicle. The constraints include precursor missions, cargo landers, robotic operations, *in situ* fuel generation compatibility, crew fitness, and mission abort to the surface of Mars.

The safety philosophy addresses the criticality of mission functions and their failure paths. It correlates mission objectives to risks for life-critical, mission-critical, and mission-discretionary functions. These criticali-

---

[*] Advanced Projects Branch, Space Projects Division, Mail Stop 244-14, NASA Ames Research Center, Moffett Field, California 94035-1000.

ties apply across the major technology areas, including hazard protection, consumable generation, life support, automation and extravehicular activity.

The habitation strategy develops a generic set of requirements to support the science objective in the working environment and the habitation objective in the living environment. This strategy focuses on "environment-human" interactions. It correlates the three criticalities to habitation functions including activation, access, EVA, and architecture.

INTRODUCTION

In 1992, NASA performed a First Mars Outpost study under the leadership of Dr. Michael Duke, Chief of the Lunar & Planetary Exploration Division at NASA's Johnson Space Center. Mars exploration naturally decomposes into two logical activities: "Getting There," which involves interplanetary transport between Earth and Mars, and "Being There" which involves activities on the Martian surface. While many previous studies of human Mars missions had been performed which focused on solving the interplanetary transportation problem, no previous study had really addressed the surface mission in depth. A study team at NASA Ames Research Center focused on the Martian surface mission and this paper represents a product of that work.

In approaching the challenge of defining the strategy for Mars surface missions, the Ames working group asked several questions that shaped their process:

Do we know what we want to do on Mars?
Do we know how to do it?
Do we have the roots of the technologies necessary to carry out these missions?
When can Mars mission planners be ready to go to Mars?

The answer to the first three questions proved to be "yes!" This report elaborates on this yes. Figure 1 explicates this process of inquiry that led to the surface mission requirements. In seeking the answer to the last question, the working group chose the date 2008 for two reasons.

The first reason concerns the 2008 launch opportunity. Historically, NASA has tried to design planetary missions around optimal launch windows that promise the minimum requirements for launch loads and thus the lowest costs. However, NASA has not enjoyed uniform success taking advantage of optimal launch windows. Therefore, the working group began from the precept that the mission strategy should be sufficiently robust to be independent of more favorable or less favorable launch windows. The launch opportunity in 2008 for a conjunction class mission to Mars is the most difficult and unfavorable for many decades to come. If we can plan a human mission to Mars in 2008, we should be able to sustain Mars exploration for any other launch opportunities.

The second reason concerns technology readiness. The participants from all the disciplines asked when they see their respective technologies will be ready. The consensus was that a crash program could make all the necessary technologies ready in 10 years. This approach assumes two technology development cycles of 3 years each plus 4 years to build flight hardware. A program at a more business-as-usual pace would take about 16 years, with two cycles of focused technology development at 5 years each and 6 years to build flight hardware. Interestingly, this 16 year program would come to fruition in 2008.

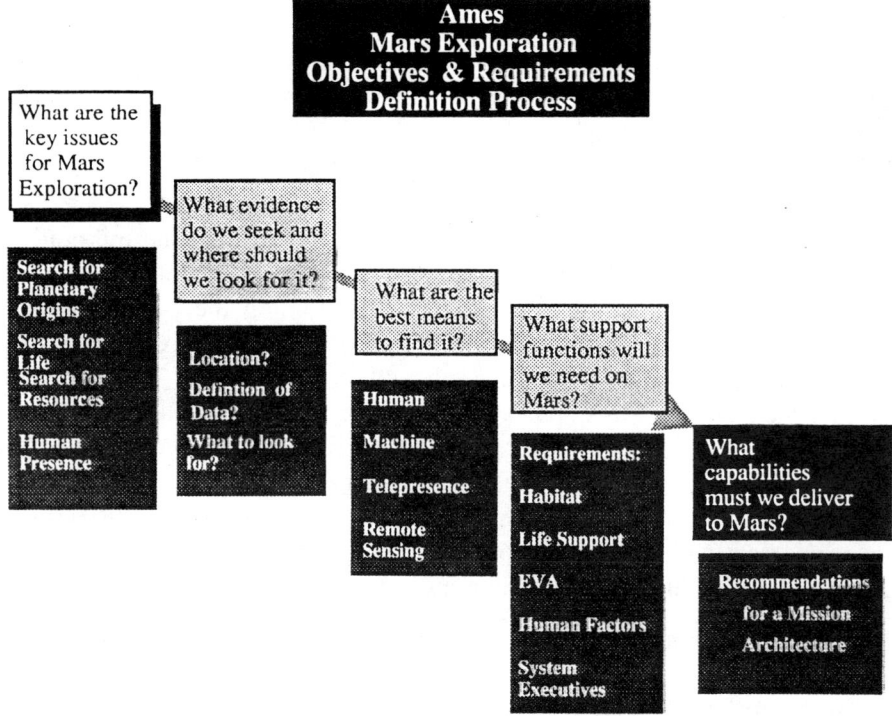

Figure 1 The process of inquiry that led to developing the Mars Surface Mission requirements.

Some of this technology development may correspond to the technology for the First Lunar Outpost (FLO). However, this Mars Surface Mission Study did not assume any direct commonalty with FLO hardware, although the Mars program may benefit from hardware developed for the moon.

The Mars Surface Mission Study began with the assumption that it was somebody else's responsibility to deliver the surface mission crew and all its hardware to Mars. This decomposition of the problem proved serendipitous because it relieved the surface mission of the great burden of system engineering and integration that an Apollo-type program would impose. There was no assumption that the lander would serve as the exploration habitat or that the habitat would serve as part of the ascent vehicle. This problem structure leads to a remarkably small set of "interface requirements" between the two mission elements: "Getting There" and "Being There."

This problem structure means that the design criteria and standards for a Mars habitat differ from a propulsive vehicle in crucial respects. It can be more forgiving of failures. The surface habitat can take advantage of opportunities for resupply of parts and maintenance that would not be available on a propulsive vehicle.

Another important precept was that this Mars Surface Mission Study was not a design study in the sense of providing specific or "point design" solutions. Rather, this undertaking explores the design problem definition, and uses specific designs only insofar as they help explore the problem definition. The purpose of this paper is to articulate the *principles, requirements and strategies* that the surface mission study team believed to be crucial to a successful human exploration of Mars *within our lifetime*. Design examples serve only to illustrate these ideas.

MISSION OBJECTIVES

The design of a Mars surface mission must take into account the mission's objectives. The study team decided that the surface mission would focus on two major objectives: (1) scientific exploration of Mars and (2) determining the potential for human habitation of Mars. Thus, the science and habitation objectives received equal emphasis because the study participants viewed them as equal in importance. The science objective involves understanding Mars' potential to support life in the past, present and future. The scientific approach to meeting this objective is discussed in the paper by Stoker [Chapter 23, this volume]. The habitation and science objectives are mutually reinforcing. Most scientific investigations will demand people on the surface to apply their unique judgment and skills. It is possible to support people on the surface to carry out the science objective effectively only by meeting the habitation objective.

Supporting habitation requires two major elements: the habitat environment and support systems. The habitat environment encompasses the facilities for the crew to live and work on the surface of Mars. The support systems take an integrated approach to life support, power, and *in situ* resource utilization. The crucial common factor for the support systems is the extraction of consumables from the environment of Mars, and stockpiling them into a cache of life support and power generation supplies.

The study team adopted a safety philosophy that approaches the First Mars Outpost more like an Earth-based facility than a spacecraft. The level of risk and required reliability vary with the activities that the crew members undertake and the capabilities necessary to support those activities. The three levels of safety criticality and risk are life-critical, mission-critical and mission-discretionary. The safety philosophy includes an analysis of the mission objectives in relation to the safety criticality levels.

HABITATION OBJECTIVE

The habitation objective is to develop and demonstrate the capability for humans to live on the surface of Mars. Mars habitation devolves into a set of component objectives, beginning with articulating the living and the working environment within the habitat and outside it. This articulation requires NASA to develop and demonstrate the capability of people to work effectively on Mars for long mission durations of up to approximately 600 days. This sustained performance implies a division of labor between tasks that can be performed inside and outside of the primary surface habitat, including the effective use of autonomous, robotic and telerobotic systems. Sustained performance also implies that the crew has effective capabilities to perform both intra-habitat and extra-habitat tasks. At a top mission design level, it is thus necessary to identify locations or situations on Mars where the crew can perform suitable demonstrations of these capabilities.

Assuring sustained crew performance during the surface mission requires the ability to support people on the surface of Mars for up to 2 years. This long duration means that supporting systems must function with high safety, availability, maintainability, and reliability in the Martian environment.

Insuring sustained crew performance also involves demonstrating that the mission risk is acceptable, and that the risk to long-term human health from a single Mars mission is commensurate with other mission risks. This means the risk of loss of a crew member during the mission should be no greater than that expected from comparable, Earth-based hazardous activities. Mission planners must demonstrate that the risks of the surface environment of Mars are within acceptable ranges or that they can develop and deliver suitable precautionary measures, design solutions, or countermeasures.

In Situ Resource Utilization (ISRU) comprises the most critical support system. The mission design must use Mars resources to reduce the dependence on Earth supply so that the mission is neither too expensive nor too dependent on Earth resupply to survive failures. Mission planning should include a demonstration of effective system designs and processes for utilizing *in situ* material and resources to replace products that otherwise would require additional deliveries from Earth. In particular, the ISRU system must extract, process, and use hydrogen, oxygen, buffer gases and water locally at habitation sites. It would extract these life support consumables from local resources, primarily the atmosphere of Mars, and store them. In a similar manner, the ISRU system would produce and store propellants for surface and ascent vehicles, including the ability to transfer propellants to those vehicles. The ISRU development effort may lead to the discovery of other aspects of the environment of Mars that may be important to Mars operations and settlement.

Using the consumables from the ISRU system, the life support systems provide a habitable environment that can sustain human life and productivity in a reasonably Earth-like environment. This robust life support capability assures human survival and success on the surface of Mars for the full mission duration and beyond it, if for some reason the crew is unable to return to Earth as planned. This idea consists of life critical, mission critical, and mission-discretionary capabilities. *Life-Critical* systems ensure crew survival if major systems fail. *Mission-Critical* systems enable the crew to conduct all the baseline mission activities. *Mission-Discretionary* systems enable the crew to go beyond their minimum exploration capabilities and take serendipitous advantage of all the opportunities that being on Mars will afford them. Achieving and sustaining mission-discretionary activities is the true payoff for the Mars-habitation strategy.

A key concept for the Mars surface mission is to create reserves of consumables in a cache on the Martian surface. Establishing an ISRU and life support system encompasses a life support, fuel, and energy cache. To provide life-critical and mission-critical reserves of consumables in a cache on the surface of Mars, the study team recommended sending the ISRU and cache plant to Mars ahead of the first crew. Before the first crew departs the Earth for Mars, the ISRU system would collect and verify a sufficient supply of life support consumables to support the crew on Mars for the entire surface stay-time. After the first crew arrives on Mars, the goal would be to operate the ISRU extraction, consumables cache and the life support systems in a steady state with the crew's requirements for atmosphere, water, food, fuel, and power.

The life support system provides safety, reliability and robustness through functional redundancy. The redundant design operates in several modes. First, it provides an open loop system, drawing consumables from the cache. Second, it provides physical/chemical systems that can revitalize and control the atmosphere, recycle water and control contaminants. Third, it provides closed ecological life support systems that use plant-based units to provide all the functions of the physical/chemical systems plus processing solid waste to provide nutrients for plants and growing food for the crew.

MARS SURFACE MISSION OVERVIEW

This section describes the essential milestones of the surface mission concept: precursors, establishing the first outpost, the arrival of the crew, conducting the exploration mission and departing the Martian surface. It outlines the major hardware elements of the surface mission, including the habitat and its support systems. It describes the major operational steps for the crew to arrive at the First Mars Outpost, activate it, live and work in it for the mission duration, deactivate it, and depart.

Precursor Missions

Before committing to a human Mars surface mission, there will be a series of robotic precursor missions to explore the surface of Mars and to determine environmental conditions. Precursors will also help to select both a landing site and the site for the First Mars Outpost (FMO) habitat. Before sending the FMO hardware to Mars, it should be tested extensively under severe conditions on Earth such as in Antarctica and in Mars atmospheric chambers. It may also be possible to test hardware on the Moon taking advantage of facilities at the First Lunar Outpost.

Mars Surface Science & Site Survey. The precursor missions will provide all the data necessary for selecting the site for the FMO and designing the equipment to prepare it for human arrival. Selecting a region and a site for the FMO will demand an exacting set of decisions that will depend on data sent back by several generations of precursor missions. These autonomous, teleoperated and remote sensing missions will enable NASA to select and survey candidate sites for the FMO as well as, local, regional and global scientific excursions from the outpost. Precursors may include preliminary soil sampling to determine appropriateness of a site for a Mars base or *in situ* resource production.

The site of the FMO will be selected based on an optimization of both regional—general, and local—specific criteria. The general requirements for selecting the site include: proximity to geographic, topographic and geological features of major scientific interest, and proximity to any resources of special value in the Mars regolith. The specific site selection requirements include: a reasonably flat terrain free of large obstacles, lander flight approach and ascent vehicle departure paths that are unobstructed by large features, sufficient flat area to locate the landing zone approximately 1 to 2 kilometers from the habitat core, and soil bearing strength, granularity, angle of repose, etc., at both the landing zone and the habitat zone sufficient to support a 60 to 65 MT lander (wet) without uneven settlement.

The time frame for the precursor missions begins with the launch of the currently planned missions, and follows up with small Mars landers and rovers dedicated to supporting the Mars base development. An exact site selection decision will precede the final design of the Mars base infrastructure to respond most precisely to the local conditions. This decision will determine the need for a habitat foundation, grading or other site preparation.

Mars Surface Technology Experiments. Mars surface precursors present a vital opportunity to understand the salient characteristics of the environment to better plan and design the FMO hardware. These environmental factors may affect surface mission design and costs significantly, especially if the Mars mission team cannot or does not take them into account during the design process. Perhaps the most significant technology precursor needs include a long duration exposure facility to measure the full effects of the Mars environment on space-rated materials, refined characterizations of Martian dust abrasion and deposition effects, the Mars radiation environment, and the potential threat of meteoroids or micrometeoroids. Other technology-related precursor issues include biological and chemical contamination, both forward contamination from a spacecraft to Mars and back contamination from Mars to Earth.

Technology subsystems tests may comprise another class of Mars precursors. It would be highly useful to deploy a prototype or model ISRU plant on Mars to test the ability to extract consumables from the atmosphere or soil of Mars. An experimental plant growth chamber may provide another key technology experiment, to ascertain that a robotically operated system can start seeds and grow plants in the very remote, partial gravity environment. Similarly, it may prove cost-effective to first test robotic construction equipment, rovers, and automated life support systems on Mars prior to designing the final flight articles.

Site Selection: Generic Mars Base Site Plan

Figure 2 shows a generic site plan for the First Mars Outpost. It is generic in the sense that it is not specific to any chosen location, but rather illustrates the typical topographic features of a candidate location on the surface of Mars. To the extent that this generic site plan represents a favorable combination of site factors, it shows an "ideal" site, with the best characteristics for a Mars base.

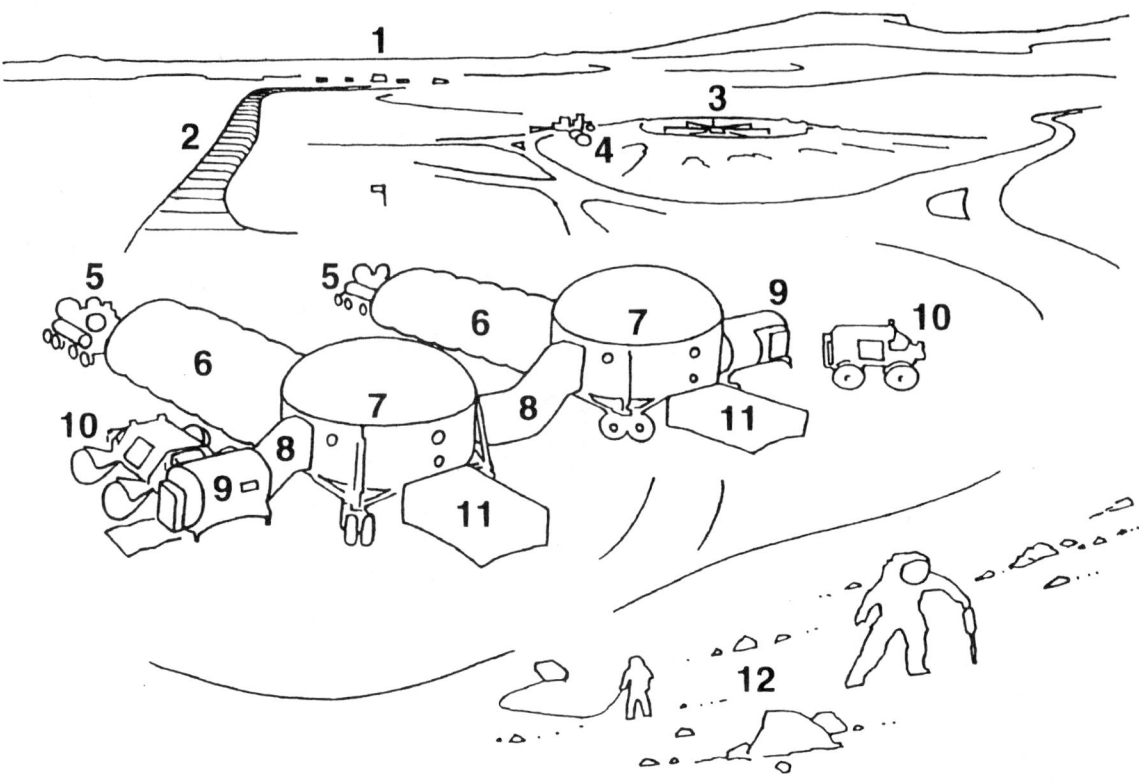

Figure 2 Perspective view of a generic site layout for the First Mars Outpost. Preliminary design by Marc M. Cohen. Artist's rendering by Carter Emmart.

Global Factors. The first criterion is that the site should be "close" to a scientifically interesting area. This closeness may range from tens of meters to tens of kilometers, depending on the availability of a suitable base site near the scientific sites of highest priority.

The location of the site will affect system design in a number of ways. The tilt of Mars's axis of rotation is about 24°, which compares closely to Earth's tilt of 23.5°. The Martian day of about 25 hours is also similar to the Earth's day, so the diurnal cycles of heating and cooling, expansion and contraction of materials are somewhat similar. However, because the Martian atmosphere is only about 1/100 as dense as the Earth's, the mean temperature is much colder, rarely rising above the freezing point of water. If the site is located away from the equatorial zone, so that the solar exposure presents an angle of incidence, the design of a photo-voltaic collector or any other device that makes use of direct sunlight must take this into account. This solar orientation also presents implications for exposure to ionizing radiation. No doubt there are some microclimates in local areas with favorable Sun exposure, with reflecting vertical rock surfaces that also shelter the site. However, it will be necessary to trade any thermal advantage of solar exposure to the disadvantage of direct radiation exposure. There are two major components to the incident ionizing radiation: Cosmic and solar in origin. Cosmic radiation is omni-directional. Solar radiation is directional, including solar protons which are the major problem in a solar storm. If the habitat is always in shadow from a cliff, it cuts out about 50% of the cosmic and nearly 100% of the solar direct. The trade-off is that a habitat always in shadow will be *very cold*, never warming up much during the Mars day. There is one advantage to being cold all the time, which is to minimize thermal cycling, but there may be other penalties we don't fully understand, associated with constant very low temperatures. The alternative approach is to locate the habitat for maximum solar exposure, maximizing reflection from surrounding landforms, and shelter from winds. This approach of making the habitat and Mars base as warm as possible offers advantages for human comfort and performance, while achieving the 50% cut in cosmic radiation exposure. However, it means that the protection against solar storms comes entirely from shielding in and on the habitats.

Regional Factors. The regional factors that make a particular location attractive for a Mars base are primarily the topography and its suitability for the diverse functions of the base. The primary consideration would be to find two level areas, each several hundred meters across, located from 2 to 4 km apart. One site should be free of any large obstacles or obstructions to provide a clear and safe landing zone. The other site should nestle against a cliff to shelter it from cosmic radiation and to maximize solar heat gain. The soil at both these sites should be fairly well consolidated, with a minimum of fine dust accumulation. These two sites do not need to be at the same altitude, but if they are not, there should be a smooth stretch of ground between them to allow hauling the habitats and other large freight from the landing zone to the habitat site. If the landing zone and the habitat site are not at the same altitude, the slope from one to the other probably should not exceed about 1/50, so as not to impose excessive demands for horsepower on the traction engines that haul the freight. For example, if the landing zone and the habitat site are 2 km apart, there should be no more than a 40 m difference in altitude. This altitude difference may present an advantage if the habitat is 40 m higher than the landing zone. A Mars lander is likely to kick up a quantity of dust and debris, and if the habitat site is somewhat higher than the landing zone, it provides a measure of protection against ejecta from a landing.

Local Factors - Landing Zone. According to the mission profile used in this study, as many as four landers of approximately 40 to 50 mTons each would fly to Mars on each launch opportunity of approximately 60 days, every 26 months. This flight pattern suggests short periods of relatively intensive use of the landing zone, followed by long periods of inactivity. The major installation at the landing zone that would always operate is the *in situ* production plant for ascent vehicle fuel. Although the exact formulation of this fuel is open to discussion, the idea is to deliver the ascent vehicle "dry" to the surface of Mars, and then to fill its tanks with fuel processed from the Martian atmosphere. It is possible that each ascent vehicle will carry its own ISRU fuel plant. However, it may turn out more efficient to have a larger capacity permanently on the ground, consisting of one large plant or several smaller plants, making use of the empty tanks on used landing descent stages to store the propellant as they accumulate it. Having this "tank farm" filled and in place would make it possible to fuel an ascent vehicle soon after it landed, without having to wait months or years for it to process enough of its own fuel to take off again. This operational scenario implies that after every lander arrives at the landing zone, a traction engine will move it away from the primary landing area to a perimeter zone. If the ISRU propellant plant requires an external source of power, it may be beneficial to set up solar photo-voltaic arrays or a Topaz 2-type power generator, connected by a cable to the ISRU plant. It would be necessary in both cases to protect the plants from dust and ejecta. It would be possible to install the Topaz 2 nearby in a deep crater. However, the only way to protect the photovoltaic arrays (which will be very sensitive to dust diminishing their capacity to generate power) will be to locate them at a substantially greater distance from the landing zone, with a longer cable.

Habitat Site Selection - Local Factors. The habitat site must accommodate the habitat core, the habitat support functions and all the activities that occur around them. The habitat core consists of a cluster of elements. Following the safety criteria described in this paper, generally there will be two of everything, and more than two in the case of certain life-critical components. Like the landing zone, the habitat system may comprise a nuclear reactor such as the Russian Topaz 2 reactors or photo-voltaic arrays or both. The best location for the reactors again would be in a small crater to provide radiation shielding. Protecting the photo-voltaics from dust will not be as critical as at the landing zone which has the problem of ejecta. Like the landing zone, it may be advantageous to locate a fuel production ISRU plant close to the power supplies, in this case to produce fuel for the pressurized roving vehicles. The combination of the ISRU plant plus Topaz 2 should be at least 100 m from the habitats to protect them against possible mishap or explosion. The power generator provides electricity to the habitat through a cable that may extend several hundred meters, all of which robots emplace. The design of the base must route this cable to avoid the rover and traction engine operation areas to protect it from being run over and damaged. It is probably not necessary to bury the cable in a trench *if* it is possible to safely route it away from traffic, but it would be beneficial to bury it under a small mound or berm of soil to protect it from ultraviolet radiation or accidental damage.

The site plan would place the habitat cluster close to the base of a cliff or escarpment but not so close as to be in danger of falling rock. The area immediately around the habitat core, to a radius of about 50 m, should be clear of all obstacles and have well consolidated soil that can bear the loads that the habitat imposes without excessive or uneven settlement. It may be practical to use the traction engine to help compact loose soil before emplacing the habitats on it. If the habitat modules weigh 40 mT dry, with their water and atmosphere tanks filled, they may weigh twice as much, imposing perhaps 80 mT each on the soil. This ground should be strong enough to support the habitats without needing a foundation or any specialized site preparation beyond rolling with a traction engine.

Habitat Core. Figure 3 shows a schematic plan of the habitat core with ancillary functions. In this scheme, each habitat module has four access ports on the ground level. The allocation and disposition of these pressure ports involves a set of crucial design and operational decisions. The inter-module connector takes up one port access on each habitat module. The habitat core must provide two separate, isolatable pressure volumes. In the event of fire, decompression or contamination in one volume, the crew can evacuate from one to the other volume and close the hatch.

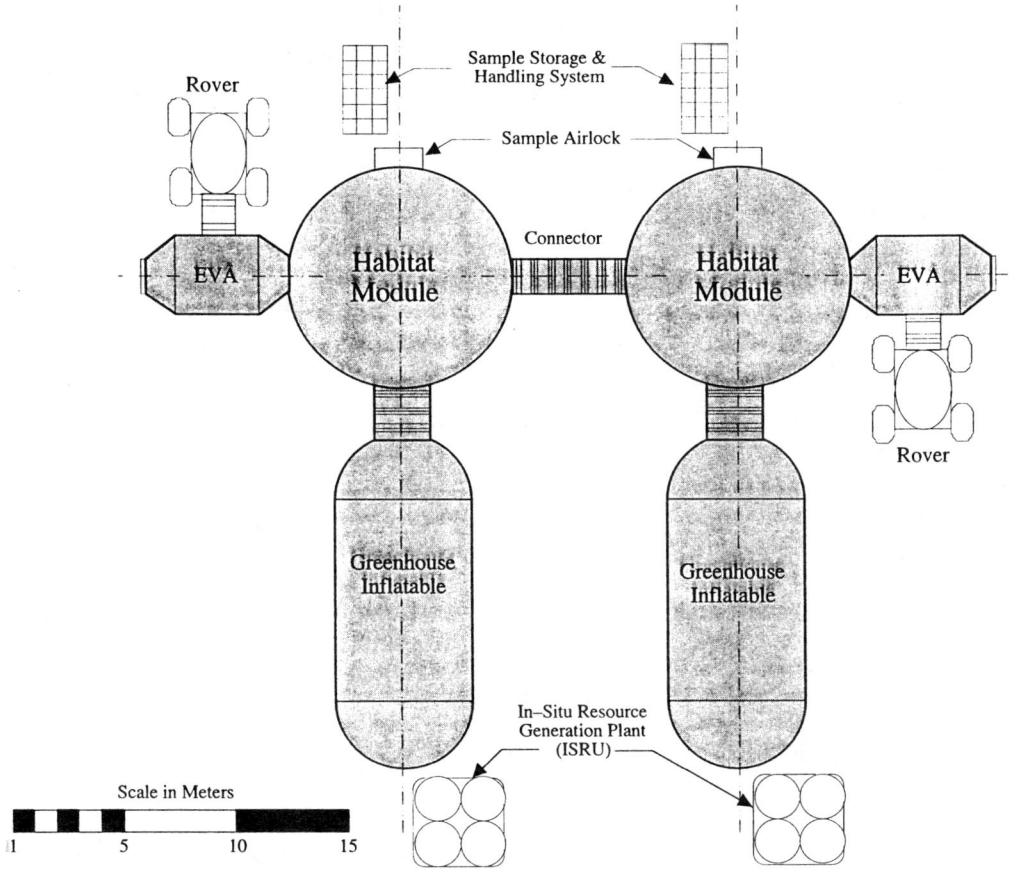

Figure 3 Mars base habitat core with cluster of supporting elements. Dimensions are in meters.

One port on each habitat module provides access to a greenhouse module, probably an inflatable that robots set up before people arrive. The human crew will perform the final set-up and start the plants growing. This greenhouse will provide the growing environment for plants that supply the bioregenerative portion of any ecologically-based life support system. The plants in the greenhouse supplement the physical/chemical life support system. They provide CO_2 removal and O_2 resupply. They process waste water into potable water and consume some solid waste as food. The unique function of the plants, which physical/chemical systems cannot duplicate, is to grow food for the crew.

One port on each habitat gives access to a special EVA (extravehicular activity)+Rover access module, with a specialized airlock to support the crew in donning space suits to explore the surface of Mars. This module also provides a port for the pressurized rover. The rovers can dock to any available pressure port, but there are advantages to the ante-chamber function that the EVA+Rover access module provides. This EVA+Rover module is separate from the habitat core because it is one of the heaviest pieces of equipment, consisting as it does of its own pressure vessel with pumping and compressing equipment. There is likely to be substantial commonality between this access module spacesuit support system and the rover systems.

Figure 4 shows one of the most promising concepts for a Mars surface airlock: the Suitport EVA Access Facility [Cohen, 1989]. The suitport provides pressurized access to a spacesuit through a rear-entry hatch. It saves on pump-down time, power, cooling, and atmosphere loss because the volume to pump down is very small—just the interstitial space between the suit rear hatch and the airlock hatch. An independent Boeing assessment rated the suitport highest for minimizing consumables and for dust mitigation [Capps and Case, 1993]. A NASA Langley/Department of Energy study for a lunar rover proposed installing two suitports in its rear bulkhead [Williams *et al.*, 1993].

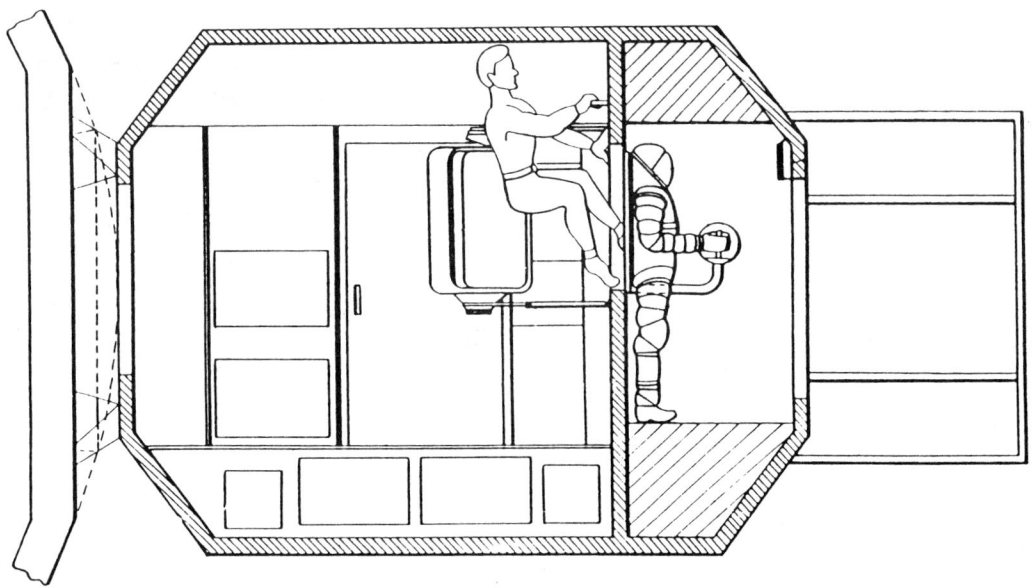

Figure 4 Longitudinal section through a dedicated module for the Suitport Extravehicular Access Facility, US Patent 4,842,224.

The fourth access port in the perimeter of the habitat module accommodates a sample airlock. The crew will collect a large number of samples, and it will be neither practical nor desirable to bring them all into the habitat laboratories at one time. This airlock provides the opportunity for crew members to pass samples into the habitat and back out again without passing themselves in and out of the pressurized environment. A sample storage area will occur near the sample airlock. It would not be very difficult for the general purpose robots to retrieve samples and place them in the airlock or return them to storage, thereby relieving the crew of the need to perform an EVA just to bring in some rocks.

Figure 5 shows a schematic cross section-elevation through a habitat module. This figure shows two crew decks with a half-height equipment deck in between. The shell geometry consists of oblate ellipsoids above and below a fairly short straight cylindrical section. In this variation, the pressure ports occur at the mezzanine level, with flexible tunnels providing the connection to the other functions. This sketch shows the earth bermed up under the end of the tunnel, at the level at which it might connect to a greenhouse or an EVA+Rover access module.

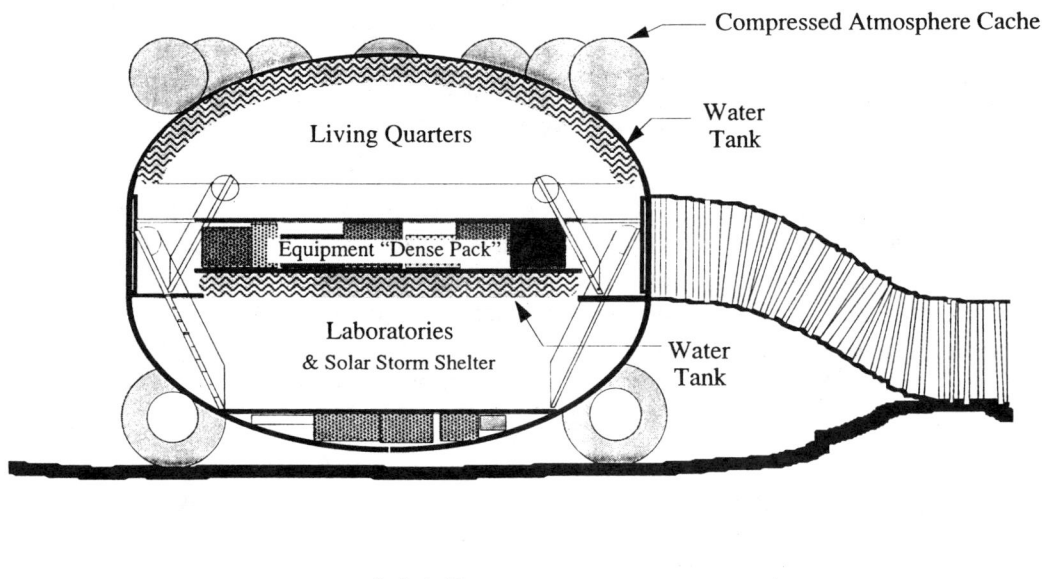

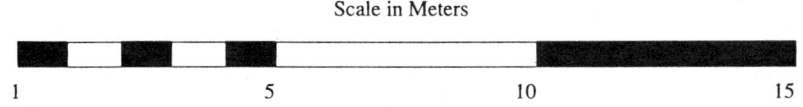

Figure 5 Section elevation through a Mars habitat module.

The location of fixed equipment in an interstitial deck offers advantages over installing it in a ring around each floor deck, as in the Space Station *Freedom* paradigm. The frequent justification for the perimeter scheme is to provide as much radiation protection for the crew as possible. While a perimeter-packing scheme may offer some radiation protection to the crew in a trans-Mars injection habitat (where radiation comes from all directions), it offers no advantage on the surface of Mars, where most radiation comes in at approximately normal (vertical) incidence. This equipment will be much more effective as a radiation shield on the Martian surface if it all occurs in a solid mass, in a position to block normal incident radiation. Removing the equipment from the perimeter also frees up this prime floor area for more valuable, active uses.

Pre-Human Arrival Activities

The pre-human arrival assets will arrive on Mars on four landers during the 2008 launch opportunity. Each lander will carry a payload of 40 MT, which may cluster in unloader modules or units of about 5 or 10 MT. Robotic systems will unload the payloads and move them to their precise destinations where they will install, erect and activate each payload. The principal exception to this set of assumptions is the habitat core, which will land in the exact location where it will stand and operate.

Site Preparation. The purpose of site preparation is to ready the site for habitat construction, activation, and mission-long operations (for at least three crew visits). The site preparation will include a detailed local survey of soil conditions, including borings to determine bearing strength, granularity, angle of repose, etc. On the basis of this data, NASA will select the exact site locations for a landing zone, habitat core, *in situ* production units and resource storage, including tankage for the consumables cache. Site preparation includes soil moving such as grading to remove loose soil at all the erection sites, grading and possibly compaction of a road between the landing zone and the habitat area.

The ideal site would consist of two very large (football field size), flat outcroppings of bedrock—one for the landing zone and one for the habitat zone—to which the robots could anchor the structures, including the habitat core, directly with the equivalent of anchor bolts. If such ideal conditions are not available in the vicinity of features of scientific interest, a greater degree of site preparation becomes necessary, up to and including grading, excavation, and even the installation of prefabricated foundations.

The power systems represent special cases. A solar power array system will require precise orientation. A nuclear reactor such as the U.S. SP-100 or Russian Topaz 2 will need a relatively remote location, ideally with some local shielding between the reactor and the habitat. The best way to provide this installation site may be the "low tech" expedient of explosives that will provide both a suitable hole in the regolith and additional shielding from the ejecta. A chemical/fuel cell type energy system will benefit from proximity to an ISRU production unit and its associated cache storage.

Initiate In Situ Resource Production. The installation of *in situ* resource production plants for O_2, N_2, and H_2O production for life support and energy will be one of the earliest steps to assure sufficient time for resource accumulation. Storing this production will depend on the lander tankage available to store the output. Once the robots completely unload each lander, and remove the descent engines, they will move the lander chassis with the propellant tanks to the ISRU location where the production plant will use the tanks to store its consumables. Following the safety strategy that calls for redundancy in all life-critical systems, there will be two ISRU locations proximate to the habitat core. The output from the ISRUs will flow into the life-critical and mission-critical cache of consumables.

Mars Base Construction. The need for pre-emplacement of infrastructure is becoming an acknowledged stipulation for both the Moon and for Mars. The more that robots or landers pre-emplace, the more critical landing accuracy and/or repositioning capability becomes. Another accepted stipulation is that some construction is required under any long duration lunar/planetary habitat. The real issue is the degree and type of pre-emplacement versus construction.

Habitat Core. In order to meet the life-critical safety design requirements, the habitat core must land completely pre-integrated and ready for activation "at the flip of a switch." This habitat core will consist of a complete lander, fully loaded with a single payload module. Most of the Mars base payload units will be relatively light—in the range moved by a medium, off-road truck on Earth. But the habitat core will be far larger than anything moved by a single vehicle anywhere except on a prepared highway. Therefore, it demands particular attention as a benchmark of the largest task on the surface of Mars. The habitat core lander will home in on a radio beacon that a robot will pre-position, with a goal of landing within 1 m of the target.

Establishing the First Mars Outpost

Robots will deploy, erect, and assemble structures on Mars. The Mars Surface Mission Control (MSMC) on Earth activates and checks out the various portions of the surface system. Any robotic elements remaining in operational condition from the precursor phase will become part of the "inventory" of the robotic surface equipment to support base operations. Until MSMC completes all the necessary testing and certification, the habitat systems operate in the remote, Earth-controlled mode. As the robots complete each phase of construction, MSMC prepares discrepancy reports (if any) and begins to plan remedial actions for the human flight or other flights after completion of the base. MSMC places the surface systems in autonomous operation mode until the crew is ready to depart. Immediately before departure, MSMC will again check all the surface systems. While the crew is en route to Mars, MSMC will monitor the surface systems continuously, providing status updates, and additional plans and training as required for the crew to remedy any discrepancies that may arise.

Human Occupancy

The FMO habitat will support all the human exploration activities on the Martian surface. It comprises a base that includes surface transportation, power supply, consumables cache and the pressurized living and working environment.

Crew Arrival on the Surface of Mars. After arrival in Mars orbit, the crew prepares to descend to the surface. The crew executes a final engineering interrogation of the surface systems. They place the interplanetary transit vehicle into a dormant mode. They enter the Mars Excursion Vehicle (MEV) lander, go through the full check-out and verification, then descend to the surface. No crew members remain on the interplanetary "mother ship."

On arrival at the surface, the crew will immediately place the Mars Excursion Vehicle (MEV) in the habitat mode, to support their initial surface activities. In a contingency situation where the surface system requires remedial action before it is habitable, the crew will execute an extravehicular activity (EVA) to perform the work necessary to bring the surface system to life-critical habitation mode. After transferring operations to the FMO habitat and certifying the surface system in the "mission-critical" habitation mode, they convert the MEV into its storage mode. The MEV will provide sufficient consumables on board to support several EVAs on the surface of Mars, without need to resort to the cached consumables.

The crew performs all maintenance and repair actions identified in the engineering certification tests. Then they complete the operational readiness inspection. The crew begins to experiment, evaluate, test, and certify the engineering and operational capabilities. These procedures apply to all aspects of the Mars surface system.

Surface Exploration Activity. On completing these tests, the crew is ready to begin pursuing the mission-critical and mission-discretionary activities during their surface stay time of 600 days. The crew will carry out these activities over a period of approximately 540 days. The surface activities schedule allows time to convert the FMO to autonomous storage mode, after which the crew prepares for ascent from the surface.

Departing the Surface of Mars. At the completion of the surface mission, the crew reactivates the MEV and converts it to the habitat mode The crew or MSMC reactivates the "mother ship" for the trans-Earth injection. The crew transfers their operations to the MEV. They check out the MEV, review their proficiency training and certify the MEV for the ascent to orbit. After ascent and transfer to the Earth-return vehicle, the crew or MSMC complete the transition of the FMO surface systems to the autonomous mode.

Conducting the Exploration Mission

Once the crew settles into the habitat and verifies that all systems are in order, they will be ready to conduct the surface exploration mission. Figure 6 illustrates the crew time allocation over the 600 days on the Martian surface. The periods called "reserve" periods provide flexible time and a buffer for contingencies/emergencies. This chart derives in part from a database of archetypal capabilities and equipment ideas [Planet Surface Systems Office, 1991]. The horizontal axis illustrates how the crew would spend their days. It assumes that during each day, each crew member has 7 hours available for "production," to carry out the mission objectives (The Mars day lasts about 25 Earth hours, but this allocation does not assume any additional useful time—thus the 24 hour cycle). "Overhead" functions consume 3 hours each day to maintain the habitat and crew health status. The balance of time—14 hours—remains for personal crew use, including eating, sleeping, recreation, time off and personal communications back to Earth. The vertical axis illustrates the sequence of events over the 600 days on the surface of Mars. The crew uses the first 90 days to set up and verify the Mars base and habitat. Similarly, toward the end of the 600 days, the crew devotes about 60 days to shutting down the base and surface exploration systems and preparing for their departure from the surface.

Once the crew completes the base set-up procedures, after about 90 days, they begin a cycle of exploration excursions. A subset of the crew—perhaps two to four—would journey out over the surface of Mars in one or more pressurized surface roving vehicles. This schedule suggests that as many as eight excursions may be possible, in some combination of "local" and "distant" expeditions. A local excursion traverses up to

100 km from the base. A distant or regional excursion traverses up to 500 km from the base. Excursion time includes preparation. Excursions vary from 5 to 20 days, depending on the distance and the time at the exploration sites (although this schedule averages them all to 10 days). After completing the excursion, the crew members return to the habitat for sample analysis and data reduction lasting about 40 days.

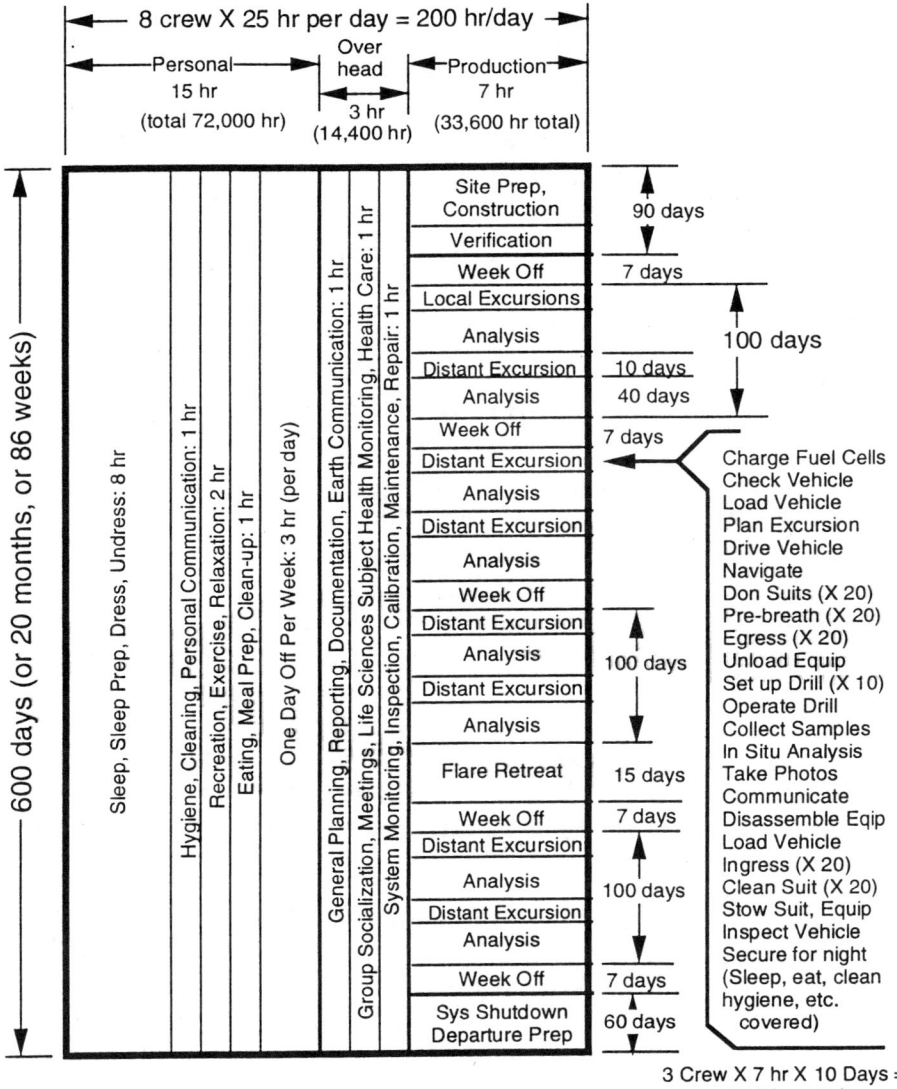

Figure 6 Example of Mars surface mission crew time allocation, courtesy of Roger Arno.

Besides the active base set-up, shut-down, excursion and analysis cycles, this schedule includes 5 weeks of uncommitted "reserve." It also reserves 2 weeks for solar flares when the crew retreats into a radiation storm shelter. This time allocation presents a very conservative approach to time allocation as a baseline for minimum performance. With allocation as a baseline, it creates many opportunities for the automation and robotics, human factors, life science, and life support disciplines first to verify that this baseline is feasible and second to improve on it to achieve higher levels of human productivity.

Surface Mission Habitation Elements

The surface mission will consist of several major elements operating at the FMO site. The ensemble of these elements will provide a potent capability to support human exploration. The major elements include the power system, the *in situ* resource utilization (ISRU) extraction plant and cache, the pressurized habitat (including "greenhouses") and pressurized rovers.

Power Supply. The power system will be the essential enabling factor. Power must be plentiful, which suggests a nuclear reactor as the primary power source. Some other sources such as solar collectors may also prove useful for special requirements or backup. No good estimate of power requirements is available yet.

In Situ Resource Utilization (ISRU). The FMO will depend on technologies to "live off the land" on the planet surface. The ISRU plant will extract valuable life support consumable resources from the atmosphere of Mars including water, oxygen, and nitrogen for a buffer gas [Meyer and McKay, 1981]. Oxygen is relatively abundant as an element in the carbon dioxide, CO_2, that comprises 95 percent of the atmosphere of Mars. Using the Sabatier reaction process, it is possible to reduce CO_2 through the addition of hydrogen, H_2 to generate methane, water and excess heat [Wydeven, 1988, pp. 5-6]. Water vapor also occurs naturally in the atmosphere of Mars and may become accessible through a direct (although energy intensive) compression and dehumidification process [Clapp, 1984]. Then, taking the water produced from one or more of these processes, it is possible to use water electrolysis processes to liberate oxygen, O_2, and hydrogen, H_2, from the water [Wydeven, 1988, pp. 7-9]. The technology is also fairly straightforward to reduce the CO_2 in the atmosphere of Mars directly to extract H_2 [Ash, 1987].

The ISRU system can also produce the fuels methanol CH_3OH for the rovers [Clark *et al.*, 1991], or methane CH_4, for the rovers and the ascent vehicles [French, 1984 Zubrin, 1991, p. 13]. These physical/chemical processes tend to produce an excess of methane compared to likely needs. The ISRU system may recover valuable hydrogen from excess methane through a pyrolysis process that yields elemental carbon and H_2 [Zubrin, 1991]. These simple chemical reactions can produce the compounds essential to supporting life in the FMO.

These consumables flow to the storage cache, which supplies the habitat and its life support systems. The habitat and greenhouse life support systems may also send consumables, including waste products, to the cache.

Consumables Cache. The cache is a central idea in planning for a Mars outpost. The ability to extract raw materials from the Martian atmosphere and store them implies substantial savings in launch mass from Earth and the potential to use launch capacity more productively than shipping water and air to Mars. The cache should store primarily food, oxygen, water, fuel for the rovers, and buffer gases for use by the crew in the habitat. The cache should store sufficient consumables to support the crew throughout their surface stay time or at least 600 days. Figure 7 illustrates the central cache paradigm for life support. The system startup elements include an initial supply of hydrogen from Earth. The cache strategy would include an analogous set of elements for energy and fuel production and storage. The interaction between the energy and life support caches occurs in the conversion of matter to energy and the chemical storage of energy in matter. One example of such a system is a fuel cell that reacts hydrogen and oxygen to produce water and electrolyzes water to liberate hydrogen and oxygen.

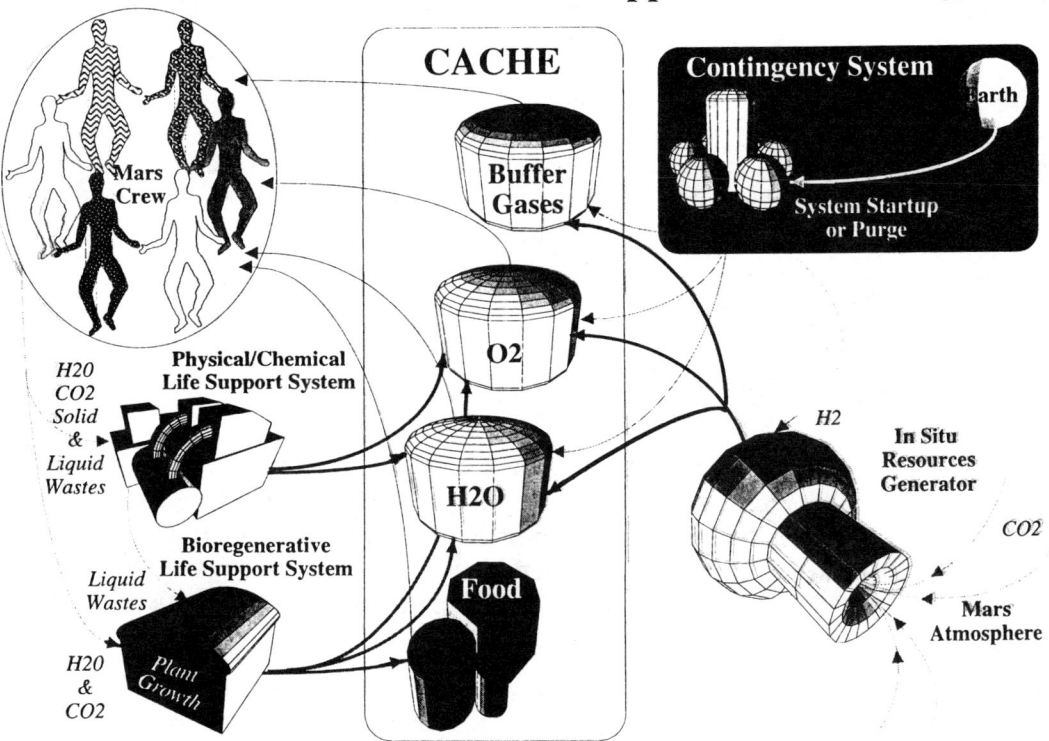

Figure 7 Diagram of the life support cache strategy.

Life Support Systems. The life support system will integrate closely with the cache system. It can run in an open-loop mode, drawing consumables from the cache without recycling. The FMO will incorporate two modes of regenerative life support: physical/chemical (P/C) and bioregenerative, each of which backs up the other mode as an implementation of functional redundancy.

Habitat. The habitat is a pressurized environment that provides the long term living and working environment for the crew members. It provides all the essentials and amenities to support the crew for a healthy and productive sojourn of 600 days. The habitat also houses the laboratories in which the crew scientists analyze and study the samples they collect on the Martian surface. It incorporates or attaches to the EVA airlocks. It provides docking ports for shirt-sleeve entry into the pressurized rovers.

Greenhouses. The Mars outpost ensemble will include one or more greenhouses for plant growth environments dedicated to growing food, restoring oxygen to the atmosphere and recycling liquid and solid wastes. The greenhouses are a special but integral part of the regenerative life support system.

Rovers. Surface exploration will require one or more pressurized roving vehicles that can make traverses of up to 20 days. The rovers should cover a range of 500 km from the Mars outpost. A rover should accommodate two to three crew members for long exploration excursions and the entire crew for a short traverse, as from the habitat to the landing and ascent zone. The rovers must be compatible with the EVA system.

EVA System. Both the habitat and the rovers must support the extravehicular activity (EVA) system. They include airlocks or airlock functions to allow a space suited crewman to transfer easily and quickly from the cabin pressure to the low pressure of the Martian atmosphere. The Mars EVA suit will accommodate the particular attributes of Mars gravity (0.38 g) and the needs of the crew members to use the appropriate tools, bend over to pick up samples and conduct their exploration tasks.

MISSION DESIGN LOGIC

This section attempts to gather and integrate the parameters for designing the Mars surface mission. In organizing these objectives into a coherent whole, it is necessary to make clear the assumptions that underlie them.

For a Mars surface mission, the crew size and cargo mass prove to be major determinants of the mission characteristics. However, the size, mass, and time should not be the sole design drivers. Instead, the scientific and habitation objectives should provide primary guidance to determine the mission design. This section attempts to lay out that philosophy and to illustrate how it may correlate to the mission objectives and thus affect design, implementation and operations.

1. *The first mission departure date is 2008:* 2008 is a difficult launch window for a conjunction class mission to Mars. If NASA develops a mission architecture—including launch capability—that is sufficiently robust to sustain the exploration program through even the toughest launch opportunities, it will greatly enhance the continuity and safety of the Mars exploration program. 2008 is also a reasonable departure date for technology development and building flight hardware.

2. *The crew size is eight people:* The baseline crew consists of eight people, as determined by a task analysis. The minimum crew number is six people. Eight crew members provide a contingency margin over the minimum of six. The crew may consist of any combination of men and women, from all male to all female and may consist of four married couples. This number of crew members provides the essential capabilities to carry out the mission objectives, but it does not consider the potential need for skill redundancy and backup among the crew members.

3. *The baseline surface stay time is 600 days:* The Mars surface mission will take advantage of the relative orbital mechanics of the Earth and Mars to sojourn on Mars for 600 days. The conjunction class mission affords the contingency option of returning to Earth after 30 days [Clark, 1988, AIAA 88-0064]. Using a conjunction class mission, the total mission time is about 1,000 days, with a flight opportunity to Mars at 26 month intervals [Babb *et al.*, 1986].

    Mars exploration enthusiasts advocate a wide variety of mission architectures as "point designs." The "flag and footsteps" 30 to 60 day mission and the "First Mars Outpost" 600 day mission comprise two alternatives within the conjunction class that often appear as opposite choices. Most other options, except for the opposition class and cycler class fall between these two extremes [Rand Corp., 1991]. However, even the fastest and shortest-duration mission must provide for the possibility of the crew becoming stranded on the surface of Mars for 600 days if they miss the return launch window.

    Given the complete set of constraints and capabilities, preliminary estimates suggest that the cost of staying 600 days may be only marginally greater than staying for 30 days with a 600 day abort on the Mars surface backup capability. However the benefit in increased scientific and exploration opportunities will vastly outweigh the difference in cost between a 30 day and a 600 day surface stay time. In this context, the FMO approach does not constitute a "point design." Thus, we use 600 days as a baseline for parametric analysis.

4. *The Mars lander, habitat and ascent vehicle may all be separate designs:* The Mars lander and ascent vehicle are not necessarily the same vehicle. The Mars habitat is separate from both the descent and ascent stages, and its design will focus on supporting the surface mission. This approach seeks to avoid any penalties from forcing commonalty or integration where it is not advantageous. In particular, the habitat does not demand commonality with the trans-Mars or trans-Earth vehicles or the Mars lander. This "decoupling" means that the crew would not need to land in the Mars surface habitat. The habitat may arrive on the previous launch opportunity to take advantage of robotic preparation capabilities *or* it may arrive simultaneously but separately from the crew.

    This example illustrates this mission architecture for the interplanetary vehicle, crew lander, and surface habitat:
    1. The trans-Mars vehicle (TMV) and the trans-Earth vehicle (TEV) are common to each other.

2. The Mars crew ascent vehicle and descent vehicle are common, and may comprise a single stage to orbit and return system.
3. The Mars surface habitat uniquely suits the Mars surface conditions.

5. *Energy will be plentiful:* This assumption means that energy abundance or availability will not limit the mission significantly. This assumption probably means that a solar electric system will not suffice. The FMO most likely requires a nuclear reactor. The Russian Topaz 2 represents a possible prototype of a reactor adapted to operation in the Mars environment. The big issues associated with this reactor include installing it for safe operation and decommissioning it safely afterward.

6. *Technology exists or will exist for this mission:* This assumption should be self-explanatory, but it demands one caveat: technology development does not happen in a vacuum. The Mars surface mission will require a technology development program that commits people and resources over a substantial time. This program must focus on well-articulated goals to support the Mars surface mission to avoid being diverted into other short-term, quick-payback ways of thinking.

Interface Constraints

Mission planners typically describe the interfaces between design elements as *requirements* that one element levies on another as a positive characterization or action. However, the *s*urface mission does not demand commonality with the MEV lander or Earth-to-Mars vehicle. Thus, the FMO approach to element interfaces is more passive than explicit requirements. Instead, *constraint* is a better term to describe these top level element relationships. These interface constraints place limitations on the other elements of the mission, but do not actively demand a particular way in which they should accomplish their objectives.

1. *Autonomous, remote sensing precursor missions—beginning with Mars Surveyor—will provide sufficient data for an informed FMO site selection:* The state of the art for remote sensing satellites and surface landers or rovers, allows a sufficiently thorough survey of the entire surface of Mars to provide an informed site selection process. In the areas that present the greatest interest for the FMO objectives, particularly science, the technology of telerobotics and remote control of surface rovers will facilitate a detailed site investigation directed by explorers and mission planners from the Earth.

2. *The delivery of cargo and crew landers to Mars shall conform to the capacity to launch HLLVs from Kennedy Space Center during single Mars launch window opportunities:* The launch throughput capacity at the Cape is four HLLVs for each launch opportunity (which lasts about 60 days) once every 26 months. The surface manifest shall distribute its resources among four Earth-to-Mars payloads of 40 MT each—that are removable from the lander—plus the mass of the MEV lander. In the case of the habitat core, it will remain on the lander, and may take advantage of the lander mass.

The first launch shall be a cargo mission to deliver the four cargo packets to provide the initial infrastructure for the FMO, including the habitat core and a MEV stage that remains in Mars orbit. The second launch opportunity in 2010 shall include the first Mars surface crew who use the MEV lander and three cargo packets of 40 MT each, one of which is an MEV. The third and fourth launch opportunities in 2012 and 2014 will resemble the 2010 launch configuration.

3. *Between the first and second launch opportunities, an ensemble of autonomous, robotic and telerobotic systems shall set up the FMO and prepare it for human arrival:* It is necessary to send a great bulk of the cargo for the FMO about 2.5 years before the first surface crew arrives for several reasons: There will not be sufficient HLLV throughput capacity at the Cape to send everything on one launch opportunity. The ISRU consumables production plants will operate slowly, and so will need considerable time to build up the consumables cache to a 600 day supply. The packages of equipment that will comprise the FMO will be fairly massive, requiring large machines to unload them. These machines will require a substantial degree of autonomy due to the communication delay to Earth.

Surface crew time is the most precious commodity, and if the crew devoted a large share of their time to construction activities it would divert their attention from the mission-critical objectives—particularly science—and they would need to make extensive use of robotics anyway. Thus, robots will execute as much FMO site preparation, base construction and assembly as possible before the crew arrives. These robotic systems will remain available for the crew to use for a diverse range of tasks.

4. *The fuel tankage on the MEV landers shall be reusable and plumbed to serve as storage capacity for ISRU generated consumables, without need for cleaning, flushing or maintenance:* The gas and liquid consumables that the ISRU plants will most need to produce by extracting and compressing them from the atmosphere of Mars include: H_2O, O_2, N_2, CH_4, and CO_2. This requirement implies that the lander descent engines cannot run on any fuel that is toxic or otherwise incompatible with these essential renewables and consumables. To conserve launch mass, it will be desirable to send the ascent vehicle with empty tanks that the crew will fill from fuel generated and cached on the surface of Mars. Methane (CH_4) is a candidate fuel for ISRU production, using Sabatier and water electrolysis techniques.

5. *The Crew will arrive on Mars in good physical and mental condition, fit and ready to begin productive work:* The crew will also depart the surface of Mars in good condition to ascend to orbit and make the long return trip to Earth [Hurtt, 1992]. It will become essential to define clearly what "good physical and mental condition" means across the entire mission so that the specific context of the surface mission is clear. To serve effectively as a mission design guideline, the definition and requirements for crew health must identify the measures for every biomedical and psychological parameter:
   1. The crew's Earth-normal condition upon launch from Earth.
   2. The crew's status upon arrival at Mars +125-200 days).

3. The crew's status upon ascent from the surface of Mars (+725-800 days).
4. The crew's status upon return to Earth (+925-1,100 days).

It becomes the responsibility of the team designing the interplanetary vehicle to ensure that the system design meets the health requirements at milestone 2 and 4. These two milestones involve the special problem of high gravitational stress possibly coming after an extended period of reduced or micro-gravity while in transit between the Earth and Mars. Lyne [1992] concludes that this situation constrains the deceleration of aerocapture to 5 g's in both the Mars and Earth atmosphere. This deceleration limit, in turn, constrains the aerocapture maneuvers and entry corridor width. In a similar manner, the surface habitation designers must find and implement research-based design parameters to maintain and enhance crew health from milestone 2 to milestone 3.

6. *Abort to Mars surface:* If an emergency occurs on Mars or in Mars orbit, staging a rescue mission from Earth will be difficult and excruciatingly slow because of the narrow launch windows and long transit times. If the crew encounters an emergency that prevents them from returning to the Earth as scheduled, they will abort to the surface—either staying on the surface for another Mars orbital cycle or returning there from the mother ship. The primary abort strategy is to the life-critical capabilities in the FMO habitat.

Problem Decomposition and Design Logic

The mission design logic follows the problem decomposition of *getting there* versus *being there*. Thus, mission elements are decoupled at the highest level of the mission design. Clark [1991] defines decoupling as

. . . an approach to mission hardware architectural design that divides operational functionality into as many discrete hardware units as reasonable, with the objective of minimizing inter-unit dependency. . .

The FMO mission design takes a modular approach in the sense that it avoids highly integrated systems in which a small change in one part can create huge ripple effects throughout the whole. However, it seeks to avoid the downside of decoupling by providing functional redundancy as appropriate for major systems. For example, in the crucial life support domain, the three systems—open loop, physical/chemical regenerative and bioregenerative—can back each other up, but they have different designs.

Availability—Maintainability—Reliability

The FMO approach to safety emphasizes the robustness and survival of the surface systems. We use the following definitions where "probability" is synonymous with "degree of confidence" or "degree of certainty"[Kaplan, 1991]:

Availability.

1. Ready-to-hand alternatives or backups for failed components or systems.

2. Availability (operational): The probability that a system or equipment when used under stated conditions and in an actual supply environment shall operate satisfactorily at any given time [Goldman and Slattery, 1964, p. 26].

Maintainability.

1. Prevention and repair of failures.
2. The probability that an item will conform to specified conditions within a given period of time when maintenance action is performed in accordance with prescribed procedures and resources [Goldman and Slattery, 1964, p. 34].
3. Characteristics of design and installation of an item which enable it to be retained in or restored to a specified operational condition by using prescribed resources and procedures [Space Station Program Office, 1987].

Reliability.

1. Control of failures.
2. The probability that an item will perform its intended function under defined conditions within specified limits without failure [Space Station Program Office, 1987, p. xi].

Safety. Control the loss or hazard resulting from failures.

Traditional spacecraft design methods seek the Holy Grail of near-absolute reliability that is the present watchword for crewed spacecraft. Reliability is vital for some elements for which redundancy is not feasible, such as a pressure vessel shell primary structure. However, the conventional framework of reliability versus redundancy—used to the exclusion of maintainability and resupply capabilities—may stifle creative and cost-effective mission design.

Availability offers a less expensive but equally safe alternative for Mars surface mission design. Availability includes the consideration of reliability. Goldman and Slattery [1964, p. 28] explicate availability as a function of reliability, maintainability and supply effectiveness interacting together. Over a long period (approximately 10 years) at a great distance from Earth, design for *availability* will prove more feasible, cost-effective and ultimately safer than design for reliability alone.

The problem decomposition of *getting there* versus *being there* means that the surface habitat is fundamentally distinct from the interplanetary vehicle habitat. The interplanetary vehicle habitat is part of a propulsive vehicle that requires an extremely high degree of reliability—the proverbial 0.99999—because, during critical maneuvers, there will be little opportunity to make repairs. By contrast, the surface habitat will be a fixed facility that can benefit from a maintainability strategy taking advantage of the resupply capability for spare parts. Whereas the design of the interplanetary vehicle and all its components must optimize to protect against catastrophic failure, the First Mars

Outpost (FMO) surface habitat design can optimize to allow graceful, non-catastrophic degradation until the crew can make repairs or replacements.

Commonality

These distinctions between *availability, maintainability,* and *reliability* shape the commonality approach. It may not be beneficial to design one habitat to serve both the interplanetary vehicle and the FMO because it would impose severe penalties in attempting to accommodate the vastly different and potentially conflicting reliability and maintainability requirements. A comprehensive and integrated lunar/planetary exploration program could benefit from a common architecture for the First Lunar Outpost (FLO) and FMO habitats. A strong case exists also for commonality between the trans-Mars vehicle habitat and the trans-Earth vehicle, which may turn out to be the same piece of hardware. However, because of the very different distances and transit times, there does not appear to be any case for commonality between a trans-lunar vehicle and the Earth-to-Mars vehicle.

SAFETY PHILOSOPHY

The vital question is how to ensure the crew's safety during their stay of nearly 2 years on the Martian surface. The answer to this question demands that the mission planners make their values explicit in formulating their approach to ensure safety. These values include a deep appreciation of availability, maintainability, reliability, and safety, and of the relationships between these disciplines. This safety philosophy "jumps off" from Rockwell International's Space Station Crew Safety Alternatives Study, in particular their precept:

> Cause no damage to space station or injury to crew which would result in a suspension of operations [Peercy *et al.*, 1985, p. 21].

Table 1 shows how we apply this philosophy to define the scope of the safety strategy options that are available for a Mars mission. Generally, the cost increases with the assurance against damage or injury. However, there are numerous scenarios in which it is impossible to ensure *no injury* or *no damage*. The measures necessary to protect against all injury and all damage may preclude carrying out many or most of the surface mission objectives. Thus it is not cost-effective to over-design the safety system. At the other end of the spectrum, it is possible to design the FMO so that the crew can survive any failure by escaping, either with the loss of the FMO (or its key capabilities), or with a significant amount of repair required before it can become operational again. Thus it is not cost-effective either to under-design the safety system, thereby putting the entire Mars mission at risk.

The major gradient in Table 1 concerns the damage to the FMO systems and the necessity and ability to repair those systems. At the top, there is no damage ever, from any "credible failure." At the bottom, there is a recognition that some "credible failures" may cause the loss of the FMO as an operational facility provided the crew can survive by escape or rescue. Both extremes pose potential flaws.

Table 1
FIRST MARS OUTPOST SAFETY PHILOSOPHY PRECEDENCE
(Adapted from Peercy 1985 Table of Space Station Safety Philosophy Precedence)

OPTIONS	COMMENTS
• Cause no damage whatsoever to First Mars Outpost (FMO) and no injury to crew.	Most desirable but most costly
• Cause no damage to First Mars Outpost beyond routine maintenance capability.	Cost trade. All failures or damage fall within routine maintenance
• Cause no damage to First Mars Outpost or injury to crew that will result in a suspension of operations.	**Selected Philosophy** -- repairs possible while operational
• First Mars Outpost is repairable and operational within a specified period of time (depending upon crew status).	May require escape/rescue, Suspension of operations
• Crew survival at expense (or loss) of the First Mars Outpost (depending on crew status).	Implies evacuation and rescue, as a minimum

The "cause no damage or injury" option at the top is extremely expensive—and may not prove practicable—without providing a robust capability to respond to "incredible" failures. Engineering a system to never fail would also limit its ability to conduct any meaningful operations or work.

The "loss of FMO" option at the bottom emphasizes crew survival over the survival of the FMO itself. However, given the unique circumstances of a Mars surface mission, it may not be possible to divorce the survival of the crew from the survival of the FMO habitat and its support systems. This option is also extremely expensive if it means writing off the entire FMO investment because of a single failure that mission designers should anticipate.

The selected philosophy, in the middle, recognizes that crew survival and FMO habitat survival will be interdependent. The second gradient in the table, "injury to crew" reflects this interdependence. The table refers to crew illness or injury only in the top, middle, and bottom boxes. The top option, "cause no injury" is virtually inconceivable because of the potential adventures on the surface of Mars that are difficult, if not impossible, to anticipate. The bottom option, crew survival through evacuation or rescue, does not address the medical care necessary for an injured crew member who is being moved or who must experience a high g lift-off from the surface of Mars.

The middle option allows the FMO to continue operations while a crew member recovers from illness or injury. It offers the most robust alternative. The habitat protects the crew and the crew protects the habitat. The habitat provides a facility for the crew to heal any illness or injury and the crew provides the capability to repair and restore the habitat.

Understanding this choice of philosophy is vital to understanding the whole purpose of this discussion. Different philosophies may yield different approaches, methodologies, and implementations. In distinguishing between the criticality levels, this philosophy allows for some items, components, subsystems or systems to fail and result in a suspension of "mission-discretionary" activities that generally occur outside the habitat. However, this philosophy calls for the ability to repair any failed system without the suspension of life-critical operations or capacity in the habitat core. It also implies the ability to repair any failed system outside the habitat without suspending mission-critical operations.

From this FMO safety philosophy, we extract two top level safety criteria:

1. *The health and safety of the crew is the top priority for all mission architecture, infrastructure, hardware, software and operations.*

2. *The mission design shall protect the crew against any hazard that the mission planners can predict or prevent.*

These two criteria signify that system engineers should not devote a great deal of time to excessively sophisticated but speculative trade studies to justify minimizing or "optimizing" protection from fairly well understood threats such as radiation. These two criteria imply that designers and engineers should spend the least time possible "playing God" with the risks to the crew's health and safety. Instead, design engineers should devote their attention to finding cost-effective and reliable means of protecting the crew. Designers should restrict their use of speculative methods to potential hazards that are difficult to predict, prevent, or perhaps even to understand.

Risk

Understanding risk is central to safety. While some risks apply globally to any mission, all risks stem from specific causes. Following Kaplan [1991] we take the approach, not of defining risk in the abstract, but rather of asking "What is the risk, from a given process or activity?" Then we decompose this question into three others that provide a "'set of triplets' definition of risk":

1. What can go wrong? (scenario)

2. How likely is that to happen? (likelihood)

3. If it does happen, what are the consequences? (damage).

Risk and uncertainty affect almost all human decision making. Design decisions entail risks. The key is how to assess and ultimately to manage the risks involved in these decisions.

The habitation objective addresses the mission risk in stating that "the risk of loss of a crew member during the mission shall be no more than that expected from comparable, Earth-based hazardous activities" [Cohen, 1993]. To meet this objective, it is first necessary to comprehend the codes and standards by which humans evaluate and handle hazards on Earth. For example, most building codes in the United States derive their egress requirements directly or indirectly from the *Life Safety Code*. This code offers guidance on the characteristics, number, and functional and performance requirements of egress-ways to escape from a fire hazard in one occupancy to a safe location in another occupancy or outside the building [National Fire Protection Association, 1991]. In this example, egress becomes a measure *by* which to analyze safety and *to* which to design a safe living environment.

Soon after the last Moon landing, William Thornton MD, an astronaut, described his vision of safety responsibility [Thornton, 1974]:

Safety has to cut across the whole program itself to literally the last man in the chain.

Thornton's "chain" assumes the utmost importance. In this chain, safety is not a separate discipline. Neither is safety the last link—and only the last link. Each link in the chain must stand for safety.

Annick Carino describes an attitude of this sort as a *Safety Culture*. Carino's safety culture "is a behavioral attitude, individual and collective, whereby a special attention is paid to all safety matters." [Carino, 1991]. It is vital that these concerns pervade the design project organization. Guy Freeman describes this safety-conscious design organization as a "Design Culture" [Freeman, 1991].

This First Mars Outpost approach anticipates and requires a safety-conscious design culture. Safety should serve as a driver for mission success that enjoys equal importance with the mission objectives of science and habitation. Only in this way will the mission designers and planners devote that "special attention" from the first to the last "link in the chain."

Cost of Safety

A fundamental assumption for implementing design safety is that when designing a safe mission, spacecraft, habitat, or *anything*, the initial cost of designing it to be safe is not significantly different from designing it to be unsafe. The huge costs sometimes attributed to safety programs come far downstream in the design process. These costly corrections occur when the project management discovers overlooked conflicts, errors, faulty assumptions, inadequacies, omissions, or other unsafe design features. Huge cost overruns do not arise from doing it right the first time. This discussion attempts to set the stage for going to Mars right on the first attempt.

Safety Criticality

At the same time that FMO attempts to be highly safety-conscious, it departs from previous safety philosophies or strategies that espouse a single, uniform level of risk-tolerance. It recognizes that it is neither practical nor cost-effective to provide a uniform level of risk tolerance for all portions of a Mars surface mission. This recognition led to a new approach to the application of criticality, failure mode and redundancy analyses.

It is vital to reinforce the precept that all crew members are essential and equally deserving of protection from all hazards. However, some activities inherently carry greater risk than other activities. Similarly, the facilities that support these activities imply varying degrees of risk with corresponding needs for reliability and redundancy. These dual variations of risk and reward come together in the idea of criticality.

Criticality is a key idea in evaluating the safety characteristics of a space system or component design. Traditionally, criticality applies to individual items, which combine in a bottoms-up manner ("critical items list") to shape a safety baseline. Criticality has several dimensions:

1. The effect of a malfunction of an item on the performance of a system [Goldman and Slattery, 1964, p. 27].

2. The relative measure of the consequences of a failure mode. Criticality assumes the loss of all redundant hardware items [Space Station Program Office, 1987, p. ix].

3. The criticality categorization of a hardware item shall be made on the basis of the worst-case potential failure effect. Criticality shall be determined by the categorization of the failure mode effect on the system/user/mission support/crew/. . . Program Element assuming loss of all redundancy (like and/or unlike) for performing the function [Space Station Program Office, 1987, p. 3-2].

The mission architecture subdivides *functionally* into three categories of risk per reward: *Life-Critical, Mission-Critical,* and *Mission-Discretionary.*

Life-Critical capability ensures the survival of the crew for the duration of the mission and their safe ascent from the surface of Mars. It does not meet either the science or habitation objectives.

Mission-Critical capabilities ensure that the crew can meet the fundamental science and habitation objectives for the Mars surface mission in the near vicinity of the First Mars Outpost.

Mission-Discretionary capabilities enable the crew to take advantage of the all opportunities for discovery and exploration on the surface of Mars, pursuing the science objective to some considerable distance from the FMO and to meet the full range of habitation objectives.

These criticality categories apply both to the FMO architecture, mobility capabilities, and to the activities that the crew uses to carry out the mission objectives. The relationship among these three categories is a hierarchy of needs and capabilities. It dictates that the crew does not undertake mission-discretionary activities unless the mission-critical and life-critical capabilities are both nominal. The crew does not undertake any mission-critical activities unless the life-critical capabilities are nominal. These categories of functionality provide systemic backup to each other to meet a range of potential system failures and safety threats.

Failure Path

Another perspective on risk and safety comes from the notion of failure path and system redundancy to protect against the effects of failure. The terms fail-op, fail-safe, and fail-restorable are among the currency of space system designers.

Fail-Operational (Fail-Op): If one level fails, the system can continue to operate normally and still have a backup [Cerimele, 1991].

Fail-Safe: If a second level fails, the system has become "fail-safe," because it can still operate safely, but it no longer has a back-up. The mission would then be aborted [Cerimele, 1991] (or that part of the mission would be aborted).

Fail-Restorable: As an alternative to aborting the mission or some part of the mission, it is possible to restore the back-up capability that makes the system fail-operational, without loss of normal operations.

For the FMO safety philosophy, these system redundancy definitions act in parallel with the criticality definitions. By combining these two analytical approaches, it became possible to articulate a comprehensive safety strategy that ultimately drives the architectural design of the FMO habitat ensemble.

Life-Critical Fail-Op/Fail-Safe/Fail-Restorable. Life-critical elements and capabilities ensure the basic—but bare—crew survival if there is a severe failure in mission-critical systems, but not to do any science work. All life-critical items shall benefit from the fail-op/fail-safe capability implicit in the mission-critical systems. In addition, the life-critical systems shall have the capability for the crew to repair or restore the back-up systems while in a multi-failure "safe haven" mode, without loss of life-critical systems or operations.

The life support system for the life-critical capability shall depend upon "'rock-solid' technology" to use the 600 day cached consumables. "Rock-solid" means that there is no need to regenerate consumables, which the system draws in open-loop mode from the cache. Life-critical capabilities include proximity EVA so that crew members can make repairs on systems outside the habitat.

Mission-Critical Fail-Op/Fail-Safe/Recoverable. Mission-critical capabilities provide the essential ability to carry out the objectives for science, habitation, and resource utilization, and to support the crew in their normal, daily routines. All mission-critical elements shall follow the dual redundant model of the essential widget: One widget in service, one in scheduled maintenance, one widget in excellent condition on standby as a system backup. In the event of any component or subsystem failure, there is a backup ready and waiting to take over. The backup will kick into service automatically or be "replaceable on line," depending on the specific item and situation. This backup means that the crew will not need to drop everything they are doing whenever a backup system kicks in automatically. However, in a two-level failure, the crew would need to stop their mission-critical activities and

1. Deal with the problem,

2. Retreat to a life-critical status *or*

3. Drop that mission-critical activity.

Mission-critical functions would be recoverable, but not "fail-restorable." This distinction means that the crew would be able to restore a mission-critical capability after a sequence of failures, but would not be able to maintain that capability in operational status at the same time. The nominal time to restore a mission-critical function that fails completely would be 2 weeks, the same period as mission-critical start-up.

The retreat from mission-critical to a life-critical status corresponds to "Cerimele's 'Category III'" for a system that qualifies for man-rating, where "the system will have some means of escape or safe haven" [Cerimele, 1991, p. 4]. this life-critical safe haven strategy differs significantly from the Rockwell Crew Safety Alternatives Study that considered retreat to an internal safe haven as one of five options for escape and rescue [Peercy et al., 1985, p. 11]. For the much more distant and isolated FMO habitat, the life-critical safe haven appears to be virtually the only option, although where to locate the safe haven remains an important design question. The design of mission-critical systems, including the habitat itself, must consider this scenario as an operational approach to dealing with contingencies and emergencies.

The life support system baselined for the mission-critical capability is a regenerative physical/chemical system that buffers its consumables through the 600 day cache. The caution and warning system will enunciate alarms that provide an intelligent analysis of the system status in the case of a component or subsystem failure that requires the substitution of a backup component, subsystem or system.

Mission-Discretionary Fail-Safe/Recoverable. Mission-discretionary exploration and experimentation activities promise the greatest potential for scientific reward but they also entail the greatest potential for risk. Mission-discretionary capabilities depend on the crew completing mission-critical tasks and demonstrating mission-critical elements. Mission-discretionary elements provide the systems that may not be initially available on crew arrival. Mission-discretionary items include growing plants for CELSS, uses that depend on successes in demonstration in the mission-critical tasks—such as starting a regional (up to 500 km) rover excursion only after the crew successfully accomplishes a local (up to 100 km) rover excursion.

Like the mission-critical functions, the mission-discretionary capabilities should all be recoverable, but not fail-restorable. The nominal time to recover a mission-discretionary capability is 2 months, the same as mission-discretionary start-up. All mission-discretionary items shall have at least one backup system or replacement. The failure or loss of a subsystem in a discretionary capability or subsystem means that the crew needs to interrupt what they are doing—to assess the problem or to evacuate to another location to better assess the problem. Because the mission-discretionary capabilities are all fail-safe, the repair or replacement of the discretionary items can wait until the crew can schedule it. However, the crew will not be able to recommence the mission-discretionary activity until they accomplish the repair.

The long-range rover excursions on the surface of Mars offer an example of a mission-discretionary fail-safe system. Clark proposes dual long-range rovers to provide backup and safe return [Clark, 1988, AIAA 88-0354]. Two rovers traveling together create a fail-safe situation. Each rover backs up the other on the excursion. If one of the rovers fails in an unrecoverable manner, its crew may transfer to the other pressurized rover. The crew ceases all mission-discretionary activities because they no longer have a back-up, and then return immediately to their base. Once they return safely to the FMO, the crew may consider whether to return to the failed rover to fix it and plan how to do it. Thus, they may recover the failed rover, but only with the suspension of mission-discretionary activities.

If three rovers traveled together on a long traversal excursion, it might be possible to upgrade the safety criticality from mission-discretionary to mission-critical. This upgrade bears the caveat that any one rover must be capable of accommodating the crews of all three rovers in the event that two rovers fail. This example of a three-rover caravan demonstrates that there are potential limits to cost-effective redundancy. The requirement that all rovers must accommodate three crews to qualify as fail-op/fail-safe may prove an unacceptable penalty for the rover design.

The life support system associated with the mission-discretionary capability is a bio-regenerative system that uses plants to maintain a normal atmosphere, process solid and liquid wastes and to grow some of the crew's food, providing fresh vegetables, grains, and legumes and possibly fruit. In the case of the CELSS system, it will probably share a common system of ductwork and sensors with the mission-critical and life-critical life support systems, but add a completely separate set of regeneration components.

Functionality and Risk

The most significant aspect of the relationship among these three criticalities is the way it distributes the specific activities into risk categories. For example, in the life-critical mode, the crew performs no science work and pursues no mission objectives except to ensure their own survival. At the opposite end of the spectrum, the mission-discretionary capabilities promise the greatest pay-off for the Mars surface mission, such as visiting widely dispersed and very different exploration sites for comparison of data. In terms of the mission-discretionary consumables, successful operation of a self-sustaining CELSS that provides all needs to the crew, including all the basic food stuffs would be the Holy Grail for all life support system designers. The mission-critical capabilities occupy the middle ground, on which the crew accomplishes the prescribed set of goals, but not so easily that the decision is automatic to move on to mission-discretionary activities or enhancements.

Philosophy of Design for Risks versus Objectives

Table 2 illustrates the three risk levels of life-critical, mission-critical and mission-discretionary functions. These functions back each other up in a distributed strategy. This strategy articulates a hierarchy of needs such that the crew does not do mission-discretionary activities unless the mission-critical capabilities are nominal and they do not do mission-critical activities unless the life-critical capabilities are functioning. In the event of a problem, the crew would "fall back" from the mission-discretionary capabilities to the mission-critical capabilities to the life-critical capabilities. However, a fallback does not imply total loss of the resources from evacuated facilities or capabilities. Thus, the network operates with the mission-discretionary resources backing up the mission-critical resources and both of them backing up the life-critical capabilities.

Table 2
MARS SURFACE MISSION SAFETY DESIGN PHILOSOPHY

	LIFE-CRITICAL "SAFE HAVEN"	MISSION-CRITICAL	MISSION-DISCRETIONARY
Failure Path	*Fail-Op / Fail-Safe / Fail-Restorable*	*Fail-Operational / Fail-Safe / Recoverable*	*Fail-Safe / Recoverable*
Hazard Protection	Solar Storm & Ambient Internal Radiation, Dust storms, Fire, Contamination, Decompression.	Ambient Internal Radiation, Dust Storms, Fire, Contamination, Decompression of hard pressure vessels. Soil conditions on Rover excursions	Ambient Internal & Higher Risk External Radiation Exposure, Decompression of inflatables, Fire, Dust Storms, Contamination.
Consumables Cache	*600 Days* stored O_2, N_2, H_2O & Energy to ensure survival in Life-Critical mode.	*200 Days* stored O_2, N_2, H_2O, Energy & Fuel (for Rovers) to support Mission-Critical activities.	*TBD Days* stored O_2, N_2, H_2O, Energy, Fuel (for Rovers) & Food to support Mission–Discretionary activities.
Life Support Consumables	Open Loop: ISRU Produced & Stored Cache.	Primary: Physical / Chemical Regenerative with In-Situ Cache Support.	Primary: Bioregenerative with Physical/Chemical backup. Includes growing food.
Technology Reliability	"Rock Solid" stand-alone plus redundant backup from other parts of habitat. Accept no losses.	Redundancy and Robustness. Accept loss of one unit of each system or of one complete module.	Redundancy and Robustness. Accept loss of a complete system or all inflatables.
Automation	Full automation with complete manual backup.	Automation to support and enhance human productivity, human-machine interfaces.	Choices of automation vs. human control
Egress / Access	1 to pressurized volume 1 to EVA airlock 1 to shirtsleeves rover	1 to pressurized volume 1 to EVA airlock 1 to shirtsleeves rover	1 or more to greenhouse 1 to rover garage

Criticality: Risk versus Mission Objectives

The three criticality levels present profound implications in how the two mission objectives—science and habitation—interact with the risks that they entail. Table 3 illustrates this interaction as it affects a wide range of activities. The FMO design introduces new activities at the mission-discretionary level where the crew sets them up, tests, demonstrates and verifies them before incorporating them into the mission-critical level of activities. The crew will not do mission-discretionary activities unless the mission-critical capabilities are nominal, nor mission-critical activities unless the life-critical capabilities are nominal.

Table 3
600 DAY MARS SURFACE MISSION 1 (2010) - ASSESSMENT OF RISK VS. OBJECTIVES

MISSION 1 RISK LEVELS	OBJECTIVES		
	Science	Habitation	
		Living & Working Environment	Life Support & ISRU
Life-Critical	• None	• Survival in Habitat • Mental and Physical Health • Proximity EVA (<2km)	• Open-loop Consumables from cache for 600 days. • Energy generation. • Fuel Production for Ascent Vehicle
Mission-Critical	• Proximity Operations (<20Km) Rovers & EVA	• Productivity • Sustained, Reliable Human Performance	• Regenerative Life Support (Physical Chemical or Bio) • Cache Restoration • Energy Production
Mission-Discretionary	• Local Operations (<100Km) Rovers & EVA • Regional Operations (<500Km) Rovers & EVA • Global Operations (Tele-Op)	• Recreation • Extended Time Off Duty • Access to greenhouses • Access to inflatable "garage"	• CELSS: Waste, CO_2, O_2, & Food • Inflatables • Fuel Production for Rovers

Note: This table presents a "Dependency Matrix." The activities and elements at each risk level depend upon the ability to perform the entries in the risk level above it.

Table 3 describes the activities on the first crewed Mars surface mission. This mission focuses on mission-critical activities. It limits the mission-critical science excursions to a proximity not to exceed 20 km, a distance from which it may be possible for the crew to walk back to the FMO (it assumes two rovers). In the life support domain, the crew activates the ecological systems. Thus all CELSS systems are mission-discretionary for this first mission.

The next step for Mars surface mission safety planning is to develop a safety methodology that covers several key points:

1. "How much safety is necessary (i.e., cost effective)?"
2. "What are the threats that the program is prepared to deal with?"
3. "What are the strategies and interdependence of these strategies to meet the criteria developed to deal with the threats?" [Peercy et al., 1985, p. 4].

The identification of threats to safety involves analyzing and understanding the unique conditions on the surface of Mars. The development of strategies depends largely upon the activities to pursue the scientific research and habitation, comparing these activities to each other. From the threat and strategy assessments, it should become possible to decide "how much safety is necessary and cost-effective."

HABITATION STRATEGY

A human mission to Mars will involve a total mission duration of 900 to 1,100 days that will change the familiar character of space exploration both quantitatively and qualitatively. Mission duration [Cohen, 1991] drives the human factors, life support, systems autonomy, safety/reliability and health maintenance issues of a space mission more than any other single factor. Long mission duration compounds and magnifies all the critical aspects of isolation, confinement, social organization, training and decision making. Human factors and human performance are central to the relationships, interactions and mutual dependencies of these technical capabilities.

This section addresses the facility requirements, the implementation plan to set them up, and the operations plan to operate them. These requirements derive from both the science and habitation objectives, framed by the habitation strategy.

The single most important key to mission success is human habitation on Mars. To assure this success, the crew will live in a habitat designed for human health, productivity and safety and for economical native resource utilization. The habitat includes the living and working quarters for the crew as well as the facilities that will enable the crew to handle virtually any emergency or problem that may arise. The pressurized rover capabilities will augment the habitat capabilities to provide a robust surface habitation system. This section explores the functional and performance requirements for the FMO habitat for the First Mars Outpost.

The psychologist Abraham Maslow developed a model of human needs as "an attempt to formulate a positive theory of [human] motivation" [Maslow, 1970, p. 35]. It represents an attempt to create a synthesis of the diverse physiological, social, emotional, perceptual and cognitive bases of human motivation. The habitable environment influences human motivation and behavior, through gratification or deprivation, or a host of other perceptions or conditions [Maslow, 1970, p. 46]. Figure 8 illustrates Maslow's model as a hierarchical pyramid having five levels, characterized from the bottom up as: physiological needs, safety, belonging, self-esteem and self-actualization. Each level serves as a prerequisite to support the levels above it.

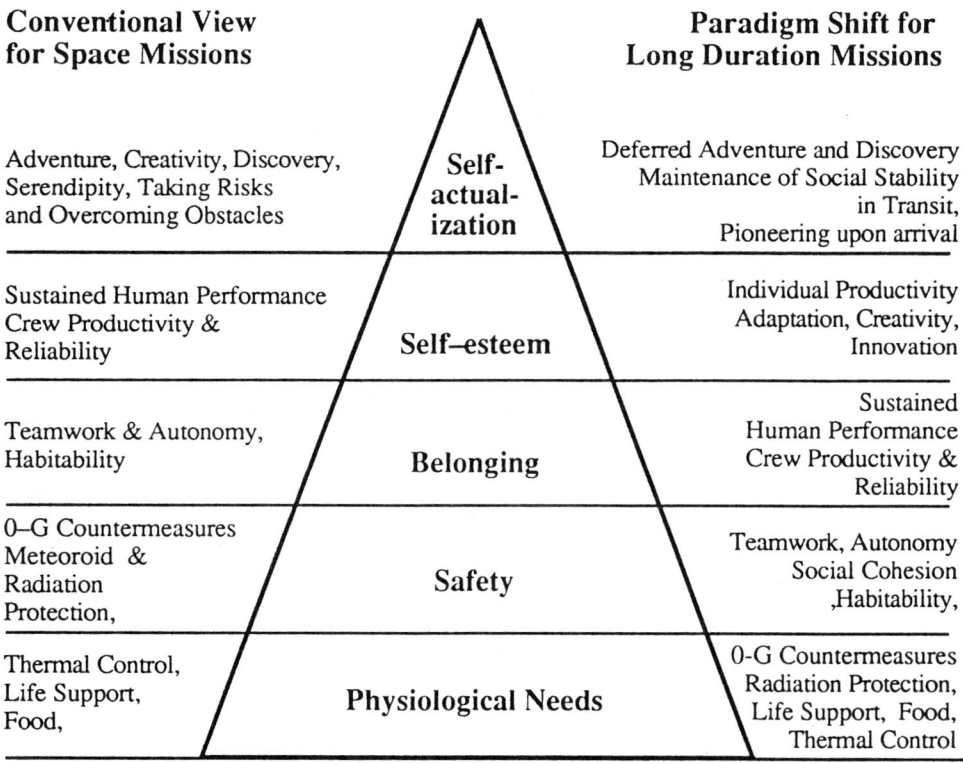

Figure 8 Maslow's model of human motivation as a hierarchy of human needs showing the paradigm shift from current, short-term space missions (e.g., Space Station) to long-duration exploration missions (e.g., 600 Day Mars Surface Mission).

This diagram of Maslow's model suggests the possibilities of interaction between both different levels of support and the diverse, complex technologies on which human space flight depends. Health problems related to zero gravity or radiation could undermine crew productivity, reliability, safety and capability for sustained performance, thus reducing the effectiveness of teamwork, which in turn could compromise the monitoring and maintenance of thermal control and life support. These "'cascading' system effects" are characteristic of human error-caused disasters in aviation and nuclear power plants.

Maslow's theory has far-reaching implications for space habitat architecture. It matches up with issues in the current space station program shown to the left of the

501

pyramid. While there appears to be fundamental agreement on physiological needs such as air, water, food and thermal comfort, less agreement exists concerning the higher level, human productivity needs. Connors *et al.* [1985] describe the baseline human requirements for long duration missions. As one moves up the pyramid, the program management tends to treat these issues more and more as expendable options. However, for Mars missions, the engineering temptation to "'trade-off' cost for comfort" would be a "major mistake" from the human factors point of view [Clearwater and Harrison, 1990, pp. 513-518].

Human Motivation and Needs Paradigm Shift. The paradigm of space systems engineering holds that every component of a space program has features of cost and benefit that are subject to manipulation "trade-offs." For a successful interplanetary journey (and perhaps for most other, more near-term long duration missions) this paradigm must change to recognize that some elements are essential to crew performance beyond just keeping them alive and working long shifts. The alignment of mission system engineering values and decision making will need to shift downward against the hierarchy of human motivations and needs in Maslow's model.

Countermeasures to weightlessness and radiation are good examples of how this paradigm shift will occur. Presently, both the Soviet and American space programs are contemplating missions to Mars using drug and exercise countermeasures to counteract bone demineralization and muscle atrophy. They consider some degree of deterioration (and recovery after return to Earth) as acceptable [Newkirk, 1990, p. 303]. However, for a "permanent human presence in space," in the form of a lunar or Mars base, providing artificial gravity countermeasures shifts from a safety trade-off option to an absolute physiological requirement. Similarly for radiation protection, the traditionally allowable—and realistically measurable—exposure is defined by the month, 90 days or the year, but not for a lifetime. The National Council on Radiation Protection also promulgates a "career" or "lifetime" limit on radiation exposure which NASA has adopted [NASA STD-3000, 1989, Vol. 1, Rev. A, p. 5-98]. However, it is difficult to imagine NASA designing a Mars mission around a *lifetime* radiation exposure limit. There is good reason to reject the "design to the exposure limit" approach—even an annual exposure—limit for a mission that may last 3 years. Raasch, Peercy and Rockoff state:

> The time is coming when the astronaut population will need to be considered as part of the general population and not a small and separate group with separate standards or radiation exposure levels [Raasch, Peercy & Rockoff, 1985, p. 99].

This precept holds true across all the participating disciplines—that project could not regard the space crew subjects as "a small and separate group with separate standards." Rather, this theme emerged gradually as an underlying agreement that a purpose of this project was to make living and working in space—if not routine—then at least predictable, reliable, safe and sustainable.

Habitation Assumptions

These assumptions provide a point of departure for defining the habitation requirements.

Crew Health & Fitness: The crew will arrive in good physical and mental condition. They will be able to ambulate safely by EVA to the habitat immediately on arrival on the surface.

Getting There versus *Being There:* Clearly, the surface facility is distinct from the interplanetary transportation system. The Mars surface crew does not "fly the first habitat down to the surface," because that habitat is not primarily a propulsive vehicle. It is a payload on an automated, remote controlled lander. The habitat should not conform to the severe reliability requirements that characterized a propulsive vehicle.

Mars Lander versus *Habitat:* Neither the Mars crew lander (MEV) nor the crew ascent vehicle is part of the habitat core. The lander and habitat core may arrive on Mars during separate launch opportunities, without prejudice to the capabilities of either one.

Mars Base Buildup Sequence

This section describes an approach to building a Mars base and setting up a crew habitat that supports the two objectives science and habitation. This approach involves four steps, each of which prepares the ground both literally and figuratively for the next step:

Step 1 — Data from Precursor Missions: describes the knowledge that the MSM will need to prepare the final architectural and engineering designs. There are two types of precursors: Mars-simulated gravity data for biomedical countermeasures and systems performance and Mars surface data on the local environment, site, and soil conditions.

Step 2 — Robotic Base Preparation and Set-up: occurs after the first launch opportunity in 2008, when the first four landers arrive on Mars, including the habitat core. Because the mission plan calls for the crew to fly on the next launch opportunity in 2010, automation and robotics will play an essential role in initiating ISRU, preparing the site, setting up the base, and checking out, verifying and activating the habitat.

Step 3 — Human Occupancy: presents the Mars surface mission scenario of how the crew enters the habitat, activates it, prepares to carry out their mission-critical and mission-discretionary activities. It involves an assessment of how the living and working requirements for the mission affects the habitat design parameters. This section includes a set of preliminary habitat design requirements. Human occupancy subsumes several of the other sections including human factors, crew factors, autonomous systems and Mars gravity countermeasures.

Step 4 — Growth Path: describes how the Mars base must follow a careful strategy to achieve clearly defined capability enhancements at specific milestones, within a rigorous geometry, rather than an on-going, incremental growth path (like SSF). Thus the growth and expansion of the Mars base would act more like a step function—which may not correspond to each mission—but to specific levels of operations and self-sufficiency.

Table 4
FIRST MARS OUTPOST - ARCHITECTURAL HABITATION DESIGN, IMPLEMENTATION AND OPERATION PARAMETERS

	LIFE-CRITICAL (SAFE HAVEN)	MISSION-CRITICAL	MISSION-DISCRETIONARY
Role	Stay alive and healthy until Habitat is fixed or safe, OR crew departure or rescue.	Begin productive work & normative living	Enhance mission capabilities and versatility
Time to Access / Activation	Ensure survival -- transfer from lander in *2 hours*.	Move in -- start science work in *2 weeks*.	Begin agriculture -- mission enhancements in *2 months*.
Access Conditions	Total Pre-Integration allows "turn the switch" entry and activation.	Check-out, some assembly, connect some utilities, etc.	Move plant growth chambers into the inflatables & set up.
Egress Options	1 Egress to each: Rover, pressure, airlock	2 Remote Egresses to each: pressure, airlock(s), rovers	Multiple remote egresses to pressure & rovers
Habitation Functions	Camp-out survival, with medical support, repair & escape capabilities.	Private quarters, cooking, dining, Labs & workshops	Recreation & rover repairs in inflatables, inflatables, CELSS & plants for food.
Architecture	Hardened core, EVA airlock, medical clinic, repair shop, solar storm shelter.	2 or more pressure volumes, EVA airlock, living and working environments.	1 or more inflatable domes -- pressurized rover, greenhouses, "garage"
Integration and Assembly	Completely pre-integrated and assembled.	Largely pre-integrated. May require some assembly.	Partially pre-integrated. Will require equipment installation.
Approx. % of Pressurized Volume	10%	30%	60%

Each of these steps involves an implementation concept and an operations concept. Table 4 states the performance requirements for step 1, the data from Mars precursor missions, focuses in detail on steps 2 and 3—the base architecture and habitat, elaborating their living and working support system, and provides a general guideline for step 4, the growth path.

Human Occupancy Implementation Concept

The Mars surface mission habitat implementation concept rests on a set of assumptions, interface constraints and design criteria.

Habitat Design Assumptions. Several key assumptions shape the habitat design. Table 4 illustrates the context and philosophy of these FMO habitat design assumptions. Perhaps the most critical set of assumptions is the time to access/activation, which derives from the human occupancy operations concept. The approximation of final pressurized volume refers to the estimate for FMO base buildup over three missions.

1. *Single Pressure Atmosphere:* There will be a single atmospheric pressure that obviates the need for internal airlocks. This pressure regime will optimize for EVA operations, human health and for plant growth. It will be acceptable to vary the local gas mix within specific volumes so long as the overall pressure is uniform. For example, it may be acceptable to maintain a higher partial pressure of CO_2 in the greenhouses to accelerate plant growth *or* a higher partial pressure of O_2 in the EVA airlock antechamber to enable a zero-prebreathe EVA and ensure against aerospace bends. However, the total pressure should be the same throughout the habitat and FMO to assure safe and easy egress from one pressure volume to another.

2. *Redundancy:* The key to reliability and safety in critical habitat systems is to provide "like redundancy" of the same capability, "unlike redundancy" of multiple capabilities that perform the same function, or distribution of essential capabilities in multiple locations.

3. *Multiple Volumes:* There will be at least two, separate, isolatable pressurized volumes within the habitat core to allow egress from one volume to the other in contingency situations. These contingencies include fire, contamination, loss of pressure, or another event that compels the crew to evacuate one of the volumes. It will be possible for evacuating crew members to close the hatches between volumes quickly and easily.

4. *The Habitat Core:* The habitat provides all the basic life-critical habitation functions (safe haven), enhanced by the pressurized rover and inflatables capabilities. It also provides most of the mission-critical capabilities for both habitation and IVA science activities. Most of the mission-discretionary capabilities reside outside the habitat core, in the inflatables (greenhouses and rover garage) and in the pressurized rovers.

Table 4 presents the FMO habitat design, implementation and operation parameters. This table attempts to quantify two of the most significant time limits: time to activation—measured in hours, weeks and months, and quantity of stored consumables parameters—measured in hundreds of days.

Human Occupancy Operations Concept

Crew Arrival and Entry to Habitat. The crew will arrive in a lander approximately 1 to 2 km from the habitat core. Once they make the lander safe and place it in surface/standby mode, the crew will begin to activate, check out, verify, and finally transfer to the habitat. The crew powers up the life-critical systems. They certify that all life-critical functions are operational and sufficiently robust. If any repairs or maintenance are required, the crew performs a contingency EVA. The crew transfers themselves and some equipment to the habitat within a few hours after landing. There will be a redundant set of options for the crew to reach the habitat:

Nominal Transfer: A pressurized rover will automatically approach the crew lander and dock to the berthing port. Three or four crew members will enter the rover in a "shirtsleeve" environment. They will drive to the habitat where they will dock the rover to the habitat rover-berthing port, through which they will enter the habitat shirtsleeve environment.

Contingency 1 Transfer: If the pressurized rover is not available for any reason, two or three crew members will don EVA suits and make a "contingency EVA." They will have the option of trying to enter the pressurized rover and making it work *or* using an unpressurized rover to travel to the habitat. Once there, they enter through a rover-berthing port or an airlock.

Contingency 2 Transfer: If neither of the rovers are available, two crew members will make a "contingency EVA" and walk the 1 km to 2 km distance to the habitat, which they will enter through the airlock.

Habitat Activation Sequence. The crew places the lander in stand-by mode. Two or more crew members conduct an operational readiness inspection (ORI) on all the life-critical systems and elements. Once they successfully complete this inspection, the crew transfers all crew operations to the habitat. They power down the lander and place it in storage mode. Then the crew begins the ORI for mission critical systems and accomplishes any assembly, construction or set-up necessary to initiate mission critical activities. Finally they perform the same sequence of procedures for mission-discretionary activities. The habitat now operates in a nominal activity and maintenance mode.

1. *Life-Critical - Activate Fully in 2 Hours:* From the time the first crew EVA of two persons leaves the lander, they should be able to travel to the habitat, enter it, and activate all the systems that comprise the life-critical capability, including the safe haven, in 2 hours. The life-critical elements include radiation storm protection, fire detection, protection and suppression, medical support, repair shop and emergency supplies. The emergency supplies consist of food and consumables supply for 600 days (stored at the habitat), including a chemical energy storage cache. Once the crew activates the life-critical systems, they can survive in the habitat core for the mission duration, even without activating the mission critical elements.

2. *Mission-Critical - Activate Fully in 2 Weeks:* Once the full crew transfers from the lander to the habitat, they will devote up to 2 weeks to "moving in": activating and

unpacking all parts of the living quarters, including the galley, dining and recreation areas. They will also unpack and prepare all the scientific equipment to make the laboratories fully functional in 2 weeks. The engineers will inspect all the external and excursion equipment, including *in situ* resource production, power generation to verify that it is in good enough condition to support the inflatables that are essential to a CELSS system and agriculture. They will test drive the rovers, and test the scientific EVA equipment.

3. *Mission-Discretionary - Activate in 2 Months:* The primary support structure for the mission-discretionary activities shall be inflatables, the first one of which the robots will emplace, deploy, inflate, and check out before the first crew arrives. The crew will install the plant growth chambers in inflatable "greenhouses" and plant seeds to commence food production and bioregenerative life support within 2 months of arrival on Mars. The first fresh vegetable should be ready in about 4 months after the crew arrives. The Mars base will phase over gradually from a physical/chemical system to partial CELSS and then full CELSS life support by the third mission.

4. *Nominal Operations - 520 to 540 Days:* The crew operates on one shift for most activities—waking up and going to sleep on the same diurnal, Mars-normal schedule. The crew divides the operational responsibilities on a real-time, autonomous basis, with advice but without command from Earth (except on unusual occasions). Mission support on Earth will provide recommended schedules that the crew will adapt to immediate, local conditions. Scientists pursue science investigations. The engineers operate and maintain the base and its associated equipment.

5. *Stand-Down and Departure:* The crew will devote about 20 to 30 days to preparing the habitat for dormancy until the next crew arrives. They will perform a shut-down readiness inspection and inventory of tools, equipment and consumables. They make final recommendations to mission support on Earth about what additional items to add to the accompanying payload lander(s) for the following Mars crew. The crew secures, in a careful sequence, all the mission-discretionary, mission-critical and life-critical systems (not necessarily in this order):

They reconfigure the Mars base, habitat and associated systems for autonomous or Earth-control mode to await the next crew. The crew inspects and checks out the ascent vehicle. They power-up the ascent vehicle and certify all its systems. They prepare and practice for return to orbit. They conduct a final flight readiness review. All the crew members transfer to the ascent vehicle, and leave the rover in storage mode. The crew ascends. After a successful ascent, the crew deactivates all the life-critical systems in the habitat, placing it in storage mode.

HABITAT REQUIREMENTS

The habitation system level requirements consist of several interrelated precepts. The functional requirements are fundamental. They describe the basic capabilities that the habitat must provide. The general performance requirements describe a small set of indices that all the capabilities under functional requirements must meet. A follow-on to this study would add performance requirements for individual functional elements.

The habitation system shall provide a pressurized environment that ensures a healthy, safe and productive crew. This environment shall be compatible with the crew IVA activities as well as the biota in the plant growth chambers and a no-pre-breathe EVA. Supporting these functions, the caution, warning, and fault detection system shall connect throughout all elements of the FMO, including the habitat.

The *life-critical* functions will be fail-op/fail-safe/fail restorable and will provide survival-level living conditions for the balance of the mission or until the crew can either fix the problem in the habitat that put them in life critical mode or ascend to Mars orbit. The life-critical functions provide: radiation detection and protection from all incident radiation, life support and a pressurized living volume, medical care (emergency and minimum maintenance), command, control and communications (Mars to Earth and local surface-to-surface), access to interfacing systems (power, EVA, ascent vehicle, repair and maintenance), emergency detection and response (fire, contamination, decompression). Life-critical functions include proximity EVA.

The *mission-critical* functions will be fail-op/fail-safe/recoverable. Except for the power supply system and consumables cache they reside in the habitat core. Mission-critical functions will include the life-critical functions above, plus working areas and volumes including laboratory capabilities, and increased quality of life features (recreation, human factors, privacy, group activities, exercise). Greenhouse facilities for food and life support would come on line by approximately +4 months into the first mission. Mission-critical functions include local EVA and mobility systems

The *mission-discretionary* functions will be fail-safe/recoverable and will provide access to discretionary systems and elements. They include working areas and volumes in inflatables connected to the habitat core. Initially all functions, activities, and structures associated with the inflatables are mission-discretionary. On the later missions, the greenhouses mature to become a necessary part of the mission-critical capability. Initially, local and regional EVA and mobility systems are mission-discretionary. Eventually these systems become mission-critical. Global telepresence exploration systems, operated by MSMC or by the crew from the FMO are mission-discretionary. Most local resources experiments are mission-discretionary. Examples include an experiment to grow plants in Mars soil, an experiment to mix and test Mars regolith concrete, an aquaculture experiment to raise fish on Mars for protein in water from Mars resources, an experiment to produce life support reagents from Mars soil, and possible research in Mars partial-gravity gravitational biology and countermeasures.

FUTURE GROWTH PATH

Mars base growth must follow a careful strategy to achieve clearly defined capability enhancements at specific milestones, rather than an on-going, incremental growth path (like SSF). Thus the growth and expansion of the Mars base would act more like a step function—which may not correspond to each mission—but to specific levels of operations and self-sufficiency.

The infrastructure will make the base fully operational for the first three missions. It should also be possible to design the base to provide a clear growth path to double or triple in size for subsequent mission sets. However, it would be a serious mistake to try to design a base that is infinitely expandable—an attempt to create an infinite growth path will impose severe penalties on any useful base geometry. Instead, the base design should present a clear completion point after some future set of missions. A strong geometry is a good way to achieve this structured growth. A strong and clear geometry is also an advantage for robotic construction techniques.

If the Mars base is such a success that NASA wants to expand it beyond 300%, it will be time to go back to develop a new geometry, modularity, delivery system, automation and robotics, probably heavy soil excavators or borers, and build a new, second-generation base. This new base may be next to the first or it may suit our successors to select a different site.

SUMMARY AND CONCLUSIONS

Previous mission studies focused primarily on interplanetary transportation and did not look in detail at how to perform a surface mission. This First Mars Outpost study divided the Mars mission into two separate problems domains: "getting there" and "being there." This decomposition shows that there is little overlap or commonality between the two problem definitions. It leads to sensible and practical systems engineering opportunities because it does not compound the surface mission with all the challenges of launch to low Earth orbit, trans-Mars injection, etc. This study looked primarily at the Mars surface systems to develop a strategy to achieve the mission goals for exploration and habitation.

Using an appropriate design philosophy, the First Mars Outpost would safely support a crew of eight people on the Martian surface for 600 days. The design philosophy includes using Martian resources to create a consumables cache that would always be full, that is, always be able to support the crew in a contingency for 600 days. This study defined the surface systems for the First Mars Outpost, for both the living and working environments. Robotic landers would pre-emplace most of these systems, particularly the habitat, and ISRU facilities before the crew arrives. Thus, the crew arrives at an operational Mars base, which is ready for human occupancy.

The scientific, engineering, and operational mission requirements suggest an optimal crew size of eight, which includes a built-in 33% margin of safety. This figure derived from consideration of how to apportion the crew's time, and the mix of crew duties between operations and science.

The safety philosophy separated the surface mission functions into three categories by increasingly higher levels of acceptable risk: life critical, mission critical, and mission discretionary. The strategy for performing each of these functions varies with the level of acceptable risk.

This study found that a First Mars Outpost program which begins to establish surface infrastructure robotically in 2008, and begins human occupancy in 2010, seems feasible without requiring a "crash program" to Mars. It does, however, require a deliberate and sustained effort in technology development.

REFERENCES

Ash, R.L., J.A. Werne, M.B. Haywood, "Design of a Mars Oxygen Processor," AAS 87-263, in *The Case for Mars III: Strategies for Exploration—Technical*, C.R. Stoker, ed., American Astronautical Society, Science and Technology Series, 75, pp. 479-487, San Diego, CA: Univelt Incorporated, 1989, Proceedings of the Third Case for Mars Conference, July 17-22, 1987.

Babb, G.R. and W.R. Stump, "Comparison of mission design options for manned Mars missions," in *Manned Mars Missions: Working Group Papers, Vol. 1.*, NASA TM-89320, June 1986, p. 38.

Capps, S. and C. Case, "A new approach for a lunar airlock structure," AIAA 93-0994, Aerospace Design Conference, Irvine, CA, February 16-19, 1993.

Carino, A., "An integrated approach to safety, including the human factors element," in *RISK MANAGEMENT: Expanding Horizons in Nuclear Power and Oher Industries,* symposium Sept 6-7, 1989, sponsored by GPU Nuclear Corporation in Whippany, NJ Knief, R.A., ed., New York NY,: Hemisphere Publishing Corp, 1991, pp. 125-126.

Cerimele, M.P., *November Guidelines for Man-Rating Space Systems*, JSC-23211, NASA-Johnson Space Center, Houston, TX, 1991 pp. 4 and 17.

Clapp, W.M., "Water supply for a manned Mars base," AAS 84-181, in *The Case for Mars II,* C.P. McKay, ed., American Astronautical Society, Science and Technology Series, 62, pp. 557-566, San Diego CA: Univelt Incorporated, 1985, Proceedings of the Second Case for Mars Conference, Boulder, CO, July 10-14, 1984.

Clark, B.C., "Human exploration of Mars," *AIAA-88-0064,* AIAA 26th Aerospace Sciences Meeting, Reno NV, Washington DC: American Institute of Aeronautics and Astronautics, January 11-14, 1988, pp. 1-2.

Clark, B.C., "Long-range exploration by humans on the surface of Mars," *AIAA-88-0354,* AIAA 26th Aerospace Sciences Meeting, Reno Nevada, Washington DC: American Institute of Aeronautics and Astronautics, January 11-14, 1988.

Clark, B.C., "Human planetary exploration strategy featuring highly decoupled elements and conservative practices," *AIAA-91-0327,* 29th Aerospace Sciences Meeting, Reno, Nevada, Washington DC: American Institute of Aeronautics and Astronautics, January 7-10, 1991, pp. 2-3.

Clark, B.C., J.L. Kalkstein, and S. Meyer, "The case for the methanol powered planetary rover," *IAF-91-447,* 42nd Congress of the International Astronautical Federation, Montreal Canada, Paris: International Astronautical Federation, October 5-11, 1991.

Clearwater, Y.A., and A.A. Harrison, "Crew support for an initial Mars expedition," *Journal of the British Interplanetary Society*, Vol. 43, 1990, pp. 513-518.

Cohen, M.M., *Suitport Extra-Vehicular Access Facility*, US Patent 4,842,224, June 27, 1989.

Cohen, M.M., "Human factors issues for interstellar spacecraft," Proceedings of the 28th Space Congress, Cocoa Beach, Florida, Cape Canaveral, FL: Canaveral Council of Technical Societies, April 23-25, 1991, pp. 1-19 to 1-29.

Cohen, M.M., "Objectives and overview," *Mars 2008: Surface Habitation Study, Center for Mars Exploration Position Paper*, Moffett Field: NASA-Ames Research Center, May 1993, pp. 4-5.

Connors, M.M., A.A. Harrison, F.R. Akins, *Living Aloft: Human Requirements for Extended Spaceflight,* NASA SP-483, Washington DC: NASA Scientific and Technical Information Branch, 1985.

Freeman, G., "Integrated use of probabilistic safety assessment," in *RISK MANAGEMENT: Expanding Horizons in Nuclear Power and Oher Industries,* symposium Sept 6-7, 1989, sponsored by GPU Nuclear Corporation in Whippany, NJ Knief, R.A., ed., New York NY,: Hemisphere Publishing Corp, 1991, p. 187.

French, J.R., "The impact of Martian propellant manufacturing on early manned exploration," AAS 84-178, in *The Case for Mars II*, C.P. McKay, ed., American Astronautical Society, Science and Technology Series, 62, pp. 521-524, San Diego, CA: Univelt Incorporated, 1985, Proceedings of the Second Case for Mars Conference, Boulder, CO, July 10-14, 1984.

Goldman, A.S., and T.B. Slattery, *MAINTAINABILITY: A Major element of System Effectiveness*, New York: John Wiley & Sons, Inc., 1964, pp. 26, 28, 34.

Hurtt, C.B.(ed.), Strategic considerations for support of humans in space and moon/Mars exploration missions, *Volume 1, NASA TM-107983 and Volume 2, NASA TM-107984*, NASA Advisory Council, Aerospace Medicine Advisory Committee, 1992.

Kaplan, S., "Risk assessment and risk management - basic concepts and terminology," in *RISK MANAGEMENT: Expanding Horizons in Nuclear Power and Other Industries*, Knief, R.A., ed., symposium September 6-7, 1989, sponsored by GPU Nuclear Corporation in Whippany NJ, New York: Hemisphere Publishing Corp, 1991.

Lyne, J.E., "Physiologically constrained aerocapture for manned Mars missions," *NASA TM 103954*, Moffett Field CA: NASA-Ames Research Center, August 1992.

Maslow, Abraham H., *Motivation and Personality*, 2nd edition, New York: Harper & Row, Publishers, 1970, pp. 35, 46.

Meyer, T.R. and C.P. McKay, "The atmosphere of Mars - resources for the exploration and settlement of Mars," AAS 81-244, in *The Case for Mars*, P.J. Boston, ed., American Astronautical Society, Science and Technology Series, 57, pp. 218-229, San Diego, CA: Univelt Incorporated, 1984, Proceedings of the First Case for Mars Conference, Boulder, CO, April 29 - May 2, 1981.

NASA, Man-Systems Integration Standard, NASA STD 3000, Vol. I, Rev. A, 1989, p. 5-98, Figure 5.7.2.2.1-1

National Fire Protection Association, *Code for Safety to Life from Fire in Buildings and Structures*, Quincy, MA, 1991.

Newkirk, D., Almanac of Soviet Manned Space Flight, Houston, TX: Gulf Publishing Company, 1990, p. 303.

Peercy, R.L. Jr., R.F. Raasch and L.A. Rockoff, Space station crew safety alternatives study—final report: volume i—final summary report, *NASA CR-3854*, Washington DC: NASA Scientific and Technical Information Branch, June 1985, pp. 4, 11, 21.

Planet Surface Systems Office, *Element/Systems Data Base (ESDB)* Release 91.1, JSC-45107, NASA-Johnson Space Center, Houston, TX, 1991.

Rand Corporation, "Architectures/missions for the space exploration initiative: results from project outreach," Santa Monica: Rand Publication Series, unpublished report to NASA and the USAF, USAF Contract F49620-86-C-0008, 1991.

Raasch, R.F., R.L. Peercy, Jr., and L.A. Rockoff, Space Station Crew Safety Alternative Study - Final Report: Volume II - Threat Development, *NASA CR-3855*. Washington DC: NASA Scientific and Technical Information Branch, June, 1985, p. 99.

Space Station Program Office, Instructions for preparation of failure modes and effects analysis and critical items list for space station, JSC 30238, Houston TX: NASA-Johnson Space Center, 1987, pp. x-xi.

Thornton, W.E., "Safety and the American space program," *Aerospace Medicine*, August 1974, p. 925.

Williams, M.D., R.J. De Young, G.L. Schuster, S.H., Choi, J.E. Dagle, E.P. Coomes, Z.I. Antoniak, J.A. Bamberger, J.M. Bates, M.A. Chiu, R.E. Dodge, and J.A. Wise, "Power transmission by laser beam from lunar-synchronous satellite," *NASA TM 4496*, pp. 19-20, November 1993.

Wydeven, T., "A survey of some regenerative physical-chemical life support technology," *NASA TM-101004*, Moffett Field CA: NASA-Ames Research Center, November 1988, pp. 5-9.

Zubrin, R.M., "In-situ propellant production, the key technology required for the realization of a coherent and cost-effective space exploration initiative," *IAF 91-668*, 42nd Congress of the International Astronautical Federation, Montreal Canada, Paris: International Astronautical Federation, October 5-11, 1991, p. 13.

Section V

Science on Mars

Humans enhance the ability to do science directly, with hands, brains, hammers and instruments.

Chapter 22

SCIENTIFIC OBJECTIVES OF HUMAN EXPLORATION OF MARS

Michael H. Carr[*]

While human exploration of Mars is unlikely to be undertaken for science reasons alone, science will be the main beneficiary. A wide range of science problems can be addressed at Mars. The planet formed in a different part of the solar system from the Earth and retains clues concerning compositional and environmental conditions in that part of the solar system when the planets formed. Mars has had a long and complex history that has involved almost as wide a range of processes as occurred on Earth. Elucidation of this history will require a comprehensive program of field mapping, geophysical sounding, *in situ* analyses, and return of samples to Earth that are representative of the planet's diversity. The origin and evolution of the Mars' atmosphere are very different from the Earth's, Mars having experienced major secular and cyclical changes in climate. Clues as to precisely how the atmosphere has evolved are embedded in its present chemistry, possibly in surface sinks of former atmosphere-forming volatiles, and in the various products of interaction between the atmosphere and surface. The present atmosphere also provides a means of testing general circulation models applicable to all planets. Although life is unlikely to be still extant on Mars, life may have started early in the planet's history. A major goal of any future exploration will, therefore, be to search for evidence of indigenous life.

INTRODUCTION

Mars will inevitably be the first planet, beyond the Earth-Moon system to be explored by humans. The crushing pressures and oven-like temperatures of Venus will surely prevent any kind of human presence there in the foreseeable future, and the enormous daytime heat of Mercury and permanent cold of the outer planets and their satellites will guarantee that their exploration will follow long after that of Mars. The rationale for any form of exploration is complex. People explore to satisfy their desire to know the unknown, to gain political power, to demonstrate their technological prowess, to gain economic advantage, and to escape intolerable political and economic conditions. Whatever the mix of reasons that ultimately drives us to explore Mars, science is likely to be the major beneficiary. Political imperatives gave birth to the Apollo pro-

---

[*] U.S. Geological Survey, Menlo Park, California 94025.

gram, but the permanent legacy that now remains long after the original political pressures have dissipated, is scientific. Our vastly improved knowledge of the origin and evolution of the Moon gained by Apollo is permanent. So it is likely to be for Mars. The purpose of this short paper it to outline some of the science issues that would be addressed by humans on Mars. The science goals will be discussed in the context of three broad categories: the origin and evolution of the solid planet, the origin and evolution of the atmosphere, and the potential of the planet for harboring life. The discussion draws heavily upon previous documents outlining the goals of Mars exploration [COMPLEX, 1978; SSEC, 1986; TBSWG, 1977; TGPLEX, 1985].

Which science problems are best done by unmanned vehicles and which might best be accomplished by humans will not be discussed. It should, however, be emphasized that the long communication time between Mars and Earth makes robotic exploration of Mars difficult. Robotic systems are not currently on the technological horizon that could sense the surroundings, interpret the scene from a science and safety viewpoint, draw on experience as to how best to proceed next, and then implement decisions. Any robotic exploration of Mars will require a large amount of human intervention and will necessarily be painfully hesitant and inefficient, until humans are at the planet.

THE ORIGIN AND EVOLUTION OF THE SOLID PLANET

All solid planets retain a subtle and incomplete record of the processes that ultimately resulted in the planet's present state and the sequence in which those processes operated. The record is partly in the configuration of rocks near the surface, partly in the chemistry and mineralogy of those rocks, partly in the chemistry of the atmosphere, and partly in the internal structure of the planet. The task of a planetary explorer is to gather the data necessary to reconstruct from this record how the planet formed and how it evolved into what we observe.

Much of the current theory about the formation of planets, particularly those in the inner solar system, is derived from the study of meteorites. These provide us with samples of the materials from which the the planets formed and give an indication of what conditions were like in the solar system as the planets formed. The bulk composition of the Earth can be deduced from the dimensions and compositions of the core, mantle, and crust. Comparisons of the bulk composition with those of different meteorites show what mix of meteorite types accumulated to form the Earth [Ringwood, 1979; Wänke, 1981]. In addition, detailed comparison of the chemistry of the Earth's mantle with the chemistry of meteorites gives an indication of the timing of accretion with respect to other global events. The evidence suggests that global differentiation occurred at the tail end of accretion, but before accretion was complete. Comparisons between the bulk composition of the Moon, meteorites and the Earth's mantle strongly support the supposition that the Moon was derived from the infant Earth by an extremely large impact. Thus, knowledge of the dimensions of the major components of the Earth and Moon, derived largely from seismic studies, and knowledge of the chemistry of these components, derived largely from detailed geochemical studies, have led to an understanding of events very early in the planet's history.

We can expect similar success in understanding how Mars formed and when its major components separated. Seismic studies, coupled with knowledge of the moment of inertia will tell us if Mars has a core, and, if so, of what size and density, from which its composition can be inferred. Magnetization of surface rocks will indicate whether the core ever generated a significant magnet field, and if so, when. The composition of the core coupled with detailed study of the chemistry of volcanic rocks, particularly samples of the mantle brought to the surface, will demonstrate what mix of materials accumulated to form Mars. At present we have this information only for the Earth-Moon system. Having comparable information for Mars will enable us to infer compositional gradients in the solar system during accretion, and to determine more precisely the "feeding zones" for the different planets. Detailed knowledge of the chemistry of the Mars mantle will also demonstrate whether accretion was essentially complete at the time of global differentiation, as has been suggested [Dreibus and Wänke, 1984] or whether significant amounts of volatile-rich material was added late in accretion, after global differentiation, as apparently was the case with the Earth [Anders, 1968; Ringwood, 1979]. Increased understanding of these early events at Mars will contribute to a general understanding of how planets form, particularly since Mars appears to have formed without the major perturbation experienced by the Earth as a result of formation of its large moon.

Highlands and Plains

Some clues as to how Mars evolved after its formation are revealed in the present configuration of the surface (for general summaries of Mars geology see Carr [1981], Baker [1982]). One of the most striking features of Mars is its asymmetry. The surface of the planet can be divided into two main components: an ancient cratered highlands, covering most of the southern hemisphere, and low-lying plains, that are mostly at high northern latitudes. Superimposed on this two-fold division are the high-standing volcanic provinces of Tharsis and Elysium. The cratered highlands cover almost two thirds of the planet. They are mostly at elevations of 1-3 km above the datum, in contrast to the high-latitude, northern plains, which are mostly 1-2 km below the datum. The cause of the division between the highlands and plains is unclear but it may be the result of a very large impact at the end of accretion [Wilhelms and Squyres, 1984]. The density of impact craters in the Martian highlands is comparable to the lunar highlands (see Figure 1). The surface clearly dates back to the very earliest history of the planet when impact rates were high. On the Moon the transition from very high impact rates to rates comparable to the present took place around 3.8 billion years ago. On Mars the transition probably took place at at a similar time. The Martian highlands differ from the lunar highlands in three main ways. First, sparsely cratered plains are more common between the larger craters in the Martian highlands than on the Moon. Many of these plains are presumed to be volcanic because of the presence of occasional flow fronts and wrinkle ridges that resemble those on the lunar maria (see Figure 2). They suggest that relatively high rates of volcanism prevailed during, and immediately after, the period of heavy impact. The second difference between the lunar and Martian highlands concerns the impact ejecta. The ejecta around lunar craters generally has a coarse, hummocky texture at the rim crest and grades outwards into a finer, randomly hummocky or radial texture. This is also true for the largest Martian craters, but most well preserved Martian craters with diameters between 5 and 100 km have ejecta arrayed in discrete lobes, each lobe being outlined by a low ridge. This is true irrespective of location. Two reasons have been suggested for the characteristic Martian ejecta patterns. The first is that impact craters above a certain size penetrate the permafrost zone and eject water-laden, or ice-laden materials that tend to flow across the surface following ejection from the crater [Carr *et al.*, 1977]. This will be amplified on later. The second suggestion, based on wind-tunnel experiments, is that the interaction between the ejecta and the atmosphere causes the flow like patterns [Schultz and Gault, 1984]. The third difference between the lunar and Martian highlands, is the presence, within the Martian highlands, of numerous branching valley networks. At low latitudes they are almost everywhere within the highlands. They superficially resemble terrestrial river valleys and are believed to be formed by running water (see Figure 3). They will be discussed more fully below under water erosion.

Figure 1 Typical cratered upland at low latitude. The scene is 60 km across.

Figure 2 Ridged plains of Syria Planum. The northeast southwest trending ridges resemble mare ridges on the Moon, and probably form by compression as a result of the presence of the Tharsis bulge to the northeast. A 30-km diameter crater in the center of the picture has the typical lobed ejecta pattern of fresh-appearing Martian craters.

The plains are located mostly in the northern hemisphere. The number of superimposed craters on them varies substantially, indicating that they continued to form throughout the history of the planet. The plains are diverse in origin. The most unambiguous in origin are those on which numerous flow fronts are visible. They are clearly formed from lava flows superimposed one on another, and are most common around the volcanic centers of Tharsis and Elysium. On other plains, such as Lunae Planum, flows are rare but wrinkle ridges like those on the Moon are common. These are generally assumed to be volcanic, although wrinkle ridges merely imply deformation, not primary origin. But the vast majority of the low-lying northern plains lack obvious volcanic features. Instead they are curiously textured and fractured. Many of their characteristics have been attributed to the action of ground ice, or to their location at the ends of large flood features, where lakes must have formed and sediments been deposited. In some areas, particularly around the north pole, dune fields are visible. In yet other areas, are features that have been attributed to the interaction of volcanism and ground-ice. Thus the plains appear to be complex in origin, having variously formed by volcanism and different forms of sedimentation, and then subsequently been modified by tectonism and by wind, water and ice.

Figure 3 Branch valley networks in the cratered upland at 43°S, 92°W. The large crater a the bottom of the picture is 50 km across.

Figure 4 Olympus Mons. The main edifice, outlined by a cliff that is in planes 6 km high, is 550 km across. At the summit, 27 km above the surrounding plains, is a complex caldera.

Volcanism

The most prominent volcanoes are in two regions, Tharsis and Elysium. Tharsis is at the center of a bulge in the planet's surface, the bulge being over 4,000 km across and 10 km high at the center. A similar bulge centered on Elysium is around 2,000 km across and 5 km high. Three large volcanoes are close to the summit of the bulge and Olympus Mons, the tallest volcano on the planet, is on the northwest flank (see Figure 4). All these volcanoes are enormous by terrestrial standards. Olympus Mons is 550 km across and 27 km high, and the three others have comparable dimensions. Lava flows and lava channels are clearly visible on their flanks, and each has a large summit caldera. They all appear to have formed by eruption of fluid lava with very little pyroclastic activity. The large size of the volcanoes has been attributed to the lack of plate tectonics on Mars and the greater depths to their magma source as compared to terrestrial volcanoes [Carr, 1973]. The small number of superimposed impact craters on their flanks indicates that their surfaces are relatively young, although the volcanoes could have been active for most of the planet's history. To the north of Tharsis is Alba Patera, the largest volcano on the planet in areal extent. It is 1,500 km across but only a few km high. Flows are visible on parts of its flank, but elsewhere the volcano flanks are dissected by numerous branching channels. The easily eroded, channeled deposits have been interpreted as ash [Wilson and Mouginis-Mark, 1987]. Densely dissected deposits on other volcanoes such as Ceraunius Tholus in Tharsis, Hecates Tholus in Elysium, and Tyrrhena Patera in the southern highlands have also been interpreted as ash. Thus Mars appears to have experienced both Hawaiian style of volcanism, involving mostly relatively quiet effusion of fluid lava, and more violent, pyroclastic eruptions, that result in deposition of extensive ash deposits.

Elysium Mons appears to be a shield volcano formed largely of fluid lava. However, huge channels start at the periphery of the volcano and extend northwest, down the regional slope for over a thousand kilometers. The channels have a streamlined forms and enclose tear-drop shaped islands. Similar large channels start adjacent to Hadriaca Patera on the rim of the large impact basin, Hellas. They start at the volcano and extend, for hundreds of kilometers down into the Hellas basin. All these channels have the characteristics of large floods. They are thought to have formed by massive release of water following melting of ground ice by the volcanoes. Numerous other features in the Elysium area and elsewhere have also been interpreted as a result of volcano-ice interactions [Squyres *et al.*, 1987].

Tectonism

The most widespread indicators of surface deformation are normal faults, indicating extension, and wrinkle ridges indicating compression. The most obvious deformational features are those associated with the Tharsis bulge. Around the bulge is a vast system of radial graben that affects about a third of the planet's surface (see Figure 5). The graben are particularly prominent north of Tharsis where they are diverted around the volcano Alba Patera, to form a fracture ring. Circumferential wrinkle ridges are also present in places, particularly on the east side of the bulge in Lunae Planum. Both the fractures and the compressional ridges are believed to be the result of stresses in the lithosphere caused by the presence of the Tharsis bulge. No comparable system of deformational features occur around Elysium, but fractures occur in other places where the crust has been differentially loaded, as around large impact basins, such as Hellas and Isidis, or around large volcanoes, such as Elysium Mons and Pavonis Mons.

The vast canyons on the eastern flanks of the Tharsis bulge are the most spectacular result of crustal deformation (see Figure 6). The canyons extend from the summit of the Tharsis bulge, eastward for 4,000 km until they merge with chaotic terrain and large channels south of the Chryse basin. In the central section, where several canyons merge, they form a depression 600 km across and several kilometers deep. Although the origin of the canyons is poorly understood, faulting clearly played a major role. The canyons are aligned along the Tharsis radial faults, and many of the canyon walls are straight cliffs, or have triangular faceted spurs, clearly indicating faulting. In one location a large impact crater is present on the canyon floor, as though the floor were a downfaulted section of the surrounding surface. Other processes were also involved in formation of the canyons. Parts of the walls have collapsed in huge landslides, other sections of the walls are deeply gullied. Fluvial sculpting is particularly common in the eastern sections. Faulting may have created most of the initial relief, but the relief then enabled other processes such as mass wasting, and fluvial action to occur. Creation of massive fault scarps may also have allowed groundwater to leak into the canyons, thereby creating temporary lakes.

Figure 5 Graben northeast of Tharsis. The faulting is believed to be the result of tension in the crust caused by the presence of the Tharis bulge. Lines of pits, whose origin is poorly understood, are commonly associated with the graben. The scene is 100 km across.

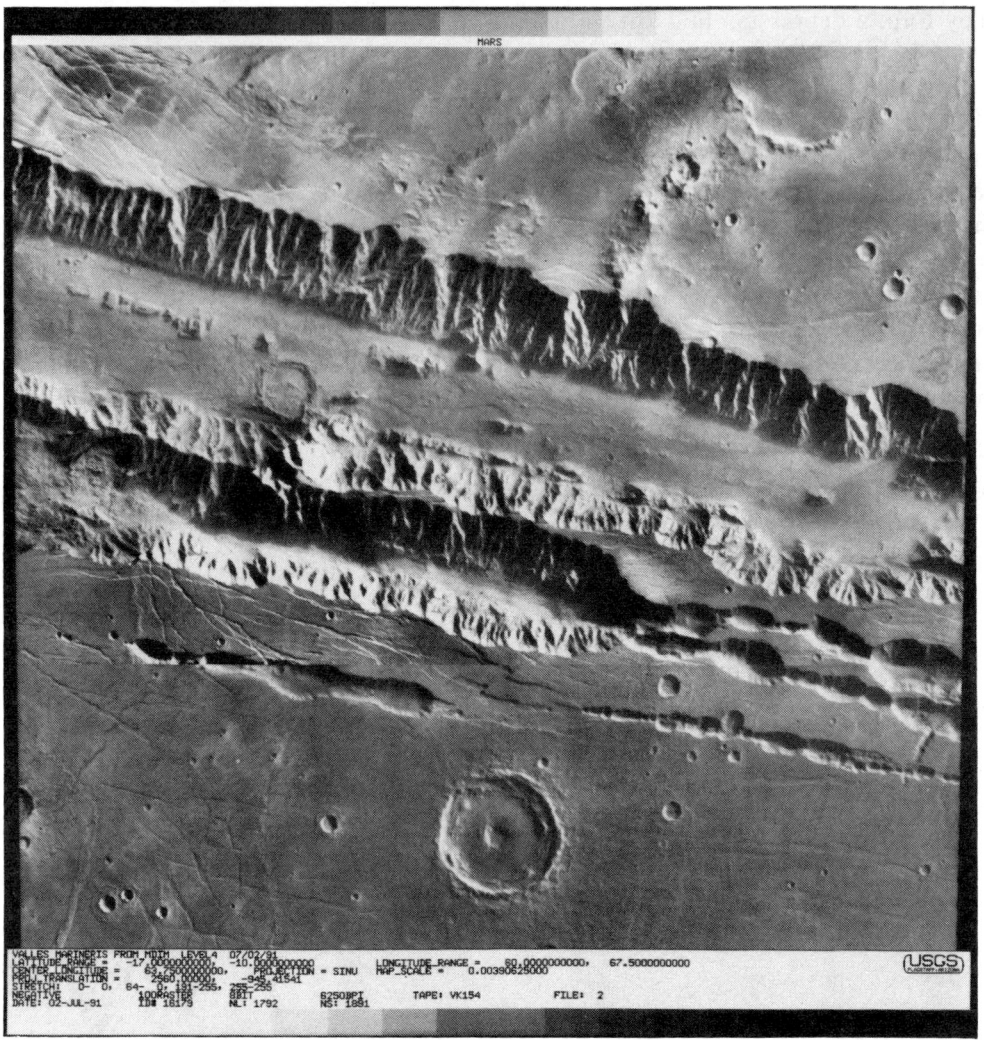

Figure 6 Section of Valles Marineris. The upper canyon, Coprates Chasma is 200 km across and 4 km deep.

Exploration Objectives

Major exploration objectives with respect to the solid planet are, to determine, as from seismic and heat flow measurements, the present thickness of the lithosphere and how it varies with location. We also need to know how the lithosphere thickness has changed with time. This can be deduced from a variety of investigations such as the dating and detailed petrology of volcanic rocks, from the degree of isostatic compensation revealed by measurements of the surface gravity and topography, and from heat flow measurements.

The composition and thickness of the crust almost certainly vary with location. The oceanic and continental components of the Earth's crust are the result of two totally different kinds of processes: igneous activity at divergent plate boundaries and welding of volcanics and metasediments to the upper plate at convergent boundaries. The anorthositic crust of the Moon appears to be the result of dry fractionation. The composition of the Martian crust is at present unknown, but its nature and distribution should similarly reveal the nature of processes that took place very early in the planet's history, processes that include both those that formed the crustal components and those that resulted in their present uneven distribution. The composition of the upland crust can be readily determined by direct sampling of the deeply gardened materials that constitute the uplands. More indirect means, such as examination of ejecta from large impacts, may be necessary to determine the composition of the lowland crust if it is different from the sediments and mantle-derived volcanics exposed at the surface. Regional variations in the thickness of the crust and its different components can be determined seismically.

While the near-surface materials constitute only a small fraction of the bulk of a planet, they are the materials that are most accessible and the ones on which we must rely for reconstruction of the planet's history. We know from previous missions that the configuration and nature of rocks at the Martian surface are the result of a variety of processes, including volcanism, global and local tectonics, impacts, weathering, and erosion by wind, water, and ice. We must be able to identify, separate, and place in chronological sequence the effects of the different processes that have operated at the surface if we are to successfully reconstruct what conditions were like at various times in the planet's history and how the planet arrived at its present state. In the next few paragraphs some of the questions that must be asked and some of the measurements that must be done at Mars to answer them are indicated. Because of the variety of processes that have operated on Mars, the Mars explorer's task is comparable in complexity to the tasks that confront geologists, geochemists, and geophysicists trying to interpret the Earth.

Among the questions that need to be addressed concerning volcanism are the following: Is the planet now volcanically active and where is current activity located? How have the location and type of activity changed with time? What do the volcanic products tell us about depth of origin of the magmas, the composition of the source materials, and nature of the materials through which the magmas pass on the way to the surface? Is there any systematic change in these characteristics with time and location? These and similar questions can be answered only with the kind of intensive mapping, sampling, and geophysical monitoring, supported by a wide range of laboratory and theoretical work, that is done in support of volcanic studies on the Earth. Success is also dependent on return to Earth of a representative set of samples.

While plate tectonics appear not to have operated on Mars, the surface has undergone considerable deformation both on global and local scales. Key questions are: What is the nature of the boundary between the uplands and plains? How tectonically active is the planet now? When and in what way did the Tharsis and Elysium bulges originate? When, how and in what sequence did the fault systems around Tharsis and Elysium develop? To what extent is the present surface relief isostatically adjusted, and to what extent is this dependent on the age and origin of the relief? Is the equatorial canyon system largely tectonic or erosional in origin? What is the age of the large faults within the canyons and are they still active? Answers to these questions are going to depend on dating, determination of intersection relations, monitoring the location of earthquakes and documenting anisotropies in mantle temperatures and lithosphere thicknesses.

Impacts have clearly played a major role in Mars' history. Impact craters have their own intrinsic interest, but they are also useful in that they drill through the near-surface materials and excavate materials from great depths. They therefore provide the means for determining vertical profiles through the crust. One characteristic of Martian impact craters that appears unique among the inner planets is that the ejecta of most craters appears to have been very fluid when emplaced, an attribute that has been ascribed to entrapment of water or ice in the ejecta. Key questions concerning impact are as follows: How has the impact rate changed with time? Are the fluidized ejecta patterns caused by entrainment of water and ice? Are mantle and deep crustal materials exposed in the central peaks of large craters?

The role of water in the evolution of the Martian surface is one of the most controversial issues concerning the geologic evolution of the planet. Water plays a crucial role in many geologic processes including both deep-seated magmatic processes and surficial changes such as weathering and erosion. Because most of the water is believed to have outgassed early, the near-surface inventory provides clues about the accretion of the planet, the thermal and fractionation history of the interior, and the formation of the atmosphere. As discussed below, the amount of water also has climatological implications which in turn affect the probability that life could have started on the planet. There may even be practical implications, for water is a potential resource that explorers could use for fuel and sustenance. Because the volatile history is so closely related to the climatic history, the numerous questions related to water and volatiles will be discussed in the next section.

Reconstruction of the origin and evolution of the solid planet thus requires a broadly based exploration program, including geophysical monitoring, general field studies in different terrains to document the configuration of the surface materials, and a comprehensive sampling program, so that the full analytical capability of the terrestrial science community can be utilized to understand the planet. While the Apollo program serves as an excellent model, the diversity of Mars, and the complexity and longevity of its geologic history, dictate that the exploration should not be narrowly localized to a few discrete sites, but should be carefully planned to return information representative both of the variety of processes that have operated on the planet and the different epochs during which activity took place.

THE ORIGIN AND EVOLUTION OF THE ATMOSPHERE

By analogy with the Earth, most of the atmosphere-forming materials probably outgassed from the planet during the segregation of the planet into crust mantle and core at the tail end of accretion. Major uncertainties are the amounts of the various volatiles outgassed at this time and their subsequent history. Estimates of the volatiles outgassed currently range over two orders of magnitude [Carr, 1987], thereby making reconstruction of the climatic history of the planet very uncertain. A question of comparable importance to the total amounts of volatiles initially outgassed is the extent to which the volatiles were retained by the planet. Were major amounts lost by impact erosion and other processes very early in the planet's history? If major amounts of the initial inventory of volatiles were retained, where are they now, and what has been the history of their fixation in the surface? Are there now major sinks of water, carbonates and nitrates on the planet, and if so, where? To what extent has outgassing continued throughout the history of the planet? The answers to these questions lie embedded in the detailed chemistry of the atmosphere and in the history of the interaction of the atmosphere with the surface as recorded in the surface materials.

Water Erosion and Climate History

The climatic history of the planet is of crucial importance with respect to whether life could have started on the planet. The preponderance of the current evidence is that early in the history of the planet the atmosphere was considerably thicker, and that surface temperatures were higher than at present [Pollack *et al.*, 1987]. The evidence is the almost ubiquitous presence of small valleys on ancient heavily cratered surfaces. The scarcity of similar channels on younger surfaces and the minute amount of erosion that has occurred on these surfaces suggest that similarly clement climatic conditions were rare throughout most of the planet's subsequent history. Branching valley networks resemble terrestrial river valleys in that they have tributaries and increase in size downstream, although only rarely can a channel be observed within the valley. The valleys are generally short compared with terrestrial river systems, most being less than a few hundred kilometers in length so rarely does one valley system dominate drainage over a large area. The most plausible explanation for the valleys is that they formed by erosion of running water. The open nature of some networks, alcove like terminations of tributaries, and the range of junction angles between branches is suggestive of groundwater sapping [Pieri, 1980]. Other networks, however, lack these characteristics, and more resemble valleys formed by surface runoff.

The origin of the valley networks is controversial. Their branching pattern and small size suggest that they are formed by slow erosion of running water, but the precise mode of formation and the climatic conditions required are both unknown. Under present climatic conditions small streams on Mars would rapidly freeze, so that the source of water for the larger rivers downstream would be cut off. Most of the valley networks are in the cratered highlands, the oldest part of the planet's surface. The valley networks suggest, therefore, that early Mars was much warmer than at present, so that liquid water could flow across the surface. Greenhouse calculations indicate that for surface temperatures on Mars to be above freezing, a CO_2 atmosphere 1-3 bars thick is required [Pollack 1979]. If such a thick atmosphere were present, it would be inherently unstable since the CO_2 would react with the surface rocks in the presence of liquid water to form carbonates. It has been suggested, therefore, that Mars had an early thick atmosphere that kept Mars warm, but that early in the planet's history, this atmosphere largely dissipated as a result of carbonate formation so that the planet assumed its present cold, hostile conditions. This simple model is, however, questionable in view of the occasional younger valley networks, such as those on the volcano Alba Patera. Some of these younger valleys have been attributed to local hydrothermal recycling of water but Baker *et al.* [1990] claim that the younger valleys are also the result of climate change and that Mars has experienced repeated massive changes in global climates, these changes being triggered by the large floods. Another problem is that massive carbonate deposits have not been detected. If the valley networks are the result of global climate change, then the 1-3 bars of CO_2 should be in carbonate deposits younger than the valley networks themselves, and remote sensing has failed to reveal them.

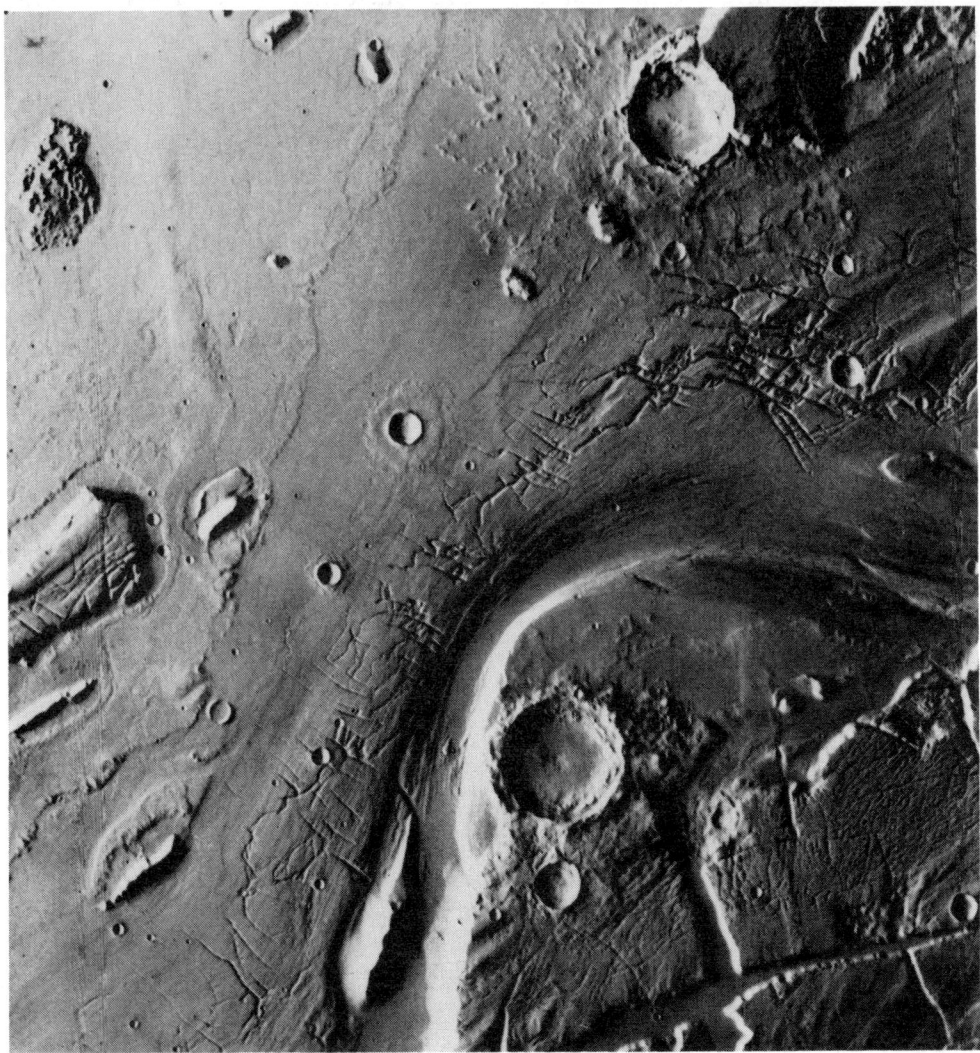

Figure 7 Part of the large outflow channel Kasei Vallis. Scouring of the ground over much of the image area indicates that the flood was not confined to the central channel. The alrge crater in the lower half of the picture is 27 km across.

The large dry valleys represent a very different type of fluvial feature, interpreted as having formed by large floods (see Figure 7). Many of the valleys start full-size, with few, if any, tributaries in areas of what has been termed chaotic terrain in which the ground has seemingly collapsed to form a surface of jostled and tilted blocks, 1-2 km below the surrounding terrain. The areas of chaotic terrain range in size up to several hundred kilometers across. The largest areas are in the Margaritifer Sinus region east of the canyons and south of the Chryse basin. Several large dry valleys emerge from this area and extend northward down the regional slope for several hundred kilometers. Other large channels to the north and east of the canyons converge on the Chryse basin and then continue further north, where they merge into the low-lying northern plains. The channels have streamlined walls, scoured floors, and commonly contain tear drop shaped islands. All these characteristics suggest that they are the result of large floods, rather than the slow erosion of running water. Although most of the flood features are around the Chryse basin, they are found elsewhere. Those near Elysium and Hellas have already been mentioned. Others occur in Memnonia and western Amazonis. Impact craters superimposed on the flood channels suggest that they formed throughout the history of the planet.

The floods were enormous. The largest known terrestrial floods are those that cut the Channeled Scablands in eastern Washington State in the late Pleistocene. These are estimated to have had peak discharges of about 10^7 m^3 s^{-1} as compared with the average discharge of 10^5 m^3 s^{-1} for the Mississippi. Discharges for the Martian floods are estimated to have been as high as 10^9 m^3 s^{-1} [Baker, 1982]. The amount of water that flowed through the channels around the Chryse basin is estimated to be the equivalent of 45 m spread over the whole planet [Carr, 1987]. The cause of the large floods is unclear, and they may not all be of the same origin. We have already seen that floods in Elysium and Hellas are, in some way, connected with volcanism. Volcanism may have resulted in melting of ground ice, migration and groundwater beneath the permafrost, and ultimately, catastrophic breakout. Similarly, volcanism in Tharsis may have resulted in outward migration of groundwater to lower-lying regions around the edge of the Tharsis bulge, where repeated breakouts occurred. Sediments within the large equatorial canyons suggest that the canyons at one time contained lakes, probably as a result of groundwater flow out from under the surrounding plateau. Catastrophic release of water from these lakes may have caused some of the large channels that connect with the canyons to the east. After the floods were over, large lakes must have been left at the ends of the channels. The fate of this water would have depended on the latitude of the lakes. Many of the Chryse channels, and those in Elysium and Hellas end at high latitudes. If climatic conditions were similar to present conditions, these lakes would have frozen and possibly formed permanent ice deposits, which may account for some of the peculiarities of the low-lying, high latitude plains. Other channels, such as Mangala Vallis end at low latitudes. The water in the terminal lakes in these cases, would have frozen and then sublimed into the atmosphere, to be ultimately frozen out at the poles. Because of their enormous size the floods probably could form under present climatic conditions, so, unlike the branching networks, have no climatic implications other than to indicate the presence, at the Martian surface, of large amounts of water.

Figure 8 Debris flows at 46°N, 311°W. High standing remnants of upland are surrounded by smooth-surfaced debris flows. The flows are believed to form from ice-rich talus shed from the high ground. The scene is 120 km across.

Ground Ice

Several features of the Martian surface are most plausibly explained as due to the presence of ground-ice [Kuzmin, 1983; Lucchitta, 1981]. We have seen that ice is unstable at all depths below the surface at low latitudes, but stable at depths deeper than a few meters at high latitude. Although ice is unstable at low latitudes, it may be present at depths of a few to several hundred meters because of the slow rate of diffusion of water vapor away from the ice, through the overlying materials, into the atmosphere. Kuzmin [1980] attempted to measure the depth to the top of the ground ice layer from the morphology of impact craters, on the assumption that petal-like ejecta patterns were due to the presence of ground ice or groundwater. He found that, in any given region, the petal-like patterns were found only around craters larger than a certain minimum size. The size decreased from 6-8 km diameter at the equator to 2-4 km at 40° latitude.

From this he inferred that ice was common at depths greater than about 300 m at the equator and at depths greater than about 100 m at 40° latitude.

Debris flows may also indicate the presence of ground ice (see Figure 8). In the 35-50° latitude band in both hemispheres debris flows commonly occur at the base of cliffs. These are convex upward flows that extend about 20 km away from the base of the cliff. Such features are rare, if present at all, at low latitudes. The simplest explanation is that at low latitudes, where cliffs form, talus simply accumulates on the cliff slope, and so inhibits further erosion. At high latitudes, however, because of the presence of ground ice, ice becomes incorporated into talus, thereby enabling it to flow away from the cliff, exposing the slope to further erosion. Debris flows are particularly common in regions of what has been termed fretted terrain, in which flat floored valleys, filled with debris flows, extend from low lying plains far into the cratered uplands. Formation of these valleys appears to be connected in some way with the formation of the debris flows. A general softening of terrain at high latitudes has also been attributed to ground ice. At low latitudes many features, such as crater rim crests and wrinkly ridges, are crisply preserved. However, poleward of about 40° latitude features tend become rounded or softened in appearance. Squyres and Carr [1986] attributed the general softening at high latitudes to ice-abetted creep of the near materials.

Poles

At each pole, and extending outward to about the 80° latitude circle, is a thick stack of layered sediments (see Figure 9). They are at least 1-2 km thick in the north and at least 4-6 km thick in the south. Incised into the smooth upper surface of the deposits are numerous valleys and low escarpments. These curl out from the pole in a counterclockwise direction in the north and a predominantly clockwise direction in the south. Between the valleys, which are roughly equally spaced and 50 km apart, the surface of the deposits is very smooth and almost crater free. For most of the year the layered deposits are covered with CO_2 frost, but in summer they become partly defrosted. The layering is best seen as a fine horizontal banding on defrosted slopes. The deposits are believed to be composed of dust and ice, with the layering caused by different proportions of the two components. The scarcity of impact craters indicates a relatively young age, although the age of the deposits could still be in the order of hundreds of millions of years.

The layering suggests cyclic sedimentation, and the origin of the deposits may be connected in some way with the obliquity cycle. The obliquity is the angle between the equatorial plane of a planet and the orbital plane. The obliquity of Mars oscillates between 15° and 35° on a 1.2 million year cycle. At the highest obliquity twice as much insolation falls on the poles as at the lowest obliquity. As a consequence, the capacity of the high latitude regolith to hold adsorbed CO_2, the size of the CO_2 cap, and the atmospheric pressure may all change with the obliquity cycle. These changes could affect global wind regimes, dust storm activity and sedimentation rates at the poles, thereby causing cyclic sedimentation. Layering would also be caused by any event that resulted in large amounts of water vapor being introduced into the atmosphere. Possibilities are volcanic eruptions, large floods, and cometary impacts. These would all result in deposition of an ice rich layer at the poles.

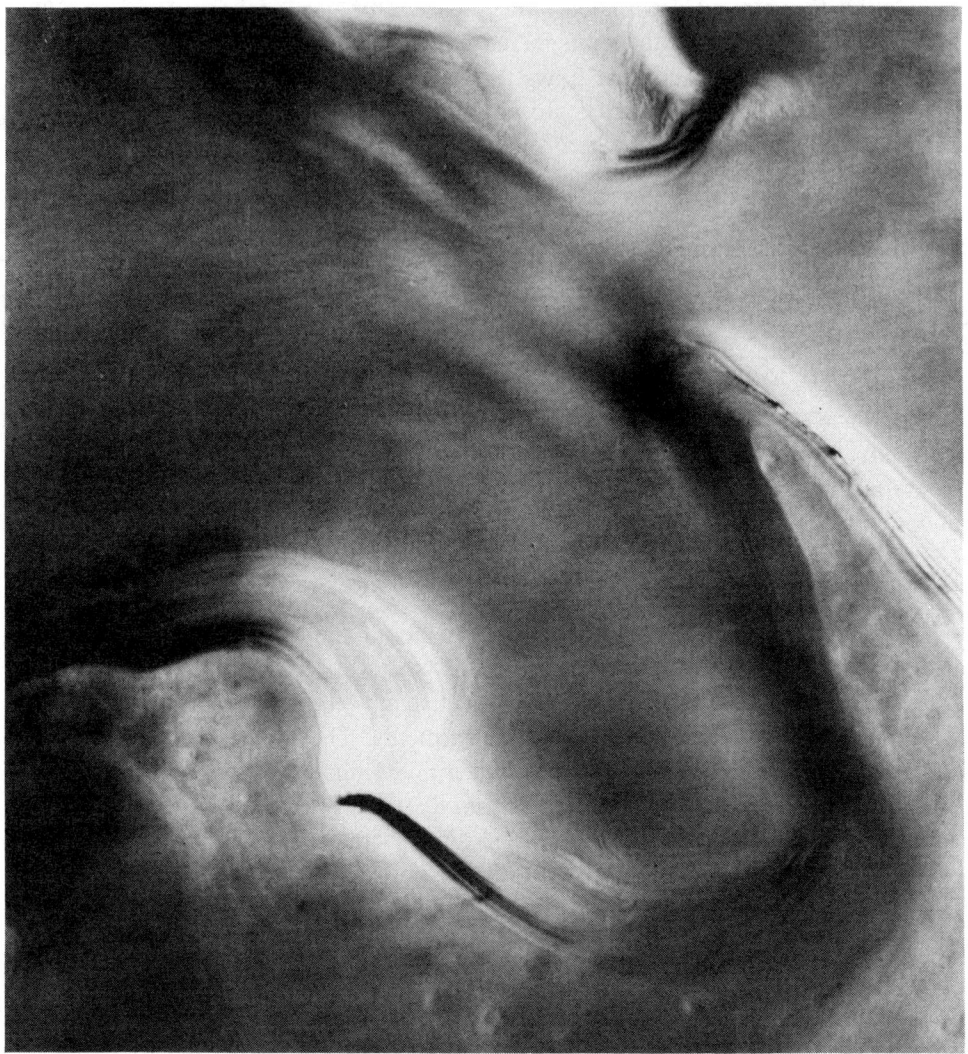

Figure 9 The edge of the north polar layered terrain. The layers are clearly visible in the scarp at the bottom of the picture. The upper surface of the layered terrain is sparsely cratered, indicating a young age.

Exploration Objectives

Following the end of the postulated early, relatively clement, epoch, climatic conditions probably changed to become similar to the present, although there may have been occasional major excursions [Baker *et al.*, 1990] and almost certainly modest cyclical variations in response to variations in obliquity. Major questions related to the climate are as follows: What were climatic conditions like at the time the geologic record emerged at the end of heavy bombardment? What was the subsequent climatic history? What were climatic conditions like during the 600 million years that preceded the end of heavy bombardment? The era between formation of the planet and termination of heavy bombardment (around 3.8 billion years ago) is of special biologic interest in that life may have started on Earth during or immediately after this time. It is crucial therefore to understand what climatic conditions were like during this time.

The climatic history of the planet is strongly linked to the role of water in the evolution of the surface. Liquid water has reacted with the surface, eroded it, and deposited sediments, and possibly evaporites, upon it. Important questions concerning water are: What is the total amount of water outgassed from the planet, and what fraction currently remains? How did the water inventory change with time? Were conditions ever such that liquid water was stable at the surface, and if so, when? Were there ever long-lived standing bodies of water at the surface? What are the relative roles of precipitation (rainfall or snow) and release of groundwater in formation of the channels? What has been the history of fluvial activity on the planet? What is the present distribution of ground-ice? When did the materials that form the layered terrains accumulate? How have weathering rates and the nature of the weathering products changed with time? Again, answers to these questions will only be found though a coordinated program of remote sensing, *in situ* mapping and geochemical measurements, and geophysical sounding coupled with extensive sampling.

Independent of its origin and history, the Martian atmosphere is of interest because of its contrasting dynamics as compared to the Earth. As ever more complex models of the circulation of the Earth's atmosphere are constructed, the models are better able to predict the behavior of the atmosphere here on Earth. However, on Earth the models can be tested only against the rather narrow range of variations that annually occur here. The validity of the models in predicting effects of secular changes, such as a rise in the global temperatures of the carbon dioxide content of the atmosphere, are very uncertain. One way of testing the generality of the models is to use them in environments that differ considerably from those that occur here on Earth. Mars is particularly suitable as an alternative test candidate for atmospheric circulation models because, while its atmosphere differs greatly from the Earth's atmosphere, both atmospheres have a variety of characteristics in common. A major goal of future Mars exploration should therefore be to monitor the atmosphere in order to further refine global circulation models.

BIOLOGICAL OBJECTIVES

Biologic objective of Mars exploration are examined more fully elsewhere in this volume. Researchers have long entertained the possibility that life might have started on Mars. Indeed, until the 1950's it was thought that climatic conditions on Mars were similar to those on the Earth, and it was widely assumed that some of the seasonal changes observed from Earth were the result of changing vegetation patterns. With recognition that the atmosphere is very thin and that surface temperatures are such that liquid water is unstable at or close to the surface, the prospects for present day life faded. Prospects dimmed further when the Viking spacecraft failed to detect any life forms and, more importantly, showed that the surface materials are devoid of organic molecules down to levels of a part per billion. The organic materials are apparently destroyed by a combination of the UV radiation and the highly oxidizing nature of the soil. While no experiment can demonstrate definitively that no life presently exists on Mars, the prospects are very poor.

The chances that life may have started early in the planet's history are, however, quite different [McKay, 1986]. Life forms have been detected on Earth in 3.6 billion year old rocks. Conditions on Earth and Mars may have been quite similar at these very early times. We have abundant evidence that liquid water was present on the Martian surface. This implies that the atmosphere was considerably thicker and that temperatures were much higher than on present-day Mars. In addition, both planets appear to have had CO_2-N_2 atmospheres. Mars may even have had a magnetic field at this time to deflect the potentially lethal solar wind and solar flares. Thus a question of major importance that must be addressed by all future missions to the planet is "Did life start on Mars?." If the answer to this question is yes, then a number of additional questions follow: What was the nature of this primitive life? What were the conditions under which it evolved? Did it become extinct, and if so, when? In what habitats did the life exist? If life did not ever start on Mars, then the question arises as to what conditions were like on early Mars and how did they differ from those conditions on Earth that brought forth life. Many of the questions just posed are difficult of answer. They will require both a search for fossils and a search for chemical signatures of former life. The search will have to be strongly coupled with geologic studies to identify in the rock materials that were either deposited organically or deposited in environments, such as lakes, that were likely to be hospitable to life.

The ability of Mars to support terrestrial life is another important life science issue. Our ability to stay on the surface and ultimately establish long lived outposts will depend on the extent that food can be grown on the planet and that water and appropriate nutrients can be found there. Any expedition to Mars should, therefore, include experiments designed to utilize the planet's resources for survival, with the ultimate aim of self-sufficiency at the planet.

CONCLUSIONS

We are fortunate that Mars, the most accessible of the planets to human exploration, is also one of the most scientifically interesting. While science may not be the main motivation for sending humans to Mars, enormous scientific returns will result. Many of the questions posed above are of fundamental importance, affecting our perception of how the planets form, how they evolve and how life started. Whether any of the questions can be answered definitively is uncertain, but what is certain is that in trying to answer them, we are going to learn much more about the solar system, and how it evolved to what we observe today.

REFERENCES

Baker, V.R., *The channels of Mars*, 198pp., University of Texas Press, Austin, Texas, 1982.

Baker, V.R., R.G. Strom, S.K. Croft, V.C. Gulick, J.S. Kargel, G. Komatsu, "Ancient ocean-land-atmosphere interactions on Mars: Global model and geological evidence," *Lunar and Planet. Sci., XXI*, pp. 40-41, 1990.

Carr, M.H., "Volcanism on Mars," *J. Geophys. Res.*, 78, pp. 4049-4062, 1973.

Carr, M.H., *The surface of Mars*, 232pp., Yale University Press, New Haven, Conn., 1981.

Carr, M.H., "Water on Mars," *Nature*, 326, pp. 30-35, 1987.

Carr, M.H., L.S. Crumpler, J.A. Cutts, R. Greeley, J.E. Guest, H. Masursky, "Martian impact craters and emplacement of ejecta by surface flow," *J. Geophys. Res.*, 82, pp. 4055-4065, 1977.

COMPLEX, *Strategy for exploration of the inner planets: 1977-1987*, National Academy of Science, Washington, D.C., 1978.

Dreibus, G., and H. Wänke, "Accretion of the Earth and inner planets," *Proc. 27th Int. Geol. Congr.*, Moscow, 11, pp. 1-11, 1984.

Kuzmin, R.O., "Morphology of fresh Martian craters as an indicator of the depth of the upper boundary of the ice-bearing permafrost; a photogeologic study," *Lunar and Planet. Sci. XI*, pp. 595-586, 1980.

Kuzmin, R.O., *Kriolitosfera Marsa*, 142pp. Nauka, Moscow, 1983.

Lucchitta, B.L., "Mars and Earth: Comparison of cold-climate features," *Icarus*, 45, pp. 264-303, 1981.

McKay, C.P., "Exobiology and future Mars missions: the search for Mars' earliest biosphere," *Adv. Space Sci.*, 6, pp. 220-285, 1986.

Pieri, D.C., "Martian valleys: Morphology, distribution, age and origin," *Science*, 210, pp. 895-897, 1980.

Pollack, J.B., "Climate change on the terrestrial planets," *Icarus*, 37, pp. 479-553, 1979.

Pollack, J.B., J.F. Kasting, S.M. Richardson, and K. Poiliakoff, "The case for a wet, warm climate on early Mars," *Icarus*, 71, pp. 203-224, 1987.

Ringwood, A.E., "Terrestrial origin of the Moon," *Nature*, 322, pp. 323-328, 1986.

Ringwood, A.E., *Origin of the Earth and Moon*, 295pp., Springer-Verlag, New York, 1979.

Schultz, P.H., and D.E. Gault, "On the formation of contiguous ramparts around Martian impact craters," *Lunar and Planet. Sci. XV*, pp. 732-733, 1984.

Squyres, S.W., and M.H. Carr, "Geomorphic evidence for the distribution of ground ice on Mars," *Science*, 231, pp. 249-252, 1986.

Squyres, S.W., D.E. Wilhelms, and A.C. Moosman, "Large scale volcano-ground ice interactions on Mars," *Icarus*, 70, pp. 385-408, 1987.

SSSEC, *Planetary exploration through the year 2000: An augmented program*, NASA, Washington, D.C., 1986.

TBSWG, *Report of the Terrestrial Bodies Science Working Group*, JPL Publication 77-51, 1977.

TGPLEX, *Major directions for Space Science*, National Academy of Science, Washington, D.C., 1985.

Wänke, H., "Constitution of the terrestrial planets," *Phil. Trans. Roy. Soc., London, Ser A*, 303, pp. 287-302, 1981.

Wilhelms, D.E., and S.W. Squyres, "The Martian hemispheric dichotomy may be due to a giant impact," *Nature*, 309, pp. 138-140, 1984.

Wilson, L., and P.J. Mouginis-Mark, "Volcanic input into the atmosphere from Alba Patera, Mars," *Nature*, 330, pp. 354-357, 1987.

A balloon carries stereo views and data to a driver in unfamiliar territory.

AAS 95-493

Chapter 23

SCIENCE STRATEGY FOR HUMAN EXPLORATION OF MARS

Carol R. Stoker[*]

This paper addresses the scientific strategy for human exploration of Mars. A scientific theme *"Exploring Mars as another home for life: past, present and future"* is proposed to provide a focus for the science strategy. There are two major science goals within this theme: (1) determine the relationship between planetary evolution, climate change and life, and (2) determine the habitability of Mars and its suitability for future human settlement.

The space and time scales required for Mars exploration are discussed. Human exploration of Mars will be effective only if we can perform detailed science investigations of a variety of globally-distributed sites over a period of decades. A strategy for achieving this range of mobility consists of at least three parts: (1) local mobility supported by spacesuits and low-mass "dune-buggy" style piloted rovers, (2) regional mobility in field excursions supported by live-in "motor-home" style piloted rovers, and (3) global mobility supported by teleoperated robotic rovers placed in widely-distributed areas on Mars. The time scales of Mars exploration suggest that missions with long surface stay times are needed, and that science can best be performed when staged from a permanent Mars base. Telepresence operation of remotely operated robotic vehicles will be used in all phases of exploration to expand the range and the capabilities of human crews on Mars, and to provide access to information and the exploration experience to scientists, students and the general public on Earth.

Site selection for a Mars base must be viewed in a regional context. The Coprates Quadrangle and adjacent areas is proposed as a geographical area of Mars which could be the focus of Mars exploration activities and provide a rich scientific return over a period of decades.

The science questions and activities motivated by the science theme in the areas of geoscience, atmospheric science, exobiology, and habitability are described in detail for the precursor and human exploration phases of the program. Finally, I describe how scientists at a Mars base would spend their time.

---

[*] Space Science Division, NASA Ames Research Center, Moffett Field, California 94035.

INTRODUCTION

Human exploration of Mars is a vast undertaking of unprecedented complexity. The scientific potential of Mars exploration is enormous, but the strategy for accomplishing the science has not been worked out in detail. A human presence on Mars would open up new possibilities in terms of the breadth and detail of scientific exploration of that planet. This paper proposes a strategy for human exploration of Mars which begins with robotic precursor missions and proceeds through the first human landings and the establishment of a permanent research base.

Science Theme - Life: Past, Present, and Future

The scientific exploration of Mars is nothing less than opening another world to scientific study. If Earth were another planet, how would we go about exploring it? Exploring Mars is a comparable challenge. Carr [Chapter 22, this volume] has reviewed the major science questions for Mars exploration, which are based on the current state of knowledge about Mars. However, knowing the scientific issues provides little guidance in determining the relative priorities, the pace, the strategy, or the approach needed to explore Mars.

To help provide a focus for human exploration of Mars, it is important to consider the rationale for Mars exploration and the scientific goals that support this rationale. Based on these considerations, [Stoker et al., 1989; Lemke et al, 1990], a consistent theme has emerged which provides a focus for scientific activities on Mars over the course of several decades. This theme is: **Exploring Mars as another home for life, both for evidence of past or present life on Mars, and as a potential future home for life.** From this theme, two key areas of research emerge: (1) understand the relationship between planetary evolution, climate change and life on Mars and (2) determine the potential of Mars to support future human settlement.

Research on the relationship between planetary evolution, climate change and life deals with understanding the evolution of the terrestrial planets and determining the role that the origin of life plays in that evolution. Clearly, Mars and Earth have had very different biological histories. If Mars is without life, present or past, then it represents a stark contrast to the Earth which has been influenced by life from very early in its history. If the early environment of Mars was conducive to the origin of life, then the comparison of this early period with the early Earth would be compelling. In addition, understanding the cause of climate change on Mars may help us understand global climate change on Earth. In order to understand planetary evolution, with or without the presence of life, the climatic history and geological record of Mars must be investigated. Mars provides a natural laboratory for the study of planetary evolution, both in terms of its present processes and the record of its past environments.

Early Mars is thought to have been warm and wet compared to its current state [McKay and Stoker, 1989]. This warmer climate was probably sustained by a greenhouse effect produced by an atmosphere composed of carbon dioxide along with other greenhouse gases [Kasting, 1991]. This thick atmosphere did not last very long on Mars, however, as atmospheric CO_2 was depleted by the formation of carbonate rocks. Initially, when early Mars was volcanically active, crustal rocks were recycled with accompanying outgassing of CO_2 [Pollack et al., 1987]. However, as Mars' interior cooled, the intense volcanism subsided and the atmospheric pressure gradually decreased to its current value. The formation of carbonates from atmospheric CO_2 requires the presence of liquid water which is unstable on Mars at present because the mean atmospheric pressure is close to the triple point pressure of water vapor. Once the atmospheric pressure on Mars dropped so low that liquid water ceased to form, the processes which depleted the atmospheric pressure stopped, leaving Mars with its current thin atmosphere. Another probable consequence of the decline and eventual disappearance of liquid water on Mars was the gradual decline of life. The existence of liquid water is an absolute requirement for life, at least on Earth. Since Mars is clearly not teeming with life, it is unlikely that life on Mars was able to successfully adapt to the dry, frozen conditions there. It is, however, still possible that life has survived on Mars in some protected ecological niches. The discovery of such life would be of profound scientific importance and considerable public interest. However, it is unlikely that extant life will ever be found without detailed exploration of the Martian surface by human crews.

Even if life *never* evolved on Mars, it clearly has a strong potential for supporting human settlement in the future. Evaluating this potential requires understanding what resources are available on Mars and how they can be used to support a human presence. The fact that Mars appears to have the resources needed to support human settlement, and that it may be possible to locate, extract, and utilize them, makes this a viable area of scientific inquiry. In addition, a substantial body of research must be performed to determine whether people can adapt to living in the Martian environment. There are compelling scientific questions dealing with the medical and psychological factors associated with living and working on Mars. Thus, evaluating Mars as another home for life provides a focus for Martian exploration which will require sustained human presence on the surface of Mars over a long period of time.

The convergence of science interests within the theme of *"Life: Past, Present and Future"* can be illustrated as follows. There is considerable evidence of hydrothermal activity on Mars throughout its history [Walter and Desmarais, 1993]. There is also evidence that geologically recent volcanism has occurred on Mars [Lucchitta, 1990] which suggests that volcanic or even hydrothermal activity could still be occurring on Mars today. Water is a key resource needed to sustain humans on Mars. Reliable energy sources are also an important consideration. Both these key resources would be available if a site of current hydrothermal activity were located on Mars. Furthermore, hydrothermal activity is associated with the formation of ore bodies, which will be of interest in evaluating Mars for future human settlement [Cordell, 1985]. A site of current hydrothermal activity would also be very important for exobiology since it would offer the greatest potential for finding extant life on Mars. Microbial life is widespread in hydrothermal sources on Earth and life may even have originated in submarine hydrothermal vents [Holm, 1992]. Conditions in hydrothermal environments also lead to high rates of microbial fossilization [Walter and Des Marais, 1993]. Thus, ancient hydrothermal environments may be good places to look for a record of past life on Mars. Therefore, both ancient and modern hydrothermal sources are very high priority sites for exobiology investigations. In addition, hydrothermal sites would be of prime interest for understanding the geological and climatic evolution of Mars. Thus, determining the mechanism, geographic distribution, evolution, and current state of hydrothermal activity on Mars should be given a high science priority from the perspective of the science theme.

TEMPORAL AND SPATIAL SCALES FOR MARS EXPLORATION

The science return from human exploration of Mars varies widely depending on the type of Mars mission strategy used. Human exploration of Mars will be an expensive endeavor, regardless of how it is performed. However, it will be more cost-effective if the mission strategy is selected based on the scientific return on the investment, as well as on the engineering constraints. The engineering issues associated with mission strategy are discussed by Lecompte and Stets [Chapter 9, this volume] and Neihoff and Hoffman [Chapter 8, this volume]. Fundamentally, missions fall into two categories: those with brief surface stay times (1 month or less) and those with long surface stay times (1 year or more). To understand the impact that mission strategy has on science, we must first consider the spatial and temporal scales required for Mars exploration.

Mars has a surface area comparable to the land area of Earth. The most interesting features on Mars are vast by terrestrial standards and, indeed, are planetary in scale. For example, if the Valles Marineris, a volcanic rift canyon, were placed on Earth, it would stretch from the east to the west coast of the North American continent (Figure 1). To understand features on the surface of Mars, it will be necessary to study them over dimensions comparable to the size of the features. Since features of interest are hundreds to thousands of kilometers in size, this is the required spatial scale of exploration. In addition, features of interest are located all across the surface of Mars. Figure 2 is a simplified geologic map of Mars showing the major geologic units. To really understand the geologic and climatic history of Mars, it will be necessary to perform detailed exploration of a range of features on the Martian surface including the ancient cratered terrains, the northern plains, young volcanic terrains, ancient valley networks, and paleolake terrains. Furthermore, the record of cyclic climatic change associated with variations in Mars' orbital parameters is likely stored in the Polar Layered terrains [Plaut *et al.*, 1988]. Thus, achieving a level of exploration accomplishment that would justify the significant expenditure of putting people on Mars requires them to have access to most or all of the Martian surface.

Scientific field work is an iterative process. Field geologists typically visit a site repeatedly over a period of years. At each stage of the research, hypotheses are formulated and tested. A good example of the conduct of scientific field work is the United States Antarctic Research Program, which has maintained an active science program in Antarctica for over forty years. In this program, key areas of scientific inquiry and science priorities are identified at a high level. Scientists propose individual investigations which usually result in field teams visiting a particular site for several months of each year, for three to ten years. By analogy, scientific field work on Mars will require investigators returning repeatedly to field sites of special interest over a period of years. Furthermore, science interests are likely to evolve into new areas of study resulting from knowledge gained during the field work. Thus, human exploration of Mars should yield a rich scientific harvest for decades.

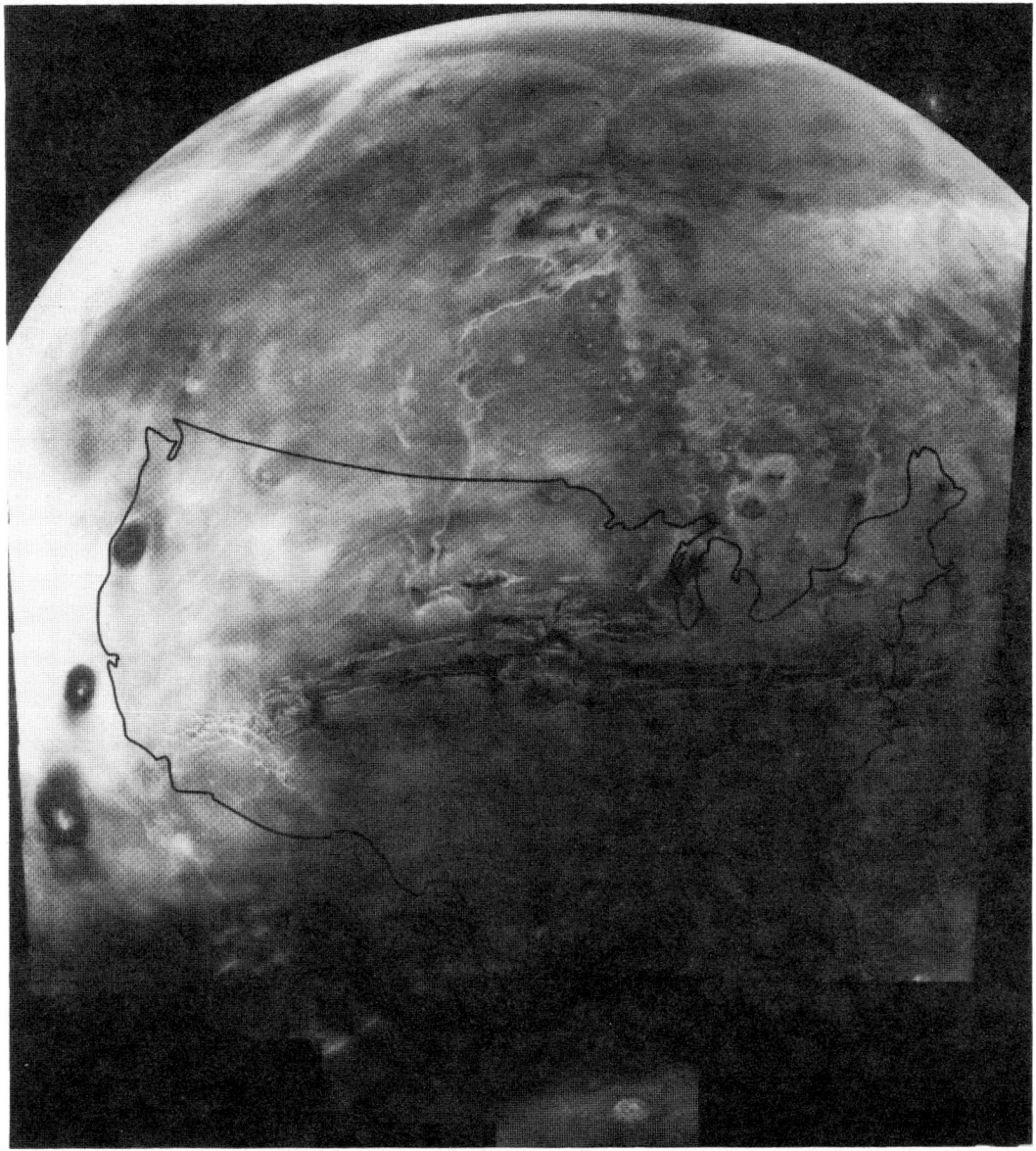

Figure 1 Image of Mars showing the Valles Marineris canyon system. Overlaid on the image is an outline of the North American continent to illustrate the relative size of the canyon system.

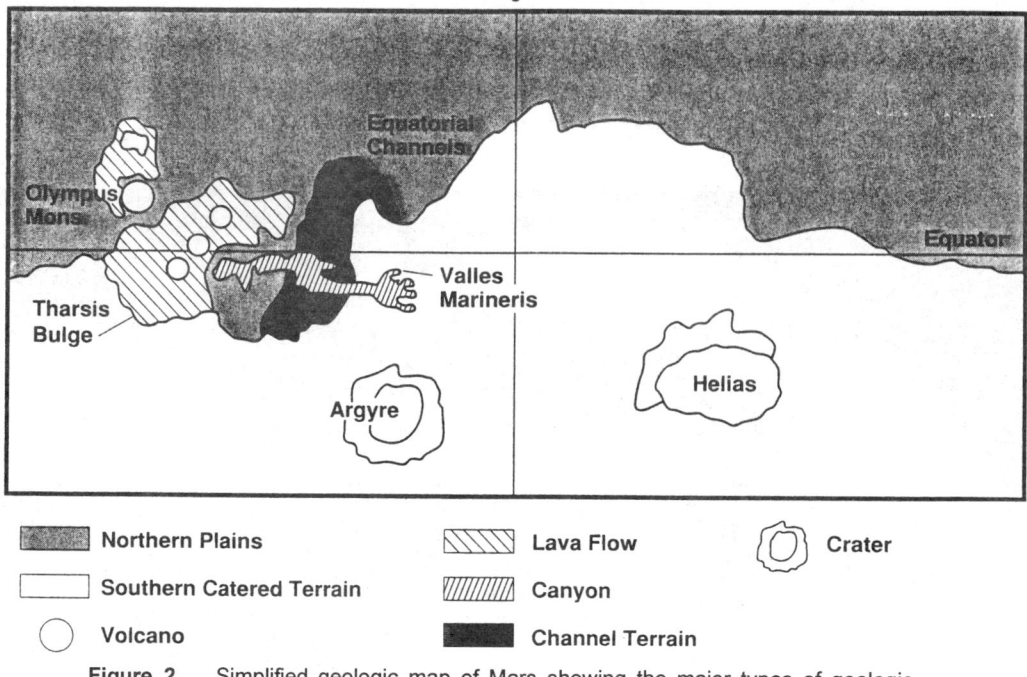

Figure 2 Simplified geologic map of Mars showing the major types of geologic units.

Science Impact on Mission Strategy

The space and time scales required to do science on Mars impact mission strategy. There is a level and scope of scientific activity on Mars which would comprise a valid rationale for the expenditures necessary to accomplish a program of human exploration of Mars. However, missions which are often referred to as "Sprint" missions, that is, missions which have the objective of placing a small crew on Mars for a short period of time and returning them safely to Earth, would not reach the necessary level of scientific achievement. Such missions establish little infrastructure on the Martian surface, have typical stay times at Mars of 30 days or less, and provide little surface mobility. While such missions still can accomplish good science, they are roughly equivalent to what could be accomplished robotically, probably at reduced cost. A "Sprint" mission might be viewed as an initial mission in a long-term program of human exploration which could lead to the eventual settlement of Mars. If cost were no object, it might be attractive to have several brief missions to different locations on Mars before selecting a final location for a Mars base. Such a strategy might lead to a more optimal choice of base location. However, the history of the Apollo lunar missions offer an important lesson about the impact of mission strategy on science. Apollo planners had the long-term goal of establishing a permanent lunar base, and the program performed detailed exploration of six sites on the Moon in preparation for that goal. However, public and political support for the program evaporated once the goal that was publicly stated by President Kennedy ("To land a man on the Moon and return him safely to Earth") had been achieved. Once a single "Sprint" mission to Mars has been performed, a continued program will probably be vulnerable to cancellation regardless of the compelling science yet to be achieved. Furthermore, "Sprint" missions are unlikely to be significantly less expensive than missions with longer surface stay times. Therefore, science considerations argue for missions which have long surface stay times.

Mobility is another key requirement for scientific exploration of Mars. Missions with long surface stay times but a limited range of mobility can accomplish significant science in the short run, provided that the landing site is properly chosen. However, to accomplish first rate science over the long term requires the ability to explore a large portion of the Martian surface. In the next section, I will discuss the strategy for exploring Mars from a Mars base where regional exploration is supported by piloted rovers capable of traversing distances of hundreds of kilometers over periods of several weeks and global-scale exploration will be performed by teleoperating a suite of automated rovers at various locations on the Martian surface.

THE ROLE OF SURFACE MOBILITY IN MARS EXPLORATION

The spatial scale required for the exploration of Mars (Figures 1 and 2) is in stark contrast to the range of mobility which the early Mars explorers are likely to achieve. That range is limited by the type of mobility that can be provided, which, in turn, depends on the mass that can be transported from Earth, as well as issues of life support and power sources. It is logical to expect that initial human missions will have relatively limited mobility and that later missions will be able to provide increasingly greater capabilities leading eventually to the desired global range. However, this view is based on the assumption of a long-term program of human missions which establish considerable infrastructure and capability on the surface of Mars. This, in turn, depends on a philosophy of Mars exploration which involves a continued human presence on Mars, a substantial commitment to build and sustain infrastructure on Mars, and use of Martian resources to support the increased capability of crews on the surface.

Figure 3 shows the expected range provided by various types of human mobility. The minimum requirement for surface mobility in any mission would be a small, low mass rover vehicle similar to the Apollo Lunar Rover. The range of such a "dune-buggy" style vehicle will be limited by the life support capabilities of an astronaut's Personal Life Support System—an integral part of a spacesuit. Thus, the suited astronaut with a rover has a range of approximately 50 km.

Clark [Chapter 20, this volume] describes a rover concept to provide human mobility over regional distances on Mars—a pressurized all-terrain vehicle which incorporates a portable life support system providing a shirt-sleeve environment inside the rover. This type of "motor-home" style vehicle could prove critical for activities in and around a Mars base. In some mission strategies [e.g. Baker and Zubrin, 1990] this type of rover is a critical component of the mission because it provides the means to get the crew from the landing vehicle to a pre-emplaced surface habitat. A piloted rover would also provide crucial mobility for surface science and could support field trips of several weeks duration. Depending on the difficulty of the terrain, such a vehicle might allow field trips in a range of up to 1,000 km from a central base. Still, this is substantially less capability than is desirable for global scientific access to the surface of Mars.

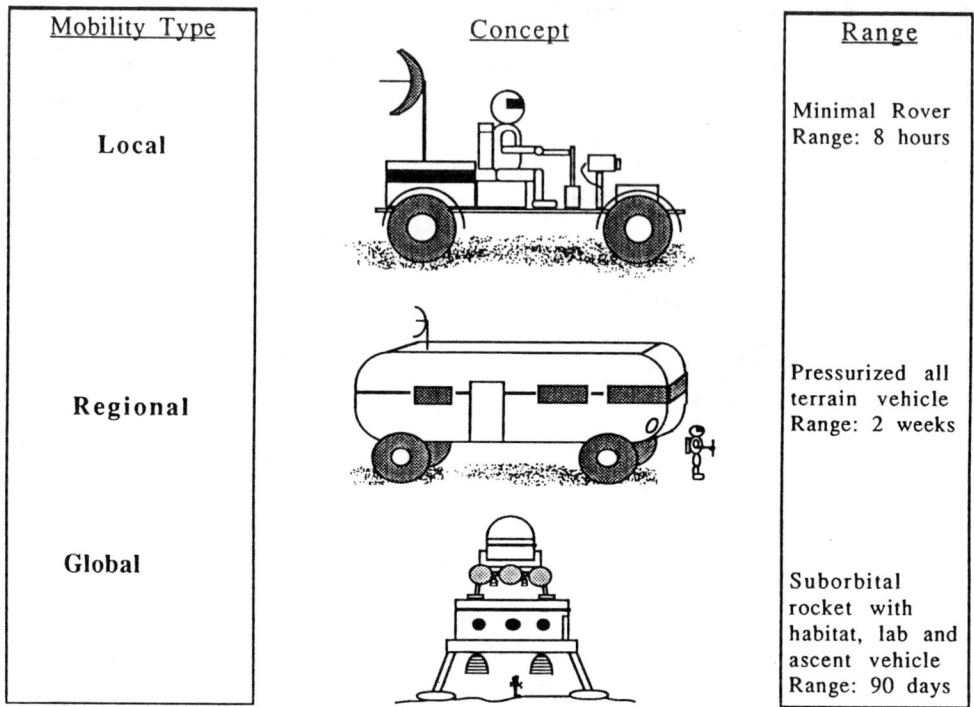

Figure 3 Three types of human mobility that could support science for Mars exploration. The first two types of mobility could be provided with the first human mission, but the third type would await the development of a mature Mars base.

Eventually, one can envision a global capability for human exploration being provided in the form of a suborbital rocket vehicle capable of taking a crew on an expedition of several months duration to any point on the surface of Mars. Zubrin [1989] proposed such an exploration vehicle which uses propellants derived from the Martian atmosphere. Expeditions would originate from a central Mars base facility and would provide the crew with a similar level of capability that could be provided by the surface rover, but without the limitation in range. Such an approach to global mobility is attractive, but it would necessarily await the development of substantial infrastructure on Mars.

Telepresence - a Shortcut to Surface Mobility

An alternative to providing mobility to take astronauts to scientific field sites would be to provide them with teleoperated robots that could be placed virtually anywhere on Mars. Teleoperated robotic rover systems would significantly augment human mobility. In fact, the provision of telepresence as an exploration tool is potentially so powerful that it could establish a new paradigm of exploration where humans and robots interact synergistically to achieve exploration goals.

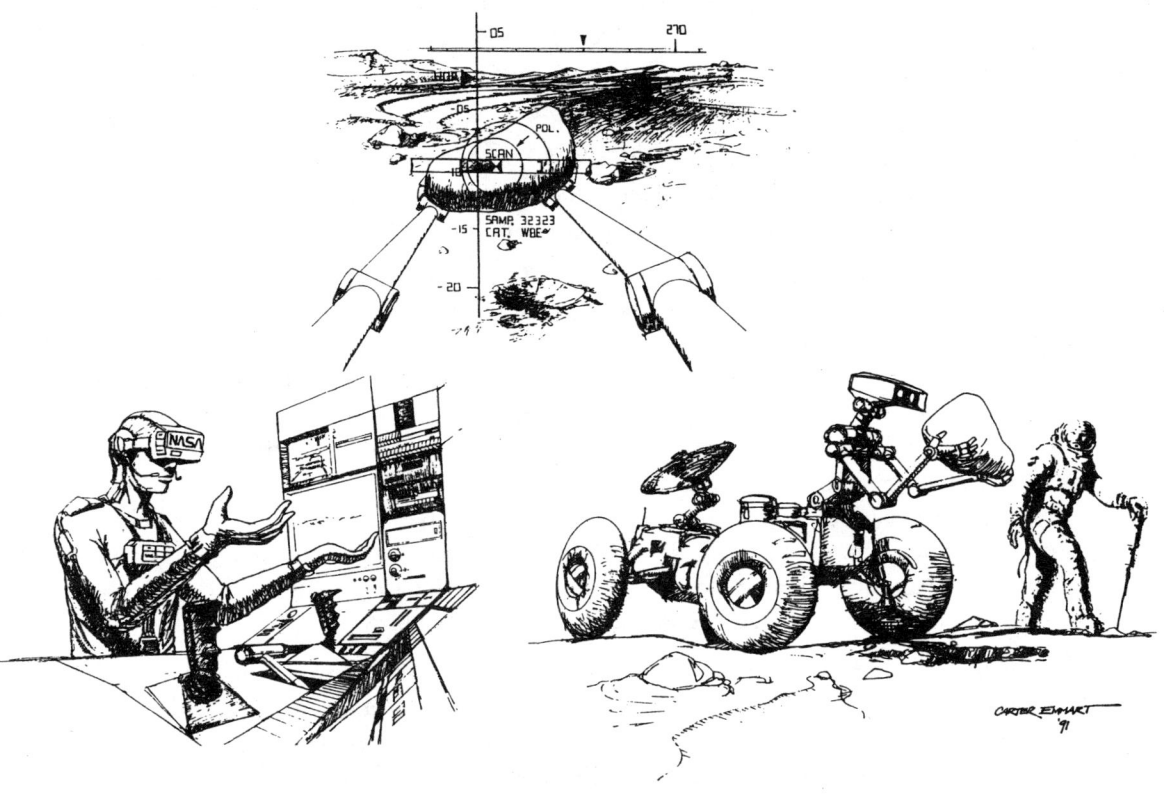

Figure 4 Artists concept of the telepresent operation of a robot on Mars (right panel) from a virtual reality workstation at the Mars base (left panel). The center panel represents the virtual reality data interface between the human and the robot.

Telepresence is a high fidelity form of remote control which projects the senses of the human operator into a robot at a distant work site. Telepresence represents the marriage of two important technologies: virtual reality and advanced robotics. Using telepresence, a robot can be operated in such a way as to give the user a strong sense of presence in the remote environment. The images from the robot cameras can be directly projected into the eyes of the operator while they control the motions of the robot. Figure 4 schematically illustrates the telepresent operation of a robot on Mars. Telepresent operation of robots is an alternative to strict Artificial Intelligence because human operation can be mixed with automated control of low-level functions. Thus, the complexity of the computer control and software involved is much lower. Teleoperated vehicles have long been used in undersea exploration. Telepresence operation from a remote site has been demonstrated in underwater exploration in the Antarctic [Stoker *et al.*, 1994; Hine *et al.*, 1994] and shown to be an effective alternative to human field work in extreme environments. In principle, robots can be controlled on Mars from the Earth, and they are a component of planned robotic Mars missions including the Mars Surveyor missions. However, the most effective and time efficient use of teleoperated robots requires real time, high bandwidth communication. Thus, part of the science strategy for Mars exploration would involve scientists at a Mars base teleoperating remote robots deployed at distant locations on the Martian surface.

The obvious advantage of telepresent operation of remote vehicles is that they could be landed anywhere on the surface of Mars and then operated from a Mars base. Thus, the range that humans can explore using telepresence depends on the communication range and the practicality of deploying the remote vehicles.

Teleoperation of remote vehicles also offers the potential for substantially leveraging crew time. Teleoperated rovers can be designed so that they can function in a variety of ways. For example, one need only consider the stages of a typical geology field trip to see how to partition the work in a telepresent field trip. Such a field trip starts out with transport over relatively uninteresting terrain to reach the study site. On arrival at the study sight, researchers first perform a general survey, with the aid of maps and aerial photographs, to get the broad overview of the area. After surveying the region on a broad scale, various areas are picked out for intensive investigation. The field geologist at this stage is on hands and knees, touching and tasting, breaking off fresh rock faces and examining them with a hand lens, and collecting samples for later analysis. All of this requires extensive documentation with field notes.

A telepresence/robotic field trip might function as follows. The rover might be given a ground command set of instructions to go to a location of interest. If the terrain were relatively simple and trafficable, the robot might navigate autonomously through it, perhaps with occasional supervisory control from Earth to help avoid obstacles and danger. There would probably be times during the traverse when the terrain is sufficiently complex or unforeseen problems arise that require real-time control of the robot from the Mars base. Once the robot reached the area of interest, it could perform a general survey of the region. Again, this activity could be controlled from Earth since it would be tolerant of long time delays (the round trip light time between Mars and Earth is up to 40 minutes depending on the relative positions of the two planets). The data obtained could be sent back to both Earth and the Mars base and could be accessed and analyzed to select the locations of greatest interest. During the intensive stage of field work, detailed real-time control of the robot is crucial. Then, the scientist on Mars would use the telepresence mode of operation of the robot to perform detailed scientific field work in selected sites. This mode would use the full range of telepresence capability to give the scientist a strong sense of presence in the remote environment.

Telepresence offers other advantages for exploration besides increasing the range of access to the Martian surface. Because the telerobot can record and transmit all the information it collects, it frees the scientist/operator from the drudgery of keeping field notes. The operator can instead keep a running verbal record of what s/he is thinking while operating the telerobot. The entire experience is recorded and so it can be replayed. Another obvious advantage of having this high-fidelity record is that the community of access to the exploration experience can be vastly expanded. One can "relive" the exploration experience in one of two ways. The record from the original operator could be passively replayed so that the user could literally hear and see everything the original explorer did. Alternatively, the data could be used to create high fidelity virtual reality models of the field sites. Thus, in virtual reality, one could roam around in the model terrain at will and each person could have a unique experience of the area. Thus, a wide scientific community, as well as the general public, could access the information. The power of this technological capability for education is stupendous. Students at every level could have the thrill of participating in the exploration experience while learning in a very first hand way the mental processes and techniques used in field science. Thus, telepresence offers a tremendous enhancement of the capabilities available to perform field science on Mars.

Even with the aid of teleoperated rovers, there will still be a desire to perform detailed field work with human crews. However, sending a crew out on a traverse to a new area in a piloted rover will be a dangerous undertaking, and considerable crew time will be required for these excursions, so they must be planned very carefully. Teleoperated rovers could be used for this reconnaissance. The entire traverse and the preliminary survey of the field site could be performed with a teleoperated rover to prepare for the human excursion, and to familiarize the rover pilot with the traverse route and any hazards along the way. The advantage of having detailed reconnaissance is so great, and crew time on Mars will be so limited, that any site visited by a human crew would probably be first visited by a teleoperated rover.

SITE SELECTION FOR HUMAN BASE

As discussed above, we have framed the science objectives for the human exploration of Mars within the context of a long-term program of exploration, lasting at least several decades. Thus, site selection for human landings and for the location of a permanent base must consider two scientific requirements: (1) the landing site must have proximity to areas of scientific interest suitable for human exploration during the first mission; (2) the regional area of the base site must offer compelling and continuing scientific issues that can be addressed over the long term. Areas close to many types of materials, terrains, and land forms with diverse geology would qualify. For example, an area where a deep water-carved channel has cut through older terrain would be both geologically and biologically interesting. Also, interesting areas should be relatively easy to reach and in the range of human mobility.

There are many sites that can satisfy the first of the above scientific requirements, as evidenced by the long list of interesting sites identified in the Mars Landing Site Catalog [Greeley and Thomas, 1994]. Human exploration introduces a new dimension to site selection with the requirement that important science activities must be maintained during the period when human access is regional but not yet global. This phase could last for many decades. Thus, selection of a site for the permanent human base must be viewed regionally. We expect that the crew at a Mars base will have access, either directly or via teleoperated robots, to sites separated by distances of several thousand kilometers.

With these constraints in mind, we suggest that the region containing the Coprates Quadrangle and adjacent areas should be the site of the first human base on Mars. We propose this region (shown in Figure 5) as the primary science region, too. Included in this region are: (1) Tharsis volcanoes; (2) Valles Marineris; (3) the ancient cratered terrain of the Margaritifer Sinus region, containing several paleolake sites; (4) Kasei Vallis, site of runoff channels; (5) the region northwest of Valles Marineris containing numerous outflow channels; (6) The northern plains in the vicinity of Chryse; and last but not least, (7) The Viking 1 landing site. This collection of fundamentally interesting areas (plus any other discoveries of interesting sites) would provide a compelling and continued harvest of scientific return from Mars that would befit a human exploration effort.

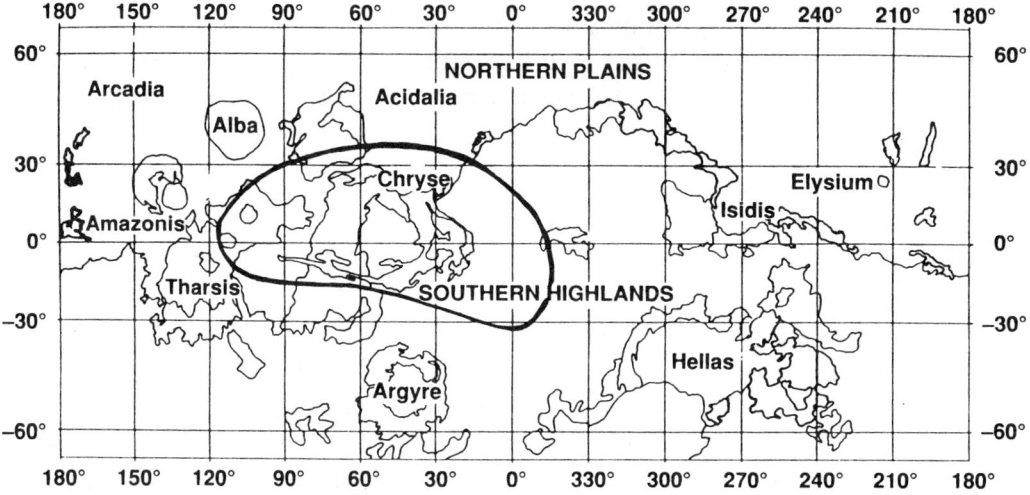

Figure 5 Outline of the major features on Mars. The bold outline surrounds the Coprates Quadrangle and the adjacent regions and is the suggested regional area for establishing a Mars base.

SCIENCE STRATEGY ON MARS

The strategy for exploring Mars is described in this section. As discussed above, the overall strategy is focused by the goal of exploring Mars as a home for life, both for evidence of past or currently indigenous life on Mars, and as a potential future home for human life. Investigating this theme requires understanding the climatological and geological history of Mars, determining whether life ever evolved and its eventual fate, determining whether Mars has the resources to support eventual settlement and determining whether the environment of Mars could be modified to make it more habitable. A scientific approach to answering these questions is developed below for the topics of geologic science, atmospheric science, exobiology, and resource assessment.

Geoscience and Geologic History

As noted previously, the theme "Life: Past, Present, Future" encompasses two major issues: (1) The relationship between planetary evolution, climate change, and life on

Mars, and (2) determining the habitability of Mars. The study of the geologic history of Mars plays a key role in both areas. The major geoscience goals of Mars exploration are listed in Table 1. The sequence of Mars exploration, from precursor to human exploration phases, will help to develop and verify, at progressively finer levels of detail, a geological model of Mars. In addition to focusing our scientific understanding of the planet, the geologic model can be used in a predictive way to help identify the resources necessary to support human presence, and to understand what might be done to make Mars more habitable.

The key issues to be addressed concern understanding the present state of Mars, its formation, and how it evolved over time. Many of the geologic processes that have operated on Earth have also operated on Mars. Like the Earth, Mars is differentiated into crust, mantle, and core. Both planets have had a long history of volcanic activity, and their surfaces have been sculpted by wind, water, and ice. Yet the differences between the two planets are profound. We need to understand the present configuration of Mars, the geologic, geophysical and atmospheric processes presently operating and those that have operated on the planet in the past, the history of their action, and why Earth and Mars are at present so geologically distinct. The activities in each major phase of exploration are summarized below.

Table 1
GEOSCIENCE GOALS AND ISSUES

Geologic goals

1. Develop and test a geological model of the current surface and interior of Mars.

2. Use the geological model to identify resources which are necessary or useful to support a human presence on Mars.

3. Identify and understand the internal and surface processes which have been active throughout Mars history.

4. Establish a time frame for the occurrence of events in Mars history.

Key issues to be addressed:

1. **What is Mars like now?**
   - Dimensions and composition of crust, mantle and core
   - Distribution, type and age of different rock exposed at the surface
   - Sites of present volcanic and tectonic activity

2. **How did Mars form, and how did its mode of formation compare with the Earth's?**
   - Materials from which Mars formed
   - Accretional history
   - Timing and nature of differentiation into crust, mantle and core

3. **How did Mars evolve to its present state?**
   - Impact history
   - Volcanic history
   - Deformational history and contrast with plate tectonics
   - Erosion and sedimentation history, with special reference to the action of water and ice, atmospheric composition and variation

Precursor Phase. The main goal of this phase is to collect as much geologic data as possible in order to understand Mars' geology so that sites can be chosen for human exploration that satisfy both engineering requirements, to safely deliver and sustain humans on Mars, and science requirements, to send humans to locations where the most critical science issues can be addressed. From this data base, a global geologic model of how Mars formed and evolved will be developed. Initially, the data base will be largely remote sensing data. Subsequently, the model will be refined and either validated or changed by acquiring ground truth. This will be accomplished by progressively more complex missions such as network science and rover/sample return missions.

Earth-based studies form a part of the precursor program and would include observations of dust storms to determine where dust might be deposited to prevent building the base in an area doomed to be buried by dust, and ground-based spectroscopy of Mars to provide useful information about surface mineralogy and hydrology over larger spatial scales.

Orbital data will help to identify geological and topographical features as well as determine mineralogy and the state and distribution of water on smaller spatial scales. This information will help to select landing and sampling sites, assess landing site hazards, recognize and interpret the origin of surface processes, and identify resources.

Landers and/or penetrators can be used to understand the composition of the Martian surface and subsurface including vertical extent and uniformity of deposits, understand soil chemistries, and look for hydrated minerals and permafrost.

Rovers can yield detailed information about the Martian surface including a close-up view of different terrains, the physical and chemical properties of soils, the trafficability of the region, and can perform subsurface sounding. Returned samples are needed to aid in understanding the formation and history of Mars, to determine ages associated with units, to provide ground truth to the orbital spectroscopic data, to understand the environmental and volcanic histories, weathering dynamics, changes in atmospheric composition over time, and the possibility that life once existed on Mars. Samples obtained from areas with a well characterized crater size frequency distribution will be used to calibrate age dating schemes based on the cratering record [Barlow, 1988]. These samples would include crustal rocks from orbitally well-characterized geologic units such as intrusive, extrusive, and metamorphic rocks, impact breccia, sediments and surface deposits including soils, aeolian material, glacial till, volatile deposits, and fresh unoxidized rocks.

Human Missions-Early. In the early stage of human exploration, the most effort should be directed toward collecting a diverse sample set for Earth-return. In order to select and characterize samples best suited to answer the pertinent scientific and engineering questions, the crew must establish an understanding of the local geological history and field relationships. Therefore, the crew should first survey and collect samples from the local area (approximately 10 km diameter circle) surrounding the landing area and teleoperate rovers to survey and collect samples from the surrounding region. Sample analysis equipment will be used primarily to select the best samples to return to

Earth for detailed analysis. Shallow rock cores may be used to obtain subsurface or unweathered samples. Active seismic techniques and electromagnetic sounding will be used to probe the subsurface to look for water-rich permafrost and other useful resources.

As time goes on, crews will collect samples and survey regions previously investigated only by teleoperated rovers. Field crews will expand the knowledge of local and regional geology using field mapping, stratigraphic analysis, geomorphologic analysis, and shallow drilling in sediments. The field expeditions supported by piloted rovers will require basic sample analysis capability to help select the best samples for return to the base. Teleoperated rovers will be used to explore interesting remote areas. A mature base will need extensive sample analysis capability including petrography, mineralogy, major and minor element chemistry, scanning electron microscopy, volatile analysis, and spectroscopy. However, the main intent will still be selection of optimum samples for return to Earth for comprehensive analysis. Field activities will also expand the knowledge of resources including assessing ore bodies and subsurface water deposits.

Mature Base. At this stage, crews will have the capability to conduct field trips to virtually any location on the planet and thus will investigate previously inaccessible but interesting places such as: potential exobiology sites, the polar regions, polar layered terrains, the bottoms of craters, the calderas of volcanoes, the bottoms of canyons, dune fields, and lava flows. The equipment necessary for deep drilling should be available, so drilling and mining for a variety of purposes (such as resource gathering) will be possible. During this phase, questions raised during previous phases can be answered and sites can be revisited for further observation.

Atmospheric Science and Climate History

The key science questions to be addressed on Mars in the field of atmospheric science and climate history are shown in Table 2. An important objective of atmospheric sciences research on Mars is to study the evolution of the Martian atmosphere over time, and the physical processes responsible for climate change, both past and present.

Table 2
ATMOSPHERIC SCIENCE KEY QUESTIONS

1. Was there an early dense atmosphere on Mars?
• What was the nature and composition of the early atmosphere? • For how long and by what mechanisms was any early dense atmosphere maintained? • Where did those volatiles go?
2. What processes have been responsible for atmospheric evolution over time?
• What is the nature and distribution of the present volatile reservoirs on Mars? • What processes have controlled the state and distribution of near-surface reservoirs? • What mechanisms dominate the depletion of atmospheric gases?
3. What aspects of the current climate of Mars can be useful as a key to the past? • Has there been cyclic climatic change on Mars? • What are the present cycles of water, dust, and CO_2 on Mars? • What exogenic and endogenic processes control the current climate of Mars?

Major questions include the early state of the atmosphere as well as recent climate cycles and annual transport of water, sediment and CO_2. Considerable geologic evidence indicates major climate changes have occurred on Mars. The occurrence of valley networks and other fluvial features on ancient terrains suggests Mars may have had a dense atmosphere early in its history. The layered terrains that characterize the polar regions could be the result of a succession of ice ages during more recent times. When viewed in conjunction with Earth, Mars offers an opportunity to help separate exogenic processes (such as variations in solar radiation) from endogenic ones (like volcanic eruptions) as factors that affected climate through Martian history. Once the mean state of the Mars atmosphere, climate, and its interannual variation are understood, along with real-time physical processes, we can generate models that reconstruct how processes operated in the past to moderate and change Martian climate. The activities which would occur in each phase of exploration are summarized below.

Precursor Phase. This phase concentrates on improving our understanding of the Martian surface and atmosphere. The goal is to establish a climate baseline before human exploration begins, and to optimize human landing site selection to best meet science objectives. Therefore, a planetary observing system for monitoring Mars weather and climate should be built up during this phase which includes surface meteorological stations and continuous monitoring of weather information from orbit. The integration of orbiter, network, balloon, and rover observations should be achieved during this phase. Interdisciplinary studies oriented toward understanding the dust, water, and CO_2 cycles, as well as the general circulation of the atmosphere, should be undertaken.

The strategy for constraining climate history on Mars starts with obtaining data on the current state of the Martian atmosphere. Each mission in the precursor phase will guide its follow-on. Remote sensing of the atmosphere from orbit can establish the basic structure of atmospheric circulation patterns. Orbital instruments (imaging and infrared spectroscopy) can be used to map the surface composition and geologic units that can help identify samples of interest, for example carbonate sediments and volatile reservoirs. A network of surface stations can then provide *in situ* measurements needed for ground truth of the remote sensing data, and further narrow the search for suitable sample return sites. A global meteorological network would also provide the basis for weather forecasting in support of human landings. Rovers on the surface would be capable of more sophisticated analyses and could obtain a range of samples to help establish evolutionary processes. Returned samples would be necessary to establish more precise mineralogy and age dating.

Human Missions, Early. Once people land on Mars, local meteorological networks will be deployed at the landing/base site. To help define site climatology and broaden the global observational data base, meteorological stations will measure wind, temperature, pressure, atmospheric dust loading and humidity. The crew can monitor global weather activity by tying into the existing weather network established during the precursor phase. This will enhance their operational capability by providing real-time weather reports.

A major goal of this phase is the acquisition of suitable samples for return to Earth. This could be done directly by the crew or through teleoperated vehicles. Several types of samples could yield clues about Mars' past climate: waterlain sediments may contain carbonates or other chemicals indicative of a dense early atmosphere; old impact breccias may contain actual samples of the early Martian atmosphere; certain rocks may provide information on the time evolution of the Martian atmosphere through analysis of their cosmic ray exposure record; others may yield clues about the early Martian environment through their weathering products. Samples of polar layered terrain would provide information about cyclic climate change. A better understanding of the current composition of the Martian atmosphere could be obtained by returning samples of the atmosphere itself.

As time progresses, the diversity of samples collected would be broadened and more sophisticated on-site analysis capability will become available on Mars (e.g. elemental abundance, mineralogy, isotopic composition, etc.). However, samples should still be returned to Earth for detailed analysis. The density of meteorological stations would continue to be increased by deploying them during field trips away from base or by teleoperated vehicles.

Mature Base. Global mobility will permit the collection of samples from the extensive layered deposits in the polar regions. These deposits are believed to be the result of cyclic climate changes. Obtaining and preserving good cores or examining naturally exposed facies directly may yield the Rosetta Stone of the Martian "Ice Ages." The composition and age of these terrains should be determined as a function of depth. Global mobility will also enable field trips to study regions previously found from orbiter and teleoperated rover studies to be good sites of waterlain sediments and evaporites.

A variety of meteorological phenomena can be examined through field studies. Dust storm source regions can be instrumented to learn more about the conditions necessary for storm development and decay. The composition and behavior of the polar caps can also be studied and monitored.

Exobiology: Search for Life on Mars (Past or Present)

Table 3 shows the principle objectives for exobiology, which are prioritized by the probability of success in achieving those goals.

Present Life: Currently there is no evidence for the existence of living organisms on Mars, nor does the current Martian climate appear hospitable towards life [see McKay and Stoker, 1989 for review]. However, the scientific question of extant life on Mars is closely associated with the policy issues of forward contamination of Mars by terrestrial organisms, and back contamination of Earth by Martian organisms [see Nealson *et al.*, 1992]. Therefore it is expected that precursor activities dealing with planetary protection issues will address the existence of present life on Mars. If definite evidence for life is found, this will have a significant impact on the planning for human exploration. If the precursor program supports what current scientific indications suggest, i.e. no indication that there is any extant life on Mars, then it is unlikely that the search for extant life will be a major science driver of the human exploration mission. However, it will be difficult to rule out completely the possibility that there are living organisms hidden in oases or refugia somewhere on Mars and there will be continued interest in testing samples and examining sites for the presence of living forms, however low the probability of success.

Table 3
PRIORITIZATION OF EXOBIOLOGY STRATEGY

Objective	Probability of success	Level on interest	Priority
Search for evidence of chemical evolution and any organic material	High, follows from evidence of liquid water in early epoch	Strong scientific interest	High
Search for fossil and chemical evidence of past life	Possible, depending on the time over which liquid water persisted on Mars	Strong scientific and general interest	High
Search for present life in specialized habitats	Speculative, no evidence at present	Fundamental and broad scientific and public interest	Low

Past Life: As discussed above, there is considerable geological evidence that Mars had liquid water on its surface in the past. This warm, wet, early period may have been conducive to the origin of life [see McKay and Stoker, 1989 for review]. However, searching for evidence of ancient life on Mars will be difficult. It will first be necessary to identify appropriate sites to search for such a fossil record. The key requirement is finding sites where there may have been biological activity in the past and where a chemical or morphological record of such activity has been recorded. This implies the frequent or persistent existence of liquid water. Possible candidate sites include: sites of past bodies of standing water (paleolakes), hydrothermal sites and/or springs, and sites of periodic liquid water activity, such as runoff channels. Of this list, the first two represent sites of significant potential in terms of supporting biological activity and leaving a fossil record of it. No active hydrothermal or spring sites on Mars have been identified but geological features suggests that these could have been prevalent in the past [Walter and Des Marais, 1993].

The strategy for searching for past life in the various phases of exploration is as follows.

Precursor Phase. First, sites of biological interest must be located using remote sensing data. Such interesting sites would include paleolakes, hydrothermal sites and springs, or sites of episodic water activity. The hypothesis that the target sites had liquid water habitats suitable for life should then be verified via *in situ* measurements. Next, a robotic rover vehicle should search for organic material below the surface. Subsurface searching is required because the evidence from the Viking experiment suggests that processes occurring on the surface of Mars destroy organic compounds [e.g. Klein, 1978]. Rovers are preferred over landers or penetrators because exobiologically relevant deposits are likely to occur on small spatial scales and mobility will be required to find them.

Human Missions, Early Phase. During this phase, scientists would intensively investigate biologically-interesting sites both directly (for sites near the landing region) and using telepresence for promising but distant sites. Biologically relevant samples would be collected for return to Earth for detailed analysis although some limited analysis would be performed on Mars to ensure selection of the best samples.

As time progresses, biological investigations relevant to both past or present life would expand to include more sites, and sites located further away from the base location. Extensive sample analysis would be performed on Mars which would include microscopy, elemental analysis, and organic and mineralogical analysis. Experiments to detect the presence of living organisms (e.g. culturing, wet chemistry, RNA sequencing) would also be performed. Sample collection and analysis would be used to guide further field work and selection of the most important samples for detailed analysis at Earth.

Mature Base. From a mature Mars base, interesting sites could be investigated during field trips of several months duration. Sites which could possibly harbor present life in confined ecological niches would be visited and samples returned to the Mars base. Most sample analysis would be performed on Mars using a fully equipped analytical laboratory.

Human Habitability

The key science focus of this topic is to assess the resource potential of Mars. The identification of Martian resources to support human activities begins in the precursor phase and must include Earth-based research to test resource extraction concepts. The importance of this can be illustrated by the case of water. Water has been identified in the atmosphere and polar caps of Mars and is believed to exist in the soil in the form of hydrated minerals [Stoker *et al.*, 1993] and possibly as permafrost at high latitudes [Fanale *et al.*, 1986]. Assessing the availability of water will be an important part of planning for a Mars base in the near term and, ultimately, the availability of water may be the limiting resource in the settlement of Mars. Determining the best way to obtain water is an engineering question which must be balanced against other constraints. Atmospheric water may be technically difficult and energy intensive to extract but is ubiquitous. Soil water of hydration may be abundant but may pose problems of handling feedstock and tailings. Permafrost water and polar ice may place undesired restrictions on the latitude range of the Mars base while still presenting problems with materials handling. Earth-based research will help provide the information needed to choose the best strategy for using Martian resources.

Another important part of the Mars habitability theme is understanding of how living organisms can survive on Mars, both as part of enclosed life support and agriculture systems and exposed to the harsh environment. Further into the future, it may be possible to alter the environment of Mars to be more habitable to Earth life [McKay et al., 1991].

Mars has many, if not all, of the resources in some accessible form that will be needed to sustain humans on the surface. By using local resources, operations on Mars will become independent from Earth-based supply lines. Meyer and McKay [Chapter 19, this volume] review the strategies for utilizing Martian resources. A brief list of useful resources of Mars would include: water, volatiles needed for life support consumables and fuels, suitable construction and insulating materials, useful topographic features, energy supplies, and accessible sources of ores bearing useful amounts of K, Ca, N, S, Na, P, Si, Al, Fe, Mg, Ti.

Table 4
PRECURSOR MISSION REQUIREMENTS FOR RESOURCES

Resource	Mission Type
Water-Atmospheric	
Assess variability of atmospheric source (spatial, temporal, sources and sinks)	Orbiter
Design and test atmospheric extraction technology	Earth-based studies
Water-Soil	
Assess global distribution, amount and variation of soil water	Lander
Assess energy required to extract water	Lander, Sample return
Design and test soil extraction technology	Earth-based studies
Water- Permafrost	
Assess distribution and depth	Orbiter, Penetrator
Design and test permafrost mining strategy	Earth-based studies
Water-Polar caps	
Assess thickness of polar caps	Orbiter, Lander
Assess existence of subpolar liquid aquifer	Lander
Water- subsurface liquid	
Search for evidence of current hydrothermal activity	Orbiter, Rover
Insulating and Building Materials	
Locate potential base sites with accessible sources of loose soils	Orbiter, Rover
Locate potentially useful topographic features	Orbiter
Engineering design and test of habitat construction using local materials	Earth-based studies
Energy Sources	
Global search for evidence of hydrothermal activity	Orbiter, Rover
Determine near surface wind regime in base location	Landed vehicle
Design and test energy use technologies	Earth-based studies
Metal Ores	
Global elemental abundance and mineralogy of surface	Orbiter, Lander/Rover
Surface chemistry and mineralogy of candidate base sites	Lander, sample return

Precursor. Much of the needed resource survey could be accomplished in the precursor phase of Mars exploration. A thorough resource assessment and technology development program to use Martian resources would allow more capable initial missions which make use of Martian resources. Furthermore, the use of Martian resources has the potential to substantially reduce Mars mission costs. Table 4 shows the types of precursor missions that would allow a thorough assessment of Martian resources.

Human Missions, Early. Once people land on Mars, an immediate objective will be to verify the existence and location of local resources inferred from precursor missions. Crews would deploy and test prototype resource extraction units and will begin to build stockpiles of useful resources. Such stockpiles will provide safety backup to reserves the early crews bring from Earth. Furthermore, these stockpiles can be relied upon by later crews. Prospecting for useful resources will be an important activity of human missions. The use of Martian-derived propellants will be a key resource enabling regional exploration of Mars by providing fuel for piloted rovers. Searching for useful sources of water, volatiles, and ore bodies will be a key objective of such regional field work.

Mature Base. During this phase, the global scale exploration of Mars for useful resources will take place. Verifying the state and distribution of subsurface water on Mars will be a high priority during this phase. Water is believed to occur at high latitudes in the form of permafrost [Fanale *et al.*, 1986], and there is a theoretical basis for the occurrence of subsurface aquifers, at least in the polar regions [Clifford, 1987]. Deep subsurface drilling and seismic sounding will help to provide an assessment of these water reservoirs.

Global scale prospecting for useful minerals and ore bodies will be another important area for habitability research, since the settlement of Mars will eventually require materials fabrication.

SCIENCE ON MARS: A DAY IN THE LIFE OF A BASE SCIENTIST

This section describes the duties of a scientist at a Mars base. Scientific activities would divide into three basic categories: (1) scientific field work, (2) deploying and monitoring scientific instruments and equipment and (3) laboratory analysis of samples collected during field work.

Scientific field work would be done in the proximity of the base on day trips in EVA mode, which means the field excursion is performed by either walking in a spacesuit or driving in an open-air rover. Regional field work would be done supported by a "motor-home" style piloted rover (Figure 3). A small crew of scientists would go out on a field expedition for several weeks and travel up to 1,000 km from the base. Along the traverse of the rover, and at the field site, the crew would perform tasks in an EVA mode to accomplish scientific field work goals. Prior to performing a regional expedition of several weeks duration, the scientists would teleoperate a rover to reconnoiter the field site as well as to verify that the site could be reached safely. Samples returned to the base by the teleoperated rover might be analyzed to help prepare for the field expedition. Samples would also be collected on the field expedition for analysis at the base, or for return to Earth for analysis.

Global scientific field work would be performed inside the Mars base habitat by the scientists controlling teleoperated rovers placed in a variety of locations on Mars. Samples might be collected by these rovers for return to the base, where that is practical, but most analysis by the rovers would, of necessity, be done on location. The tasks of operating these rovers would be divided between personnel on Earth and Mars. Earth-based operators would control the rovers for routine tasks that would be tolerant of having a long time delay. Mars-based operators would teleoperate the rovers to perform intricate tasks like detailed scientific investigations that would be much more difficult to perform from Earth.

Laboratory analysis of samples in the Mars base lab would involve cutting and sectioning samples and using various analytical instruments. For geological samples, standard techniques for determining mineralogy, petrology, grain size, elemental composition, age dating, isotopic composition, and trapped volatile analysis could be used. For samples of biological interest, macro and micro-scale inspection of any prospective fossils would be performed as well as organic analysis, biological culturing, and wet chemistry.

In summary, a scientist on Mars would divide his or her time at the base between local field work, teleoperating distant rovers, laboratory analysis of samples, and communication with colleagues on Earth. Regional field work would require the dedicated and full time effort of the scientists involved, both to plan and to perform the excursion, and to analyze and communicate the results to Earth.

SUMMARY AND CONCLUSIONS

The major science theme for the exploration of Mars is determining its role as another home for life, in the past, present and future. Within this theme, two major science questions must be addressed which provide a focus for science activities on Mars. These are: (1) determine the relationship between planetary evolution, climate change and life, and (2) determine the habitability of Mars and its suitability for future human settlement. Thus, understanding the evolution of the Mars environment and how it relates to the search for life will be a major science driver. In addition, determining the long-term habitability of Mars will be a key focus of Mars exploration.

Three major phases of exploration were identified. These include the precursor phase, prior to human landings, early human missions which include the first few human landings, and exploration from a mature Mars base. The activities during each phase and the extent of exploration will be determined by the mobility range of human-occupied and teleoperated rovers. Since the limitations on human mobility are the key defining feature of human exploration of Mars, it is important to begin now to develop rover and spacesuit capabilities to expand the range of human mobility on Mars. It is equally important to develop the technology for teleoperated rover vehicles on Mars to expand the science capabilities of human explorers during all phases.

The precursor science program must be formulated so that it supports the focus and capabilities of human exploration on the surface of Mars. There is no doubt that a well-structured science program, which involves humans on the surface of Mars can do better, more in-depth science than the best possible robotic exploration program. Thus, some science objectives which could be accomplished robotically might be delayed until human crews arrive on Mars.

When selecting a site for human landings, the multi-decade time-frame of the human exploration program, as well as the expanding range of mobility of human crews, must be considered. Thus, site selection must be viewed in a regional context and a regional area must be selected for the exploration which will provide rich scientific return over this time scale. The Coprates Quadrangle and adjacent areas represent a geographical area of Mars which could be the focus of Mars exploration activities and provide a rich scientific return over a period of decades. Detailed studies are needed to select sites for human landings within this geographical area.

Human exploration of Mars which maximizes the science return will involve a mix of direct human involvement and teleoperation using telepresence. Telepresence operation of remotely operated robotic vehicles will be used in all phases of exploration to expand the range and the capabilities of human crews on Mars, and to provide access to information and the exploration experience to scientists, students and the general public on Earth. Further definition of the role of teleoperation/telepresence in human exploration is needed to determine how robotic presence can aid human exploration. In particular, telepresence and other field techniques need to be developed and tested in Mars-analog terrestrial environments. To maximize the science capabilities of these systems, it is important to gain field experience by using them to perform scientific exploration on Earth.

REFERENCES

Baker, D. and R. Zubrin, "Mars direct: combining near-term technologies to achieve a two-launch manned Mars mission," *J. British Interplanet. Soc.*, 43, pp. 519-525, 1990.

Barlow, N.G., "Crater size-frequency distribution and a revised Martian relative chronology," *Icarus*, 75, pp. 285-305, 1988.

Clifford, S.M., "Polar basal melting on Mars," *J. Geophys. Res.*, 92, pp. 9135-9152, 1987.

Cordell, B., "A preliminary assessment of Martian natural resource potential," in *The Case for Mars II*, C.P. McKay, ed., American Astronautical Society, Science and Technology Series, 62, pp. 627-639, San Diego, CA, 1985.

Fanale, F.P., J.R. Salvail, A.P. Zent, and S. Postawko, "Global distribution and migration of subsurface ice on Mars," *Icarus*, 67, pp. 1-18, 1986.

Greeley, R. and P.E. Thomas, Mars landing site catalog, NASA Reference Publication 1238, 392pp., 1994.

Klein, H.P., "The Viking biology experiments on Mars," *Icarus*, 34, pp. 666-674, 1978.

Kasting, J.F., "CO_2 condensation and the climate of early Mars," *Icarus*, 94, pp. 1-13, 1991.

Hine, B., C. Stoker, *et al.*, "The Application of Telepresence and Virtual Reality to Subsea Exploration," in *Proceedings of ROV '94, The 2nd Workshop on: Mobile Robots for Subsea Environments*, Monterey CA, May, 1994.

Holm, N., ed., "Marine hydrothermal systems and the origin of life," *Origin Life Evol. Biosphere*, 22, pp. 1-242, 1992.

Lemke, L.G., C.R. Stoker, O. Gwynne and C.P. McKay, "Mission strategy for human exploration of Mars: summary of workshops from the Case for Mars IV conference," Boulder CO, June 1990.

Lucchitta, B.K., "Young volcanic deposits in the Valles Marineris, Mars?" *Icarus*, 86, pp. 476-509, 1990.

McKay, C.P. and C.R. Stoker, "The early environment and its evolution on Mars: implications for life," *Rev. Geophys.*, 27, pp. 189-214, 1989.

McKay, C.P., O.B. Toon, and J.F. Kasting, "Making Mars habitable," *Nature*, 352, pp. 489-496, 1991.

Nealson, *et al.*, *Biological Contamination of Mars: Issues and Recommendations*, Report of the Space Studies Board, National Research Council, Washington D.C., 1992.

Plaut, J.J., R. Kahn, E.A. Guiness and R.E. Arvidson, "Accumulation of sedimentary debris in the south polar region of Mars and implications for climate history," *Icarus*, 75, pp. 357-377, 1988.

Pollack, J.B., J.F. Kasting, S.M. Richardson, and K. Poliakoff, "The case for a wet, warm climate on early Mars," *Icarus*, 71, pp. 203-224, 1987.

Stoker, C., J. Barry, D. Barch and B. Hine, "Use of telepresence and virtual reality in undersea exploration: 1993 Antarctic telepresence experiment," *Proc. AAAI workshop AI Technologies in Environmental Applications*, Seattle WA, July, 1994.

Stoker, C.R., J.L. Gooding, T. Roush, A. Banin, B. Clark, D. Burt, G. Flynn and O. Gwynne, "The physical and chemical properties and resource potential of Martian surface soils," *Resources of Near Earth Space*, J. Lewis, ed., University of Arizona Press, pp. 659-708, 1993.

Stoker, C.R., C.P. McKay, R.M. Haberle, and D.T. Andersen, "Mars human exploration science strategy," Report of Mars science workshop, NASA Ames Research Center, 30-31 Aug. 1989.

Walter, M.R. and D.J. Des Marais, "Preservation and biological information in thermal spring deposits: developing a strategy for the search for fossil life on Mars," *Icarus*, 101, pp. 129-143, 1993.

Zubrin, R., "Nuclear rockets using indigenous Martian propellants," in *The Case for Mars III*, C.R. Stoker, ed., American Astronautical Society, Science and Technology Series, 75, pp. 495-504, San Diego, CA, 1989.

Section VI

Costs and Benefits of Mars Exploration

AAS 95-494

Chapter 24

THE COST OF SENDING HUMANS TO MARS

Humboldt C. Mandell, Jr.[*]

The costs of human missions to Mars, estimated with conventional aerospace industry cost estimation methods, are more than any realistic estimate of funding availability. And this will be so for the foreseeable future, even with realistic levels of international participation.

A program can not cost more than the resources available to it. Therefore, if the resources will not increase to the levels required to support existing ways of doing business (as reflected in the cost models), it is axiomatic that the ways of doing business must change to meet available resources, *or the venture will simply not happen.*

Changing the ways of doing business to the degree required and in the time required could be impossible; or it could be relatively easy. To try to lower costs within an existing organizational management paradigm is all but impossible in the short term, particularly in the absence of a threat or an imperative. On the other hand, NASA studies have found benchmarks in high technology industries which produce high quality products for a fraction of what NASA has historically spent for comparable products. The difficulty will be in creating the requisite management culture (*not just changing an existing culture*) to insure that the low cost conditions are met.

This chapter presents the analytical relationship between management paradigms of several industries and the costs of comparable products produced by those industries. It further suggests an economic strategy, perhaps the *only* strategy, which will produce human exploration missions to Mars in this generation.

BACKGROUND

The greatest barrier to American exploration of the planet Mars is not the development of the technology needed to deliver humans and return them safely to Earth. Neither is it the cost of such an undertaking, as has been previously suggested, although, certainly, such a venture may not be inexpensive by some measures.

---

[*] Ph.D., Deputy Manager, New Initiatives Office, NASA Johnson Space Center, Houston, Texas 77058.

There is, rather, a high cost of *not* sending Americans to Mars. Just as the Chinese once made a conscious decision to remain a regional power, and thus limited themselves for a millennium to such a role, is now the United States able to afford the cost of being an Earth-bound culture?

The venture can be very *inexpensive*, and, in fact, can probably be accomplished with no increases to the currently-projected NASA budget, and could possibly even be accomplished within a decade.

If not technology, and if not cost, then what is the greatest barrier to American exploration? It will be demonstrated that the *perception* of the costs of such a venture, and the cultural responses to the challenges involved, are the primary factors inhibiting the exploration of Mars.

Most of what has been written since the presidential announcement of a space exploration initiative in July of 1989 has either implicitly or explicitly stated that the costs of such a venture would be prohibitive, at least in the near term, at least for the remainder of the 20th century. Pessimism on a national scale perhaps prompted President Bush to set a "latest" date for the human landing on Mars to be in the 50th year from the first lunar landing in 1969, or in the year 2019.

One of the barriers to be overcome is the very fact that positions have been taken on the cost of human exploration of Mars. The NASA "Ninety Day Study" produced development cost estimates, predicated on current NASA business practices, of approximately $200 billion 1991 dollars. And, indeed, such an undertaking could very well cost that much, if resources were available, and if NASA were to allow it to cost that much.

However, no one has asked the question of how much it *would* cost to send humans to Mars and return them safely to Earth. The question asked was "how much will it cost to explore the Moon and Mars, set up bases in both places, send precursor missions (some not required for exploration), and maintain the NASA institution, all the while doing business as we have chosen to do it for current and past space programs. The two questions are literally hundreds of billions of dollars apart in their answers.

The answer to the first question *could* be that it would cost very little in addition to the currently projected NASA budget, certainly not requiring the doubling of the budget predicated by the answer to the second question. Human exploration missions to Mars of $65 billion *or less* (1992 dollars) are probably feasible, by current calculations, and, in fact, could require even less in new money, with some reallocation of the existing NASA budgetary resources (approximately $14 billion per year, 1993 dollars). *To put that into perspective, we could have Americans walking on Mars at a cost of about five years of the current NASA budget.*

But positions have been taken, papers written, and perceptions created which will be very difficult to reverse. Perception is a very powerful force in a political society, and perception can become self-fulfilling reality.

Research by NASA has shown that the most powerful influence on development cost is the culture of the developing organization, more powerful, even, than the technical and technological drivers on a particular venture. It will be demonstrated that ways of doing business exist, in fact are well known and widely practiced, which can reduce the costs of spacecraft development by 50% or more. Such business practices are being used more and more by other American industries.

There are, therefore, three factors working against low cost planetary exploration. The first is *expectation*; expectation has been based a great deal on perception, in the case of human exploration of the Moon and Mars. The second is *culture*, another qualitative influence which is very hard for typical engineers to deal with, and extremely difficult to change, particularly in the absence of a perceived threat.

And the third is *politics*, a product of culture; any major new undertaking is often expected to serve the existing institutional structure, whether it be in the U.S. automobile industry or in the space exploration industry. Thus, NASA plans for human exploration have been based on past roles and missions, future hopes and aspirations of regional and center interests, and, understandably, on preservation and growth of the existing institution. As other industries have learned (notably the U.S. auto industry), preservation of existing institutions can cause reductions in efficiency which result inevitably in a loss of competitiveness.

Recently, NASA held a celebration of the tenth anniversary of the first shuttle flight. A forum was held for some of those prominent in the shuttle program to share their views on where NASA has been and where it is going.

On the subject of the American exploration initiative, it was interesting that the managers of the shuttle program were very pessimistic about American exploration of Mars. Only John Young, the original Space Shuttle commander, and professional risk taker, expressed the view that the United States should explore Mars through "a crash effort, like Apollo." He alone held the opinion that a great deal of money would be saved by doing the exploration initiative quickly and efficiently: "We would save so much money that you couldn't carry it out of here in a wheelbarrow."

John Young is undoubtedly right.

ANOTHER VIEW OF HOW MUCH HUMAN MARS EXPLORATION CAN COST

If the hypotheses expressed here are true, the stated costs of human exploration of the Moon and Mars are a matter more of perception than of analysis to this point in time. The opportunity therefore exists for a true analytical approach to the estimation of the costs of human space exploration.

One major influence on how much human space exploration will cost is how much it *can* cost. The major design driver in the Space Shuttle program is generally acknowledged to have been the peak annual funding requirement (of approximately $1.2 billion 1971 dollars). Designs for the shuttle had evolved to the point of entering into the design and development phase of the program with a two stage fully reusable configuration. Early cost estimates showed that this configuration could not be developed within the expected budget (doing business as NASA then did business).

This unexpected finding caused a rapid search for a configuration which would fit the OMB-mandated funding profile. The search ended, after dozens of iterative design studies, with the current partially-reusable configuration.

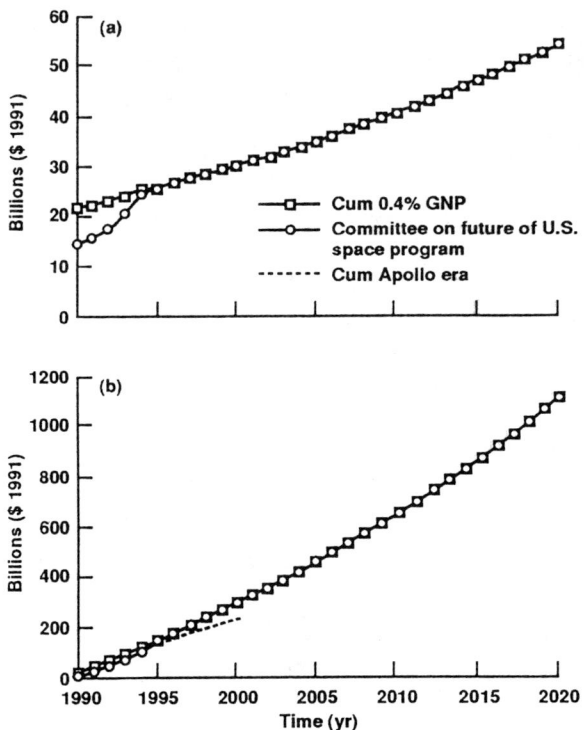

Figure 1 (a) Budget model of the Committee on the Future of the U. S. Space Program; (b) Comparison of Apollo era vs. growth NASA budget.

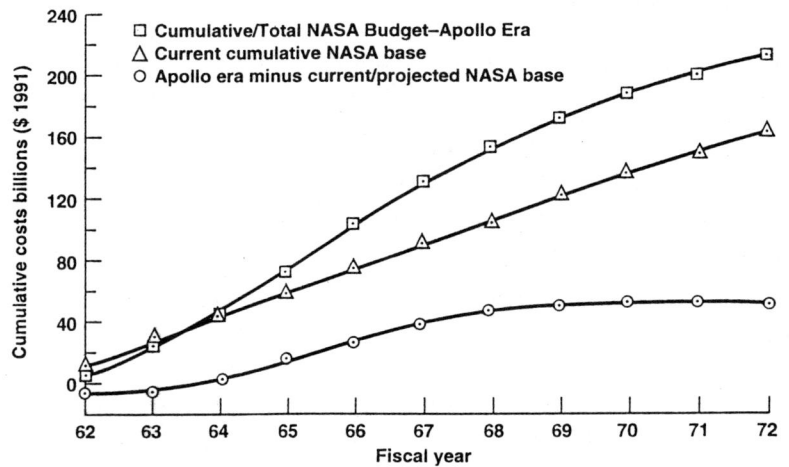

Figure 2 Returning to Apollo budget levels is not the solution for Mars exploration.

Likewise, in the case of exploration, funding constraints will almost certainly determine the pace, if not the content, of the activity.

Early estimates were based on current management practices and very ambitious "architectures" (architecture is the word used by NASA to connote the scenario, or top level flight schedule, of a particular approach to the initiative). These architectures were assembled in a very logical, sequential, building-block fashion, but were not tailored to any budget availability predictions.

Since the content of the various architectures was unconstrained by any logical budget, and was not determined by the accomplishment of any single stated mission, other than a very general presidential announcement, the makeup of the architectures was determined by an accumulation of missions proposed from throughout NASA. These were combinations of spacecraft and science which were, as could have been expected, very costly when compared to any logical prediction of available budget. Time did not permit these early mission proposals to be iterated to fit a realistic funding constraint.

One super ordinate constraint on human space exploration is therefore the available budget; exploration can not cost *more* money than is available. Several attempts have been made to effectively model budget availability for new exploration ventures. Perhaps the most analytical was performed by the Advisory Committee on the Future of the U.S. Space Program (the "Augustine" committee, or ACFUSSP).

This committee proposed a budget for NASA which would increase from the current $15 billion level by a real growth of 10% per year until it reached 0.4% of the GNP, a level commensurate with levels reached during the Apollo lunar landing program (see Figures 1a and 1b). However, when the costs of the current NASA base are removed from Apollo-level budgets, only approximately $5 billion (1991 dollars) per year would remain, still not enough to fund a $200 billion initiative (Figure 2).

An additional problem is the sustaining of budget support for the long periods of time envisioned by human space exploration planners. As shown in Figure 2, Apollo levels of funding, after the costs of the current NASA base are removed, would only yield some $50 billion (1991 dollars) in new money over a decade.

Since any budget level at this time is purely conjectural, NASA has therefore used three arbitrarily-chosen alternative levels for study (see Figure 3). For a given cumulative program cost, the possible first mission dates can be approximated from that chart. The example shown is that of a $200 billion program. It can be seen that at least $8 billion additional dollars would have to be made available to achieve a 2019 Mars landing date, using none of the existing $15 billion budget.

The answer to the question "what *can* human space exploration cost?" is then a complex one, involving at least the following variables:

1. The mission selected.
2. The content of the architectures chosen to perform the missions.

3. The amount of money made available by NASA from its existing budget.

4. The budget available from the Administration and the Congress (itself a function of the chosen architecture: more exciting architectures will perhaps stimulate the availability of more resources, and a higher annual rate of expenditure).

5. The development culture chosen by NASA, or, more precisely, the degree of cultural change made by NASA.

Using Figure 3, anyone can make his own predictions of what exploration *can* cost. The constraints, however, are difficult to predict with precision. Certainly, a doubling of the NASA budget in the current national budgetary climate is unlikely.

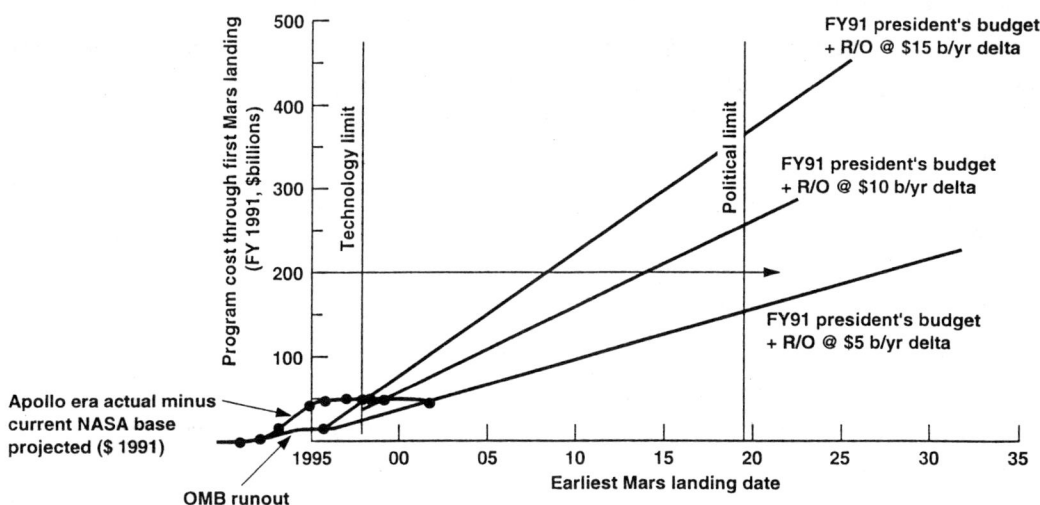

Figure 3   The date of the first Mars landing is a function of available funding.

HOW CAN THE COSTS OF HUMAN MARS EXPLORATION BE ESTIMATED?

At this point in its evolution, any cost estimate will be heavily assumption laden. To the five major variables shown above must be added another list which describes the characteristics of the space vehicles to be employed. The remainder of this discussion will focus on architectures which assume the following:

1. A very straightforward mission statement: The current mission of space exploration is to place Americans on the Moon and Mars and return them safely to Earth as early as possible within national resource constraints.

2. The simplest and least expensive architectures available to perform the missions.

3. The use of about 1/3 of the current NASA budget for exploration purposes.

4. Little or no addition to the current NASA budget (less than or equal to a 3% real growth).

5. A development culture consistent with that of many other major new government ventures, from the Manhattan project, to the Apollo program, to the DOD ballistic missile programs.

Estimation Principles

Cost estimation is prediction. Prediction can be done in a number of ways. Physical laws and models are helpful to engineers in predicting masses, sizes, and performance of spacecraft. These, in turn, are useful in predicting costs.

Much estimation is done by analogy. The more precise the analogy, the more accurate the prediction. Thus, to predict the cost of an automobile battery on one day, one only has to know what a similar battery cost the previous day to make an accurate prediction. Estimating uncertainties increase as the type of battery changes, the car model varies, or the time elapsed increases, giving rise to probable cost estimation variances.

Another effective tool of the estimator is statistics, although one observes much misuse of statistics. The primary reasons for misuse are twofold. First of all, in the spacecraft development industry, very few analogous data points are usually available. Analysts often fit curves to two or three data points, which is statistically very risky.

The second misuse stems from the omission of organizational culture as a variable, i.e., the assumption that cultures of the past will continue into the future, and that historical data will therefore be predictive of the future. This will be further discussed, below.

Cost Models

A cost model is an algebraic accumulation of estimating methods. This accumulation is based on a logical, hierarchic, multi-level arrangement of the tasks to be performed (called a work breakdown structure, or WBS) tailored to the project or program being costed. Typically, a program is broken down into systems, systems into subsystems, and subsystems into components. In a typical model, costs are estimated at the subsystem or system level of a program utilizing a set of "Cost Estimating Relationships" (CER's), often regression equations based on costs and technical histories from analogous systems or subsystems. A dependent variable (cost or manpower) is often predicted based on performance- or size-related parameters, such as weight or thrust.

Systems or subsystems costs are combined algebraically to higher work breakdown levels, and integration costs added as each level is combined to form the next higher level. Management and overhead costs are generally added as percentages of the sum of system and integration costs.

Cost models are only as good as the logic used in their assembly. Often, models are created for a specific purpose (e.g., a launch vehicle cost model, or a crewed spacecraft cost model). In concept, these models, based on historical analogs, are assumed to have predictive power for future systems. But there are at least four sources of error in the estimation process.

Sources of Error

This discussion will be limited to errors not of the mathematics of cost models, but of methodology. Anyone interested in the discussion of statistical errors should consult any of the many good texts on statistics.

First, most parametric cost models are based on mathematical transformations of the historical data set from Cartesian to logarithmic space, which tends to graphically obscure divergences in the historical data, and make correlations appear greater than they are in actuality.

Second, historical data is often not homogeneous, for a number of reasons, most of them associated with the lack of rigor in data collection, and the absence of standards, such as accounting standards, which would make McDonnell Douglas data comparable to Rockwell data, and to Northrop data, etc.

Third, historical data are seldom purged of cost growth. The theory is that all programs will experience cost growth proportional to that in the data base of the model, in this case, a very weak assumption. In fact, in executing human space exploration missions, the scarcity of resources will demand that cost growth be controlled much more closely than has been the norm in spacecraft development programs, which will invalidate this assumption.

The final source of errors, that associated with the culture of the developing organization, is so powerful that it will be dealt with separately.

The Influence of Organizational Culture on Program Cost

NASA has known for decades that it is a high-cost developer. At first, this was assumed to be a natural consequence of the high risks, both technological and engineering, of space flight.

As NASA budgets began to decline at the end of the Apollo lunar landing program, the understanding of the causes of high cost became a priority. The first thing discovered was that NASA technologies, while highly advanced, were not unique. For example, at approximately the same time NASA's contractors were developing the rather conventional aluminum and steel structures for the Apollo lunar vehicles, Lockheed was developing the all-titanium SR-71 aircraft for the U.S. Air Force (and doing it in 22 months).

Components of spacecraft environmental control and life support systems were discovered to be outgrowths of aircraft technology in many cases. Where new components were developed, very often the technologies were not highly advanced. Of course there were a great many exceptions, but when added together, the exceptions did not dominate the development cost picture.

Gone, then, was the most simple explanation for high costs. At this point, NASA brought in RCA Price Systems Division to perform an analysis designed to determine the real causes. RCA had the advantage that they had developed similar devices for a wide variety of customers, from commercial aircraft companies, to the spacecraft industry, to the consumer market.

It was hypothesized that perhaps the NASA requirements for parts traceability (NASA has a standard requirement that every component of every spacecraft flown must be traceable to the basic materials from which it was produced) was one reason; another guess was that high safety, quality assurance, and reliability requirements added to the costs; and another guess was made that the heavy documentation requirements placed on all contractors contributed to NASA costs being higher than those in comparable industries, if indeed there was a comparable industry.

The RCA analysis did find correlations with all of the hypothesized reasons for higher cost. However, a large residual error remained in the regression equations. Only after a variable was introduced to identify the developing agent did high correlations emerge.

The final report [RCA Price Systems Division, 1971] makes the following statement: "Organizational work habits do not readily change completely as a result of imposing a different specification level. Residual costs were attributable only to the organizational manner of doing business [culture]."

These results have been recently verified independently in the Advanced Mission Cost Model research effort which is ongoing at the NASA Johnson Space Center [Cyr, 1990].

The latter research has produced quantified correlations between development culture and program cost (Figure 4). This chart is presented not to suggest that spacecraft programs can be produced with the same management cultures as aircraft programs, but to illustrate the large "leverage" available from slight variations in the culture dimension.

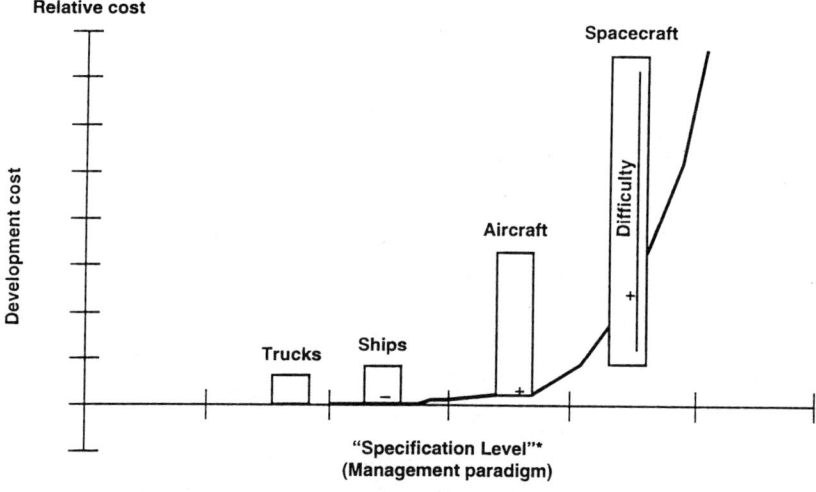

Figure 4   The management paradigm (culture) employed is the largest single influence on cost. *Expressed mathematically, this is the residual regression error after all technical parameters are removed from the data [Cyr, 1990].

Industry Interviews

Once these results had been obtained, NASA set about to find what, if anything, could be done to change "work habits" toward the reduction of costs, particularly development costs. An extensive interview program was designed and conducted with managers of many successful (mostly aerospace) programs.

It was soon discovered that a common message was being sent from the private sector. The ingredients of low cost, successful, high technology programs were well known in the industry, and in fact had been widely practiced outside of NASA, primarily in parts of the DOD, and in the competitive commercial aircraft sector.

The primary message was that NASA had become too involved in telling the private sector *how* to build and fly spacecraft. And through their "phased program planning" approach to program execution, NASA had institutionalized a process recognized by those interviewed to introduce higher levels of cost, without commensurate additions to the benefits achieved, or the reduction of risks.

At the height of the development program of the Space Shuttle, several oversight groups were established to review NASA business practices. One group, chaired by Donald Hearth of NASA, investigated reasons for high costs and cost growth in robotic spacecraft development programs. Hearth's recommendations included requiring NASA to complete demonstration of all key technologies prior to the start of full scale development, which would have introduced a major change into the Phased Program Planning process. The recommendation has not yet been fully implemented into the development of spacecraft for human missions.

The interview process had revealed the identical finding five years earlier. NASA was advised by many of the successful program managers to complete technology demonstration while active competition still remained in effect. Competitions based on paper studies were pointed to as major contributors to cost growth.

As to the NASA involvement in the development process itself, observations came from a number of program managers that the high degree of interaction between NASA and the developing contractor was not only contributing to high change rates, but was compromising the award fee process.

It was recommended that NASA explore the fee arrangements to place more emphasis on good performance of *products*, as opposed to the more subjective evaluation involved in the performance of the humans involved in producing the products. In other words, NASA was advised to reward contractors for good results, as demonstrated in space flight, not for running good meetings and writing good reports.

The interview group also observed that it took NASA far too long to develop new products. Once technological uncertainties are resolved, several told NASA, almost anything can be built in two to three years. NASA had evolved to decade-long development programs, which has produced an expectation in its current culture that development of human exploration spacecraft is a decade-long undertaking.

These "lessons learned" have received a great deal of discussion within and without NASA, and the more important ones have been documented to provide direction to current cultural change processes (e.g., the TQM program recently initiated by NASA).

THE EFFECT OF CULTURE ON COST: THE MAJOR COST DRIVER

Research has therefore revealed the pervasive nature of development culture on the costs of products, not only in the spacecraft industry, but in many sectors of the economy. But simply knowing a cause-and-effect relationship, when it comes to dealing with human cultures, is not sufficient.

At the beginning of a typical high technology program in the United States today, all of these lessons are typically recognized by program management in every industry involved, and program planners have subscribed to the need for cultural change. However, other, more powerful influences, are at work.

First of all, there is the inertia of the culture itself. Behavioral science has dealt with this problem for decades, without producing prescriptive remedies. It is known that cultures change only in response to their environments, and that threatening environments produce more rapid change rates than unthreatened environments. Rapid change, in the absence of major threat, will not happen.

Xerox Corporation has received the Malcolm Baldridge award for the successful cultural changes which led to the healing of that Corporation. David Kearnes, Xerox executive, tells the story of how, in the process of "benchmarking" a competing Japanese copier, it was discovered that Xerox could not even manufacture the copier for the selling price of the competitor. This very real threat provided all the stimulus required to force that successful change experience.

But in government agencies like NASA there often is no such perceived threat. The government customer is often a monopsony, i.e., a sole customer able to dictate the behavior patterns of its suppliers, which all but rules out unilateral cultural change on the part of the private sector suppliers. And monopsonies, by definition, have no competition, and hence threats are harder to recognize.

Instead, the influences which were present at the beginning of the typical government high-technology program are more institutional. As national and agency budgets decline, each government installation struggles to maintain the maximum share of the budget. The primary influence can become the competition for funds. As a consequence, the efficiency of managing a particular program can become secondary to the quest for funding.

Another lesson learned in the interview process was never to begin a program until the requirements for funds are matched by the available resources. In the NASA culture, the uncertainties and political influences involved in initiating a program have sometimes overwhelmed efforts to match program cost requirements with available resources.

Planning of human space exploration missions must therefore be commensurate with *realistic* predictions of program budgets.

A final lesson taught by the culture is that cultures themselves resist change. The individual who violates a norm is identified by the group quickly as being away from the norm, or "ab-norm-al": the individual either conforms to the norm eventually, or is purged by the group.

Cultures are changed by changing their norms, i.e., setting new standards for the behavior of the group. It is therefore essential that any attempt to lower NASA development costs by cultural change be accompanied by willing acceptance by the group of each change; the alternative would be the failure of the change, and the eventual purging of those involved in making the change.

HOW MUCH *WILL* MARS EXPLORATION COST?

If it is then known that programs can cost whatever they are allowed to cost, and that cost is highly driven by the culture, or management paradigm of the developing organization, what then should the nation expect to have to spend to explore the planet Mars with humans?

Cost in the Context of the Current NASA Paradigm

Many Americans today are convinced that a successful exploration program, through the first American landing on the planet Mars, will cost at least $200 billion *new* (current year) dollars, *above the current NASA budget.* Those who hold this belief are probably wrong, not because the initiative could not cost that much, but simply because that amount of money will probably not be made available within a politically-viable time horizon. It is clear that there is a relationship between how much exploration *can* cost and how much it *will* cost. *It can not cost more than is available.*

Linkage of Accomplishments and Budgets

The subject of how much exploration can cost was discussed above. But the subject of how much it *will* cost is much more complex, if only because of the strong linkage between what it *can* cost and the accomplishments promised by the venture itself. A question which needs to be answered here, then, is what can be done by advocates of the program to incorporate into program "architectures" features which are attractive enough to the Administration and the Congress to elicit the maximum support of those who must supply the resources.

Toward that end, it is believed that early accomplishments must be a major feature of any future NASA proposal to perform human space exploration missions. *Early accomplishments must not be stunts*, but must be real, meaningful, elemental parts of a deliberate, well-planned program designed to land Americans on the Moon and Mars. Any flight hardware produced should have a direct planetary exploration mission, and make a direct and indispensable contribution to human exploration of the Moon and Mars.

While "early" is a rather subjective measure, it is believed that timing can be measured in terms of political horizons which are meaningful to the decision makers who will influence the availability of funds. It is submitted that milestones much after the beginning of the 21st century do not satisfy the criterion of "early."

Cost in the Context of the Cultural Change Paradigm

The case has already been made that three things must be done to make human space exploration missions economically viable. First, the available budget must be maximized, both by offering meaningful, early, results and by focusing resources from within the existing NASA $15 billion budget on human space exploration.

Second, realistic "architectures" must be found, which are mission specific, i.e., accomplish the mission with the minimum expenditure of time and money. And third, some measure of cultural change must occur within NASA, possibly in the context of the current NASA TQM initiative, but one which is tailored specifically to the needs of human space exploration missions, and which takes advantage of the large cultural leverage available (Figure 4).

If Mars exploration will not cost $200 billion (through the first American landing on Mars), because it can not, what then will it cost? What cost will provide a safe, economically and technologically viable American exploration achievement?

Cost models designed to explore the sensitivity of the cultural variable currently suggest that there are realistic architectures costing less than $80 billion (1993 dollars) through the first American landing on Mars (*much less, if other resources, like international participation might be utilized*). These amounts of funding *pass all of the criteria* previously mentioned. If these early results prove to be true, American space exploration will have a very bright future.

International Participation

Many have suggested that one way, if not the *only* way to achieve a human landing on Mars would be to perform the venture in cooperation with other space faring nations. While international cooperation is attractive, international partnerships should be scrutinized very carefully.

Lessons learned in previous NASA space programs, currently the Space Station *Freedom* program, point to the dangers of multiple entity developments. If international partners are used, clear and clean interfaces must be established early in the venture and observed scrupulously throughout. From past experience (e.g., the Apollo program) it is clear that a spacecraft can be developed independently from its Earth to orbit launch system. Thus, use of, say, a Soviet Energia launch vehicle could be viable, resulting in the savings of some 5 to 10 billion dollars in development funds, which could be utilized in other parts of the venture, or as the means to accelerate the mission accomplishment dates.

But the use of international partners to develop subsystems for a U.S. spacecraft could actually *increase* total program cost to the U.S., if the principle of clean interfaces is not applied scrupulously.

The main point to be made is that, if early evaluations of mission architectures prove to be correct, *the use of international partners will NOT be necessary for an early American landing on the planet Mars.* Certainly, however, the content of any mission can be enriched, and/or the dates of accomplishment accelerated, by the availability of the additional resources which could be made available by the participation of international partners.

Keys to Making it Happen

Program cost estimates are probabilistic variables. There is not a deterministic model which will predict costs with absolute certainly. Thus, cost models only predict costs *subject to the exact reproduction of the assumptions made* in the estimation process. If assumptions are made as to the numbers of development test articles, mass properties, technological requirements, and development culture variables, all of these assumptions must be reproduced in the execution of the program for the estimates to have any chance of being realized.

If NASA bases the costs of American planetary exploration on the introduction of a certain amount of cultural change (e.g., more delegation of development risk to the private sector), it must be willing to satisfy the assumptions made.

The conditions necessary to produce lowest costs have been previously discussed. However, each must be analyzed fully prior to committing NASA to making them come to pass. Some might prove to be, after further analysis, inconsistent with mission requirements such as crew safety.

CONCLUSIONS

The usual processes utilized by the aerospace industry will not be sufficient to estimate the costs of American human space exploration, and, in particular, human missions to Mars. In fact, use of these methods can prevent the venture from ever happening. The undertaking is too massive in scale to submit to traditional methods. The usual cost prediction parameters of size, technological difficulty, and scale, have been superseded by interactions between the mission and the political environment, which will determine the resources available (which will become the costs), the timing of events, and the development culture employed. Once these parameters and their cost influences have been defined, then the more traditional estimation techniques can be called into play.

In economic terms, the question of how much human exploration of space will cost is a supply-side issue. Supply will be determined by the benefits promised, as well as by the demands of competing ventures on the national scene.

The costs are, then, not only a function of the content of the mission architectures chosen, but are a function of how much the nation is willing to spend. A tourist class flight to London, it could be argued, produces the same result as a first class flight, although the latter costs roughly 400% of the former. But promised the benefits of early, meaningful, accomplishments, the policy makers might well decide to upgrade the class of fare. Without early significant accomplishments, however, the nation may very well decide not to go at all in this generation.

Early estimates have been made of the supply expectations, and these can support, by early analysis, several very attractive architectural approaches to planetary exploration.

However, to make this possible, NASA must dedicate a significant part of its existing resource base to human exploration; mission plans must focus strongly and exclusively on the human exploration objectives, and some measure of cultural change must be introduced to lower program development costs.

Under the proper circumstances, and given the willingness to focus on the missions, to the exclusion of other influences, NASA should be fully capable of fulfilling the dream of planetary exploration within a relatively short time horizon, perhaps even within a decade.

But many barriers must be overcome. The first is the set of perceptions that has been created about the possibility, duration, and costs of human space exploration. The second is the ability of NASA to set a mission statement giving exploration a high internal priority, and to focus on that mission, to the exclusion of other influences. The third is the ability and motivation of NASA to make cultural change in the midst of many ongoing activities. And finally, there is the barrier of convincing the nation and itself that there are worthwhile early accomplishments which can be made by refocusing NASA on the human exploration of space.

There should be little doubt that NASA is up to the challenge.

REFERENCES

Cyr, Kelley J., *The Role of Cost Analysis in Manned Spacecraft Development.*, SAE International (SAE Paper 901863), Warrendale, PA, October, 1990.

RCA Price Systems Division, *Equipment Specification Cost Effect Study* (Phase 2, final report). Cherry Hill, NJ, RCA, November 30, 1971.

A mature base as envisioned, and dictated in 1990, by Tom Paine.

AAS 95-495

Chapter 25

MARS COLONIZATION: TECHNICALLY FEASIBLE, AFFORDABLE, AND A UNIVERSAL HUMAN DRIVE

Thomas O. Paine[*]

In the 21st century men and women from Planet Earth will establish a new human society on the resource-rich Red Planet. Several years ago I presented a 100 year "Timeline for Martian Pioneers" at the Case for Mars II Conference at the University of Colorado [Paine, 1984]. Revisiting the rationale and schedule for Mars colonization, I find prospects are now brighter than ever for international investment in the settlement and economic development of the Inner Solar System. This universal human drive will spur continuing technological advance, global cooperation and economic growth. The settlement of Mars on the proposed schedule can eliminate Malthusian limits to the aspirations of humanity, and open our way to the stars.

INTRODUCTION

The future of mankind will be decided by the race between two competing human drives, one unleashing military power to compete for Planet Earth's finite resources, the other organizing international cooperation to provide access to unlimited extraterrestrial resources. The outcome of this fateful competition will be decided within our lifetimes. I believe that rational statesmen will choose peace and the extension of humanity to other worlds. To establish the required transportation and communication network linking bases throughout the inner solar system we must:

- Avoid major wars through increased international cooperation and armament reductions;

- Reduce global tensions by raising living standards on all continents; and

- Apply a small fraction of the vast capital, technological, and educated human resources thereby liberated to advance spaceflight technologies and learn to live off the land in self-sustaining Martian communities.

This paper looks ahead ten decades to consider briefly: (1) The prospects for international cooperation to achieve peace and prosperity that will make Mars settlement

---

[*] Administrator - NASA, 1968-70; Chairman - U.S. National Commission on Space, 1984-86; Member - Advisory Committee for the Future of the U.S. Space Program, 1989-91. Dr. Thomas Paine is now deceased.

technically feasible and affordable; (2) The rationale that will make Mars settlement a universal human drive; and (3) A potential decade-by-decade timeline for creating an independent, high-technology Martian society.

INTERNATIONAL COOPERATION FOR PEACE AND PROSPERITY

The Stone Age may return on the gleaming wings of science.

— Winston Churchill

This short paper cannot rigorously address geopolitical issues of war and peace, yet the avoidance of major wars and catastrophic nuclear exchange is an essential precondition for the settlement of Mars, as is global economic and technical advance to promote stability and prosperity on Earth. These conditions will activate Martian settlement in the 21st century. Brief comments therefore follow on the chances of achieving these prerequisites.

Avoiding Major Wars and Nuclear Armageddon

This is clearly imperative! Since World War II nuclear weapons have increased in numbers and destructive power by more than a factor of a thousand, and are already in the possession of at least seven nations. Their awesome power effectively blocked their use in the sporadic conflicts of the post-colonial era, and deterred the nuclear superpowers from escalating warfare. In the dark shadow of the hydrogen bomb responsible leaders of the United States and the Soviet Union avoided direct armed conflict despite Cold War tensions, while well-disciplined military controls prevented an accidental firing which could trigger Armageddon. The threat of nuclear holocaust is diminishing in the new geopolitical environment of the 1990s. With good luck and responsible statesmanship, there appears to be a good chance that non-proliferation treaties and international inspection can avert nuclear war. Diplomacy based upon enlightened self interest must minimize armed conflicts while a vigorous Martian civilization becomes self-sustaining. This will represent a great victory of life over death.

Enough fissionable material exists in the world's nuclear warheads to provide fuel to power Martian civilization for centuries. Beating terrestrial nuclear swords into extra-terrestrial plowshares provides statesmen with a compelling rationale for mutual contribution of nuclear arms to provide power for Mars, thus removing these destructive weapons from Earth. The conquest of Mars, unlike terrestrial military invasions, provides a unique unifying force around the Earth that reaches out to all people. A vigorous international program to settle Mars can provide an alternative to destructive wars that could decimate high-tech civilization on Earth, and humanity's chance to reach the stars.

Prospects for Global Technical and Economic Advance

If major conflicts can be avoided through wise diplomacy and effective international institutions, can we achieve the positive prerequisite of global technical and economic progress on a scale that makes sustained Martian investment affordable? There are many alternative scenarios for future world economic development, from neomalthusian doom and gloom to technocratic optimism. In his classic *The Two Cultures*, Lord C. P. Snow [1959] observed that non-technical writers often project a pessimistic view of the future, based on their fears of debased human nature manipulated by despotic bureaucracy (e.g. *Brave New World* [Huxley, 1932] and *1984* [Orwell, 1949]) while technically-oriented writers often paint euphoric pictures of advancing science and technology rationally applied for human benefit in global democratic enterprise (e.g. *2001, A Space Odyssey* [Clarke, 1968], and *The World, the Flesh and the Devil* [Bernal, 1929]). This paper assumes that basic human nature changes very little, but that exponential technological advances initiate revolutionary social change. For the purposes of this paper I have selected as most likely my late friend Herman Kahn's "Guarded Optimism" scenario, which sees the steady economic growth of the last two hundred years continuing as the industrial revolution continues to spread. His provocative book, *The Next 200 Years* [Kahn, 1976], is based in large part on an analysis of the technical, economic, and social changes that have taken place in industrializing countries since the start of the Industrial Revolution in late 18th century Europe, and the implications for future advances.

One quarter of Earth's population now lives in the 80 percent urbanized and industrialized nations of Europe, North America, Australasia, and Japan. This group enjoys a per capita Gross Domestic Product exceeding $10,000 and a life expectancy 50 percent greater than that prevailing in the least developed nations. This uneven progress of economic development is reflected in a roughly 50:1 disparity in productivity and wealth between modern industrialized and traditional agrarian economies. In spite of this gap, the impact of science and technology is felt in every nation. Witness, for example, the unprecedented global medical campaign that completely eradicated smallpox from our planet, and the proliferating "Green Revolution" that in the space of two decades made possible a grain surplus for a third of the human race in Asia.

Comparing different eras and economies involves complex problems of concept and methodology, so results can be ambiguous. With this limitation in mind, Table 1, from *Pioneering the Space Frontier* [National Commission on Space, 1986], summarizes two-centuries of steady economic growth in the United States.

Table 1
U.S. REAL ECONOMIC GROWTH - 1800-1985

Mid-year	Population (1000s)	Total GDP (1985 $ billions)	GDP per person (1985 $ billions)
1800	5,297	4.9	923
1855	27,386	48.3	1,762
1885	56,658	156.4	2,761
1900	76,094	277.7	3,650
1933	125,579	517.3	4,120
1950	152,271	1,245.7	8,181
1965	194,303	2,157.6	11,104
1985	238,631	3,769.4	15,796

In the past 185 years, America's population has multiplied about 45 times, Gross Domestic Product about 770 times, and per capita Gross Domestic Product about 17 times. This experience is paralleled throughout the industrialized world.

In 1800, Planet Earth's estimated total population was about 900 million people, 10 percent urbanized, producing a Gross World Product (GWP) of about $350 billion, or $390 per capita (all monetary units are in 1985 U.S. dollars). Today's world population is estimated to be 5,320 million people, about 40 percent urbanized, producing a GWP of about $20,000 billion, or $3,750 per capita (about the U.S. per capita output of 1900). In other words, the first 200 years of the Industrial Revolution saw the population of the world multiply about 5.9 times, Gross World Product about 57 times, and GWP per capita about 10 times.

This momentum appears likely to continue through the next 100 years, but with uneven progress in differently managed economies. In the 21st century an increasing percentage of humanity will possess advanced technologies, growing capital resources, trained scientists and engineers, and future-oriented leaders. This combination will encourage many nations to participate in pioneering the space frontier.

Because most of the 100 new post-colonial nations are of subcritical size, free world trade is essential for their economic growth. Less developed nations need ready access to the industrialized world's technology, capital and markets. With good management, sustained growth can then be achieved, as has been demonstrated by Japan and the four neoconfucian economies of East Asia: South Korea, Taiwan, Hong Kong and Singapore. Between 1974 and 1983 the percent growth in GNP (in constant dollars) of these economies was: U.S. 22%, Japan 47%, South Korea 89%, Taiwan 104%, Hong Kong 108%, and Singapore 124%. The South Korean farm family who received on average $747 in 1970 earned $4,800 in 1983; the factory worker who was paid $0.10 an hour in 1970 got $1.69 in 1983. Beijing's emphasis on economic development for world markets has impelled China to follow the path of its East Asian neighbors, improving prospects for global economic growth by bringing another quarter of the human race into productive global trading and investment relationships. These successes are consistent with the observations condensed in Table 2, a one-page commentary on the organization of productive economic institutions at seven levels of aggregation based upon my personal observations in international business. Others may disagree, but I believe that this pragmatic view is basically correct, and that similar considerations will apply to super-high-productivity industry on Mars.

The point here is that the development of advanced industrial organizational skills and the attainment of much higher productivity and wealth on every continent of Earth appear well within reach in the next 100 years. New technologies, like biological engineering, optical computer networks, robotics and artificial intelligence, will accelerate this trend. There are many uncertainties, but I believe that productivity and living standards will continue to advance in much of the world. These demographic and economic trends could lead to a 2090 world population on the order of 20 billion people producing a Gross World Product around $100 trillion, for a per capita GWP of $5,000. This is half of the $10,000 per capita GNPs of developed nations today, but will require a challenging eightfold increase in GWP.

Table 2
ONE-PAGE COMMENTARY ON ORGANIZING FOR PRODUCTIVITY

The most productive economic institutions at seven levels of human organizations are:

Level I	A trained, well-equipped team of healthy, motivated workers; within
Level II	A well-managed, internationally competitive industrial enterprise that balances the interests of customers, investors, and workers; within
Level III	A decentralized modern urban complex with excellent financial, educational, and cultural institutions linked to the world through efficient communication and transportation; within
Level IV	A 99% literate, 90% urbanized electrified nation with an effective legal system, equal opportunity, minimal gap between rich and poor, and globally-oriented, incorruptible leaders; within
Level V	An actively cooperating region with ready access to world markets, capital, technology and resources; within
Level VI	A freely trading world at peace; within
Level VII	A solar system rich in energy and materials under cooperative human exploration and development.

(Note: The world's *least* productive worker is an illiterate, unhealthy, low cast peasant working without incentives for a bureaucratic, xenophobic, military dictatorship in a corrupt, war-ravaged, agrarian nation.)

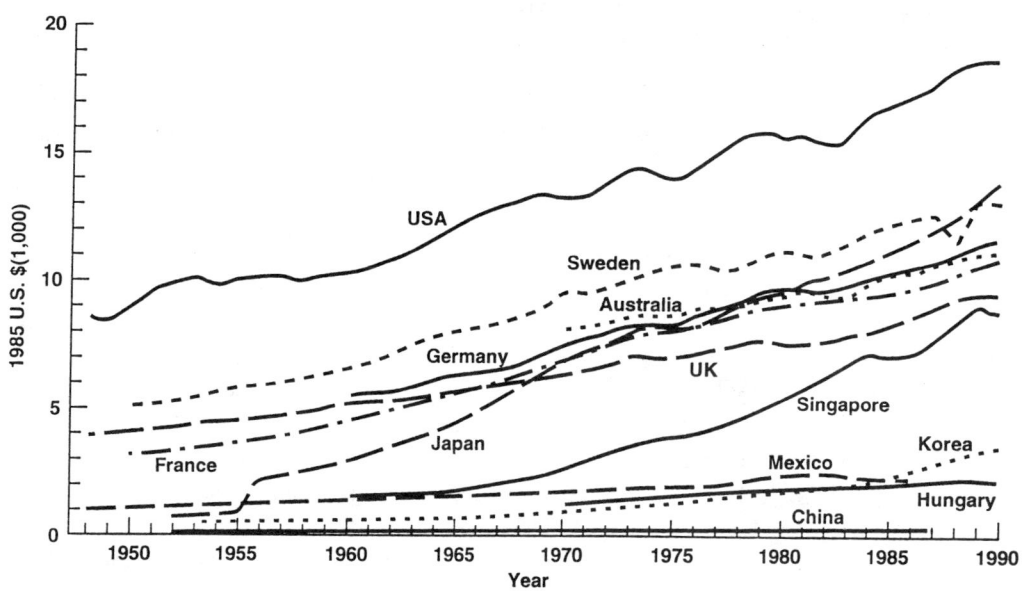

Figure 1 Estimates of economic growth in terms of Gross Domestic Product per capita.

Figure 1 provides a sense of historical perspective by depicting estimates of comparative real economic growth, in 1985 dollars and exchange rates, for 12 nations across the development spectrum in terms of Gross Domestic Product per capita. As the figure shows, some countries, such as Germany, Japan, and Singapore, are achieving relatively rapid rates of economic growth, surpassing the per capita incomes of other countries with long histories of higher ranking positions. Virtually no one looks on such statistics as ends in themselves, but they do correlate with other social measures to indicate nations whose citizens enjoy greater opportunities and higher standards of living.

Adjusting to technological change will bring many social problems as the broadening industrial revolution continues to transform our planet. Strengthened international trade and environmental institutions will be required to guide wise economic growth, but five fundamental factors suggest favorable prospects for international cooperation in the 21st century:

- 45 years of Cold War ending without a nuclear Armageddon;

- The prospect of declining military expenditures freeing capital to update the world's industrial, agricultural, communication, and transportation infrastructure (without economic distortions due to superpower competition);

- Accelerating technical progress across a broad spectrum that will improve public health, education, and spur further widespread productivity gains;

- Demonstrated rapid growth of national economies in well-managed, global-market-oriented nations; and

- Increasing capabilities of effective international organizations like Intelsat to bring nations together in challenging high-tech enterprises.

It therefore appears likely that the substantial economic, human and technical resources required for the settlement of Mars in the 21st century will be broadly available on an increasingly affluent Earth.

This brings us to our basic question: Will Mars settlement be affordable within the probable 21st century Gross World Product? Until we have a settlement plan it is premature to discuss costs, but I estimate that the U.S. space budget required would grow to about twice NASA's $20 billion annual expenditures in the Apollo lunar landing era. With a comparable sum spent by other technically-oriented nations, the Earth's investment in Mars would reach around 80 billion 1992 dollars per year (about the annual output of an automobile company). Assuming that the expenditure level will rise gradually over the century (although with peaks and valleys), this would represent a cumulative total Earth investment in a self sustaining Mars settlement of about $5 trillion. I believe that this price tag is not unreasonable. If the annual GWP in this period rises from today's $20 trillion to $100 trillion in 2085, this $5 trillion investment in Mars would represent less than a tenth of a percent of the cumulative GWP over the century. A prospering Earth can readily afford major investment in the extension of terrestrial life to Mars. Will it wish to do so?

DRIVING FORCES FOR MARS SETTLEMENT

Men need the mystery and romance of new horizons almost as badly as they need food and shelter. In the difficult years ahead, we should remember that the Snows of Olympus lie silently beneath the stars, waiting for our grandchildren.

— Arthur C. Clarke, Author, Consultant to NASA, and former President of The British Interplanetary Society.

Once Europeans mastered deep sea navigation, settlements were rapidly established on other continents for many reasons, attracting pioneers with equally diverse motivations. Men and women endured great hardships crossing stormy seas in sailing vessels, or opening the American West in Conestoga Wagons, impelled both by dissatisfaction with conditions behind them and visions of a better life ahead. Families undertook hazardous six to nine month trips as a matter of course. When space navigation and closed-ecology life support technologies have been mastered in the 21st century, Mars will be settled for a similar wide range of national and human motivations, including:

- *Economic Development* - Long term investment in a creative new growth economy, in which super-high-productivity industry utilizes unique extraterrestrial materials and energy to initiate the economic development of the inner solar system;

- *Limitless Growth Potential* - Access to virgin continents with vast untapped resources, opening an endless frontier that eliminates Malthusian limits to human aspirations;

- *National Pride and Leadership* - The desire of far-seeing leadership and citizens of affluent nations to participate in history's greatest high-tech human venture;

- *Religious, Ideological, and Humanistic Values* - The basic desire to preserve life and to expand and transmit deeply treasured human beliefs and cultural heritages to an expanding posterity;

- *Martian Descendants* - Opportunities for the transplantation and growth of genes, families, races and ethnic groups on new worlds;

- *A Fresh Start* - The attraction of an advanced, intellectually-based civilization working toward a limitless future for mankind free of the old world's diseases, ignorance, fears and outworn prejudices;

- *Technical Pilgrim's Haven* - The chance to build forward-looking, technically-oriented, frontier societies with advanced social compacts that place a high value on scientific vision, technical competence, individual responsibility and cooperative enterprise;

- *High-Productivity Systems Driver* - Initial labor costs on Mars will be a hundred times those on Earth, stimulating the development of reliable robotic production systems with a hundred times the productivity of terrestrial factories;

- *Research and Exploration* - A unique R&D base and technical society ideally located for astronomy, planetary research, spaceflight development and the exploration of the asteroids and outer solar system; and

- *Prototype Extraterrestrial Community* - A representative low-gravity, low-atmospheric-pressure, lifeless planet with rich resources (including water and carbon) for advanced engineering of self-replicating robotic mining, agriculture, industry and closed-ecology life support systems that future generations will employ throughout our solar system - and, eventually, on an expanding new frontier of remote planets circling other stars.

I am convinced that these driving forces will appeal to many individuals and national leaders, attracting more volunteers than can be initially accommodated. How then is the exploration and settlement of Mars likely to proceed over the next 100 years?

A TIMELINE FOR MARTIAN EXPLORATION AND SETTLEMENT

> *Only a miraculous insight could have enabled the scientists of the eighteenth century to foresee the birth of electrical engineering in the nineteenth. It would have required a revelation of equal inspiration for a scientist of the nineteenth century to foresee the nuclear power plants of the twentieth. No doubt, the twenty-first century will hold equal surprises, and more of them. But not everything will be a surprise. It seems certain that the twenty-first century will be the century of scientific and commercial activities in outer space, of manned interplanetary flight and the beginnings of the establishment of permanent human footholds outside the mother planet Earth.*
>
> — *Wernher von Braun*

Let's now project the achievements that we should expect to see in space exploration over the next ten decades if our three assumptions are correct:

- That major wars and nuclear Armageddon on Earth can be avoided,

- That steadily growing world capital, technological, and educated human resources will spur a major Mars settlement project, and

- That for a variety of reasons increasingly affluent nations will direct their space efforts toward Mars, attracting talented space pioneers willing to undergo hardships to participate in this inspiring adventure on the cutting edge of humanity.

The 1957-1992 Technical Foundation

We are considering Mars settlement today in the fourth decade of the Space Age. Let's begin by summarizing some significant achievements that provide the technical foundation on which the Mars venture will be built:

- Large rocket engines and heavy-lift launch vehicles (HLLVs) with capacities up to 120 tons to Low Earth Orbit (LEO), the U.S. Space Shuttle for round trips to LEO, European and Chinese commercial launchers, and Russian Energia HLLVs and automated LEO resupply freighters;

- High energy solid fuel and liquid hydrogen-liquid oxygen upper stages capable of propelling spacecraft from LEO into interstellar space, with advanced R&D on alternative high-specific-impulse spacecraft propulsion systems, from electric ion thrusters to thermal nuclear stages;

- Global tracking networks with large diameter antennas and tracking and data relay satellites capable of high-data-rate communication with spacecraft beyond Neptune and Pluto;

- Crewed orbital flights of many months duration in Salyut, Mir and Skylab space stations that demonstrate adaptation to weightlessness by cosmonauts and astronauts adequate for the long round trip to Mars;

- Life science advances, including closed-ecology life-support experiments by NASA, Biosphere II, and Russian laboratories;

- Robotic reconnaissance of the solar system from Mercury to Neptune, including fly-bys, orbiters, soft landers on three worlds, surface rovers, automated sample returns, and years of climate observations from two Viking landers on Mars;

- Earth observation satellite systems studying oceans, land masses, ice fields, the atmosphere, and their complex interactions;

- Reconnaissance satellite systems capable of monitoring arms control and peace keeping treaties;

- Orbiting astrophysical observatories scanning the universe across the spectrum from infrared through gamma rays; and

- Six manned lunar expeditions that demonstrated 23 years ago surface exploration with lunar landers, self-contained spacesuits, electric jeeps, automated lunar orbit rendezvous, and other systems and techniques usable on Mars.

In addition to this solid foundation of applicable space experience, we have many untapped sources of valuable information in the history of European exploration, in Magellan's and Drake's circumnavigations, in the colonization of "The New Worlds" of America and Australia, in the long voyages of New England Whalers, in stressful wartime submarine patrols lasting up to a year at sea, and in Antarctic research stations. Historical research to develop data on demonstrated human endurance, the settlement of new continents, and the theory of the American frontier can contribute greatly to planning the Mars program.

This is our heritage. What lies ahead as we build on our 35 years of spaceflight experience to pioneer the space frontier? Here is my proposed decade-by-decade future history of Mars settlement.

The 1990-2000 Decade

NASA's new management decides that in the fourth decade of the Space Age U.S. industry knows how to build space systems, and U.S. university teams know how to conduct space research. New contracting procedures are therefore instituted that give aerospace firms and academia new freedom and responsibility to carry out faster-paced programs at lower cost. Detailed specifications of every nut and bolt, followed by micromanagement of every operation, are eliminated in favor of performance contracts. This results in new launch vehicles and spacecraft being deployed on accelerated schedules, from families of expendable launch vehicles and direct broadcast satellites, to man-rated HLLVs and Earth observation systems. Fresh information on the planets is streaming in from Magellan at Venus, the U.S. Mars Observer and Russian Mars 4 spacecraft at Mars, asteroid fly-by missions, Galileo at Jupiter, Cassini at Saturn and other robotic platforms. NASA's Space Station *Freedom* is canceled in favor of a later HLLV-launched station designed to support the Space Exploration Initiative. The result is a new orbital maintenance and operations center, with life science laboratories, spacecraft maintenance, check-out, refueling and launch capabilities, experimental space construction test bed, tether experiments, and other 21st century-oriented capabilities.

The 2000-2010 Decade

The major spaceflight program of the decade is a return to the Moon 30 years after Apollo to establish permanent international lunar bases. Aerospace planes and man-rated HLLVs with crew-recovery capability from launch pad to orbit are in initial operation, greatly lowering the cost and risk of transporting men and women between Earth and LEO. The shuttle fleet is retired after another tragic accident. Additional international lunar base construction is supporting higher levels of activity in space, including high-resolution, multi-spectral, robotic surveys of the Moon and Mars. Proposed Martian habitats and workshops are being tested on the Moon. A crash program has been instituted to send prospector probes with penetrators to Phobos, Deimos and the Martian surface. Prototype systems under development and test include low-cost electric orbit transfer vehicles using large electric ion thrusters and aerobraking, closed-ecology life support habitations, robotic mining and primary metallurgy and ceramics industrial systems designed to process lunar and Martian resources, and 1-10 rpm artificial gravity spacecraft. A number of new university departments have received NASA contracts to step up work in such fields as planetary sciences, exo-agriculture, closed-ecology, robotics, artificial intelligence and exo-sociology. For example, NASA dedicates a new permafrost laboratory at the University of Alaska.

A man-rated space transportation system is extended to Geosynchronous Earth Orbit (GEO) and the Moon; man-tended orbiting bases constructed in LEO are deployed to GEO and lunar orbit using economical electric ion propulsion. Martian rovers with automated sample-return capability are helping to select the most promising sites for resource development. Meanwhile, back on Earth, the supporting R&D in academia is advancing rapidly under the stimulus of the accelerating Mars programs.

The 2010-2020 Decade

In this decade the first U.S.-led international Mars expedition shoves off from an Earth-orbital base to rendezvous with Phobos and Deimos and check out predeployed surface habitats. The first humans then land on Mars to activate bases and initiate exploration. Prototype extraterrestrial communities based on robotic resource development begin operation. A new *International Space Agency* is organized to manage extraterrestrial operations. All of these advances are eclipsed in public interest, however, by the dramatic announcement from Cal Tech, quickly confirmed by the Russian Academy of Sciences, that the first signals have been detected indicating the presence of an advanced extraterrestrial civilization in a remote arm of our galaxy.

The 2020-2030 Decade

The first automated five gigawatt nuclear electric generating station goes on-line to power Martian bases. Its energy is used to separate hydrogen and oxygen from permafrost to fuel surface transport and surface-to-orbit shuttles. Economical electric ion thrusters are providing propulsion for robotic freighters delivering heavy cargoes from Earth, and to emplace the first multipurpose space base in low Martian orbit. This facility serves as Mars Mission Control Center, orbital laboratory, transportation node, space maintenance base, and refueling station. One of its first missions is to experiment with advanced tether techniques to lower orbit transfer costs. Automated aerobraking freighters are using standardized nuclear generators to provide power for their electric thrusters; these one-way transports deliver structural materials and their automated electric power equipment for use on Mars. The first large cycling spaceships are emplaced into permanent orbits linking Earth and Mars. A number of planets are detected in orbit around nearby stars; one of them has an atmosphere rich in oxygen and water vapor, suggesting the presence of life. Simultaneous global holidays on Earth and Mars mark the birth of the first Martian baby.

The 2030-2040 Decade

The population of the Moon now numbers several thousand people, and the first extraterrestrial children are thriving. Closed-cycle lunar communities are now increasingly supported by robotic mining, metallurgy, ceramic processing, factory production and agriculture using indigenous resources, but hydrogen and carbon are still costly imports. Next-generation systems designed for future deployment on Mars are being tested at lunar test sites. Martian research bases are experimenting with high-productivity resource development and construction techniques adapted to Martian conditions. They are manned by several hundred men and women scientists, engineers and technicians who have volunteered for seven year tours of duty. First priority is given to launch pads, air strips and shelter. Additional cargoes to build Martian infrastructure continue to arrive from LEO during favorable Earth-Mars conjunctions.

The 2040-2050 Decade

Technologically this decade is advanced as far beyond 1992 as the 1940s are behind 1992. Closed-ecology lunar prototype communities and resource development systems now support 10,000 lunarians, but more advanced extraterrestrial life support systems utilizing indigenous resources are being emplaced on Mars. By 2045 there are a thousand people on Mars completing installation and initial operation of Martian infrastructure, with emphasis on lunar-demonstrated robotic agricultural and industrial facilities to reduce dependence on costly shipments from Earth.

The 2050-2060 Decade

Despite advancing technologies, teleportation has not been invented in the classic science fiction sense of *Beam me up, Scotty, there's no intelligent life here.* Because optical data communication between planets is so much cheaper than the transport of mass, however, robotic factories on Mars with direct software links to twin factories on Earth are achieving an effective "teleportation." Equipment like hydrogen-oxygen fueled jeeps with artificial intelligence guidance produced on Earth can be simultaneously duplicated on Mars, and vice versa. By 2055 this system of "teleportation via robotic factory software," in which bits are transmitted instead of objects, is creating a wealth of new opportunities for import-export and balance of trade among planets. The population of Mars has grown to 5,000 people through continuing immigration and strong encouragement of multiple births. Robotic Martian factories are working 24 hours and 37 minutes a day, 686 days a year, to turn out robotic Martian factories. The Arthur C. Clarke Observatory is dedicated atop towering Mons Olympus.

The 2060-2070 Decade

The population of Mars is passing 10,000 humans and 100,000 equivalent robots. Two-way Mars-Earth trade is beginning to move toward a balanced exchange and joint scientific research programs are underway. The space programs of Earth and Mars are cooperatively exploring the asteroids and the giant outer planets with their resource-rich moons; materials that are scarce on Mars and Earth's Moon are being sought on low-gravity worlds to reduce dependence on terrestrial resources. Mars-Earth astronomical interferometers have pinpointed the location of eleven other civilizations in the galaxy, but all are too distant to permit communication within a lifetime. The industrial policy of Mars is to reduce dependence on imported goods from Earth and attain self-sufficiency at the earliest possible date. On Earth fusion power is hailed as the space propulsion drive of the future.

The 2070-2080 Decade

As robotic mining, manufacture and construction extends Martian community facilities, immigration from Earth is stepped up and the human population of Mars nears 50,000 people. Trade with Earth is now nearing balance, with artistic products, engineering designs, scientific research results and other software products predominating. Specialized high value goods continue to be imported and exported, and tourism is becoming an important source of foreign exchange for the Martian economy. On Mars matter-antimatter drives are hailed as the space propulsion drive of the future.

The 2080-2090 Decade

The vigorous young Mars settlement contains 100,000 people and close to a million equivalent robots. Martian immigration policy is promoting a rich gene pool from every continent on Earth, while biological laboratories, zoos, gene libraries, museums, art galleries, and book and image libraries house extensive collections of terrestrial life. Although a lively trade is underway, the continuation of life on Mars no longer depends upon imports from Earth; the same is true of the Moon. Martians, lunarians and earthlings are now evolving independently on three resource-rich worlds.

Genetically engineered bacteria are under test to initiate the terraforming of Venus and Mars. Young Martians are pressing for additional settlements beyond the asteroid belt, and in response promising sites on Pluto, Charon, Triton, Titan Ganymede, Europa and other moons of Uranus, Saturn and Jupiter are being surveyed. Joint Mars-Earth-Moon research on matter-antimatter and photon propulsion to drive spacecraft at speeds approaching ten percent of the speed of light shows promise of opening the challenging "stellar frontier"; long range plans now call for launching robotic probes to temperate planets circling nearby stars before the end of the century.

ACKNOWLEDGMENTS

I am grateful to Carol Stoker for asking me to revisit this topic, and to Fred Paroutaud and Dr. Mark Fisher for invaluable assistance.

REFERENCES

Bernal, J.D., *The World, the Flesh and the Devil; an Inquiry into the Future of the Three enemies of the Rational Soul*, Paul, Trench, Tribner and Co., London, 1929.

Clarke, A.C., *2001: A Space Odyssey*, New American Library, New York, 1968.

Dyson, F., "The world, the flesh and the devil," Third J.D. Bernal Lecture, delivered at Birkbeck College, London, 16 May 1972; reprinted as Appendix D in *Communication with Extraterrestrial Intelligence*, Carl Sagan, ed., MIT Press, Cambridge, MA, 1973.

Huxley, A., *Brave New World*, Harpers, New York 1932.

Kahn, H., W. Brown, and L. Martel, (with the assistance of the Staff of the Hudson Institute): *The Next 200 Years: A Scenario for America and the World*, W.M. Morrow & Co., New York, 1976.

The National Commission on Space: *Pioneering the Space Frontier*, Bantam Books, New York, 1986.

Orwell, G., *1984*, Secker & Warburg, London, 1949, and Harcourt Brace Jovanovich, New York, 1949.

Paine, T.O., "A timeline for Martian pioneers," AAS 84-150, in C.P. McKay, ed., *The Case for Mars II*, American Astronautical Society, Science and Technology Series, 62, pp. 3-21, (Proceedings of the Second Case for Mars Conference held July 10-14, 1984), Univelt, Inc., San Diego, CA, 1985.

Snow, Lord C.P., *The two cultures and the Scientific Revolution*, Cambridge University Press, Cambridge, 1959.

Epilogue

Just a beginning.

AAS 95-496

Chapter 26

BEYOND MARS . . . INTO THE UNIVERSE AT LARGE

Leonard David[*]

The distant reddish shores of Mars stands today as a metaphor for human exploration. In future decades, such will not be the case.

The Earth . . . the Moon . . . and Mars comprise a triad of worlds that will support even grander leaps of exploratory zeal. In the scheme of things yet to come, we as a species represent the "third worlders" of our solar system. Third planet thinking prompted the survival instincts of *Homo erectus* and use of crude tools . . . the Neanderthal and the discovery of fire . . . and the overall inventiveness of *Homo sapiens*. Our coupling of three worlds—the Earth, the Moon, and Mars—will signify a coming of age for *Homo exploratorias*.

Humankind has long been a curious animal, enough so to set sail from safe harbors to probe beyond the immediate horizon. Advances in technology—be it binding logs together to make the first raft or the building of a hypersonic airliner—have opened up remote corners of the Earth to exploration, scientific inquiry, as well as commerce and trade.

Likewise, humankind's early steps into the continent of space mimic in spirit and purpose the early missions of exploration that took place centuries ago. But even those early sojourns of geographical expansion led to a retrogression of exploration by the European civilization, later called the "Dark Ages." Throughout history, in fact, bold and daring periods of exploration are followed by intervals of inactivity. The stagnant period of the Dark Ages ended with the explosive Age of Discovery, spawning the epic voyages of Cook, Columbus, Magellan, and others.

Exploration, settlement, and enterprise—this progression is at work here on Earth, and this framework will support humankind's emergence into the solar system. The horizonal expansion of the pioneering spirit has moved to all points on Earth and is now in transition to the vertical.

Perhaps it would be a fitting historical twist for future chroniclers to consider humanity's plunge into the lightless, vacuum void of outer space as the "Dark Ages"—a period of time when brave solar sailors pulled up anchor near Earth and moved into the

---

[*] Editor, *Final Frontier Magazine*, 306 4th Street, S.E., Washington D.C. 20003-2044.

darkness of space, plying the gravitational trade routes between Earth . . . the Moon . . . Mars . . . and then outward.

One can easily consider the solar system in which we reside as a giant swimming pool, its inner and outer distances separated by the asteroid belt lying between Mars and Jupiter. Just like children learning how to swim in the shallow part of the pool, eventually they will journey into the deep end. Initially clinging to the buoyed rope denoting shallow from deep, children will eventually let go, voyaging into unfamiliar territory.

Similarly, short forays of humans beyond low Earth orbit to establish small encampments on the Moon and Mars will give way to longer stay times away from our planet and larger outposts. To do so, however, will require a reduction in our dependance on water, food, propellant, air, construction materials, and other necessities supplied from Earth. From our first human missions to the Moon and robotic studies of Mars, we know that indigenous resources can be helpful in this cause. So too will the material found in comets, near-Earth approaching, and Mars-crossing asteroids will foster the settling of the inner solar system. Over time, the quality of life for humans distant from Earth can be expected to increase. Learning to live and work within the confines of the inner solar system will also serve as an incubator for new technologies of radiation protection, energy storage, gravity management, and closed life support systems. The settlers of the inner space frontier will be a melding of ethnic background, nationality, and personal philosophy; they themselves a crucible for what could be called an evolving "space race."

Husbanding local available resources in the Earth-Moon-Mars triad, the first human travel into the asteroid belt can be guided from Mars. Indeed, Mars' two moons—Phobos and Deimos—provide a convenient locale to testbed human and mechanical technigues of asteroid exploration. While doing so, the salvation of Deimos can also be undertaken. The tiny moon, threatened by tidal forces that are eventually expected to break the natural satellite into fragments, can be hauled to higher orbit around Mars and placed into a neutral Singer zone of safety.

For the most part, asteroids are huddled in a zone between the orbits of Mars and Jupiter. Believed to be the outcome of planetary accretion occurring early in the solar system's history, these bodies are, in a real sense, time capsules; encapsulated specimens of primordial matter awaiting scientific scrutiny, perhaps holding clues to the genesis of life itself.

In addition, these treasure troves are caches of raw material for fueling expeditions ever-deeper into the solar system. Alternatively, mining operations on these motherlodes of minerals could hollow out residences for humans. Asteroids can yield through advanced processing techniques quantities of hydrogen and oxygen to power cislunar traffic, as well as provide yet another staging base for exploratory treks toward the outer planets.

Through "island hopping" and "gravitational field harvesting," humankind can move expeditiously throughout the outer planets. Tapping the resource-rich gaseous worlds of Jupiter, Saturn, Uranus, and Neptune, as well as the collective entourage of moons circling those planets, humanity will find itself on the outskirts of the known family of planets. Most certainly, Pluto and its moon, and the yet-to-be-detected planet X, can help support humankind's stretch into the deeper of the solar system. If hypothetical turns to reality, the immense shell of comets surrounding the entire solar system—called the Oort Cloud—can energize to an even greater degree the emergence of humanity outside its home to new worlds beyond the light.

In all likelihood, our ability to rapidly traverse inner and outer solar system mileage, and then interstellar distances, will be progressively enhanced over the next several decades by new technologies. To speculate on what these breakthroughs might be is difficult, given the state of research already underway, such as work in nuclear, laser, fusion, and anti-matter. "Doesn't matter" propulsion will have a profound effect on the movement of men, women, and cargo across what we now view as the great expanse of space between the known worlds and the unknown worlds circling other stars.

At some point in the far future, suspected to be a comfortable five billion years hence, our Sun will expand to cover an area of space out beyond the orbit of Venus. The Sun then shrinks to a burnt out ember of its former self. One wonders whether orbiting observatories, such as the Hubble Space Telescope, may find a smattering of burnt out Sun-like stars with ages far shorter than now predicted, thus throwing into question the true lifetime of our own star. So goes the Sun . . . so goes life on Earth.

If we as a species are fortunate, our evolution into the cosmos at large will have started by migrating from Earth to the highlands of the Moon and to the plains of Mars. The task ahead is not without danger. Opening up the solar system to human habitation will mean sacrifices, failures, accidents and deaths. These earliest of space pioneers who chart the uncharted are homesteading new frontiers.

And from these modest beginnings . . . onward to new frontiers and new beginnings.

Appendices

PUBLICATIONS OF THE AMERICAN ASTRONAUTICAL SOCIETY

Following are the principal publications of the American Astronautical Society:

JOURNAL OF THE ASTRONAUTICAL SCIENCES (1954 -   )
Published quarterly and distributed by AAS Business Office, 6352 Rolling Mill Place, Suite #102, Springfield, Virginia 22152. Back issues available from Univelt, Inc., P.O. Box 28130, San Diego, California 92198.

SPACE TIMES (1986 -   )
Published bi-monthly and distributed by AAS Business Office, 6352 Rolling Mill Place, Suite #102, Springfield, Virginia 22152.

AAS NEWSLETTER (1962 - 1985)
Incorporated in *Space Times*. Back issues available from AAS Business Office, 6352 Rolling Mill Place, Suite #102, Springfield, Virginia 22152.

ASTRONAUTICAL SCIENCES REVIEW (1959 - 1962)
Incorporated in *Space Times*. Back issues still available from Univelt, Inc., P.O. Box 28130, San Diego, California 92198.

ADVANCES IN THE ASTRONAUTICAL SCIENCES (1957 -   )
Proceedings of major AAS technical meetings. Published and distributed for the American Astronautical Society by Univelt, Inc., P.O. Box 28130, San Diego, California 92198.

SCIENCE AND TECHNOLOGY SERIES (1964 -   )
Supplement to *Advances in the Astronautical Sciences*. Proceedings and monographs, most of them based on AAS technical meetings. Published and distributed for the American Astronautical Society by Univelt, Inc., P.O. Box 28130, San Diego, California 92198.

AAS HISTORY SERIES (1977 -   )
Supplement to *Advances in the Astronautical Sciences*. Selected works in the field of aerospace history under the editorship of R. Cargill Hall. Published and distributed for the American Astronautical Society by Univelt, Inc., P.O. Box 28130, San Diego, California 92198.

AAS MICROFICHE SERIES (1968 -   )
Supplement to *Advances in the Astronautical Sciences*. Consists principally of technical papers not included in the hard-copy volume. Published and distributed for the American Astronautical Society by Univelt, Inc., P.O. Box 28130, San Diego, California 92198.

Subscriptions to the *Journal of the Astronautical Sciences* and the *Space Times* should be ordered from the AAS Business Office. Back issues of the *Journal* and all books and microfiche should be ordered from Univelt, Incorporated.

ADVANCES IN THE ASTRONAUTICAL SCIENCES SERIES
(1957-)

ISSN 0065-3438,

LIBRARY OF CONGRESS CARD NO. 57-43769

Proceedings of Major AAS Technical Meetings

Vol. 1 Third Annual AAS Meeting, Dec. 6-7, 1956, New York, NY, 1957, 184p., ed. Norman V. Petersen, Microfiche only, $20 (ISBN 0-87703-002-2)

Vol. 2 Fourth Annual AAS Meeting, Jan. 29-31, 1958, New York, NY, 1958, 440p., eds. Norman V. Petersen, Horace Jacobs, Microfiche only, $20 (ISBN 0-87703-003-0)

Vol. 3 First Western National AAS Meeting, Aug. 18-19, 1958 530p., eds. Norman V. Petersen, Horace Jacobs, Microfiche only, $20 (ISBN 0-87703-004-9)

Vol. 4 Fifth Annual AAS Meeting, Dec. 27-31, 1958, Washington, D.C., 1959, 462p., ed. Horace Jacobs, Microfiche only, $20 (ISBN 0-87703-005-7)

Vol. 5 Second Western National AAS Meeting, Aug. 4-5, 1959, Los Angeles, CA, 1960, 364p., ed. Horace Jacobs, Microfiche only, $20 (ISBN 0-87703-006-3)

Vol. 6 Sixth Annual AAS Meeting, Jan. 18-21, 1960, New York, NY, 1961, 968p., eds. Horace Jacobs and Eric Burgess, Hard Cover $45 (ISBN 0-87703-007-3)

Vol. 7 Third Western National AAS Meeting, Aug. 4-5, 1960, Seattle, WA, 1961, 464p., eds. Horace Jacobs and Eric Burgess, Microfiche only, $20 (ISBN 0-87703-008-1)

Vol. 8 Seventh Annual AAS Meeting, Jan. 16-18, 1961, Dallas, TX, 1963, 602p., ed. Horace Jacobs, Microfiche only, $20 (ISBN 0-87703-009-X)

Vol. 9 Fourth Western Regional AAS Meeting, Aug. 1-3, 1961, San Francisco, CA, 1963, 910p., ed. Eric Burgess, Hard Cover $45 (ISBN 0-87703-010-3)

Vol. 10 Manned Lunar Flight (AAS/AAAS Symposium) Dec. 19, 1961, Denver, CO, 1963, 310p., eds. George W. Morgenthaler and Horace Jacobs, Hard Cover $35 (ISBN 0-87703-011-1)

Vol. 11 Eighth Annual AAS Meeting, Jan. 16-18, 1962, Washington, D.C., 1963, 808p., ed. Horace Jacobs, Hard Cover $45 (ISBN 0-87703-012-X)

Vol. 12 Scientific Satellites - Mission and Design (AAS/AAAS Symposium), Dec. 27, 1962, Philadelphia, PA, 1963, 262p., ed. Irving E. Jeter, Hard Cover $25 (ISBN 0-87703-013-8)

Vol. 13 Interplanetary Missions, 9th Annual AAS Meeting, Jan. 15-17, 1963, Los Angeles, CA, 1963, 690p., ed. Eric Burgess, Hard Cover $45 (ISBN 0-87703-014-6)

Vol. 14 Second AAS Symposium on Physical and Biological Phenomena under Zero G Conditions, Jan. 18, 1963, Los Angeles, CA, 1963, 382p., eds. Elliot T. Benedikt and Robert W. Halliburton, Hard Cover $30 (ISBN 0-87703-015-4)

Vol. 15 Exploration of Mars Symposium, Jun. 6-7, 1963, Denver, CO, 1963, 634p., ed. George W. Morgenthaler, Hard Cover $45 (ISBN 0-87703-016-2)

Vol. 16 Space Rendezvous, Rescue, and Recovery Symposium, Sept. 10-12, 1963, Edwards, CA 1963, 1408p., ed. Norman V. Petersen, Hard Cover, Part 1, 1028p., $45 (ISBN 0-87703-017-0); Part 2, 380p., $30 (ISBN 0-87703-018-9)

Vol. 17 Bioastronautics - Fundamental and Practical Problems (AAS/AAAS Symposium), Dec. 30, 1963, Cleveland, OH, 1964, 128p., ed. William C. Kaufman, Microfiche only, $10 (ISBN 0-87703-019-7)

Vol. 18 Lunar Flight Programs, 10th Annual AAS Meeting, May 4-7, 1964, New York, NY, 1964, 630p., ed. Ross Fleisig, Hard Cover $45 (ISBN 0-87703-020-0)

Vol. 19 Unmanned Exploration of the Solar System Symposium, Feb. 8-10, 1965, Denver, CO, 1965, 1000p., eds. George W. Morgenthaler, Robert G. Morra, Hard Cover $45 (ISBN 0-87703-021-9)

Vol. 20 Post Apollo Exploration, 11th Annual AAS Meeting, May 3-6, 1965, Chicago, IL, 1966, 1220p., ed. Francis Narin, Microfiche only, Part I, 572p., $30 (ISBN 0-87703-022-7); Part 2, 648p., $35 (ISBN 0-87703-023-5)

Vol. 21 Practical Space Applications Symposium, Feb. 21-23, 1966, San Diego, CA, 1967, 508p., ed. Lawrence L. Kavanau, Microfiche Only $40 (ISBN 0-87703-024-3)

Vol. 22 The Search for Extraterrestrial Life, 12th Annual AAS Meeting, May 23-25, 1966, Anaheim, CA, 1967, 388p., ed. James S. Hanrahan, Microfiche only $30 (ISBN 0-87703-025-1); Microfiche Suppl. (Vol. 1 AAS Microfiche Series) $12 (ISBN 0-87703-132-0)

Vol. 23 Commercial Utilization of Space, 13th Annual AAS Meeting, May 1-3, 1967, Dallas, TX, 1968, 512p., eds. J. Ray Gilmer, Alfred M. Mayo, Ross C. Peavey, Hard Cover (ISBN 0-87703-026-X); plus Microfiche Suppl. (Vol. 3 AAS Microfiche Series) $60 (ISBN 0-87703-216-5)

Vol. 24 Exploitation of Space for Experimental Research, 14th Annual AAS Meeting, May 13-15, 1968, Dedham, MA, 1968, 363p., ed. Harry Zuckerberg, Hard Cover $30 (ISBN 0-87703-027-8)

Vol. 25 Advanced Space Experiments, Sept. 16-18, 1968, Ann Arbor, MI, 1969, 530p., eds. O. Lyle Tiffany and Eugene M. Zaitzeff, Hard Cover $40 (ISBN 0-87703-028-6)

Vol. 26 Planning Challenges of the 70's in Space, 15th Annual AAS Meeting, Jun. 17-20, 1969, Denver, CO, 1970, 470p., eds. George W. Morgenthaler and Robert G. Morra, Hard Cover $35 (ISBN 0-87703-053-7); Microfiche Suppl. (Vol. 14 AAS Microfiche Series) $20 (ISBN 0-87703-130-4)

Vol. 27/28 Space Stations (v27) and Space Shuttles and Interplanetary Missions (v28), 16th Annual AAS Meeting, Jun. 8-10, 1970, Anaheim, CA, 1970, Vol. 27, eds. Lewis Larmore and Robert L. Gervais, 606p., Hard Cover $45 (ISBN 0-87703-054-5); Vol. 28, eds. Lewis Larmore and Robert L. Gervais, 488p., Hard Cover $35 (ISBN 0-87703-055-3)

Vol. 29 The Outer Solar System, 17th Annual AAS Meeting, Jun. 28-30, 1971, Seattle, WA, 1971, 1358p., $85; ed. Juris Vagners, Hard Cover, Part 1, 618p., $40 (ISBN 0-87703-059-6); Part 2, 740p., $45 (ISBN 0-87703-060-X)

Vol. 30 International Congress of Space Benefits, 19th Annual AAS Meeting, Jun. 19-21, 1973, Dallas, TX, 1974, 528p., ed. Francis S. Johnson, Hard Cover $40 (ISBN 0-87703-065-0)

Vol. 31 The Skylab Results, 20th Annual AAS Meeting, Aug. 20-22, 1974, Los Angeles, CA, 1975, 1174p., eds. William C. Schneider and Thomas E. Hanes, Microfiche only (ISBN 0-87703-072-3); Plus Microfiche Suppl. (Vol. 22 AAS Microfiche Series) $60 (ISBN 0-87703-043-6)

Vol. 32 Space Shuttle Missions of the 80's, 21st Annual AAS Meeting, Aug. 26-28, 1975, Denver, CO, 1977, 1364p., eds. William J. Bursnall, George W. Morgenthaler, Gerald E. Simonson, Hard Cover, Part 1, 598p., $40 (ISBN 0-87703-078-2); Hard Cover, Part 2, 766p., $55 (ISBN 0-87703-087-1); Microfiche Suppl. (Vol. 25 AAS Microfiche Series $65 (ISBN 0-87703-133-9)

Vol. 33 AAS/AIAA Astrodynamics Conference, July 28-30, 1975, Nassau, Bahamas, 1976, 390p., eds. William F. Powers, Herbert E. Rauch, Byron D. Tapley, Carmelo E. Velez, Hard Cover $35 (ISBN 0-87703-079-0); Microfiche Suppl. (Vol. 26 AAS Microfiche Series) $40 (ISBN 0-87703-142-8)

Vol. 34 Apollo Soyuz Mission Report, 1977, 336p., ed. Chester M. Lee, Hard Cover $35 (ISBN 0-87703-089-8)

Vol. 35 The Bicentennial Space Symposium - New Themes for Space: Mankind's Future Needs and Aspirations, 22nd AAS Meeting, Oct. 6-8, 1976, Washington, D.C., 1977, 242p., ed. William C. Schneider, Hard Cover $25 (ISBN 0-87703-090-1)

Vol. 36 The Industrialization of Space, 23rd Annual AAS Meeting, Oct. 18-20, 1977, San Francisco, CA, 1978, 1160p., eds. Richard A. Van Patten, Paul Siegler, Edward V.B. Stearns,

Hard Cover, Part 1, 610p., $55 (ISBN 0-87703-094-4); Hard Cover, Part 2, 550p., $45 (ISBN 0-87703-095-2); Microfiche Suppl. (Vol. 28 AAS Microfiche Series) $15 (ISBN 0-87703-121-5)

Vol. 37 Space Shuttle and Spacelab Utilization, What are the Near-Term and Long-Term Benefits for Mankind?, 16th Goddard Memorial Symposium, 24th Annual AAS Meeting, March 8-10, 1978, Washington, D.C., 1978, 865p., eds. George W. Morgenthaler and Manfred Hollstein, Hard Cover, Part 1, 400p., $40 (ISBN 0-87703-096-0); Hard Cover, Part 2, 465p., $45 (ISBN 0-87703-097-9)

Vol. 38 The Future U.S. Space Program, 25th Anniversary Conference, Oct. 20 - Nov. 2, 1978, Houston, TX, 1979, 880p., eds. Richard S. Johnston, Albert Naumann, Jr., Clay W. G. Fulcher, Hard Cover, Part 1, 444p., $45 (ISBN 0-87703-098-7); Hard Cover, Part 2, 436p., $40 (ISBN 0-87703-099-5); Microfiche Suppl. (Vol. 30 AAS Microfiche Series) $15 (ISBN 0-87703-129-0)

Vol. 39 Guidance and Control 1979, Feb. 24-28, 1979, Keystone, CO, 1979, 492p., ed. Robert D. Culp, Hard Cover $45 (ISBN 0-87703-100-2); Microfiche Suppl. (Vol. 31 AAS Microfiche Series) $10 (ISBN 0-87703-128-2)

Vol. 40 AAS/AIAA Astrodynamics Conference, Jun. 25-27, 1979, Provincetown, MA, 1980, 996p., eds. Paul A. Penzo, Bernard Kaufman, Louis Friedman, Richard Battin, Hard Cover, Part 1, 494p., $45 (ISBN 0-87703-107-X); Soft Cover $35 (ISBN 0-87703-108-8); Hard Cover, Part 2, 502p., $45 (ISBN 0-87703-109-6); Soft Cover $35 (ISBN 0-87703-110-X); Microfiche Suppl. (Vol. 32 AAS Microfiche Series) $20 (ISBN 0-87703-139-8)

Vol. 41 Space Shuttle: Dawn of an Era, 26th Annual AAS Meeting, Oct. 29-Nov. 1, 1979, Los Angeles, CA, 1980, 980p., eds. William F. Rector, III and Paul A. Penzo, Hard Cover, Part 1, 452p., $45 (ISBN 0-87703-111-8); Soft Cover $35 (ISBN 0-87703-112-6); Hard Cover, Part 2, 528p., $55 (ISBN 0-87703-113-4); Soft Cover $40 (ISBN 0-87703-114-2); Microfiche Suppl. (Vol. 33 AAS Microfiche Series) $10 (ISBN 0-87703-136-3)

Vol. 42 Guidance and Control 1980, Feb. 17-21, 1980, Keystone, CO, 1980, 738p., ed. Louis A. Morine, Hard Cover $60 (ISBN 0-87703-137-1); Soft Cover $45 (ISBN 0-87703-138-X)

Vol. 43 Shuttle/Spacelab - The New Transportation System and its Utilization, (3rd DGLR/AAS Symposium), Apr. 28-30, 1980, Hannover, Germany, 1981, 342p., eds. Dietrich E. Koelle and George V. Butler, Hard Cover $45 (ISBN 0-87703-144-4); Soft Cover $35 (ISBN 0-87703-146-0)

Vol. 44 Space-Enhancing Technological Leadership, 27th Annual AAS Meeting, Oct. 20-23, 1980, Boston, MA, 1981, 580p., ed. Lawrence P. Greene, Hard Cover $65 (ISBN 0-87703-147-9); Soft Cover $50 (ISBN 0-87703-148-7); Microfiche Suppl. (Vol. 35 AAS Microfiche Series) $10 (ISBN 0-87703-164-9)

Vol. 45 Guidance and Control 1981, Jan. 31- Feb. 4, 1981, Keystone, CO, 1981, 506p., ed. Edward J. Bauman, Hard Cover $60 (ISBN 0-87703-150-9); Soft Cover $50 (ISBN 0-87703-151-7); Microfiche Suppl. (Vol. 36 AAS Microfiche Series) $15 (ISBN 0-87703-156-8)

Vol. 46 AAS/AIAA Astrodynamics Conference, Aug. 3-5, 1981, North Lake Tahoe, NV, 1982, 1124p., eds. Alan L. Friedlander, Paul J. Cefola, Bernard Kaufman, Walt Williamson, G.T. Tseng, Hard Cover, Part 1, 552p., $55 (ISBN 0-87703-159-2); Soft Cover $45 (ISBN 0-87703-160-6); Hard Cover, Part 2, 572p., $55 (ISBN 0-87703-161-4); Soft Cover $45 (ISBN 0-87703-162-2); Microfiche Suppl. (Vol. 37 AAS Microfiche Series) $40 (ISBN 0-87703-163-0)

Vol. 47 Leadership in Space - For Benefits on Earth, 28th Annual AAS Meeting, Oct. 26-29, 1981, San Diego, CA, 1982, 310p., ed. William F. Rector, III, Hard Cover $45 (ISBN 0-87703-168-1); Soft Cover $35 (ISBN 0-87703-169-X)

Vol. 48 Guidance And Control 1982, Jan. 30 - Feb. 3, 1982, Keystone, CO, 1982, 558p., eds. Robert D. Culp, Edward J. Bauman, W. E. Dorroh, Jr., Hard Cover $65 (ISBN 0-87703-170-3); Soft Cover $50 (ISBN 0-87703-171-1); Microfiche Suppl. (Vol. 38 AAS Michrofiche Series) $10 (ISBN 0-87703-180-0)

Vol. 49 Spacelab, Space Platforms, and the Future, Fourth AAS/DGLR Symposium and 20th Goddard Memorial Symposium, Mar. 17-19, 1982, Greenbelt, MD, 1982, 502p., eds. Peter M. Bainum, Dietrich E. Koelle, Hard Cover $55 (ISBN 0-87703-174-6); Soft Cover $45 (ISBN 0-87703-175-4); Microfiche Suppl. (Vol. 42 AAS Microfiche Series) $15 (ISBN 0-87703-181-9)

Vol. 50 Proceedings on an International Symposium on Engineering Sciences and Mechanics, Dec. 29-31, Tainan, Taiwan, 1983, two parts, 1570p., eds. Han-Min Hsia, Richard W. Longman,

You-Li Chou, Hard Cover $120 (ISBN 0-87703-176-2); Microfiche Suppl. (Vol. 43 AAS Microfiche Series) $10 (ISBN 0-87703-215-7)

Vol. 51 Guidance and Control 1983, Feb. 5-9, 1983, Keystone, CO, 1983, 494p., eds. Edward J. Bauman, Zubin W. Emsley, Hard Cover $60 (ISBN 0-87703-182-7); Soft Cover $50 (ISBN 0-87703-183-5); Microfiche Suppl. (Vol. 44 AAS Microfiche Series) $10 (ISBN 0-87703-214-9)

Vol. 52 Developing the Space Frontier, 29th Annual AAS Meeting, Oct. 25-27, 1982, Houston, TX, 1983, 436p., eds. Albert Naumann, Grover Alexander, Microfiche Only $45 (ISBN 0-87703-189-4)

Vol. 53 Space Manufacturing 1983, May 9-12, 1983, Princeton, NJ, 1983, 496p., eds. James D. Burke, April S. Whitt, On Microfiche Only $50 (ISBN 0-87703-188-6)

Vol. 54 AAS/AIAA Astrodynamics Conference, Aug. 22-25, 1983, Lake Placid, NY, 1984, two parts, 1370p., eds. G.T. Tseng, Paul J. Cefola, Peter M. Bainum, David A. Levinson, Hard Cover $120 (ISBN 0-87703-190-8); Soft Cover $90 (ISBN 0-87703-191-6); Microfiche Suppl. (Vol. 45 AAS Microfiche Series) $40 (ISBN 0-87703-192-4)

Vol. 55 Guidance and Control 1984, Feb. 4-8, 1984, Keystone, CO, 1984, 500p., eds. Robert D. Culp, Parker S. Stafford, Hard Cover $60 (ISBN 0-87703-199-1); Soft Cover $50 (ISBN 0-87703-200-9); Microfiche Suppl. (Vol. 48 AAS Microfiche Series $15 (ISBN 0-87703-201-7)

Vol. 56 From Spacelab to Space Station, Fifth DGLR/AAS Symposium, Oct. 3-5, 1984, Hamburg, Germany, 1985, 270p., eds. H. Stoewer, Peter M. Bainum, Microfiche Only $30 (ISBN 0-87703-209-2)

Vol. 57 Guidance and Control 1985, Feb. 2-6, 1985, Keystone, CO, 1985, 618p., eds. Robert D. Culp, Edward J. Bauman, Charles A. Cullian, Hard Cover $65 (ISBN 0-87703-211-4); Soft Cover $50 (ISBN 0-87703-212-2); Microfiche Suppl. (Vol. 50 AAS Microfiche Series) $15 (ISBN 0-87703-213-0)

Vol. 58 AAS/AIAA Astrodynamics Conference, Aug. 12-15, 1985, Vail, CO, 1986, two parts, 1556p., eds. Bernard Kaufman, Joseph J.F. Liu, Robert A. Calico, Felix R. Hoots, Hard Cover $140 (ISBN 0-87703-245-9); Soft Cover $110 (ISBN 0-87703-246-7); Microfiche Suppl. (Vol. 51 AAS Microfiche Series); $60 (ISBN 0-87703-247-5)

Vol. 59 Space Station Beyond IOC, 32nd Annual AAS Meeting, Nov 6-7, 1985, Los Angeles, CA, 1986, 188p., ed. M. Jack Friedenthal, Hard Cover $40 (ISBN 0-87703-252-1); Soft Cover $30 (ISBN 0-87703-253-X)

Vol. 60 Space Exploitation and Utilization, First AAS/JRS Symposium, Dec. 15-19, 1985, Honolulu, HI, 1986, 740p., eds. Gayle L. May, Peter M. Bainum, Kenji Ikeda, Tamiya Nomura, Tatsuo Yamanaka, Ryojiro Akiba, Hard Cover $70 (ISBN 0-87703-254-8); Soft Cover $55 (ISBN 0-87703-255-6); Microfiche Suppl. (Vol. 52 AAS Microfiche Series) $10 (ISBN 0-87703-256-4)

Vol. 61 Guidance and Control 1986, Feb. 1-5, 1986, Keystone, CO, 1986, 460p., eds. Robert D. Culp, John C. Durrett, Hard Cover $60 (ISBN 0-87703-257-2); Soft Cover $50 (ISBN 0-87703-258-0); Microfiche Suppl. (Vol. 53 AAS Microfiche Series) $10 (ISBN 0-87703-259-9)

Vol. 62 Tethers in Space, Proceedings of First International Conference on Tethers in Space (NASA & PSN Sponsors; AIAA, AAS, & AIDAA Co-Sponsors), Sept. 17-19, 1986, Arlington, VA, 1987, 784p., eds. Peter M. Bainum, Ivan Bekey, Luciano Guerriero, Paul A. Penzo, Hard Cover $80 (ISBN 0-87703-264-5); Soft Cover $70 (ISBN 0-87703-265-3)

Vol. 63 Guidance and Control 1987, Jan. 31 - Feb. 4, 1987, Keystone, CO, 1987, 638p., eds. Robert D. Culp, Terry J. Kelly, Hard Cover $75 (ISBN 0-87703-268-8); Soft Cover $60 (ISBN 0-87703-269-6)

Vol. 64 Aerospace Century XXI, 33rd AAS Annual Meeting, Oct. 26-29, 1986, Boulder, CO, 1987, all three parts, Hard Cover $225 (ISBN 0-87703-276-9); Soft Cover $180 (ISBN 0-87703-277-7); Part I, Space Missions and Policy, 686p., eds. George W. Morgenthaler, Gayle L. May, Hard Cover $75 (ISBN 0-87703-279-3); Soft Cover $60 (ISBN 0-87703-282-3); Part II, Space Flight Technologies, 608p., eds. George W. Morgenthaler, W. Kent Tobiska, Hard Cover $75 (ISBN 0-87703-280-7); Soft Cover $60 (ISBN 0-87703-283-1); Part III, Space Sciences, Applications, and Commercial Developments, 724p., eds. George W. Morgenthaler, Jean N. Koster, Hard Cover $75 (ISBN 0-87703-281-5); Soft Cover $60 (ISBN 0-87703-284-X); Microfiche Suppl. (Vol. 54 AAS Microfiche Series) $25 (ISBN 0-87703-278-5)

Vol. 65 AAS/AIAA Astrodynamics Conference, Aug. 10-13, 1987, Kalispell, MT, 1988, two parts, 1774p., eds. John K. Soldner, Arun K. Misra, Robert E. Lindberg, Walton Williamson, Hard Cover $180 (ISBN 0-87703-285-8); Soft Cover $150 (ISBN 0-87703-286-6); Microfiche Suppl. (Vol. 55 AAS Microfiche Series); $70 (ISBN 0-87703-287-4)

Vol. 66 Guidance and Control 1988, Jan. 30 - Feb. 3, 1988, Keystone, CO, 1988, 576p., eds. Robert D. Culp, Paul L. Shattuck, Hard Cover $75 (ISBN 0-87703-288-2); Soft Cover $60 (ISBN 0-87703-289-0); Microfiche Suppl. (Vol. 56 AAS Microfiche Series) $10 (ISBN 0-87703-290-4)

Vol. 67 Space - A New Community of Opportunity, 34th AAS Annual Meeting, Nov. 3-5, 1987, Houston, TX, 1989, 472p., eds. William G. Straight, Henry N. Bowes, Hard Cover $70 (ISBN 0-87703-297-1); Soft Cover $55 (ISBN 0-87703-298-X)

Vol. 68 Guidance and Control 1989, Feb. 4-8, 1989, Keystone, CO, 1989, 708p., eds. Robert D. Culp, Robert A. Lewis, Hard Cover $85 (ISBN 0-87703-299-8); Soft Cover $70 (ISBN 0-87703-300-5)

Vol. 69 Orbital Mechanics and Mission Design, Apr. 24-27, 1989, Greenbelt, MD, 1989, 862p., ed. Jerome Teles, Hard Cover $95 (ISBN 0-87703-311-0); Soft Cover $80 (ISBN 0-87703-312-9); Microfiche Suppl. (Vol. 57 AAS Microfiche Series) $10 (ISBN 0-87703-313-7)

Vol. 70 The 21st Century in Space, 35th AAS Annual Meeting, Oct. 24-26, 1988, St. Louis, MO, 1990, 446p., ed. George V. Butler, Hard Cover $90 (ISBN 0-87703-314-5); Soft Cover $75 (ISBN 0-87703-315-3); Microfiche Suppl. (Vol. 58 AAS Microfiche Series) $10 (ISBN 0-87703-316-1)

Vol. 71 AAS/AIAA Astrodynamics Conference, Aug. 7-10, 1989, Stowe, VT, 1990, two parts, 1472p., eds. Catherine L. Thornton, Ronald J. Proulx, John E. Prussing, Felix R. Hoots, Hard Cover $200 (ISBN 0-87703-317-X); Soft Cover $170 (ISBN 0-87703-318-8); Microfiche Suppl. (Vol. 59 AAS Microfiche Series) $50 (ISBN 0-87703-319-6)

Vol. 72 Guidance and Control 1990, Feb. 3-7, 1990, Keystone, CO, 1990, 676p., eds. Robert D. Culp, Arlo D. Gravseth, Hard Cover $95 (ISBN 0-87703-320-X); Soft Cover $80 (ISBN 0-87703-321-8)

Vol. 73 Space Utilization and Applications in the Pacific, Third (PISSTA) AAS/JRS/CSA Symposium, Nov. 6-8, 1989, Los Angeles, CA, 1990, 764p., eds. Peter M. Bainum, Gayle L. May, Tatsuo Yamanaka, Yang Jiachi, Hard Cover $95 (ISBN 0-87703-325-0); Soft Cover $80 (ISBN 0-87703-326-9)

Vol. 74 Guidance and Control 1991, Feb. 2-6, 1991, Keystone, CO, 1991, 730p., eds. Robert D. Culp, James P. McQuerry, Hard Cover $120 (ISBN 0-87703-334-X); Soft Cover $90 (ISBN 0-87703-335-8)

Vol. 75 AAS/AIAA Spaceflight Mechanics Meeting, Feb. 11-13, 1991, Houston, TX, 1991, two parts, 1354p., eds. John K. Soldner, Arun K. Misra, Lester L. Sackett, Richard Holdaway, Hard Cover $220 (ISBN 0-87703-338-2); Soft Cover $190 (ISBN 0-87703-339-0); Microfiche Suppl. (Vol. 62 AAS Microfiche Series) $50 (ISBN 0-87703-340-4)

Vol. 76 AAS/AIAA Astrodynamics Conference, Aug. 19-22, 1991, Durango, CO, 1992, three parts, 2590p., eds. Bernard Kaufman, Kyle T. Alfriend, Ronald L. Roehrich, Robert R. Dasenbrock, Hard Cover $390 (ISBN 0-87703-347-1); Microfiche Suppl. (Vol. 63 AAS Microfiche Series) $75 (ISBN 0-87703-348-X)

Vol. 77 International Space Year (ISY) in the Pacific Basin, Fourth ISCOPS (formerly PISSTA) AAS/JRS/CSA Symposium, Nov. 17-20, 1991, Kyoto, Japan, 1992, 798p., eds. Peter M. Bainum, Gayle L. May, Makoto Nagatomo, Yoshiaiki Ohkami, Yang Jiachi, Hard Cover $120 (ISBN 0-87703-351-X); Soft Cover $90 (ISBN 0-87703-352-8)

Vol. 78 Guidance and Control 1992, Feb. 2-6, 1992, Keystone, CO, 1992, 754p., eds. Robert D. Culp, Richard P. Zietz, Hard Cover $120 (ISBN 0-87703-353-6); Soft Cover $90 (ISBN 0-87703-354-4); Microfiche Suppl. (Vol. 64 AAS Microfiche Series) $20 (ISBN 0-87703-355-2)

Vol. 79 AAS/AIAA Spaceflight Mechanics Meeting, Feb. 24-26, 1992, Colorado Springs, CO, 1992, two parts, 1312p., eds. Roger E. Diehl, Ralph G. Schinnerer, Walton E. Williamson, Daryl G. Boden, Hard Cover $240 (ISBN 0-87703-358-7); Microfiche Suppl. (Vol. 65 AAS Microfiche Series) $40 (ISBN 0-87703-359-5)

Vol. 80 Space Business Opportunities, 37th and 38th AAS Annual Meetings, Nov. 5-7, 1990, Dec. 3-5, 1991, Los Angeles, CA, 1992, 380p., eds. Wayne J. Esser, Don K. Tomajan, Hard

Cover $90 (ISBN 0-87703-360-9); Soft Cover $70 (ISBN 0-87703-361-7); Microfiche Suppl. (Vol. 61 AAS Microfiche Series) $50 (ISBN 0-87703-331-5); (Vol. 66 AAS Microfiche Series) $60 (ISBN 0-87703-362-5)

Vol. 81 Guidance and Control 1993, Feb. 6-10, 1993, Keystone, CO, 1993, 648p., eds. Robert D. Culp, George Bickley, Hard Cover $120 (ISBN 0-87703-365-X); Soft Cover $90 (ISBN 0-87703-366-8); Microfiche Suppl. (Vol. 67 AAS Microfiche Series) $20 (ISBN 0-87703-367-6)

Vol. 82 AAS/AIAA Spaceflight Mechanics Meeting, Feb. 22-24, 1993, Pasadena, CA, 1993, two parts, 1454p., eds. Robert G. Melton, Lincoln J. Wood, Roger C. Thompson, Stuart J. Kerridge, Hard Cover $240 (ISBN 0-87703-368-4); Microfiche Suppl. (Vol. 68 AAS Microfiche Series) $15 (ISBN 0-87703-369-2)

Vol. 83 Dynamics of Space Tether Systems (English Language Edition), 1993, 508p., by Vladimir V. Beletsky, Evgenii M. Levin, Hard Cover $120 (ISBN 0-87703-370-6); Soft Cover $90 (ISBN 0-87703-371-4)

Vol. 84 AAS/GSFC International Symposium on Spaceflight Dynamics, Apr. 26-30, 1993, Greenbelt, MD, 1993, two parts, 1450p., eds. Jerome Teles, Mina V. Samii, Hard Cover $240 (ISBN 0-87703-378-1); Microfiche Suppl. (Vol. 69 AAS Microfiche Series) $10 (ISBN 0-87703-379-X)

Vol. 85 AAS/AIAA Astrodynamics Conference, Aug. 16-19, 1993, Victoria, British Columbia, Canada, 1994, three parts, 2750p., eds. Arun K. Misra, Vinod J. Modi, Richard Holdaway, Peter M. Bainum, Hard Cover $390 (ISBN 0-87703-380-3); Microfiche Suppl. (Vol. 70 AAS Microfiche Series) $30 (ISBN 0-87703-381-1)

Vol. 86 Guidance and Control 1994, Feb. 2-6, 1994, Keystone, CO, 1994, 700p., eds. Robert D. Culp, Ronald D. Rausch, Hard Cover $120 (ISBN 0-87703-384-6); Soft Cover $90 (ISBN 0-87703-385-4)

Vol. 87 AAS/AIAA Spaceflight Mechanics Meeting, Feb. 14-16, 1994, Cocoa Beach, FL, 1994, two parts, 1272p., eds. John E. Cochran, Jr., Charles D. Edwards, Jr., Stephen J. Hoffman, Richard Holdaway, Hard Cover $240 (ISBN 0-87703-386-2)

Vol. 88 Guidance and Control 1995, Feb. 1-5, 1995, Keystone, CO, 1995, 600p. eds. Robert D. Culp, James D. Medbery, Hard Cover $120 (ISBN 0-87703-399-4); Soft Cover $90 (ISBN 0-87703-400-1)

Vol. 89 AAS/AIAA Spaceflight Mechanics Meeting, Feb. 13-16, 1995, Albuquerque, NM, 1995, two parts, 1774p., eds. Ronald J. Proulx, Joseph J. F. Liu, P. Kenneth Seidelmann, Salvatore Alfano, Hard Cover $280 (ISBN 0-87703-401-X); Microfiche Suppl. (Vol. 71 AAS Microfiche Series) $15 (ISBN 0-87703-402-8)

Vol. 90 AAS/AIAA Astrodynamics Conference, Aug. 14-17, 1995, Halifax, Nova Scotia, Canada, 1996, two parts, 2270p., eds. K. Terry Alfriend, I. Michael Ross, Arun K. Misra, C. Fred Peters, Hard Cover $290 (ISBN 0-87703-407-9); Microfiche Suppl. (Vol. 72 AAS Microfiche Series) $15 (ISBN 0-87703-408-7)

Order from Univelt, Inc., P.O. Box 28130, San Diego, California 92198
STANDING ORDERS ACCEPTED

SCIENCE AND TECHNOLOGY SERIES (1964-   )

ISSN 0278-4017

A Supplement to *Advances in the Astronautical Sciences*. Proceedings and monographs, most of them based on AAS technical meetings.

Vol. 1 Manned Space Reliability Symposium, Jun. 9, 1964, Anaheim, CA, 1964, 112p., ed. Paul Horowitz, Hard Cover $20 (ISBN 0-87703-029-4)

Vol. 2 Towards Deeper Space Penetration (AAS/AAAS Symposium), Dec. 29, 1964, Montreal, Canada, 1964, 182p., ed. Edward R. Van Driest, Hard cover $20 (ISBN 0-87703-030-8)

Vol. 3 Orbital Hodograph Analysis, 1965, 150p., ed. Samuel P. Altman, Hard Cover $20 (ISBN 0-87703-031-6)

Vol. 4 Scientific Experiments for Manned Orbital Flight, 3rd Goddard Memorial Symposium, Mar. 18-19, 1965, Washington, D.C., 1965, 372p., ed. Peter C. Badgley, Hard Cover $30 (ISBN 0-87703-032-4)

Vol. 5 Physiological and Performance Determinants in Manned Space Systems (AAS/HFS Symposium, Apr. 14-15, 1965, Northridge, CA, 1965, 220p., ed. Paul Horowitz, Microfiche Only $20 (ISBN 0-87703-033-2)

Vol. 6 Space Electronics Symposium (AAS/AES Meeting), May 25-27, 1965, Los Angeles, CA, 1965, 404p., ed. Chung-Ming Wong, Hard Cover $30 (ISBN 0-87703-034-0)

Vol. 7 Theodore von Karman Memorial Seminar, May 12, 1965, Los Angeles, CA, 1966, 140p., ed. Shirley Thomas, Hard Cover $30 (ISBN 0-87703-035-9)

Vol. 8 Impact of Space Exploration on Society, Aug. 18-20, 1965, San Francisco, CA, 1966, 382p., ed. William E. Frye, Hard Cover $30 (ISBN 0-87703-O36-7)

Vol. 9 Recent Developments in Space Flight Mechanics, (AAS/AAAS Symposium), Dec. 29, 1965, Berkeley, CA, 1966, 280p., ed. Paul B. Richards, Hard Cover $25 (ISBN 0-87703-037-5)

Vol. 10 Space in the Fiscal Year 2001, 4th Goddard Memorial Symposium, Mar. 15-16, 1966, Washington, D.C., 1967, 458p., eds. Eugene B. Konecci, Maxwell W. Hunter, II, Robert F. Trapp, Hard Cover $35 (ISBN 0-87703-038-3)

Vol. 11 Space Flight Specialist Conference, Jul. 6-8, 1966, Denver, CO, 1967, 618p., ed. Maurice L. Anthony, Microfiche Only (ISBN 0-87703-039-1); Plus Microfiche Suppl. (Vol. 2 AAS Microfiche Series) $60 (ISBN 0-87703-221-1)

Vol. 12 Management of Aerospace Programs Conference, Nov. 16-18, 1966, Columbia, MO, 1967, 392p., ed. Walter K. Johnson, Hard Cover $30 (ISBN 0-87703-040-5)

Vol. 13 Physics of the Moon (AAS/AAAS Symposium), Dec. 29, 1966, Washington, D.C., 1967, 260p., ed. S. Fred Singer, Hard Cover $25 (ISBN 0-87703-041-3)

Vol. 14 Interpretation of Lunar Probe Data, Sept. 17, 1966, Huntington Beach, CA, 1967, 270p., ed. Jack Green, Hard Cover $25 (ISBN 0-87703-042-1)

Vol. 15 Future Space Program and Impact on Range and Network Development Symposium, Mar. 22-24, 1967, Las Cruces, NM, 1967, 588p., ed. George W. Morgenthaler, Hard Cover $40 (ISBN 0-87703-043-X)

Vol. 16 Voyage to the Planets, 5th Goddard Memorial Symposium, Mar. 14-15, 1967, Washington, D.C., 1968, 184p., ed. S. Fred Singer, Hard Cover $20 (0-87703-044-8)

Vol. 17 Use of Space Systems for Planetary Geology and Geophysics Symposium, May 25-27, 1967, Boston, MA, 1968, 623p., ed. Robert D. Enzmann, Hard Cover $45 (ISBN 0-87703-045-6); Microfiche Suppl. (Vol. 5 AAS Microfiche Series) $15 (ISBN 0-87703-135-5)

Vol. 18 Technology and Social Progress, 6th Goddard Memorial Symposium, Mar. 12-13, 1968, Washington, D.C., 1969, 170p., ed. Philip K. Eckman, Hard Cover $20 (ISBN 0-87703-046-4)

Vol. 19 Exobiology - The Search for Extraterrestrial Life (AAS/AAAS Symposium) Dec. 30, 1967, New York, NY, 1969, 184p., eds. Martin M. Freundlich, Bernard W. Wagner, Hard Cover $20 (ISBN 0-87703-047-2)

Vol. 20 Bioengineering and Cabin Ecology (AAS/AAAS Symposium) Dec. 30, 1968, Dallas, TX, 1969, 162p., ed. William Cassidy, Hard Cover $20 (ISBN 0-87703-048-0)

Vol. 21 Reducing the Cost of Space Transportation, 7th Goddard Memorial Symposium, Mar. 4-5, 1969, Washington, D.C., 1969, 264p., ed. George K. Chacko, Microfiche only $25 (ISBN 0-87703-049-9)

Vol. 22 Planning Challenges of the 70's in the Public Domain, 15th Annual AAS Meeting, Jun. 17-20, 1969, Denver, CO, 1970, 504p., eds. William J. Burnsnall, George K. Chacko, George W. Morgenthaler, Hard Cover $40 (ISBN 0-87703-050-2); Microfiche Suppl. (Vol. 13 AAS Microfiche Series) $20 (ISBN 0-87703-131-2); See also Vols. 15-17, AAS Microfiche Series

Vol. 23 Space Technology and Earth Problems Symposium, Oct. 23-25, 1969, Las Cruces, NM, 1970, 418p., ed. C. Quentin Ford, Hard Cover $35 (ISBN 0-87703-051-0); Microfiche Suppl. (Vol. 12 AAS Microfiche Series) $20 (ISBN 0-87703-134-7)

Vol. 24 Aerospace Research and Development, Jul. 14, 1966, Holloman AFB, NM, 1970, 500p., ed. Ernst A. Steinhoff, Microfiche Only $40 (ISBN 0-87703-052-9)

Vol. 25 Geological Problems in Lunar and Planetary Research, Feb. 17-18, 1969, Huntington Beach, CA, 1971, 750p., ed. Jack Green, Hard Cover $45 (ISBN 0-87703-056-1)

Vol. 26 Technology Utilization Ideas for the 70s and Beyond, Oct. 30, 1970, Winrock, AR, 1971, 312p., eds. Fred W. Forbes, Paul Dergarabedian, Microfiche only $30 (ISBN 0-87703-057-X)

Vol. 27 International Cooperation in Space Operations and Exploration, 9th Goddard Memorial Symposium, Mar. 11, 1971, Washington, D.C. 1971, 194p., ed. Michael Cutler, Hard Cover $20 (ISBN 0-88703-058-8)

Vol. 28 Astronomy from a Space Platform (AAS/AAAS Symposium) Dec. 27-28, 1971, Philadelphia, PA, 1972, 416p., eds. George W. Morgenthaler, Howard D. Greyber, Hard Cover $35 (ISBN 0-87703-061-8)

Vol. 29 Space Technology Transfer to Community and Industry, 10th Goddard Memorial Symposium, 18th Annual AAS Meeting, Mar. 13-14, 1972, Washington, D.C., 1972, 196p., eds. Ralph H. Tripp, John K. Stotz, Jr., Hard Cover $20 (ISBN 0-87703-062-6); on Microfiche $15

Vol. 30 Space Shuttle Payloads (AAS/AAAS Symposium) Dec. 27-28, 1972, Washington, D.C., 1973, 532p., eds. George W. Morgenthaler, William J. Bursnall, Hard Cover $40 (ISBN 0-87703-063-4)

Vol. 31 The Second Fifteen Years in Space, 11th Goddard Memorial Symposium, Mar. 8-9, 1973, Washington, D.C., 1973, 212p., ed. Saul Ferdman, Hard Cover $25 (ISBN 0-87703-064-2)

Vol. 32 Health Care Systems Conference, Nov. 21-22, 1972, Dallas, TX, 1974, 265p., ed. Eugene B. Konecci, Hard Cover $25 (ISBN 0-87703-067-7)

Vol. 33 Orbital International Laboratory, 3rd and 4th IAF/OIL Symposia, Oct. 5-6, 1970, Constance, Germany, Sept. 24-25, 1971, Brussels, Belgium, 1974, 322p., ed. Ernst A. Steinhoff, Hard Cover $30 (ISBN 0-87703-068-5)

Vol. 34 Management and Design of Long-Life Systems, Apr. 24-26, 1973, Denver, CO, 1974, 198p., ed. Harris M. Schurmeier, Hard Cover $20 (ISBN 0-87703-069-3)

Vol. 35 Energy Delta, Supply vs. Demand, (AAS/AAAS Symposium) Feb. 25-27, 1974, San Francisco, CA, 1975, 2nd Printing 1976, 604p., eds. George W. Morgenthaler, Aaron N. Silver, Hard Cover $35 (ISBN 0-87703-070-7); Soft Cover $25 (ISBN 0-87703-082-0); on Microfiche $20

Vol. 36 Skylab and Pioneer Report, 12th Goddard Memorial Symposium, Mar. 8, 1974, Washington, D.C., 1975, 160p., eds. Philip H. Bolger, Paul B. Richards, Hard Cover $20 (ISBN 0-87703-071-5)

Vol. 37 Space Rescue and Safety 1974, 7th International IAA Symposium, Sept. 30 - Oct. 5, 1974, Amsterdam, Netherlands, 1975, 294p., ed. Philip H. Bolger, Microfiche Only $25 (ISBN 0-87703-073-1)

Vol. 38 Skylab Science Experiments, (AAS/AAAS Symposium) Feb. 28, 1974, San Francisco, CA, 1976, 274p., eds. George W. Morgenthaler, Gerald E. Simonson, Microfiche only $20 (ISBN 0-87703-074-X)

Vol. 39 Environmental Control and Agri-Technology, 1976, 346p., ed. Eugene B. Konecci, Microfiche only $20 (ISBN 0-87703-075-8)

Vol. 40 Future Space Activities, 13th Goddard Memorial Symposium, Apr. 11, 1975, Washington, D.C., 1976, 182p., ed. Carl H. Tross, Microfiche only $20 (ISBN 0-87703-076-6)

Vol. 41 Space Rescue and Safety 1975, 8th International IAA Symposium, Sept. 21-27, 1975, Lisbon, Portugal, 1976, 230p., ed. Philip H. Bolger, Hard Cover $25 (ISBN 0-87703-077-4)

Vol. 42 The End of an Era in Space Exploration, From International Rivalry to International Cooperation, 1976, 216p., by J.C.D. Blaine, Hard Cover $25 (ISBN 0-87703-084-7); without volume number (ISBN 0-87703-080-4)

Vol. 43 The Eagle Has Returned, Part I, International Space Hall of Fame Dedication Conference, Oct. 5-9, 1976, Alamogordo, NM, 1976, 370p., ed. Ernst. A. Steinhoff, Hard Cover $30 (ISBN 0-87703-086-3)

Vol. 44 Satellite Communications in the Next Decade, 14th Goddard Memorial Symposium, Mar. 12, 1976, Washington, D.C., 1977, 188p., ed. Leonard Jaffe, Hard Cover $20 (ISBN 0-87703-088-X)

Vol. 45 The Eagle Has Returned, Part 2, International Space Hall of Fame Dedication Conference, Oct. 5-9, 1976, Alamogordo, NM, 1977, 454p., ed. Ernst A. Steinhoff, Hard Cover $35 (ISBN 0-87703-092-8)

Vol. 46 Export of Aerospace Technology, 15th Goddard Memorial Symposium, Mar. 31 - Apr. 1, 1977, Washington, D.C., 1978, 174p., ed. Carl H. Tross, Hard Cover $20 (ISBN 0-87703-093-6)

Vol. 47 Handbook of Soviet Lunar and Planetary Exploration, 1979, 276p., by Nicholas L. Johnson, Microfiche Only $25 (ISBN 0-87703-105-3)

Vol. 48 Handbook of Soviet Manned Space Flight, 2nd Edition, 1988, 474p., by Nicholas L. Johnson, Hard Cover $60 (ISBN 0-87703-115-0); Soft Cover $45 (ISBN 0-87703-116-9)

Vol. 49 Space - New Opportunities for International Ventures, 17th Goddard Memorial Symposium, Mar. 28-30, 1979, Washington, D.C., 1980, 300p., ed. William C. Hayes, Jr., Hard Cover $35 (ISBN 0-87703-124-X); Soft Cover $25 (ISBN 0-87703-125-8); see also Vol. 2 AAS History Series

Vol. 50 Remember the Future - The Apollo Legacy, Jul. 20-21, 1979, San Francisco, CA, 1980, 218p., ed. Stan Kent, Hard Cover $25 (ISBN 0-87703-126-6); Soft Cover $15 (ISBN 0-87703-127-4)

Vol. 51 Commercial Operations in Space 1980-2000, 18th Goddard Memorial Symposium, Mar. 27-28, 1980, Washington, D.C., 1981, 214p., eds. John L. McLucas, Charles Sheffield, Hard Cover $30 (ISBN 0-87703-140-1); Soft Cover $20 (ISBN 0-87703-141-X); Microfiche Suppl. (Vol. 34 AAS Microfiche Series) $10 (ISBN 0-87703-165-7); see also Vols. 2 and 3, AAS History Series

Vol. 52 International Space Technical Applications, 19th Goddard Memorial Symposium, Mar. 26-27, 1981, Washington, D.C., 1981, 186p., eds. Andrew Adelman, Peter M. Bainum, Hard Cover $30 (ISBN 0-87703-152-5); Soft Cover $20 (ISBN 0-87703-153-3); see also Vol. 5, AAS History Series

Vol. 53 Space in the 1980's and Beyond, 17th European Space Symposium, Jun. 4-6, 1980, London, England, 1981, 302p., ed. Peter M. Bainum, Hard Cover $40 (ISBN 0-87703-154-1); Soft Cover $30 (ISBN 0-87703-155-X)

Vol. 54 Space Safety and Rescue 1979-1981 (with abstracts 1976-1978), Proceedings of symposia of the International Academy of Astronautics held in conjunction with the 30th, 31st, and 32nd International Astronautical Federation Congresses, Munich, Germany, 1979, Tokyo, Japan, 1980, and Rome, Italy, 1981, 1983, 456p., ed. Jeri W. Brown, Hard Cover $45 (ISBN

0-87703-177-0); Soft Cover $35 (ISBN 0-87703-178-9); Microfiche Suppl. (Vols. 39-41 AAS Microfiche Series) $39 (ISBN 0-87703-222-X); (ISBN 0-87703-223-8); (ISBN 0-87703-224-6)

Vol. 55 Space Applications at the Crossroads, 21st Goddard Memorial Symposium, Mar. 24-25, 1983, Greenbelt, MD, 1983, 308p., eds. John H. McElroy, E. Larry Heacock, Hard Cover $45 (ISBN 0-87703-186-X); Soft Cover $35 (ISBN 0-87703-187-8)

Vol. 56 Space: A Developing Role for Europe, 18th European Space Symposium, Jun. 6-9, 1983, London, England, 1984, 278p., eds. Len J. Carter, Peter M. Bainum, Hard Cover $45 (ISBN 0-87703-193-2); Soft Cover $35 (ISBN 0-87703-194-0); Microfiche Suppl. (Vol. 46 AAS Microfiche Series) $15 (ISBN 0-87703-195-9)

Vol. 57 The Case for Mars, Apr. 29 - May 2, 1981, Boulder, CO, 1984, Second Printing 1987, 348p., ed. Penelope J. Boston, Hard Cover $45 (ISBN 0-87703-197-5); on Microfiche $25 (ISBN 0-87703-198-3)

Vol. 58 Space Safety and Rescue 1982-1983, Proceedings of the International Academy of Astronautics held in conjunction with the 33rd and 34th International Astronautical Congresses, Paris, France, Sept. 27 - Oct. 2, 1982, and Budapest, Hungary, Oct. 10-15, 1983, 1984, 378p., ed. Gloria W. Heath, Hard Cover $50 (ISBN 0-87703-202-5); Soft Cover $40 (ISBN 0-87703-203-3)

Vol. 59 Space and Society - Challenges and Choices, April 14-16, 1982, University of Texas at Austin, 1984, 442p., eds. Paul Anaejionu, Nathan C. Goldman, Philip J. Meeks, Hard Cover $55 (ISBN 0-87703-204-1); Soft Cover $35 (ISBN 0-87703-205-X)

Vol. 60 Permanent Presence - Making It Work, 22nd Goddard Memorial Symposium, Mar. 15-16, 1984, Greenbelt, MD, 1985, 190p., ed. Ivan Bekey, Hard Cover $40 (ISBN 0-87703-207-6); Soft Cover $30 (ISBN 0-87703-208-4)

Vol. 61 Europe/United States Space Activities - With a Space Propulsion Supplement, 23rd Goddard Memorial Symposium/19th European Space Symposium, Mar. 27-29, 1985, Greenbelt, MD, 31st Annual AAS Meeting, Oct. 22-24, 1984, Palo Alto, CA, 1985, 442p., eds. Peter M. Bainum, Friedrich von Bun, Hard Cover $55 (ISBN 0-87703-217-3); Soft Cover $45 (ISBN 0-87703-218-1)

Vol. 62 The Case for Mars II, July 10-14, 1984, Boulder, CO, 1985, 730p., ed. Christopher P. McKay, Hard Cover $60 (ISBN 0-87703-219-1); Soft Cover $40 (ISBN 0-87703-220-3)

Vol. 63 Proceedings of 4th International Conference on Applied Numerical Modeling, Dec. 27-29, 1984, Tainan, Taiwan, 1986, 800p., ed. Han-Min Hsia, You-Li Chou, Shu-Yi Wang, Sheng-Jii Hsieh, Hard Cover $70 (ISBN 0-87703-242-4)

Vol. 64 Space Safety and Rescue 1984-1985, Proceedings of the International Academy of Astronautics held in conjunction with the 35th and 36th International Astronautical Congresses, Lausanne, Switzerland, Oct. 7-13, 1984, and Stockholm, Sweden, Oct. 7-12, 1985, 1986, 400p., ed. Gloria W. Heath, Hard Cover $55 (ISBN 0-87703-248-3); Soft Cover $45 (ISBN 0-87703-249-1)

Vol. 65 The Human Quest in Space, 24th Goddard Memorial Symposium, Mar. 20-21, 1986, Greenbelt, MD, 1987, 312p., ed. Gerald L. Burdett, Gerald A. Soffen, Hard Cover $55 (ISBN 0-87703-262-9); Soft Cover $45 (ISBN 0-87703-263-7)

Vol. 66 Soviet Space Programs 1980-1985, 1987, 298p., by Nicholas L. Johnson, Microfiche Only $45 (ISBN 0-87703-266-1)

Vol. 67 Low-Gravity Sciences, Seminar Series 1986, University of Colorado at Boulder, 290p., ed. Jean N. Koster, Hard Cover $55 (ISBN 0-87703-270-X); Soft Cover $45 (ISBN 0-87703-271-8)

Vol. 68 Proceedings of the Fourth Annual L5 Space Development Conference, Apr. 25-28, 1985, Washington, D.C., 1987, 268p., ed. Frank Hecker, Hard Cover $50 (ISBN 0-87703-272-6); Soft Cover $35 (ISBN 0-87703-273-4)

Vol. 69 Visions of Tomorrow: A Focus on National Space Transportation Issues, 25th Goddard Memorial Symposium, Mar. 18-20, 1987, Greenbelt, MD, 1987, 338p., ed. Gerald A. Soffen, Hard Cover $55 (ISBN 0-87703-274-2); Soft Cover $45 (ISBN 0-87703-275-0)

Vol. 70 Space Safety and Rescue 1986-1987, Proceedings of the International Academy of Astronautics held in conjunction with the 37th and 38th International Astronautical Congresses,

Innsbruck, Austria, Oct. 4-11, 1986, and Brighton, England, Oct. 11-16, 1987, 1988, 360p., ed. Gloria W. Heath, Hard Cover $55 (ISBN 0-87703-291-2); Soft Cover $45 (ISBN 0-87703-292-0)

Vol. 71 The NASA Mars Conference, Jul. 21-23, 1986, Washington, D.C., 1988, 570p., ed. Duke B. Reiber, Hard Cover $50 (ISBN 0-87703-293-9); Soft Cover $30 (ISBN 0-87703-294-7)

Vol. 72 Working in Orbit and Beyond: The Challenges for Space Medicine, Jun. 20-21, 1987, Washington, D.C., 1989, 188p., ed. David Lorr, Victoria Garshnek, Hard Cover $45 (ISBN 0-87703-295-5); Soft Cover $35 (ISBN 0-87703-296-3)

Vol. 73 Technology and the Civil Future in Space, 26th Goddard Memorial Symposium, Mar. 16-18, 1988, Greenbelt, MD, 1989, 246p., ed. Leonard A. Harris, Hard Cover $50 (ISBN 0-87703-301-3); Soft Cover $35 (ISBN 0-87703-302-1)

Vol. 74 The Case for Mars III: Strategies for Exploration - General Interest and Overview, July 18-22, 1987, Boulder, CO, 1989, 744p., ed. Carol Stoker, Hard Cover $75 (ISBN 0-87703-303-X); Soft Cover $55 (0-87703-304-8)

Vol. 75 The Case for Mars III: Strategies for Exploration - Technical, July 18-22, 1987, Boulder, CO, 1989, 646p., ed. Carol Stoker, Hard Cover $70 (ISBN 0-87703-305-6); Soft Cover $50 (ISBN 0-87703-306-4)

Vol. 76 Global Environmental Change: The Role of Space in Understanding Earth, 27th Goddard Memorial Symposium, Mar. 8-10, 1989, Washington, D.C., 1990, 178p., ed. Richard G. Johnson, Hard Cover $50 (ISBN 0-87703-322-6); Soft Cover $40 (ISBN 0-87703-323-4); Microfiche Suppl. (Vol. 60 AAS Microfiche Series) $10 (ISBN 0-87703-324-2)

Vol. 77 Space Safety and Rescue 1988 - 1989, Proceedings of the International Academy of Astronautics held in conjunction with the 39th and 40th International Astronautical Congresses, Bangalore, India, Oct. 8-15, 1988, and Málaga, Spain, Oct. 7-12, 1989, 1990, 500p., ed. Gloria W. Heath, Hard Cover $70 (ISBN 0-87703-327-7); Soft Cover $55 (ISBN 0-87703-328-5)

Vol. 78 Leaving the Cradle: Human Exploration of Space in the 21st Century, 28th Goddard Memorial Symposium, Mar. 14-16, 1990, Washington, D.C., 1991, 348p., ed. Thomas O. Paine, Hard Cover $70 (ISBN 0-87703-336-6); Soft Cover $55 (ISBN 0-87703-337-4)

Vol. 79 Space Safety and Rescue 1990, Proceedings of the International Academy of Astronautics held in conjunction with the 41st International Astronautical Congress, Dresden, Germany, Oct. 6-12, 1990, 1991, 232p., ed. Gloria W. Heath, Hard Cover $65 (ISBN 0-87703-341-2); Soft Cover $50 (ISBN 0-87703-342-0)

Vol. 80 Prospects for Interstellar Travel, 1992, 390p., by John H. Mauldin, Hard Cover $50 (ISBN 0-87703-344-7); Soft Cover $27 (ISBN 0-87703-345-5)

Vol. 81 Humans and Machines in Space: The Vision, The Challenge, The Payoff, 29th Goddard Memorial Symposium, Mar. 14-15, 1991, Washington, D.C., 1992, 204p., ed. Bradley Johnson, Gayle L. May, Paula Korn, Hard Cover $50 (ISBN 0-87703-356-0); Soft Cover $35 (ISBN 0-87703-357-9)

Vol. 82 Space Safety and Rescue 1991, Proceedings of the International Academy of Astronautics held in conjunction with the 42nd International Astronautical Congress, Montreal, Canada, Oct. 5-11, 1991, 1993, 270p., ed. Gloria W. Heath, Hard Cover $65 (ISBN 0-87703-372-2); Soft Cover $50 (ISBN 0-87703-373-0)

Vol. 83 Space: A Vital Stimulus to Our National Well-Being, 31st Goddard Memorial Symposium, March 9-10, 1993, Arlington, Virginia, and World Space Programs and Fiscal Reality, 30th Goddard Memorial Symposium, April 9-10, 1992, Alexandria, Virginia, 1994, 334p., ed. Gayle L. May, Saunders B. Kramer, Paula Korn, Leonard David, Barbara Sprungman, Hard Cover $70 (ISBN 0-87703-389-7); Soft Cover $50 (ISBN 0-87703-390-0)

Vol. 84 Space Safety and Rescue 1992, Proceedings of the International Academy of Astronautics held in conjunction with the World Space Congress, Washington, D.C., Aug. 28 to Sept. 5, 1992, 1994, 372p., ed. Gloria W. Heath, Hard Cover $70 (ISBN 0-87703-391-9); Soft Cover $55 (ISBN 0-87703-392-7)

Vol. 85 Civil Space in the Clinton Era, 32nd Goddard Memorial Symposium, March 1-2, 1994, Crystal City, Virginia, and Partners in Space . . . 2001, 41st Annual Meeting, November 14-16, 1994, Crystal City, Virginia, 1995, 292p., ed. Donald R. McConathy, Paula Korn, Hard Cover $70 (ISBN 0-87703-397-8); Soft Cover $50 (ISBN 0-87703-398-6)

Vol. 86 Strategies for Mars: A Guide to Human Exploration, 1996, 644p, ed. Carol R. Stoker, Carter Emmart, Hard Cover $70 (ISBN 0-87703-405-2); Soft Cover $45 (ISBN 0-87703-406-0

Order from Univelt, Inc., P.O. Box 28130, San Diego, California 92198
STANDING ORDERS ACCEPTED

AAS HISTORY SERIES

Vol. 1 Two Hundred Years of Flight in America: A Bicentennial Survey, Edited by Eugene M. Emme, 1977, 326p, Third Printing 1981, Hard Cover $35 (ISBN 0-87703-091-X); Soft Cover $25 (ISBN 0-87703-101-0); special price for classroom text or bulk purchase.

Vol. 2 Twenty-Five Years of the American Astronautical Society: Historical Reflections and Projections, 1954-1979, Edited by Eugene M. Emme, 1980, 248p, Hard Cover $25 (ISBN 0-87703-117-7); Soft Cover $15 (ISBN 0-87703-118-5).

Vol. 3 Between Sputnik and the Shuttle: New Perspectives on American Astronautics, 1957-1980, Edited by Frederick C. Durant, III, 1981, 350p, Hard Cover $40 (ISBN 0-87703-145-2); Soft Cover $30 (ISBN 0-87703-149-9).

Vol. 4 The Endless Space Frontier: A History of the House Committee on Science and Astronautics, By Ken Hechler, Abridged and edited by Albert E. Eastman, 1982, 460p, Hard Cover $45 (ISBN 0-87703-157-6); Soft Cover $35 (ISBN 0-87703-158-4).

Vol. 5 Science Fiction and Space Futures: Past and Present, Edited by Eugene M. Emme, 1982, 278p, Hard Cover $35 (ISBN 0-87703-172-X); Soft Cover $25 (ISBN 0-87703-173-8).

Vol. 6 First Steps Toward Space, Edited by Frederick C. Durant, III and George S. James, 1986, 318p, Hard Cover $45 (ISBN 0-87703-243-2); Soft Cover $35 (ISBN 0-87703-244-0).

Vol. 7 History of Rocketry and Astronautics, Edited by R. Cargill Hall, 1986, Part I, 250p, Part II, 502p, sold as a set, Hard Cover $100 (ISBN 0-87703-260-2); Soft Cover $80 (ISBN 0-87703-261-0).

Vol. 8 History of Rocketry and Astronautics, Edited by Kristan R. Lattu, 1989, 368p, Hard Cover $50 (ISBN 0-87703-307-2); Soft Cover $35 (ISBN 0-87703-308-0).

Vol. 9 History of Rocketry and Astronautics, Edited by Frederick I. Ordway, III, 1989, 330p, Hard Cover $50 (ISBN 0-87703-309-9); Soft Cover $35 (ISBN 0-87703-310-2).

Vol. 10 History of Rocketry and Astronautics, Edited by Å. Ingemar Skoog, 1990, 330p, Hard Cover $60 (ISBN 0-87703-329-3); Soft Cover $40 (ISBN 0-87703-330-7)

Vol. 11 History of Rocketry and Astronautics, Edited by Roger D. Launius, 1994, 236p, Hard Cover $60 (ISBN 0-87703-382-X); Soft Cover $40 (ISBN 0-87703-383-8).

Vol. 12 History of Rocketry and Astronautics, Edited by John L. Sloop, 1991, 252p, Hard Cover $60 (ISBN 0-87703-332-3); Soft Cover $40 (ISBN 0-87703-333-1).

Vol. 13 History of Liquid Rocket Engine Development in the United States 1955-1980, Edited by Stephen E. Doyle, 1992, 176p, Hard Cover $50 (ISBN 0-87703-349-8); Soft Cover $35 (ISBN 0-87703-350-1).

Vol. 14 History of Rocketry and Astronautics, Edited by Tom D. Crouch, Alex M. Spencer, 1993, 222p, Hard Cover $50 (ISBN 0-87703-374-9); Soft Cover $35 (ISBN 0-87703-375-7).

Vol. 15 History of Rocketry and Astronautics, Edited by Lloyd H. Cornett, Jr., 1993, 452p, Hard Cover $60 (ISBN 0-87703-376-5); Soft Cover $40 (ISBN 0-87703-377-3).

Vol. 16 Out From Behind the Eight-Ball: A History of Project Echo, by Donald C. Elder, 1995, 176p, Hard Cover $50 (ISBN 0-87703-387-0); Soft Cover $30 (ISBN 0-87703-388-9).

Vol. 17 History of Rocketry and Astronautics, Edited by John Becklake, 1995, 480p, Hard Cover $60 (ISBN 0-87703-395-1); Soft Cover $40 (ISBN 0-87703-396-X).

Vol. 18 Organizing for the Use of Space: Historical Perspectives on a Persistent Issue, Edited by Roger D. Launius, 1995, 234p, Hard Cover $60; Soft Cover $40

Order from Univelt, Incorporated, P.O. Box 28130, San Diego, California 92198
STANDING ORDERS ACCEPTED

Index

INDEX TO ALL AMERICAN ASTRONAUTICAL SOCIETY PAPERS AND ARTICLES 1954 - 1992

This index is a numerical/chronological index (which also serves as a citation index) and an author index. (A subject index volume will be forthcoming.)

It covers all articles that appear in the following:
- *Advances in the Astronautical Sciences* (1957 - 1992)
- *Science and Technology Series* (1964 - 1992)
- *AAS History Series* (1977 - 1992)
- AAS Microfiche Series (1968 - 1992)
- Journal of the Astronautical Sciences (1954 - September 1992)
- Astronautical Sciences Review (1959 - 1962)

If you are in aerospace you will want this excellent reference tool which covers the first 35 years of the Space Age.

Numerical/Chronological/Author Index in three volumes,

Ordered as a set:
- Library Binding (all three volumes) $120.00;
- Soft Cover (all three volumes) $90.00.

Ordered by individual volume:
- Volume I (1954 - 1978) Library Binding $40.00; Soft Cover $30.00;
- Volume II (1979 - 1985/86) Library Binding $60.00; Soft Cover $45.00;
- Volume III (1986 - 1992) Library Binding $70.00; Soft Cover $50.00.

Order from Univelt, Inc., P.O. Box 28130, San Diego, California 92198.

NUMERICAL INDEX

VOLUME 86	**SCIENCE AND TECHNOLOGY SERIES**, *STRATEGIES FOR MARS: A GUIDE TO HUMAN EXPLORATION* (1996)
AAS 95-471	Why Should Humans Explore Space?, Lawrence G. Lemke
AAS 95-472	The Significance of the Martian Frontier, Robert M. Zubrin
AAS 95-473	The Millennium Project, Harrison H. Schmitt
AAS 95-474	Mars: The Media... the Masses... and the Message, Leonard David
AAS 95-475	Strategic Communications Planning and the Case for Mars, Frank White
AAS 95-476	Managing the Exploration of the Moon and Mars, Michael D. Griffin
AAS 95-477	Mars Mission Concepts: The von Braun Era, Frederick I. Ordway III
AAS 95-478	Pathways to Mars: An Overview of Flight Profiles and Staging Options for Mars Missions, John C. Niehoff and Stephen J. Hoffman
AAS 95-479	Mars Mission Designs: Comparing the Near Term Options, Malcolm A. LeCompt and Julie P. Stets
AAS 95-480	Artificial Gravity: Design Implications for Mars Vehicles, Lawrence G. Lemke
AAS 95-481	Nuclear Rockets: High-Performance Propulsion for Mars, Clayton W. Watson
AAS 95-482	Nuclear Electric Propulsion for Human Mars Missions, Ernst Stuhlinger
AAS 95-483	Biomedical Issues in the Exploration of Mars, Rosalind A. Grymes, Charles E. Wade and Joan Vernikos
AAS 95-484	The Human Side of Mars Flight: A Review of Human Factors Issues, Mary M. Connors and Albert A. Harrison
AAS 95-485	From the Great Voyages of Exploration to Missions to Mars, Ben Finny
AAS 95-486	The Interplanetary Radiation Environment and Methods to Shield from it, Lawrence W. Townsend and John W. Wilson
AAS 95-487	Moving in on Mars: The Hitchhiker's Guide to Martian Life Support, Penelope J. Boston
AAS 95-488	Living in Space: Results from Biosphere 2's Initial Closure, an Early Testbed for Closed Ecological Systems on Mars, Mark Nelson and William F. Dempster
AAS 95-489	Using the Resources of Mars for Human Settlement, Thomas R. Meyer and Christopher P. McKay
AAS 95-490	Mars Rovers, Benton C. Clark

AAS 95-491	First Mars Outpost Habitation Strategy, Marc M. Cohen
AAS 95-492	Scientific Objectives of Human Exploration of Mars, Michael H. Carr
AAS 95-493	Science Strategy for Human Exploration of Mars, Carol R. Stoker
AAS 95-494	The Cost of Sending Humans to Mars, Humbolt C. Mandell, Jr.
AAS 95-495	Mars Colonization: Technically Feasible, Affordable, and a Universal Human Drive, Thomas O. Paine
AAS 95-496	Beyond Mars. . . Into the Universe at Large, Leonard David
AAS 95-497	Steps to Mars, Daniel S. Goldin
AAS 95-498 to -500	Not Assigned

AUTHOR INDEX*

Boston, P. J., AAS 95-487, S&T v86, pp327-361

Carr, M. H., AAS 95-492, S&T v86, pp515-535

Clark, B. C., AAS 95-490, S&T v86, pp445-462

Cohen, M. M., AAS 95-491, S&T v86, pp465-512

Connors, M. M., AAS 95-484, S&T v86, pp241-264

David, L., AAS 95-474, S&T v86, pp41-48; AAS 95-496, S&T v86, pp595-597

Dempster, W. F., AAS 95-488, S&T v86, pp363-390

Finny, B., AAS 95-485, S&T v86, pp267-281

Goldin, D. S., AAS 95-497, S&T v86, pp xi-xix

Griffin, M. D., AAS 95-476, S&T v86, pp59-66

Grymes, R. A., AAS 95-483, S&T v86, pp225-239

Harrison, A. A., AAS 95-484, S&T v86, pp241-264

Hoffman, S. J., AAS 95-478, S&T v86, pp99-125

LeCompt, M. A., AAS 95-479, S&T v86, pp127-150

Lemke, L. G., AAS 95-471, S&T v86, pp3-10; AAS 95-480, S&T v86, pp153-166

Mandell, H. C., Jr., AAS 95-494, S&T v86, pp563-577

McKay, C. P., AAS 95-489, S&T v86, pp393-442

Meyer, T. R., AAS 95-489, S&T v86, pp393-442

Nelson, M., AAS 95-488, S&T v86, pp363-390

Niehoff, J. C., AAS 95-478, S&T v86, pp99-125

Ordway, F. I., III, AAS 95-477, S&T v86, pp69-96

Paine, T. O., AAS 95-495, S&T v86, pp579-591

Schmitt, H. H., AAS 95-473, S&T v86, pp27-39

Stets, J. P., AAS 95-479, S&T v86, pp127-150

Stoker, C. R., AAS 95-493, S&T v86, pp537-560

Stuhlinger, E., AAS 95-482, S&T v86, pp193-221

Townsend, L. W., AAS 95-486, S&T v86, pp283-323

Vernikos, J., AAS 95-483, S&T v86, pp225-239

Wade, C. E., AAS 95-483, S&T v86, pp225-239

Watson, C. W., AAS 95-481, S&T v86, pp167-190

White, F., AAS 95-475, S&T v86, pp51-58

Wilson, J. W., AAS 95-486, S&T v86, pp283-323

Zubrin, R. M., AAS 95-472, S&T v86, pp13-24

---

* For each author the paper number is given. The page numbers refer to Volume 86, AAS *Science and Technology Series.*

QB 641 .S77 1996

STRATEGIES FOR MARS